Y0-BCR-518

Springer Series on Signals and Communication Technology

Signals and Communication Technology

Satellite Communications and Navigation Systems
E. Del Re and M. Ruggieri
ISBN: 0-387-47522-2

Wireless Ad Hoc and Sensor Networks
A Cross-Layer Design Perspective
R. Jurdak
ISBN 0-387-39022-7

Cryptographic Algorithms on Reconfigurable Hardware
F. Rodriguez-Henriquez, N.A. Saqib,
A. Díaz Pérez, and C.K. Koc
ISBN 0-387-33956-6

Multimedia Database Retrieval
A Human-Centered Approach
P. Muneesawang and L. Guan
ISBN 0-387-25627-X

Broadband Fixed Wireless Access
A System Perspective
M. Engels and F. Petre
ISBN 0-387-33956-6

Distributed Cooperative Laboratories
Networking, Instrumentation, and Measurements
F. Davoli, S. Palazzo and S. Zappatore (Eds.)
ISBN 0-387-29811-8

The Variational Bayes Method in Signal Processing
V. Šmídl and A. Quinn
ISBN 3-540-28819-8

Topics in Acoustic Echo and Noise Control
Selected Methods for the Cancellation of Acoustical Echoes, the Reduction of Background Noise, and Speech Processing
E. Hänsler and G. Schmidt (Eds.)
ISBN 3-540-33212-x

EM Modeling of Antennas and RF Components for Wireless Communication Systems
F. Gustrau, D. Manteuffel
ISBN 3-540-28614-4

Interactive Video
Methods and Applications
R. I Hammoud (Ed.)
ISBN 3-540-33214-6

ContinuousTime Signals
Y. Shmaliy
ISBN 1-4020-4817-3

Voice and Speech Quality Perception
Assessment and Evaluation
U. Jekosch
ISBN 3-540-24095-0

Advanced ManMachine Interaction
Fundamentals and Implementation
K.-F. Kraiss
ISBN 3-540-30618-8

Orthogonal Frequency Division Multiplexing for Wireless Communications
Y. (Geoffrey) Li and G.L. Stüber (Eds.)
ISBN 0-387-29095-8

Circuits and Systems
Based on Delta Modulation
Linear, Nonlinear and Mixed Mode Processing
D.G. Zrilic ISBN 3-540-23751-8

Functional Structures in Networks
AMLn—A Language for Model Driven Development of Telecom Systems
T. Muth ISBN 3-540-22545-5

RadioWave Propagation for Telecommunication Applications
H. Sizun ISBN 3-540-40758-8

Electronic Noise and Interfering Signals
Principles and Applications
G. Vasilescu ISBN 3-540-40741-3

DVB
The Family of International Standards for Digital Video Broadcasting, 2nd ed.
U. Reimers ISBN 3-540-43545-X

Digital Interactive TV and Metadata
Future Broadcast Multimedia
A. Lugmayr, S. Niiranen, and S. Kalli
ISBN 3-387-20843-7

Adaptive Antenna Arrays
Trends and Applications
S. Chandran (Ed.) ISBN 3-540-20199-8

Digital Signal Processing with Field Programmable Gate Arrays
U. Meyer-Baese ISBN 3-540-21119-5

Neuro-Fuzzy and Fuzzy Neural Applications in Telecommunications
P. Stavroulakis (Ed.) ISBN 3-540-40759-6

SDMA for Multipath Wireless Channels
Limiting Characteristics and Stochastic Models
I.P. Kovalyov ISBN 3-540-40225-X

continued after index

Satellite Communications and Navigation Systems

Edited by:

Enrico Del Re

Marina Ruggieri

Edited by:

Enrico Del Re
University of Florence
Italy

Marina Ruggieri
University of Tor Vergata, Rome
Italy

Satellite Communication and Navigation Systems

Library of Congress Control Number: 2007921308

ISBN 978-0-387-47522-6 e-ISBN 978-0-387-47524-0

Printed on acid-free paper.

9 8 7 6 5 4 3 2 1

springer.com

Preface

Globalisation of network and services is stimulating a new awareness about the role of satellites and related applications. Even in case "*becoming global*" is "just" seen as a convergence of technologies, it implies the effective exploitation of all components (terrestrial, air and space-based) and media (wired, wireless) in a fully integrated and, in perspective, seamless way to the end-users.

A new and important integration strategy concerns *Navigation* and *Communications* architectures and services. The vision involves an "active" integration, proposing services, applications, integrated business opportunity able to merge two worlds – communications and navigation - that have been considered apart for years.

The *2006 Tyrrhenian International Workshop on Digital Communications (TIWDC'06)* was purposely devoted to the topic of *Satellite Navigation and Communications Systems*, addressing specifically their integration in the satellite scenario.

TIWDC'06 offered to the international satellite navigation and communications community an opportunity of exchanging results and perspectives towards the implementation of the global integrated vision. The workshop activities have been developed under the technical co-sponsorship umbrella of the IEEE AESS (Aerospace and Electronic Systems Society) and the ComSoC (Communication Society), that are gratefully acknowledged for their trust and support.

This volume, that gathers the contributions presented at *TIWDC'06*, includes the state-of-the-art of system concepts, envisaged services and applications as well as enabling technologies for future satellite integrated navigation and communications systems. The contributions come from leading international experts and researchers in the field.

Chapter I *Trends of Communications and Navigation System Integration* deals with the vision of the integration concept, including dual use, based on current and foreseen satellite systems.

Chapter II *Navigation Satellite Technologies* addresses the enabling technologies for future navigation systems.

Chapter III *Satellite Navigation: Perspectives and Applications* describes the envisaged applications and proposals of new navigation services.

Chapter IV *Advanced Satellite Communications Systems & Services* deals with architecture and technologies for near future communication systems.

Chapter V *Perspectives in Satellite Communications* addresses the medium-to-long-term trends in satellite communications.

The Editors would like to express their sincere and grateful appreciation to the session organisers, whose dedicated and enthusiastic effort has rendered the *TIWDC'06* an event of highly scientific value and importance, to the Technical Programme Committee chair Prof. G. Galati and valuable members for their support and to all authors for their state-of-the-art contributions.

Finally, the Editors would also like to thank the members of the Organising Committee for their highly appreciated and dedicated work, that gave a deep contribution to the success of *TIWDC'06*.

Enrico Del Re
Marina Ruggieri

2006 TYRRHENIAN INTERNATIONAL WORKSHOP ON DIGITAL COMMUNICATIONS (TIWDC'06) SATELLITE NAVIGATION AND COMMUNICATIONS SYSTEMS

General Chairs:
Marina Ruggieri, University of Tor Vergata, Rome, Italy
Enrico Del Re, University of Florence, Italy

Technical Program Chair:
Gaspare Galati, University of Tor Vergata, Rome, Italy

Technical Program Committee Members
Antonio Arcidiacono, EUTELSAT, France
Vidal Ashkenazi, Nottingham Scientific, UK
Giovanni Barontini, Finmeccanica, Italy
Paolo Binelli, Telespazio, Italy
Saverio Cacopardi, University of Perugia, Italy
Massimo Comparini, Alcatel Alenia Spazio, Italy
Franco Davoli, University of Genoa, Italy
Patrizio De Marco, Selex SI, Italy
Giuseppe Di Massa, University of Calabria, Italy
Barry G. Evans, University of Surrey, UK
Romano Fantacci, University of Florence, Italy
Pietro Finocchio, Teledife, Italy
Paul Gartz, Boeing, USA
Giuliano Gatti, ESA/ESTEC, The Netherlands
Giordano Giannantoni, OCI, Italy
Filippo Graziani, University of La Sapienza, Rome, Italy
Sergio Greco, Alcatel Alenia Spazio Italia, Italy
Ram Gopal Gupta, Ministry of Communications and Information Technology, India
Guenter Hein, University FAF Munich, Germany
Abbas Jamalipour, University of Sydney, Australia
Shuzo Kato, Pacific Star Comm and NICT, Japan
Letizia Lo Presti, Polytechnic of Turin, Italy
Eric Lutz. DLR, Germany
William F. Lyons, Boeing, Australia
Mario Marchese, University of Genoa, Italy
Franco Marconicchio, ASI, Italy
Francesco Martinino, Alcatel Alenia Spazio Italia, Italy
Takis Mathiopoulos, NOA, Greece
Sergio Palazzo, University of Catania, Italy
Aldo Paraboni, Polytechnic of Milan, Italy
Jorge Pereira, European Commission
Ramjee Prasad, University of Aalborg, Denmark
Luca Ronga, CNIT University of Florence, Italy
Enrico Saggese, Finmeccanica, Italy

Acknowledgements

The *2006 Tyrrhenian International Workshop on Digital Communications (TIWDC'06) - Satellite Navigation and Communications Systems* has been supported by the following sponsors, whose contributions is gratefully acknowledged:

Table of Contents

Chapter II. Navigation Satellite Technologies

Chapter III. Satellite Navigation: Perspectives and Applications

Chapter IV. Advanced Satellite Communications Systems & Services

Chapter V. Perspectives in Satellite Communications

Chapter I

Trends of Communications and Navigation System Integration

Network Centric Operations: The Role of Satellite Communications

Gen. Isp. GA.r.n. Finocchio ing. Pietro

TELEDIFE/AFCEA,
Viale Dell' Università 4, 00100 Roma, Italy
Phone: +39 06 4440204, Fax: +39-06-4986 3658,
e-mail: pietro.finocchio@tiscali.it

Abstract. Nowadays there are new common challenges and objectives for the Defence and Security communities: new forms of conflicts, new players, new tasks, augmented speed of technological innovation.

The NCO concept requires to implement the so called Network Enabled Capabilities (NEC): this means to better exploit the different assets (already operational or in acquisition) and to make them interoperable to allow an efficient information exchange for the interconnection of every node (People, Weapons, Sensor or C2) wherever located in the world. In this context, the SatCom segment (e.g. the Italian Sicral System), represents a valuable asset and the adequate solution to interconnect in a flexible manner the mentioned nodes.

Today, transparent repeaters, similar to the ones available on Sicrall satellite, are available, with some limitations in terms of network flexibility and traffic handling.

The paper proposes the adoption, in the near future, of the Processed EHF/KA Satellites (e.g. the Fidus Mission Satellite), and as long term solution, the Processed and IP Routing EHF/KA Satellites, to implement flexible network configurations as well as IP routing for maximum resource utilization; indeed the processing of the IP packets on board, as usually performed by IP routers on ground, allows to better support delay sensitive traffic and services.

1 Introduction and Aim

Plotting the future is always a taught duty, also because future is never waiting for you and none has a crystal ball to avoid mistakes.

Nevertheless this is what a General Directorate of Ministry of Defence has to manage, building it up consistently with plans made by NATO and allied nations, taking into account international threats.

In plotting the future, we look around, catching news in technology and seeing what others are doing, in order to identify the target and choose the road. The subject of this paper is the Network Centric Operations and the satellite role.

The evolution represents an historical characteristic, and nowadays the changes have characteristics of a revolution more than an evolution. The actual instability period should continue and could be identified as an "instable transient".

The current scenario represents a major modification of the old logic of the "cold war" with the counter position of two major assets; globalisation is nowadays a key word with a consequent revolution in the context of security concept.

The causes of such revolution are not of military nature, but have to be searched in the events that are changing the world and in their effects:

- The growing development gap between countries, that produces major differences in prospective and expectations.
- The technology proliferation, especially information technology, that leads to a "digital" division between those that have access to digital data and those that do not.
- The globalisation and the general growing information interconnection, that are levelling the world.
- Lastly, the loss of countries sovereign that evolves into two directions:
  - The first, positive, that tries to modify the sovereign characteristics of countries towards new kind of extra-national form of aggregation.
  - The second, negative, that leads to difficulties in governance with frequently dramatic consequences (such as ethnic-social-religious conflicts or terrorism).

These causes tend to form, on one side, an interconnected set (core) of countries (having access to technology) that are active part in the globalisation process and, on the other side, a residual "disconnected" and isolated part (gap), which have more and more the perception of possible opportunities but less and less the possibility to access them.

The first set includes: North America countries, European (founder and new) countries, Russia, China, India and Australia.

The residual part includes: the Balkans and Persian Gulf areas, a part of Asia, Korea, almost the entire Africa (with the exception of South Africa) the Centre-Western part of South America.

It can be noticed that it is just on the border between the core and gap areas that instability and crisis were more evident in the past. They are actually in progress and are most probably to turn up in the future.

This border goes through the European-Asiatic continent, crossing strategic areas of primary importance (due to the presence of energy sources) up to the Pacific Ocean, affecting important sites for the maritime trade (see Fig. 1).

Faced with the causes that determine the revolution of the scenario and the security globalisation, the courage to change drastically the security approach is needed.

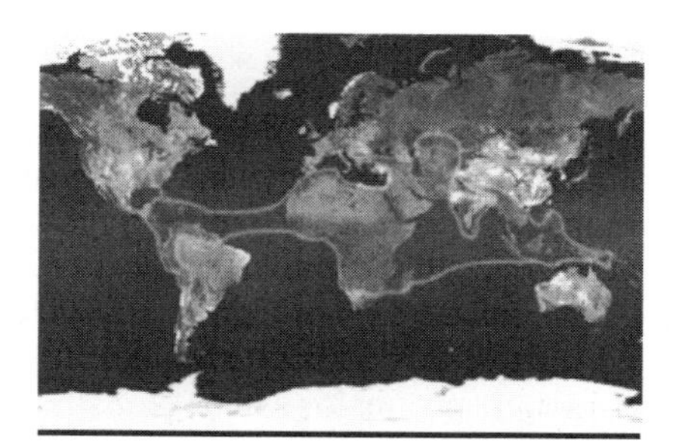

Fig. 1. World instability areas.

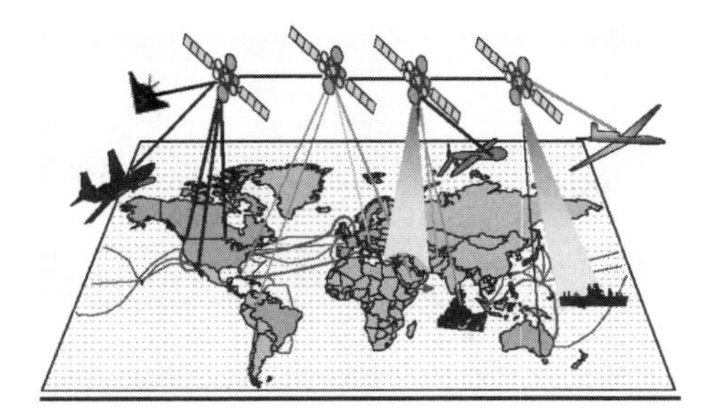

Fig. 2. Synergic utilization of resources, capacity, methodology and instruments, matching the World Instability Areas.

First of all, it has to be understood that the national government holds decreasing power in favour of an increasing support to the international organization for a multilateral approach.

Secondary, taking into account the characteristics of the risks, it has to be noticed that the security concept can be viewed as a continuum, where internal and external security are not separated.

Finally it shall be taken into account that the globalise security concept have to be considered in its more general mean, going over the classic geostrategic one.

The deduction of what has been said is that multilateralism and internal – external continuum, must be the founding elements for a new holistic approach, requiring a synergic development and utilization of capacity, methodology and instruments (see Fig. 2).

2 Requirements and Constraints

Nowadays there are new common challenges and objectives for Defence and Security communities: new forms of conflicts, new players, new tasks, augmented speed of technological innovation. The capability to conduct Network Centric Operations (NCO) is considered the most appropriate solution, to respond to the new challenges of the so called "Information Age" and the asymmetric conflicts.

The Network Centric Operations (NCO) concept requires to implement the so called Network Enabled Capabilities (NEC). In a first phase this means to put together different assets (already operational or in acquisition) and to make them interoperable to allow an efficient information exchange. NCO allows to distribute information only to "need to know" users and to delegate some decision to the very last and remote organizational entity (fighting users).

These assets are usually classified in three different segments: Finders (sensors), Deciders and Effectors, so making a 3D system (Detect-Decide-Destroy).

The **Finders** shall provide Intelligence, Surveillance, Target Acquisition and Reconnaissance (ISTAR) information to the Deciders increasing the "situational awareness".

The **Deciders** at all levels shall be connected with the **Effectors** by mean of integrated Command and Control chain in order to speed up the sensor to operator cycle (3D Cycle) and to increase the mission effectiveness.

The space assets contribute to all these three segments and provide a very flexible service.

Furthermore in a secure framework environments three main requirements have to be highlighted:

- First of all the **bandwidth**, whose wideness cannot be chosen; it is simply required as much bandwidth as technology makes available, and without any differences among the needs of strategically, deployable or tactical network nor among physical layers. That means same bandwidth on fiber optics as well on satellite channels.
- The second requirement is the **IPv6 convergence**, as it seems to fit the multi-services scenarios, allowing effective use of unique interface for all users, to serve people and computers, communicating data and applications in national or multi-national operational environments. That means the development of new software, more collaborative applications and enabling the diffusion of Voice over IP, safe-guarding the quality of service in the meanwhile keeping on using the existing equipments and networks.
- Last but not least, there is a need for **node switching** which must be able to adapt channel to information, which means they should be aware of the semantic of data. As consequence an effective and shared tag information system has to be settled.

That is enough to understand what we are looking for: a really new kind of equipments, networks and systems, in which it will be impossible to distinguish between computing and communicating: even the crypto functions will be implemented in the same hardware and utilizing the same software which will replace the nowadays assets.

This kind of new assets (which will be the structure of the new world of information called "info-structure") will update what we use today and force us to revise the systems already in service in order to define the new programs.

The implementation approach for a NATO Network Enabling Capability (NNEC) is to build a Networking and Information Infrastructure as a "Federation of Systems" (FoS) in services in different nations. In this way it is not required a heavy NATO infrastructure. Assets from different countries will contribute for NII (Networking and Information Infrastructure) remaining under autonomous control of each nation. Of course this idea relays on trusted security and strong interoperability. In this context the IP protocol results to be the most suitable to interconnect heterogeneous networks and to support different services (telephony, multimedia services) by means of various physical layers.

3 Evolution, Not Revolution

The replacement of existing network is really a serious matter, since we don't have yet the new equipments we need, whereas operation are every day performed and no interruption is allowed.

Can you imagine how huge is to reengineer the set of sensors, decisors and actuators that must be in such an infrastructure?

How must they be changed? And what happens when we change one but not the others? Even more which should change first? And what about the improvement we look for?

And of course, there is a financial aspect that is not the least to be mentioned among the constraints.

To find out the correct answer for transforming while operating, Italy decided to have a national study done by Finmeccanica and to participate in conducting two international feasibility studies, one in NATO and the other in EU context.

4 The Italian Milsatcom Programme

As regards the Italian MILSATCOM Programme, in the following it will be described the architecture and some technical solutions implemented in the SICRAL Project and the standards that have been adopted in order to grant interoperability with Allied and Partner Nations.

4.1 "The Present": The SICRAL 1 Project – Operational Effectiveness

The need for a military satellite communication system arose, in Italy, around the end of the '80s. In a global scenario that rapidly evolved towards the radical changes that were characteristic to the end of the past century, the requirement for the capability to have real time access to the information became more and more pressing.

In the above scenario, the need for an adequate C3 instrument was self-evident. It had to be able to provide high levels of mobility, flexibility and deployability, and, at the same time, to be robust, jam-resistant, reliable and performing.

Considering that the national industry was already in possession of the adequate know-how and experience, it was considered that the time was ripe to provide the Italian Armed Forces with such a capability. As a result, the SICRAL 1 satellite was launched in 2001 and is now a reality (see Fig. 3).

Fig. 3. SICRAL 1 satellite.

According to the military requirements, the primary mission of SICRAL is to provide communications for national Forces, both in real world operations and in exercises, in particular when forces are deployed abroad. That function can also be extended to allied or coalition forces involved in the same activities or for their national use.

In addition, it can provide communications in case of "Disaster Relief", when Armed Forces are called to intervene in areas where usual communication media are jeopardized by natural phenomena, such as an earthquake or a flood.

As a secondary mission, the system can integrate the infrastructural communication networks, enhancing their capacity and providing a gap filler capability in case of failure of the fixed networks.

The architecture was therefore designed as follows:

- A space segment, based on a spacecraft with a payload operating in UHF, SHF and EHF bands, fitted with anti-jamming Satellite Control systems;
- A ground segment based on:
  - Main Satellite Control Centre, protected against direct threats (Physical, electronic and nuclear) and configured to be easily improved, in order to perform multi-satellite control functions;
  - Network Management Centre, performing the configuration of satellite services, acting, at the same time, as entry point to terrestrial networks;
  - Back-up Satellite Control Centre;
  - User terminals: fixed, transportable and mobile;
  - An integrated logistic support component is completing the architecture.

The areas covered by the satellite change according to the band in use. UHF covers the totality of the visible hemisphere, which means from the east coast of the US and Brazil to the west coast of India, considering that the satellite's position is 16°2 longitude East.

SHF fixed coverage is directed to the main areas of interest, such as Europe, the Mediterranean and the Middle East, including the Red Sea and the Persian Gulf. A mobile SHF spot, which can be directed wherever required in the visible hemisphere, completes satellite fittings.

EHF coverage is limited to homeland and its proximities.

Ground terminals are, basically, of three different types: Fixed, Transportable and Mobile.

Transportable terminals are composed by two shelters (one is for power supply) and an antenna trailer. They can also be transported on aircraft and will operate in EHF and SHF bands.

Mobile terminals are smaller in size and in performance. However, some installations, such as SHF ship borne, can provide the same throughput of transportable terminals.

Such kind of equipments is installed on board ships (SHF and UHF), aircrafts and vehicles (UHF).

In addition, the flexibility of the system is provided by man-pack terminals, operating in SHF and in UHF.

The system was designed in compliance with international and military standards, as well as with NATO STANAGs.

4.2 "The Near Future": "SICRAL 1 B" & NATO SATCOM Post 2000

Italy is participating in the NATO SATCOM Post 2000 project, managed by the NATO C3 Agency acting as Host Nation on behalf of the Alliance, for which a national contribution has been integrated in a Consortium together with France and U.K., furnishing the required Capability in UHF and SHF through the provision of Allocated Capacity from their SICRAL, SYRACUSE and SKYNET satellites.

For such reason a new communication satellite called SICRAL 1B, to be launched in the year 2007, has been conceived. SICRAL 1B is also a bridge between SICRAL 1 and SICRAL 2, the satellite that will replace the SICRAL 1, which is expected to expire by 2011, at the end of its operational life. For the development of the 2nd SICRAL constellation satellite, SICRAL 1B, some consideration has been taken into account, such as the need to exploit the investments already made within SICRAL 1 project and the adoption of advanced technologies for flexible resource management.

The Italian Defence intends to develop a self–sustained, from the financial point of view, military SATCOM programme. The SICRAL 1 has some exceeding resources that are currently made available to foreign Forces involved in multinational or allied operations and to allied Forces for their national use, producing some incomes.

With SICRAL 1B, it will be possible to greatly increase the income, satisfying the communication services required by NATO, in order to finance future development programmes, as SICRAL 2.

Moreover, the new satellite SICRAL 1B, will take advantage of an existing spacecraft adapted to the new technical requirements, of the existing Italian Control and Management Centres (Master and Back-up) and of the existing management organization and infrastructures.

The equipment has to comply with NATO STANAGs and requirements. For such purpose, anti-jamming and physical protection have to be taken into consideration in accordance with NATO requirements.

With Sicral1B therefore advantages and economics will be brought either to the National and to the NATO programmes.

The main technical characteristics of Sicral1B are the following:

- High number of possible connections;
- Interoperability with Allied users;
- Security of Satellite Control system;
- Security of communications and transmission.

The satellite is controlled by a fully redundant control system with 2 separate Control Centres. The TT&C *(Tracking, Telemetry and Control)* function is designed to operate in EHF during the nominal lifetime operations and in S-band during the launch, the LEOP *(Launch and Early Orbit Phase)* and the contingency lifetime periods. The Command function is encrypted.

The orbital position is scheduled at 11.8 degrees East.

4.3 "The Future": SICRAL 2

The effectiveness of SICRAL 1 demonstrated that SATCOM is an absolutely essential strategic asset.

However, even if outsourcing options might appear more cost-effective, it should be acknowledged that military requirements call for peculiar systems, which should be able to provide a high level of flexibility, which is currently not available on the market.

Two simple considerations may provide a better clarification of the concept:

- Communications for highly mobile users, such as ships, aircraft and small patrol units, can be provided only in UHF. However, UHF assets are not available on commercial markets, although they can be provided on military satellites;
- Communications between ships in Enduring Freedom Operation and Italy have been provided with a 1 Mbps trunk for ITS Garibaldi and with a dedicated 128 Kbps INMARSAT connection for the rest of the naval forces. The latter was obtained within one month from the request and turned out to be particularly expensive.

It is therefore necessary that the Armed Forces maintain or, better, increase their SATCOM capability, in order to ensure the coherent development of military capabilities.

It should also be mentioned that communication capabilities such as only a SATCOM system can provide, are a powerful force multiplier, not only because they allow the effective conduct of operations, but also, and more importantly, because they allow a better control of the crisis, thus representing a vital factor for the safety of forces deployed in the theatre. Taking into account the time necessary to design, develop and procure a new system, Italy has already started a feasibility study for the continuation of the programme, in order to launch a follow-on satellite, named SICRAL 2, not later than 2011.

With regard to EHF, the processed EHF is foreseen in the context of near term satellite missions (Athena/Fidus mission), as later on discussed.

5 The Italian Satellite Core Infrastructure

As already mentioned, the Network Centric Operations (NCO) concept requires to implement the Network Enabled Capabilities. The existing assets have to operate together for a well-organized information exchange and have to be configured to work efficiently with future resources.

The space assets is an important and unique assets to enable mobile communications in remote areas, as well as for providing imagery, navigation, precise position and weather information. The present and the future high performance capabilities of the satellite segment naturally brings to the integration in an overall NEC architecture.

There is one point that is never stressed enough: nowadays homeland defence and international security cannot be separated and must be guaranteed making full use of all the available resources, both military and civilian. In this context the dual use characteristic of the assests becomes even more important.

Italy is fully committed in developing and exploiting space assets for Command and Control Intelligence, Surveillance, Target Acquisition and Reconnaissance (C4 ISTAR)

In this scenario the core of the Italian Satellite Infrastructure is useful to give an overview of both the already existent facilities and the planned developments for the near future of ground and space segment.

Regarding the ground segment the Italian Infrastructure is composed of "Communication Control Centre" and "Satellite Control System" placed in Vigna di Valle and a secondary "Satellite Control System" placed in Fucino.

Both the Communication Control Centre and the Satellite Control System exchange information with the NMAC (NATO Mission Access Centre) to monitor satellite resources assigned by every nation to the NATO mission.

Also shown are the Italian military terrestrial networks connected to the Satellite Communication Control Centre.

Regarding the space segment the present constellation is composed by Sicral 1 satellite. Skyplex and Leased Bentpipe transponders, on the other hand, have the functionality of backup/gapfillers.

The evolution of the present constellation foresees the introduction of:

- Sicral 1B satellite, planned to be launched in 2007
- Sicral 2 satellite, planned to be launched in 2011
- Athena/Fidus planned to be launched in 2011, supporting processing on board with DVB-RCS standard in EHF/Ka band.

6 Network Actually Connected to CGC SICRAL

Military terrestrial networks already connected to the SICRAL Control Centre (CGC) are listed below:

- RIFOR (Roma Area Joint-Forces Optical Network): with overall Data Rate of 620 Mbps
- RIFON (National Joint-Forces Optical Network)
- RNI (Joint-Forces Digital Network) based on:
  - A Radio Bridge National Backbone
  - Several Joint-Forces Local Networks connected to RNI
- DIFENET (Internet Defence)
- SOTRIN (Integrated Transmissions Subsystem)
- ROID (Integrated Operating Network for Defence)
  - Switched voice/data network
- RINAM (National Integrated Network, Air Force)

Sicral 1 Control Centre represents an interconnection node between satellite and the military terrestrial networks listed above. Moreover by means of a secure IP network connected to the Centre it is possible to manage satellite resources assigned to NATO from different sites in Europe.

7 C4I Systems Interconnected Via SatCom, to Support the NEC

A list of Command and Control Systems that would be a useful upgrading in NEC context, for the following C4I assets. For each asset the benefits of the integration with Sicral SatCom links is reported:

- **SIACCON** (Automatic Control and Command System for the Italian Army)
  - SatCom needs: could be useful to support the SiCCAM system interaction
- **SICCONA** (**National** Command and Control System)
  - SatCom needs: it shall integrate all military units and systems in various operative scenarios
- **ACCS** (Air Command and Control System)
  - Integrated Command and Control System for Air Forces
  - SatCom needs: it will interact with SICCAM, National/NATO systems (also Army and Air Force)
- **C4I Difesa** (High Command and Control System)
  - SatCom needs: requested meshed wide band links to C2I main systems
- **SiCCAM** (Air Force Command and Control System)
  - C2I System for avionic operations (on development phase)
  - It will include previously existent systems and it will interact with NATO ACCS (Air Command and Control System), ACE ACCIS (Allied Command Europe Automated Command and Control Information System) and other C2I national system.
  - SatCom needs: It could interact with SIACCON and National/NATO C2I systems
- **C2M** (Air Force Command and Control Mobile System)
  - Command and Control Mobile System of Italian Air Force operative since 1998 and now on optimization phase: further elements will be included such as MATRA and DCE
  - SatCom needs: it could be used to advantage Sicral mobile transportable SHF/UHF stations, particularly on theatre application
- **MIDS-**LVT (Multifunctional Information Distribution System-Low Volume Terminal)
  - Integrated system for tactical, identification and navigation information distribution, applicable with small dimension terminals according to STANAG 4175
  - SatCom needs: It could be integrated to Sicral Satellite System

8 Platforms Interconnected Via SatCom, to Support the NEC

In the following, the platforms that can be interconnected via SatCom to support the NEC are listed, for each one the benefits of the integration with Sicral SatCom links is reported:

- TETRA e TETRA-TAC TETRA-CAMPALE
  - short-range communication platforms
  - SatCom needs: Necessary an Access Point to Sicral System

- UAV PREDATOR, MIRACH 26, FALCO e NIBBIO
  - Unmanned Airbone Vehicle platforms for surveillance/reconnaissance operations
  - SatCom needs: high capacity satellite link
- EF2000
  - Combat aircraft
  - SatCom needs: Possible use of Sicral UHF terminal
- FUTURE SOLDIER
  - R&S program for the development of ITC future soldier equipments
  - Necessary an integration to C2I Network (SICCONA e SIACCON)
  - SatCom needs: Necessary wireless Access Point via SatCom (Man-pack UHF/EHF)
- NEW NAVAL UNITS
  - SatCom needs: new EHF/SHF/UHF terminals
- BMD (TBMD, NATO-BMD, ALT-BMD, THAAD)
  - Ballistic Missile Defense:
  - SatCom needs: meshed wide band links (with on board processing)

9 SatCom Requirements in the NEC Context

NEC allows to facilitate the interconnection of every nodes (People, Weapons, Sensor or C2) positioned in continental areas. In this context the SaTCom segment should represent an adequate solution to interconnect several nodes in the same network.

To reach this goal, it is possible to use three satellite payload solutions:

- Bentpipe (Transparent) Repeater
- Processing Repeater
- Processing and IP Routing Repeater

Nevertheless, today it is possible to use only Bentpipe Repeater, by means SICRAL 1 payload, that is characterized by fast connectivity to deployed forces in star topology. This system presents some limitations in term of network flexibility (point to point connection) and traffic priority handling.

10 Evolution Toward SatCom Processed Systems

As discussed above at present satellite communication mainly rely on transparent configurations, but the prospective and the service requirements for the near future scenario lead toward "processed" systems. In particular a Near Future Scenario (year 2011) is represented by the Processed EHF/Ka Satellite FIDUS, characterized by:

- Flexible network configuration by means point to point and meshed two ways connectivity for all network terminals
- Optimized satellite capacity management
- Small terminals with reduced power
- Optimal use of the satellite transmission power

A Long Term Scenario (year 2014) is represented by Processed and IP Routing EHF/Ka Satellite (SICRAL 3), that will have the same features of a "Processed EHF/Ka Satellite" and in addition:

- On Board IP Routing for maximum resource utilization, using the IP Protocol, as usually implemented in Internet
- Priority Policy for Delay Sensitive Services.

11 Implementation of NEC Requirements

NEC implementation requires that every node can exchange data with other nodes. Obviously with Transparent Satellite (Present Scenario) there are many limitations such as:

- Mainly Star network configuration (Hub centric) supported
- Meshed connectivity possible only by means of double hop between nodes
- No support of alternative IP routes, as regards the routing path.

On the other hand, "Processed and IP Routing Satellite" will provide several benefits; the major ones are: Possible use of both Star and Mesh (Figs. 4–5) communication network configuration

- Meshed connectivity by means of single hop between nodes; the meshed solution is the one that maximize the connectivity between nodes giving the opportunity to

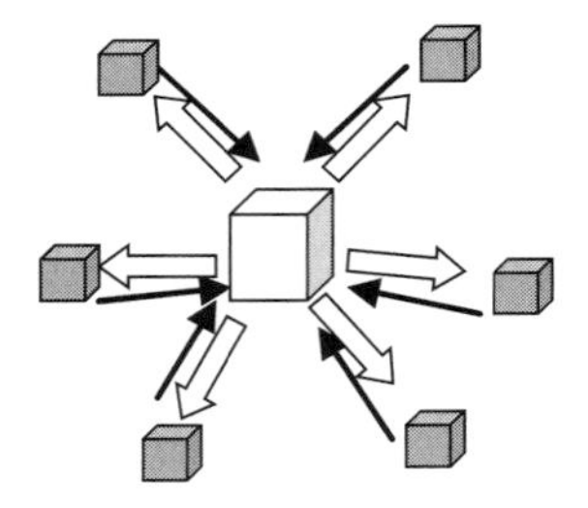

Fig. 4. Star topology.

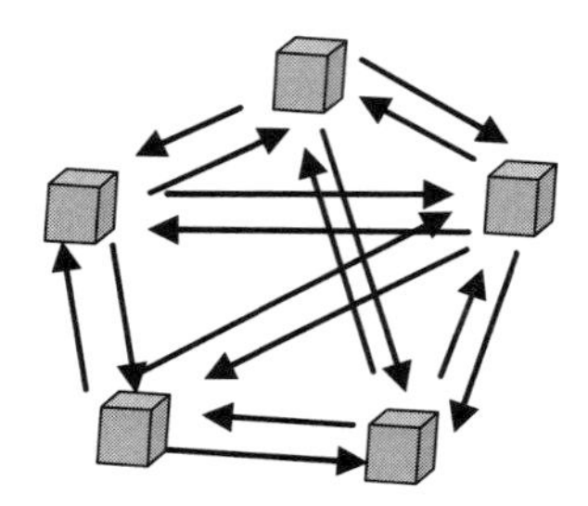

Fig. 5. Meshed topology.

every node to exchange data towards all other nodes, Network Operation Centre or Headquarter.
- On board IP routing, allowing more alternative IP routes (composed of terrestrial and satellite paths)
- The regenerative approach allows also an optimal use of the satellite transmitting power, avoiding noise, jammers, interference, intermodulation retransmission and signal unbalances.

Figure 6 shows a meshed network via processing and IP Routing satellite.

Other significant difference, between Transparent and Processed/IP Routing Satellites, consists in the dynamic bandwidth allocation, that allows different Information exchange profiles for every mission phase and for every node. Requirements concerning the information exchange are reported below:

- high capacity links (several Mbps)
- high level of asymmetry allowed (forward versus return link)
- end users easily upgradeable as Content Sources (bandwidth availability)

Using "Not Processed Satellite" there are the following limitations:

- Difficult Channel Bandwidth Adaptation (without IP priority processing)
- Long times for channel bandwidth adaptation (several double hop time intervals due to star communication)

"Processed and IP Routing Satellite" will carries several benefits in terms of Channel Bandwidth Adaptation:

- Dynamic Channel Bandwidth Allocation based on the IP priority scheme
- Improved Reactivity Time (about 300 ms. to modify the used bandwidth).

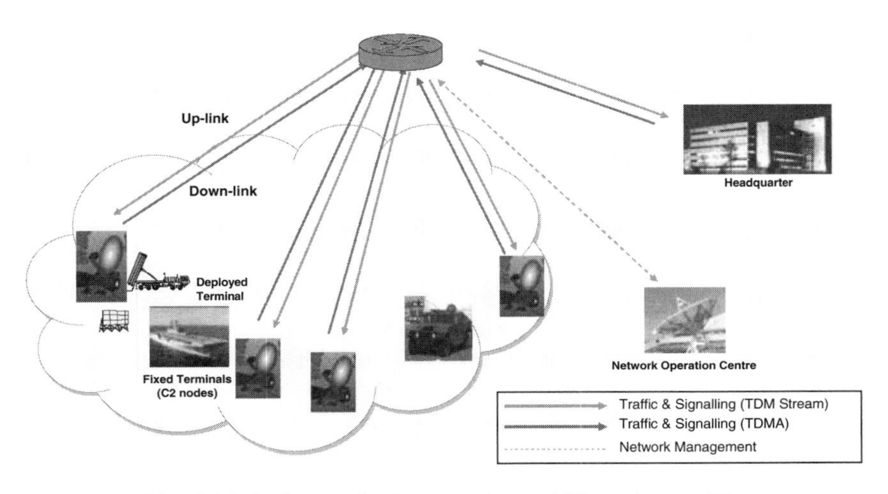

Fig. 6. Meshed network via processing and IP routing satellite.

Link robustness is an other important requirement for NEC. It is characterised by:

- Anti-jamming features, implemented by means of Spread Spectrum Modulations
- Backup bearers given by multiple routing paths
- Quality of Service (in terms of Delay), implemented with differentiated traffic handling (priority policy)

Using "Not Processed Satellite" there are the following limitations in terms of link robustness:

- Anti jamming capability reduction, caused to the absence of on board dispreading
- Low efficiency of capacity utilisation adopting multiple-routing paths
- Efficiency reduction for delay sensitive applications, due to the relevant round trip time delay (several double-hop time intervals needed in the star topology)

On the contrary, "Processed and IP Routing Satellite" guarantees link robustness as:

- High anti-jamming performance is obtained by means of On Board Spread Spectrum Dispreading
- Unwanted and jamming signals are blocked before transmission on the down link
- IP Routing in the sky is implemented, supporting more alternative IP routes (composed of terrestrial and satellite paths)
- Single hop round trip time ameliorates the TCP/IP performance and Quality of Service.

The service area for NEC should be related to a word wide coverage. In order to guarantee such a coverage, with the link dimensioning constraints (adopting very small aperture terminals), a multi Beam Antenna on board the satellite is needed.

In this case, adopting "Not Processed Satellite" there are limitations in terms of Beam to Beam connectivity, as the fixed frequency association scheme, between up and down link of different beams, is too rigid for NEC context. "Processed and IP Routing Satellite" instead will carry several benefits such as:

- Inter beam connectivity managed at IP Packet Routing level
- Inter beam connectivity implemented with single hop (minimum time delay).

12 On Board Processing and IP Traffic Routing

The IP-over-Satcom network target is to cope with the growing request for new Network Centric Infrastructures providing packet and circuit switched connections with flexible resource allocation on request by user. This process requires an on board routing, which *performs on the Satellite the same functionality, usually implemented within an Internet Terrestrial Router*:

- IP routing of the up linked IP packets toward downlink destination address (Address Field in Fig. 7)

- Priority policy to privilege Delay-Sensitive Traffic, for immediate transmission on the downlink (Priority Field in Fig. 7)

Figure 8 shows the conceptual scheme of the On board Processing with IP Traffic Routing with priority policy.

In the picture the service areas transmitting packets toward the satellite with three priority levels (High, Medium and Low) are shown, with three channel addresses (Ch1, Ch2 and Ch3). The on board processor manages each received IP packet and forwards it to the channel indicated in the IP header address field. Each IP Packet is stored in a Buffering Stage and Queue matched to its own level of priority. Finally, the priority server controls the output packets sequence, according to a delay sensitive traffic policy, privileging the high priority packets.

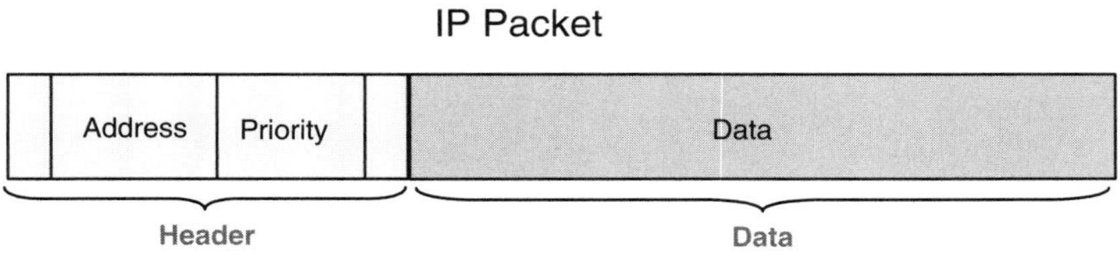

IP Packet

	Address	Priority		Data

Header — Data

Fig. 7. IP packet.

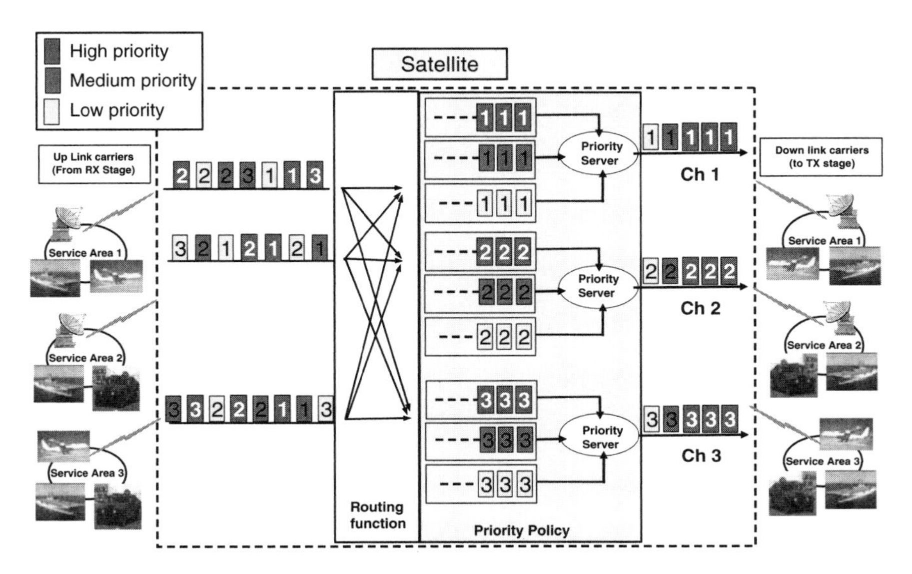

Fig. 8. On board processing and IP traffic routing.

13 Conclusions

The implementation of the Network Enabled Capabilities represents a challenge and an opportunity for technological innovation.

The aim is to make full use of the assets already in inventory following the "Transforming while operating" concept and, in this frame, the Italian SatCom Infrastructure, has been conceived to cope with such an evolving scenario.

NCO/NEC concept system leads to special communications requirements in terms of interconnectivity, flexibility, robustness, dynamic bandwidth assignment and coverage.

In particular the interoperability and interconnection issues, which represent the main obstacles in the path to a fully integrated Defence System, can be solved thanks to the IP based common protocol transport layer. Processing and IP Routing SatCom represents indeed the ideal solution to deploy NEC, although it requires to proceed with standardization agreements (at NATO level) and with a technology program for the relevant development activities.

Comparison and Integration of GPS and SAR Data

M. Calamia[1], G. Franceschetti[2,3], R. Lanari[4], F. Casu[4,5], M. Manzo[4,6]

[1] Dipartimento Elettronica e Telecomunicazioni, Università di Firenze, Via di Santa Marta 3, 50139 Firenze, Italy
[2] Dipartimento di Ingegneria Elettronica e delle Telecomunicazioni, Università degli studi di Napoli Federico II, Via Claudio 21, 80125 Napoli, Italy
[3] University of California at Los Angeles (UCLA), USA
[4] Istituto per il Rilevamento Elettromagnetico dell'Ambiente (IREA), National Research Council of Italy (CNR), Via Diocleziano 328, 80124 Napoli, Italy
[5] Dipartimento di Ingegneria Elettrica ed Elettronica, Università degli studi di Cagliari, Piazza d'Armi, 09123 Cagliari, Italy
[6] Dipartimento di Ingegneria e Fisica dell'Ambiente, Università degli Studi della Basilicata, Viale dell'Ateneo Lucano 10, 85100 Potenza, Italy

Abstract. We compare the surface deformation measurement capability of the Differential Synthetic Aperture Radar Interferometry (DIFSAR) technique, referred to as the Small BAseline Subset (SBAS) approach, and of the continuous Global Positioning System (GPS). The analysis is focused on the Los Angeles (California) test area where different deformation phenomena are present and a large amount of SAR data, acquired by the European Remote Sensing Satellite (ERS) sensors, and of continuous GPS measurements is available. The carried out analysis shows that the SBAS technique allows to achieve an estimate of the single displacement measurements, in the radar line of sight (LOS), with a standard deviation of about 5mm, which is comparable with the LOS-projected GPS data accuracy. Final remarks on the complementariness and integration of the SAR and GPS measurements are also provided.

1 SAR and GPS Data Comparison in the Los Angeles (California) Area

The Global Positioning System, usually referred to with the acronym GPS [1], is a fully-operational satellite navigation system based on a constellation of more than 24 GPS satellites; they broadcast precise timing signals via radio to the GPS receivers, allowing them to accurately determine their location (longitude, latitude, and altitude) with any weather, during day or night and everywhere on the Earth. GPS has become a global utility, indispensable for modern navigation on land, sea, and air around the world, as well as an important tool for map-making and land surveying. GPS also provides an extremely precise time reference, required for telecommunications and some scientific research. Among all these possible uses, the GPS data are also widely employed in geophysical applications to detect and follow the deformations of the Earth surface on a millimeter/centimeter scale, via the differential operational mode [1]. In the near future, the advanced European GA LILEO Positioning System will be operational.

More recently, the remote sensing technique referred to as Differential Synthetic Aperture Radar Interferometry (DIFSAR) has been developed [2], that also allows to investigate surface deformation phenomena on a millimeter/centimeter scale. In this case, it is exploited the phase difference (usually referred to as the interferogram) of two SAR images relevant to temporally separated observations of an investigated area. An effective procedure to detect and follow the temporal evolution of deformations is via the generation of time series; to achieve this task the information available from each interferometric data pair must be properly related to those included in the other acquisitions by generating an appropriate sequence of DIFSAR interferograms.

Several approaches aimed at the DIFSAR time series generation have been proposed. In this work, we focus on the technique referred to as the Small BAseline Subset (SBAS) algorithm [3], that has been originally developed to investigate large spatial scale displacements with relatively low resolution (typically of the order of 100×100 m). The SBAS approach relies on an appropriate combination of differential SAR interferograms characterized by small spatial and temporal separations between the orbits (baseline). As a consequence, the SAR data involved in the interferograms generation are usually grouped in several independent small baseline subsets, separated by large baselines. A way to easily "link" such subsets is the application of the Singular Value Decomposition (SVD) method.

The capability of the SBAS approach to generate deformation maps and time series from data acquired by the European Remote Sensing Satellite (ERS) sensors have been already shown in different applications [4, 5, 6]: an analysis on the quality of the DIFSAR measurements and comparison with geometric leveling and GPS techniques has been provided [7]. Accordingly, we focus in this presentation to the comparison between the SBAS-DIFSAR results and the measurements available from continuous GPS data. In particular, the investigated test site is the Los Angeles metropolitan zone (Southern California, USA), a tectonically active region with surface deformations caused by a variety of natural and anthropogenic actions. We remark that a key element for the seismic surveillance on the whole area is represented by the Southern California Integrated GPS Network (SCIGN) [8], which is an array of 250 GPS stations spread out across Southern California and northern Baja California, Mexico.

For what concerns the presented DIFSAR analysis, the SBAS algorithm has been applied to a set of 42 SAR data (track: 170, frame: 2925), acquired by the ERS satellites during the 1995–2002 time interval and coupled to 102 interferograms. Each interferometric SAR image pair has been chosen with a perpendicular baseline value smaller than 300 m and with a maximum time interval of 4 years; precise satellite orbital information and a Shuttle Radar Topography Mission (SRTM) Digital Elevation Model (DEM) [9] of the area have also been used. All the DIFSAR products have been obtained following a complex multilook operation [10] with 20 looks in the azimuth direction and 4 looks in the range one, with a resulting pixel dimension of the order of 100×100 m.

As a first result of the SBAS algorithm analysis, we present in Fig. 1a the geocoded SAR amplitude image relevant to the investigated area, with superimposed the retrieved line of sight (LOS) mean displacement velocity map and the locations of the GPS SCIGN sites (black and white squares) within the region.

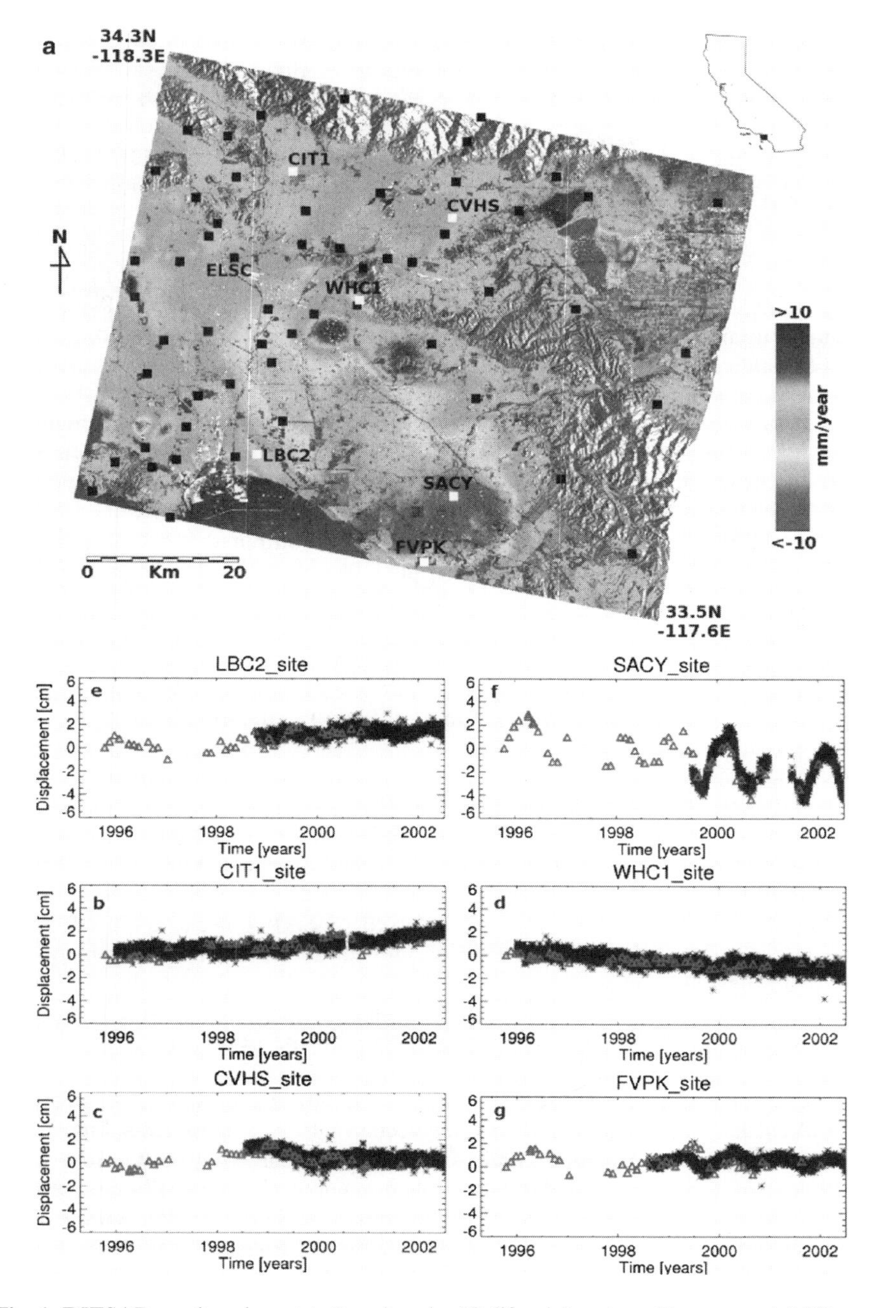

Fig. 1. DIFSAR results relevant to Los Angeles (California) metropolitan area. a) LOS mean deformation velocity map with superimposed the locations of the investigated GPS stations (black and white squares). White squares mark the selected stations relevant to the plots shown in Figs. 1b-g; the label ELSC identifies the reference point for both DIFSAR and geodetic measurements. b-g) Comparison between the DIFSAR LOS deformation time series (triangles) and the corresponding GPS measurements projected on the radar LOS (black stars), for the pixels labeled in Fig. 1a as CIT1, CVHS, WHC1, LBC2, SACY and FVPK, respectively.

Note that, we selected in our analysis the GPS stations located in coherent areas, and for which the measurements are relevant to a time interval starting before the year 2000 (in order to ensure at least two years of overlap with the available SAR data).

Following the mentioned GPS selection, we have compared the DIFSAR time series with the corresponding LOS-projected GPS measurements, the latter obtained through the SCIGN web site [11]. As first result of this comparison we show in Figs. 1b-g the plots relevant to the DIFSAR and GPS measurements for six sample GPS stations identified in Fig. 1a by the white squares and labeled as CIT1, CVHS, WHC1, LBC2, SACY and FVPK, respectively. Note that both DIFSAR and geodetic measurements have been referred to the same pixel located in correspondence of the ELSC station (see Fig. 1a). The presented results clearly show the good agreement between these two measurements.

Let us now move from a qualitative to a quantitative analysis. Accordingly, we have computed the standard devition values of the differences between the DIFSAR and the LOS-projected GPS time series for each GPS site identified by black and white squares in Fig. 1a (see Table 1) and the average of all these values; in this case, we obtained $\sigma_{d_{SAR}}$ = 6.9 mm. Subsequently, we computed the LOS-projected GPS errors available from the SCIGN web site [10] and removed the bias due to the estimated errors relevant to the geodetic measurements. Following this bias removal, we finally achieved the mean value $\sigma_{d_{SAR}}$ = 5.6 mm for the standard deviation of the difference between DIGSAR and GPS data. We remark that this value is rather close to the standard deviation of LOS-projected GPS errors whose mean value, computed from the measurements of the right column of Table 1, corresponds to $\sigma_{d_{GPS}}$ = 4 mm.

2 Conclusions and Future Developments

We have compared the surface deformation measurement capability of the SBAS-DIFSAR technique and of continuous GPS. The analysis has been focused on the Los Angeles (California) test area and involved 42 SAR images, acquired by the ERS sensors in the 1995–2002 time interval, and continuous GPS measurements relevant to 38 sites. The carried out analysis has shown that the SBAS technique allows to achieve an estimate of the single displacement measurements with a standard deviation of about 5 mm. which is comparable with the LOS-projected GPS data accuracy.

As additional remark we underline that the integration of DIFSAR and GPS (and GALILEO soon) measurements is foreseen in future developments. Indeed, the former technique may provide spatially dense measurements but limited to a single component of the detected deformation phenomenon. On the contrary, the GPS (GALILEO) data are relevant to single points but allow to retrieve temporally dense time series with fully 3D information. The complementariness of the two techniques is evident.

A convenient possibility is the use of GPS (GALILEO) data as "tie points" at the DIFSAR phase-unwrapping stage. This renders the procedure overdetermined and appropriate processing techniques may be implemented to improve the accuracy of the solution. Forthcoming constellation SAR satellites will drastically reduce the

Table 1. Results of the compariosn between SAR and LOS-projected GPS deformation time series. The values shown in the third column have been obtained by projecting along the radar line of sight the information relevant to the errors available form the SCIGN web site (http://www.scign.org/).

GPS stations	Standard deviation of the difference between SAR and los-projected GPS measurements [cm]	Los-projected GPS errors [cm]
AZU1	0.71	0.43
BGIS	0.48	0.38
BRAN	0.85	0.42
CCCO	0.68	0.37
CCCS	0.73	0.37
CIT1	0.70	0.38
CLAR	1.09	0.37
CRHS	0.41	0.37
CSDH	0.58	0.36
CVHS	0.53	0.38
DSHS	0.54	0.43
DYHS	0.49	0.37
ECCO	0.48	0.38
EWPP	0.61	0.37
FVPK	0.82	0.37
JPLM	0.93	0.41
LASC	0.76	0.37
LBC1	0.79	0.42
LBC2	0.59	0.37
LONG	0.90	0.43
LORS	0.81	0.39
LPHS	0.47	0.40
MHMS	1.13	0.50
NOPK	0.47	0.44
PMHS	0.50	0.38
PVHS	0.77	0.44
PVRS	0.77	0.40
RHCL	0.49	0.39
SACY	0.85	0.43
SNHS	0.72	0.39
SPMS	0.65	0.37
TORP	0.69	0.38
USC1	0.73	0.42
VTIS	0.80	0.40
VYAS	0.48	0.39
WCHS	0.49	0.39
WHC1	0.38	0.39
WHI1	0.20	0.39

revisiting time, thus allowing statistical analysis to possibly detect systematic errors. In addition, knowledge of the full deformation map, as provided by DIFSAR, allows to optimize the GPS (GALILEO) sensors location providing also appropriate hints to the development of geological models.

Acknowledgments

This work has been partially sponsored by the CRdC-AMRA, ASI and the (Italian) GNV. We thank ESA which had provided the ERS data of the Los Angeles zone through the WInSAR data archive in collaboration with dr. P. Lundgren, JPL, Caltech. The GPS measurements relevant to the SCIGN network have been achieved through the SCIGN web site (http://www.scign.org/). Moreover, the DEM of the investigated zone has been achieved through the SRTM archive while precise ERS-1/2 satellite orbit state vectors are courtesy of the TU-Delft, The Netherlands.

References

[1] Kaplan, E., Hegarty, C., *Understanding GPS: Principles and Applications, 2nd Edition*, Artech House, Boston, MA, 644 pages, ISBN 1580538940, 2005.

[2] Gabriel, A. K., Goldstein, R. M., Zebker, H. A., Mapping small elevation changes over large areas: Differential interferometry, *J. Geophys. Res.*, 94, pp. 9183–9191, 1989.

[3] Berardino, P., Fornaro, G., Lanari, R., Sansosti, E., A new Algorithm for Surface Deformation Monitoring based on Samll Baseline Differential SAR Interferograms, *IEEE Trans. Geosci. Rem. Sens.*, 40, 11, pp. 2375–2383, 2002.

[4] Lundgren, P., Casu, F., Manzo, M., Pepe, A., Berardino, P., Sansosti, E., Lanari, R., Gravity and magma induced spreading of Mount Etna volcano revealed by satellite radar interferometry, *Geoph. Res. Lett.*, 31, L04602, doi: 10.1029/2003GL018736, 2004.

[5] Lanari, R., Lundgren, P., Manzo, M., Casu, F., Satellite radar interferometry time series analysis of surface deformation for Los Angeles, California, *Geoph. Res. Lett.*, 31, L23613, doi:10.1029/2004GL021294, 2004.

[6] Borgia, A., Tizzani, P., Solaro, G., Manzo, M., Casu F., Luongo, G., Pepe, A., Berardino, P., Fornaro, G., Sansosti, E., Ricciardi, G. P., Fusi, N., Di Donna, G., Lanari, R., Volcanic spreading of Vesuvius, a new paradigm for interpreting is volcanic activity, *Geoph. Res. Lett.*, 32, L03303, doi:10.1029/2004GL022155, 2005.

[7] Casu, F., Manzo, M., Lanari, R., A Quantitative Assessment of the SBAS Algorithm Performance for Surface Deformation Retrieval from DIFSAR Data, in press on *Remote Sensing of Environment*, doi: 10.1016/j.rse.2006.01.023, 2006.

[8] Hudnut, K. W., Bock, Y., Galetzka, J. E., Webb, F. H., Young W. H., The Southern California Integrated GPS Network (SCIGN). *Proc. Of the The 10th FIG International Symposium on Deformation Measurements*, 19–22 March 2001 Orange, California, USA, pp. 129–148, 2001.

[9] Rosen, P. A., Hensley, S., Gurrola, E., Rogez, F., Chan, S., Marthin, J., SRTM C-band topographic data quality assessment and calibration activities, *Proc, of IGARSS'01*, pp. 739–741, 2001.

[10] Rosen, P. A., Hensley, S., Joughin, I. R., Li, F. K., Madsen, S. N., Rodriguez, E., Goldstein, R., Synthetic Aperture Radar Interferometry, *IEEE Proc.*, 88, pp. 333–376, 2000.

[11] SCIGN web site: http://www.scign.org/.

Integration of Navigation and Communication for Location and Context Aware RRM

Ernestina Cianca[1], Mauro De Sanctis[1], Giuseppe Araniti[2], Antonella Molinaro[2], Antonio Iera[2], Marco Torrisi[1], Marina Ruggieri[1]

[1] University of Rome Tor Vergata, Department of Electronics Engineering, via politecnico 1, 00133 Rome, Italy
[2] University "Mediterranea" of Reggio Calabria, D.I.M.E.T., via Graziella lo. Feo di Vito, 89100, Reggio Calabria, Italy
Phone: 1 +390672597284, Fax: +390672597455, e-mail: cianca@david.eln.uniroma2.it

Abstract. Next Generation Wireless Networks (NGWNs) will allow the user to roam over different access networks, such as UMTS, Wi-Fi, satellite-based networks. Currently, these networks are integrated/assisted by more and more accurate navigation systems, which can make available the information on the location of the mobile terminal. This information is then typically used to provide location based services. This paper addresses a novel way of jointly using navigation and communication systems: the information on location together with other information on the situation of the user/network nodes is used in order to optimize the mobility and resource management over satellite/terrestrial heterogeneous networks.

1 Introduction

A heterogeneous communication network is foreseen for Next Generation Wireless Networks (NGWNs) where different Radio Access Networks (RANs) such as Terrestrial Universal Mobile Telecommunications System (T-UMTS), Satellite UMTS (S-UMTS) and Wireless LANs (WLANs) can be offered to the user in order to access the same core network. The efficient integration of different access systems is one of the main challenges for the scientific and industrial telecommunications community. In this framework, the new concept that is addressed by this paper is that the deployment of NGWNs may benefit from the capability of exploiting location/situation information for:

- more efficient "seamless" integration of heterogeneous radio access networks;
- improved resource allocation both within one single network and in the integrated heterogeneous scenario.

Some works have already shown that this information can be used to improve radio resource management or mobility management (i.e., horizontal handover) by properly designed mechanisms [1–5]. In this paper, this concept will be further developed. We claim that location information, together with the knowledge of the user situation or context, could become the most important enabling function in order to provide efficient integration of different access technologies.

The paper first presents an architecture able to collect the location/situation information, process it and distribute it to the network nodes that will exploit it to successfully implement location/situation aware radio resource management (RRM) or mobility management mechanisms. Later on, the agent-based middleware implemented in some components of the proposed communication platform is described in details. Finally, possible location/situation aware RRM and mobility management mechanisms are proposed in a heterogeneous network which includes HAPs and satellites. Advantages offered by these enhanced RRM mechanisms and also the limits due to the attainable accuracy of currently available radio-location techniques are discussed.

2 Overview of Radiolocation Techniques

In this paper, the term *location* refers to the geographical co-ordinates of the mobile users and, in some cases, also to the speed, direction and orientation of the users movements. A network control center can process the information about the mobile user location with the aim of computing:

- user location with respect to the cell of coverage;
- user distance from the access nodes;
- path and next location of the user/node.

In satellite-based networks, HAP-based networks and ad-hoc networks, nodes are mobile in nature. Therefore, in such networks, the term *location* can be also referred to the geographical co-ordinates of the *network nodes*. The knowledge of this information could be used to more efficiently manage the dynamics of the network topology, the coverage and also the resource allocation.

Location systems can be classified into physical or symbolic. Physical information provides the position of a location on a physical coordinate system (x,y,z), for example the Electronics Engineering Department is at (x1,y1) coordinates. Symbolic location information provides a description of the location, for example the Radar laboratory at the Electronics Engineering Department. Furthermore, the Symbolic location is related to abstract ideas; physical location information can be derived by symbolic position with additional information. Using only symbolic location information can yield very coarse grained physical positions [6].

An absolute location system uses a shared reference grid for all located objects, while in a relative system, each object can have its own reference frame. An absolute location can be transformed into a relative location.

In indoor location architectures, there are, in general, two different types of mobile devices: active and passive. In an active mobile architecture, an active transmitter on each mobile device periodically broadcasts a message on a wireless channel. On the other hand, in a passive mobile architecture, fixed nodes at known positions periodically transmit their location (or identity) on a wireless channel, and passive receivers on mobile devices listen to each beacon [7].

In the following, we will compare several indoor/outdoor radio-location systems in terms of some parameters of interest for the application considered in this paper.

2.1 Indoor/Outdoor Sensing Systems

Active Badge: it uses cellular proximity system that employ diffuse Infrared technology. The system can locate every person that wear a small infrared badge. The badge emits a globally unique identifier every 10 seconds or on demand [6]. These periodic signals are picked up by a network of sensors placed around the host building. A master station, also connected to the network, polls the sensors for badge 'sightings', processes the data, and then makes it available to clients that may display it in a useful visual form. An active badge signal is transmitted to a sensor through an optical path. This path may be found indirectly through a surface reflection, for example, from a wall [8].

Active Bat: this location system uses an ultra sound time-of-flight lateration technique to provide more accurate physical positioning than active badges. Users and objects carry active bat tags. It combines a 3D ultrasonic location system with a pervasive wireless network [6].

A short pulse of ultrasound is emitted from a transmitter (a bat) attached to the object to be located, and the time-of-flight of the pulse to receivers mounted at known points on the ceiling is measured. The speed of sound in air is known, so we can calculate the distance from the bat to each receiver - given three or more such distances, we have enough information to determine the 3D position of the bat (and hence that of the object on which it is mounted) [9].

Cricket: the system is decentralized so that each component of the system whether fixed or mobile is configured independently, no central entity is used to register or synchronize elements. This architecture uses beacons to disseminate information about a geographic space to listeners. A beacon is a small device attached to some location within the geographic space it advertises. To obtain information about a space, every mobile and static node has a listener attached to it. A listener is a small device that listens to messages from beacons, and uses these messages to infer the space it is currently in [10].

Enhanced 911 (E 911): it is used to determine cellular phones location and can be used in applications that need to find the nearest gas station, post office etc. [6].

RFID: the basic premise behind RFID systems is that you mark items with tags. These tags contain transponders that emit messages readable by specialized RFID readers. A reader retrieves information about the ID number from a database, and acts upon it accordingly. RFID tags fall into two general categories, active and passive, depending on their source of electrical power [11].

Most of the applications of RFID technology, however, assume that the readers are stationary and only the tags that are attached to objects or persons move. The main focus is to trigger events if a tag is detected by a reader or entering the field of range [12].

Cell-ID: the cellular based location system is overlaid on the existing cellular communication system. The common geolocation techniques used in cellular-based location systems are signal strength measurements, time of arrival, time difference and angle of arrival [13].

The cell ID only has to be associated with location, i.e. the coordinates of the BSs must be known. In this method, no calculations are needed by the mobile unit to obtain location information [14].

Table 1. Properties of the location sensing systems.

Technology	Technique	Physical location	Symbol location	Absolute location	Relative location
Active badges	Diffuse Infrared Cellular Proximity	No	Yes	Yes	No
Active bats	Ultrasound, of lateration	Yes	No	Yes	No
E 911	Triangulation	Yes	No	Yes	No
Cricket	Proximity, Lateration	No	Yes	Yes	Yes
Cell-ID	Signal strength measurements, time of arrival, time difference and angle of arrival	Yes	No	Yes	No
RFID	To mark the items with tags.	No	Yes	No	Yes
GPS	Radio TOF lateration	Yes	No	Yes	No

2.2 Outdoor Only Sensing Systems

GPS (Global Positioning System): it uses multiple synchronized sources with known locations (satellites) and a single receiver with unknown location to determine a position. Each satellite of a constellation transmits a unique code, a copy of which is created in real time in the user-set receiver by the internal electronics [15]. GPS provides physical position and absolute locations, inexpensive GPS receivers can even determine and locate positions to within 10 meters for approximately 95% of measurements. A minimum of four satellites must be visible for most applications.

DGPS (Differential GPS): the precision of the GPS can be enhanced by means of DGPS which uses a network of fixed ground based reference stations to broadcast the difference between the positions indicated by the satellite systems and the known fixed positions.

Table 1 shows the main features of some location sensing systems where the location system properties are defined.

In Table 2, a comparison of the location sensing systems in terms of accuracy and precision, scale, cost and limitations is provided.

3 Context Managements Techniques

The location information described above can be considered as a part of the user context. The term *context* is referred to a set of parameters that can be used to describe the environment in which the user is embedded, the devices, and the access networks with which the user interacts. To better characterize a user context, we add to the *geographical* user location, which has been previously defined, the *environmental* user location, which includes more information on the specific environment a user is currently located (e.g., stadium, hospital, ambulance, city centre, etc.). Sensor devices can be used to get environmental contextual information. The collection and forwarding of context parameters to the access network is responsibility of the *Context Detector* (CD) entity that will be described in Section V.

Table 2. Comparison of the location sensing systems.

Technology	Accuracy and precision	Scale	Cost	Limitations
Active badges	Room Size, Active architecture	1 Base Station per room, badge per Base Station per 10 sec	Inexpensive	Sunlight, fluorescent interference with infrared, and unsuccessful for many applications that require fine-grained 3D location and orientation information.
Active bats	9cm (95 %), Active architecture	1 Base Station per 100 m, 25 Computation per room per sec	Cheap tags and sensors, inexpensive and low-power.	Required Ceiling grids
E 911	150m–300m, (95 %) of calls	Density of cellular, infrastructure	Cell Infrastructure, Expensive	Only whee cell, Coverage exists
Radar	3 to 4.5 m, (50 %)	3 BSs per floor	802.11 network, installation, Expensive	Wireless NICs required
Cricket	4*4 ft region, (100 %)	1 beacon per 16, square-foot regions inside a room	Expensive	No central, management, receiver
Cell-ID	Depend by cell's topology	Depend by cellular's infrastructure	Expensive for installation the cell's network	Cover based
RFID	Depend by the power usage and used frequency	The read range of RFID is larger than that of a bar code reader	Convergence of Lower cost	The two categories, active or passive, depending on their source of electrical power, it does not require line-of-sight access to read the tag
GPS	10 to 30 meters (95–99 %)	24 satellites world-wide	Too expensive infrastructure	Stricted to outdoors only
DGPS	<1 m	24 satellites world-wide	Too expensive infrastructure	Stricted to outdoors only

The *user profile* (UP) can be used to provide some personal information and preferences about the user and its devices, such as gender, age, and type of user (e.g., business, traveler, private), type of terminal and its status (e.g., supported media, computational capability, battery level/energy resources, preferred screen color, font type). UP can include other information and preferences related to the service level agreement, such as the maximum cost that the user is willing to pay for that service, the residual credit, the target QoS level (hard, soft), the preferred level of security, etc.

During the exploitation of a service, the user is allowed to change his/her UP, which can be stored into one of the user's devices (e.g. in a smart card), or even partly stored in some network databases.

The term *situation* is referred to the interpretation of the physical, social or environmental contextual information that can be referred to the user and/or to the access network. This interpretation requires a set of rules defined (personalized) by the user. These rules can be defined by the user profile. Of course, *situation* is an evolving concept; therefore its description must be continuously updated. The distributed information that characterizes a specific situation (e.g,, network resources, class of users, devices, applications, etc.) must be collected and *dynamically, autonomously*, and *proactively* handled to create a logical representation of the current user's working environment.

4 Architecture Description

In NGWNs the user terminal(s) allows the connection of the user with at least one of the RANs available in a given area. The user terminal can be either an integrated device, or a set of different devices in a Wireless Personal Area Network (WPAN), or a completely reconfigurable terminal. Figure 1 shows the proposed reference scenario, where the user can get connected to the heterogeneous network with the support of three entities: the Location Enabler (LE), the Context Detector (CD) and the User Profile (UP).

The LE is a device with localization capabilities. It can be a GPS receiver or any other device that is able to get information about the position of the user (see section III) and it is able to communicate this information to other terminals by using the Bluetooth technology or any future WPAN standards. In our architecture the LE is represented as a separated device, however, it could be also integrated into one of the other user's devices. Seeing it as a separated device helps to logically

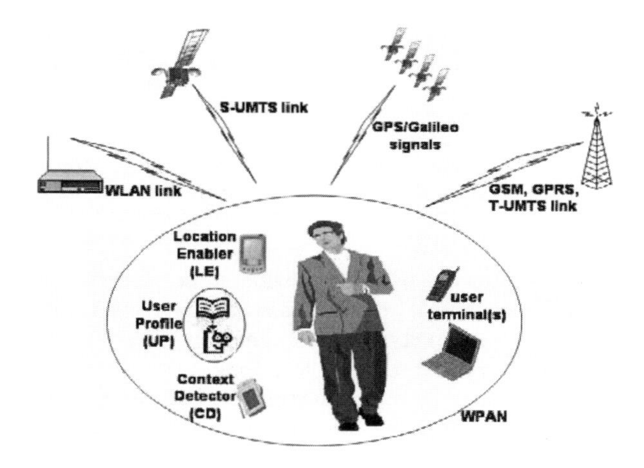

Fig. 1. Reference scenario.

separate its functionality from the communication functionality of other devices in the architecture and to focus on the central role of the location information in this framework.

The CD is an entity responsible of collecting contextual information from the environment surrounding the user. This could mean that the CD has to be able to collect the information from physical sensors (e.g. temperature, humidity or light sensor) or to detect information by itself.

The UP represents a set of personal information which can be partly stored in one of the user devices and is useful for the definition of the situation given a certain location and context. Some variables included in the user profile are almost static and can be stored in some network databases, some other variables are dynamic and can be continuously updated by the user through interaction with the UP module.

4.1 Communication Architecture

This Section defines the main components of the architecture that are able to collect, process and distribute the location/situation information to the user and/or network nodes, in order to use them through properly designed algorithms.

The proposed architecture is shown in Fig. 2 and it consists of different access networks, referred as Slave Networks, which interact through a common platform called Roaming Provider (RP).

Figure 3 shows the components of the RP, which consists of:

- *AAA database*, which stores the agreements of the user with each access network.
- *User profile database*, which stores profiles of the users.

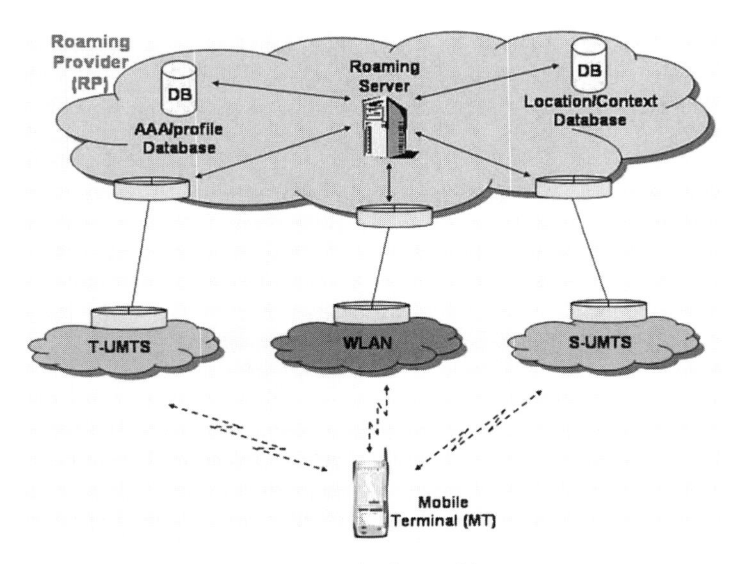

Fig. 2. Communication architecture.

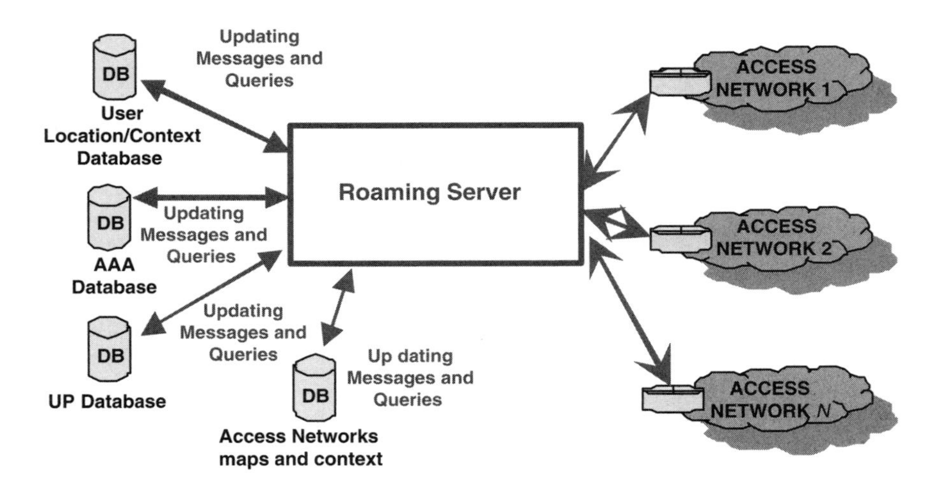

Fig. 3. Roaming provider architecture.

- *User location/context database*, which collects the relevant location/context information of the users in the area covered by the underlying access network.
- *Access network map and context database*, which stores the map and the current context state of the access networks under the control of the RP.

The Roaming Server (RS) processes the information stored in the four databases and sends the results of this processing back to the users/nodes, with two different purposes, which are:

- to achieve the key parameters in input to specifically designed location/situation-based algorithms of resource and mobility management.
- to start vertical handover procedure involving some of the underlying access networks.

The RP must process a number of information that is directly proportional with the number of users and networks available in a given service area. Therefore, it is better to distribute the RP functionality over several interconnected servers. Thanks to its features, a HAP could provide a privileged host for the RP functionality.

4.2 Middleware

We are convinced that in 4G networks, under heterogeneous RANs, several providers, and manifold terminal devices and applications, a multiagent-based middleware platform is a powerful solution to provide global area service management. In this section we describe the agent-based middleware we propose to be implemented in the communication architecture of a Roaming Provider. Intelligent agents have the task to support the user during access network selection, resource and handover management procedures, as well as in service discovery and QoS parameters adaptation procedures.

Our architecture includes five agent typologies: User Agent (UA); Accounting Agent (AA); Network-side User Agent (NUA); Radio Resource Agent (RRA); and Service Agent (SA). They are distributed in the overall architecture as illustrated in Fig. 4. The Databases shown in Fig. 4 represent the repositories illustrated in Fig. 3 for user profiles (*Profile Database – PDB*), authentication information (*AAA Database*), location and context information (*User Location/Context Database – LCDB*), and *access network maps and context database* (ANDB).

The Roaming Provider could either belong to a single operator owning several RANs (e.g., WLAN, UTRAN, satellite master control stations, etc.), or it could be owned by a "third-party" establishing agreements with both the user and a number of operators in order to offer ubiquitous service access. Whatever the choice, the middleware platform will be deployed in a NGWN composed of different RANs and a backbone equipped with a Roaming Server for each Roaming Provider.

The RS functionalities described in Section IV-B can be split in two parts:

- On the network side, there are the functionalities relevant to the *Network Resource Manager* (NRM)
- On the middleware side, there are the functionalities relevant to Roaming Decision Maker (RDM); Information Collector (IC); Location Tracer (LT), and Situation Tracer (ST), that represent some of most important functionalities of the NUA agent.

The proposed middleware architecture is *situation and location aware*, since it reflects the instantaneous changes in the observed scenario. As already mentioned, instantaneous positions are available at the user terminal through the LE device; they will

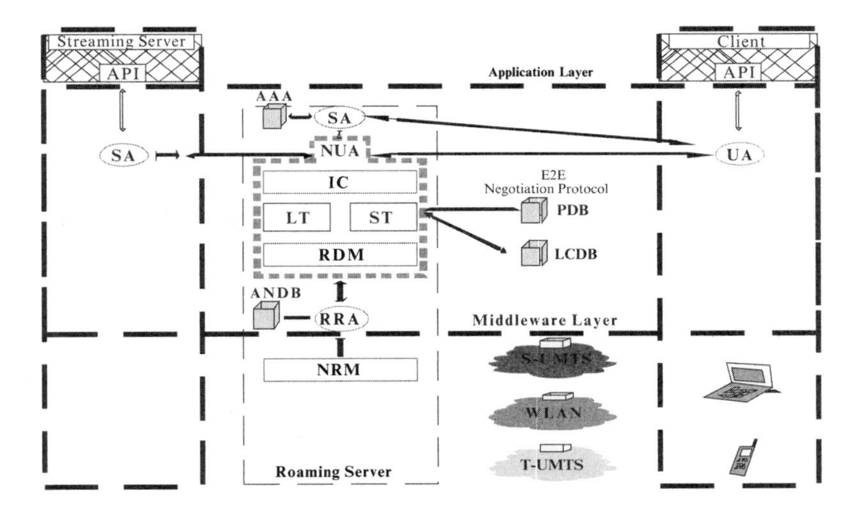

Fig. 4. Middleware architecture.

be periodically signaled to the middleware platform through interaction with the User Agent.

The UA is implemented in the user terminal; it follows the user when he/she is roaming and continuously monitors his/her local context. The UA interacts with:

- the SA, which is involved in a service discovery procedure within the RP's platform;
- the AA, which performs user identification and accounting (AAA functions), and triggers the instantiation of the NUA for the new user in the RP's platform
- the User, in order to manage his/her preferences and to forward them to NUA, which has the task to store them in the PDB. In addition, the UA notifies the user of the decisions taken on behalf of him/her by the NUA after negotiations with other agents in the provider's platform.
- the Application layer in order to individuate the requisites of the application, in terms of both network (QoS) and elaboration (hardware and software) resources, and to communicate them to the NUA. This latter will, then, request the allocation of the necessary radio resources and telecommunication services in the provider's platform.
- the Device, to obtain the information about available network interfaces and their features (e.g., their current status, the received signal strength, etc.)
- the NUA, which represents the user counterpart in the provider's network. Periodically, the UA updates its counterpart with new location information. The update frequency depends on the user speed, the precision and the rapidity the offered service requires this information.

The NUA, typically placed at the middleware layer in the RS, has the task to interact with the UA and with other network-side agents (RRAs and SAs). An NUA will be instantiated for each user who wants to access the provider's platform. The NUA migrates with the user from one roaming server to another each time the user roams from a domain managed by a provider to another domain belonging to a different provider.

The NUA implements the algorithms for selection of both the access network and the available services for the user. Furthermore, as already mentioned, the NUA includes different functional blocks: the IC module is responsible of the reception of location, context and profile information from users and access networks and it has the task to store, update and query this information from the databases. The location information of users and mobile nodes managed by the IC is then sent to the RDM through the LT module.

Moreover, it is the NUA that interprets the context information by using the UP to get the situation information about the user or network nodes by using the ST module.

The input parameters which help the NUA in taking decisions include both historic (almost static) information and more dynamic information related to the user, and also network-related information. They are the following:

- the *historic user profile*, stored in the PDB;
- the *running user profile*, notified by the UA, including current device and requested application characteristics;

- the *network profiles*, signaled by the RRA, which includes information on available access networks status and resources availability obtained by the NRM;
- the *service profiles*, received from the SAs, which include information of the currently available services through the various RANs of the provider's platform.

The SA is the agent with the task of performing service discovery. It delivers the NUA the list of the available services offered by the service providers, which have competence in that location and in that situation. The NUA selects the service for the user based on his/her profile, on the device he/she is currently using, and on the access network situation. According to the NUA choice, the SA has to take care of the service delivery at the right QoS level. This means that it has to take care of the determination and adaptation of the right service configuration both during setup and at runtime.

The RRA is implemented in the middleware layer of the RS and has the task to verify the resources availability at the network layer, across multiple access networks managed by the provider. To this aim, it periodically contacts the NRM devices to verify the resources availability in the various RANs.

5 Location-Aware Mobility Management

In this Section we show an example of location-aware handover mechanism. In particular, we consider the vertical handover between T-UMTS and WLAN: the user is making a call using VoIP with T-UMTS and is approaching a WLAN hot spot area. The handover procedure consists of two phases:

1) detection of a WLAN coverage and activation of the WLAN air interface (activation phase);
2) decision phase about the opportunity of making or not the handover (decision phase).

Both phases exploit the information on the location of the user.

5.1 Activation Phase

Let us consider a user moving at speed v along the x-axes of a reference plane centered in the centre of the hot spot area of the WLAN. Let us assume that the time needed to "wake up" the WLAN air interface is a couple of seconds. We can then fix a minimum threshold S_{Dmin}, which defines the minimum distance between the mobile terminal and the border of the WLAN spot area at which the Roaming provider should send the message to activate the WLAN air interface. This information can be made available at the NUA through interaction with the UA. This threshold will depend on:

- "wake up" time t_{ris}, which is the minimum time needed by the air interface before the terminal is able to detect the beacon signal of a WLAN. During this time the terminal will move of a distance:

$$D_{ris} = v \cdot t_{ris} \tag{1}$$

- time interval between two consecutive updates of the information on the user location (position, speed and orientation of the movement), denoted as t_{update}. For instance, in case of GPS, t_{update} = 1s. In this time interval the user moves ahead of:

$$D_{update} = v \cdot t_{update} \tag{2}$$

If we do not consider D_{update}, it could happen that the user has already moved into the WLAN area without activating the air interface.

Finally S_{Dmin} depends on the accuracy on the user location information. If we denote with ε the maximum error on the user position, S_{Dmin} can be written:

$$S_{Dmin} = D_{ris} + D_{update} + \varepsilon \tag{3}$$

In Fig. 5 it is shown for different levels of accuracy the value of S_{Dmin} vs. the user's speed. We can observe that by using a GPS with a typical accuracy of 30–10m, S_{Dmin} is already very high with low-average speed (20m/s) with respect to the case of a DGPS, which has a range of 3–0.5m.

Table 3 shows the value of S_{Dmin} for different environments in case of GPS or DGPS localization system. From Table 3 we can conclude that for this application it is important to use some improved GPS, such as the DGPS. On the other hand, Table 3 also shows that for high speeds (i.e., 120km/h) the high value of S_{Dmin} discourages the use of this "wake up" procedure.

In a multimode device enabled to the handover process, the two air interfaces are always waken up, and, hence, a classical handover procedure does not include an activation phase with its activation rule. The advantage of this activation phase is that only one air interface at a time is active, thus decreasing the energy consumption of the multimode device.

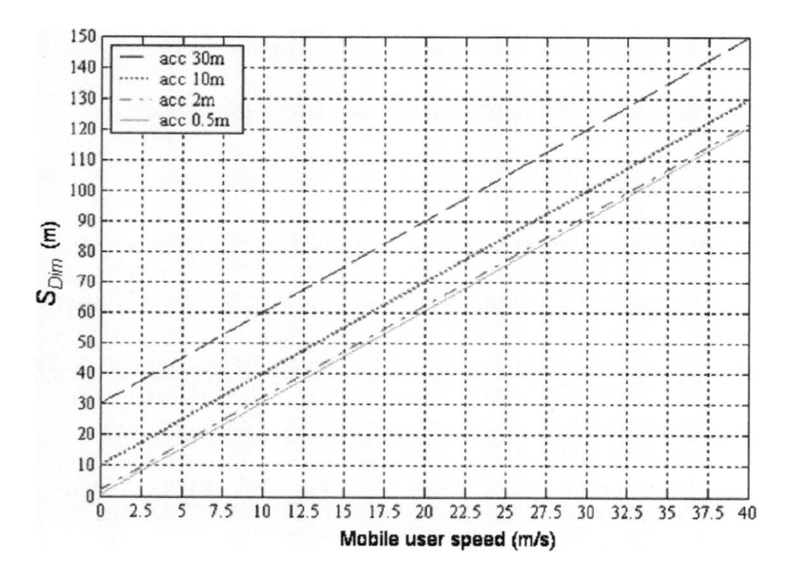

Fig. 5. Minimum threshold vs mobile user speed.

Table 3. Minimum threshold in case of GPS and DGPS for different environments.

	Speed		S_{Dmin}, [m]			
Mobile user speed	km/h	m/s	acc. 30m	acc. 10m	acc. 2m	acc. 0,5m
Pedestrian/indoor	3,6	1	33	13	5	3,5
Urban area	50	13,88	71,64 (72)	51,64 (52)	43,64 (44)	42,14 (42,5)
Urban street at high speed	80	22,22	96,66 (97)	76,66 (77)	68,66 (69)	67,16 (67,5)
Highways	120	33,33	129,99 (130)	109,99 (110)	101,99 (102)	100,49 (100,5)
			GPS		DGPS	

5.2 Decision Phase

The decision phase starts when the WLAN air interface is active and when the mobile terminal probes a WLAN signal with enough strength. Then it sends a message to the Roaming Server (specifically the NUA module), which is responsible for deciding about starting or not the handover procedure. The decision will be made according to a prediction of the average time the user will remain in the hot spot WLAN area. In fact, the handover to WLAN will be effective only if the user will be able to experience the higher transfer rate in the WLAN. However, taking into account the time needed to conclude the handover procedure, which is a period of time with no transmission of data, if the user will stay in the WLAN area for too short, the amount of transferred data could be even less than in the case of not handover at all. Therefore, the decision criteria will be based on the evaluation of the The minimum Required Visit Duration (RVD), which is the minimum time that a user must remain within the same WLAN coverage area to ensure the successful completion of the handoff procedure and the transfer of a sufficient (configured) amount of data over the WLAN network. In other words, it is the amount of time that the user must remain within the same WLAN to allow the application to benefit from the higher data rates and compensate for the handoff-related delays. The RVD can be evaluated by:

$$RVD_{\min} = L_a + (L_c + L_{MIP} + L_s)\,\tau \tag{4}$$

where:

- L_a is the latency due to one single iteration of the decision algorithm, from the arrival time of the request to the time notifying that the decision has been made;
- L_c is the latency associated to procedure for configuring the address;
- L_{MIP} is the latency related to the Mobile IP operations and the registration phase;
- L_S is the latency associated to the stabilization phase of the connection;
- τ is the period of time needed to receive with the WLAN the same amount of data that would be received with the UMTS connection in the period of time equivalent to the handover delays. τ can be evaluated as follows:

$$\tau = (L_c + L_{MIP} + L_s) \times (R_c / R_t) \tag{5}$$

where:

- R_c is the maximum data rate serving the user before the handover;
- R_t is the maximum data rate in the network towards the user is moving.

Once the RVD has been evaluated by the NUA, the algorithm must estimate the Predicted Path Length (PPL), which is the distance covered by the user moving at speed ν in the time RVD:

$$PPL = RVD \times \nu \qquad (6)$$

If the user starts to move from a point of coordinates (x_1,x_2), and covers a distance of PPL in the direction θ, he/she will move to the point (x_2,y_2) which are:

$$\begin{aligned} x_2 &= x_1 + \mathrm{b} \\ y_2 &= y_1 + \mathrm{a} \end{aligned} \qquad (7)$$

where:

$$\begin{aligned} a &= PPL \times \sin(\theta) \\ b &= PPL \times \cos(\theta) \end{aligned} \qquad (8)$$

Therefore, the NUA queries the *access network maps and context* database containing the map of the hot spot area, and checks if the new coordinates fall into the WLAN area. If so, then the NUA will activate the handover procedure, otherwise the handover will not take place.

One of the main limitation of this algorithm is related to the value of L_a, which must be kept as lowest as possible. A high value of L_a results in a high value of RVD, which could mean that the handover towards the WLAN has a low probability to happen, even if it is beneficial to the user. The value of L_a is mainly related to the localization information acquisition time. A high value of L_a could be due to the obstruction of the satellite in a GPS localization system. However, if the value of L_a is updated each time a new position is detected, any wrong decision in the algorithm will not persist for a long time.

6 Radio Resource Management

This Section provides some examples of the use of the location/situation information about the user and/or network nodes in RRM mechanisms.

The knowledge of the location of users and access network nodes can be important for the optimization of the RRM when the access network nodes are mobile and the distance between the user and the access node varies in a wide range, which is the case of satellite and HAP networks.

In such networks, the knowledge of the users and nodes location can be used to aid handover, scheduling, power control, call admission control, etc. For instance, scheduling mechanisms could give priority to users in specific geographical positions with respect to other users for different objectives (reducing the experienced delay, reducing the congestions in some areas etc.). More in general, location/situation awareness can be used to guide a mobile user from a bandwidth-impoverished to a bandwidth-rich environment. In a heterogeneous scenario, where more than one access network is available, the user can choose the best solution according to its position, its preferences (user profile) and the specific context. Assuming that each Roaming Provider controls different RANs, the task of efficiently control the assignment of resources in the various RANs can be performed in a distributed and cooperative way by the RRA agent, the NRM and RDM modules in the Roaming Server. Areas managed by an RRA can coincide with one access network, or be a part of a network, or even they can include more networks. Thereby, the resource management functionality in the Roaming Provider's platform can be performed by a unique RRA agent or more agents (typically one for each access network).

7 Conclusions

This paper proposed a middleware-based architecture to perform efficient location/situation aware mobility and resource management mechanisms in future heterogeneous networks. In particular, one location/aware vertical handover mechanism is presented where the location information plays a key role in both the phases of the handover algorithm: "wake up" of the air interface and the decision phase. The example shows that the accuracy and the update time of the information are of key importance. The accuracy and the update time of current localization systems already made possible the use of the proposed algorithms in many cases (i.e., user speeds). However, to make cost-effective the implementation of the proposed location/situation aware mechanisms, they should be further improved.

References

[1] C.A. Patterson, R.R. Muntz, C.M. Pancake, "Challenges in Location Aware Computing", IEEE Pervasive Computing, pp. 80–89, April-June 2003.
[2] S. Sharma, A.R. Nix, S. Olafsson, "Situation Aware Wireless Networks", IEEE Communications Magazine, pp. 44–50, July 2003.
[3] D. Jeong, Y.G. Kim, H.P. In, SA-RFID, "Situation-Aware RFID Architecture Analysis in Ubiquitous Computing", Proceedings of the 11th Asia-Pacific Software Engineering Conference (ASPEC 04).
[4] Jian Ye, Jiongkuan Hou, S. Papavassiliou, "A Comprehensive Resource Management Framework for Next Generation Wireless Networks", IEEE Transactions on Mobile Computing, vol. 1, no. 4, pp. 249–264, Oct.-Dec. 2002.
[5] Siamak Naghian, "Location-Sensitive Radio Resource Management in Future Mobile Systems", WWRF Results of NG4 May 10–11, 2001 of NG4 May 10–11, 2001.
[6] J. Hightower, G. Borriello; "Location systems for ubiquitous computing.", IEEE Computer, Volume 34, Issue 8, pp. 57–66, Aug. 2001.
[7] A. Smith, H. Balakrishnan, M. Goraczko, and N. Priyantha; "Tracking Moving Devices with the Cricket Location System.", MIT Computer Science and Artificial Intelligence Laboratory, http://nms.csail.mit.edu/cricket/
[8] R. Want, A. Hopper, V. Falcão and J. Gibbons; "The Active Badge Location System.", Olivetti Research Ltd. (ORL), Cambridge, England, 1992.

[9] Cambridge University Computer Laboratori, "The Bat ultrasonic location system", http://www.cl.cam.ac.uk/Research/DTG/attarchive/bat/
[10] N.B. Priyantha, A. Chakraborty, and H. Balakrishnan, "The Cricket Location-Support System", Proceedings of the 6th Annual {ACM} International Conference on Mobile Computing and Networking (ACM MOBICOM), Boston, MA, August 2000.
[11] Ron Weinstein, "RFID: A Technical overview and its applications to enterprise.", IEEE IT Professional, Volume 7, Issue 3, pp. 27–33, May-June 2005.
[12] D. Hahnel, W. Burgard, D. Fox, K. Fishkin, M. Philipose, "Mapping and localization with RFID technology", Proceedings of the 2004 IEEE International Conference on Robotics & Automation, Volume 1, pp. 1015–1020 Vol.1, 2004.
[13] S.S. Manapure, H. Darabi, V. Patel, P. Banerjee, "A Comparative Study of Radio Frequency-Based Indoor Location Sensing Systems.", IEEE International Conference on Networking, Sensing and Control 2004, Volume 2, pp. 1265–1270 Vol.2, 2004.
[14] Heikki Laitinen (editor), Suvi Ahonen, Sofoklis Kyriazakos, Jaakko Lähteenmäki, Raffaele Menolascino, Seppo Parkkila, "Cellular location technology.", IST CELLO Project Deliverable, Document Id: CELLO-WP2-VTT-D03-007-Int, Nov. 2001.
[15] N. Bulusu, J. Heidemann, D. Estrin, "GPS-less low-cost outdoor localization for Very Small Devices.", IEEE Personal Communications, Volume 7, Issue 5, pp. 28–34, Oct. 2000.

Convergence of Networks: An Aerospace-Friendly Strategic Vision

Ramjee Prasad[1], Marina Ruggieri[2]

[1] Center for Teleinfrastruktur (CTIF), Aalborg University,
Niels Jernes Vej 12, 9220 Aalborg Øst, DENMARK
e-mail: prasad@kom.aau.dk
[2] Dpt. of Electronics Engineering, University of Roma Tor Vergata,
Via del Politecnico, 1 - 00133 Roma, ITALY
e-mail: ruggieri@uniroma2.it

Abstract. The paper focuses on the revolutionary changes that could characterise the future of networks. Those changes involve many aspects in the conceivement and exploitation of networks: architecture, services, technologies and modeling. The convergence of wired and wireless technologies along with the integration of system components and the convergence of services (e.g. communications and navigation) are only some of the elements that shape the perpsected mosaic. Authors delineate this vision, highlighting the presence of the space and stratospheric components and the related services as building block of the future.

Keywords: Convergence, Integration, Communications, Satellite Navigation, Layerless

1 Introduction

The future of systems is *globalisation* and the future of networks is *convergence*. The full deployment of this challenging evolution brings through some cross-linked revolutionary changes in the conceivement and design of systems and services:

- Convergence (*C-approach*) of wired and wireless;
- Integration (*I-concept*) of components (terrestrial, stratospheric and space);
- Convergence of services (*S-convergence*) (e.g. Navigation and Communications, *NavCom*; Earth Observation and Communications, *ObsCom*, Earth Observation, Navigation and Communications, *NavObs Com*);
- Layereless design.

Those changes are taking place in a variegated technological scenario, where consolidated and emerging technologies need a continuous harmonization for the effective deployment of complex system architectures.

The diversity of technologies is due to many aspects, such as technical problems in different domains, ranging from the physical to the application layer, various market players and their interests, etc. A set of examples in the differences that can be found, for instance, in the wireless standards today are: coverage, data-rates, services, medium access control protocols, Quality of Service methods, network architecture, mobility solutions, security methods (authentication, key-management, encryption schemes).

At the terminal level, which is the first and closest experience of any user with technology, convergence can be seen as an invisible and seamless service provisioning, in the sense that apparently to the user, devices can easily interact with each other and offer services to the user according to his/her needs under given circumstances. An example, that is provided in Fig. 1, is the so called *flying screen* [1]. This concept involve the user need for a display service, which can be either the TV, laptop screen or the mobile phone.

The vision is that any content can be shown at any time and anywhere. However, according to present technology, it is not a simple matter to show pictures taken by the mobile phone camera on the TV. With the flying screen, pictures can be, instead, easily shown on either one of the available displays, without constraining the user to perform boring - and not always successful – technical setups and software installations for interacting with the service-providing device.

In the above frame, the present paper provides authors' vision on the four listed cross-linked revolutionary changes in system and service deployment – *C-approach, I-concept, S-convergence* and *layereless design* – moving from the activity of growing intensity that is being developed around each of the above topics.

The paper is organised as follows: in Section 2 the convergence of networks and technology, particularly in terms of wired-wireless convergence is faced; in Section 3 the integration of components for the deployment of integrated networks is dealt with: in Section 4 the focus in on service convergence; Section 5 is devoted to the visionary concept of layerless networks; finally in Section 6 conclusions and further perspectives are drawn.

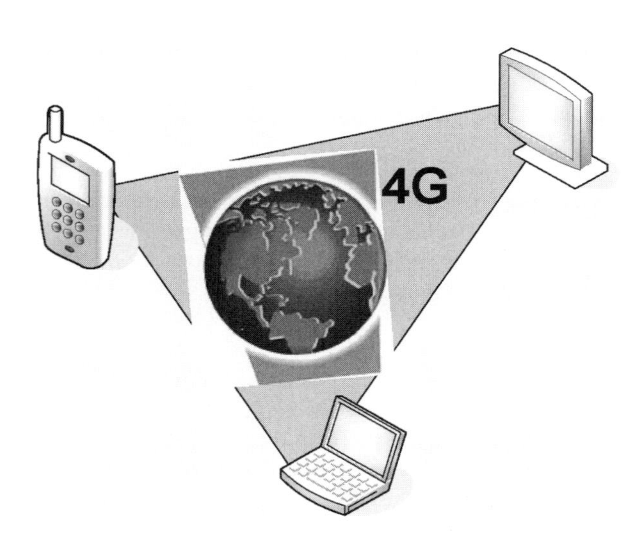

Fig. 1. The flying screen concept [1].

2 Convergence of Wired and Wireless

The convergence of wireless and wired technology (*C-approach*) – along with the *I-concept*, that will be dealt with in the following Section – brings to the conceivement of systems and applications without any polarisation in terms of technology and medium. In the development of a communications system, this approach would bring to the effective exploitation, in particular, of wireless or wired connections, in order to provide the optimal benefit to the user in terms of performance, variety of services, terminal technology and associated costs.

The *C-approach* has been the core, in 2004 and 2005, of closed-door Strategic Workshops on "Wire-(d)/-(less) Convergence towards 4G" and "Future Convergence of Wired and Wireless Network", respectively, that gathered managers and experts from European, USA and Asian manufacturers and operators. The aim of the events was to highlights the issues deriving from the wired-wireless convergence and to shape a medium and long term vision for the communications world [2–6].

The *C-approach* frames in the eMobility European Technology Platform's vision: in 2015 the wireless and wired communications networks will bring to reality "individual's quality of life improvement by providing an environment for instant connectivity to relevant multi-sensory information and content" [7].

This vision calls for user-centricity coupled with a secure communications environment, where users will experience:

- convenience
- usability
- trust
- privacy

Users will be able to meet all their communications needs, as well as get timely access to their personal data anywhere, anytime, and by the means of any device and perceived with multiple senses. The vision necessitates the co-design of security technologies and the communications infrastructure.

The security features need to be designed from the end-to-end point of view, considering also the technologies to enable authentication of the biological user of the system, whereas currently the user equipment are authenticated and it is mostly assumed that the biological user is the authorised user of the equipment.

An effective implementation of the *C-approach* can be obatianed once the major open points in the wireless and wired components are solved. In particular, there are some technological challenges to be overcome in the wireless networks. Positioning technologies will be improved with the advent of the European GALILEO satellite navigation system [8]. Other positioning technologies could be coupled with satellite positioning to ensure continuous service regardless of location – also indoors. Integration of communications and positioning technologies is a future main application enabler. It is also important to take into account the possibility of extremely fast moving terminals taking advantage of the network services.

When designing the networks, user-centricity and application usability need to be considered from the beginning. Involving real users and case study analysis in the

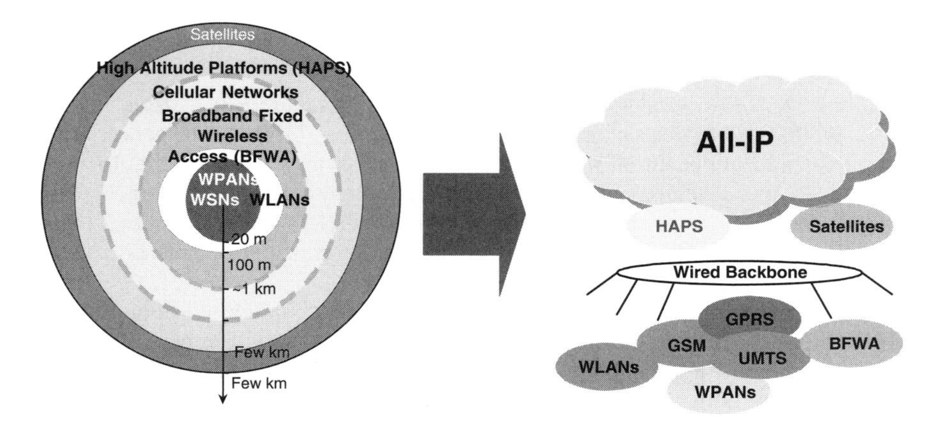

Fig. 2. Convergence in network technology [9].

development will enable to derive user requirements. The latter need to be converted into technical system requirements that, in turn, specify the system. In order to achieve a global roaming, large scale coordination and collaboration has to take place to align the future communications technologies views around the world.

The *full* network *convergence* will be achieved once the wireless transmission speeds and network latencies can be improved close to those in the wired networks. Then the wealth of the applications currently available in the fixed networks will be easily migrated to take advantage of the added convenience offered by the wireless networks.

In the deployment of the full convergence, a trend is being delineating at network level. Today a wide range of network technologies already exists on the market, some are IP based and others are not (e.g. Bluetooth, Zigbee). Some network are local, such as Wireles Local Area Networks (WLAN), Wireless Personal Area Networks (WPAN), while others are global, e.g. the Internet, satellite networks. From a user's point of view, a common technology is beneficial, since this basically enables communications between various devices, over short as well as long distances through whatever communication means that is available. The trend at network level is that an IP based solution will become the major technology for future network, as depicted in Fig. 2 [9].

Furthermore, a common platform as IP makes software development much easier, both for new network and application components, but also for services. This is a key issue that eventually will benefit the user, as a standardised interface will make interaction between applications much easier, and ultimately will make the convergence on terminal level happen.

3 Integration of Components

As the *C-approach*, the integration concept (*I-concept*) brings to the conceivement of systems and applications with the effective exploitation of satellite, terrestrial and stratospheric technologies to provide optimal performance (cost-to-benefit), without feeling constrained by any "lack-of-trust" in specific media or specific technologies.

The *I-approach* is the focus of various initiatives at local or international level. In particular, research activities have been undertaken in Italy since 1998, co-funded by the Italian Space Agency (ASI) and the Italian Ministry of University and Research (MIUR), addressing the integration of terrestrial, satellite and eventually stratospheric components to build advanced networks for mobile and fixed communications. Among the deployed programs it is worth mentioning: *DAVID* (*DAta and Video Interactive Distribution*) developed in 1998–2003 and funded by ASI, *CABIS* (*CDMA for Broadband mobile terrestrial-satellite Integrated Systems*), developed in 2000–2002, *SHINES* (*Satellite and HAP Integrated NEtworks and Services*), developed in 2002–2004, co-funded by MIUR and *WAVE* (*W-band Analysis and VErification*), 2004-on-going, funded by ASI [10–15].

Moving from the results achieved by the abovementioned activities, a new program, named *ICONA* (*I*ntegrated *CO*mmunications and *NA*vigation) and co-funded by the Italian MIUR, has been recently started in early 2006 [10].

A more detailed analysis if the *I-concept* brings to distinguish between *interoperability* and *pure integration*. In fact, interoperability implies proper interfaces at system and user terminal level, that integrate systems mostly conceived independently each other. The pure integration, instead, envisages that the components share common parts, according to an integrated-oriented design followed when initially conceiving the system. This approach obviously involve also the user terminal.

It is interesting to frame the *dual use* in the *I-concept*. Dual use brings to systems that from the beginning are conceived in an integrated mode, being able to meet the requirements of two very different class of users: military and civilian. If the sytem is composed of terrestrial, satellite and/or aerial components – according to the network centric vision – a multiple integration level is reached.

The I-concept translates into networks composed of existing or dedicated parts (*NoN*, Networks of Networks). An *I*ntegrated *Net*work (*I-Net*) is an *NoN* where all elements cooperate effectively in terms of:

- performance
- security
- seamless behavior to the user.

In an *I-Net*, terrestrial facilities are envisaged in terms of network nodes and connections; satellite and aerial components can be usefully considered to complement coverage, provide a back-up to the terrestrial section, provide additional services, strengthen the services mostly provided by the terrestrial facilities. On the other end, the satellite or the aerial component can be the main one, supported by a terrestrial section.

The aerospace component is structured in various layers: the HAP (High Altitude Platform) layer, where manned or unmanned stratospheric vehicles are located at about 20 km height, and the layers where satellites can be mainly located (low, medium, geostationary and highly elliptical orbits) [3,6,10].

The behaviour of an *I-Net* is related to that of its components and to the effectiveness of the integration. In terms of interoperability, the components can be meant as the various systems which work jointly. In the true integration, the components can be, for instance, identified in terms of height from the Earth (i.e.

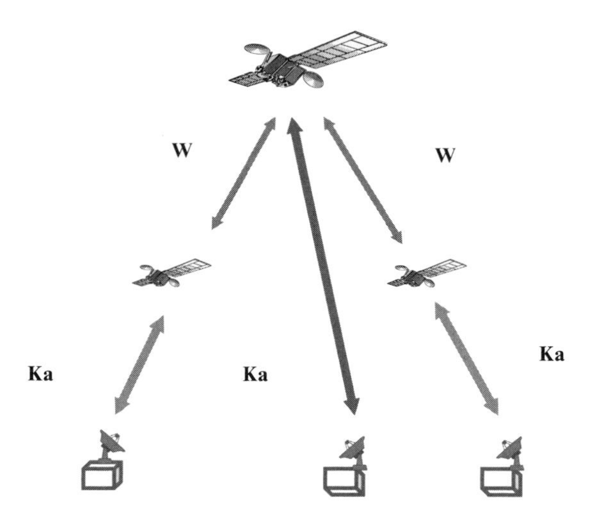

Fig. 3. An example of multi-orbit satellite *I-Net*.

terrestrial, aerial, space), service type (e.g. navigation, communications, Earth observation, data handling, etc), user type (e.g. military, civilian), technology (e.g. optical, radio, digital), access (e.g. wired, wireless), etc.

In both cases, the *I-Net* performance can be expressed in terms of *Quality of Integration* (*QoI*) as:

$$QoI = \left\{ \sum_j w_j P_j \right\} \eta_I \quad (1)$$

where P_j is the performance of the j-th component that is shared in the integrated system; w_j is the weight of the j-th component performance; the sum is extended to all *I-Net* components; η_I is the *integration efficiency*, i.e. the capability of exploiting effectively the various components and their performance of interest in the integrated system.

Figures 3 and 4 provide examples of *I-Nets* that integrate satellites in different orbits and the stratospheric component with the terrestrial part. The user segment in Fig. 4 is both aerial and terrestrial. The Figures point out also that different frequency ranges can be exploited for the various links and, in particular, the use of the innovative W-band (75–110 GHz) is advised for most of them [13–15].

4 Convergence of Services

The convergence of services (*S-convergence*) offered by different systems or by different parts of the same system is becaming a key aim for the exploitation of system features, the improvement of market potential and the protection of nations at civil, military or dual level.

In particular the development of second generation of Global Navigation Satellite Systems (GNSS) is stimulating the effective convergence between navigation

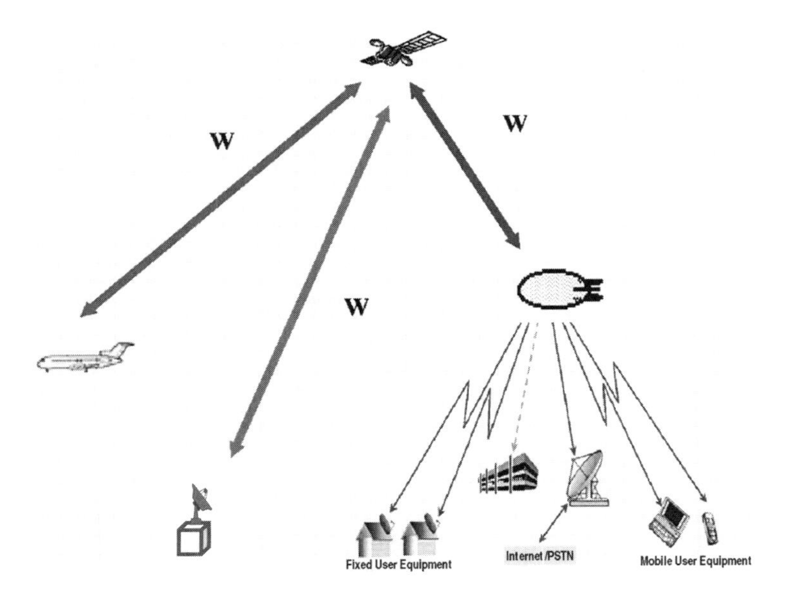

Fig. 4. An example of *I-Net* with satellite and stratospheric components.

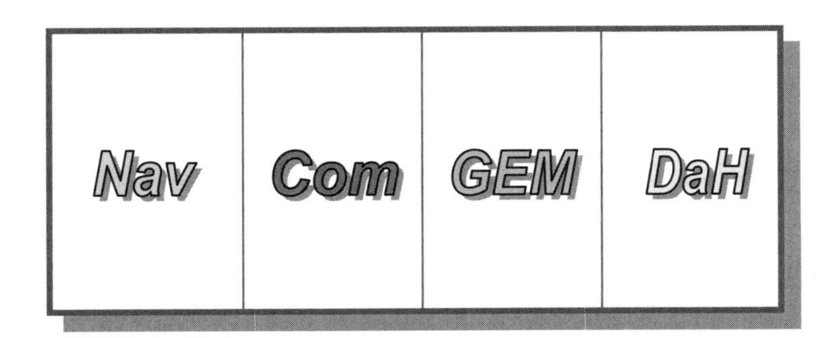

Fig. 5. An *S-convergence* package.

and communications services, bringing to a new class of *NavCom* (Navigation & Communication) users and services [6, 8, 10, 16–18].

The convergence is also being investigated and experimented among Earth observation, navigation and communications services for the effective deployment of the "Global System" concept. The Nav, Com and Global Earth Monitoring (GEM) component together with the supporting Data Handling (DaH) and user interface section could converge in global multiservice systems suitable to an impressive number of civilian, military and dual applications (Fig. 5).

The user is the core in the universe of service convergence: "personalisation", "personalized services" are the key-words for the user centric vision that is characterising the passage towards future systems.

This revolution has been started by the mobile communications world, in its transition from the present third generation systems (3G) to the next generation (4G). In authors' vision, 4G will be obtained by adding to B3G (i.e. beyond third generation), which is defined as the integration of existing systems to interwork with each other and with the new interface, the *personalisation*, which is the key issue in personal networks [19].

Even in today's market, the end user plays an important role, but in 4G the user will be the centre point and not the technology as it has been previously. In this sense, user centricity means applications and services will be developed with the end user as a person and not some anonymous entity that will have to use whatever the technology is capable of offering. This person centric approach means that applications and services will need to adapt to who the user is, to his/her needs and to current situations. From a users standpoint, some high level requirements can be set, for which the technology simply have to adapt to use a service or application anywhere at anytime; to do it in a cheap and efficient way; to not need a heavy technical parameter setup; to account for user's current situation and interests.

In fact, meeting all these requirements is not only a matter of ensuring a high coverage or high data rates but of taking into account *user profiles, user context* and *adaptation* towards service and applications that the user will be using.

Taking into account these matters enables the technology to adapt its behaviour by changing system parameters, bringing to a context aware service discovery The envisioned personalisation will have a potential impact on our society and the way we communicate, as it will assists our everyday of interacting with our work, family, friends and so on. It is also this concept that acts as a trigger to develop a new network paradigm such as Personal Network.

5 The Layerless Approach

Another aspect of the future mosaic concerns the network structure. The trend today in wireless communication is to utilise information from various layers in the protocol stack to adapt and optimise the behavior of the protocol to the given circumstances. This is typically referred to as cross layer optimisation and is prune to continuously research. In fact, many information found in the lower layer, such as the physical and medium access control layers may be beneficial also for the networking and application layer and viceversa.

Looking at the way the information is used, i.e. in particular the vertical movement of information, blurs the picture of how the typical networking model (OSI) represents today's network.

The OSI model refers to clean and well defined interfaces between the different layers, where none of the layers needs to care of the communication either below nor above the layer itself. This is not the case when discussing cross layer optimisation, since the natural phenomena – such as fading, interference, device mobility – that are influencing the bit and frame error rates, ultimately put limits on the user's experience of bandwidth and delay.

Different techniques can be used to remedy this, but in most cases usage of information above and below one specific layer is useful to adapt to circumstances.

Getting all information necessary up and down in the various layers of a protocol stack necessarily introduces a more complex interface structure which, in turn, may even delete the benefits of structuring the control software in layers.

Instead, one could define a *layerless communication model*, based on a well defined communication platform, i.e. IP. In such a model, all information are available to all control mechanisms that are potentially involved in the optimisation process when doing wireless communication. The main benefit from taking such a drastic direction in designing software, is simply to keep the interfaces very simple.

Needless to say that security, as well as the privacy of the user, have to be maintained as key objects in this vision. Security solutions should be all tailored for securing the information exchanged between user devices and access points. Under all circumstances, security will need to be an integrated part of any network solution in order to meet the user's requirement of end to end security.

6 Conclusions

The paper has depicted a challenging picture of the future, where the user is the center and all components (terrestrial, satellite, stratospheric), technologies (wired, wireless) and services (communications, navigation, Earth observation, etc) will participate to the deployment of global system where "convergence" and "integration" are the implementation guidelines.

Some of the major expected achievements within the coming years is the convergence, layerless communication, a sensing and reacting communication environment as well as security and privacy.

Convergence toward an all-IP based world wide platform is one of the key issues, but to reach that goal a long range of challenges must be solved whereas standardisation for interoperability and ensuring sufficient user capacity on a global scale are examples on some of the most important ones.

Cross layer optimisation is a key issue in wireless communication, but the introduction of this concept requires a lot of information exchange in the traditionally layered software architecture. Hence future technology may introduce a concept where the layered structure is not visible as in today's technology.

Building intelligent software, which can sense and react on the user's environment and contextual situation, is also a key point wherein many technical challenges are still to be solved in order to achieve an operational system.

References

[1] Y. K. Kim and R. Prasad, "4G Roadmap and Emerging Communication Technologies", Artech House, 2006

[2] R.Prasad, M. Ruggieri, Editorial – "Wire-(d)/-(less) Convergence towards 4G", *Springer Wireless Personal Communications*, Vol.33 Nos.3–4, June 2005, pp.213–216.

[3] E. Cianca, M. De Sanctis, M. Ruggieri, "Convergence towards 4G: a Novel View of Integration", *Springer Wireless Personal Communications*, Special Issue on 'Wire-(d)/-(less) Convergence towards 4G', Vol.33 Nos.3–4, June 2005, pp.327–336.

[4] Ole Brun Madsen, M. Ruggieri, Editorial – "Future Convergence of Wired and Wireless Networks", *Springer Wireless Personal Communications*, 2006.
[5] R.Prasad, R.Lovenstein Olsen, J.Saarnio, "Strategic Vision on Convergence of Wired and Wireless Networks", Special Issue on "Future Convergence of Wired and Wireless Networks" (Selected Topics from the Strategic Workshop 2005), 2006
[6] M. Ruggieri, "Next Generation of Wired and Wireless Networks: the NavCom Integration", *Springer Wireless Personal Communications*, Special Issue on "Future Convergence of Wired and Wireless Networks" (Selected Topics from the Strategic Workshop 2005), 2006.
[7] R.Prasad, R. Lovenstein Olsen, J.Sarnio, "Strategic Vision on Convergence of Wired and Wireless Networks, *Springer Wireless Personal Communications*, Special Issue on "Future Convergence of Wired and Wireless Networks" (Selected Topics from the Strategic Workshop 2005), 2006.
[8] R.Prasad, M.Ruggieri, "Applied Satellite Navigation Using GPS, GALILEO, and Augmentation Systems", *Artech House*, Boston, 2005.
[9] IST-2000-26459, PRODEMIS Project: "Global Technology Road-map", August 2004, http://www.prodemis-ist.org
[10] M. Ruggieri, "Satellite Navigation and Communications: an Integrated Vision", *Springer Wireless Personal Communications*, Special Issue dedicated to the 60th birthday of Ramjee Prasad, Vol. 37 Nos.3–4, May 2006, pp.261–269.
[11] E. Cianca, M. Ruggieri, "SHINES: a Research Program for the Efficient Integration of Satellites and HAPs in Future Mobile/Multimedia Systems", *Proceed. WPMC*, Japan, October 2003, pp.478–482.
[12] G. Maral, S. Ohmori, M. Ruggieri, Editorial – "Broadband Mobile Terrestrial-Satellite Integrated Systems", *Kluwer Wireless Personal Communications*, Vol. 24 n. 2, pp. 97–98, 2003.
[13] S. De Fina, M. Ruggieri, A. V. Bosisio, "Exploitation of the W-Band for High-Capacity Satellite Communications", *IEEE Transactions on AES*, Vol. 39 n.1, pp. 82–93, January 2003.
[14] M. Antonini, A.De Luise, M. Ruggieri, D. Teotino, *Data Collection and Satellite Forwarding Evolution*, IEEE Aerospace&Electronic System Magazine, September 2005, Vol. 20 N.9, pp. 25–29.
[15] A. Jebril, C. Fragale, M. Lucente, M.Ruggieri, T.Rossi, *WAVE – A New Satellite Mission in W-band*, Proceed. IEEE Aerospace Conference, Big. Sky, USA, March 2005, paper n. 1007.
[16] R. Prasad, M. Ruggieri, "Technology Trends in Wireless Communications", *Artech House*, Boston, 2003.
[17] M.Antonini, M.Ruggieri, R.Prasad, U.Guida, G. F. Corini, *Vehicular Remote Tolling Services Using EGNOS*, IEEE Aerospace&Electronic System Magazine, Vol. 20, N.10, pp.3–8, October 2005.
[18] M.Ruggieri, E. Cianca, "HAP-Based Integrated Architectures for NavCom", *Workshop on Broadband Access via High-Altitude Platforms, BBEurope 2005*, Bordeaux, December 2005, www.bbeurope.org
[19] R. Prasad, L. Deneire, "From WPANs to Personal Networks – Technologies and Applications", ISBN-10:1-58053-826-6, Artech House, 2006

The Monitor Project: A GNSS Based Platform for Land Monitoring and Civil Engineering Applications

G. Graglia*, R. Muscinelli*, G. Manzoni**, M. Barbarella***, W. Roberts****

* Alcatel Alenia Space Italia SpA Navigation
and Integrated Comms Directorate
e-mail: gianluca.graglia@alcatelaleniaspazio.com, roberto.muscinelli@aleniaspazio.it
** GEONETLAB, University of Trieste. e-mail: manzoni@units.it
*** DISTART, University of Bologna. e-mail: maurizio.barbarella@mail.ing.unibo.it
**** Nottingham Scientific Ltd william. e-mail: Roberts@NSL.EU.COM

Abstract. For nearly 50 years, geodesists and surveyors developed deformation monitoring techniques based on traditional land surveying equipment, such as theodolites, Electronic Distance Measuring Devices, both absolute at sub-millimetric accuracy (Mekometer, derived by a prototype of Physical National Laboratory, Teddington, UK), and interferometric (long arm Michelson laser interferometers) and sub-millimetric levels. From the 1980's onwards, these techniques were first supplemented and later completely replaced by the more easy satellite positioning technology based on GPS. The early GPS deformation monitoring techniques were developed for the benefit of geophysicists, who were involved in the monitoring of crustal dynamics and in plate tectonics. Deformation monitoring of engineering structures was to follow suit. Although GPS suffers from a number of limitations which affect the coverage, accuracy and reliability of the satellite measurements, GNSS systems allow continuous nearly-real-time monitoring of the small movements of points.

The development of Galileo, its proposed interoperability with GPS, and the use of EGNOS, will contribute substantially to the quantity and quality of the satellite measurements thereby improving the quality of the deformation monitoring process. Moreover, the availability of signals from two different satellite systems is likely to reduce the price of GNSS receivers and sensors and thus enabling a wider spatial coverage with an increase in the number of monitoring points. In particular Galileo will increase the integrity of the GNSS measurements, which is very important for such applications affecting Safety of Life (SoL), and therefore involving legal and economic consequences.

In this context the European Galileo Project managed on behalf of the European Union (EU) and the European Space Agency (ESA), by Galileo Joint Undertaking (GJU), has opened new era in Satellite Navigation. One of the missions of the GJU, through its business development initiatives, is to develop future markets for Galileo and the European satellite based augmentation system, EGNOS, addressing a large number of user communities including Location Based Services (LBS), Road, Rail, Maritime, Aviation, and a Special Sector to which the Land and Civil Engineering community belongs.

The GJU 2nd Call was launched in June 2004, with bids submitted in October 2004. After the bids evaluation and negotiations, the MONITOR projects Consortium was successfully established and the Consortium was awarded the contract to address the Land and the Civil Engineering Community.

The partners within the Monitor Project Consortium include universities, companies and organisation based in Italy, Portugal, United Kingdom, Romania and Greece, whose combined capabilities cover all aspects of high precision monitoring of land and engineering of structural deformations, current satellites positioning techniques, engineering applications of GNSS and the potential benefits of the Galileo system.

This paper aims to describe the current status of MONITOR Project (officially kicked off in November 2005 and lasting till the mid summer of 2007) and the purposes to which, this 18 months long project, is focused on. A particular attention will be devoted to the description of the three Pilot Projects experimentation (par. 3, 4, 5), representative of the priority applications identified in a preliminary phase (par. 2), and to their relation with a technological and operative platform, represented by the Monitor Control Centre (par. 6), mandatory to provide a wide group of users, that span from the professional users (surveyors, researchers, national and local institutions, organizations, etc.) to citizenship, useful responses to their needs.

1 The Monitor Project: Overview

Performed by the adoption of high precision measurements techniques and by the identification of clear procedure that can be tested and then refined to drive future certifications and standardisation processes, the MONITOR Project is focused on demonstration of the use of satellite navigation in existing and new application areas, such as the Land Monitoring and Civil Engineering, the description and promotion of the added value to be brought by Galileo and EGNOS, and finally to pave the way for the acceptance of these added technologies and tools into our industry and to a wider user community.

The MONITOR Project is structured in three phases:

I. Critical analysis: that concerns the analysis of the land monitoring and civil engineering frame versus several enablers: (technology, market, standards, regulatory and legal aspects, etc..) with the final objective of the selection of the applications identified representative of certain priorities for the community addressed, elements that under a deep analysis provide the selection of the Pilot Projects to be implemented.
II. Pilot Project definition and execution: that concerns the definition of a Plan to be developed for the Pilot Projects (addressing the aspects of technology, market, regulation, standardisation and training), the implementation and the analysis of the results with the respect of the short term objective identified by the Plan.
III. Results analysis: that concerns the identification of all the elements of criticalities, by a deep analysis of data results, that should be improved to guarantee that each application will reach its own objectives.

The most significant phase of the project concerns the demonstration of the use of GNSS through the selection of some applications considered representative of the areas included into the land monitoring and civil engineering community, applications experimented within the Pilot Projects, and connected to each other through a technological and operative platform represented by the MONITOR Control Centre. The Pilot Projects will help to demonstrate the ideas and techniques proposed, in order to inform the specialist as well as the general public of the potential value of these techniques, that would lead to some very significant cost reductions of technology and implementation, in the near future, and pave the way to a full exploitation of Galileo within the Land and Civil Engineering community.

2 The Priority Application Selection

The MONITOR Consortium was committed to develop the definition of the Priority Application to be experimented during the second phase of the project. This process, started since the beginning of the proposal, derives from the analysis of the whole domain and on the basis of the Consortium Participant experience and awareness. The results was the identification of some applications retained priority versus the needs and the goals expressed by the Galileo Join Undertaking (GJU) in the [2].

MONITOR aims to be a technological and operational platform able to provide real answer to several territory monitoring and control criticalities, including the safeguard of anthropic infrastructures and activities adopting an higher level of innovation pushed by the satellite based technology and in particular the satellite navigation.

The assessment process has been performed in two major steps:

- The first considering different parameters such as number of users, potential revenues for GOC, private interests, public benefits, level of maturity (technological and commercial) and of innovation and potentiality for future market evolution, whose consideration has restricted the number of application relevant to the MONITOR Project from those described in a preliminary state of the art study.
- The second considering that MONITOR also aims to constitute and test a technological and operational platform able to provide real-time monitoring of different natural phenomena, civil infrastructure and to control all the activities related to their construction and maintenance, of which an elementary cell is represented by the MONITOR Control Centre. From this point of view the priority application selected for the experimentation had to be considered also in order to provide the main requirements for the Control Centre design.

Following these criteria, through the assessment process, the priority applications to be experimented are the following:

- Certified procedures for monitoring deformation of buildings at risk for landslides, in flooding areas or in subsidence area;
- Bridge deformation monitoring for real-time alert systems;
- Optimisation and safety for site management and logistics based on precise survey of the site and risks constraints.

3 Certified Procedures for Monitoring Land Movements and the Deformation of Buildings at Risk for Landslides, in Flooding Areas or in Subsidence Areas

Several natural causes can damage buildings and represent a risk to population and cultural heritage.

Landslides are the 7th most deadly natural hazard in the world, after droughts, windstorms, floods, earthquakes, volcano and extreme temperature [1]. Mass movement and landslides contribute to major disasters every year on a global scale, and their frequency is on an upward trend.

Number of deaths caused by landslides is likely underestimated, since they are usually masked by the broad disaster statistics of earthquakes and floods, but it can be said that about 1000 people died every year because of landslides, especially in Asia. In Europe the situation is certainly better.

In terms of economic damage, Japan reports annual losses of US$ 4–6 billion whether India, Italy and USA reports annual losses that are about US$ 1–2 billion.

The presence of landslides affect highly urban and infrastructural development of local communities also without having victims.

In Emilia-Romagna region (Italy), for instance, have been registered more than 70.000 landslides, luckily most of them with slow movements, that have caused damages spread over the area, but no victims.

The monitoring of buildings movements and displacements in landslides danger prone area, allows to have an operative control on potential risks over such area: some of the Monitor Project partners are Public Administrations sensitive to this subject.

GNSS surveys are the most suitable tools to perform the monitoring, either in continuous mode or by means of repeated surveys.

Considering the spread of this phenomenon on the territory, it is necessary to tune affordable and low-cost GNSS control systems to allow a wide diffusion over the territory.

GNSS Monitoring systems must come out from the scientific environment to be used on large scale and so to be adopted by specialized companies able to operate on demand raised by public agencies or private clients.

It is necessary to define not only the operating methodologies that should be applied, but also which kind of instruments should be used and finally the operating standards and control criteria. Assuming these considerations, every commitment will be able to evaluate the quality of every action performed by the nominated company.

Considering that the area under control could be wide, a goal could be identified in the implementation of a network of Continuously Operating Reference Stations in order to have instruments positioned on a potentially wide number of buildings, that can be monitored by a small number of reference stations providing measured data that could be processed by a single Control Centre.

From the operational point of view, the main goals to be achieved are:

- setup of a test field in a realistic environment aimed at study of possibilities of GNSS systems to detect movements of structures like buildings;

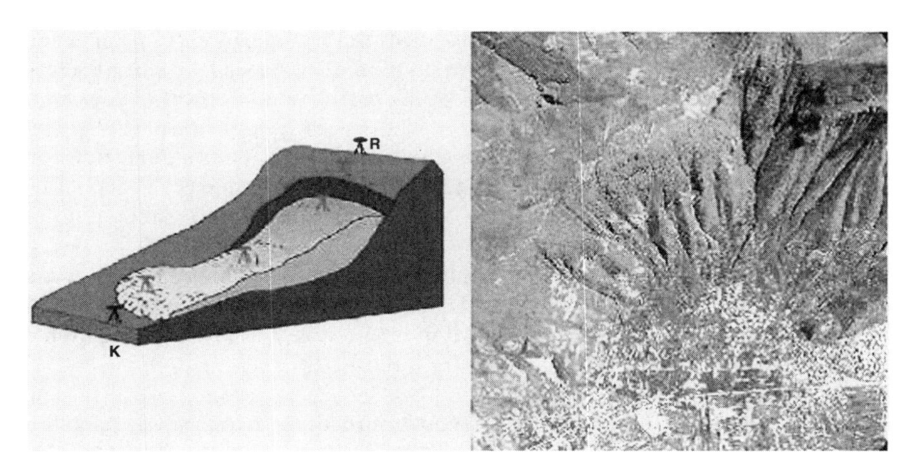

Fig. 1. Landslide scenario monitoring.

- to define a survey methodology using different types of instruments and of data transmission procedures;
- to detect improvements provided by the growth of the satellites constellations, the interoperability between different GNSS systems, and most of all, derived from integrity availability;
- to define a procedure in accordance with the Partners of the Public Administrations (Provincia di Bologna, Regione Emilia Romagna) to allow a private firm that could be entrusted with the surveys in a large scale approach.

Field tests will be executed on soils within a landslides area (Fig. 1), and the analysis will be performed paying attention on application of similar methodologies to other scenarios in which is important to detect movements and deformation with high precision such as monitoring of dams, boundary walls, buildings in problematic areas (i.e. structure liable to vibrations).

In the short term action plan, will be extremely important that field tests will be checked in any way concerning geodetic problems and boundary features such as communication link and the geological characteristics of the selected landslide. These control checks must be done for field tests as soon as the first Galileo Satellites will be ready to transmit data. In this manner, it will be possible to check the real accuracy and precision improvements for high precision real time positioning due to the Galileo system.

4 Bridge Deformation Monitoring

Bridges are part of a country's transportation infrastructure and are typically assessed and maintained by the authorities responsible for the appropriate transportation sector (road or rail). Bridge monitoring is necessary to ensure the safety of those who either use, or are affected by, the structure itself and is usually part of the legislature governing the maintenance of the sector.

Recently, the deterioration of bridge structures has become a serious problem due to issues related to modern society; reliance on the car, increased bridge traffic, environmental pollution, and the use of potentially corrosive substances (e.g. cleaning and de-icing).

Regular bridge inspection is typically a statutory duty of the bridge operator for reasons of public safety. Inspection is conducted via manual means although a few systems are deployed using fixed sensors, such as fibre optic cables laid within the structure, and, on occasions, GPS.

Bridge monitoring is an ideal application for the use of combined GNSS, incorporating GPS, Galileo, EGNOS and GLONASS. Although not necessarily a safety issue, the potential benefits of including MONITOR-style GNSS within a combined sensor system are listed below:

- Low cost equipment to provide part of the statutory duties of the bridge operators;
- Continuous monitoring should lower the cost of current routine and specialists inspections;
- Possibility of shared Local Elements between bridge operators, further reducing cost.
- The MONITOR Service Centre is the ideal platform for the dissemination of the real-time status of the structure;
- No specialised core skills are required by the bridge personnel;
- The system will provide added value, including safety to the bridge users (the general public);
- The system will provide operation benefits to the bridge operator. This includes better information for decision making, such as whether or not to close the bridge to transport or pedestrians.

At the time of writing this paper, a UK-based bridge operator have agreed to a pilot demonstration using one of their structures and an action plan for bridge monitoring within MONITOR has been developed covering technological, market and regulatory issues.

The short term strategy is aimed at promoting the use of GPS for bridge monitoring and demonstrating, though simulation, the improvements that will be gained through the adoption of Galileo and EGNOS. The actions required to achieve this goal are computer simulations involving the use of GNSS simulation software that can imitate performance of different GNSS configurations along with 3D (Fig. 2) models of the bridge, so that satellite masking can be determined. This simulation analysis is combined with a demonstration activity which involves an actual measurement and analysis campaign using the SEPA GPS-based monitoring equipment and the MONITOR service centre. The purposes of the demonstration activity are two-fold; firstly, to verify the simulation results and secondly to promote the technology to bridge operators and to infuse their interest so that they will be susceptible to the uptake of Galileo when this becomes available. For uptake of the technology, and inclusion into existing bridge monitoring systems, it is important that the activities are disseminated to a wider audience than the immediate pilot study bridge operating company, and that the appropriate authorities are urged to adapt, or develop, legislation that accommodates GNSS.

Fig. 2. Photograph and a computer simulation of the Severn Suspension Bridge, UK.

The medium and long term actions are non-MONITOR specific. These concentrate on GNSS companies and the suppliers of bridge monitoring equipment and outline the steps that are required to specify, develop and verify a commercially viable multi-sensor (including GNSS) bridge monitoring system.

5 Optimisation and Safety for Site Management and Logistic Based on Precise Survey of the Site and Risk Constraints

Accidents in yards are quite frequent an often the workers are not protected by insurances since they could be foreigner without work license. Accidents have to be prevented by adequate passive protections, with barriers between different areas, in which dangerous interactions could occur.

Active protection is the goal of MONITOR, by determining the position of the workers outside and inside the dangerous area together with the excavators, elevators, trucks with which the workers could accidentally interact.

The workers will be positioned by a single GNSS receiver, while, in general, the excavators etc. will need position and arm or movement directions. A third receiver (or a roll sensor or integrated with a roll sensor) will be useful to alert potentially dangerous positions along the yard trails.

Accuracy should be in general the best one obtainable with RTK: nevertheless a very frequent interruption of satellite signals and necessity of new initialisation are expected, which is in conflict with the safety requirements. Hence Differential Code Phase GNSS will be the initial target of the Pilot project.

GALILEO will increase the number of satellites so that the availability of positions inside a yard will increase. Moreover, since a lack of safety could be under penal responsibility the guarantee of the GALILEO SIS will have a high impact on the stakeholders with respect to GPS. At present GPS cannot be proposed in a yard, as everybody who is used to survey urban areas knows very well. Hence, the Pilot project should be carried on in

- high satellite visibility environment, just to show the correct positioning of workers and machines, anytime requested by the Control Centre and, when required, with continuity. The radio transmission coverage will be controlled both for Differential method and, when required, for RTK one. Such kind of environment has been already fixed in a permanent yard connected with the water channel system in agriculture: this is anyway a very important case for workers since they are often alone with their machines, quite fare away from any possible immediate assistance; they are already equipped with mobile phones so to be able to ask rescue. GNSS positioning will be of great help in the rescue service and the DGNSS accuracy should guarantee the side of a channel, which means to avoid misunderstandings and wrong rescue itinerary;
- poor satellite visibility, more as a sky plot simulation than as a field test, even if the highest number of GPS satellites window will be chosen on field.

It is quite clear that at present the use of GPS only is NOT feasible in a severe safety environment in relation to the evident guarantee that workers and controllers expect from such systems.

No one could be persuaded to adopt GPS as safety tool, unless in the particular case of wide area yards as for the mentioned water channel for agriculture as well as for agriculture process themselves.

The most outstanding open issue in this kind of experimentation, is the absence of rules: in general active and topographic safety tools for a yard is not carefully considered in terms of safety, so that an intensive dissemination of the GNSS possibilities will be necessary.

Yards are expected to be quite conservative and not open to innovations, unless they give strong economical benefits (Fig. 3). Therefore the dissemination needs to continue for years, particularly introducing the matter in the University classes so to make also Civil Engineers and Architects familiar with satellite positioning, in particular with the expected characteristics of GALILEO. With this dissemination on

Fig. 3. A typical construction yard scenario.

the young generations a general use of satellite positioning and integrated sensors is expected within 10 years.

Contemporarily a strong action will be carried on the Governmental Agencies for Workers Safety and Insurance including Inabilities and Hospitals. No doubt at all that these Agencies will be sensible to the methodology, provided that it works 100% and it is guaranteed. NO DOUBT that this cannot be achieved by GPS, but only by GPS + GALILEO.

6 The Monitor Control Centre

The typical functional scheme for the MONITOR Platform is depicted in the following Fig. 4.

As a fundamental part of the proposed Pilot Projects and as a "common platform" for the whole project an operational centre is implemented assuming to perform the following functionalities:

- *Real Time monitoring*: Quick, real-time results collected from the field provide alert conditions and situation assessment and improved awareness in case of natural disasters hitting the area (earthquakes, floods, etc.); this may improve the dispatching of rescue teams as well as providing a first assessment of the situation;
- *Operations Benchmark*: To provide an operational assessment in ordinary situation or to cope with disasters, the centre will be equipped with suitable capability to display the data according to the scenario;
- *Post processing*: The Centre will post-process some Pilot project data in order to provide a preliminary set of additional information to the users;
- *Pilot Projects Data collection and Archiving*: All the relevant data for monitoring of the territory and structures/buildings will be collected for possible post-processing and archiving in a single facility, from which the data can be retrieved at will;
- *Dissemination*: The centre will collect all the Pilot Project data and results organizing them into a Data Base making easier the following dissemination.

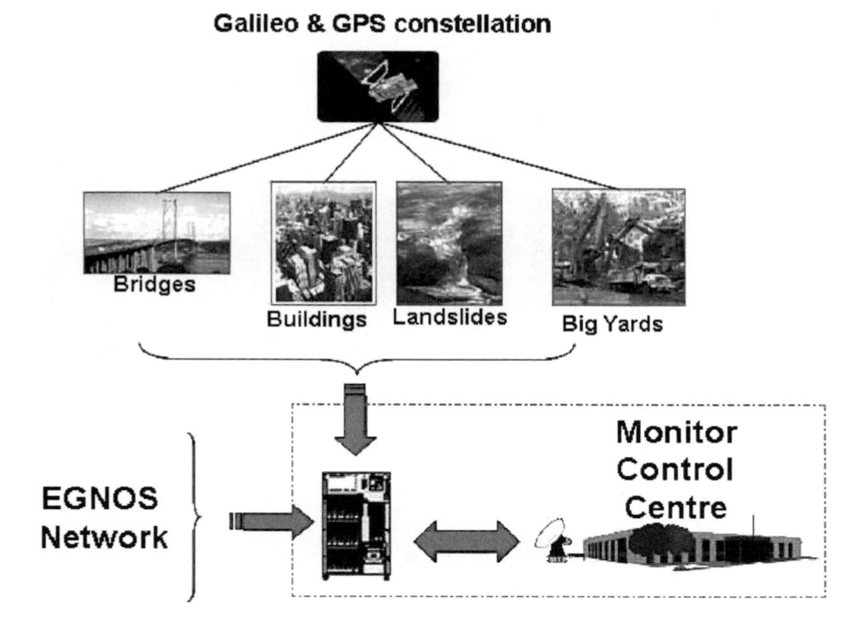

Fig. 4. The MONITOR platform concept.

All the three experiments and the MONITOR Platform, will demonstrate the potential benefits of the GNSS techniques through the implementation of a measurement and analysis campaign, sharing the following common elements:

- Low cost sensors which allow for the mass deployment of fixed monitoring networks (it is important that these sensors do not trade cost against performance);
- High data rate measurements;
- Real-time processing, at the same rate as the measurements;
- Real-time communications, allowing for data collection and distribution.

As depicted in Fig. 5, the platform is composed by three main architectural elements:

- A Network of sensors deployed on the interested site for the real time raw data acquisition;
- the *Monitoring Centre* which is the responsible of the data acquisition, processing and storage;
- the *Data Dispatcher* that is the interface aimed to the dissemination and acquisition of from/to other Monitoring Centres and also from/to other networks in order to gain data, relevant to the alarm event generation.

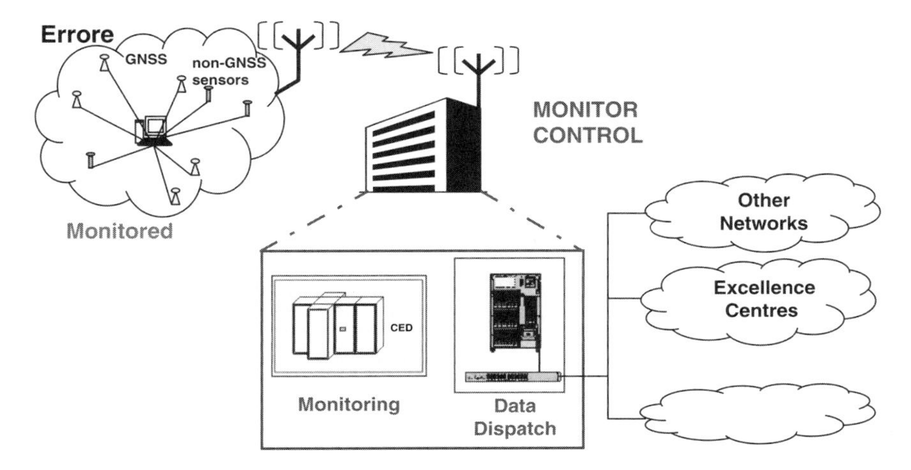

Fig. 5. MONITOR Control Centre schematic.

The platform is based on the data fusion concept, applied to a wider and heterogeneous kind of data gained from different kind of monitoring systems, involving different kind of monitoring technologies, through different kind of networks suitably connected through a proper interface. From this point of view this kind of control centre belongs to the concept and to the need of an over-national monitoring network with the aim of the convergence of technologies in a fully integrated and seamless way to end-users.

An important feature of the platform is the presence of a data base gaining information from the different monitoring campaigns to build an historical data base and to provide a source of data for the generation and validation of predictive models (for example for the validation of a FEM model of a structure). The collection and the storage of GNSS data from different sites is considered a key element to really experiment the data fusion concept using also data from those scenarios that seems to be, apparently, not correlated to the land surveying (e.g. big yards monitoring) but could be instead a valid source of information, ideas and methodology useful for other applications experimentation and a starting point for the development of future GNSS applications.

MONITOR is now at a critical stage of the project with the commencement of the demonstrations and the promotion and dissemination of the advantages brought through the early adoption of Galileo and EGNOS. Once completed, it is expected that this will fulfil its main purpose, namely to bring about the wider usage of satellite technologies within all aspects of civil engineering and surveying and to achieve higher levels of accuracy, reliability, safety, security and economy.

Acknowledgement

A special thank to Galileo Joint Undertaking represented by Dr Mario Musmeci (Business Development Division) for his precious suggestions and support for the MONITOR Project.

The authors would like also to thank all the partners of the MONITOR Projects Consortium for the valid contributions provided during the current MONITOR Project development that have allowed the issue of the present paper:

Alcatel Alenia Space Italia, GEONETLAB (University of Trieste), ELSACOM, SEPA, SOGEI, NSL, Edisoft, Provincia di Bologna, Regione Emilia Romagna, DISTART(University of Bologna Topography & Geodesy Dpt), University of Bologna II, Value Partners, Pagnanelli Risk Solution, Sistematica, IDS Italia srl, Aristotle University of Thessaloniki.

References

[1] Source: http://www.unesco.org/water/
[2] GCJ Call for Tender, GJU/04/2412-C1/NV, May 2004.
[3] State of art, TNO/MON/0001/VAL, issue2.0, April 2004

The Galileo C-Band Uplink for Integrity and Navigation Data

L. Castellano[1], S. Bouchired[1], M. Marinelli[1], I. Walters[1], E. Yau[2]

[1]Galileo Industries
Via G.V.Bona 85, 00156, Roma, Italy
Phone: +39 06 41799 529, Fax: +39 06 41799 506,
email: lucio.castellano@galileo-industries.net
[2]European Space Agency, Keplerlaan 1,
NL 2201, AZ Noordwijk, The Netherlands
Phone: +31 71 565 3158, Fax: +31 71 565 4369, email: edrich.yau@esa.int

Abstract. Galileo is a global, European-controlled, satellite-based navigation system. It will have a constellation of 30 satellites and a Ground Segment made by a Ground Control Segment (GCS) providing satellite monitoring and control, and a Ground Mission Segment (GMS) that is in charge of generating the navigation message (Navigation Function) and detecting system malfunctions (integrity function).

The System is designed to detect and broadcast real-time warnings to the users of satellite or system malfunctions regarding:

- ranging signal generation
- navigation message generation
- satellite orbit and attitude
- atomic frequency standard clocks and signals
- loss of sensor stations (GSS)

These warnings are transformed into integrity information that is transmitted to the user receivers where it is elaborated by complex integrity algorithms whose final result is the provision of the following information:

- stop using the Galileo system
- complete the current operation (i.e. landing) but then stop using the Galileo system
- use the Galileo system

The maximum latency between the time at which a malfunction is detected and the time at which the user integrity algorithm produces the "stop using the system" condition (Time-to-Alert) is 6 seconds. This latency includes the time needed to generate, uplink, downlink and process in the user receiver the integrity information produced by the GMS

This paper presents an overview of the Galileo System Architecture and describes the way the C-band uplink has been designed and dimensioned to fit to the stringent requirements of navigation data and integrity information dissemination.

1 The Galileo System Architecture

The Galileo System is designed to provide a high accurate global positioning service and information regarding the quality of this service (system integrity) at the same time. The integrity information is updated every second. Thus the system is designed to be synchronous modulo 1 second.

The system is composed by four segments:

- The Space Segment (30 satellites)
- The Ground Control Segment (GCS)
- The Ground Mission Segment (GMS)
- The Test User Segment (TUS) which includes the Test User Receivers

The space segment consists of a Walker constellation of 30 spacecraft in MEO orbits, expandable in the future to up to 36 satellites. The 27 operational satellites are distributed uniformly in 3 different orbit planes separated by 40° with the ascending nodes of the orbital planes separated by 120°. There is 1 spare satellite on each orbital plane, positioned between two operational satellites.

The constellation of satellites is controlled and commanded by two Ground Control Centres (GCC), each one made up by one GCS and one GMS, and a network of ground stations, namely 5 TTC stations operating in S-band (2.048GHz for uplink and 2.225GHz for downlink) and 9 ULS stations operating in C-band (5.005 GHz uplink only).

The S-band stations are used for satellite and constellation control. Telecommands are uploaded by the ground stations at 2.048GHz and telemetry downloaded by satellites at 2.225GHz. During nominal operations this link is operated in spread spectrum mode although a PM mode is foreseen for initial satellites deployment (LEOP/IOT) and contingencies. In this case the spacecraft are operated in FDMA mode and the frequencies are selectable among 8 different couples for uplink/downlink.

The C-band stations are used for uploading navigation and integrity data that are subsequently broadcast through the navigation signals to the users. These transmitting stations operate at 5.005GHz in spread spectrum.

The Galileo satellites carry 6-channels CDMA receivers to allow simultaneous access to 1 ULS and to up to 5 external uplink stations.

The system provides 4 different navigation services, namely the Open, Safety-of-life, Commercial and Public Regulated servicesn via simultaneous transmission to the users of three navigation signals at frequencies between 1.1GHz and 1.6GHz, as in Tables 1 and 2.

As can be seen there are 5 types of messages;

- the "Navigation message" containing:
  - time & clock corrections
  - ephemeris, almanacs
  - ionospheric corrections

Table 1. Galileo frequency plan.

Signal	Carrier-frequency	Polarisation	Transmitted bandwidth	Modulation
E5	1191.795 MHz	Right-hand circular	92.07 MHz	AltBOC
E6	1278.750 MHz	Right-hand circular	40.92 MHz	Constant envelope modulation
L1	1575.420 MHz	Right-hand circular	40.92 MHz	Constant envelope modulation

Table 2. Mapping of services on different signal components and message content/allocation.

		Message data content				
Services	Channel	Navigation	Integrity	Search & rescue	Supplementary	Service management
OS	E5a-I	Yes	No	No	No	No
OS/CS/SOL	E5b-I & L1-B	Yes	Yes	Yes	No	Yes
CS	E6-B	No	No	No	Yes	Yes
PRS	E6-A & L1-A	Yes	Yes	No	No	Yes

  - Broadcast group delay
  - SISA (Signal-in-Space Accuracy)
- the "Integrity message" containing:
  - **Integrity tables** for the global Integrity that include flags to indicate the integrity status of each navigation data broadcast by each satellite.
  - **Alerts** to be broadcast with an alert process that shall take less than 6s. This latency constraint is the major design driver for the complete Galileo system. Alerts are timely warnings to alert the users when the system cannot be used for navigation.
- The "Search and Rescue" message. Galileo supports the S&R service acting as relay of VHS distress beacons and by providing in the L-band downlink the capability to send up to 5 messages of 120 bits every 60 seconds or up to 7 messages of 80 bits every 60 seconds to those beacons equipped with a suitable Galileo receiver.
- The "Supplementary data" provided as part of the CS message on E6. The supplementary data is expected to provide weather alerts, traffic information and accident warnings, etc. This data is almost certainly geographically related and so could be shared between satellites.
- The "Service management data used to provide key management and other information to enable controlled access to the Galileo signals and message data.

The signals are transmitted by each satellite and comprise ranging codes and timing information. The timing information is included as part of the navigation message containing additional information relating to the satellite itself, the overall constellation and the integrity of the service.

The multi-frequency signals transmitted allow improving the accuracy of the fix since user receivers can calculate the corrections for ionospheric effects.

The 40 Galileo Sensor Stations (GSS) perform constantly measurements on the L-band navigation signals to detect faulty satellites. These measurements are transmitted to both GCCs. The GSSs are located in different geographical location to perform a worldwide monitoring of the navigation signals.

A simplified overview of the Galileo architecture is shown as in Fig. 1.

The single elements (MGF, CMCF, etc.) in the GCC will be described in the next sections.

The GMS and the GCS within each GCC are linked to the remote ground stations through the MDDN (Mission Data Dissemination Network) and the SDDN (Satellite Data Dissemination Network) respectively.

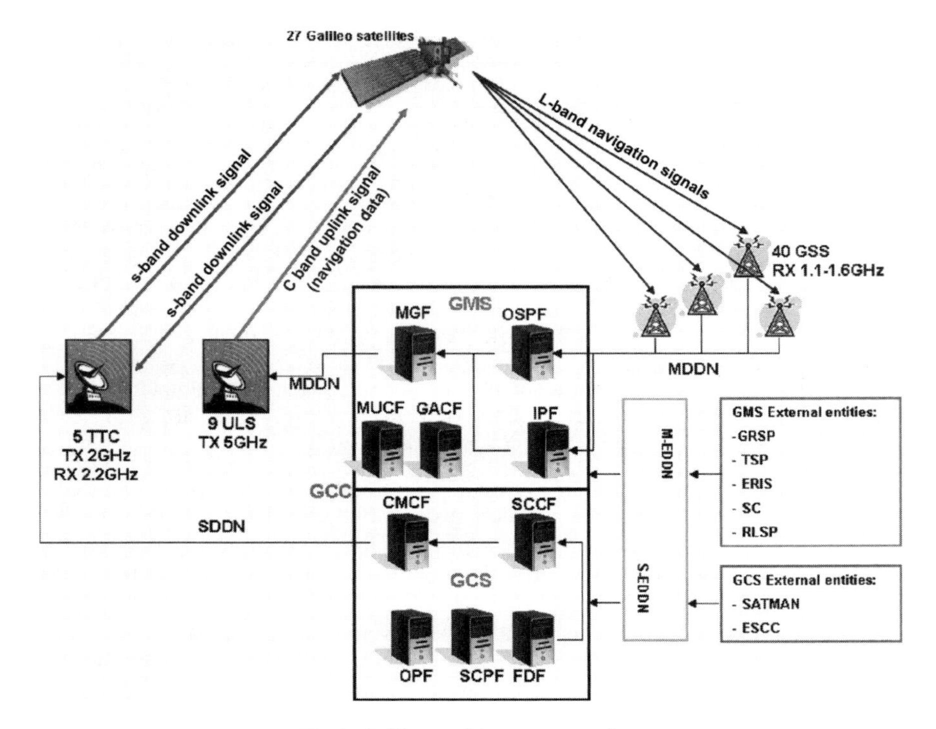

Fig.1. Galileo architecture overview.

The two ground segments are also connected to a number of entities external to Galileo. For the GMS they are namely:

- *GRSP* (Geodetic Reference Service provider). The GRSP provides the GTRF (Galileo Terrestrial Reference Frame which is linked to the ITRF
- *TSP* (Time Service provider). The TSP provides Galileo with the daily predicted value of the International atomic Time (TAI) with respect to the GST (Galileo System Time) and with frequency offset and the daily steering correction to be applied by GMS to GST
- *ERIS* (External Region Integrity System). The ERIS are external entities able to determine and distribute to users through direct (C-band) or indirect (GMS) uplink of integrity information regarding the Galileo navigation signals on a regional scale (Galileo provides indeed integrity information on a global scale).
- *SC* (Service Center). Service Centres are all those facilities implemented by the Service Providers to administrate navigation, timing, other navigation-related services and revenue-generating services offered to the users, distribute off-line Galileo data products, gather input data (commercial service data) to be disseminated by Galileo satellites, and manages CS users access control.
- *RLSP* (return Link Service Provider). The RLSP is the single interface with Galileo of the COSPAS-SARSAT, an international Search&Rescue system that processes distress signals from active beacons and interacts with the SAR community.

The GCS interface with the following external entities:

- *SATMAN* (Satellite Manufacturer). The SATMAN provides a number of satellite parameters, such as the satellite mass at the manufacturing stage, that will be used by the GCS during the satellite lifetime.
- *ESCC* (External Satellite Control Centre). The ESCC are the external leased satellite control stations that will be leased to perform LEOP (i.e. the positioning in the initial orbit) and IOT (in-orbit testing) for each satellite before hand over to the GCC control.

1.1 Description of the GMS

The main functions of the GMS are:

- to monitor the navigation signals transmitted by the Galileo satellites
- to calculate all the navigation and integrity data that need to be broadcast in the navigation signals
- to collect data coming from external entities that need to be broadcast in the navigation signals
- to multiplex these data in single sub-frames
- to define the sub-constellation of satellites to which these sub-frames have to be uploaded
- to uplink this data to the spacecraft

In addition the GMS performs a number of other functions such as monitoring of GCS assets, monitor and control of networks and global archiving.

The list of elements in the GMS and a brief description of their functions are shown as in Table 3:

Table 3. Description of the elements in the GMS.

GMS element	Name	Function
GSS	Galileo Sensor Station	Production of SIS observables for navigation & integrity algorithms
PTF (*)	Precise Time Facility	Generation of GST (Galileo System Time) steered towards TAI.
OSPF	Orbit and Synchronisation Processing Facility	Production of satellites ephemeris, clocks, almanacs and SISA predictions for OS, CS, SoL and PRS
IPF	Integrity Processing Facility	Integrity data computation for PRS and SoL (integrity tables and integrity alerts)
MGF	Message Generation Facility	Multiplex of the data streams making up the C-band Uplink Messages (including the Galileo Navigation and Integrity data generated within the GMS, the SAR data provided by the RLSP, the commercial data provided by the GOC SC, integrity data provided by ERIS, and PRS data

(Continued)

Table 3. Description of the elements in the GMS.—cont'd

GMS element	Name	Function
SPF (*)	Service Products Facility	The SPF provides the interface between the GMS elements located in the GCC and the external world.
GACF	Ground Assets Control Facility	– Technical monitoring and control of ground assets – Support to the execution of the activities planned for maintenance and the related reporting – Data archival and retrieval (including long term archiving for GCS)
MUCF	Mission and Uplink Control Facility	– Overall mission management (long-term and mid-term planning) – Short-term planning – On-line supervision of the mission performance – Mission performances prediction or short-term forecast of services performances
MKMF (*)	Mission Key Management Facility	Management of the payload security units (C-band part)
PKMF (*)	PRS Key Management Facility	Management of the payload security units (PRS navigation service)
ULS	Up-Link Station	Modulation and transmission of C-band uplink signals
MDDN	Mission Data Dissemination Network	Communications network facility to provide intersite connectivity
M-EDDN (*)	Mission External Data Dissemination Network	Provision of communication services between GMS and external entities (GRSP, TSP. SC, RLSP, etc.)

(*) These elements are not shown in Fig. 1 for sake of simplicity

1.2 Description of the GCS

The prime role of the GCS within the Galileo system is to provide satellite control and management.

This is accomplished via the generation and uplink of Telecommands (TC) and the reception and processing of spacecraft Telemetry (TM). In addition the GAS is able to perform spacecraft positioning through a two way ranging process on the S-band signals.

Aside from its primary role in spacecraft management the GCS also performs specific functions with respect to the Ground Mission Segment (GMS).

In order to achieve the primary role, the elements within the GCS provide functions that comprise of both real time and non-real time processes. These range from planning through to execution of TC, TM and Ranging (RNG). The GCS also performs its internal monitoring to ensure the efficiency and robustness of the segment elements throughout the lifetime of the programme.

Within the GCC, the GCS provides facilities for the real time monitoring and control of the Galileo satellites and ground assets (SCCF, CMCF and SKMF).

These facilities perform the routine automated operations supervised by operators, although critical operations will likely be performed manually, with the support of automated procedures.

GCS non-real time operation functions are supported by the SCPF, OPF, FDF and CSIM facilities within the GCC.

The list of elements in the GCS and a brief description of their functions is shown as in Table 4:

Table 4. Description of the elements in the GCS.

GCS element	Name	Function
TTCF	Telemetry, Tracking and Command facility	The TT&C facility support the space-ground interface for telemetry acquisition and telecommand uplink as well as the provision for 2-way ranging. Each TT&C station acquires and controls the satellites in accordance with commands received from the SCCF via the CMCF.
SDDN	Satellite Data Dissemination Network	Communications network facility to provide inter-site connectivity
S-EDDN	Satellite External Data Dissemination Network	Provision of communication services between GCS and external entities (SATMAN and ESCC)
SCCF	Spacecraft Constellation Control Facility	The SCCF is the core element that supports realtime operations for both routine and special satellite operations. This element provides for telemetry and telecommand processing status displays, manual commanding facilities together with operations automation at both schedule and procedure levels.
GCS KMF (*)	GCS Key Management facility	The GCS Key Management Facility is responsible for all TM & TC related security key management, both on the ground and on the spacecraft including the ground based encryption / decryption of TM & TC with cryptographic devices. It also provides M&C for the spacecraft security units.
CMCF	Central M&C Facility	The CMCF monitors and controls all GCS assets (only monitors the GCS KMF), both in the GCC and at each TT&C station. Each TT&C station has its own local M&C and the CMCF M&C of the TTCF is performed via the SDDN to all TT&C sites. The CMCF provides a summary of TT&C station status to the SCCF that is required to coordinate and synchronise satellite operations.
FDF	Flight Dynamics Facility	The FDF supports orbit prediction, manoeuvre planning and attitude monitoring for both individual satellites and overall constellation management. It is also capable of performing orbit determination based on S-band two-way ranging data generated by the TT&C Facility.
OPF (*)	Operations Preparation Facility	The OPF comprises the editors to develop and maintain mission operations data, including the spacecraft databases and operations procedures,

(Continued)

Table 4. Description of the elements in the GCS.—cont'd

GCS element	Name	Function
		together with tools to perform overall consistency checking of configuration data and to support configuration control.
CSIM (*)	Constellation Simulator	The CSIM provides a simulation of the whole Galileo constellation of satellites, as well as the ground control network of TT&C stations.

(*) These elements are not shown in Fig. 1 for sake of simplicity

1.3 Description of the SSeg

Galileo satellites are composed of the following subsystems:

- Payload Subsystem including the navigation payload and the SAR payload
- Structure Subsystem
- Thermal Control Subsystem (TCS)
- Electrical Power Subsystem (EPS) with the following units:
  - Solar Arrays (SA)
  - Solar Array Drive Mechanisms (SADM)
  - Battery
  - Power Conditioning and Distribution Unit (PCDU)
- Harness
- Avionics Subsystem with the
  - on-board computer (Integrated Control and Data Handling Unit, ICDU)
  - Attitude and Orbit Control System, AOCS (based on earth sensors, sun sensors, gyros, reaction wheels and magnetic torquers),
  - Software (SW)
- Telemetry, Tracking and Command (TTC) Subsystem (with S-Band Transponder and two low-gain, omni-directional antennas)
- Propulsion Subsystem (mono-propellant system with one tank and 8 thrusters)
- Laser Retro-Reflector (LRR)
- Platform Security Unit (PFSU)

The main functions of the payloads embarked on board the Galileo satellites are:

- Provision of on-board timing signals
- Receiption & storage of up-linked navigation message data
- Receiption & storage of up-linked integrity data (from 6 simultaneous uplink channels in C-band)
- Assembly of navigation message in the agreed format
- Error correction coding of navigation message
- Generation of ranging codes
- Encryption of ranging codes as required

- Generation and modulation of L-Band carrier signals
- Broadcast of navigation signals
- Relay VHF S&R distress beacon in L-band (1.544GHz)

The list of units in the Navigation Payload and a brief description of their functions is shown as in Table 5:

Table 5. Description of the payload units.

Payload unit	Name	Function
USO	Ultra Stable Oscillators: – Passive Hydrogen Maser (PHM) – Rubidium Atomic Frequency Standards (RAFS),	Provision of timing signals. This is done by high precision on-board clocks, implemented as two (cold) redundant pairs per satellite, each pair including two different technologies, the Passive Hydrogen Maser (PHM) which is the p rimary reference clock and the Rubidium Atomic Frequency Standard (RAFS), both of them being operated simultaneously.
CMCU	Clock Monitoring and Control Unit	The whole clock ensemble is under the control of a dedicated (internally cold redundant) CMCU which performs the monitoring and switching functions (selection under Ground control) and generates a highly stable on-board reference frequency of 10.23 MHz which is distributed to the other payload units.
MISANT	Mission antenna (C-band)	The C-band receiving antenna (small aperture axially corrugated circular waveguide horn)
MISREC	Mission Receiver	The MISREC includes the Mission Processor function (MISPROC) and performs the 6-channel receive function in C-band
PLSU	Payload Security Unit	The data output from the MISREC is routed to the Payload Security Unit (PLSU) which performs COMSEC treatment of the incoming signal (authentication verification)
NSGU	Navigation Signal Generator Unit	The NSGU receives the up-linked navigation data from the PLSU and uses them to generate the navigation signals in the appropriate format, performs the PRN encoding and the modulation of the 3 navigation signals (E5, E6 and L1) and passes them to the FGUU
FGUU	Frequency Generator & Up Converter Unit	The FGUU performs the up- conversion into L-band of the 3 signals
NAVHPA	Solid State Power Amplifiers	The 3 L-band navigations signals (E5, E6 and L1) are fed into three different SSPAs
OMUX	Output Multiplexer and filters	Multiplexes E5 and E6 signals at the output of the SSPAs

(*Continued*)

Table 5. Description of the payload units.—cont'd

Payload unit	Name	Function
NAVANT	Navigation antenna (L-band)	The Navigation transmit antenna is operated in RHCP polarisation and consists of a high and a low band beam-forming network and a dual band array of radiating elements which provides a global coverage iso-flux radiation pattern
RTU	Remote Terminal Unit	The payload Remote Terminal Unit performs the communication functions between the payload and the avionics subsystem via the 1553B data bus

2 The Message Generation and the Uplink Chain

The Galileo Uplink baseband signal comprises a multiplex of data streams originating from both Galileo and external origins (External Regional Integrity Systems, Search & Rescue). The multiplexing function is implemented in the Message Generation Facility (MGF) in the GMS and onboard the spacecraft.

Each second a navigation and an integrity message is generated by the MGF using data provided respectively by the OSPF and IPF which process of the measurements made by the GSSs.

These messages are uplinked every second to the satellites through a dedicated network of 9 C-band uplink stations (ULSs) and broadcast to the users, by the satellites using the L-band frequencies.

Two different types of alerts are transmitted in the integrity message depending on the failure occurred:

- Satellite failure: an alert is broadcast indicating which satellite is not ok (NOK). These alerts are coded on 1 bit per satellite.
- Ground segment failure: an alert is broadcast indicating the estimated degradation of the positioning accuracy. These alerts are coded on 4 bits per satellite.

Integrity alerts for up to 36 satellites can be uplinked and broadcast simultaneously.

A simplified description of message generation performed in the MGF and of the uplink chain is shown in Fig. 2.

The Galileo up-link message is multiplexed in the GMS and contains, as described in section 2, the following information:

- Galileo Integrity data for the SoL service (generated by the IPF)
- Galileo Integrity data for the PRS service (generated by the IPF)
- Navigation data for all the services (generated by the OSPF)
- Search-and-Rescue return link data (acknowledgment of distress signal reception provided by the RLSP)
- Commercial Service data (provided by the SC)

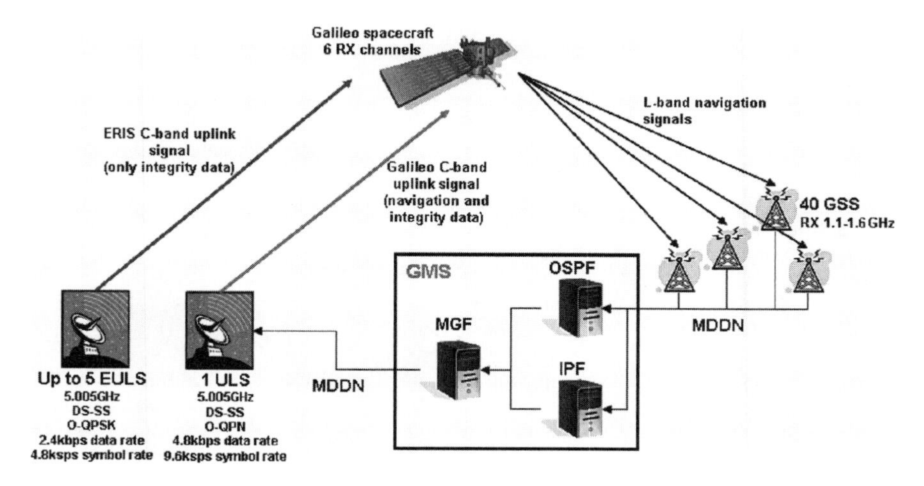

Fig. 2. The message generation and uplink chain.

The baseband message stream is sent by the MGF to the ULS where it is used to modulate the RF signal transmitted to the spacecraft.

Each Galileo satellite can be accessed simultaneously by 1 ULS to upload the messages generated by the MGF and by up to 5 EULSs (External Regional Integrity Uplink Stations) to upload regional integrity information generated outside the Galileo boundary. Therefore, an additional processing and multiplexing function is implemented in the satellites. In this way each satellite broadcasts the global integrity information generated by Galileo (GMS) and the regional integrity information generated independently by External Regions (ERIS).

This capability of allowing up to 6 simultaneous accesses to a single satellite along with the maximum time (133ms) allocated for uplinking the integrity information, have been some of the main design driver for this link.

The key requirements which have driven the design of this link are listed and discussed as in Table 6:

Table 6. C-band link: main design drivers.

Requirement	Source	Remark
The allocated band for the link is 5.000-5010 GHz	Band allocation for RNSS	The actual carrier frequency is selected taking into account that carrier, code and bit rate have to be in a fixed ration and derived by a common source. The modulation and chip shape filtering (on the TX side) are defined in such a way to have the highest possible chip rate still avoiding interference with adjacent services (radioastronomy)

(Continued)

Table 6. C-band link: main design drivers—cont'd

Requirement	Source	Remark
6 simultaneous CDMA accesses to each satellite	System requirement	Colliding requirements difficult to be met simultaneously.
ULS minimum elevation anglebetween 5° and 10° also in tropical regions where attenuation and scintillation effects are significant.	Minimization of the number of ULS sites and antennas	
No Up link power control	Simplification of system complexity	
The maximum time allocated to ULS processing and uplink (640bits of integrity information every second) is 133ms	System requirement for the maximum TTA (time-to-alarm)	Driver for the definition of the data rate for the Galileo uplink
ULS uplinked Information data twice the ERIS uplinked information data	ERISs do not uplink navigation data	Driver for the definition of the data rate and modulation scheme for the ERIS uplinks
Doppler compensation: +/– 11 KHz, step less than 2Hz	Constant data rate at receiver level	The Doppler, and its compensation affects both chip and symbol rates proportionally
Message structure and sequencing designed to allow the nominal broadcasting of NAV data to be interrupted and restored soon after integrity information transmission	System requirement for the maximum TTA (time-to-alarm)	Requires dynamic management of navigation and integrity messages.
On ground and/or on board Multiplexing of S&R and ERIS data	System requirement to support S&R service and ERIS autonomous integrity determination	

2.1 Message Design

The definition of the message structure and sequencing in both the uplink and the downlink has been one of the main challenges in the design of the Galileo system.

As described in Table 6 the two major constraints have been the management of integrity alerts and the reduction of the TTA, leading to a complex mechanism of dynamic management of alerts based on priorities. The structure of the message in the downlink allows the insertion of alerts in each 1 second interval. The C-band message is designed to allow simultaneous uplink of navigation and integrity data as described in Fig. 5.

Integrity alerts are transmitted as soon as they are received by the spacecraft, i.e. they take priority over normal integrity message transmissions. The message sequence generated by the MGF guarantees that even in case of alerts, the user will receive a complete navigation data batch in 30 seconds.

The message structure has been also optimised to allocate the S&R data and the ERIS integrity data without impacting the integrity performance.

Owing to the volume of data that must be uplinked to each satellite and the regularity of these transmissions, the uplink system is designed to be synchronous. A full data set contains thirty lines of code where each line is one second in length. This forms the uplinked page to which the receive system will be synchronised to. Synchronisation of the page is achieved by the satellite searching through the data to identify the first line within the page.

Once the uplink signal is stripped of the coding and modulation layers, it can be seen that the message is composed of one second frames per line. These frames actually contain separate Integrity and Navigation messages which are separately processed within the satellite.

Each integrity and message sub-frame then is fragmented down to multiple packets with dedicated descriptor headers identifying their purpose and application within the satellite transmitted signal. This structure allows for a simple and yet efficient provisions for each of the different services that the signal in space must provide.

The packets will thus contain updates of the ephemeris and almanac for each satellite providing the user with up-to-date information on the constellation.

2.2 The C-band Link Characteristics

The uplink transmission system adopts a phase continuous frequency compensation system to allow for the change in Doppler each satellite approaches and passes. Also, this allows to maintain a constant data rate at the C-band receiver output. The synchronism of the signal is also maintained between the carrier frequency, code rate, and symbol rate. This helps the receive system to maintain bit lock and synchronisation without the need for excessive on board buffering.

The uplink system uses commercially available 3m class Cassegrain antennas transmitting the right hand circular polarised signal to the satellites.

Compliance to the ITU requirements is met with the typical filtering processes as well as digital chip shape filtering to a 0.35 square raised root cosine.

The receive system is designed to be able to discriminate and demodulate simultaneously up to six C-band signals transmitted by ground stations accessing the satellites in CDMA mode. This also has to account for the varying ranges between each ground station (near-far range effect) as well as stringent atmospheric conditions of some of the locations (Rain attenuation and atmospheric scintillation).

The network of 9 ULS stations allows for one station to drop the uplink signal and another station to re-assume the link while still maintaining an overall seamless provision of the 2 necessary independent[1] integrity links to the end user (break-before-make system). The same design is adopted for both the Galileo and Regional integrity systems.

---

[1] "Independent path" means the each user receives the integrity information from at least two satellites connected with two different ULS sites.

In order to achieve the overall stringent requirements, each ground station is carefully coordinated and synchronised through careful scheduling.

2.3 Coding Layers and Modulation

The Galileo C-Band uplink signal employs a CDMA Direct-Sequence Spread-Spectrum together with an Offset-Quadrature Phase Shift Keying (O-QPSK) of the carrier. The gold code sequences have been selected providing excellent orthogonality and thus cross correlation isolation between signals.

This modulation has been selected to optimise signal spectral quality through easing the achievement of ground station transmit chain linearity.

The link budget is reinforced by a combination of inner and outer encoding. As with typical communications systems a standard ½ rate convolution encoder is used on the uplink signal with the corresponding Viterbi decoder (with a 3-bit soft decision algorithm as per [2]) within the receive modules.

The different Reed Solomon codes chosen for each sub-frame are designed such that the frame error rate probability will be sufficiently low that it can effectively be ignored while minimising the processing overheads of the satellites. This is required to minimise the processing time and stringent transit times for the integrity signal such that the necessary Time To Alarm can be achieved. The strength of the Reed Solomon coding is also reinforced by bit randomisation in order to ensure a sufficient number of transitions, whatever is the data to be uplinked.

The order of the coding and modulation operations is described in Fig. 3.

At the transmission side (ULS) the application data (Navigation and integrity sub-frames) after randomization and inner Reed-Solomon coding are NRZ-M encoded.. The data stream NRZMi(t) at the output of the differential encoder (at the data rate fd = 4,826 bits/s) is input into the convolutional encoder as per CCSDS recommendation [1]. The convolutional encoder produces two branch streams (carrying different data) classically denoted G1 & G2, each at rate fd. The two streams, with G2 is logically inverted, are exclusive-ored with two different pseudo-random sequences (C1, C2) at rate fc = 1023 * 4826 chips/s having the origins aligned. One of the two sequences (Y'(t) in the figure) is delayed by one-half chip. The spread branch signals modulate the in-phase and in-quadrature carriers at nominal frequency $f^0 = \omega/2\pi$ = 5.0050275 GHz yielding the I and Q signals the sum of which is the modulated signal M = X'(t)* cos(ωt) + Y"(t) * sin(ωt). The process is described in Fig. 4.

The ERIS C-band uplink is very similar to the Galileo one except that the convolution encoder outputs only one data stream which is exclusive-ored with a single pseudo-random sequence, i.e. the ERIS data rate is ½ that of Galileo.

Fig. 3. Order of the coding and modulation operations.

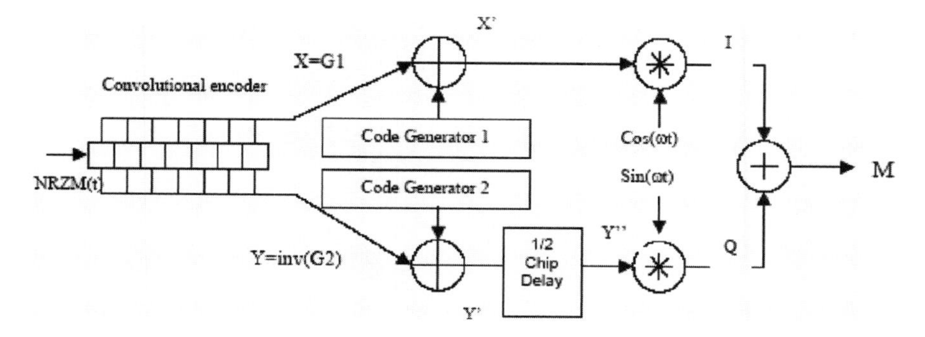

Fig. 4. Generation of the RF signal.

Table 7. Signal to interference ratio expected in some geographical locations candidate to hosting ULS stations.

	Signal / Interference [dB] (signal in nominal conditions at 10°)		Signal / Interference [dB] (signal in nominal conditions at 5°)	
	Power ratio with 5 interferers	Power ratio with a single interferer	Power ratio with 5 interferers	Power ratio with a single interferer
Kiruna	−9.17	−5.1	−12.71	−7.08
Kourou	−14.92	−10.8	−23.79	−18.16
New Norcia	−9.83	−5.7	−14.34	−8.71
Noumea	−11.60	−7.5	−17.82	−12.19
Papeete	−13.69	−9.6	−21.83	−16.26
Reunion	−12.29	−8.2	−19.15	−13.52
Santiago	−8.63	−4.5	−11.77	−6.14
Trivandrum	−15.01	−10.9	−23.06	−18.43
Vancouver	−9.97	−5.9	−14	−8.37

2.4 The Link Budget

The C-band link is dominated by self-interference. The useful signal is interfered by up to 5 signals transmitted by the other ULS/EULS. The presence of Galileo's intra-system interferers is indeed a special feature of the system. The total power of these unwanted signals far exceeds the thermal noise power at the receiver input because ULSs are operated without uplink power control.

The situation is made worse by the fact that the 9 ULS are located in geographical regions in which the effects of ionosphere, troposphere and rain can be largely different, as shown in Table 7.

The analysis of the simulations made for the different geographical locations has suggested that the EIRP transmitted by the ULSs had to be maintained at the minimum possible level. For this reason the EIRP has been defined at two different levels, 52.8dBW for ULSs located in tropical regions where atmospheric attenuations can be really significant, and 50.8dBW for those ULS located at temperate latitudes or in very dry environments.

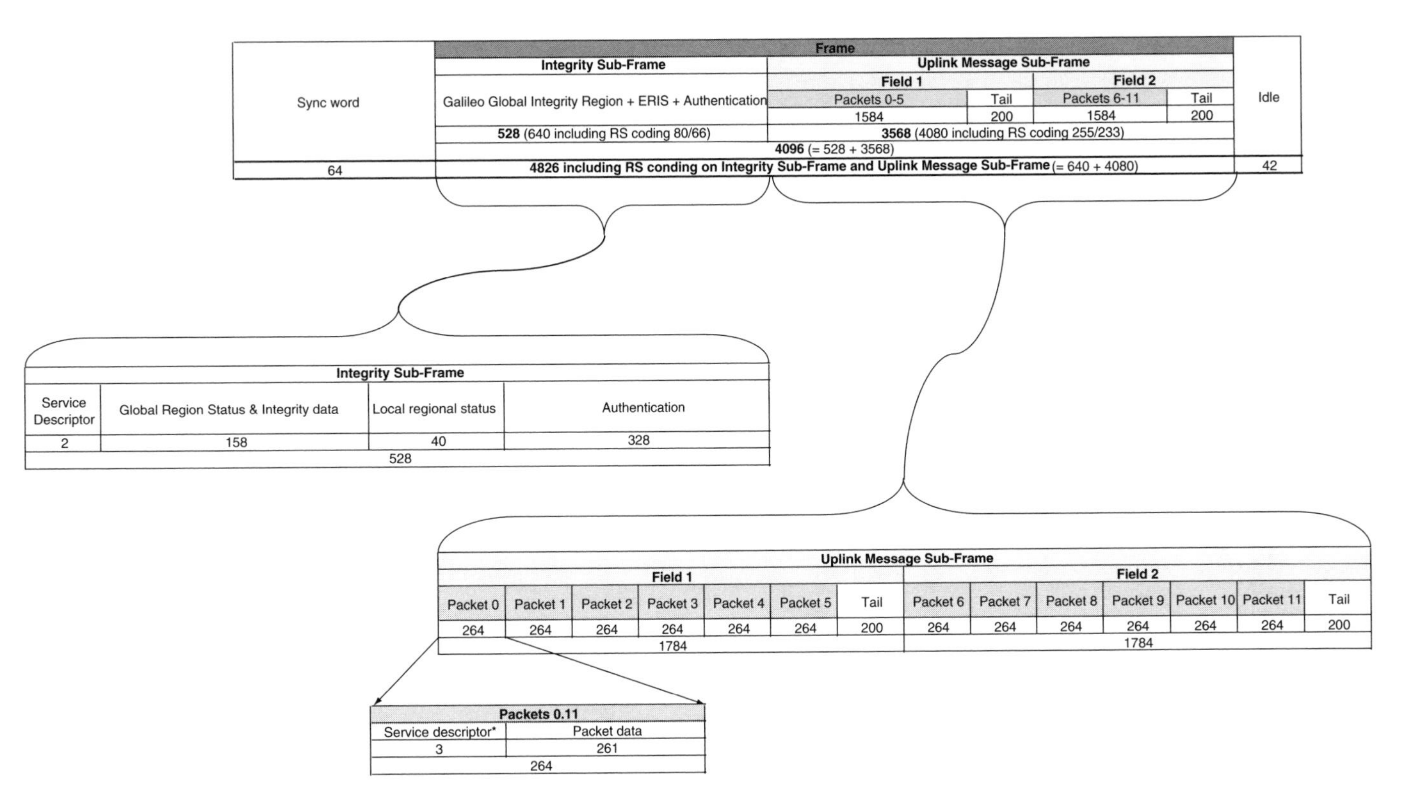

Fig. 5. Structure of the uplink message frame.

The EIRP for the EIRS uplink station has been set at 47.8dBW, taking into account that these station will not be operated below 10° elevation and that the ERIS uplink has 1/2 the data rate of Galileo.

3 Conclusions

Galileo is a satellite navigation system thought to provide global and regional integrity information to the users with extremely low latencies between the time when a failure/problem occur and the time at which the user is informed (receiver output). This capability has been the main system design driver and has led to a very complex mechanism to generate and disseminate integrity information. In particular, the requested integrity performances are based on a complex data dissemination concept of which the C-band uplink is a key part.

This link has been designed to: fulfill simultaneously all the constraints and to simplify at the maximum extend the complexity in the system.

References

[1] "*Radio Frequency and Modulation Systems - Part 1 Earth Stations and Spacecraft*", CCSDS 401.0-B, June 2001

[2] "*Channel Coding and Synchronization. Part 1: Synchronous*", Draft Recommendation for Space Data System Standards, CCSDS 131.0-R-1. Red Book. Issue 1, June 2000

A GPS/EGNOS Local Element Integrated with the VHF Communication Infrastructure Under Development in the POP-ART Project

F. Dominici[1], A. Defina[1], P. Mulassano[1], E. Loehnert[2], V. Bruneti[3], E. Guyader[4]

[1]Istituto Superiore Mario Boella
Via P.C. Boggio 61, 10138 Torino, Italy
Phone: +39 011 2276414, e-mail: paolo.mulassano@ismb.it
[2]IfEN GmbH
Alte Gruber Strasse 6, 85586 POING, Germany
Phone: +49 8121 223817, e-mail: E.Loehnert@ifen.com
[3]Sist&matica s.r.l.
via S. Pertini 17 12030 Manta (Cuneo) – Italy
Phone: +39 0175 255700, e-mail: valentina.brunetti@sistematica.it
[4]GJU – Galileo Joint Undertaking
Rue du Luxembourg, 3, B-1000 Brussels
Phone: +32 25078045, e-mail: Eric.Guyader@galileoju.com

Abstract. The paper aims at introducing the on-going activities around the design and development of a network-based positioning system exploiting the advantages of both the European Geostationary Navigation Overlay System (EGNOS) and the VHF communication infrastructure with the goal to support the Alpine Rescue Teams in the management of search and rescue operations.

Such an innovative integration strategy among analog communication channels and EGNOS is the key topic of a project named Precise Operation Positioning for Alpine Rescue Teams (POP-ART) co-funded by the Galileo Joint Undertaking as specific action toward the support of innovative ideas around EGNOS and Galileo proposed by Small Medium Enterprises (SME).

The positioning system is based on the raw GPS measurements collected by the users and transmitted to a Local Element by VHF radio channels (or GPRS when available).

The POP-ART activities are based on pre-existing works already presented in [1] [3].

The goal of the paper is to provide the latest information on the results already achieved within the project as well as the perspective on the expected performance thanks to several analyses on the EGNOS features.

In particular, the results demonstrate that the system fully exploits the improvement given by the EGNOS signals, ensuring a suitable level of accuracy in the positioning for a larger time with respect to standalone GPS positioning.

1 Introduction

This paper is focused on the overall description of the system architecture for an innovative integrated GPS/EGNOS/VHF Local Element as proposed in the Precise Operation Positioning for Alpine Rescue Teams (POP-ART) project [2].

The POP-ART project is part of the projects started under the 2nd call area 3 of the 6th Framework Programme for Galileo Research and Development This project is partially funded by the Galileo Joint Undertaking and coordinated by Sist&Matica that is an Italian SME. Moreover the POP-ART consortium is composed by the Institut für Erdmessung und Navigation (IfEN) that is a SME from Germany and the Istituto Superiore Mario Boella (ISMB) that is an Italian research institute.

With the goal of supporting the operations of the Alpine Rescue Teams, POP-ART aims to ease the real-time management and coordination of the on-field resources: precise positions of all the team members are calculated at the control centre level thanks to the use of the EGNOS augmentation data. Here the so called Network-Assisted Local Element will be realized. As a consequence, this architecture allows the operations coordinator to provide very precise directions to all the rescuers, allowing a sensible reduction of the intervention time, essential for the success of search and rescue operations.

The paper is organized as in the following: Section 2 provides some preliminary statistics about the Corpo Nazionale Soccorso Alpino e Speleologico (CNSAS) operations and presents the collaboration between the POP-ART consortium and the CNSAS. Such data justify the request from CNSAS to have a technological platform supporting their operations.

Section 3 provides a description of the system architecture, whilst Section 4 describes the expected performance of the final prototype on the basis of the EGNOS performance for static and dynamic conditions. Section 5 presents the overall project planning, while Section 6 proposes some possible R&D activities that will follow the POP-ART.

It has to be remarked that POP-ART kicked-off in March 2006. Therefore, some of the technical results here presented has to be considered as pre-existing know-how of the authors and then useful in the project development.

2 CNSAS Involvement in the POP-ART Requirement Definition

The POP-ART has to end with a ready to use system prototype; therefore, since the beginning of the project the consortium directly involved the CNSAS that is the Italian Alpine Rescue Team [5]. All the system requirements at both functional and services levels has been obtained through specific interview with CNSAS representatives, achieving a high level of system acceptance from the persons that will operate the prototype.

Hereafter some statistics related to the CNSAS interventions in Italy are reported. These graphs are useful in order to understand the strength and the potential advantages of a centralized monitoring system like POP-ART for search and rescuer operations.

It has to be remarked that the information sketched below are heavily used in the POP-ART requirement analysis for the definition of the system architecture.

The raising numbers of people who goes to the mountains has dramatically increased the number of interventions of the Rescue teams to help people injured or in danger situations, as highlighted by the Fig. 1 that shows the statistics about

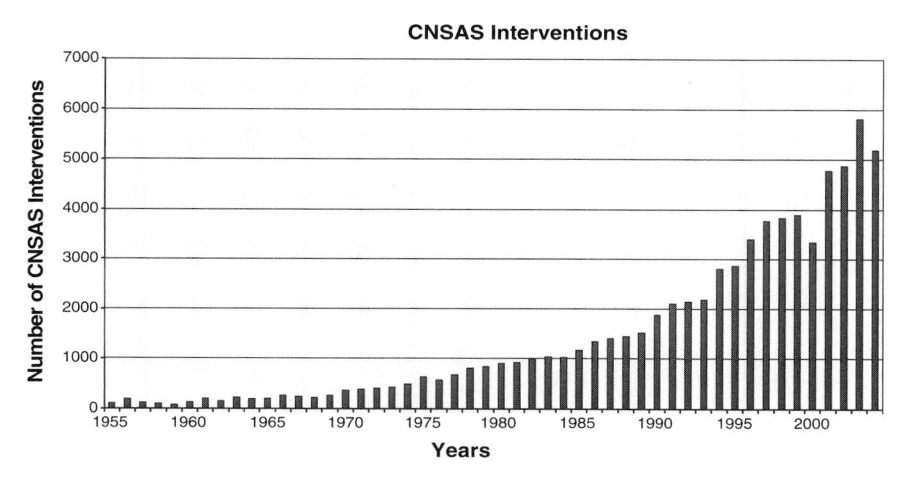

Fig. 1. CNSAS number of interventions per years from 1955 to 2004.

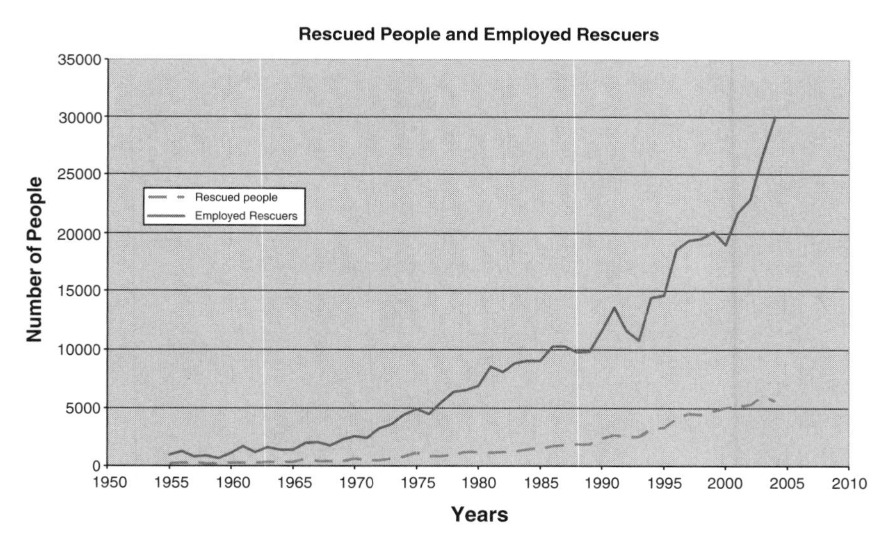

Fig. 2. Comparison between rescued people (dashed line) and number of employed rescuers (full line) per year from 1955 to 2004.

interventions from 1955 to 2004. Consequently, the number of rescuers employed in the operations has considerably increased, as shown in Fig. 2.

The trend shows in these graphs with an increasing in the CNSAS interventions number and in the employed rescuers confirms that the use of an automatic and precise monitoring system like POP-ART will became soon necessary in order to make easy the operation management and to improve the safety of both the rescued people and the rescuers.

Figure 3 shows the average number of teams' components per year; it is possible to observe that this number has always remained under 10 people. Furthermore this number has stayed around 5 people in the lasts years. This indication has to be taken into account into the POP-ART architectural design especially for the definition of

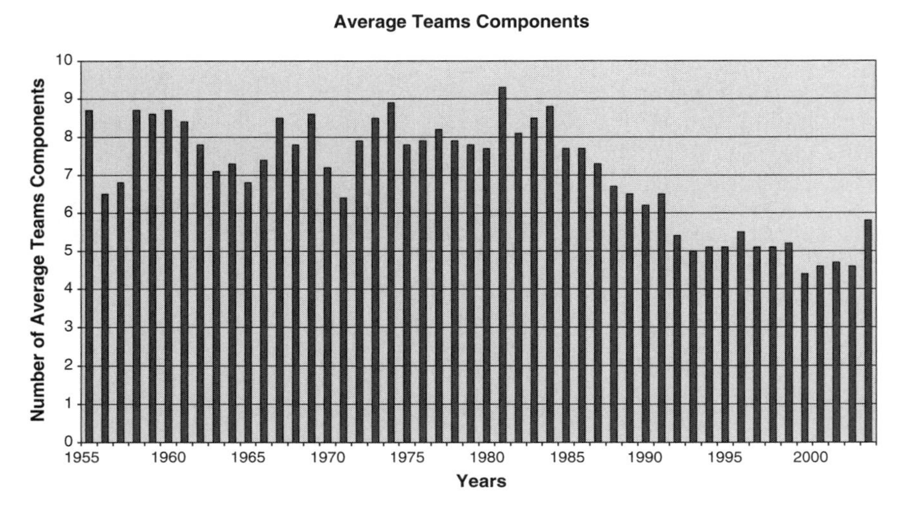

Fig. 3. Average components number of an alpine rescue team employed in search and rescue operations.

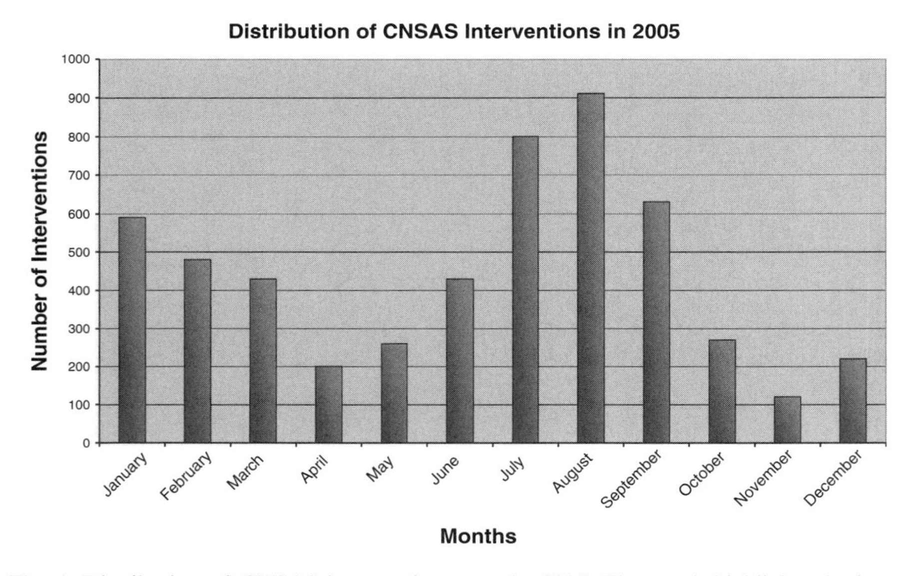

Fig. 4. Distribution of CNSAS intervention over the 2005. The graph highlights the large interventions number during the summer; it is due to the increasing of the tourist excursions.

the communication strategy that represents a key point for the success of the POP-ART system.

Figure 4 represents an example of statistic information about the CNSAS activities in the 2005, which are the most recent at the time of writing. It shows the distribution over the 2005 of the CNSAS interventions per month. The graphs highlights that most of the interventions are concentrated during the summer, especially in August, due to the strong presence of tourists in the mountains areas for holidays or for excursions.

This large number of interventions in a limited time window demonstrates the requirement to improve the management of CNSAS interventions; hence it justifies the introduction of POP-ART system for the alpine rescue teams in order to conduct the operations with higher level of efficiency.

3 POP-ART System Architecture

According to the input received by the CNSAS representatives of the Piemonte region, nowadays during standard rescue operations each rescuer is equipped with a VHF radio transceiver transmitting on a reserved and certified frequency band and in some cases with a GPS receiver. The position, when available, is communicated by voice on the radio channel to the control centre where the operations coordinator manually records on a map the received positions in order to have a complete view of all the rescuers. Such positioning information are then displayed in order to allow the operations coordination on a personal computer at the control center side using a 2D cartographic interface. All these constituent blocks are shown in Fig. 5.

The goal of POP-ART is to upgrade such system in order to provide to rescuers an automatic real-time localization infrastructure with high accuracy and availability.

In particular, the main drivers that have been taken into account are listed in the following:

- The positioning system must reach a high degree of availability and reliability, especially in mountainous environments;
- POP-ART must rely on VHF radios already in use by the Alpine Rescue Teams. GPRS data communications has to be considered as a nice to have feature due to its unreliability in critical situation; hence it cannot be use as the main COM infrastructure. It has to be remarked that CNSAS of the Piemonte region invested many financial resources in order to have a full coverage of VHF in the Alpine area. Therefore, such operative infrastructure guarantees today the best

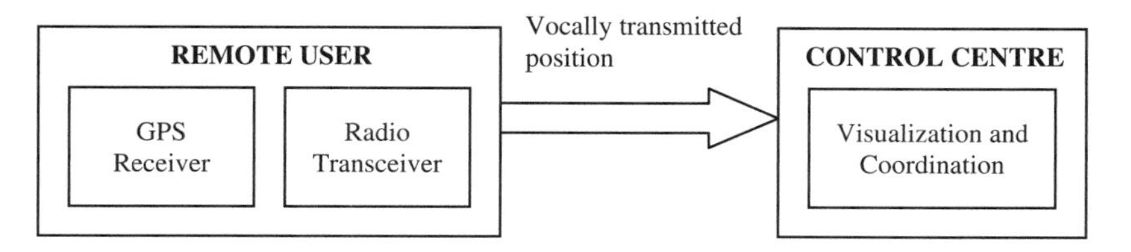

Fig. 5. System for teams management in use at the present by the CNSAS.

cost-effective solution if compared to other technical choices (i.e. with respect to satellite communication).
- The rescuers' equipment must be portable and ruggedized since it has to be used in hostile environments;
- The operation coordinator has to be able to send back waypoint or text information to each rescuer.

On the basis of such general requirements, the POP-ART prototype architecture will follow the scheme reported in Fig. 6. This basic architecture has been already presented in [1].

The POP-ART system architecture is based on the Galileo Local Element concept, i.e. a fixed system infrastructure foreseen in the Galileo architecture which will enable a centralized service provider to deliver a Galileo local positioning services within a limited, or local, geographic area to remote terminals connected through a wireless communication technology.

The idea of the POP-ART project is to equip the rescuers with Remote Terminals (RTs) that will be a portable devices able to communicate with Operation Coordinator (OC) for obtaining POP-ART precise positioning (performed thanks to the integration of EGNOS corrections) and additional location based services. The OC will be fixed station supporting the operation manager, who is the person in charge to control and manage the work force employed in the field during mission. Moreover different OCs can share useful data and information through a Remote Data Center (RDC).

The RT is composed by a Remote Terminal Device (RTD) which performs the main RT functionalities and a Personal Digital Assistant (PDA) that implement the Man Machine Interface (MMI). The RTD is endowed with a GPS chipset able to download raw GPS measures and a communication unit for the communications toward the OC. Within the POP-ART system the selected wireless communication technologies are primarily the VHF channel that is available and reliable in mountain environment, and additionally the GPRS channel, employed as backup channel.

Due to the hostile work conditions of the rescuers, the RT must be rugged in order to resist to shocks, cold temperature, water and so on. Moreover, due to its mobile nature, the RT power consumption has to be optimized.

The OC is composed by the Search and Rescue Network Precise Positioning (SAR-NPP) and the Search and Rescue Mission Control Center (SAR-MCC). The SAR-MCC is in charge to provide the graphical interface and the OC location based services as well as the interfaces between the OC and the RDC. The SAR-NPP has to manage the data communications toward the RTs by means of a communication unit and to provide the precise positioning thanks to its GPS/EGNOS professional receiver. This receiver is able to directly download EGNOS correction by the antenna or alternatively to receive the augmentation data through the internet ESA service, named Signal in Space through interNeT (SISNeT). In this way the EGNOS data should be always available at the OC side.

As sketched above, the main feature of the POP-ART system is the rescuers precise positioning demanded to the OC. The OC matches the RT data received through the active communication channel (usually VHF channel) with its own

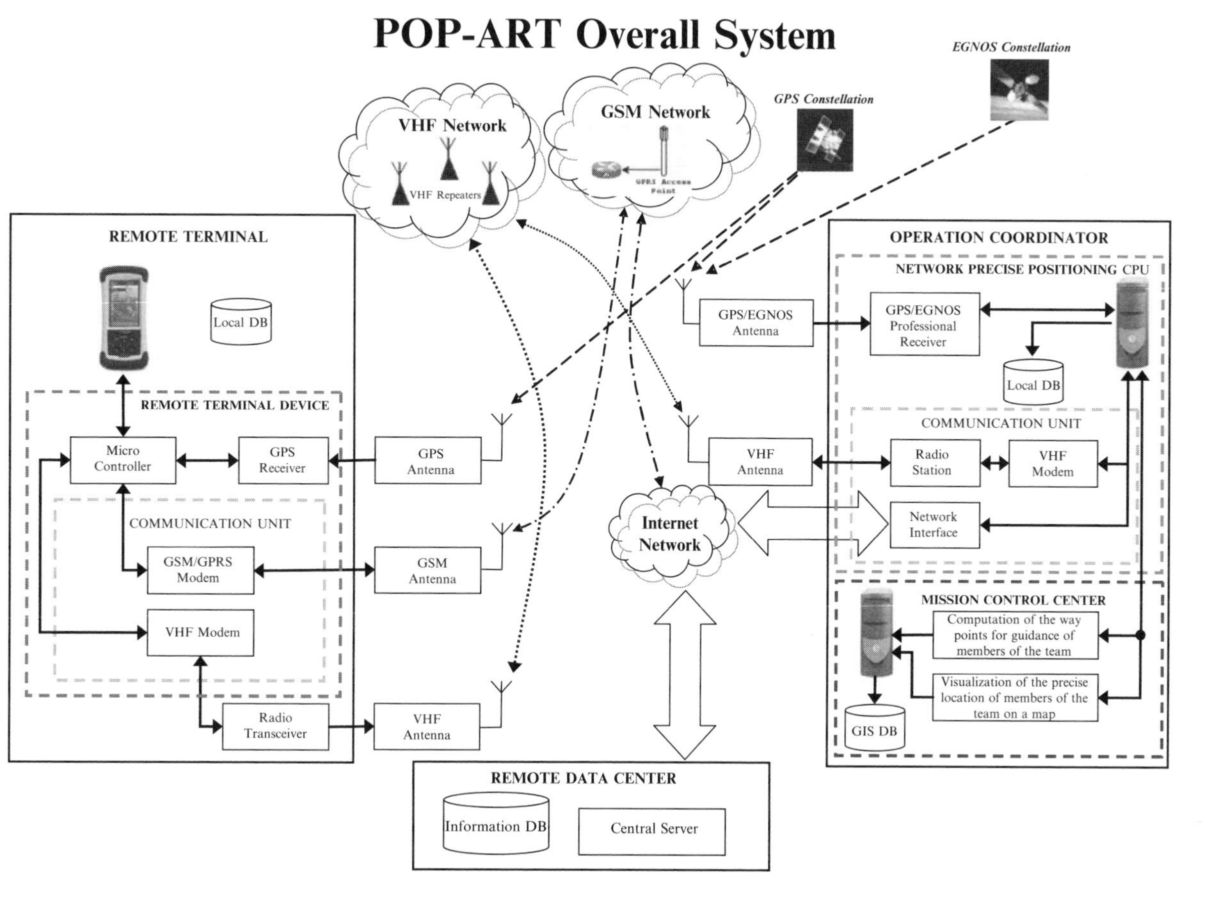

Fig. 6. The POP-ART detailed architectural design.

GNSS information, and computes the precise positioning employing EGNOS data. Then the position is sent back to the RT and immediately visualized on a digital map at the OC side. Of course also the RT, if the PDA is connected, can visualize the same position on a map.

The architecture described above permits at the operation manager level to have a real time and accurate monitoring of the rescuers employed in the field; in fact the Position Velocity Time (PVT) computation performed at the OC side is accurate (under 2 m of error, as described in Section 4) and reliable due to the continuous availability of EGNOS correction. Moreover the automatic data exchange cancels the possible mistakes causes in the rescuers position recording due to misunderstanding in the vocal communications or mistakes in recording phase.

It has to be remarked that mass-market GPS handsets enabled to apply EGNOS corrections are currently available. But the EGNOS signal is broadcast by geostationary satellites from the south direction with respect to the Alps. In Europe such satellites are seen with an elevation angle that can cause visibility problem, especially in mountain environment and, in particular, in north slopes (e.g. in the western Alps EGNOS geostationary satellite Inmarsat 3 F2, AOR-E has an elevation angle of about 30°, which could be not high enough). As the visibility of the EGNOS satellites is not assured, a network-based approach has to be followed in order to guarantee the availability, reliability and accuracy of the positioning service.

Moreover the EGNOS corrections must be continuously downloaded due to restricted validity time and applicability conditions, causing waste of power supply that is an important problem for portable devices.

The POP-ART architecture allows to rely on EGNOS features in every kind of environment, assuring also a limited (and adjustable) power consumption at the RT level.

On the top of the technical features, it is important to highlight that POP-ART allows the rescuers to receive ad-hoc LBS (Location Based Services). For example, the operation manager, can send information about point of interest such as helicopter landing or refuge, when requested.

4 Expected Performance

Regarding the accuracy that can be potentially achieved by POP-ART through the network-assisted Local Element approach; many tests have been conducted on the precise positioning using EGNOS and on the reliability of the communication infrastructure.

Such tests have been conducted under the following constraints:

- PVT computed at the network level employing raw GPS pseudorange measurements made available by standard GPS chipset (e.g. SiRF Star III) at the remote terminal level;
- EGNOS corrections downloaded from an EGNOS reference station that employs a professional GPS/EGNOS receiver and EGNOS corrections taken from the server of the SISNeT service managed by ESA;
- Static and dynamic conditions.

4.1 Static Test

Hereafter some results of a static test of about 12 hours are reported. The test has been leaded in Torino where the EGNOS satellites have an elevation angle of about 32 degrees.

Figure 7 shows the 3D positioning error with and without EGNOS corrections as well as the time percentage in which the horizontal error is lower than 2m varying the masking angle (and so varying the DOP). It has to be noted that for all the masking angles the EGNOS corrections are applied even if the EGNOS satellite results to be not in view at the terminal level. In fact, the PVT is computed at the network level where the corrections are always available as previously explained in Section 3.

4.2 Dynamic Test

Figure 8 shows the results of a test conduced in dynamic conditions within a real mountain environment in Entracque (Italy).

Such tests are needed in order to deeply understand the operational problems of non static users and of course to identify critical points, due to the hostile environment (e.g. signal availability in forests or in mountain canyons).

The selected test area is covered by the GSM service but do not give the possibility to have a stable GPRS availability, on the other hand there is a good VHF coverage.

In the Figure a specific path followed an ISMB operator has been highlighted. In such a situation the position was computed in Torino (80 km apart from Entracque) using EGNOS corrections. The rwa data where sent the VHF infrastructure. The reached accuracy demonstrates the capabilities of such a system, which relies on the integrity of the communication channel and the correct data matching between the data coming from the remote User Terminal and the data provided by Local Element.

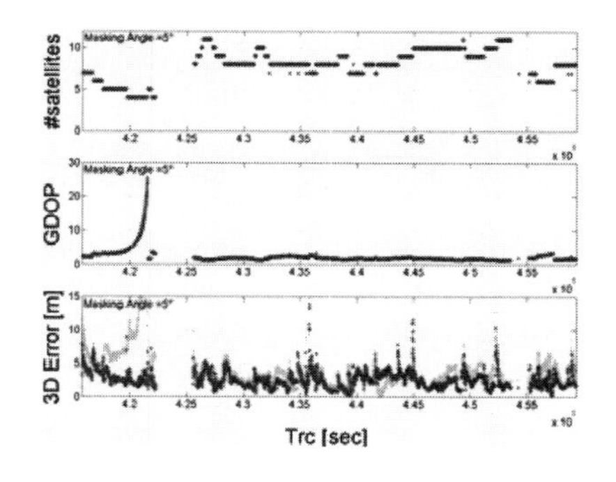

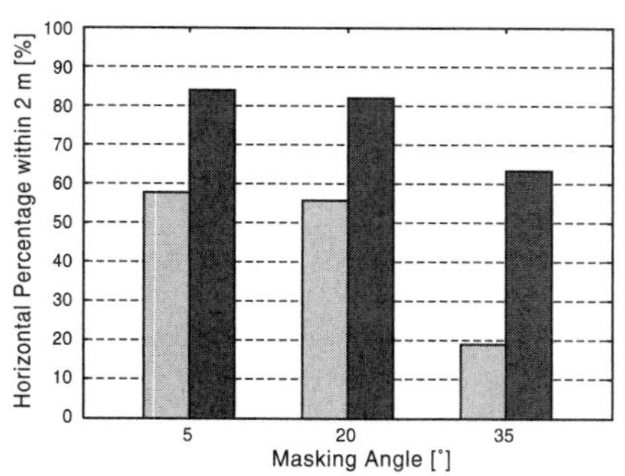

Fig. 7. Left hand-side: Comparison of the 3D positioning errors with (black) and without (grey) EGNOS corrections in the system. Right hand-side: Percentage of computed positions horizontal errors within 2 meters for different masking angles. Measurements without EGNOS corrections appear in grey while with EGNOS in black.

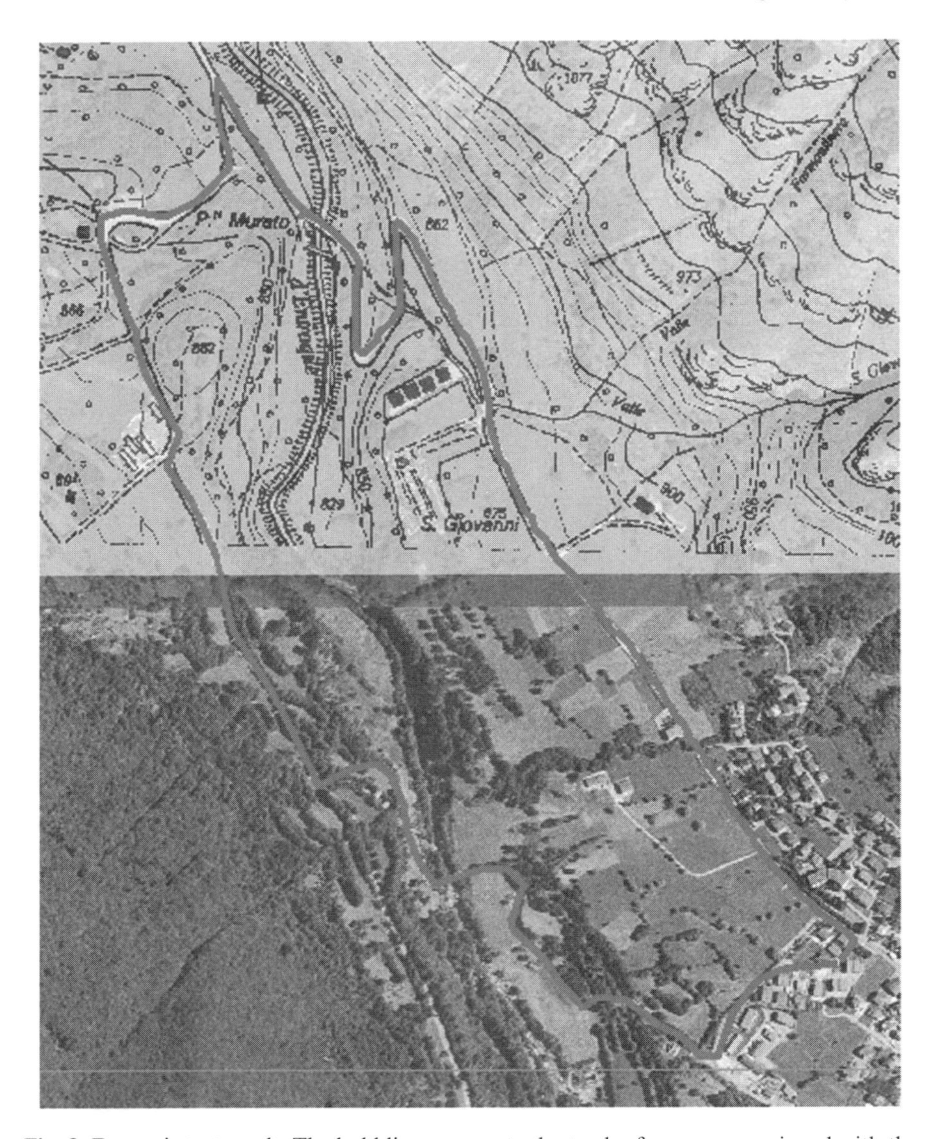

Fig. 8. Dynamic test result. The bold line represents the track of a rescuer equipped with the ISMB test prototype. This test has been conduced in a mountain environment; in particular in this area there are frequent lacks in the GPRS availability; therefore the main communication channel used for this test is VHF.

5 Development Plan

The POP-ART project has been organized following three main temporal phases (see Fig. 9). Each phase will determine a major milestone, in the project development, corresponding to the logical step of the engineering process:

- **Consolidation phase** will output a clear vision of the requirements of the specific application. As already stated this phase is conducted with a strong support of the user community. The consolidation has been ended in July 2006
- **Implementation phase** will realize a prototype of the system in all his aspects: the hardware integrated VHF/GPS/GPRS terminal, the Operations Coordination Centre, the operational software and the core of the complete service;
- **Technology transfer phase** which represents the interface towards other potential user communities (e.g. emergency management teams).

One other relevant aspect is that during this phase the steps necessary for the Galileo use will be envisaged in light of the future evolutions of the system.

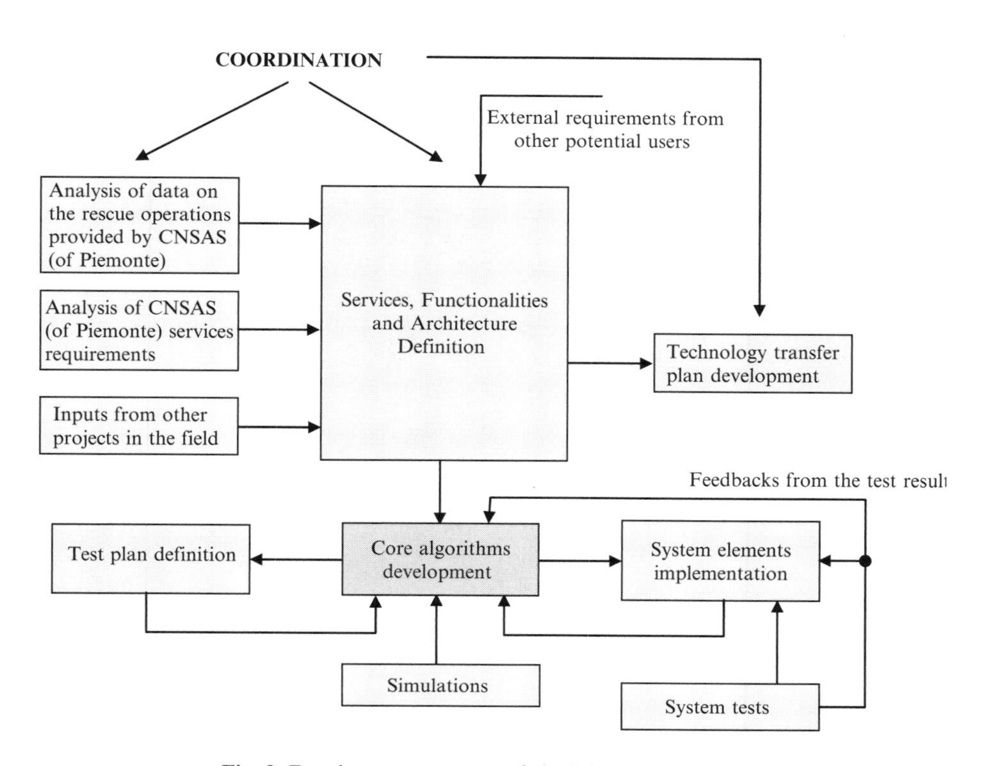

Fig. 9. Development strategy of the POP-ART project.

6 Conclusions and Future Activities

This paper presents the idea and the first results of the POP-ART project. The system adopts EGNOS corrections and VHF communications in the view of realizing a network-based PVT computation. The POP-ART system is designed taking in account the evolution of the system toward Galileo and toward the next generation of COM systems for emergency management like TETRA [6].

The presented system can improve the effectiveness of the interventions of the Alpine Rescue Teams and consequently the safety of the people involved in mountain activities.

Acknowledgements

As POP-ART is in the framework of the Galileo Joint Undertaking, the authors would like to thank all the GJU representatives for their valuable support to the work performed.

References

[1] Defina, F. Dominici, F. Dovis, M. Gianola, P. Mulassano, "An Augmented GPS/EGNOS Localization System for Alpine Rescue Teams Based on a VHF Communication Infrastructure" IEEE/ION PLANS2006, San Diego, April 2006.
[2] POP-ART web site: www.popart-project.com
[3] PARAMOUNT web site: www.paramount-tours.com
[4] P. Kovar, L. Seidl, F. Vejrazka, "Availability of the EGNOS Service for a Land Mobile User", IEEE/ION PLANS2006, San Diego, April 2006.
[5] CNSAS web site: www.cnsas.it
[6] TETRA web site: www.tetramou.com

Optical Intersatellite Links Made Easier and Affordable by Precision 3D Spacecraft Localization via GPS/GNSS Constellations

Giorgio Perrotta

IMT srl
Via Bartolomeo Piazza 8, 00161, Rome, Italy
Phone and Fax: +3944292634; email: imtsrl@imtsrl.it

Abstract. The paper focuses on the emerging space-distributed multisatellite constellations, swarms and formations, which are increasingly proposed to carry out Missions demanding tremendous intersatellite information exchange rates, thus justifying the use of optical frequencies. The paper then addresses the critical issue of how to cope with the fast initial satellite acquisition needs, considering that the satellites will increasingly be injected in orbits leading to time-variable topologies, and that communication needs increasingly require minimizing link acquisition and reacquisition times during satellites handover. Solutions based on the partial use of the GPS data available on board the satellites and broadcast to all others are described. The mix of large and powerful satellites and of less capable microsatellites demands solutions which are tailored to the planned capabilities or even capable of complementing them.

1 Facing up to an Exciting, Ever-changing Scenario

The growing interest in LEO satellite constellations, most notably for science and remote sensing applications [2] as well as for communications with mobile and fixed terminals [1], [3], [11] and new forms of space distributed assets such as satellite swarms and formations, increasingly require that optical broadband intersatellite links be reconsidered in addition to rapidly reconfigurable interconnectivity. Indeed the consolidation of the engineering methodologies for building inexpensive minisatellites [4], microsatellites and nanosatellites, will progressively lead to a shift away from a 'concentrated' architecture, characterized by very few spacecraft, to a spatially 'distributed' one where many smaller spacecraft cooperate to achieve a level of operational performance which had previously been unimaginable [5], [6]. The orbital topology of these space distributed systems will differ greatly, depending on the Mission objectives and orbit control type. For example, the satellites of swarms and constellations are conceived not to be tightly orbit controlled, therefore the distances between satellites and the line-of-sight orientation will dynamically change, while those belonging to formations are, instead, strictly controlled in terms of spacing and l.o.s. orientation . The spatial distribution of these assets brings with it new communication requirements: while some application imply quite modests intersatellite information exchange rates, others- the more interesting ones – do imply very high data rates exchanged between satellites and, what is more worrisome, a fast switching of the data flow towards different cooperating satellites.

In essence, we will witness the development of re-addressable, dynamically evolving, space data networks where each node – a satellite belonging to a constellation, formation or swarm- 'talks' sequentially with the partner satellites at very high speed and with minimum handover delay.

2 Optical Links and Related Problems

Optical technologies are ideally suited to support high data rate intersatellite link, because of the much reduced dimensions of the equipment required compared to the microwave frequencies. However the gain which is feasible with optical frequencies is paid with very narrow beamwidths, something which has two well-known consequences:

a) the necessity of disposing of very fine pointing and tracking systems to co-align the transmitter and receiver optical boresights;
b) problems with the initial spatial acquisition, and reacquisition, of the other satellite with which two way communications must be established.

The first issue is typically addressed by means of optical systems that apply well-known radar tracking techniques (e.g.: monopulse, conical scanning, step tracking) or other continuous or pulsed beacon tracking systems: but all have in common the fact that, to have a fair chance of aligning the two transceivers in a finite time, the initial angular error between their optical boresights must be in the range of a few beamwidths.

The second aspect is even more worrisome, when the initial or recurring misalignment between the optical beams boresights is unknown and of several tenth or hundredth beamwidths. Initially a brute-force, sequential, raster scan approach was proposed leading however to unpractical acquisition times. Variants were conceived, see e.g. [7], on the spatial scanning approach to decrease the acquisition times, but these remained quite high. Substantial work has been produced worldwide on the 'all optical' solutions to pointing acquisition and tracking (PAT) which often resulted in quite sophisticated [8], complex and costly systems. All this plus the fall off in the demand for commercial space communication has, 'de facto', considerably contributed, over the past few years, to discourage further development of this technology.

This author believes that combining RF and optical technologies, instead of insisting on an 'all optical' approach, may lead to solutions which are more acceptable from a technology-risk and cost viewpoints.

Indeed, the acquisition problems are bound to become even worse with the upcoming distributed space assets characterized by time-evolving spatial geometries and the requirement for very rapid handover, in the data transfer from any satellite of the group to any other in visibility, to maximize the data volume transfer per unit time.

In this scenario, system techniques are sought enabling an effective and quasi-instantaneous narrowing of the angular acquisition cell of the partner satellite with which to establish two-way wideband communications via optical ISL.

The proposed approach will be described making reference to an hypothetical constellation system, with satellites injected in multiple equal altitude orbits but different orbital planes. It is also assumed that the mission requirements would imply the sequential, but near continuous, transfer of high speed data flow between satellites in mutual visibility, and that the switching time from one satellite to the next would be minimized. The ISL will be implemented using optical transceivers and telescopes driven by a dual gimbal system – a mature technology- capable of fine pointing performance. The operational scenario implies that the distance, bearing and depression angles from one satellite to all the nearby ones will be generally different and time-varying with the satellites' orbital evolution. An optical telescope, around 10 cm in diameter, will produce approximately an 8 μrad beamwidth. The open loop initialization should bring the optical tracker boresight close to K times the optical beamwidth, with K being in the 5 to 10 range to get a fair probability of a rapid lock-on the partner satellite.

3 GPS Receivers and Low Datarate Links for Fast Open Loop Acquisition

Almost all modern satellites, both large and small, do carry GPS receivers for orbital position restitution and timing distribution Some satellites even carry a GPS receiver with four antennas placed on the tips of two interferometric arms as an additional platform attitude sensor, although this solution has been adopted with less then expected favour because ultimate performance is achieved with difficulty, due to spacecraft-induced multipath, and the increase in the receiving elements (large flare angle horns) dimensions resulting from attempts to control the dangerous multipath effects. For the proposed solution, however, we have considered satellites equipped with basic GPS receivers only.

The starting idea to solve the optical ISL initial acquisition problems is to fully exploit the GPS / GNSS navigation signal. Both satellites will carry a GPS receiver and an estimator of its spatial position in a common three-axis system, e.g. the standard inertial coordinate system. Each spacecraft will then periodically broadcast, using a near omnidirectional antenna operating at any convenient frequency, a data packet consisting of a satellite identifier followed by continuously updated three-dimensional position and speed data in the common reference system. The timing for the information exchange is proposed to be controlled in a master-slaves organization of the constellation. Thus all satellites know the three-dimensional position, in space, of all the others as well as their own position. In order to derive from these data the bearing and elevation angle, for open loop pointing the boresight of the optical communication system, the satellite must also know the orientation of the body axes w.r.t. the common coordinate system. Purely for the sake of clarity, the system geometry is shown in Fig. 1, for two satellites only.

The open loop pointing error of a gimballed optics with respect to the required direction O1–O2 results from the combination of three contributions:

- the estimate of the orbital point O1;
- the estimate of the orbital point O2, relayed from satellite #2 to satellite #1;
- the error in the attitude determination of x1,y1,z1 w.r.t. the reference coordinate system X.Y.Z

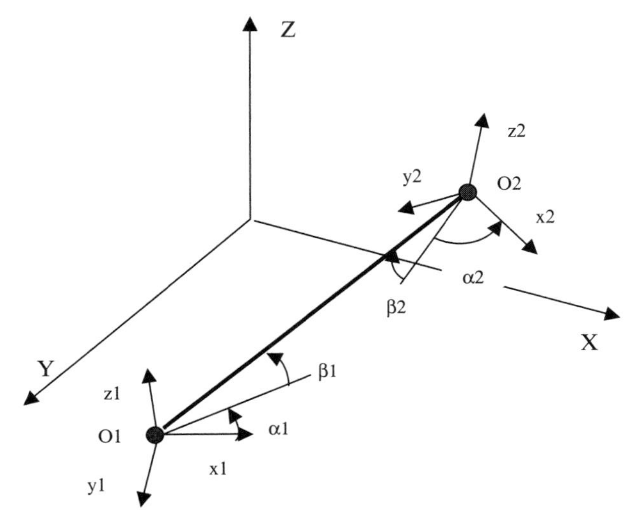

X,Y,Z: common coordinate system
x1,y1,z1; and x2,y2,z2: body coordinate axes
O1, O2: satellites' instantaneous orbital position
α1, β1; α2, β2; bearing and depression angles of O1-O2 measured in body coordinates by spacecraft #1 and #2

Fig. 1. System geometry.

The first two contributions are quite small indeed, around 10 m each, even using the C-code. This means an angular error of the order of 20 μrad for an intersatellite distance of 1000 km, which however rapidly decreases with increasing ISL distances.

The spacecraft attitude error is a cause of concern. One possibility is to use the platform' own attitude sensors (e.g. Star, Sun, Earth or a combination thereof) but the achievement of sub-mrad pointing error will require accurate and costly sensors, usually available only on larger and sophisticated spacecraft. For example state-of-the art star sensors can provide attitude measurement accuracies around 0.05 mrad. Combining this performance with the orbital position inaccuracies, one gets a boresight error estimate of around 70–80 μrads, which is 10 times the optical beamwidth. By open loop directing the optical telescope boresight according to the bearing and depression angles so computed, a fine acquistion can be performed using the optical telescope's own APT (Acquisition, Pointing and Tracking) subsystem. For ISL distances shorter than 1000 km, and down to 100 km, the error in the estimate of the vector joining the two satellites becomes significant (between 20 and 200 μrad, single measurement) but can be reduced, by one order of magnitude, applying smoothing techniques to the sequences of position, speed and attitude data of the two spacecraft.

The block diagram of the GPS-aided open loop driving of the optical telescope is shown in Fig. 2.

The specific equipment required for implementong the optical telescope open loop fast repointing, are the low data rate X_band transceiver with the associated

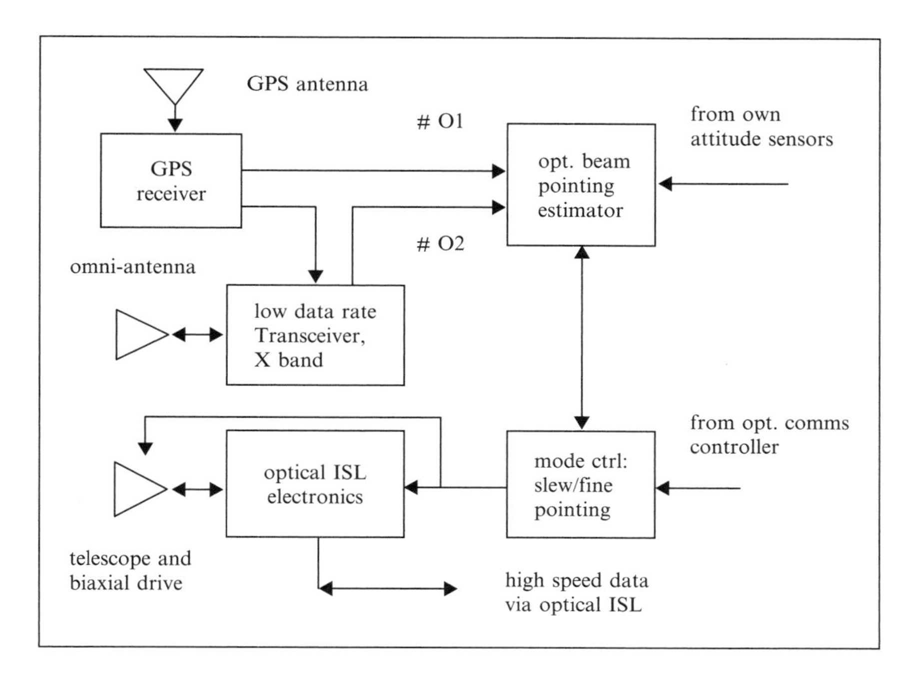

Fig. 2. Block diagram of GPS-aided open loop optical ISL pointing.

omni directional antenna and the optical beam pointing estimator. The antenna is a small bi_cone with a toroidal beam 360° × 20°; the X_band transceiver operates in simplex mode – that is transmission and reception are not simultaneous- and features a duplexer, a 1 W output power stage, a LNA and BPSK modulator and demodulator. The transceiver emits every 5 seconds, and for a duration of 1 second, a packet of about 100 bit, generated by the GPS receiver at a rate of 100 bps. The transceiver receives, instead, a sequence of 1 sec. data packets- at a rate of 100 bps- corresponding to the emission of the nearby satellites broadcasting their spatial coordinates. The timing, coordinated by the 'master satellite' takes into account the required guard times between bursts emissions.

The optical beam estimator is a simple microprocessor-based circuit that receives the position and speed data from the local GPS receiver, the remote GPS receivers hosted in the other satellites, and the locally generated data from the satellite attitude sensors. The circuit computes – by means of software routines mainly involving matrix operations, filtering and smoothing – the bearing and depression angles as imput to the optical telescope biaxal drive. For very wide repointing when passing from one satellite to the next, the circuit computes also the time sequence of the pointing coordinates during the transient to optimize the latter.

The link budgets shows that a 1 W RF power at X_band is sufficient to transmit 100 bps data rate signals in BPSK over a free space distance of up to 5000 km – the maximum practical L.O.S. between two LEO satellites. The provision of a simple coding will then provide the required margin to achieve an error rate of 10^-5 which is adequate for this application.

Obviously, the use of the auxiliary, small, X_band transceiver can be exploited for other communication functions, in which case the signal bandwidth as well as other system parameters will have to be adapted to the new requirements.

4 Interferometry Based System for Fast Acquisition and Tracking

Relying on top performance platform attitude sensors does not seem fit for the micro e minisats targeting the medium-low cost market. Indeed many of these low cost satellite, even those carrying out missions that may justify the use of optical ISL, are characterized by less than top performance Attitude Control Systems. Accordingly, other means have been considered to achieve a fast open-loop initialization of the optical APT. It is envisaged that many applications of mini, micro and nanosatellites constellations requiring very high datarates transfers might concern smaller inter-satellite distances, thus smaller telescope diameters, say in the 1 cm range or less providing some 80 μrad optical beamwidth.

One possible approach is to make resort to an interferometric system exploiting a signal broadcast by the partner satellite. The interferometric system, which has no moving parts, must have a 360° coverage in azimuth and some 20° to 30° in the elevation plane, to cope with the ambiguities in the depression angle. The sensitivity of the interferometric system depends on the arm length and operating frequency. Concerning the latter, either the X or Ku band, or even the millimeter band, are preferred to lower frequencies for the following reasons:

- greater interferometer sensitivity for a given baseline length;
- smaller antennas for a given design beamwidth.

With a baseline of 10 wavelengths at X_band (that is an interferometer arm length around 30 cm), a phase detector and associated processor capable of 0.006° sensitivity, the error in evaluating the angle of arrival of the incoming beacon signal emitted by the partner satellite will be in the range of 0.15 mrad. The error contribution due to the Doppler effect – the two spacecraft have a non-zero relative speed- is negligeable. The interferometer directly provides the bearing and depression angles of the incoming wavefront measured in body coordinates: and these can be used to drive the optical telescope dual gimbal system. Ambiguity problems, typical of long baseline interferometers, will be solved implementing a dual baseline, i.e. coarse-fine, geometry.

The potential advantage of the interferometric system is that one could eliminate the optical system pointing and tracking system by implementing a software-based optics beam pointing estimation algorithm. Indeed, the nearby satellites move with a relative velocity which can be relatively low-as in the case of a formation deployed throughout a quite limited spatial volume- or as high as twice the spacecraft orbital speed - as in the case of a constellation with satellites coarsely distributed around Earth. The relative speed justifies using an adaptive – proportional plus derivative-software-based tracker (e.g. an alpha-beta tracker or other efficient algorithms) where the individual measuremenents, performed with the interferometer, are collected and processed to extrapolate the near term relative satellite motions from actual measurements and the recent past history. This processing efficiently implements a smoothing of the data points, therefore the prediction accuracy is estimated to be substantially better than the 0.15 mrad of a single, isolated, measurement-based evaluation.

The physical implementation of the interferometric system will depend on the mini- micro or nanosatellite dimensions, most of which have a cube-like shape and while the sides are not generally available because reserved to other subsystems, the satellite wedges, or the edges of the side panels, can be- instead - cleverly used. To provide a 360° coverage in azimuth all four side panels could carry a thin dielectric strip with printed patch antennas implementing the short and long interferometer baseline. To provide the 20° or 30° coverage in the vertical plane, all four panels will also carry a printed circuit strip at 90° w.r.t. the other. A possible arrangements is given in Fig. 3, showing just one of the four side panels of the satellite.

The horizontal and vertical interferometer RF channels are kept separated. The switching between the short and long baselines of each quadrant, as well as the sequential switching of the four interferometer quadrants is made with PIN diode switches driven by a control logic. A simplified functional diagram of each interferometer arm is given in Fig. 4, which is valid for both the azimuth and elevation interferometer arms, following a highly modular approach. The beat signal resulting from the mixing of the three signals received from the three patch antennas with a carrier generated by a local oscillator are input to a phase detector a pair-at-a-time. The signals' selection is made in a SPDT PIN switch. The phase detector output is thus proportional either to the bearing angle or to the depression angle.

In a constellation environment a time-shared operation of the tranmit and receive tasks can be easily managed: a repetitive time frame is defined and divided in N slots where N is the number of satellites mutually and simultaneously visible from any satellite in the constellation. Each satellite transmits sequentially a signal during one

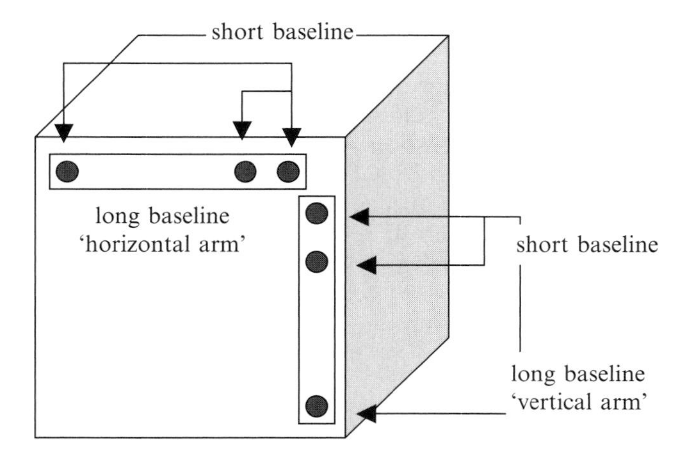

Fig. 3. Possible arrangement of long and short baseline interferometer antenna elements on the satellite side panels.

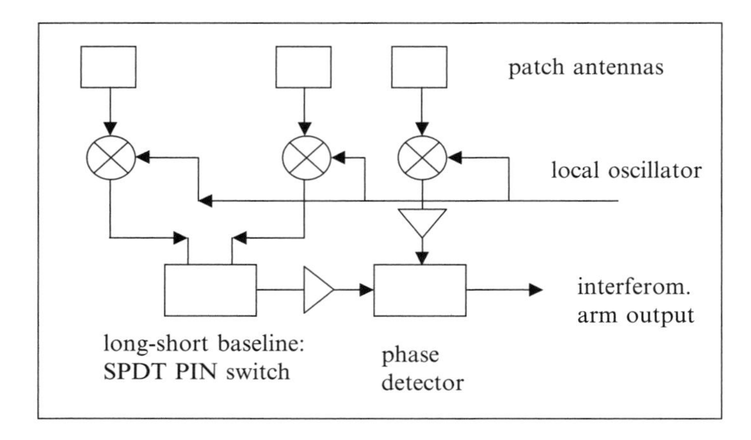

Fig. 4. Schematics of the azimuth or elevation interferometer arm.

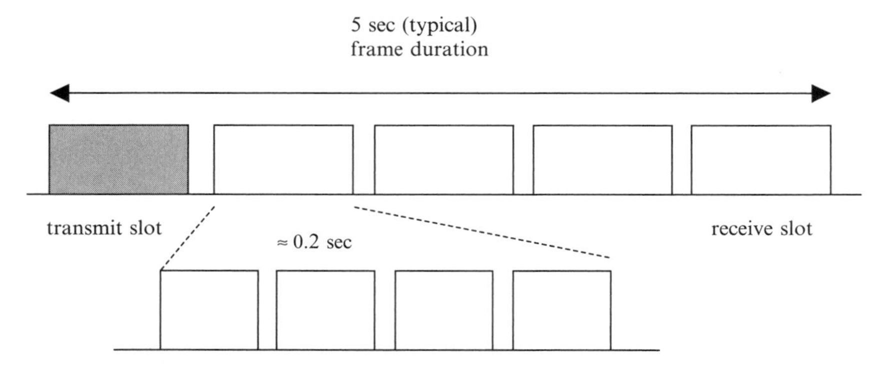

Fig. 5. Sample system timing.

time slots and can receive the other N-1 signals emitted by the other N_1 satellites in the corresponding time slots. During each 'receiving slots' the interferometer performs the measurement of the bearing and depression angles of the incoming wave front emitted by the M-th satellite. The quadrant switching is performed within each 'receive slot' and the four sequential signals at the output of the phase detectors are compared to sort out ambiguities and gross errors.

A sample system timing is shown in Fig. 5 illustrating the subdivision of each 'receiving slot' into four sub-slots for parallel evaluation of the side panels horizontal and vertical interferometer arms.

A simplified block diagram of the satellite interferometer subsystem as shown in Fig. 6.

One of the four 'horizontal arms' outputs, designated as the most likely, is then used to compute the bearing angle. A similar process is performed in parallel for the 'vertical arms', resulting in the evaluation of the depression angle. The data are temporarily stored in a bank of N_1 small memories in order to be further processed.

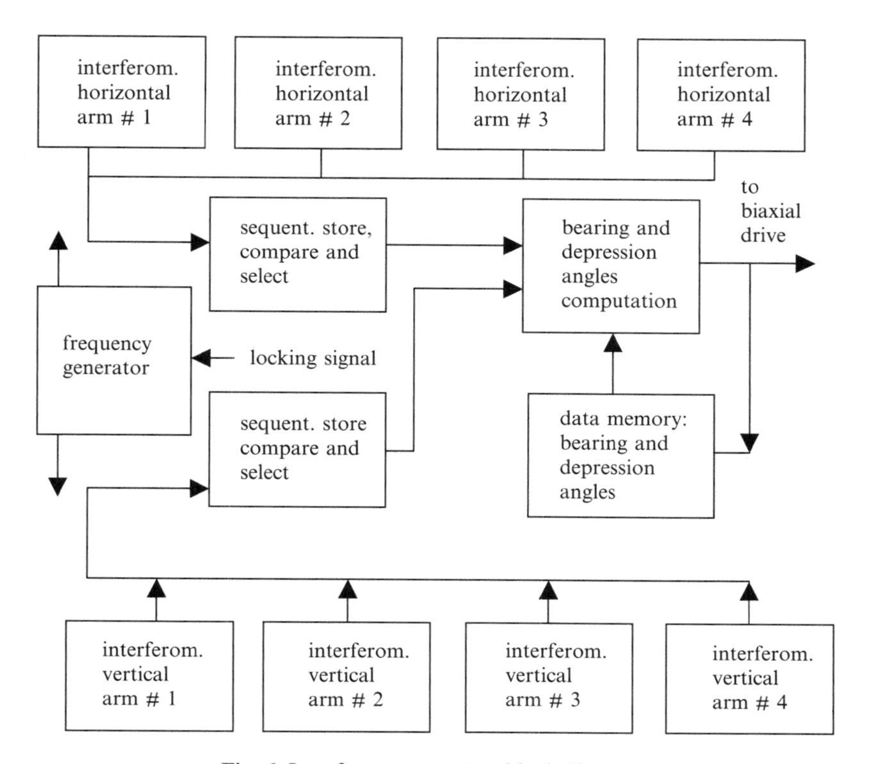

Fig. 6. Interferometer system block diagram.

The outputs from the horizontal and vertical arms of the interferometer are then input to a software-based estimator implementing advanced filtering and smoothing techniques, making use of past evaluations too, and the output is sent to the dual gimbal drive of the optical telescope.

Concerning the physical implementation of this RF wavefront angle-of-arrival sensor system, the four plus four printed circuit strip carrying the three X_band patch elements will have a width of around 10mm. The strip will also integrate the mixers, PIN switch, IF amplifiers and comparator, taking advantage of the high integration and miniaturization level achieved by RF and IF parts developed for the mass market, and that can be adapted to operate in the space environment provided suitable precautions are adopted. The arm assembly will thus take the shape of a rectangular cross-section rod, which can be either laid on the edge of the side panels or on the wedges of the satellite.

The local oscillator feeding all eight interferometer arms is phase locked to the sum signal picked-up either by a combiner of all antenna patches distributed around the satellite or by a dedicated antenna element. The carrier so generated will then be simply distributed by means of a times-eight divider without any worry as to path-length equalization since the three mixers of each arm are fed in-phase by a suitable design of the paths inside the rectangular cross-section rod.

Preliminary link budgets seem to point to the feasibility of operating the interferometer by transmitting a 0.5 W X_band carrier via the bicone antenna from the partner satellite, and assuming hemispherical beamshapes for the interferometer' arms patch elements.

5 Combined MW and Optical Technologies for Microsatellite Fast Acquisition and Tracking

It should be noted that, in this paper, we focus on existing and near-term plans for implementing formations or satellite constellations featuring augnificant intersatellite information transfers. The problem of fast satellite acquisition and beam repointing when using optical ISL in close range formations, has its own specificities which are summarized below.

- sophisticated APT systems are undesirable for cost and system integration reasons. This may indeed rule out the interferometric system described above, which seems more suitable for medium-class minisatellites;
- the platform attitude determination sensors, usually chosen to meet the essential Mission requirements, may be unsuited to support open loop pointing of the optical beam. Indeed many microsatellites are designed with platform pointing accuracies in the 0.1° to 1° range, which are orders of magnitude greater than the optical beamwidth. Indeed the latter, if designed to support – over distances in the 10–100 km range- datarates of several hundred Mbps, will have a size well under 0.01°.

A 100:1 or 10000:1 ratio in the scan cell widths to optical beamwidth ratio implies scanning, or handover times for the acquisition of the co-operating satellite, which are too long. One solution is to adopt multiple, wider, optical beams either on transmit or receive, as basically proposed in [5], [12]. However that solution penalizes the data rates that can be transmitted over the ISL, without increasing the tranmitted optical powers, something which is not compatible with the DC power budget typically available in most microsatellites.

A possible solution comes from revisiting a concept illustrated in [9]. The system configuration of that patent makes reference to a pair of satellites, neither of which knows its attitude to a high degree of accuracy but both of which are nevertheless equipped with GPS receiver. Each satellite sends its own position to the other, via a radio-link, whereupon each satellite computes the vector joining the two satellites, as described in paragraph 3. In addition, each satellite is equipped with a dual gimballed antenna which acquires and tracks the RF emissions of the other satellite. The gimbal angles are taken as a measure of the tracker boresight angles in body coordinates. From a comparison between the computed angles (body perfectly aligned with the reference axes) and the measured angles, the patent claims that it is possible to compute the spacecraft attitude error: which is indeed conceptually correct. However instead of using the system to derive the spacecraft pointing error, we propose a different embodiment.

The basic idea is to integrate – in a single multipurpose Unit- a coarse, fast, acquisition system working in the microwave bands and a fine acquisition and tracking system working in the optical bands. In must be said that the open literature reports sparse examples of Intersatellite PAT combining optical and microwave frequencies, one of which is found in [10], though applied to stratospheric platforms and mainly to combat fading.

The combined system will first search, and coarse acquire, a RF beacon emitted br the partner satellite, then will switch to the optical PAT which is in charge of the fine optical signal acquisition and tracking. To this end the optical dynamically repointable telescope will also integrate a microvave small planar antenna designed to generate both a sum beam, substantially larger than that of the optical telescope, and two 'difference' signals proportional to the error between the instantaneous l.o.s. between the two satellites and the commanded mechanical antenna boresight.

During the initial acquisition phase, the microvave system is active while the optical system is in stand-by. A nulling system is applied to the microvave antenna error outputs and to the dual gimbal drive, as in a conventional tracker; therefore the mechanical boresight will gradually move in the direction to reduce the misalignenent between the boresight and the instantaneous l.o.s. When the misalignment has been reduced to a very small value, then the microvave system goes in stand-by and the optical system is activated. The 'optical' PAT will then work within rather narrow angular boundaries, completing the fine angular acquisition then passing into the fine tracking mode. The microwave system will then be put on hold till a new wide angle acquisition in occasion of satellite handover.

It must be noted that the system operation can do without the transfer of information about the two satellites orbital position data obtained from the space-borne GPS receivers: indeed the operation of the coarse and fine nulling systems does only rely on the sequential emission, by the partner satellite, of microwave and optical beacons, either unmodulated or modulated. However, the availability of the intersatellite l.o.s. vector, if computed on board, specially in the case of under-instrumented microsatellites, can be used to improve the attitude error estimate and, ultimately, the microsatellite attitude control performance. This might be instrumental for critical Missions, such as the Earth Observation ones, implemented by means of space-distributed assets.

Figure 7 shows an Artist's view of a combined microwave-optical ISL terminal for microsatellites. The cylindrical section, with a diameter around 10 cm, includes a four-quadrant planar passive array and a beamformer providing a sum and two difference signals. The choice of an operating frequency in the 26 GHz band would give component beamswidths around 15°: which is very convenient for fast satellite acquisition. The 45° flat mirror is steered in azimuth by + – 180° and in elevation by + – 15° by means of two small drive motors which are active in all operating modes. Since the flat reflector will rotate w.r.t. the fixed array antenna, the use of circular polarization is mandatory. At the center of the flat array there is an hole for passing signals in the optical bands: see Fig. 8. The laser transmitter and the optical detector are housed in the lower part of the fixed cylindrical body: an optical prism, or dichroic filter, will separate the tramsmit from the receive optical frequencies allowing to share the same focal region.

Fig. 7. Artist's view of the combined microwave-optical ILS terminal for microsatellites.

Fig. 8. A detail of the fixed cylindrical section housing the MW flat array and the optical transceiver head.

The flat reflector, which is polished to an optical quality, reflects the optical signals establishing the correct origin to destination paths, while mirror steering, under control of the APT, allows performing both fine acquisition and tracking of the partner satellite.

The small dimensions of the assembly are compatible with micro and nanosatellites. It should be borne in mind, according to the previous discussions, that if the terminal is operated alòong with on-board GPS receivers, it can also provide attitude determination data: which would be an important by-product and microsatellite performance-enhancer.

References

[1] W.L. Morgan 'Intersatellite links'; Space Business International, 1st Quarter 1999 (mentioned for historical reasons...)
[2] D.L. Begley 'Laser Cross-Link Systems and Technology'; IEEE Communications Magazine, August 2000;
[3] V. Chan 'Optical Space Communications'; IEEE Journal on Selected Topics in Quantum Electronics; Vol. 6 N° 6, Dec. 2000;
[4] A. Wicks, A. Da Silva Curiel, J. Ward, M. Fouquet 'Advancing Small satellite Earth Observation operational spacecraft, planned missions and future concepts'; 14th Annual AUAA/USU Conference on Small Satellites
[5] W. Leeb, K. Kudielka, P. Winzel, B. Furch 'Optical cross links for microsatellite fleets'; 20th AIAA Int'l Communication Satellite Systems Conference; Montreal 2002;
[6] U.Tancredi, M. D'Errico 'Microsatellite configuration design for an Earth Observation Mission based on the distributed sensor concept'; 4th IAA Symposium on small satellites for Earth Observation, April 2003, Berlin;
[7] J. Wang, J. Kahn, K. Y.Lou 'Minimization of acquisition time in short range free space optical communications; Applied Optics Vol. 41 N° 36, Dec. 2000;
[8] M. Guelman, A. Kogan, A. Kazarian et al 'Acquisition and Pointing Control for Inter-Satellite Laser Communications'; IEEE Transactions on Aerospace and Electronic Systems, Vol. 40 N° 4, Oct. 2004;
[9] United Stated Patent 5959576 'Satellite attitude determination using GPS and intersatellite line-of-sight communications'
[10] T. Dreischer, A. Marki, B. Thieme 'Fade Tolerant Beam Acquisition and Tracking for Optical Inter_HAP crosslinks'; IST Summit 05, April 2005
[11] M. Toyoshima, 'Trends in satellite communications and the role of optical free-space communications; Journal of Optical Networking, Vol. 4, 2005
[12] W. Leeb et ali 'Optical terminal for Microsatellite Swarms'; SPIE 'Free space Laser Communication Technology'; Vol. 4635, April 2002

Chapter II

Navigation Satellite Technologies

A Satellite for the Galileo Mission

J.C. Chiarini / C. Mathew[1], H.P. Honold / D. Smith[2]

[1]Galileo Industries GmbH, Lise Meitner Strasse 2, 85521 Ottobrunn, Germany
Phone: +49 89 88 984 – 44 441, Fax: +49 89 88 984 – 44 455
e-mail: jean-claude.chiarini@Galileo-Industries.net /
e-mail: Colin.mathew@Galileo-Industries.net
[2]EADS Astrium Germany, Ludwig Bölkow Allee, D-85521 Ottobrunn, Germany
Phone: +49 89 607 28154, Fax:+49 89 607 28546
e-mail: hans.peter.honold@astrium.eads.net /
e-mail: david.smith@astrium.eads.net

Abstract. The Galileo System is based on a 30 spacecrafts constellation in MEO orbits, controlled and commanded in S-band link by a Ground Control Centre. The navigation service is achieved via transmission to the user of L-band signals comprising ranging codes and timing information.

The timing signals are provided by precision on-board atomic clocks, implemented as two redundant pairs per satellite. Two different technologies are implemented - passive hydrogen maser and rubidium.

In addition to the navigation services, a Search & Rescue service is provided, which is implemented by a dedicated payload.

The specified lifetime for the Galileo satellites is 12 years while the overall specified lifetime of the Galileo System is 20 years. This means that a full replenishment of the satellite constellation will be required. In the frame of the IOV phase, 4 satellites will be produced and deployed in two different orbit planes in two dual launches using the Soyuz.

1 General Information

The Galileo constellation comprises of 30 satellites placed in MEO orbit, with 10 satellites placed in each of 3 orbital planes distributed evenly round the equator. The active constellation comprises of 27 satellites, with each plane containing a spare satellite which can be moved to replace any failed satellite within the same plane, thereby reducing the impact of failures upon quality of service. All satellites are identical in terms of design, performance capability and fuel load. The satellite flight configuration is shown in Fig. 1. The earth-pointing face is defined along the +Z axis. In launch configuration the solar arrays are stowed on the +/– Y sides of the satellite. The volume and outer shape of the satellite is compatible with the shroud dimensions of the selected launchers. Clearly visible on the earth-facing side of the satellite are the search and rescue and navigation payload antennas.

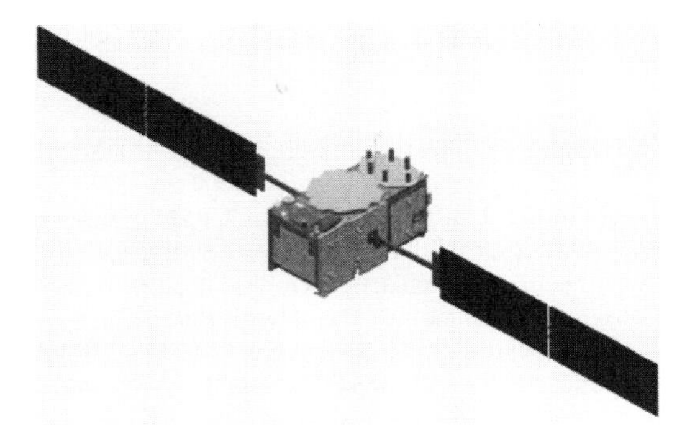

Fig. 1. Galileo satellite flight configuration.

The satellite is composed of the following subsystems:

– Payload Subsystem including the navigation payload and the SAR payload
– Structure Subsystem
– Thermal Control Subsystem (TCS)
– Electrical Power Subsystem (EPS) with the following units:
  - Solar Arrays (SA)
  - Solar Array Drive Mechanisms (SADM)
  - Battery
  - Power Conditioning and Distribution Unit (PCDU)
– Harness
– Avionics Subsystem with
  - on-board computer (Integrated Control and Data Handling Unit, ICDU)
  - Attitude and Orbit Control System, AOCS (based on earth sensors, sun sensors, gyros, reaction wheels and magnetic torquers),
  - Software (SW)
– Telemetry, Tracking and Command (TTC) Subsystem (with S-Band Transponder and two low-gain, omni-directional antennas)
– Propulsion Subsystem (mono-propellant system with one tank and 8 thrusters)
– Laser Retro-Reflector (LRR)
– Platform Security Unit (PFSU)

Launchers for IOV

For IOV Launchers the launchers already selected is Soyuz, with dual launch. The configuration under fairing is shown in Fig. 2.

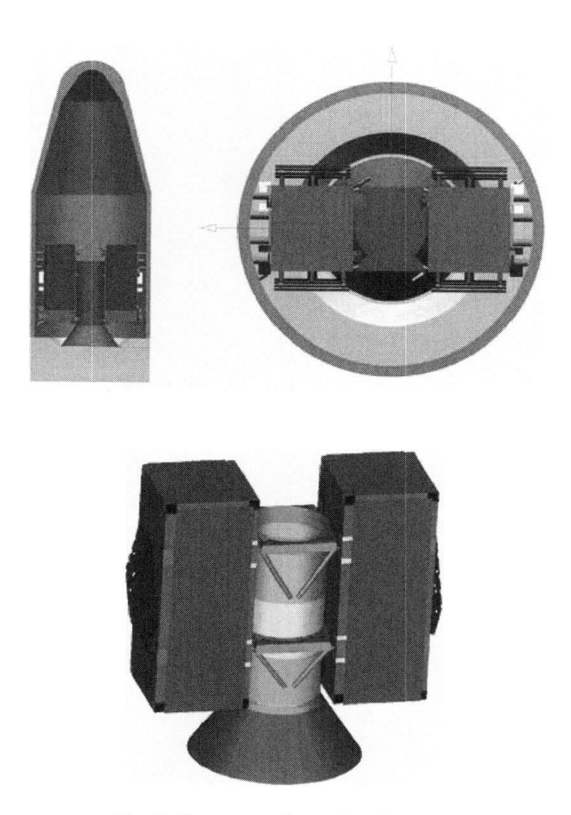

Fig. 2. Soyuz configuration for IOV.

2 Payload Architecture

The Galileo satellites include two payloads, the Navigation payload and the Search and Rescue payload.

The overall payload block diagram is presented in Fig. 3, here under.

Navigation Payload

The main functions of the navigation payload are:

- Provision of on-board timing signals
- Receipt & storage of up-linked navigation message data
- Receipt & storage of up-linked integrity data
- Assembly of navigation message in the agreed format
- Error correction coding of navigation message
- Generation of ranging codes

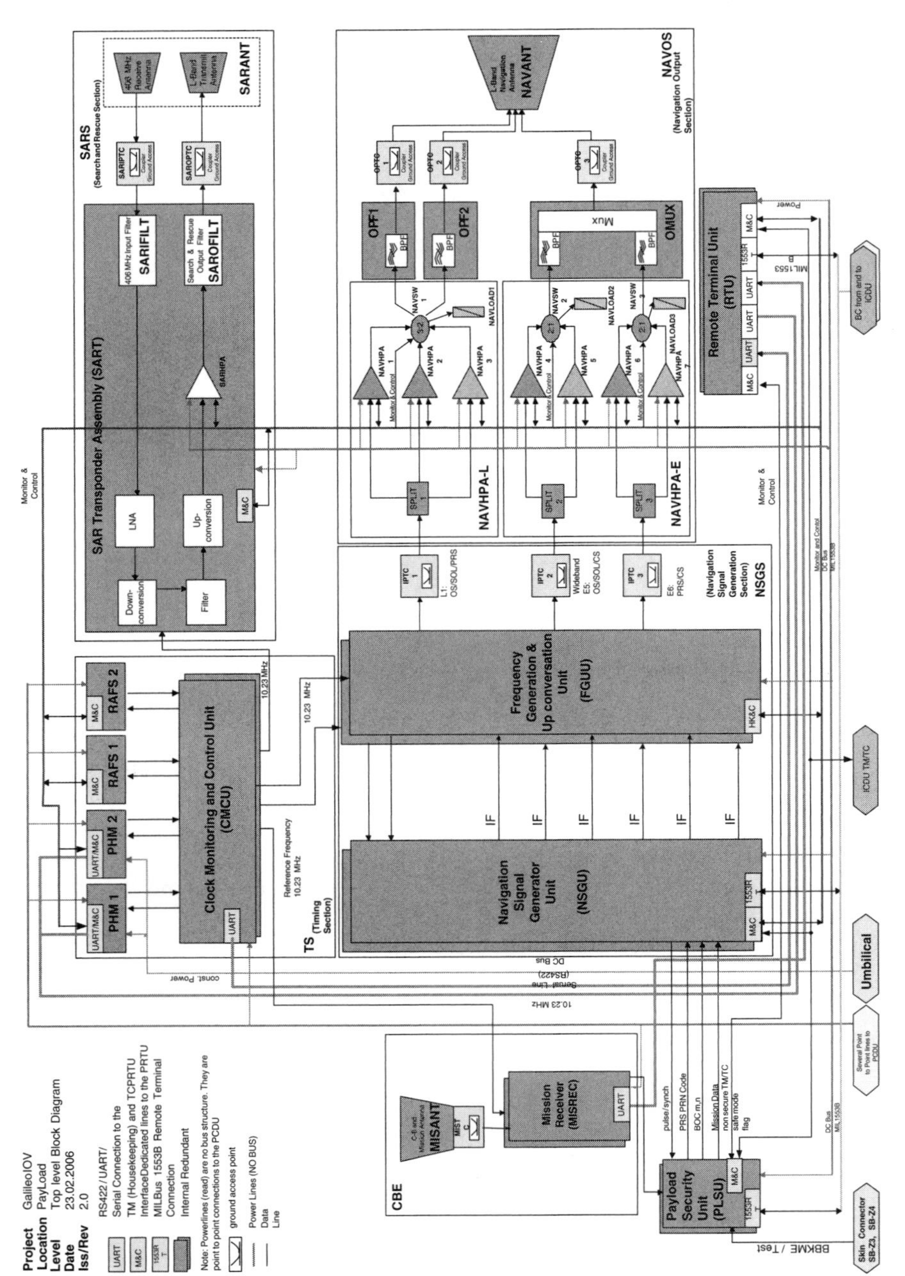

Fig. 3. Navigation payload.

- Encryption of ranging codes as required
- Generation and modulation of L-Band carrier signals
- Broadcast of navigation signals

The timing signals are provided by high precision on-board clocks, implemented as two (cold) redundant pairs per satellite, each pair including two different technologies, the Passive Hydrogen Maser (PHM) which is the primary reference clock and the Rubidium Atomic Frequency Standard (RAFS), both of them being operated simultaneously. Due to the highly stable frequency stability requirements the clocks are mounted on a separate radiator panel, which is kept facing deep space using a yaw steering law controlled by the Avionics.

The whole clock ensemble is under the control of a dedicated (internally cold redundant) Clock Monitoring and Control Unit (CMCU) which performs the monitoring and switching functions (selection under Ground control) and generates a highly stable on-board reference frequency of 10.23 MHz which is distributed to the other payload units.

The navigation data (including integrity data, Search and Rescue data and other mission data) are contained in the C-band spread spectrum uplink signal which is received via the Cband mission receive antenna operating in RHCP polarisation (baseline solution is a small aperture axially corrugated circular waveguide horn). The Mission Receiver (MISREC) which includes the Mission Processor function (MISPROC), performs the receive function, the despread and demodulation functions in order to provide a data stream which is routed to the Payload Security Unit (PLSU) which performs COMSEC treatment of the incoming signal, and passed to the Navigation Signal Generator Unit (NGSU). The MISREC is an internally cold redundant unit.

The Navigation Signal and Generator Unit (NSGU) which includes internal cold redundancy, receives the up-linked navigation data and uses them to generate the navigation signals in the appropriate format, performs the PRN encoding and the modulation of the 3 navigation signals (E5a + E5b, E6 and L1) and passes them to the Frequency Generation and Upconversion Unit (FGUU) which performs the up- conversion into L-band of the 3 signals.

The FGUU includes internal cold redundancy.

The 3 L-band navigations signals are then routed to a 2:1 (E5a + E5b and E6 with respectively about 65 W and 70 W SSPA RF output power) or 3:2 (L1 with parallel amplification at about 50 W SSPA RF output power) SSPA redundancy ring. The two amplified L1 signals are routed to the navigation antenna and combined in free space while the two other signals are multiplexed (within the NAVOMUX) before being routed to the navigation antenna.

The Navigation transmit Antenna (NAVANT) which is operated in RHCP polarisation consists of a high and a low band beam-forming network and a dual band array of radiating elements which provides a global coverage iso-flux radiation pattern.

Test couplers are included in order to allow the testing of the different navigation payload sections during the payload and satellite AIT cycle.

The payload Remote Terminal Unit performs the communication functions between the payload and the avionics subsystem via the 1553B data bus as well as the acquisition of all payload units telemetry and the distribution of all commands to the payload units except the On/Off commands and the acquisitions of the correspon-ding status of the payload units connected to the 1553B data bus (in order to allow these units to be switched off in case of anomalies occurring on the 1553B data bus).

The Navigation Payload consists of the following units:

- Ultra Stable Oscillators
  - Passive Hydrogen Maser (PHM),
  - Rubidium Atomic Frequency Standards (RAFS),
- Clock Monitoring and Control Unit (CMCU),
- MISANT antenna (C-band)
- Mission Receiver (MISREC)
- Payload Security Unit (PLSU),
- Navigation Signal Generator Unit (NSGU),
- Frequency Generator & Up Converter Unit (FGUU),
- NAVANT antenna (L-band),
- NAVHPA (SSPA),
- Output Multiplexer (OMUX) and filters,
- Remote Terminal Unit (RTU).

Search & Rescue Payload

The Search And Rescue (SAR) payload is principally based on a transparent transponder receiving the uplink signal in the 406.0–406.1 MHz band and retrans-mitting it in the 1544.05–1544.15 MHz band. The SAR payload includes no redun-dancy at satellite level.

Redundancy at mission level is obtained via the satellites of the constellation.

The SAR antenna performs both the receive function at UHF-band (RHCP polar-isation) and the transmit function at L-band (LHCP polarisation). The baseline solu-tion consists of a ring of 6 quadri-filar helix radiating elements for the UHF receive function, with a central 'splashplate' feed acting as the L-band transmit antenna.

The receive signal is routed to the Search and Rescue Transponder (SART).

The SART performs low pass filtering of the received signal, amplification via a Low Noise Amplifier (LNA), output filtering and then down-conversion into IF, close-to-band filtering using crystal filtering (switchable in order to provide a wide-band (90 kHz) or narrow band mode (50 kHz)) and up-conversion into L-band.

The signal is then amplified with a nominal RF output power of 5 W, filtered and routed to the SAR antenna transmit port for transmission.

Test couplers are included in order to allow the testing of the SAR payload dur-ing the payload and satellite AIT cycle.

The Navigation Payload consists of the following units:

- Search and Rescue (SAR) Transponder,
- SAR Antenna (at UHF for receive and L-Band for transmit).

The SAR payload uses the 10.23 MHz signal from the navigation payload for the frequency up-conversion which is the only interface between the two payloads.

3 Electrical Architecture

The overall satellite electrical architecture is presented in Fig. 4.

The satellite power is distributed to the electrical units via a 50 V fully regulated power bus delivered by the Power Conditioning Unit (PCDU).

The power sources include the solar array which delivers approximately 1700 W at EOL and a Li-ion battery (baseline is a 9s3p SAFT battery design with G5 cells).

The sizing requirement for the battery is the energy required to supply the satellite during the long duration launch sequence used for direct injection. During this phase the satellite is supplied (nearly) only by the battery. In the cruise phases where the launcher does not perform any manoeuvre, the attitude of the launch vehicle can be optimised for the satellite requirements. It is therefore foreseen to follow a barbecue type attitude law with the sun direction being normal to the rotation direction. This allows both a minimum amount of power to be provided via the external panel of the stowed array (external face is the cell face).

Most satellite units are switched off during launch. However a significant amount of heating power is required to maintain the satellite units in their applicable temperature limits (mostly non operating limits).

Most part of the radiator area is covered by the stowed solar array wings, which limits the heat loss but the clock radiators and a part of the platform radiators are uncovered.

Furthermore, all external units (sensors, thrusters, etc.) require heating. The battery capacity which is maximised at launch via a specific charge procedure (at a higher cell voltage than the one used for cycling operations) has been sized to be compliant to all launch.

However, the current baseline is based on conservative assumptions in terms of heating power and since no power is assumed to be delivered by the solar array.

The SADM performs the solar array orientation function and the transfer of the solar array power to the satellite. Due to the yaw steering law, the SADM is used in normal operation around the –X axis within an angular range which depends on the sun elevation (w.r.t. the orbit plane) and which can reach up to ± 90° for zero sun elevation. However, the SADM offers the possibility to orient the SA in any position around the Y axis in order to provide the capability to perform long duration orbit-keeping manoeuvres (which require solar array power) with the necessary flexibility (w.r.t. the sun direction) to achieve the repositioning duration requirements.

The battery management function is performed by the PCDU. The 50 V power bus is distributed to most electrical units except to the propulsion units (pressure transducer, thrusters and latch valves) which are supplied via a dedicated 28 V power supply (also provided by the PCDU) and to some of the avionics units such as the sun sensors and the magnetic torque rods, which are supplied by the ICDU.

The PCDU performs the power bus regulation function via shunt switching, the distribution and the protection functions according to a star topology using SSPC as protection devices to the permanent lines (S-band transponder Rx, PFSU, ICDU) protected by Foldback Current Limiters (FCL) and to the switchable lines

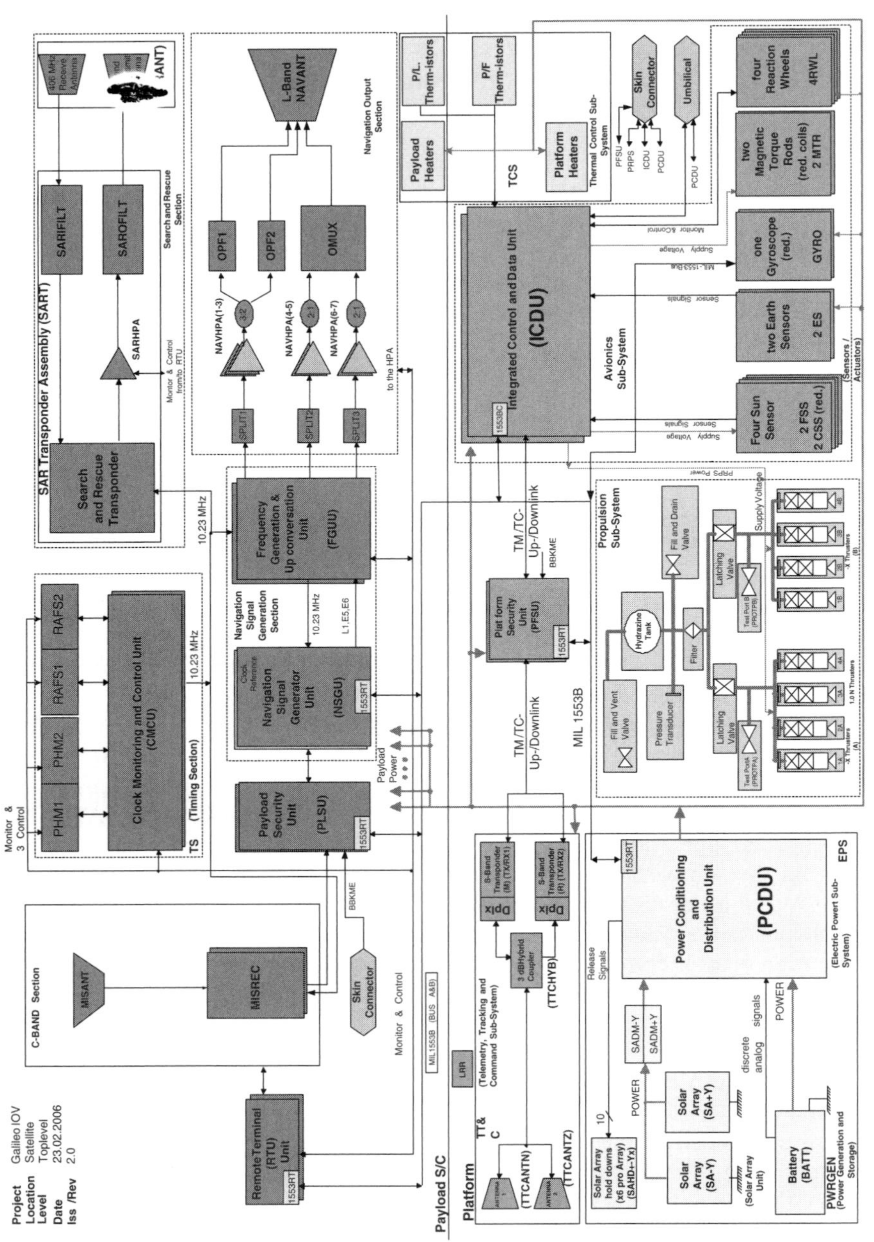

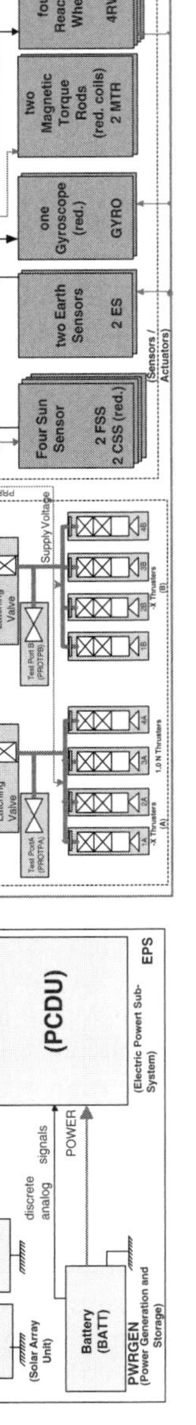

Fig. 4. Electrical architecture

(all other units) protected by Latching Current Limiters (LCL). This approach leads to a high cleanliness of the power bus supply in comparison to a fuse protection approach which allows the units converters to be optimised.

All switchable lines may be switched On or Off at PCDU level under ICDU control (in addition to the nominal switch On and Off function). The PCDU also provides the pyro and heaters interface functions.

The battery can be isolated from the PCDU via a dedicated strap located on a skin connector. The pyro commands can also be armed/disarmed via dedicated plugs to be connected on a skin connector.

The ICDU which is the core unit of the avionics subsystem includes in the baseline the 3 following modules:

- The TMTC-RM module which performs the following hot redundancy functions: TC decoding, monitoring and reconfiguration, context memory (Safe Guard Memory), distribution of HPC commands, and the OBT including the synchronization mechanism, as well as the following cold redundancy functions: TM encoding, TM storage (TM memory),
- The PM-BC modules (operated in cold redundancy) combining the ERC-32 processor with the related memory and the 1553 Bus Controller,
- The IOM modules (operated in cold redundancy) which deliver all the input/output signals necessary to control and monitor the platform equipments and some payload equipments and performs the thermistors acquisition. The IOM modules provide the interface functions to both nominal and redundant AOC units
- The last module is the Converter module, based on 3N and 3R independent DC/DC converters each of them dedicated to one module (TMTCRM, PM or IOM).

A Space-Wire network provides the interface between each module and a full cross-strap allowing each PM to communicate with both Main and Redundant other modules.

The 1553B redundant data bus is used for communication between the ICDU and the following units:

- Payload: RTU, PLSU and NGSU
- Platform: PCDU and PFSU

All RT use the long stub option, i.e. transformer coupling, and are connected to both nominal and redundant buses. Each nominal and redundant BC has the capability to control both the nominal and the redundant bus.

The RTU performs the TM/TC interface function between the payload units and the ICDU via the 1553B data bus. It distributes various types of command signals (High Level Commands, extended HLC, Bi-level Commands, serial load commands) and performs the acquisition of the various payload units TM signals (thermistor, analogue telemetry, bi-level telemetry, discrete relay / switch status, serial telemetry). It includes internal (cold) redundancy and internal cross-strapping.

The other platform subsystems and their overall electrical architecture are then presented:

The TT&C subsystem consists of:

- two S-band conical quadri-filar helix antennas accommodated symmetrically on the satellite in order to provide a quasi omni-directional coverage which perform both the receive and transmit functions. The antennas are circularly polarised, the antenna used in nominal attitude being RHPC, the other one being LHPC,
- of a 3 dB hybrid coupler,
- of two S-band transponders including diplexers (operated in hot redundancy for the receive function and in cold redundancy for the transmit function) which perform the TC receive, the TM transmit and the ranging functions and which can be operated in 2 different selectable modes, namely ESA standard mode and spread spectrum mode.

In addition, a Laser Retro Reflector used for high accuracy ranging performed with laser ranging stations is implemented on the +Z face of the satellite.

The TC signals delivered by the TC receiver function of the TTC transponder including the data, clock and data validity signals are routed to the PFSU (the carrier lock signal being routed directly to the ICDU) which performs COMSEC processing of the data signal. The base-band TC data signal is a 1 kbps NRZ-L BCH encoded signal. The signals are then generated again by the PFSU in the same format (including BCH encoding) and are routed to the ICDU.

Similarly, the PFSU receives the TM signals (data and clock) from the ICDU in the form of a 20 kbps NRZ-L RS encoded signal. The PFSU performs the COMSEC processing of the data signal. The signals are then generated again by the PFSU in the same format (including RS encoding) and are routed to the TTC transponder. The conversion into NRZ-M and the convolutional coding is performed within the transponder transmitter function.

Since the output signals delivered by the PFSU have the same format as the input signals, it is possible to by-pass the PFSU (using an appropriate test harness) in AIT and to operate the TC/TM chain without the PFSU being present. This introduces some robustness of the AIT schedule w.r.t. a late delivery of the PFSU.

The PFSU are operated in hot redundancy and are cross-strapped for the TC function with the nominal and redundant transponders. They are cross-strapped both for the TC function and for the TM function with the nominal and redundant ICDU functions.

The avionics subsystem includes, additionally to the ICDU, the following AOC units (the TM/TC interface of which is performed by the ICDU) which are all cross-strapped w.r.t. the IOM module of the ICDU:

- 2 coarse sun sensors (cold redundant)
- 1 fine sun sensors (cold redundant) with power supply delivered by the ICDU
- two earth sensors (cold redundant) connected to the power bus
- two rate integrating gyros (cold redundant) connected to the power bus
- two magnetic torque rods (redundant coils) commanded by the ICDU
- four reaction wheels (4:3 redundancy) connected to the power bus

The propulsion system includes a hydrazine propellant tank, a filter, a propulsion transducer, two latch valves, the piping to distribute the propellant to the 8 thrusters

(4 nominal and 4 redundant) via two redundant branches and the necessary fill and drain valves and test ports. A pressure transducer allows the monitoring of the pressure in the system. The pressure transducer as well as the thrusters valves and the latch valve are supplied by the PCDU via a dedicated 28 V supply.

The thruster catalytic bed heaters are supplied via the 50 V primary bus by the PCDU via four standard heater interfaces. Four heaters are connected on each heater line (two in series and two in parallel). Each of the 8 thrusters is equipped with one nominal and one redundant catbed heater.

Thruster valve commands (not the latch valve command since the latch valve needs to be operated open shortly before launch) as well as pyro commands are inhibited in the not separated launch configuration via the umbilical connector in order to fulfil the safety requirements. These commands are indeed executed after separation during the autonomous satellite initialisation sequence and for the solar array deployment following the initial sun acquisition by the software upon separation detection.

The separation detection is obtained via three separation straps distributed on both umbilical connectors and conditioned by the ICDU via three independent acquisition channels to the maximum practical extent (use of common function is only allowed if any failure of the common part can be detected in which case a reconfiguration to the redundant IOM module would be performed). The software performs then a majority voting on the three signals to detect separation. The separation detection system is designed to be tolerant to any single failure.

Furthermore, the electrical architecture includes the necessary skin connectors used during the AIT activities. In flight configuration, the flight plugs (e.g. pyro arm, thrusters arm, SADM arm, etc.) are connected to the corresponding skin connectors while EMC covers are installed on the other skin connectors.

In order to be able to perform the key loading operations into the security units via the BBKME, dedicated skin connectors (6 in total, 2 for the PFSU, 2 for the C-band function of the PLSU and 2 for the PRS function of the PLSU) are implemented. These skin connectors are directly linked to the corresponding security units via dedicated cables which follow additional design constraint and which are included in the TEMPEST test campaign performed at unit level. All these skin connectors are equipped with seals (as well as on unit side) which are periodically controlled in AIT in order to control the integrity of the corresponding function.

The PHM ion pumps require to be permanently supplied via a dedicated low power high voltage supply except for periods of limited duration (lower than 10 days). Since this supply is not Corona free, it shall be switched off at launch and during the pump-down phase of the satellite thermal vacuum test. In order to supply the PHM ion pumps when the satellite is off, dedicated skin connectors are foreseen which will be used to connect the corresponding EGSE.

4 Mechanical Architecture

The mechanical architecture is based on a parallelepiped box made of aluminum sandwich panels. The size of the structure is $2530 \times 1200 \times 1100$ [mm] with extensions of the +/– Y payload panels of 100 [mm] in + Z direction and 150 [mm] in – Z

direction. These extensions have been implemented to provide the necessary radiative area for the payload units.
The primary structure provides the interfaces to the directly mounted units or to the secondary structure components (brackets, etc.). Figure 5 presents an external view of the overall satellite architecture as well as the definition of the main reference coordinate system. The satellite is composed in a modular way, by a P/F and a P/L Module for separate integration and test activities at different locations. This modularity concept is shown in Fig. 6.
The body of the platform module consists of:

- the +/–Y P/F-panels, carrying the major part of the platform electronics units,
- the –X panel which is dedicated to the propulsion subsystem (thrusters, valves etc),
- the internal panel which supports the four wheels,
- and the 2 Shear Frames which serve as tank support, load path for the separation system and connecting element for the modules,
- the access panels (–Z).

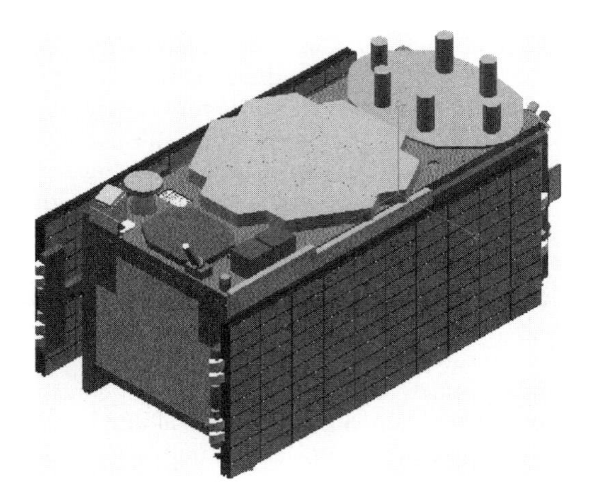

Fig. 5. External view of the overall satellite architecture.

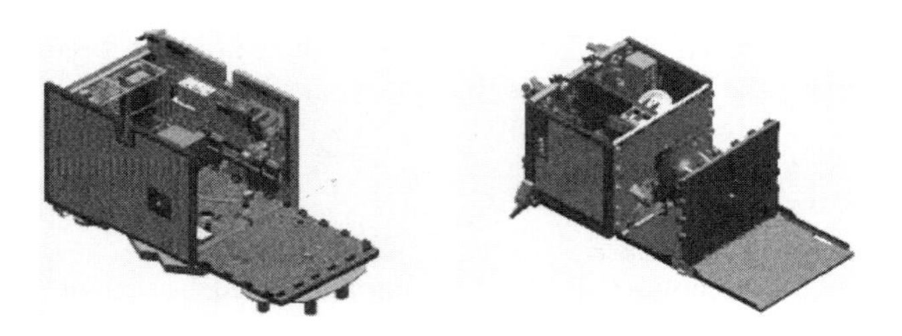

Fig. 6. PL Module and PF module.

The structure will be closed at the −Z side by the 3 removable −Z access panels which can be removed separately, allowing access to the 3 compartments (platform compartment and the two other compartments which allow the access to the tank and to the payload units).

The body of the payload module consists of:

- the +/−Y P/L-panels, carrying the high dissipating payload units,
- the +Z panel which has the length of the entire satellite body and carries the payload antennas and additional low dissipating payload equipment on its internal side,
- the +X panel, which accommodates the clocks.

Structure Design

The sandwich layout, i.e. overall thickness of the sandwich, core density and thickness of the face sheets have been optimized w.r.t. the structural needs (strength, stiffness, load carrying capability of inserts etc) in order to optimize the structure w.r.t. mass.

Where necessary, the panels have been reinforced, mainly at highly loaded areas or cutouts.

For high-load introduction into the panels, embedded brackets are foreseen, e.g. in the areas of the separation systems.

All sandwich panels are vented by use of perforated core material.

The design of the panel assembly via brackets (cleats) is based on heritage from previous programs.

The cleats are metallic profiles, reinforced by webs as necessary to achieve the required stiffness. The cleats will be made of 7475-T7352 alloy, for the attachment of the +x-panel of Ti-6Al-4V for reduced thermal conductivity.

The design of the cleats as well as the location accounts for good accessibility for integration activities, i.e. the integration flow has been considered in the attachment configuration for the cleats. In addition, it allows easy removal of panels if requested by AIT activities.

The brackets will be assembled with the panels via face-to-face inserts for M5 bolts. For the fixation of the units, the panels provide inserts which will be the space qualified ENN398M4/M5 type. These inserts contain replaceable, self-locking Helicoil threads, which allow a limited number of operations and easily can be replaced before the number of allowable operations has been exceeded.

Accessibility in Ait

Special attention has been paid to achieve the required access to the internal part of the satellite. The foreseen access area will be via the – Z panel. This panel is split into 3 parts, each part corresponding to one compartment, i.e. platform, tank + payload and payload. As these three panels belong to the primary structure and therefore

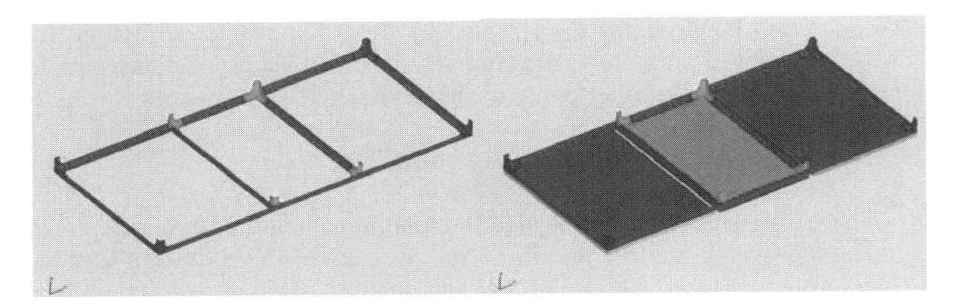

Fig. 7. Frame and access –Z panels (Centre Panel).

have a structural function, only one panel at a time may be removed to maintain the satellite mechanical integrity.

The panels are attached to a frame which is also part of the primary structure. Fig. 7 presents the removable −Z access panels as well as the −Z frame.

Separation System Interfaces

The separation system will be provided by the launcher authority. The structure will provide a "standard interface" for the attachment of such a system, i.e. threaded holes in embedded brackets. In addition the surrounded structure includes the necessary reinforcement.

Specific Interfaces

The structure shall provide specific interfaces for the the tank interfaces and the Navigation Antenna interfaces.

Tank Interface. The tank will be supported by two frames, via 2 polar mounted supports. The propellant tank has a significant impact on the dynamic behaviour of the satellite and on the design of the support structures due to its high mass (which may reach up to 85 kg when fully loaded with propellant).

It shall be noted that the structure, and more specifically the tank support interface, is specified to support the propellant tank loaded at full capacity. This allows both to provide margins w.r.t. the currently needed propellant mass and to provide the flexibility to fully load the propellant tank in launch scenarios where the launch mass is not limited by the launcher performance.

Navigation Antenna Interface. The antenna shall be attached via 6 bolts homogeneously arranged on a circle of 730 mm diameter.

5 Thermal Architecture

The satellite thermal control architecture consists of 6 main thermal zones as follows:

- The clocks zone (payload),
- The payload global thermal control zone, excluding clocks
- The platform global thermal control zone (excluding battery and propulsion)
- The battery zone,
- The propulsion zone,
- The external elements (antennas, sensors) specific thermal control

Clock Zone. The thermal control stability of the clocks is one of the major design drivers for the thermal control subsystem. In order to reach the required stability the clocks and their radiators are accommodated on the +X panel. In order to achieve the required thermal stability performance, the following measures are implemented:

- Only the clocks are accommodated on the clock panel in order to avoid any disturbance coming from other units with regard to the thermal stability.
- The clock panel is conductively decoupled from the other panels by using thermal washers.
- The clock panel is radiatively decoupled from the satellite internal environment by using two MLI tents, one for each clock pair.
- The +X clock radiator is free of any satellite protuberance as far as practicable (the only present item is the TTC antenna which impact is acceptable); in particular, no radiator extension is implemented in the +X direction of the Y payload panels.
- Each clock pair of a same technology (PHM on one hand, RAFS on the other hand) is accommodated on a doubler (one for each pair) which spreads the heat to the corresponding clocks radiator. This approach provides a homogenous heat distribution on the radiator, optimises the thermal stability (by using the thermal heat capacitance of the redundant unit) and minimises the heater consumption.
- Active thermal control of the clocks by heaters which are regulated by a PI algorithm using high accuracy thermistors (three with majority voting for failure tolerance).

Payload Global Zone. The payload zone corresponds to the ±YPL panels (Fig. 8) and the +Z panel.

All high dissipating payload units are located on the ±Y***PL*** panels. These S/C panels, which are submitted to very limited sun incidence and therefore to minimum external heat load, include the OSR radiators and are therefore the most suitable panels for the accommodation of the dissipating units.

However, these panels are submitted to the reflected sun radiation and infrared radiation from the solar array and the yoke onto the ±Y***PL*** panels. The influence of the solar array panels is minimised via the yoke which takes away the panels and thus minimises their factor of view to the radiator. The yoke shall be mechanically (a frame type design shall be used) and thermally (white painted) designed such as to minimise the radiative heat transfer between the yoke and the Y radiators.

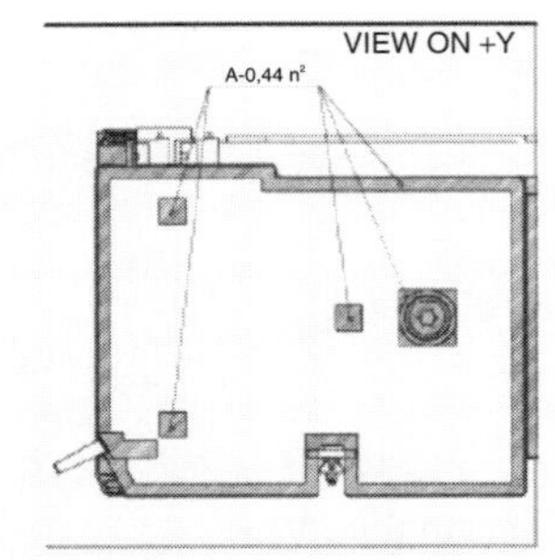

+YPL panel external radiator

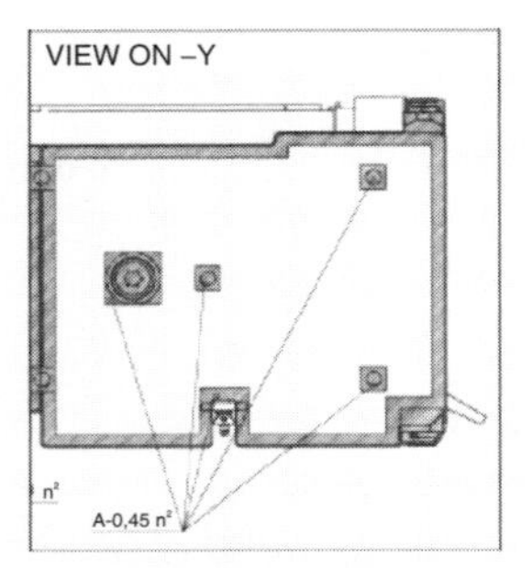

YPL panel external radiator

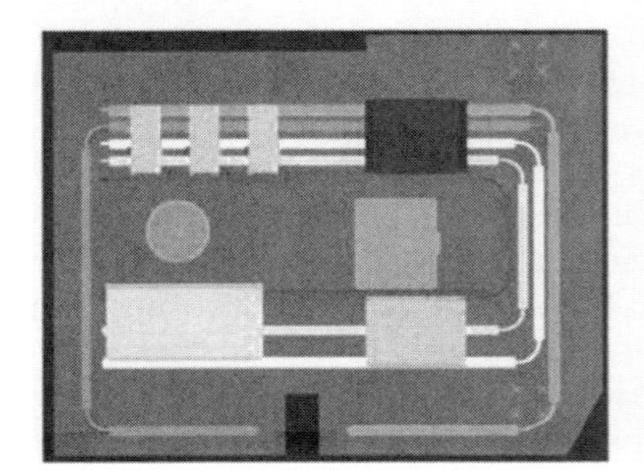

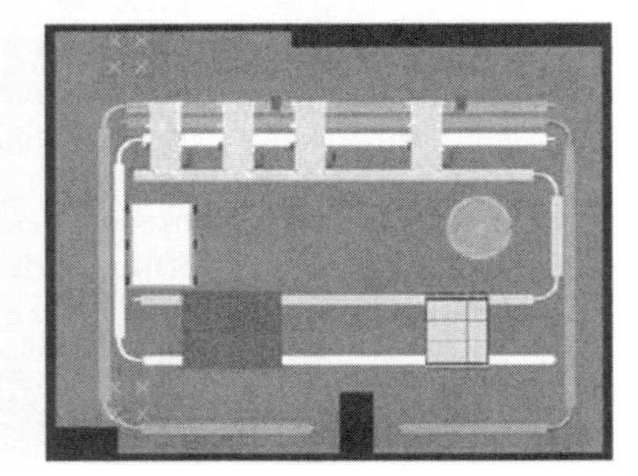

Fig. 8. Radiators & baseline heat pipe layout on the $+Y_{PL}$ and $-Y_{PL}$ panels.

The dissipating payload units are distributed on the ±Y***PL*** panels such as to balance the total dissipated power on the two walls.

Additionally the following requirements are fulfilled with the actual payload unit configuration:

- accessibility of the PLSU and PFSU from the satellite −Z side during AIT
- minimum RF-cables length
- no interferences of units und sufficient margin for unit connector

All external sides of the ±Y***PL*** panels are covered with OSR excluding the position of the SADM and the 3 solar array hold down points which are covered with MLI.

Most of the payload units are located on a heat pipe network to achieve a good conductive heat transfer from the unit via the heat pipes to the complete radiator area. Thermal fillers are implemented between the units and the heat pipes to achieve good thermal conductance.

U-shaped heat pipes are implemented on the ±Y***PL*** panels.

Dissipating units are black painted as well as the internal surfaces of the S/C panels to homogenise internal temperature and to facilitate heat rejection to the radiators.

On each Y***PL*** panel there are three heater lines. Most heaters are attached on the heat pipes.

Some units have heaters on the panels beside the units. Units will be heated via the heat pipe system with several survival heaters to guarantee that they are always within the acceptance temperature limits.

Three thermistors are measuring the temperatures on several positions and the output of the majority voter is used for the heater regulation.

Most units located on the +Z panel are low dissipative units (e.g. switches and couplers).

However, the OMUX and OPF which have significant dissipation, are also accommodated on the internal side of the +Z panel. They have been positioned as close as possible to the Y***PL*** panels to achieve a high view factor between the units and the Y***PL*** panels and to transfer the heat from the units to the panels via radiation.

Accommodation of the OMUX and OPF on the Y***PL*** panels has been considered in the architecture trade offs. But this solution has not been selected since the available area on the Y***PL*** panels was too limited to accommodate them together with the corresponding harness, switches and couplers. Such a solution would have led to an unacceptable length of the RF cables and therefore to unacceptable cable losses.

Platform and Battery Zones. The battery and the PCDU are conductively coupled to the +Y***PF*** panel to radiate its dissipated power via the external radiator to deep space.

The PCDU radiator is such that it is not covered by the stowed solar array in order to maintain the unit within its temperature limits during the launch phase until solar array deployment.

All other radiator areas of the ±Y***PL/PF*** panels are covered by the stowed solar array during launch (which allows the heating power in this phase to be minimised).

The battery zone is made of a dedicated battery radiator (conductively decoupled from the +YPF panel), a MLI including MLI support which provides radiative decoupling and the associated active thermal control.

The platform units: ICDU, PFSU, Tx/Rx S-band Transponder and Gyros are conductively coupled to the -YPF panels to radiate dissipation power via the external radiator to deep space.

A thermal doubler is used below the Tx/Rx S-band transponders to improve the thermal coupling between nominal and redundant units. Active heating with heaters on the panels will keep the units within their temperature limits. Black paint on the units and on the panels is used to harmonize the temperature within the compartment.

The reaction wheels are fixed to the M panel and on the shear frame 1. The dissipation power will be transferred radiatively to the ±YPF panels. The wheels and the panels will be black painted. Additional possible solutions to increase the heat transfer of the wheel to the ±YPF panels are:

- Thicker face sheets of the shear frame 1
- Thermal straps from the wheels to the ±YPF panels

Dedicated radiator area on the ±YPF panels where the thermal straps are connected Active heating with heaters on the panels are used to maintain the units within their temperature limits.

Propulsion Zone. All propulsion items (but the tank) are arranged on the –XPF panel.

The piping from the tank to the –XPF panel is routed on the M-panel. The propulsion items such as valves, pressure transducer, filter, etc. are arranged in the compartment/enclosure where the PCDU/Battery is located. This is favourable for minimising the heater power due to the PCDU which is dissipating in all phases.

Active heating with heaters implemented directly on the units (valves, pressure transducer, filter, tank and the piping system) allows the units to be maintained within their temperature limits. The units and the piping system will be thermally decoupled from the –XPF panel by thermal washers to minimize heater power consumption. Propulsion units such as valves, pressure transducer, filters and pipes will be covered with small MLI boxes.

A main and a redundant heater as well as three thermistors are mounted on the outer surface of the thruster valve.

The external side of the –XPF panel is completely covered with MLI except the required stayout area of the thrusters.

Skin connectors are positioned on the –XPF panel which are covered with MLI in flight configuration.

External Units. The external elements are insulated from the satellite, radiatively by MLI on their rear side and conductively by insulating washers in order to minimise mutual interaction between the satellite and the external element thermal control. This allows the heat flow between the S/C interior and the external elements to be minimised in order to achieve the required thermal stability requirements (e.g. of the payload units) and to minimise the required heating power.

All the units on the external side of the +ZPL panel will be therefore covered with MLI except the apertures of the antennas, sensors and the LRR. The remaining S/C +ZPL panel will be covered as well with MLI. The MLI of the S/C will have an overlapping with the antenna MLI blankets. Standoffs are the preferred fixation method for the purpose of several integration cycles.

Dedicated radiators are used for some external units such as the IRES and the CSS/FSS.

6 Command and Control Architecture

The C&C for the IOV-satellite is based on the implementation of the Packet Utilization Standard (PUS) tailored for Galileo IOV needs. This standard defines a set of services to monitor and control a satellite. Each service consists of subservices, which are either a telecommand or a telemetry report. The format of telemetry and telecommand packets are standardized to a certain extent to support re-use of existing onboard or ground software.

The following table gives an overview about the implemented services.

The hardware independent part of the PUS is implemented by a so called Core Data Handling System (CDHS), which is a library of data handling services for the support of PUS applications. The CDHS library has been used in different projects, like GSTB-V2, and TerraSarX. The hardware depended services (device commanding, memory management etc.) will be implemented to fit to the Galileo IOV ICDU design.

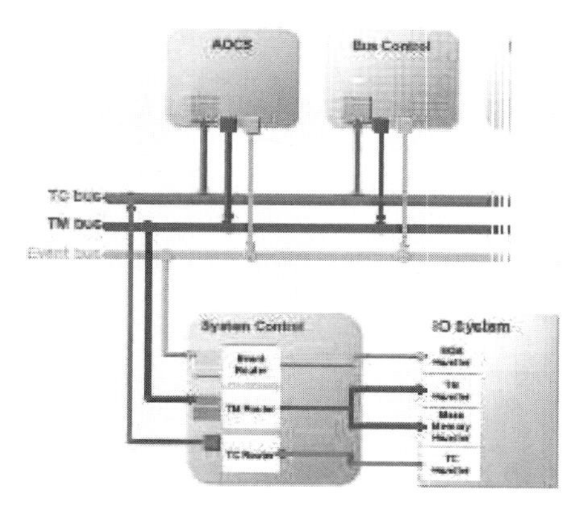

Fig. 9. Software buses.

The CDHS is based on the implementation of software busses (Fig. 9) for TM, TC and events. These busses provide a common interface which allows the different software applications to communicate in a standardized way.

Telecommands. Telecommands received by the transponder are routed to both hot redundant PFSU. If the PFSUs are in clear mode they will route the data to the TC-decoders. The TC decoder (if addressed by the VCID) will decode the CLTU, perform the frame acceptance procedure, generate the command link control word and route the TC segment to the ICDU TC interface if all checks are successfully passed. The avionics software reads the TC segment and performs the de-segmentation of a TC packet (if the command consists of more than one segment). When the telecommand packet has been reconstructed, the avionics software performs checks on the packet header and packet error control, generates the appropriate command acceptance report (TM 1,1 or TM 1,2) and routes the TC packet onto the TC-bus. The application software addressed by the application process identifier in the packet header removes the TC from the TC-bus, checks the telecommand parameters and executes the telecommand itself or routes the TC via the Mil-Std-1553B bus to the final destination to perform the TC handling. If the TC is executed by the avionics software itself, the avionics software will produce the appropriate telecommand execution report. If the telecommand is processed by an external unit (NSGU, PFSU, PLSU) this unit will generate the reports.

The following telecommand types are implemented:

- direct telecommands (DTC) are directly executed by the command pulse distribution unit (CPDU) of the TC-decoder without any software interaction
- immediate telecommands are processed and executed immediately after reception by the software
- time tagged telecommands are inserted into the master time line according to the time tag information and executed by the software when the time tag is due.

Telecommands are in addition classified according to their criticality:

- potential hazardous commands: if executed at the wrong time or in wrong configuration, this type of command could cause the loss or damage of the satellite. For this reason these telecommands are protected by a HW measures like arming by DTC or by electrical inhibits such as separation straps (for the commands which need to be performed autonomously).
- vital commands: if executed at the wrong time, this type of command could cause degradation of the mission. Therefore these commands are protected by a software arming and firing mechanism.
- non-critical commands: all other commands

Telemetry. The downlink data are transferred from the ICDU, from the satellite subsystems and from the payload through the TM encoder, the Security Unit until the TM transmitter. The different applications of the on-board software either generate PUS-packets itself or read PUS-packets from external units (NSGU, PLSU, PFSU). The avionic software forwards these packets to the TM-encoder of the ICDU or to the internal data storage for later downlink. The TM-encoder takes either real-time packets from virtual channel (VC0) or replay data via VC1.

If no TM data are available the ICDU will generate idle packets via VC7 automatically. The TM-frames generated by the ICDU are passed through the PFSU without processing if the PFSU is in clear mode or will be encrypted by the PSFU if it is in secure mode. From the PFSU the data are sent to the transponder.

The telemetry concept is driven by the services defined in the PUS. Each service generates defined telemetry reports. One report can consist of more than one packet.

The reports can be:

- a response to a telecommand (e.g. memory dump)
- generated periodically if enabled by telecommand (e.g. housekeeping reports)
- an asynchronous report due to an on-board event
- a telecommand verification report generated during the different steps of the telecommand processing.
- a special TM report (e.g. encrypted TM packets).

All telemetry packets are put on a so called TM-bus. A telemetry router as part of the system control application reads periodically the TM-bus and routes the packets according to their destination which can be defined by PUS-service 15 telecommand either to a packet store within the HK-memory or to the TM forward control (PUS service 14). The TM forward control allows the control of which TM packets are enabled to be sent to ground. All enabled packets are put by the TM manager to the virtual channel VC-0 hardware interface. Finally the HW TM-encoder takes the TM-packets from both virtual channels and generates TM-frames to be downlinked to ground. The HK-memory output is routed directly to VC-1 without any SW interaction, beside start and stop of the HK-memory output.

On-board Storage and Retrieval. The ICDU provides the capability to store 1.65 Gbits of telemetry data, which can be downlinked to ground via VC-1 on request

(TC). This allows the storage of selected telemetry data during the time between two ground contacts. It is possible to define up to 5 different packet stores, which can be managed by telecommand independently. PUS service 15 provides the necessary commands and telemetry reports to monitor and control the different packet stores.

On-board Communication. The main interface for data exchange between the central avionics system and the platform units and the payload is based on an on-board bus concept realized by the Mil-Std 1553B bus. Bus controller (BC) is located in the ICDU. The following units are connected to the bus: NSGU, PLSU, RTU, PFSU, PCDU,

Figure 10 gives an overview how the units are connected to the Mil-Std-1553 and how the redundancy concept is realized.

Bus controller and remote terminals are connected to the redundant bus lines via long stub transformer bus couplers. Each RTU has two Mil-bus remote terminal interfaces, where each interface is connected to both busses. The ICDU provides two bus controllers each of them being able to access to both busses.

On-board Time Management. In the following, the baseline on-board time management architecture is described.

As shown below five different and independent time management systems exist onboard the GALILEO satellite:

- NSGU Time: This is the highly accurate time, needed for the navigation mission. The time origin is the CMCU / Atomic Clocks. It is called Local Galileo System Time (LGST).
- PLSU Time PRS: This is a separately generated time function at PLSU side synchronized by a pulse per second (PPS) and the LGST received from the NSGU.
- PLSU Time C-Band: This is a separately generated time at PLSU side, synchronised to the Galileo Time by (second level encrypted) ground command or synchronized to the PPS and LGST provided by the NSGU. This time is mainly needed for TC time tagging and TM time stamping.
- PFSU Time: This is a separately generated time at PFSU side, synchronised to the Galileo Time by (second level encrypted) ground command, mainly needed for TC time tagging and TM time stamping at PFSU side.
- ICDU Time: The ICDU Time is the time generated by the ICDU of the avionics subsystem of the GALILEO Satellite. The time is needed on-board for correct performance of the satellite (P/F) system-, subsystem management and attitude and orbit control functions, which are required for safe and autonomous operation of the satellite during all mission phases including also non-nominal situations.

7 FDIR Architecture

The overall FDIR is organized in a hierarchical form defined by four levels with increasing complexity in terms of recovery actions. The goal of this organization is to recover from failures on the lowest possible level. In principle an additional (w.r.t. the ones listed below) level 0 exists, which is equipment internal (e.g. EDAC) and considered transparent.

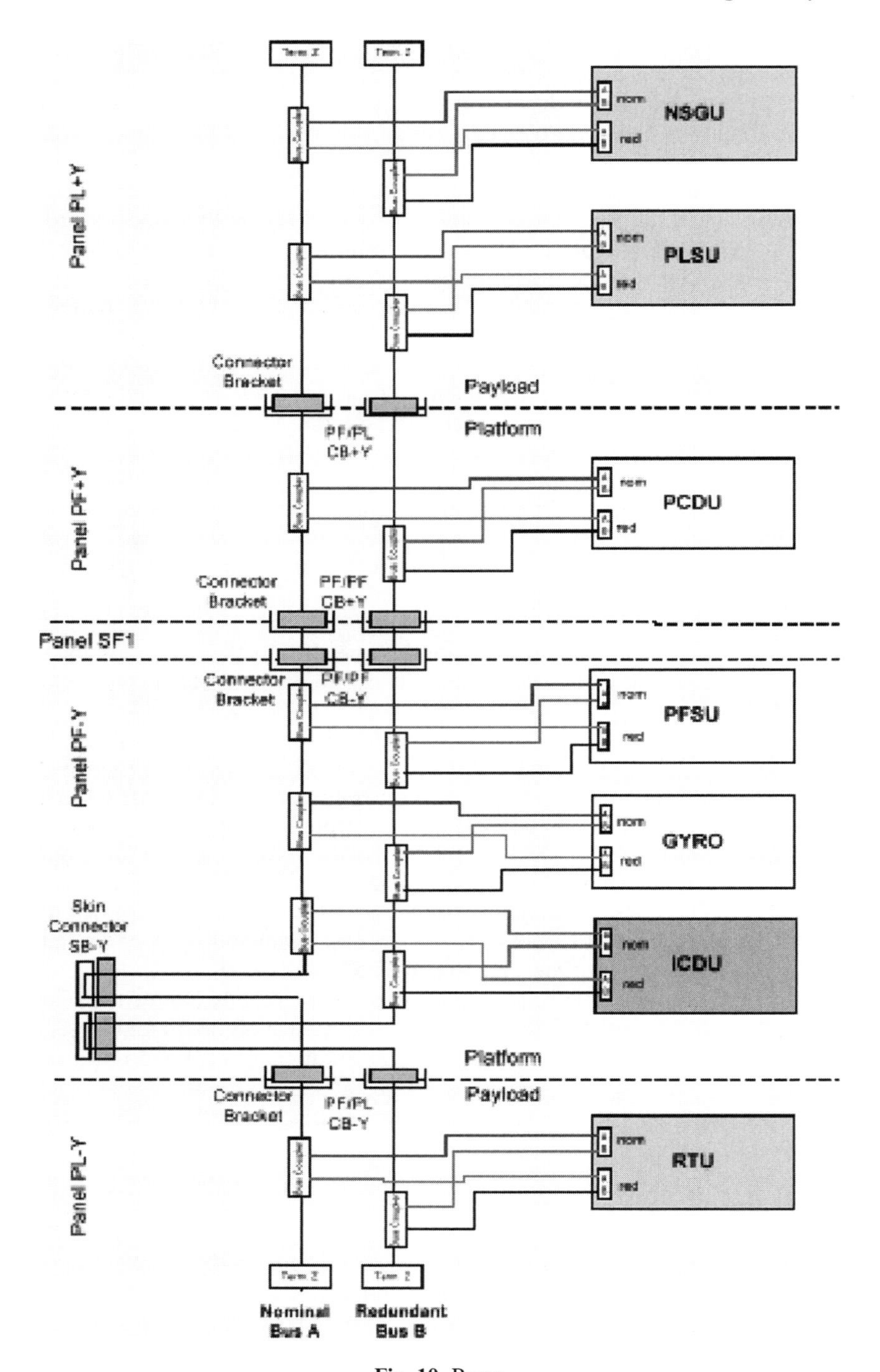

NSGU
nom
red
PLSU
nom
red
Panel PL+Y
Connector
Bracket
Payload
Platform
PF/PL
CB+Y
Panel PF+Y
PCDU
Connector
Bracket
PF/PF
CB+Y
Panel SF1
Connector
Bracket
PF/PF
CB-Y
Panel PF-Y
PFSU
GYRO
Skin
Connector
SB-Y
ICDU
Platform
Payload
Connector
Bracket
PF/PL
CB-Y
Panel PL-Y
RTU
Nominal
Bus A
Redundant
Bus B

Fig. 10. Buses.

The FDIR functions are controllable by ground (i.e. enabled and disabled) at monitoring and recovery level.

Monitored values are filtered in order not to react on a single out of limit event.

All parameters for FDIR (e.g. thresholds and filters) are modifiable by telecommand.

When in Intermediate Safe Mode or in Safe Mode the Satellite rejects on-board stored commands (e.g. in the Master Time Line) and all ground commands (but a dedicated command allowing to recover normal command operation) and relies on the ground for the transition back to Normal Mode or to Sun Acquisition Mode. This allows the execution of telecommands which are no more suited to the new situation (entry in ISM) to be avoided.

To support ground investigation and recovery actions, the observability of the failure is a major issue.

Observability is ensured through:

- On-board storage of events in the event log buffer stored in SGM as well as in ICDU TM-Memory
- Telemetry which is stored in the ICDU TM-Memory and which provides all sensors acquisition data and actuator raw commands as well as the monitored FDIR parameters (>28 hours storage).

The general approach is to implement S/S FDIR functions directly in the related SW application (decentralised approach). This has the advantage to design, develop and validate these applications as a whole with less external interfaces. The use of OBCP will be limited to real justified needs.

FDIR Levels. The FDIR concept is aimed at minimizing the impact of all kinds of failures on the system performance by implementing 4 different FDIR levels with increasing complexity. Level 1 and 2 are S/W measures whereas Level 3 and 4 are H/W based.

Level 1: Failures allow the continuation of the current Satellite Mode. The detection and the isolation of the failure is based on equipment status information, consistency checks between units and surveillance of subsystem behaviour. After failure isolation the suspected unit(s) and/or interface(s) are switched over.

Level 2: Failures imply a Satellite Mode change, possibly in combination with an AOC mode or sub-mode change. These failures become apparent by violations of certain subsystem performances which require a block switching of all involved units and interfaces to be performed and/or to use another mode. As described in the chapter "Mode Dependent FDIR" the recovery mode to be applied is different in satellite NM where priority is put on the satellite mission availability to all other modes where priority is put on the satellite safety.

Level 3: These are malfunctions of the ASW internal to the software or caused by PM failures. Their detection is based on PM specific checks like the watchdog signal, etc. Recovery actions are Warm start, Reset, Cold start, Reconfiguration of the processor (implying a cold start of the new processor)

Level 4: Hardwired alarms like sun- and earth presence checks are implemented to detect major system anomalies. If one of these alarms becomes active, a Safety Sequence (SFS) is performed which consists of:

- A switchover to redundant ICDU and redundant equipments as pre-set by ground
- A transition to Satellite Safe mode and AOC safe mode
- Disabling of S/W checks and H/W alarms

Level 0: In addition to the FDIR levels 1 to 4 described above, also a FDIR level 0 exists. The level 0 mechanisms are part of the unit function (e.g. correction of a single bit by the EDAC) and are transparent for the FDIR system.

8 Satellite Main Budgets

Mass Budgets

Overall Max. Mass:	700 kg
Including Maturity and Uncertainty	

Power Budget

Safe Mode:	
Max. Power in Sunlight:	1290 W
Max Power in Eclipse:	1345 W
Max Power Peak	1525 W
Safe Mode:	
Max. Power in Sunlight:	950 W
Max Power in Eclipse:	1115 W
Max Power Peak	530 W
Safe Mode:	
Max. Power in Sunlight:	1290 W
Max Power in Eclipse:	1345 W
Max Power Peak	1525 W

Galileo Rubidium Standard and Passive Hydrogen Maser

Current Status and New Development

Fabien Droz, Pierre Mosset, Gerald Barmaverain, Pascal Rochat, Qinghua Wang[1], Marco Belloni, Liano Mattioni[2], Ulrich Schmidt, Timothy Pike[3], Francesco Emma, Pierre Waller, Giuliano Gatti[4]

[1]Temex Neuchatel Time / Switzerland
[2]Galileo Avionica / Italy
[3]EADS Astrium / Germany
[4]European Space Agency (ESTEC) / Netherlands

Abstract. The pointing accuracy of satellite navigation systems relies to a great extent on the stability of the on-board atomic clocks.

The Passive Hydrogen Maser (PHM) and the Rubidium Atomic Frequency Standard (RAFS) constitute respectively the master and the hot-redundant clock of Galileo Satellite Navigation System. Their development has been continuously supported by ESA.

This article gives a general overview on the RAFS and the PHM current status and the new developments foreseen.

1 Introduction

GALILEO is a joint initiative of the European Commission and the European Space Agency (ESA) for a state-of-the-art global navigation satellite system, providing a highly accurate, guaranteed global positioning service under civilian control. It will probably be inter-operable with GPS and GLONASS, the two other Global Navigation Satellite Systems (GNSS) available today.

The fully deployed Galileo system will consist of 30 satellites (27 operational and 3 active spares), stationed on three circular Medium Earth Orbits (MEO) at an altitude of 23 222 km with an inclination of 56° to the equator.

Atomic clocks represent critical equipment for the satellite navigation system. The Rubidium Atomic Frequency Standard (RAFS) and Passive Hydrogen Maser (PHM) are at present the baseline clock technologies for the Galileo navigation payload. According to the present baseline, every satellite will embark two RAFSs and two PHMs. The adoption of a "dual technology" for the on-board clocks is dictated by the need to insure a sufficient degree of reliability (technology diversity) and to comply with the Galileo lifetime requirement (12 years). Both developments are based on early studies performed at the Observatory of Neuchâtel (ON) from end of 1980s and Temex Neuchâtel Time (TNT) since 1995. These studies have been continuously supported by Switzerland within ESA technological programs especially since the set-up of the European GNSS2 program. Galileo Avionica (GA) started the electronic development of the PHM already in 2000 and EADS Astrium-GmbH (AST-GmbH) jointed the RAFS development activity in 2001.

The activities related to Galileo System Test Bed (GSTB-V2) experimental satellite as well as the implementation of the In Orbit Validation phase are in progress. One experimental satellites was already launched the 28th of December 2005 (GIOVE-A) and the second one (GIOVE-B) will be launched second half 2006. The main objectives of these two satellites are to secure the Galileo frequency fillings, to test some of the critical technologies, such as the atomic clocks, to make experimentation on Galileo signals and to characterise the MEO environment. There are two RAFS on the satellite supplied by Surrey Satellite Technologies Ltd (GIOVE-A) and there will be one PHM and two RAFS on board the satellite supplied by Galileo Industries (GIOVE-B). This article gives a general overview on the space RAFS and the PHM current status and further development foreseen.

2 Current Status of On-Board Clocks

2.1 Current Activities of Rubidium Atomic Frequency Standard

The RAFS clocks on-board of the GIOVE satellites are issued from 8 years of dedicated development activities for navigation application [1] & [2].

Since 2001, a Swiss-German industrial consortium led by TNT with AST-GmbH as the subcontractor for the electronics package is set in place to develop and produce the RAFS clocks. The current model (RAFS2) includes:

- An optimised physics package with low frequency sensitivity to temperature variation (<5E-14/°C) resulting in a better short/mid term stability with a temperature & vacuum environment similar to satellite platform environment.
- A DC/DC converter and the satellite TT&C interface compatible with ESA's last requirements. Figure 4 shows the performances achieved in term of frequency & time stabilities.

Within this configuration RAFS clocks shows capabilities to perform time stability close to 1 ns over 1 day (drift removed) as reported in Fig. 2. It is the type of RAFS on-board of the GIOVE satellites.

In the frame of GSTB-V2, one Qualification Model, one Proto-Flight Model (PFM) and five Flight Model (FM) units have been delivered. The PFM and FM1 are integrated in GIOVE-B and ready for launch. The FM4 and FM5 are integrated in GIOVE-A and in orbit since 28th December 2005. In addition, the FM2 and FM3 are available as FM spare units. Table 1 lists the achieved RAFS performance for GSTB-V2. Fig. 1 shows the measured frequency stability of GSTB-V2 PFM and FM1 to FM5. Fig. 3 shows the RAFS equipment equipped with the thermally regulated base-plate & DC-DC converter.

2.2 Development & Qualification Activities of Passive Hydrogen Maser

The space hydrogen maser will be the master clock on the Galileo navigation payload. The first maser development activity tailored to navigation applications was

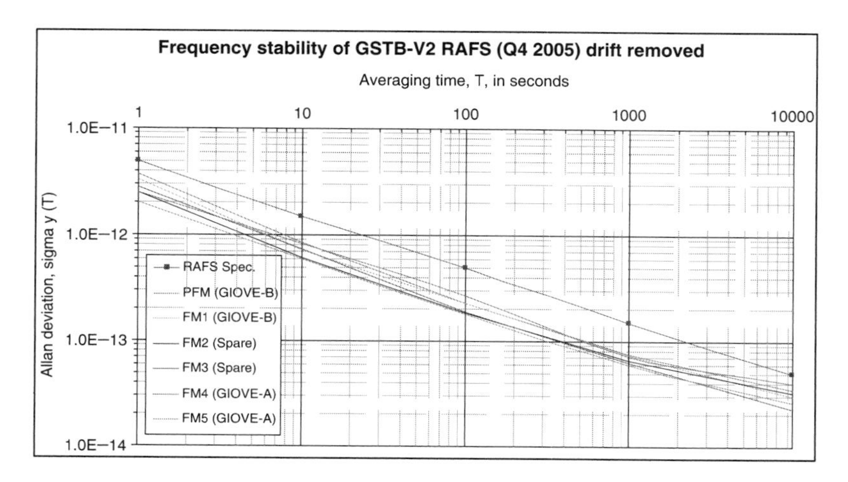

Fig. 1. GSTB-V2 RAFS2 frequency stability.

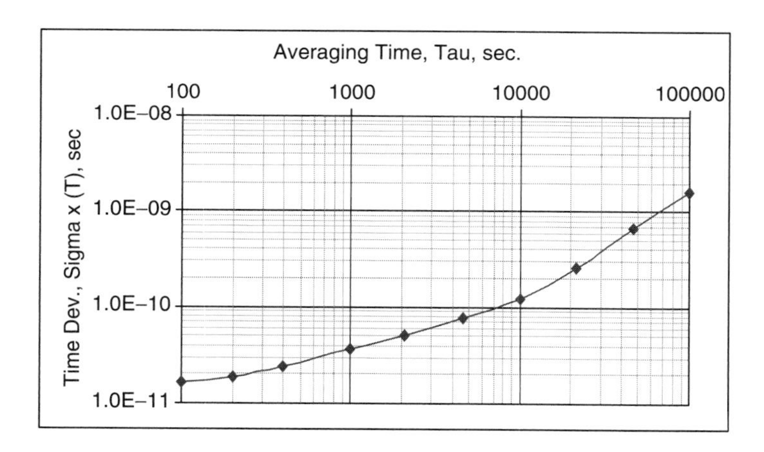

Fig. 2. RAFS stability in the time domain.

Table 1. RAFS for GSTB-V2 performance achieved.

Parameter	Measurement
Frequency stability	< 4* 10^{-14} @ 10,000 sec
Flicker floor	< 3* 10^{-14} (drift removed)
Thermal sensitivity	< 5* 10^{-14} / °C
Magnetic sensitivity	< 1* 10^{-13} / Gauss
Mass and volume	3.3 kg and 2.4 litre

Fig. 3. RAFS equipment in navigation configuration.

kicked off in 1998. It was initiated by the development of an active maser at ON. However, at the Galileo definition phase, it became clear that the accommodation of the active maser on the satellite was too penalizing in term of mass and volume, and the excellent frequency stability performances of the active maser were not required. In 2000 it was re-orientated towards the development of a PHM based on the industrial design and ON heritage on active maser studies.

The development of a prototype was completed at the beginning of 2003 [3], under the lead of ON with Galileo Avionica (GA) subcontractor for the electronics package and TNT supporting the activity in view of the future PHM industrialisation. The instrument has been under continuous test since June 2003 for assessment of long term, reliability and lifetime performances.

The industrialization activity aimed at PHM design consolidation for future flight production was started in January 2003 [4]. The industrial consortium is led by GA designing the electronics package with TNT responsible for the manufacturing of the physical package and the ON supporting the transfer of technology. The overall structure of the instrument was reviewed to increase compactness and to ease the Assembly, Integration and Test (AIT) processes on the satellite by the inclusion of an external vacuum envelope. Main efforts in the industrialization frame focused on the definition of repeatable and reliable manufacturing processes and on the development of more compact electronics. In addition to the PHM Qualification Model, four Models for life demonstration are being manufactured and will be submitted to prolonged testing on ground.

In the frame of GSTB-V2 (now GIOVE-B); one Proto-Flight Model (PFM) was submitted to proto-qualification testing and hence delivered in May 2005. One spare Flight Model (FM1) has been also delivered by Q1 2006. Figures 4 and 5 and Table 2 show the achieved performance of PHM PFM & FM1 for GSTBV2. Better performance has been achieved by improving the magnetron cavity design in the last model (FM1).

Since the beginning of the development, the PHM lifetime was a subject of discussion. The lifetime is being sized to guarantee 12 years of orbit life plus 3 years of ground storage, including the complete AIT program. The operational life is mainly limited by capacities of the hydrogen container (for H_2 supply), bulk getters (for H_2 sorption), ion pump (for pumping ungetterable background gases) and the total dose of ionising radiation. The lifetime capability has been confirmed by detailed analysis and tests of subassemblies (Fig. 6).

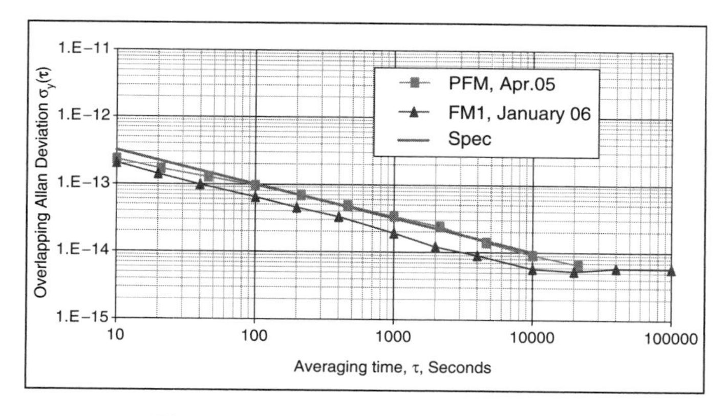

Fig. 4. GSTB-V2 PHM frequency stability.

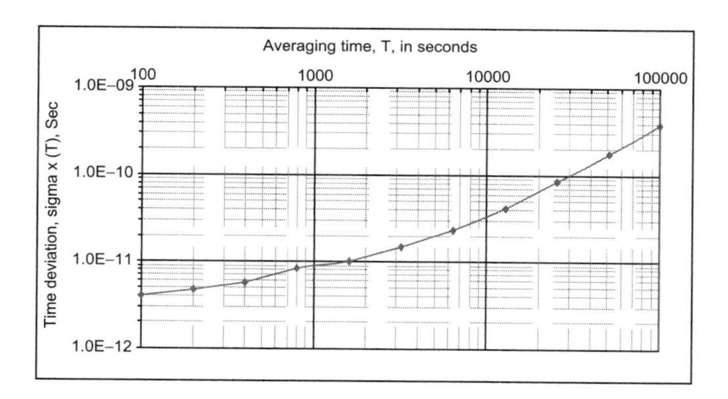

Fig. 5. PHM stability in the time domain.

Table 2. PHM for GSTB-V2 performance achieved.

Parameter	Measurement
Frequency stability	$< 1^* \ 10^{-14}$ @ 10,000 sec
Flicker floor	$< 7^* \ 10^{-15}$
Thermal sensitivity	$< 3^* \ 10^{-14}$ / °C
Magnetic sensitivity	$< 4^* \ 10^{-14}$ / Gauss
Mass and volume	18 kg and 28 liter

3 Further Development on RAFS and PHM

3.1 Further Development on RAFS

Further investigations to improve the flicker floor are under way. By improving the RF atomic interrogation signal stabilisation circuitry, RAFS has demonstrated stabilities in a range fo $7^* \ 10^{-15}$ for half of day or longer observation time. A careful

Fig. 6. PHM PFM equipment.

worst case analysis of possibles drifts of parameters has been performed and demonstates the feasibility and possible repeatability of a RAFS having short term stability over one day lower than $1^* \ 10^{-14}$.

For navigation, the further developments on the RAFS clocks will be concentrated on the improvement of performances. For telecommunication, reduction of the mass, volume and cost of the RAFS are the main drivers of the new development. The goal is to propose a clock with frequency stability of few 10^{-13} within a volume of 1 litre and a mass of 1.5 kg.

3.2 Further Development on PHM

The present PHM instrument is a master clock specifically designed for navigation applications, offering a unique stability performances, requested today by very few other scientific missions, besides navigation.

In order to increase the attractiveness of the PHM and possibly broaden its application field, further developments will be focused on improving its interface characteristics, like mass, size and power consumption, while keeping its very good stability performances and lifetime. In addition, an improvement of thermal sensitivity will be pursued.

4 Conclusions

In total, eight flight clocks were produced for GSTB-V2, which provides the first flight opportunity for Galileo clocks qualification. GIOVE-A with two RAFS onboard is in orbit since 28th December 2005. Both RAFS are fully operational with expected very good frequency stability. The first PHM will be launch very soon on-board of GIOVE-B. With more than 10 years of efforts, two clock technologies for Galileo are qualified. These clocks use reliable and mature technologies leaving room from further improvements in term of mass & performances. Based on this success, new applications could be considered in Europe like for telecommunication or science.

References

[1] A. Jeanmaire, P. Rochat, F. Emma, "Rubidium atomic clock for Galileo," 31st Precise Time and Time Interval (PTTI) Meeting, 07–09 December, 1999, California (USA), pp. 627–636.

[2] F. Droz, P. Rochat, G. Barmaverain, M. Brunet, J. Delporte, J. Dutrey, F. Emma, T. Pike, and U. Schmidt, "On-Board Galileo RAFS, current status and Performances," 2003 IEEE International Frequency Control Symposium Jointly with the 17th European Frequency and Time Forum, 05–08 May, 2003, Tampa (USA), pp. 105–108.

[3] P. Berthoud, I. Pavlenko, Q. Wang, and H. Schweda, "The engineering model of the space passive hydrogen maser for the European global navigation satellite system GalileoSat," 2003 IEEE International Frequency Control Symposium Jointly with the 17th European Frequency and Time Forum, 05–08 May, 2003, Tampa (USA), pp. 90–94.

[4] L. Mattioni, M. Belloni, P. Berthoud, I. Pavlenko, H. Schweda, Q. Wang, P. Rochat, F. Droz, P. Mosset, and H. Ruedin, "The development of a passive hydrogen maser clock for Galileo navigation system," 34th Precise Time and Time Interval (PTTI) Meeting, 03–05 December, 2002, Reston (USA), pp. 161–170.

China-Europe Co-Operation Agreements for Navigation: SART and LRR Developments

Francesco Emma (ESA), Rafael Garcia Prieto (ESA),
Juergen Franz (AST-D), D. Hurd (AST-Uk), H. Ding (CGI),
Y. Sun (NCRIEO), G. Peng, C. Janrong (CAST-XIRST)

ESA/ESTEC Navigation Department
Keplerlaan 1 2200AG Noordwijk-The Netherlands
Phone: +31 71 5653193, Fax: +31 71 5654369,
e-mail: francesco.emma@esa.int

Abstract. The European Commission and the European Space Agency through the Galileo Joint Undertaking (GJU), have launched in China a number of developments related to Navigation Applications and Infrastructurs. The programmatic framework is part of a broad agreement between the GJU and the Chinese Ministry of Science and Technology through which China is providing an important contribution to the Galileo development. The type of activities included in the co-operation agreement cover system studies and navigation applications services plus the development of both ground and onboard equipment for Galileo. Eleven contracts have already been kicked off, others are subject of ongoing tender actions which will raise the first, overall contribution of China to Galileo to an amount of about 65 M€. The space related developments are currently involving the development of the Galileo Search and Rescue Transponder (SART) and the satellite Laser Retro Reflector (LRR). An overview of the ongoing activities is provided in this article with a more detailed reporting on the SART and LRR developments.

1 China-Europe Co-Operation Framework

The European Commission and ESA have signed in 2003 a co-operation agreement with the Chinese Ministry of Science and Technology defining the contribution of China to the overall development plan of Galileo. The technical annex defining the contents of this co-operation foresees the development of on-board/ground equipment, plus activities related to system studies to promote the use of Galileo in China. The contribution of China to the Galileo development is channelled through the Galileo Joint Undertaking which is the institutional body constituted by the European Commission and the European Space Agency to steer the activities related to the validation and full deployment phase of Galileo.

2 Galileo Activities in China: Overview

Between the year 2005 and 2006 the Galileo Joint Undertaking has kicked off eleven different projects. A detailed description of the SART and LLR project is provided in the following paragraphs, while a summary description of the other activities is given here below.

China Galileo Test Range. The overall objective of this activity is the development, deployment and operation of a ground based infrastructure of Galileo pseudolites, to be used:

- As a tool to perform analysis and research on the Galileo 'Signal In Space' (SIS)
- Act as a 'Test Environment' for Galileo receiver and applications.
- Form the basis of a series of demonstration and promotion activities for Galileo services and applications.
- Act as a local augmentation system to deliver high performance positioning and navigation services.

Location Based Services Standardisation. This is activity is dedicated to the definition of the work to be carried out for the standardisation of Galileo as a system for mobile phones, location applications.

Ionosphere Studies. The primary objective of this activity is the investigation of on effective ionospheric correction for single frequency Galileo receivers on a regional basis and the study of ionospheric scintillations.

Fishery Applications. The project is addressing the implementation of a system based on GNSS to support applications in the fishery domain in the Chinese region. The activity includes an overall analysis of the technical and commercial aspects related to the system implementation, and will also consider the design and development of a system demonstrator proving the effectiveness of the system concept from a technical, operational and commercial perspective.

Medium Earth Orbit Local User Terminal. The Search and Rescue transponders on the Galileo Satellites will relay the distress signals transmitted by Cospas-Sarsat emergency beacons towards dedicated ground stations MEO Local User Terminals (MEOLUTs). These MEOLUTs will be in charge of recovering the message and locating the emergency beacon, as well as providing the relevant SAR distress data to the associated Cospas-Sarsat Mission Control Centre (MCC). The objective of this activity is the development of a MEOSAR Local Users Terminal Prototype (MEOLUT Prototype).

Search and Rescue End to End Validation. The primary objective of this activity is the end-to-end verification of the SAR/Galileo system requirements and the validation the forward link service by demonstrating the system and evaluating its technical performances.

Up-link Station development. For the control of the Galileo system the ground segment is organized in control centres, located in Europe, and remote stations, located worldwide and includes the following types of stations:

- Tracking, Telemetry & Command (TTC) stations.
- Mission Up-Link Stations (ULS).
- Galileo Sensor Stations (GSS).

The ULS stations are composed of one or more full-motion antennas (approx. 3-m diameter) for transmitting a spread spectrum signal in C-band (5000–5010 MHz), without operational downlink implementation, for uploading mission related information. This activity is addressing the design of the ULS station front end.

Satellite Laser Ranging Services. Satellite Ranging Services are essential for a precise determination of the Galileo satellites orbits. Through this activity the European Space Agency is procuring SLR services in China to for the GIOVEA and GIOVEB mission (the two Galileo experimental satellites) and will develop a dedicated station for the need of the IOV phase.

3 The SART and LRR Developments

3.1 The Search and Rescue Transponder on Galileo Satellites

The Galileo constellation is designed to provide, together with a global navigation service, also support to the COSPAS-SARSAT system for the provision of Search and Rescue, MEO services. The Galileo Satellites are, for these purposes, embarking a Search and Rescue Transponder interfaced with the Navigation Payload. The aim of the Galileo support SAR services is to relay distress signals from Cospas-Sarsat-defined beacons to specialised ground facilities and to relay messages from ground to beacons equipped with a Galileo receiver, using the constellation of Galileo satellites.

The SAR/Galileo service can be subdivided in two other sub-services, the Forward Link Service (relaying distress signals from beacons to ground stations) and the Return Link Service (relaying messages from ground to beacons equipped with a Galileo receiver) Fig. 1 shows a common scenario where both SAR/Galileo sub-services are used. The roles of the different components are:

- Galileo Space Segment: to relay distress signals transmitted by type-approved beacons to ground stations, transponding signals from 406.05 MHz to 1544.1 MHz; to disseminate Return Link Messages received from ground within the L1 navigation signal. The Galileo Space Segment is part of a single MEOSAR constellation and is interoperable with other MEOSAR systems such as DASS/GPS and SAR/Glonass.
- Intergovernmental SAR Satellite System (ISSS): to process beacon signals in order to recover the transmitted message and determine the beacon location; to distribute

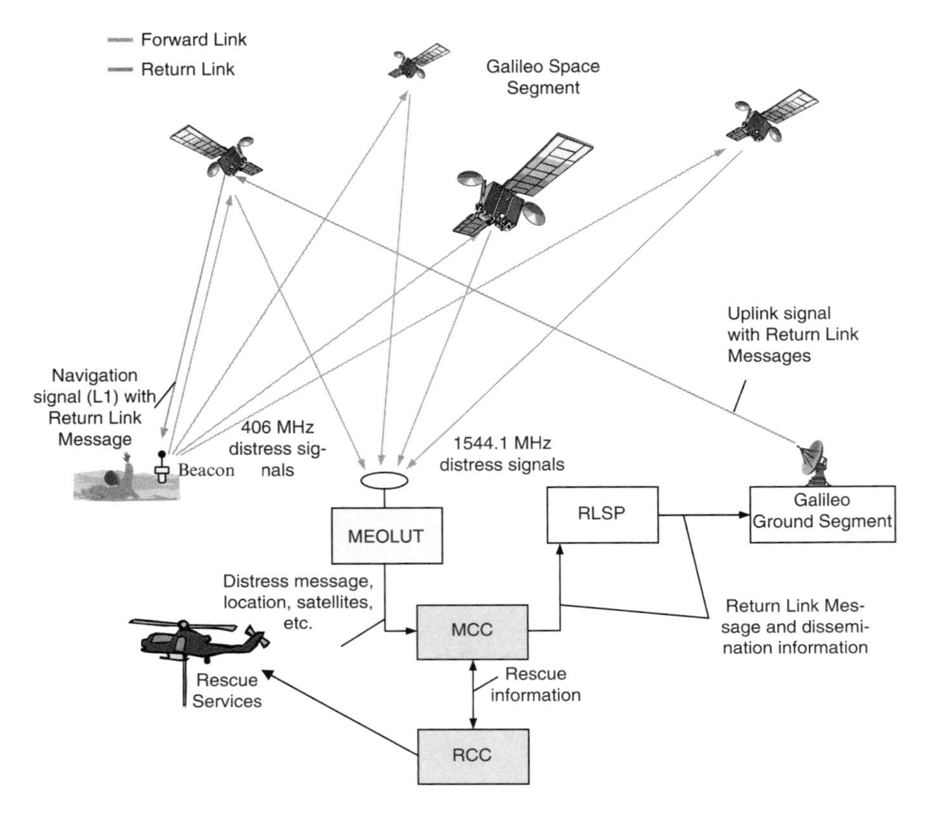

Fig. 1. Search and rescue service system architecture.

messages across the network to the Return Link Service Provider (RLSP) and appropriate Rescue Coordination Centres (RCCs). The ISSS consists of: Medium-altitude Earth Orbit Local User Terminals (MEOLUTs), Mission Control Centres (MCCs) and Nodal MCCs.

- Rescue Coordination Centres (RCCs): to launch and coordinate rescue operations.
- Return Link Service Provider (RLSP): to coordinate the requests of Return Link Messages and interface with the Galileo Ground Segment. It might be implemented as an extension/integral part of the ISSS, or as an independent centre.
- Galileo Ground Segment: to uplink the Return Link Messages to the appropriate satellites for dissemination.

3.1.1 SART Architecture and Design Driving Requirements. The SAR transponder layout and architecture are presented in Figs. 2 and 3. The transponder includes a receive chain, a down conversion section and an up conversion and amplification chain.

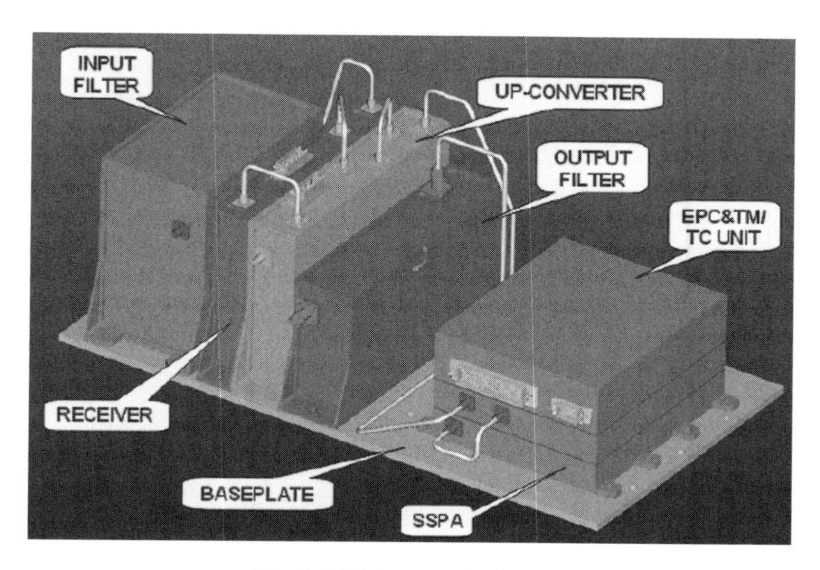

Fig. 2. SAR transponder layout.

The major design driver requirements are:

Group Delay Stability. This parameter is important as it affects directly the Time Difference of Arrival (TDOA) techniques at a MEOLUT and therefore the precise location of the distress beacon. This is a function of the BW of the input filter design.

Receiver Noise Figure. This parameter is important as it is affecting the overall G/T requirement at system level.

Frequency translation accuracy. It is an important parameter for the use of the Frequency Difference of Arrival as a distress beacon location technique. This is ensured in the SAR transponder by the use of the highly stable reference frequency from the Clock Monitoring and Control unit.

Linearity and output Power. A high linearity requirement is needed on the SART to ensure minimum occupied spectrum of the relayed beacon signals, and minimum inter-modulation or cross-products generated by the beacon signals, to avoid creating any false in-band signals. This requirement is impacting directly the design of the output power amplifier and is critical for the overall transponder power consumption. The Transponder corresponding output Power for the linearity requirement shall be equal to 5 Watt and this makes critical the design of the SSPA.

Mass and Power. Critical parameters due to the limited margins existing today at satellite level. The SART overall mass shall not exceed 8.8 kg. The overall power consumption shall be less than 45 W.

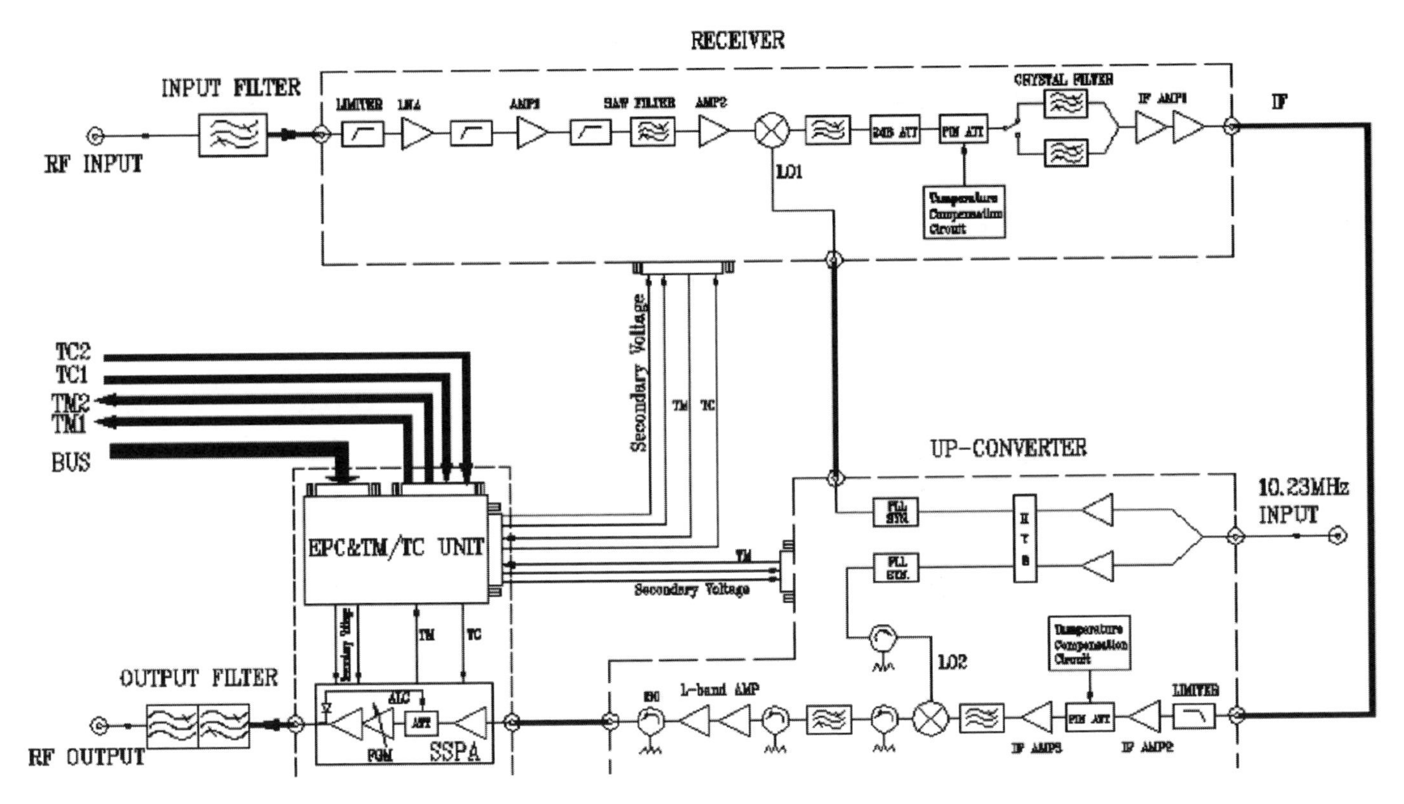

Fig. 3. SART transponder architecture.

3.1.2 Industrial consortium in China and Program Status. The SART prime contractor is China Galileo Industry with main subcontractor CAST (China Academy of Space Technology and in particular the XIRST division).

XISRT is one of the subsidiaries of CAST, located in Xi'an(China) and it is leader in payload developments in China,. Xirst has been engaged in many space related activities, including:

- Communications payloads;
- Navigation Payloads;
- Space-born communicating antenna;
- Microwave remote sensing systems;
- Space-born TT&C system and ground TT&C facilities;
- Space electronic systems and applications.

Since its foundation in 1965, it has developed a series of onboard satellite subsystems and ground applications systems for communication satellites, manned space mission and Ground TT&C stations.

The SART contract was kicked off in November 2005. It has a time span of two years and will be concluded with the delivery of 1 EQM and 4 flight models of the SART.

3.2 The Laser Retro Reflector

The Galileo Satellite will be equipped with a Laser Retro Reflector (LRR), a passive unit used for precise orbit determination. The LRR has the peculiarity to reflect Laser Pulses back to their originating source (i.e. laser ranging ground station).

The LRR consists of an Aluminium base-plate, equipped with an array of Corner Cube Reflectors (CCRs), made of fused silica and fixed in individual aluminium housings. These CCRs have the property to reflect the incoming laser pulses exactly into the direction of the laser source, independent of their incidence angle. Increased incidence angles, however, reduce the intensity of the reflected laser light. So for practical purposes the incidence angle is limited to about 15° which fully satisfies the Galileo needs.

The required LRR size (470 mm × 430 mm), respectively number of CCRs (84) is determined by the satellite orbit altitude and by the required incidence angle to receive valuable signals in the ground stations. The LRR is mounted on the +Z panel of the Galileo Satellites (Nadir orientted), lateral to the Navigation antenna, the Search and Rescue antenna is also present on this panel.

3.2.1 Major Requirements. The main requirements for the LRR are reported here below. Of particular importance for the achievement of the overall performance is the effective reflective area of the LRR that shall be grater than 660 cm^2 for a Field of View of ± 15degree. The total mass shall also be less than 5 kg as limited margins are available at satellite level for the accommodation of payload and platform equipment. The design of the LRR shall also be radiation-proof ensuring a lifetime of the equipment greater than 12 years.

LRR Main Technical Requirements

Performance	Specification
Range finding	Provide capability for range finding between laser ground station and the Galileo satellite.
Laser wavelength	532 nm
Field of View angle	Elevation: ±15°(wrt + ZLRR axis) Azimuth: 360°(about +ZLRR axis)
Ranging accuracy	To be considered to match the overall accuracy of 2 cm.
Physical Characteristics	Symmetrical, planar array, whose normal to the entrance surface coincides with the nadir axis of the Satellite.
Corner cube material	Fused silica with coated or nocoated
Total effective area	> 660 cm^2, at any point of the Earth
The mass of the LRR	< 5 kg
Lifetime	In-orbit: > 12 years

A picture of the structure of the LRR is shown in Fig. 4

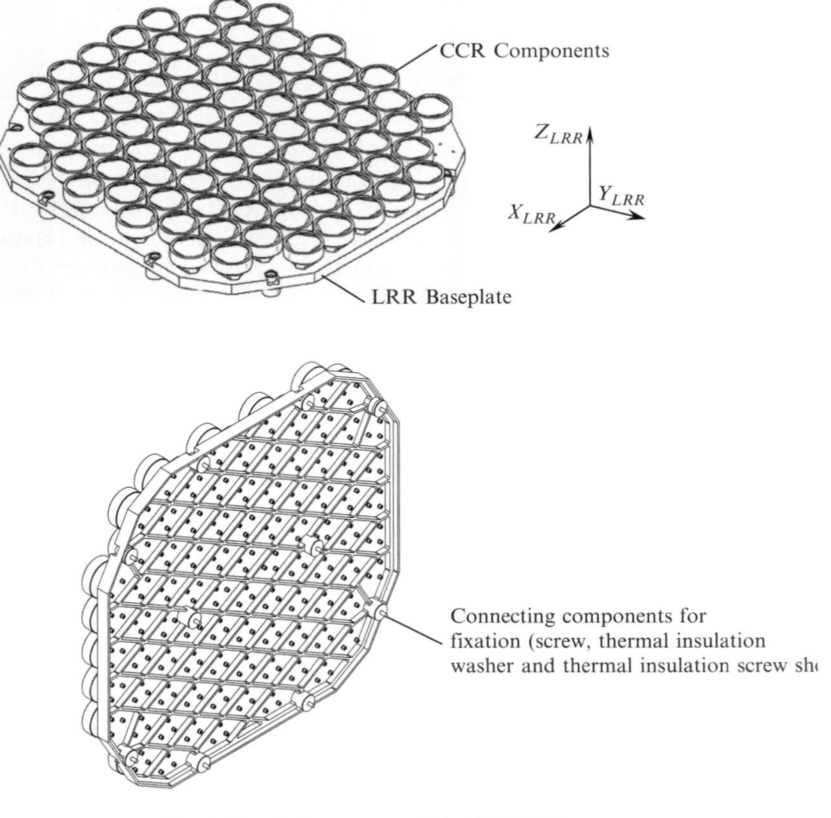

Fig. 4. Physical structure of the LRR Unit.

3.2.2 Industrial Consortium in China and Program Status. The LRR prime contractor is China Galileo Industry with main subcontractor NCRIEO. NCRIEO is a state-owned Class I institute that belongs to China Electronic Technology Group Corporation (CETC).

NCRIEO was established in 1956. As the earliest national electronic component and material institute, it undertook much important national technological projects and obtained 127 researching achievement.

NCRIEO started the R&D in infrared related technology from 1958, laser related technology from 1964. It is the only integrated institute that forms a complete set with laser and infrared material, components, devices and application system. In recent 30 years, the institute undertook and accomplished over 500 national research tasks.

The LRR contract was kicked off in November 2005. It has a time span of two years and will be concluded with the delivery of 1 QM and 4 Flight Models and 1 Flight Spare of the LRR at the beginning of 2007.

4 Conclusions

This article has provided an overview of the activities performed in China in the context of a broad co-operation agreement between the European commission, the European Space Agency and the Chinese Ministry of Science and Technology. A detailed description of the development of two on board equipment: the SART and the LRR has been given. The involvement of China in activities related to the development of Galileo and its world-wide applications is quite important and covers ground developments as well as onboard HW development.

References

[1] Search and Rescue Proposal form CAST: GAL- PRP -XIRST-SART-A-0017
[2] NCRIEO Proposal for the procurement of a Laser Retro Reflector: CGI-PM-GALILEO-LRR-A0001
[3] SAR/Galileo System Definition: GAIN Technical Note, Galileo Phase C0 contract- GAL-TNO-SAR-15-Issue 3
[4] GJU Statement of Works for China Galileo Contracts: FAS, IONO, CGTR, EGSIC, LBS, MEOLUT
[5] ESA statement of Works for: Up-Link Station Development, Satellite Laser Ranging Services, Search and Rescue Forward Link End to End Validation

The Impact of the Galileo Signal in Space in the Acquisition System

Daniele Borio[1], Maurizio Fantino[2], Letizia Lo Presti[1]

[1]Politecnico di Torino/Dipartimento di Elettronica
C.so Duca degli Abruzzi 24, 10129, Torino Italy
Phone: +39 011 5646033, Fax: +39 011 5644099, e-mail: name.surname@polito.it
[2]Istituto Superiore Mario Boella
via P.C. Boggio 61, 10138, Torino Italy
Phone: +39 011 2276431, Fax: +39 011 2276299, e-mail: name.surname@ismb.it

Abstract. This paper is about the impact of the Galileo signal in space on the Acquisition system of a GNSS receiver. The Galileo signal definition presents features allowing improvements of the acquisition and tracking performance, but these differences with respect to GPS must be taken into account in the receiver design phase. In this paper the classical acquisition blocks designed for GPS will be revisited, from a statistical point of view and in terms of performance, looking to the issues that longer codes, BOC modulations and pilot channels will introduce on the Galileo Open Service (OS).

1 Introduction

The first stage of a GNSS (Global Navigation Satellite System) receiver consists in the acquisition of the satellites in view, and in a first rough estimation of the parameters of the Signal-In-Space (SIS) transmitted by each detected satellite. This activity is performed by the so-called acquisition block, which is a system that implements some well-known results of the estimation theory, by using typical signal processing operations, such as correlation, FFT, and filtering. With the advent of the European Galileo system, some modifications in the acquisition stage have to be adopted in order to account the new characteristics of the Galileo Signal in Space (SIS). The main factors which lead to different strategies for the acquisition of the Galileo SIS: are the use of the sub-carrier BOC(1,1), the presence of the secondary code in the pilot channel and the increased rate in the data channel. These new features have been added in order to guarantee interoperability between the new Galileo and the GPS systems and better performance for indoor positioning. However some additional impairments, from the architectural point of view and in terms of performance, in the case of standard localization, have to be paid.

In this paper the typical acquisition blocks used in GPS receivers are described from a statistical signal processing point of view, in order to emphasize the impact of the Galileo SIS format on the different stages of the acquisition operations. Furthermore some modifications to be adopted in order to acquire the Galileo signals are proposed. Since the main task of the paper is to stress the main differences between GPS and Galileo, only the Open Service (OS) case will be considered. The extension to the other cases is in some cases straight forward, in some cases more complex. The main focus of the paper is on the acquisition strategies based on block processing techniques [2], that is the techniques which process simultaneously a block of L samples, as for example the FFT. In fact the presence of the secondary codes in the Galileo SIS impacts on the acquisition methods based on block processing much more than on the methods based on a sample-by-sample processing. The paper analyzes the architectural modifications that have to be implemented in the acquisition block, the additional computational load and the acquisition performance in terms of Receiver Operative Characteristics (ROC's).

The paper is organized as follows: in Section 2 a general model of GNSS signals is exposed; Section 3 provides a theoretical framework for the description of a general acquisition system; in Section 4 the modifications required by the Galileo SIS are discussed whereas in Section 5 some simulations are reported as support for the theoretical results developed in the paper. At the end some conclusions are drawn.

2 Signal Model

The signal at the input of both Galileo and GPS acquisition blocks, in one-path additive Gaussian noise environment, without data modulation, can be modeled as

$$y[n]=\sum_{i=1}^{N_s} r_i[n]+\eta[n] \tag{1}$$

that is the sum of N_s GNSS signals coming from different satellites and $\eta[n]$, the additive Gaussian noise with flat power spectral density (PSD) $N_0/2$ over the receiver band B_r and with power $\sigma_\eta^2 = \mathrm{N_0 B_r}$.

Every useful GNSS signal is of the form

$$r_i[n]=A_{IN,i}\,d(nT_s-\tau_i)\,c_i(nT_s-\tau_i)\,s_b(nT_s-\tau_i)\cos(2\pi(f_{IF}+f_{d,i})nT_s+\theta_i) \tag{2}$$

where

- $A_{IN,i}$ is the amplitude of the ith GNSS signal, whose power is given by $C_i = \mathrm{A}^2_{IN,i}/2$;
- $c_i\,(nT_s - \tau_i)$ is the ith primary spreading code (assumed to be binary) delayed by τ_i and sampled at $f_s = \frac{1}{T_s}$; in the following the dependence from the sampling interval T_s will be omitted and the primary code will be indicated by $c_i\,[n]$;
- $d(nT_s - \tau_i)$ is the secondary spreading code for the Galileo pilot channel and Data in the GPS case;
- $s_b\,(nT_s - \tau_i)$ is the subcarrier, BPSK for GPS and BOC(1,1) for Galileo[1];
- f_{IF} is the intermediate frequency of the receiver;
- $f_{d,i}$ is the unknown Doppler frequency;
- θ_i is the phase of the ith received carrier.

---

[1] Galileo code chips are further modulated by a squared sub–carrier, in this paper it is commonly referred as slot the width of the sub–carrier chip

Thanks to code orthogonality the different GNSS codes are analyzed separately by the acquisition block, thus the case of $N_s = 1$ is considered and the dependence from the index i is omitted in the rest of the paper.

3 Acquisition Concepts

The first operation performed by a GNSS receiver is the signal acquisition that decides either the presence or the absence of the satellite under test and provides a rough estimation of the code delay and of the Doppler frequency of the incoming signal. The acquisition system implements some well-known results of the detection and of the estimation theory and different logical and functional blocks take part in the process. In the GNSS literature the exact role of these disciplines and of these functional blocks is sometimes unclear. In this section a general acquisition system is described as the interaction of four functional blocks that perform four different logical operations. The framework developed by using these four elements allows to describe the majority of the acquisition systems, providing an effective tool for comparative analysis. All the acquisition systems for GNSS applications described in literature [1], [3], [4] are based on the evaluation and processing of the Cross Ambiguity Function (CAF) that in the discrete time domain can be defined as

$$R_{y,c}(\bar{\tau}, \bar{f}_d) = \sum_{n=0}^{L-1} y[n] c(nT_s - \bar{\tau}) s_b(nT_s - \bar{\tau}) e^{j2\pi (f_{IF} + \bar{f}_d) nT_s} \tag{3}$$

Ideally the CAF envelope should present a sharp peak in correspondence of the value of $\bar{\tau}$ and $\bar{f}_d$ matching the delay and the Doppler frequency of the SIS. However the phase of the incoming signal, the noise and other impairments can ruin the readability of the CAF and further processing is needed. For instance, in a non-coherent acquisition block only the envelope of the CAF is considered, avoiding the phase dependence. Moreover coherent and non-coherent integrations can be employed in order to reduce the noise impact.

When the envelope of the averaged CAF is evaluated the system can take the decision on the presence of the satellite. Different detection strategies can be employed for the decision: some strategies require only the partial knowledge of the CAF implying an interaction among the acquisition elements. The detection can be bettered by using multitrial techniques that require the use of CAF's evaluated on subsequent portions of the incoming signal.

In Fig. 1 the general scheme of an acquisition system is reported highlighting the presence of the four blocks

- CAF evaluation;
- Envelope and Average;
- Detector;
- Multitrial;

and of their possible interactions.

In the following these four blocks are discussed, with particular emphasis to the first two that are the more affected by the new Galileo SIS's.

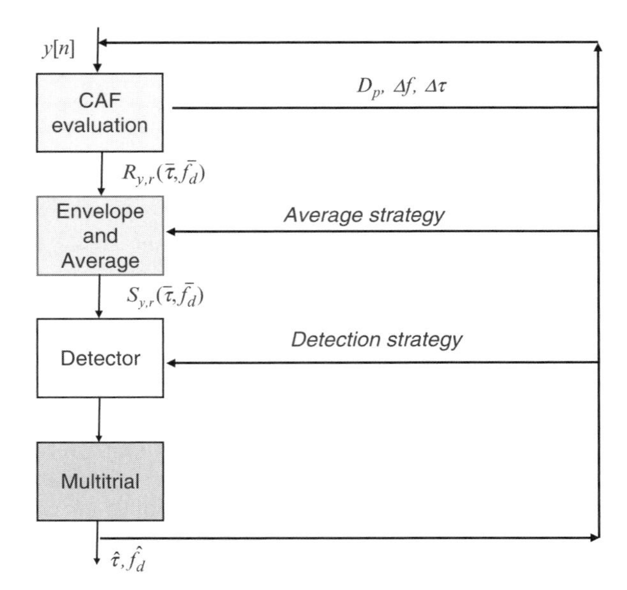

Fig. 1. Functional blocks of an acquisition system.

3.1 CAF Evaluation

In the acquisition systems described in literature different methods of evaluating the CAF are presented. They give the same (or approximately the same) results and the choice of the method mainly depends on the hardware and software tools available for the receiver implementation. The local generation of the test signal

$$\bar{r}[n] = c(nT_s - \bar{\tau})s_b(nT_s - \bar{\tau})e^{j2\pi(f_{IF}+\bar{f}_d)nT_s}$$

can be done in different ways. It is important to notice that the part of the test signal containing the code and the subcarrier, that is $c(nT_s - \bar{\tau})s_b(nT_s - \bar{\tau})$, can be obtained starting from a local code (including the subcarrier) of the type

$$c_{Loc}[n] = c(nT_s)\, s_b(nT_s)$$

with $n \in (0, L-1)$ and by applying a circular delay to the samples of $c_{Loc}[n]$. This is possible when the periodicity of the incoming code is a submultiple of the integration time L. In the actual implementation this circular delay should be done also taking into account the problem of the incommensurability constraint on the sampling frequency [6]. In fact this constraint alters the perfect periodicity of the samples of the incoming code $c(t)$, allowing the Delay Lock Loop (DLL) to work properly even if the number of samples per chip is very low. On the contrary the effect of the incommensurability on the detection and estimation operations of the acquisition block is slightly disturbing. In practice this effect is negligible since the purpose of the acquisition block is to perform only a rough estimation of the delay and Doppler frequency.

A macro classification of classical acquisition methods can be done as it follows.

- **Method 1: Serial scheme.** In this scheme a new CAF is evaluated at each n instant. The input vector **y** can be updated instant by instant by adding a new input value and by discarding the former one. To avoid ambiguity, in this case the notation $\mathbf{y}_n = [y(n)\ y(n-1) \ldots y(n-L+1)]$ will be adopted. With this approach the delay $\bar{\tau}$ moves throughout the vector $\mathbf{y}_n$ at each new instant. Therefore the local code c_{Loc} $[n]$ is always the same and the CAF is given by the expression

$$R_{y,r}(\bar{\tau}, \bar{f}_d) = \sum_{m=0}^{L-1} y[\bar{\tau} - L + m + 1]\, c(mT_s)\, s_b(mT_s)\, e^{j2\pi(f_{IF} + \bar{f}_d)mT_s} \tag{4}$$

It is quite easy to verify that this approach is equivalent to move the delay of c_{Loc} $[n]$, as the mutual delay between c_{Loc} $[n]$ and the received code is the unknown of interest. In Fig. 2 the serial scheme is reported, each value of the CAF is evaluated independently without using any block strategy. The term F_D indicates the quantity $(f_{IF} + \bar{f}_d)\, T_s$.

- **Method 2: FFT in the time domain.** In this scheme the vector **y** is extracted by the incoming SIS and multiplied by $e^{j2\pi(f_{IF} + \bar{f}_d)nT_s}$, so obtaining a sequence

$$ql[n] = y[n]\, e^{j2\pi(f_{IF} + \bar{f}_d)nT_s} \tag{5}$$

for each frequency bin. At this point the term

$$R_{y,r}(\bar{\tau}, \bar{f}_d) = \sum_{n=0}^{L-1} ql[n]\, c(nT_s - \bar{\tau})\, s_b(nT_s - \bar{\tau}) \tag{6}$$

assumes the form of a Cross-Correlation Function (CCF), which can be evaluated by means of a circular cross-correlation defined by

$$\tilde{R}_{y,r}(\bar{\tau}, \bar{f}_d) = \mathrm{IDFT}\{\mathrm{DFT}[ql[n]]\, \mathrm{DFT}[c(nT_s)\, s_b(nT_s)]^*\} \tag{7}$$

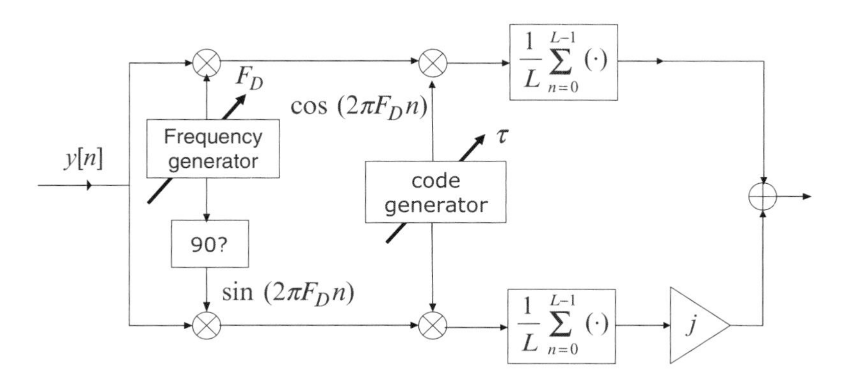

Fig. 2. The serial acquisition scheme: the CAF is evaluated independently for each value of $\bar{\tau}$ and $\bar{f}_d$.

where DFT and IDFT stand for the well-known Discrete Fourier Transform and Inverse Discrete Fourier Transform. It is easy to show that the CCF and the circular CCF coincide only in presence of periodic sequences. This is the case when $\bar{f}_d = f_d$, except for the noise contribution and a residual term due to a double frequency ($2f_d$) component contained in the term $ql[n]$. In the other frequency bins the presence of a sinusoidal component could alter the periodicity of the sequence. The proof of this is out of the scope of this paper.

- **Method 3: FFT in the Doppler domain.** In this scheme (Fig. 3) a vector y can be extracted by the incoming SIS instant by instant, as in the method 1, and multiplied by $c_{Loc}[n]$ so obtaining a sequence

$$q_i[m] = y[\bar{\tau} - L + 1 + m]\, c_{Loc}[m] \tag{8}$$

for each delay bin. A similar result can be obtained by extracting an input vector y every L samples, and multiplying it by a delayed version of the local code $c_{Loc}[n]$. As mentioned before, this delay is obtained by applying a circular shift to the samples of $c_{Loc}[n]$. At this point the term

$$R_{y,r}(\bar{\tau}, \bar{f}_d) = \sum_{m=0}^{L-1} q_i[m]\, e^{j2\pi (f_{IF} + \bar{f}_d) m T_s} \tag{9}$$

assumes the form of an inverse Discrete-Time Fourier Transform (DTFT). It is well known that a DTFT can be evaluated by using a Fast Fourier Transform (FFT) if the normalized frequency $(f_{IF} + \bar{f}_d)\, T_s$ is discretized with a frequency interval

$$\Delta f = \frac{1}{L}$$

in the frequency range (0, 1), which corresponds to the analog frequency range $(0, f_s)$. The evaluated frequency points become

$$\bar{f}_d T_s = \frac{l}{L} - f_{IF} T_s$$

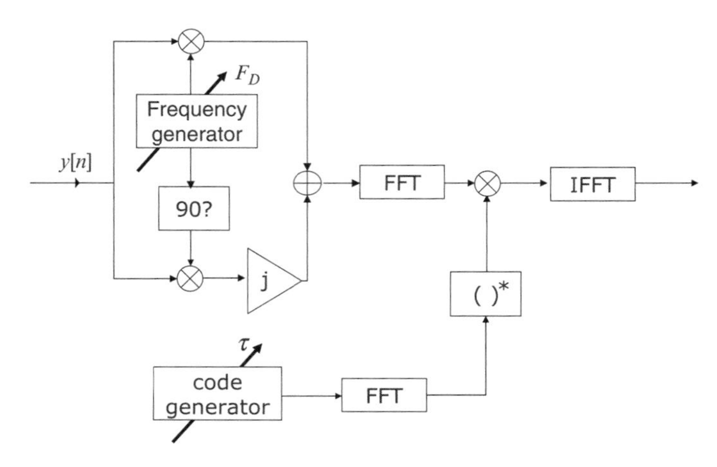

Fig. 3. The time parallel acquisition scheme: the CAF is determined by using a circular convolution employing efficient FFT's.

and the CAF can be written as

$$R_{y,r}(\bar{\tau}, \bar{f}_d) = \sum_{m=0}^{L-1} q_i[m] e^{j\frac{2\pi}{L}lm} \tag{10}$$

With this method the support of the search space along the frequency axis and the frequency bin size depend on the sampling frequency f_s and on the integration time L. If the same support and bin size used in methods 1 and 2 have to be used, the integration time has to be changed, and some decimation (with pre-filter) has to be adopted before applying the FFT. This modifies the input signal to be processed and the comparison among the methods will be affected by the signal modifications. Notice that the maximum peak loss in the Doppler frequency domain is not any more a free parameter with this method, since it is ruled by the FFT constraints, as it is shown in [2, 5]. If it is necessary to mitigate this effect some zero padding techniques can be used, at the expenses of some interpolation loss. In Fig. 4 the frequency domain acquisition block is reported. An integrate and dump block followed by a decimation unit is inserted in order to reduce the number of samples on which the FFT is evaluated. This operation reduces the computational load but introduces a loss in the CAF quality [2, 5].

3.2 Envelope and Average

After the CAF is evaluated, the acquisition system has to remove dependence on the input signal phase and to apply some noise reduction techniques. In Fig. 5 three different methodologies are reported. The phase dependence is removed by considering the CAF envelope. If different CAFs are averaged before evaluating the envelope, **coherent integrations** are employed. This kind of integrations provide the best performance in terms of noise variance reduction. In fact before the envelope the noise terms are zero mean gaussian random variables and the coherent integrations average elements that can be either positive or negative. If the average is performed after the squared envelope (Fig. 5, part b), **non-coherent integrations** are used. In this case non-negative random variables are averaged together thus a residual term, due to the noise, still remains. In [5] a deep analysis of the impact of the use of coherent and non-coherent integrations is provided. It is highlighted that the use of coherent integrations impact the Doppler frequency resolution and a greater number of Doppler bin is required for having a constant error on the determination of the Doppler frequency [5]. Thus the required computational load is greater than the one of the non-coherent integrations.

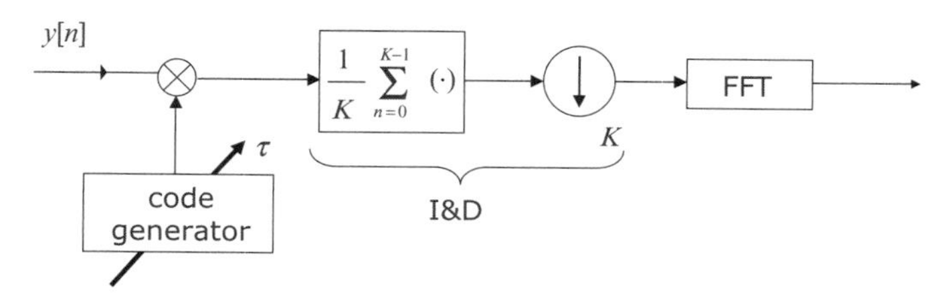

Fig. 4. The frequency parallel acquisition scheme: the CAF is evaluated by using efficient FFT.

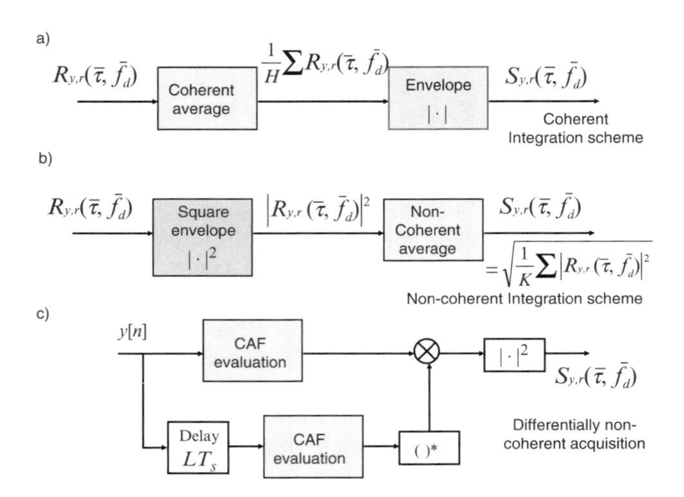

Fig. 5. Different "envelop and average" scheme.

The output of the "envelope and average" block is indicated by

$$S_{y,r}(\bar{\tau}, \bar{f}_d)$$

and two indicators of the system performance are the cell false alarm and detection probabilities associated to $S_{y,r}(\bar{\tau}, \bar{f}_d)$. Each value of $\bar{\tau}$ and of $\bar{f}_d$ defines a cell and the probability that $S_{y,r}(\bar{\tau}, \bar{f}_d)$ is greater than a fixed threshold defines

- the **false alarm probability** if $\bar{\tau}$ or $\bar{f}_d$ do not match the SIS's ones;
- the **detection probability** if the code delay and the Doppler frequency are matched.

When coherent and non-coherent integrations are used the false alarm and detection probabilities assume the following expressions:

$$P_{fa}(V_t) = \exp - \left\{ \frac{V_t^2}{2\sigma^2} \right\} \sum_{i=0}^{K-1} \frac{1}{i!} \left(\frac{V_t^2}{2\sigma^2} \right)^i$$
$$P_{det}(V_t) = Q_K \left(\frac{\sqrt{K}\,\alpha}{\sigma}, \frac{V_t}{\sigma} \right) \tag{11}$$

where V_t is the threshold, $\alpha = \frac{A_{IN}}{2}, \sigma^2 = \frac{\sigma_n^2}{2LH}$, H is the number of coherent integrations, and K the number of non-coherent integrations. Equations (11) and (12) account the fact that coherent and non-coherent integrations can be combined together. In section 5 a comparison between coherent and noncoherent integrations will be provided.

Recently a modified non-coherent detector for signal acquisition referred as differential non-coherent (DNC) has been proposed [8]. This new scheme, whose performance are analyzed in [7] can be easily described in terms of the four functional blocks reported above and its integration strategy is summarized in Fig. 5c. The analysis of this kind of system is out of the scope of this paper, however the modifications necessary for its use with the Galileo SIS can be found in [9].

3.3 Detection Strategy

Once $S_{y,r}(\bar{\tau}, \bar{f}_d)$ is evaluated the system can take the decision on the presence of the satellite. Different strategies can be employed. The detection strategies can control the previous blocks, for example, by requiring the computation of $S_{y,r}(\bar{\tau}, \bar{f}_d)$ only on a subset of the values of $\bar{\tau}$ and $\bar{f}_d$.

In [10] three different strategies are analyzed and compared in terms of system performance.

The introduction of the Galileo SIS does not essentially change the role of this block and the considerations reported in [10] still applies.

3.4 Multitrial

When a first decision about the satellite presence and a first estimation of the code delay and of the Doppler frequency are available, the system can refine these results. Thus multitrial techniques, based on the use of different $S_{y,r}(\bar{\tau}, \bar{f}_d)$, evaluated over subsequent portions of the input signal, can be employed. Two examples of these techniques are the M on N [1] and the Tong [11] methods.

Multitrial techniques generally do not require the computation of more than one complete $S_{y,r}(\bar{\tau}, \bar{f}_d)$, thus they interact with the other blocks changing the requirements for the subsequent iterations occurring in the process.

4 Galileo Impact on the Signal Acquisition

In this paper the Galileo BOC(1,1) modulation on the L1 carrier is considered. This signal is rather similar to the GPS signal, which can be derived from the C/A code by applying to it a Manchester like coding.

One difference between the GPS and Galileo signals is their bandwidth, since the application of the Manchester coding on the C/A code causes a splitting of the code power spectral density. The single–sided bandwidth, that for the GPS C/A code is B_s = 1.023 MHz, for the Galileo BOC(1,1) signal becomes B_s = 2.046 MHz, as depicted in Fig. 6.

This means that, in order to process the Galileo signal in the same way as the GPS C/A signal, the ADC antialiasing filter must have a single–sided bandwidth twice larger than the corresponding GPS ADC antialiasing filter bandwidth. Therefore, the sampling frequency must be at least twice the GPS C/A code sampling frequency.

The sampling frequency value can be derived also considering that the Galileo BOC(1,1) slot rate is twice the GPS C/A chipping rate. The final result is that every slot of the Galileo BOC(1,1) modulation has a rate twice than the chipping rate of the GPS C/A code signal, i.e. R_{BOC} = 2.046 Mslot/s. This implies that, in order to

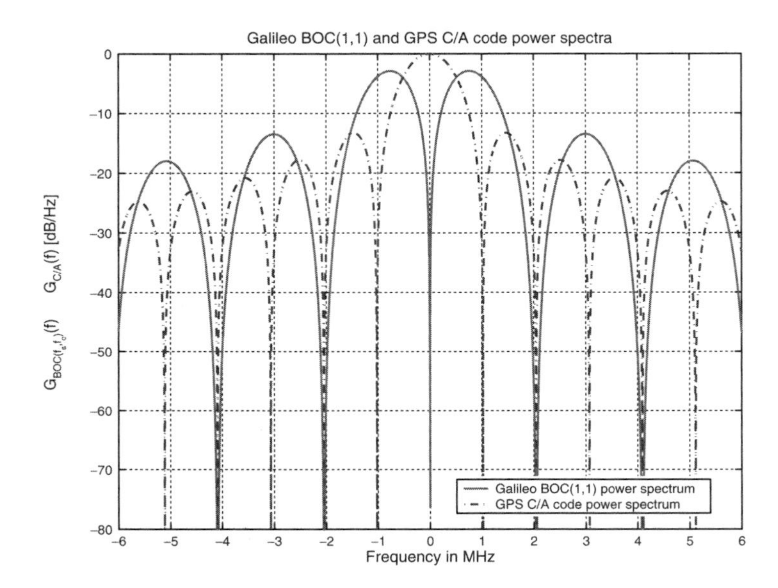

Fig. 6. Power spectral density of Galileo BOC(1,1) code (solid line) and comparison with GPS C/A code (dashed line).

obtain about two samples per BOC(1,1) slot it is necessary to sample the Galileo signal at a rate that is twice greater than the sampling rate of the GPS C/A code.

From the previous considerations it follows that a Galileo receiver can be designed on the basis of a GPS receiver, but the number of samples to process, besides, is accordingly greater than the number of samples processed by the GPS receiver in the same condition of integration time. The global complexity of a Galileo BOC(1,1) receiver is, therefore, slightly increased with respect to the GPS C/A receiver, but the same applies to the acquisition performances, as it will be pointed out in the section devoted to the performance results.

4.1 Secondary Code Transition and Single Period Integration Time

Parallel acquisition in time domain schemes, based on FFT operations, are extremely efficient, but since their intrinsic nature to process blocks of data may suffer of peak miss-detection due to the presence of the secondary code in the Galileo BOC(1,1) pilot channel. In fact, the ranging codes used for the L1F pilot channel are based on the so called tired codes. Tired codes are built modulating a short duration primary code by a long duration secondary code. The secondary code acts exactly as the data transition for the GPS signal and it can be the cause of a sign reversal in the correlation operation over the integration interval. For its natural way to look for all the code shifts over all the possible delays moving along the received signal (Fig. 7), when the correlation is performed on a single period, the serial acquisition scheme is practically insensitive to the secondary code transitions. In fact the main

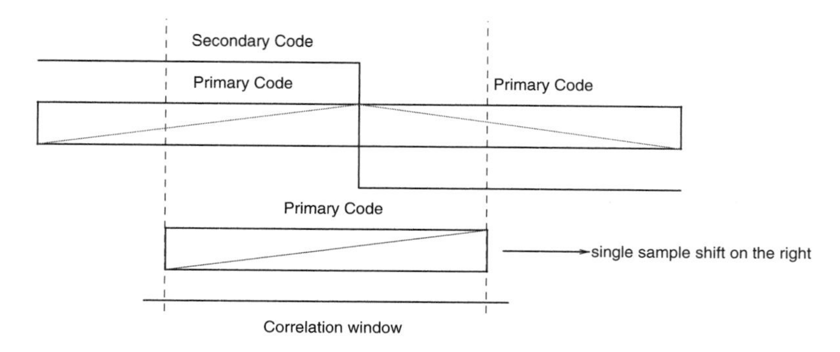

Fig. 7. Circular correlation for serial search scheme with secondary code.

lobe is identified just when the incoming and the local generated codes are perfectly aligned, hence when the secondary transitions are at the edges of the analyzed stream [12].

Since the FFT system in frequency domain performs a serial search over the code delay and a parallel search in frequency domain, it results insensitive to the code transition as well as the serial search scheme and it can be used in the acquisition of the L1 Galileo primary code.

The fast acquisition scheme computes an entire row of the search space from a block of data by means of FFT operations. Since it is not possible to know if in the data block the secondary code causes a sign reversal or not, this technique cannot be applied without changes. Figure 8 shows, how, the secondary code sign reversal within a correlation window makes it no longer periodic; by consequence, the

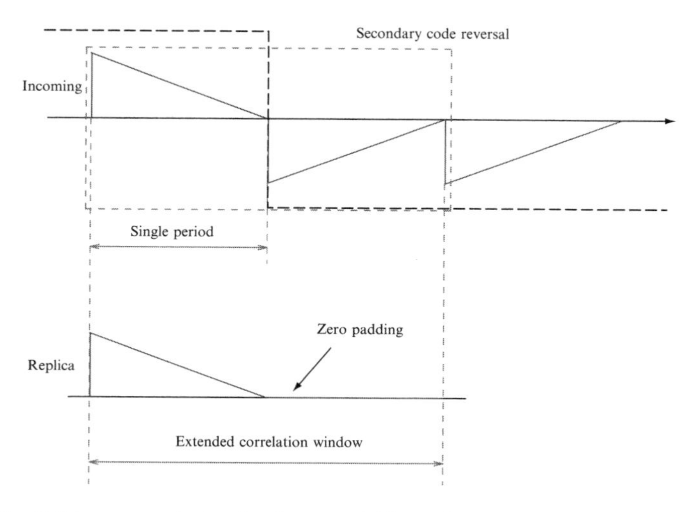

Fig. 8. Linear correlation for fast acquisition scheme with secondary code.

single period circular correlation cannot be used alone to decide the absence or the presence of a signal.

A possible solution to this problem is to conduct a linear correlation as shown in Fig. 8. Two periods of the incoming signal are correlated with a single period of the local code zero padded to fit the correlation window. The zero terms in the second period does not introduce any advantage in terms of noise reduction generally achieved with a longer integration time. The price to be paid for the fast acquisition scheme to achieve such insensitivity is to perform a linear correlation using two code periods, then to use longer FFTs.

4.2 Multiple Period Integration Time

In order to increase the detection probability for a given false alarm probability, a summation over more than one code period can be performed. In this case, the threshold value has to be increased.

The length of a data record used for the summation in an acquisition scheme is limited by two factors, the navigation data or secondary code transitions and the Doppler effect on the spreading code.

The presence of a navigation data or secondary code transitions in the data record causes a spreading effect of the output spectrum and the performances of the acquisition system are degraded. For the GPS C/A signal on the L1 carrier, the navigation data rate is 50 bit/s (see reference [1]), so that the length of a data bit is 20 ms, i.e. 20 periods of the spreading code. The maximum data record that can be used for the coherent summation is, therefore, 10 ms or 10 C/A code periods. In fact, in a 20 ms time interval only one navigation data transition can occur. Then, if there is a transition in the first 10 ms data record, the second 10 ms will be transition free. On the other hand, the sequence length cannot be less than a code period or 1 ms and even in this minimum interval a data transition can occur. In order to guarantee no data transition, the acquisition algorithm should take into account two consecutive data records of equal duration, but less than 10 ms, perform the coherent summation over these two data records and then declare the detection if one of the two envelopes or both of them exceed the threshold. Unfortunately the presence of the secondary code on the Galileo signal does not allow the possibility to perform the acquisition of consecutive pieces of signal, since every period of the primary code is modulated by the secondary short code, then it is not guaranteed the absence of a secondary code transition in the following integration period. By the way, in order to increase the detection probability, a summation over than one code period in a non-coherent way can be applied, accepting the squaring loss due to the square operation performed after the squared envelope.

5 Performance Analysis

The simulations performed to determine and analyze the acquisition systems and their optimum parameters aim to obtain the so-called *Receiver Operative Characteristic*, which will be named with the acronym ROC. This is the graph of the

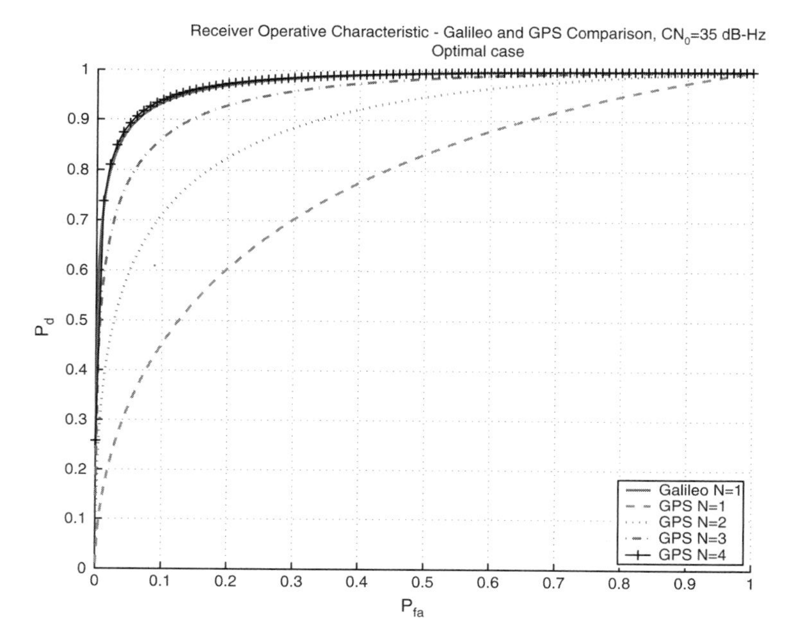

Fig. 9. Receiver Operative Characteristic Comparison, Galielo single period and from one up to five coherent GPS C/A code periods integration time under no losses hypothesis at CN_0 of 35 dB-Hz.

detection probability versus the false alarm probability, or, equivalently, of the missed detection probability versus the false alarm probability. Figure 9 depicts the comparison between the ideal optimal ROC comparison between the single period Galileo L1 signal acquisition and GPS.

Since the C/A code length is a quarter the Galileo L1 OS code, the comparison of Fig. 9 is made increasing the integration time used for the coherent autocorrelation function evaluation from 1 ms to 4 ms. As it is possible to see, considering the optimal case, better performance can be achieved increasing the integration time and without considering any correlation loss impairments GPS and Galileo are completely identical when the integration is 4 ms, value which correspond to 4 GPS C/A code periods and a single Galileo L1 OS code period.

Figure 10 shows the same comparison of Fig. 9, but considering the Acquisition impairments due to a rough code alignment in half chip/slot resolution and a coarse Doppler frequency recovery.

It is here remarked how, the correlation loss due to the Doppler shift in the bidimensional CAF evaluation does not depend on the Galileo or GPS signal structure, but only on the integration time used to perform the correlation. Different is the case of the loss due to the code misalignment, which is related to the shape of the correlation function [5]. As it possible to see in Fig. 11, where a comparison of the envelope of the GPS C/A and Galileo L1 OS correlation functions is reported, the Galileo correlation function is narrower than the GPS one. Even though, this

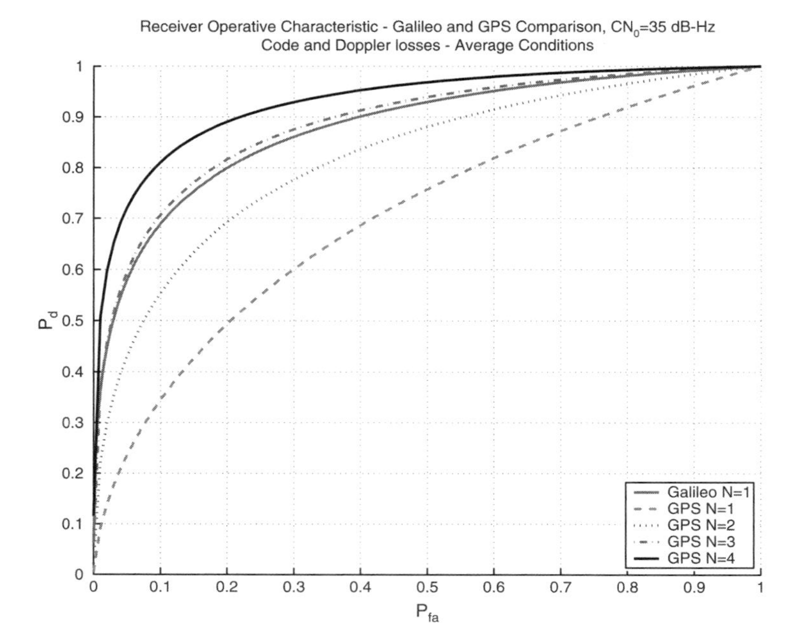

Fig. 10. Receiver operative characteristic comparison, Galielo single period and from one up to five coherent GPS C/A code periods integration time at CN_0 of 35 dB-Hz.

leads to better tracking jitter and multipath rejection performances, it makes the acquisition more challenging with respect to the classical GPS strategies. In fact it is easy to understand that for the same local code displacement the Galileo correlation functions drop faster than the GPS one, with a consequent bigger loss, as it is possible to see in Fig. 10.

Coherent integration over more than a single code period is a common strategy to increase the signal to noise ratio at the detector input in the acquisition of GNSS signals. It has been highlighted how the presence of the secondary codes in Galileo will make the coherent evaluation of the CAF quite difficult and how the required robustness in terms of signal to noise ratio can be achieved by means of non-coherent summation. This strategy, however due to its easily implementation and efficiency is already successfully used to acquire the C/A GPS signal. In Fig. 12 a comparison between the single period Galileo and multi period non-coherent GPS signal acquisition is carried out. As it is possible to outline, better performance can be achieved with the coherent approach, but this requires to reduce the Doppler bin size in order to maintain the same Doppler loss and then increasing the number of analyzed cell in the search space [5]. Moreover, a longer integration time means to increase the samples that must be processed by means of FFT operations, which is surely more cost effective than the non-coherent integration.

The Acquisition system performance can be better appreciated by means of the graph of Fig. 13, where the detection probability for a fixed false alarm probability is plotter versus the input carrier to noise ratio. Due to the long code designed for Galileo and the presence of the secondary codes, the comparison is outlined

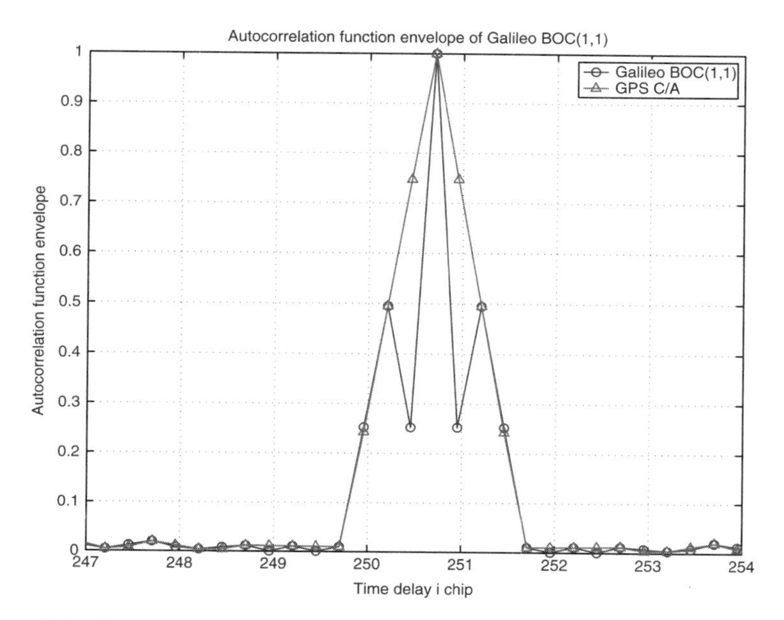

Fig. 11. GPS C/A and Galileo L1 OS envelope autocorrelation comparison for a digital sequence sampled at two samples per slot.

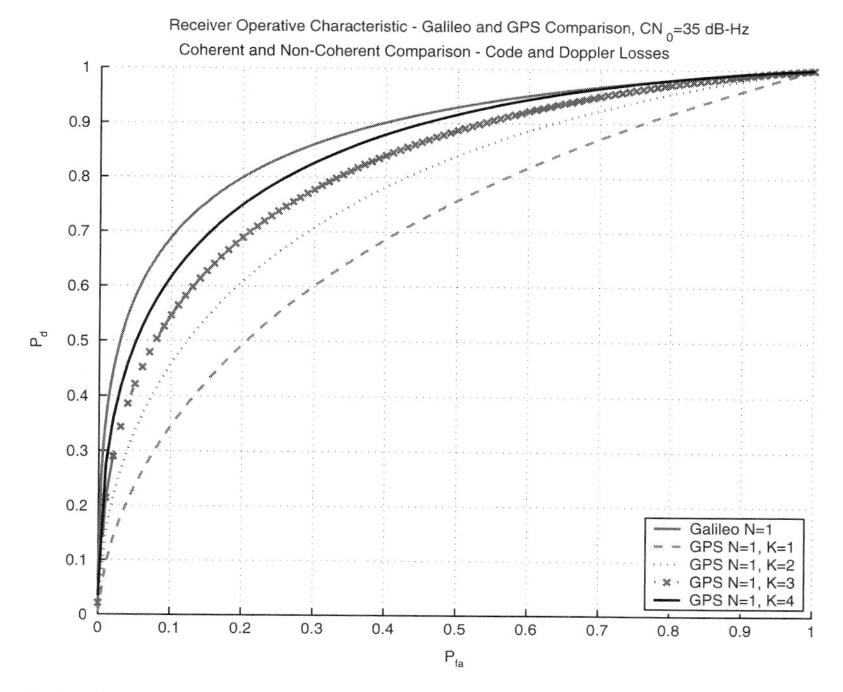

Fig. 12. Receiver operative characteristic comparison, Galielo single period and from one up to five non coherent GPS C/A code periods integration time at CN_0 of 35 dB-Hz.

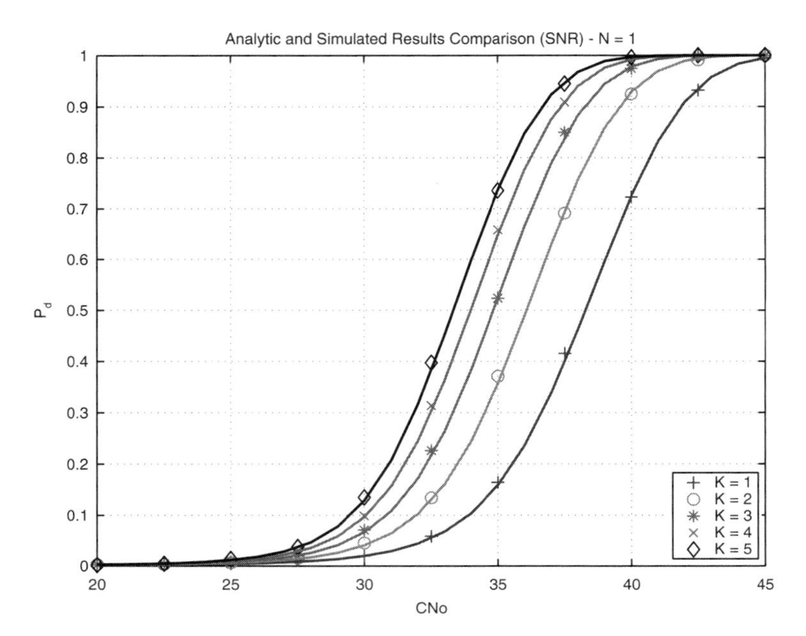

Fig. 13. Comparison for the analytic and simulated results for the Galileo BOC(1,1) signal for a desired $P_{fa} = 10^{-3}$ and from one up to five non-coherent integration times.

varying the integration time from one up to five codes period integrated in a non-coherent way.

It is possible to see how, even though this approach is less performing than the classical coherent one, it is possible to achieve good acquisition performance even at low signal to noise ratio overcoming the secondary code transition problems.

6 Conclusion

The main differences introduced by Galileo with respect to GPS are the presence of a secondary code, which acts in the same way of the GPS navigation data but with a higher rate, the particular shape of the correlation function and the presence of side lobes. The secondary code may introduce a sign reversal in the data record with a reduction of the efficiency of the circular correlation performed by the time parallel acquisition technique; the problem can be overcome by employing a linear correlation with a consequent increment of the system complexity due to the longer number of samples that have to be processed. Moreover, the sign reversal introduced by the secondary code does not allow to increase the integration time in a coherent way. In the paper these issues, coupled to the system performance are analyzed. The main factors to be accounted in the design of a Galileo acquisition block are described and the system performance are studied by means of theoretical curves and computer simulations.

References

[1] E. D. Kaplan "Understanding GPS: Principles and Applications" Norwood, MA Artech House, 1996.

[2] H. Mathis, P. Flammant and A. Thiel "An analytic way to optimize the detector of a post-correlation FFT acquisition algorithm" *ION GPS/GNSS 2003 Proceeding*, 9–12 September 2003, Portland, OR.

[3] J.B.Y Tsui, "Fundamentals of Global Positioning System Receivers. A software Approach" New York, John Wiley and Sons, 2nd edition 2005.

[4] Z. Weihua and J. Tranquilla "Modeling and analysis for the GPS pseudo-range observable", *IEEE Transaction on Aerospace and Electronic System*, 31: 739–751, April 1995.

[5] D. Borio, M. Fantino and L. Lo Presti "Acquisition Analysis for Galileo BOC Modulated Signals: Theory and Simulation" *European Navigation Conference*, Manchester, UK, 7–10 May, 2006.

[6] M. Fantino, F. Dovis and L. Lo Presti "Design of a Reconfigurable Lowcomplexity Tracking Loop for Galielo Signals" *International Symposium on Spread Spectrum Techniques and Applications, ISSSTA 2004*, Sidney, 736–740, August 2004.

[7] R. Pulikkoonattu and M. Antweiler "Analysis of Differential Non Coherent Detection Scheme for CDMA Pseudo Random (PN) code Acquisition" *Proceeding of IEEE ISSSTA*, Sydney, Australia, 30 August - 2 September, 2004.

[8] R. Pulikkoonattu, P.K. Venkataraghavan and T. Ray "A Modified Non Coherent PN Code Acquisition Scheme" *IEEE Wireless Communications and Networking Conference*, Atlanta, USA, March 2004.

[9] P. G. Matthos "Galileo L1c - Acquisition Complexity: Cross Correlation Benefits, Sensitivity Discussions on the choice of Pure Pilot, Secondary Code, or something different" *Proceeding of IEEE/ION PLANS*, San Diego, California, 25–27 April, 2006

[10] D. Borio, L. Camoriano and L. Lo Presti "Impact of the Acquisition Searching Strategy on the Detection and False Alarm Probabilities in a CDMA Receiver" *Proceeding of IEEE/ION PLANS*, San Diego, California, 25–27 April, 2006

[11] P. S. Tong "A Suboptimum Synchronization Procedure for Pseudo-Noise Communication Systems" *Proceeding of National Telecommunications Conference* 1973.

[12] M. Fantino and F. Dovis "Comparative analysis of acquisition techniques for BOC modulated signals". *The European Navigation Conference, GNSS 2005, Munich, Germany*, July 2005.

The Aalborg GPS Software Defined Radio Receiver

Kai Borre

Danish GPS Center, Aalborg University
Niels Jernes Vej 14, DK-9220 Aalborg Ø, Denmark
Phone: +4596358362, e-mail: borre@gps.aau.dk

Abstract. A receiver for the Global Positioning System (GPS) signals provides information on its position and time. The position is given in an Earth-Centered and Earth-Fixed coordinate system. This means that a static receiver keeps its coordinates over time, apart from the influence of measurement errors. The system time (GPST) counts in weeks and seconds of week starting on January 6, 1980. Each week has its own number. Time within a week is counted in seconds from the beginning at midnight between Saturday and Sunday (day 1 of the week). GPST is maintained within the system itself. Universal Time Coordinated (UTC) goes at a different rate which is connected to the actual speed of the rotation of the Earth. At present 14 seconds have to be added to UTC to get GPST.

The GPS has 6 orbital planes with at least 4 satellites. At the moment GPS consists of 29 active satellites. They complete about 2 orbits/day.

1 The Transmitted GPS Signals

Satellite positioning systems exploit ***Spread Spectrum*** (SS) techniques. SS came alive in 1980s and is popular for applications involving radio links in hostile environments. SS is an RF communications system in which the baseband signal bandwidth is intentionally spread over a larger bandwidth by injecting a higher-frequency signal. As a direct consequence, energy used in transmitting the signal is spread over a wider bandwidth and appears as noise. The ratio (in dB) between the spread baseband and the original signal is called *processing gain*. Typical SS processing gains run from 10 dB to 60 dB, see [1].

To apply an SS technique, simply inject the corresponding SS code somewhere in the transmitting chain before the antenna. That injection is called the spreading operation. The effect is to diffuse the information in a larger bandwidth. Conversely, you can remove the SS code by a despreading operation, at a point in the receive chain before data retrieval. The effect of a despreading operation is to reconstitute the information in its original bandwidth. Obviously, the same code must be known in advance at both ends of the transmission channel.

In GPS, SS modulation is applied on top of a BPSK modulation, see below.

Intentional or un-intentional interference and jamming signals are rejected because they do not contain the SS code. This characteristic is the real beauty of SS. Only the desired signal, which has the code, will be seen at the receiver when the despreading operation is exercised.

In GPS the codes are digital sequences that must be as long and as random as possible to appear as "noise-like" as possible. But in any case, they must remain reproducible. Otherwise, the receiver will be unable to extract the message that has been sent. Thus, the sequence is "nearly random". Such a code is called a ***pseudo-random number*** (PRN) or sequence. The PRN sequences applied in GPS are ***Gold sequences***. These sequences are generated by feedback shift registers, and they are inserted at the data level. This is the direct sequence form of spread spectrum (DSSS). The PRN is applied directly to data entering the carrier modulator.

All GPS satellites use the same carrier frequencies: On L_1 1575.42 MHz and L_2 1227.60 MHz. In a modernized GPS there will be a new civilian frequency L_5 (then the military might remove L_2 from civilian use).

Each satellite has two unique spreading sequences or codes. The first one is the coarse acquisition code (C/A) and the other one is the encrypted precision code (P(Y)). The C/A code is a sequence of 1 023 chips. (A chip corresponds to a bit. It is simply called a chip to emphasize that it does not hold any information.) The code is repeated each ms giving a chipping rate of 1.023 MHz. The P code is a longer code ($\approx 2.35 \cdot 10^4$ chips) with a chipping rate of 10.23 MHz. It repeats itself each week starting at the beginning of the GPS week. The C/A code is only modulated onto the L_1 carrier while the P(Y) code is modulated onto both the L_1 and the L_2 carrier.

The purpose of PRN codes is twofold: They spread the signals and they provide for measuring the travel time between satellite and receiver. The system keeps all C/A code starts aligned in all active satellites.

In the rest of this presentation we focus on the L_1 signal. Each satellite transmits a continuous signal with at least three components:

- a carrier wave with frequency $f_1 = 1575.42 = 154 \times 10.23$ MHz
- an individual PRN code which is a sequence of −1 and +1 each of length 1 millisecond
- a data bit sequence which carries information from which the satellite's position can be computed. The length of one navigation bit is 20 milliseconds.

The PRN code and the data bits are combined through modulo-2 adders. The result is modulated onto the carrier signal using the ***binary phase shift keying*** (BPSK) method: The carrier is instantaneously phase shifted by 180° at the time of a chip change. When a navigation data bit transition occurs, the phase of the resulting signal is also phase shifted 180°.

So the ***signal transmitted*** from satellite k is

$$\begin{aligned} s^k(t) = {} & \sqrt{2P_C}\left(C^k(t) \oplus D^k(t)\right)\cos(2\pi f_{L1} t) \\ & + \sqrt{2P_{PL1}}\left(P^k(t) \oplus D^k(t)\right)\sin(2\pi f_{L1} t) \\ & + \sqrt{2P_{PL2}}\left(P^k(t) \oplus D^k(t)\right)\sin(2\pi f_{L2} t). \end{aligned} \tag{1}$$

Here P_C, P_{PL1}, and P_{PL2} are the powers of signals with C/A or P code. C^k and P^k are the C/A and P(Y) code sequences assigned to satellite number k. D^k is the navigation data sequence, and f_{L1} and f_{L2} are the carrier frequencies of L1 and L2. The $\oplus$ symbol denotes the "exclusive or" operation.

2 The Received GPS Signals

Let the total received power be P, and let the transmission delay (traveling time) be τ. The carrier frequency offset is Δf (Doppler), and the received phase is θ. Then the *received* L_1 *signal* can be written as

$$S^k(t) = \text{const.} \times \sqrt{2P}\, D^k(t-\tau)\cos(2\pi(f-\Delta f)(t-\tau)+\theta). \tag{2}$$

The data coefficient D^k is a product of code sequences and navigation data for satellite k.

From the observation $s(t)$ we want to estimate τ, Δf, and θ. This is done in a two step procedure

1. find global approximate values of τ and Δf, called ***signal acquisition***
2. local search for τ, Δf and possibly the carrier phase θ:
   - If θ is estimated the search is called *coherent signal tracking*.
   - If θ is ignored, the search is called ***non-coherent signal tracking***.

The purpose of code tracking is to estimate the travel time τ. This is done by means of a *delay lock loop* (DLL). For a coherent DLL we have $\theta = 0$.

To demodulate the navigation data, a carrier wave replica must be generated.

To track a carrier wave signal, a *phase lock loop* (PLL) often is used.

3 Receiver Channels and Acquisition

The signal processing for satellite navigation systems is based on a channelized structure. Next, we provide an overview of the concept of a ***receiver channel*** and the processing that occurs.

Figure 1 gives an overview of a channel. Before allocatting a channel to a satellite, the receiver must know which satellites that are currently visible.

The received signal $s(t)$ is a combination of signals from all n visible satellites

$$s(t) = s^1(t) + s^2(t) + \cdots + s^n(t). \tag{3}$$

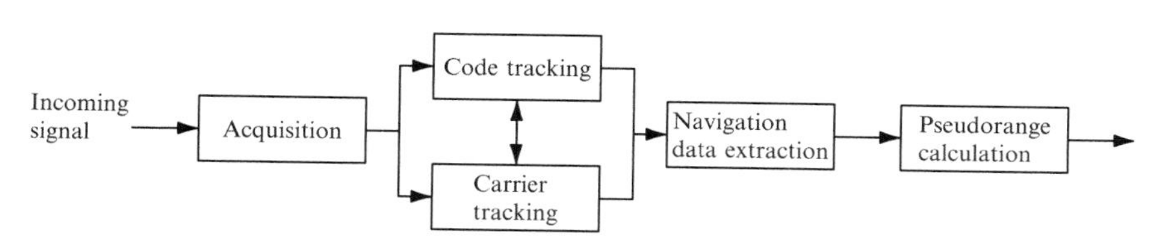

Fig. 1. One receiver channel. The acquisition gives rough estimates of signal parameters. These parameters are refined by the two tracking blocks. After tracking, the navigation data can be extracted and pseudoranges can be computed.

Before dealing specifically with satellite k we allocate a *channel* to acquiring it. This happens by using the following steps:

- the incoming signal s is multiplied with the locally generated C/A code corresponding to the satellite k
  The cross-correlation between C/A codes for different satellites implies that signals from other satellites are nearly removed by this procedure. To avoid removing the desired signal component, the locally generated C/A code must be properly aligned in time, that is have the correct code phase.
- After multiplication with the locally generated code, the signal must be mixed with a locally generated carrier wave; this removes the carrier wave of the received signal.
  In order to do this successfully the frequency of the locally generated signal must be close to the signal carrier frequency.

Next all signal components are squared and summed providing a numerical value. The acquisition procedure is a ***search procedure.***

It is sufficient to search Δf in steps of 500 Hz in the interval ±10 kHz. There are 1 023 discrete values of the code phase. A search for the maximum value over this 41 × 1 023 grid is performed either as

1. Parallel ***frequency space search*** acquisition (search 1 023 different code phases), see Fig. 2:
   a) The incoming signal is multiplied with a locally generated C/A code for satellite k
   b) the result is FFT from time domain to frequency domain
   c) absolute values of all components are computed
   d) if satellite k is present we can identify a maximum value at (code phase, frequency) = $(\tau, f - \Delta f)$. An unsuccessful search is shown in Fig. 3(a) and a successful one in Fig. 3(b),

or as

2. Parallel ***code phase search*** acquisition (search 41 carrier frequencies) (Fig. 4):
   a) Multiply incoming signal with cosine or sine, giving I and Q
   b) combine I and Q to complex input to FFT
   c) generate local C/A code for satellite k, FFT, complex conjugate and multiply with output from (b)
   d) IFFT and compute absolute values of all components
   e) maximum value at (code phase τ, frequency $f - \Delta f$) if satellite k is present. Else no distinct maximum value.

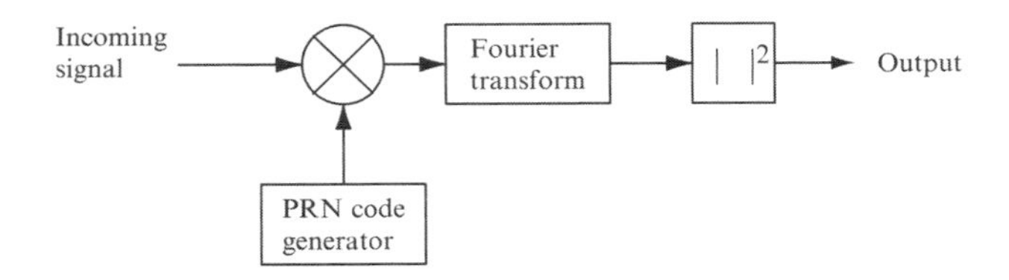

Fig. 2. Block diagram of the parallel *frequency space search* algorithm.

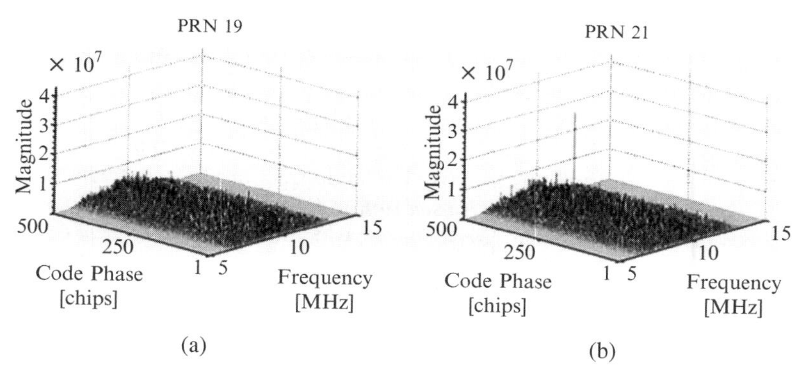

Fig. 3. Output from parallel frequency space search acquisition. The figure only includes the first 500 chip shifts and the frequency band from 5–15 MHz. (a) PRN 19 is not visible so no significant peaks are present in the spectrum. (b) PRN 21 is visible so a significant peak is present in the spectrum. The peak is situated at code phase 359 chips and frequency 9.548 MHz.

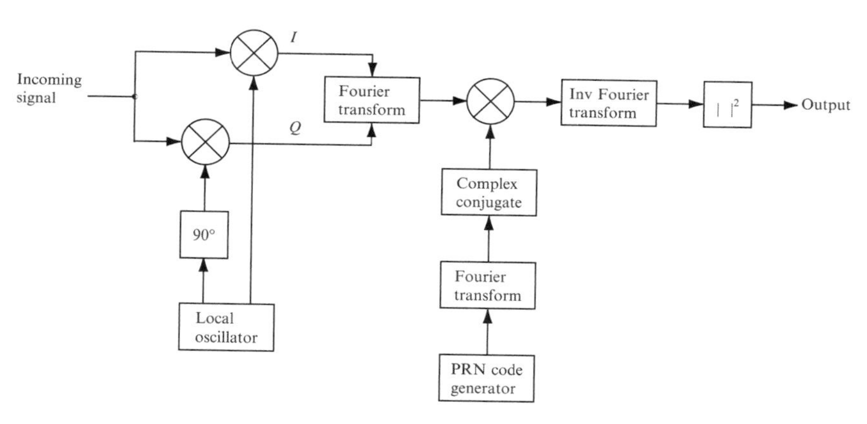

Fig. 4. Block diagram of the parallel *code phase search* algorithm.

In our implementation the code phase search is ten times faster than frequency space search.

4 Carrier and Code Tracking

The carrier tracking (phase lock loop PLL) involves the following issues:

- Improve the estimate of the carrier frequency obtained by acquisition
- generate local carrier signal
- measure phase error between incoming carrier and local carrier signal
- adjust frequency until phase and frequency become stable.

To demodulate the navigation data successfully a carrier wave replica has to be generated. To track a carrier wave signal ***Phase Lock Loops*** (PLL) or Frequency Lock Loops (FLL) are often used.

Figure 5 shows a basic block diagram for a phase lock loop. The two first multiplications wipe off the carrier and the PRN code of the input signal. To wipe off the PRN code, the I_p output from the early-late code tracking loop described above is used. ***The loop discriminator block is used to find the phase error*** on the local carrier wave replica. The output of the ***discriminator***, which is the phase error ϕ (or a function of the phase error), is then filtered and used as a feedback to the Numerically Controlled Oscillator (NCO) which adjusts the frequency of the local carrier wave. In this way the local carrier wave could be an almost precise replica of the input signal carrier wave.

The problem with using an ordinary PLL is that it is sensitive to 180° phase shifts. The PLL used in a GPS receiver has to be insensitive to 180° phase shifts due to navigation bit transitions,

The Costas loop is insensitive for 180° phase shifts. The Costas loop in Fig. 6 contains two multiplications. The first multiplication is the product between the input signal and the local carrier wave and the second multiplication is between a 90° phase shifted carrier wave and the input signal. The ***goal of the Costas loop is to try to keep all energy in the I (in-phase) arm.*** To keep the energy in the ***I*** arm some kind of feedback to the oscillator is needed. If the code replica in Fig. 6 is perfectly aligned, the multiplication in the ***I*** arm yields the following sum

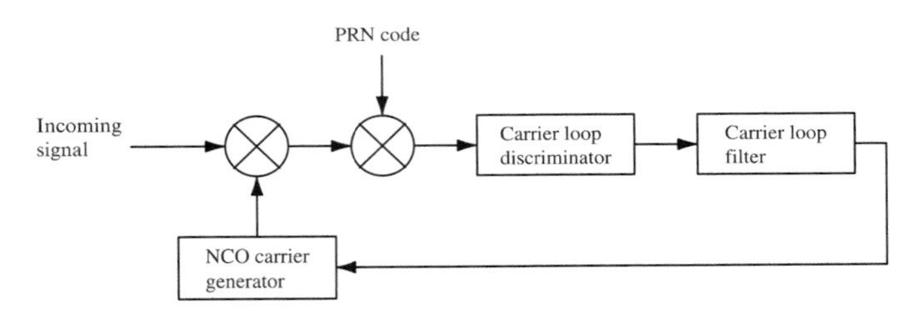

Fig. 5. Basic GPS receiver tracking loop block diagram.

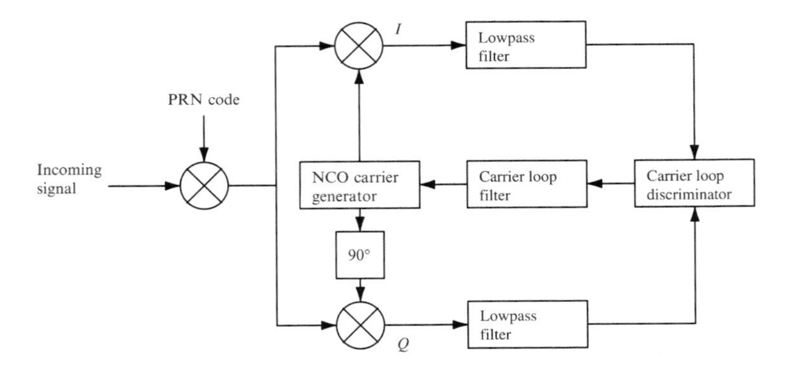

Fig. 6. Costas loop used to track the carrier wave.

$$D^k(n)\cos(\omega_{\text{IF}} n)\cos(\omega_{\text{IF}} n+\phi)=\frac{1}{2}D^k(n)\cos(\phi)+\frac{1}{2}D^k(n)\cos(2\omega_{\text{IF}} n+\phi) \quad (4)$$

where ϕ is the phase difference between the phase of the input signal and the phase of the local replica of the carrier phase. The multiplication in the quadrature arm gives the following

$$D^k(n)\cos(\omega_{\text{IF}} n)\sin(\omega_{\text{IF}} n+\phi)=\frac{1}{2}D^k(n)\sin(\phi)+\frac{1}{2}D^k(n)\sin(2\omega_{\text{IF}} n+\phi). \quad (5)$$

If the two signals are lowpass filtered after the multiplication, the two terms with $(2\omega_{\text{IF}} n + \phi)$ are eliminated and the following two signals remain

$$I^k=\frac{1}{2}D^k(n)\cos(\phi) \quad (6)$$

$$Q^k=\frac{1}{2}D^k(n)\sin(\phi). \quad (7)$$

To define a quantity to feedback to the carrier phase oscillator, the phase error ϕ of the local carrier phase replica is a good candidate which can be found as

$$\frac{Q^k}{I^k}=\frac{\frac{1}{2}D^k(n)\sin(\phi)}{\frac{1}{2}D^k(n)\cos(\phi)}=\tan(\phi) \quad (8)$$

$$\phi=\tan^{-1}\left(\frac{Q^k}{I^k}\right). \quad (9)$$

From equation (9) it can be seen that the phase error is small when the correlation in the quadrature-phase arm is close to zero and the correlation value in the in-phase arm is maximum.

The goal of a code tracking loop is to keep track of the phase of a specific code in the signal. The output of such a code tracking loop is a perfectly aligned replica of the code. The code tracking loop in the GPS receiver is a delay lock loop (DLL) called an *early-late tracking loop*. The idea behind the DLL is to correlate the input signal with three replicas of the code as seen in Fig. 8.

First step in Fig. 8: Convert the C/A code to baseband, by multiplying the incoming signal with a perfectly aligned local replica of the carrier wave. Afterwards the signal is multiplied with three code replicas. The three replicas are nominally generated with a spacing of $\pm\frac{1}{2}$ chip. After this second multiplication, the three outputs are integrated and dumped. The output of these integrations is a numerical value indicating how much the specific code replica correlates with the code in the incoming signal.

The three correlation outputs I_E, I_P, and I_L are then compared to see which one provides the highest correlation. Figure 7 shows an example of code tracking. In Fig. 7(a) the late code has the highest correlation, so the code phase must be decreased. In Fig. 7(b) the highest peak is located at the prompt replica, and the early and late replicas have equal correlation. In this case, the code phase is properly tracked.

The DLL with three correlators as in Fig. 8 is optimal when the local carrier wave is locked in phase and frequency. But when there is a phase error on the local carrier wave, the signal will be more noisy making it more difficult for the DLL to keep lock on the code. So instead the DLL in a GPS receiver is often designed as in Fig. 9.

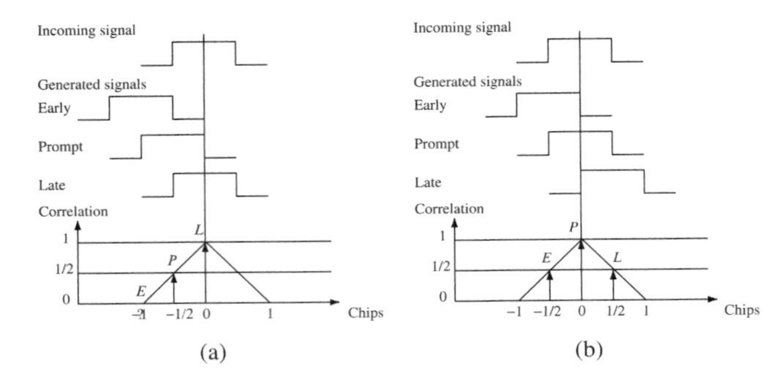

Fig. 7. Code tracking. Three local codes are generated and correlated with the incoming signal. (a) The late replica has the highest correlation so the code phase must be decreased, i.e., the code sequence must be delayed. (b) The prompt code has the highest correlation and the early and late have similar correlation. The loop is perfectly tuned in.

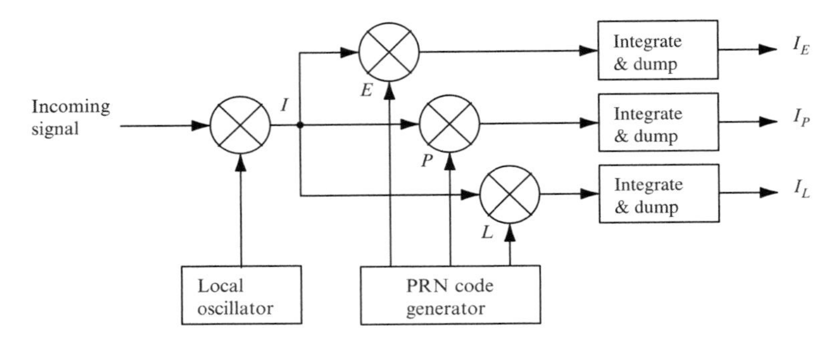

Fig. 8. Basic code tracking loop block diagram.

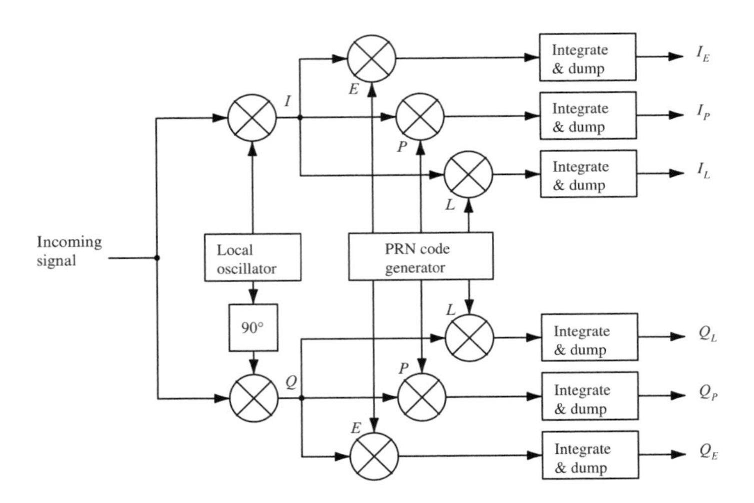

Fig. 9. DLL block diagram with six correlators.

The design in Fig. 9 has the advantage that it is independent of the phase on the local carrier wave. If the local carrier wave is in phase with the input signal, all the energy will be in the in-phase arm. But if the local carrier phase drifts compared to the input signal, the energy will switch between the in-phase and the quadrature arm. For demonstration purposes Fig. 10 shows such situation where the phase of the carrier replica drifts compared to the phase of the incoming signal. The upper plot shows the output of the three correlators in the in-phase arm and the lower plot shows the correlation output in the quadrature arm of the DLL with six correlators. This situation is a result of different frequencies for the signal and the replica; it results in a constantly changing phase difference (miss-alignment). There are a few reasons why this can happen, for example the PLL could be not in a lock state.

Figure 11 shows a case when the PLL is in a locked state. Because of the precise carrier replica from the PLL it is seen in Fig. 11 that the correlators are constant over time. This would not be the case if the carrier replica is not adjusted to match the frequency and phase of the incoming signal.

If the code tracking loop performance has to be independent of the performance of the phase lock loop, the tracking loop has to use both the in-phase and quadrature arms to track the code.

The DLL now needs a feedback to the PRN code generators if the code phase has to be adjusted. Some common DLL discriminators used for feedback are listed in Table 1.

The table shows one coherent and three non-coherent discriminators. The requirements of a DLL discriminator is dependent on the type of application and the noise in the signal. The discriminator function responses are shown in Fig. 12.

Figure 12 shows the coherent discriminator and three non-coherent discriminators using a standard correlator. The figure is produced from ideal ACFs and the space between the early, prompt, and late is $\pm\frac{1}{2}$ chip. The space between the

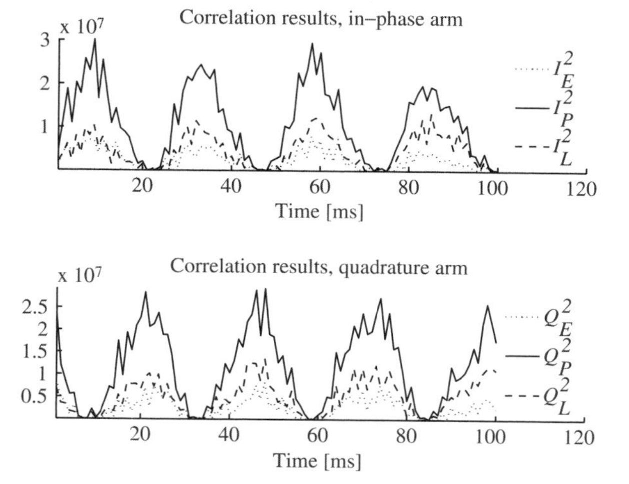

Fig. 10. Output of the six correlators in the in-phase and quadrature arms of the tracking loop. Acquisition frequency offset is 20 Hz and PLL noise bandwidth is 15 Hz (for demonstration purpose).

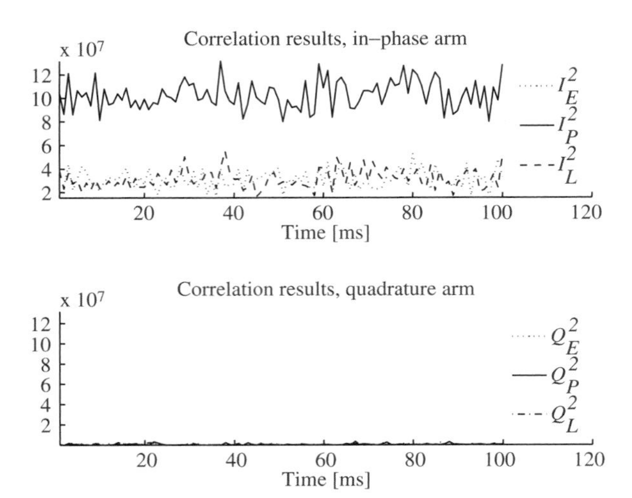

Fig. 11. Output of the six correlators in the in-phase and quadrature arms of the tracking loop. The local carrier wave is in phase with the input signal.

Table 1. Various types of delay lock loop discriminators and a description of them.

Type	Discriminator	Characteristics
Coherent	$D = I_E - I_L$	Simplest of all discriminators. Does not require the Q branch but requires a good carrier tracking loop for optimal functionality.
	$D = (I_E^2 + Q_E^2) - (I_L^2 + Q_L^2)$	Early minus late power. The discriminator response is nearly the same as the coherent discriminator inside $\pm\frac{1}{2}$ chip.
Non-coherent	$D = \frac{(I_E^2 + Q_E^2) - (I_L^2 + Q_L^2)}{(I_E^2 + Q_E^2) + (I_L^2 + Q_L^2)}$	Normalized early minus late power. The discriminator has a great property when the chip error is larger than a $\frac{1}{2}$ chip, this will help the DLL to keep track in noisy signals.
	$D = I_P(I_E - I_L) + Q_P(Q_E - Q_L)$	Dot product. This is the only DLL discriminator that uses all six correlator outputs.

early, prompt, and late codes determines the noise bandwidth in the delay lock loop. If the discriminator spacing is larger than $\frac{1}{2}$ chip, the DLL would be able to handle wider dynamics and be more noise robust, on the other hand a DLL with a smaller spacing would be more precise. In a modern GPS receiver the discriminator spacing can be adjusted while the receiver is tracking the signal. The advantage from this is that if the signal to noise ratio suddenly decreases, the receiver uses a wider spacing in the correlators to be able to handle a more noisy signal, and hereby a possible code lock loss could be avoided.

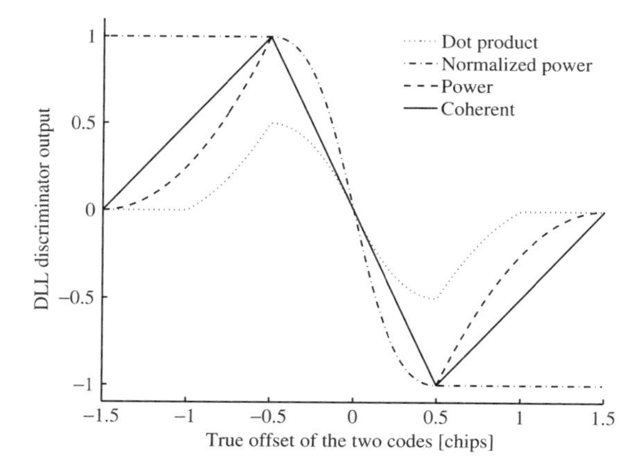

Fig. 12. Comparison between the common DLL discriminator responses.

The implemented tracking loop discriminator is the normalized early minus late power. This discriminator is described as

$$D = \frac{(I_E^2 + Q_E^2) - (I_L^2 + Q_L^2)}{(I_E^2 + Q_E^2) + (I_L^2 + Q_L^2)} \tag{10}$$

where I_E, Q_E, I_L, and Q_L are output from four of the six correlators shown in Fig. 10. The normalized early minus late power discriminator is chosen because it is independent of the performance of the PLL as it uses both the in-phase and quadrature arms. The normalization of the discriminator causes that the discriminator can be used with signals with different signal to noise ratios and different signal strengths.

The tracking loop generates three local code replicas. In this section, the chip space between the early and prompt replicas is half a chip.

As was described, the DLL can be modeled as a linear PLL and thus the performance of the loop can be predicted based on this model. In other words the loop filter design is the same, just parameter values are different.

5 Navigation Data Extraction

When the signals are properly tracked, the C/A code and carrier wave can be removed from the signal, leaving the navigation data bits. The value of a data bit is found by integration over a navigation bit period of 20 ms:

- Find start of subframe
- Decode the ephemeris data.

6 Estimation of Pseudorange

- Find common start for all satellites of a subframe. The accuracy of the pseudorange with a time resolution of 1 ms is 300 km

– The code tracking loop tells the precise start of the C/A code. Pseudorange accuracy of 8 meters. This value depends on signal sampling frequency.

Finally the receiver position is computed from the estimated pseudoranges. The next subsection describes a standard method for this computation.

7 Computation of Receiver Position

The most commonly used algorithm for position computations from pseudoranges is based on the ***least-squares method.*** This method is used when there are more observations than unknowns. This section describes how the least-squares method is used to find the receiver position from pseudoranges to four (Fig. 13) or more satellites.

Let the geometrical range between satellite k and receiver i be denoted ρ_i^k, and let c denote the speed of light. Let dt_i and dt^k be the receiver clock and satellite clock offsets. Let T_i^k be the tropospheric delay, I_i^k be the ionospheric delay, and e_i^k be the observational error of the pseudorange. Then the basic observation equation for the pseudorange P_i^k is

$$P_i^k = \rho_i^k + c(dt_i - dt^k) + T_i^k + I_i^k + e_i^k \,. \tag{11}$$

The geometrical range ρ_i^k between the satellite and the receiver is computed as

$$\rho_i^k = \sqrt{(X^k - X_i)^2 + (Y^k - Y_i)^2 + (Z^k - Z_i)^2} \,. \tag{12}$$

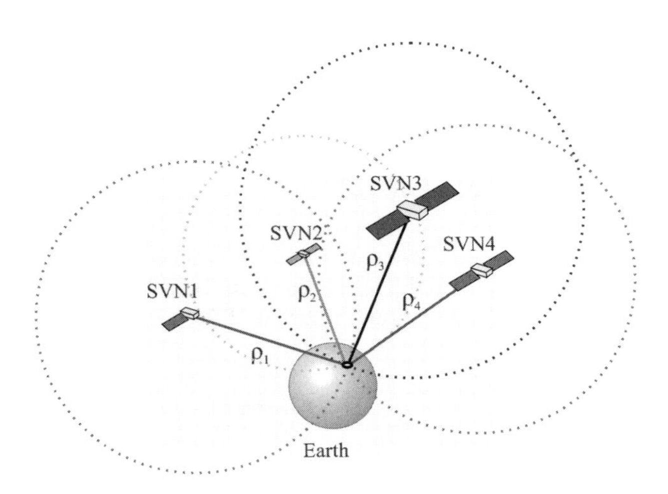

Fig. 13. The basic principle of GNSS positioning. With known position of four satellites SVNi and signal travel distance ρ_i, the user position can be computed.

Inserting (11) into (12) yields

$$P_i^k = \sqrt{(X^k - X_i)^2 + (Y^k - Y_i)^2 + (Z^k - Z_i)^2} + c\,(dt_i - dt^k) + T_i^k + I_i^k + e_i^k. \quad (13)$$

From the ephemerides—which include information on the satellite clock offset dt^k—the position of the satellite (X^k, Y^k, Z^k) can be computed. (The *M*-file **satpos** does the job.)

The tropospheric delay T_i^k is computed from an a priori model which is coded as **tropo**; the ionospheric delay I_i^k may be estimated from another a priori model, the coefficients of which are part of the broadcast ephemerides. There are four unknowns in the equation: X_i, Y_i, Z_i, and dt_i; the error term e_i^k is minimized by using the method of least squares. To compute the position of the receiver at least four pseudoranges are needed.

Equation (13) is nonlinear with respect to the receiver position (X_i, Y_i, Z_i), so the equation has to be ***linearized*** before using the least-squares method. We analyze the nonlinear range term in (13):

$$f(X_i, Y_i, Z_i) = \sqrt{(X^k - X_i)^2 + (Y^k - Y_i)^2 + (Z^k - Z_i)^2}. \quad (14)$$

Linearization starts by finding an initial position for the receiver: ($X_{i,0}$, $Y_{i,0}$, $Z_{i,0}$). This is often chosen as the center of the Earth (0,0,0).

The increments ΔX, ΔY, ΔZ are defined as

$$\begin{aligned} X_{i,1} &= X_{i,0} + \Delta X_i \\ Y_{i,1} &= Y_{i,0} + \Delta Y_i \\ Z_{i,1} &= Z_{i,0} + \Delta Z_i. \end{aligned} \quad (15)$$

These increments update the approximate receiver coordinates. So the Taylor expansion of $f(X_{i,0} + \Delta X_i, Y_{i,0} + \Delta Y_i, Z_{i,0} + \Delta Z_i)$ is

$$\begin{aligned} f(X_{i,1}, Y_{i,1}, Z_{i,1}) = f(X_{i,0}, Y_{i,0}, Z_{i,0}) &+ \frac{\partial f(X_{i,0}, Y_{i,0}, Z_{i,0})}{\partial X_{i,0}} \Delta X_i \\ &+ \frac{\partial f(X_{i,0}, Y_{i,0}, Z_{i,0})}{\partial Y_{i,0}} \Delta Y_i + \frac{\partial (X_{i,0}, Y_{i,0}, Z_{i,0})}{\partial Z_{i,0}} \Delta Z_i. \end{aligned} \quad (16)$$

Equation (16) only includes first order terms; hence the updated function f determines an approximate position. The partial derivatives in equation (16) come from (14):

$$\begin{aligned} \frac{\partial f(X_{i,0}, Y_{i,0}, Z_{i,0})}{\partial X_{i,0}} &= -\frac{X^k - X_{i,0}}{\rho_i^k} \\ \frac{\partial f(X_{i,0}, Y_{i,0}, Z_{i,0})}{\partial Y_{i,0}} &= -\frac{Y^k - Y_{i,0}}{\rho_i^k} \\ \frac{\partial f(X_{i,0}, Y_{i,0}, Z_{i,0})}{\partial Z_{i,0}} &= -\frac{Z^k - Z_{i,0}}{\rho_i^k}. \end{aligned}$$

Let $\rho_{i,0}^{k}$ be the range computed from the approximate receiver position; the first order linearized observation equation becomes

$$P_i^k = \rho_{i,0}^k - \frac{X^k - X_{i,0}}{\rho_{i,0}^k}\Delta X_i - \frac{Y^k - Y_{i,0}}{\rho_{i,0}^k}\Delta Y_i - \frac{Z^k - Z_{i,0}}{\rho_{i,0}^k}\Delta Z_i + c(dt_i - dt^k) + T_i^k + I_i^k + e_i^k \tag{17}$$

where we explicitly have

$$\rho_{i,0}^k = \sqrt{(X^k - X_{i,0})^2 + (Y^k - Y_{i,0})^2 + (Z^k - Z_{i,0})^2}\,. \tag{18}$$

A least-squares problem is given as a system $\boldsymbol{Ax} = \boldsymbol{b}$ with no exact solution. A has m rows and n columns, with $m > n$; there are more observations $b_1, \ldots, b_m$ than free parameters $x_1, \ldots, x_n$. The best choice, we will call it $\hat{\boldsymbol{x}}$, is the one that minimizes the length of the error vector $\hat{\boldsymbol{e}} = \boldsymbol{b} = A\hat{\boldsymbol{x}}$. If we measure this length in the usual way, so that $\|\boldsymbol{e}\|^2 = (\boldsymbol{b} - A\boldsymbol{x})^{\mathrm{T}}(\boldsymbol{b} - A\boldsymbol{x})$ is the sum of squares of the m separate errors, minimizing this quadratic gives the normal equations

$$A^T A\hat{\boldsymbol{x}} = A^T \boldsymbol{b} \quad or \quad \hat{\boldsymbol{x}} = (A^T A)^{-1} A^T \boldsymbol{b}. \tag{19}$$

The error vector is

$$\hat{\boldsymbol{e}} = \boldsymbol{b} - A\hat{\boldsymbol{x}}. \tag{20}$$

The covariance matrix for the parameters $\hat{\boldsymbol{x}}$is

$$\Sigma_{\hat{x}} = \hat{\sigma}_0^2 (A^T A)^{-1} \quad with \quad \hat{\sigma}_0^2 = \frac{\hat{\boldsymbol{e}}^{\mathrm{T}}\hat{\boldsymbol{e}}}{m-n}\,. \tag{21}$$

The linearized observation equation (17) can be rewritten in a vector formulation

$$P_i^k = \rho_{i,0}^k + \begin{bmatrix} -\dfrac{X^k - X_{i,0}}{\rho_{i,0}^k} & -\dfrac{Y^k - Y_{i,0}}{\rho_{i,0}^k} & -\dfrac{Z^k - Z_{i,0}}{\rho_{i,0}^k} \end{bmatrix} \begin{bmatrix} \Delta X_i \\ \Delta Y_i \\ \Delta Z_i \\ c\,dt_i \end{bmatrix} - c\,dt^k + T_i^k + I_i^k + e_i^k\,. \tag{22}$$

We rearrange this to resemble the usual formulation of a least-squares problem $\boldsymbol{Ax} = \boldsymbol{b}$

$$\begin{bmatrix} -\dfrac{X^k - X_{i,0}}{\rho_{i,0}^k} & -\dfrac{Y^k - Y_{i,0}}{\rho_{i,0}^k} & -\dfrac{Z^k - Z_{i,0}}{\rho_{i,0}^k} & 1 \end{bmatrix} \begin{bmatrix} \Delta X_i \\ \Delta Y_i \\ \Delta Z_i \\ c\,dt_i \end{bmatrix} = P_i^k - \rho_{i,0}^k + c\,dt^k - T_i^k - I_i^k - e_i^k\,. \tag{23}$$

A unique least-squares solution cannot be found until there are $m \geq 4$ equations. Let $b_i^k = P_i^k - \rho_{i,0}^k + c\,dt^k - T_i^k - I_i^k - e_i^k$ and the final solution comes from

$$Ax=\begin{bmatrix} -\frac{X^1-X_{i,0}}{\rho^1_{i,0}} & -\frac{Y^1-Y_{i,0}}{\rho^1_{i,0}} & -\frac{Z^1-Z_{i,0}}{\rho^1_{i,0}} & 1 \\ -\frac{X^2-X_{i,0}}{\rho^2_{i,0}} & -\frac{Y^2-Y_{i,0}}{\rho^2_{i,0}} & -\frac{Z^2-Z_{i,0}}{\rho^2_{i,0}} & 1 \\ -\frac{X^3-X_{i,0}}{\rho^3_{i,0}} & -\frac{Y^3-Y_{i,0}}{\rho^3_{i,0}} & -\frac{Z^3-Z_{i,0}}{\rho^3_{i,0}} & 1 \\ \vdots & \vdots & \vdots & \vdots \\ -\frac{X^m-X_{i,0}}{\rho^k_{i,0}} & -\frac{Y^m-Y_{i,0}}{\rho^m_{i,0}} & -\frac{Z^m-Z_{i,0}}{\rho^m_{i,0}} & 1 \end{bmatrix} \begin{bmatrix} \Delta X_{i,1} \\ \Delta Y_{i,1} \\ \Delta Z_{i,1} \\ c\,dt_{i,1} \end{bmatrix} = \boldsymbol{b}-\boldsymbol{e}. \tag{24}$$

The solution $\Delta X_{i,1}$, $\Delta Y_{i,1}$, $\Delta Z_{i,1}$, is added to the approximate receiver position to get the next approximate position:

$$\begin{aligned} X_{i,1} &= X_{i,0} + \Delta X_{i,1} \\ Y_{i,1} &= Y_{i,0} + \Delta Y_{i,1} \\ Z_{i,1} &= Z_{i,0} + \Delta Z_{i,1}. \end{aligned} \tag{25}$$

The next ***iteration*** restarts from (22) to (25) with $_{i,0}$ replaced by $_{i,1}$. These iterations continue until the solution $\Delta X_{i,1}$, $\Delta Y_{i,1}$, $\Delta Z_{i,1}$ is at meter level. Often 2–3 iterations are sufficient to obtain that goal, are discussed in [2].

The present description of the software-defined GPS receiver is based on [3]. Further developments in the project can be followed at **gps.aau.dk/softgps.**

References

[1] Anonymous, "An introduction to direct-sequence spread-spectrum communications," 2003, http://www.maxim-ic.com/appnotes.cfm/appnote_number/1890.

[2] G. Strang and K. Borre, *Linear Algebra, Geodesy, and GPS*. Wellesley, MA: Wellesley-Cambridge Press, 1997.

[3] K. Borre, D. Akos, N. Bertelsen, P. Rinder, and S. H. Jensen, *A Software-Defined GPS and Galileo Receiver: Single-Frequency Approach*. Boston Basel Berlin: Birkhäuser, 2006.

Ephemeris Interpolation Techniques for Assisted GNSS Services

Matteo Iubatti*, Marco Villanti*, Alessandro Vanelli-Coralli*,
Giovanni E. Corazza*, Stephane Corazza**

* DEIS/ARCES - University of Bologna - Viale Risorgimento 2 - 40136 Bologna, Italy,
Tel: +39 051 2093056, Fax: +39 051 2093053,
e-mail: {miubatti, mvillanti, avanelli,gecorazza} @ deis.unibo.it
** Alcatel Alenia Space France - Avenue J.F. Champollion 26, 31037 Toulouse, France,
Tel: +33 (0)534 354281, Fax: +33 (0)534 354943,
e-mail: Stephane.Corazza@alcatelaleniaspace.com
Ephemeris, Interpolation techniques, A-GNSS, assistance services.

Abstract. Assisted-GNSS (A-GNSS) is an interesting technology that can consistently improve positioning performance and reduce terminal complexity. One of the possible data that the A-GNSS can provide is the satellite orbit (satellite ephemeris) necessary to determine the user position starting from pseudorange measurements. In this paper, we propose two novel data structures for the transmission of ephemeris through the assistance network. It is shown that, adopting different interpolation techniques, it is possible to consistently reduce the amount of data to be transmitted and extend the ephemeris validity to 24 hours.

1 Introduction

The European Galileo program is devoted to the development of a satellite based positioning system that will provide enhanced accuracy and continuous reliability to civilian users. To enhance performance and compatibility, Galileo has been designed to be possibly interoperable with the American Global Positioning System (GPS), so that the two systems can cooperate in synergy and exploit a larger satellite constellation. Further, in order to guarantee efficient positioning and navigation services in critical environments, both Galileo and GPS can be improved by assistance services supported by terrestrial networks. With this strategy, identified as Assisted Global Navigation Satellite Services (A-GNSS), users can benefit of assistance data to perform positioning and navigation more efficiently and quickly, for example by reducing the time-to-first-fix (TTFF) in code acquisition. The potential benefits of A-GNSS on terminal complexity reduction and performance improvement have stimulated the research community towards the definition and standardization of different data aiding sets. Among many others, the transmission of ephemeris (i.e. the satellite orbit description parameters) through the terrestrial aiding network can be very effective to reduce time-to-first-fix onto satellite signals because the receiver is released from the task of receiving it from the satellite GNSS network, which has low bitrate and suffer from adverse propagation environments, such as in urban areas.

Because the accurate computation of the satellite position is a fundamental prerequisite in all GNSS positioning systems, particular attention must be taken in defining the ephemeris assistance field. A possible solution is to adopt for the assistance ephemeris field the same structure of broadcast ephemeris, which are described in the navigation message by a set of parameters to be employed in an accurate equation describing the satellite orbit, as detailed in the Galileo navigation model reported in [1]. In order to reduce the amount of data to be transmitted while preserving high accuracy, ephemeris data are valid for a time interval, which however is limited so that they need to refreshed frequently. For GPS, ephemeris are updated by the control center every 2 hours, to allow an accuracy of the computed coordinates in the order of about 3 meters, while for Galileo it is foreseen that the ephemeris will be updated every 3 hours. Also, an overlapping period is foreseen in order to prevent possible gaps, as depicted in Fig. 1.

Note that this very accurate orbit description is necessary for high-precision services, while less stringent constraints are adequate for mass market services. In fact, in low cost terminals the pseudorange estimation can be performed with a limited accuracy so that a satellite position error in the order of a few meters becomes negligible. For example, an error in the satellite coordinates of 20m is equivalent to having pseudorange estimation errors in the order of 0.1μsecond, which is a reasonable precision for mass market terminals.

Starting from this observation, the aim of this paper is to investigate new efficient solutions for ephemeris description, in order to extend the data validity with respect to the standard GNSS approach and/or to reduce the amount of data to be

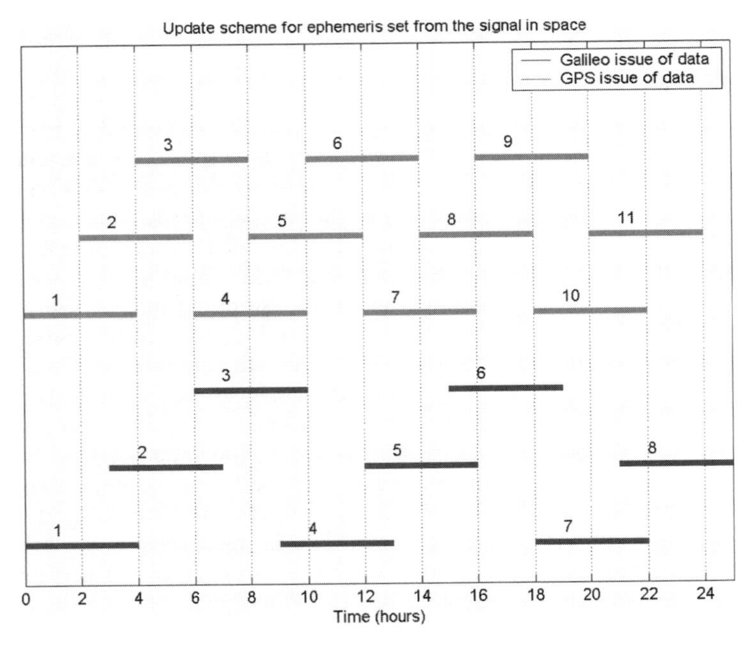

Fig. 1. Galileo and GPS ephemeris issues over 24h.

transmitted, under the constraint of preserving a precision for the satellite orbit description in the order of 20m. Two alternatives are identified in this paper: the first foresees transmitting the samples the satellite coordinates derived at a low sampling rate, and then interpolating these in the receiver to obtain the satellite position with the desired accuracy; the second envisages the transmission of the coefficients of a simplified function describing the satellite orbit. In particular, the idea is to find a polynomial function that properly describes the ephemeris in order to transmit the polynomial coefficients instead of the satellite position points. Notably, both techniques are based on interpolation, so that it is essential to find the best interpolation techniques to make these alternative approaches effective. To this aim, different alternatives are considered to find the best trade-off between accuracy and bit requirement.

2 Ephemeris Interpolation Techniques

Interpolation is a method of constructing new data points from a discrete set of known data points. Typically, this is achieved by sampling a function which closely fits the known data set. The process of identifying the approximating function is generally called curve fitting, and interpolation is a specific case of curve fitting in which the function must go exactly through the known data values.

The simplest way of interpolation is the linear interpolation, in which the known points are connected by segments. Linear interpolation has a very limited complexity but does not provide high accuracy, so that alternative methods have been developed. A generalization of linear interpolation is given by polynomial interpolation, which encompass a family of different strategies. An example of polynomial interpolation is the Lagrange interpolation, which is one of the most widely adopted interpolation techniques and considers a polynomial of degree $n - 1$ going through all the n known data values. The interpolation error is proportional to the distance between the data points to the power n. Although polynomial interpolation consistently improves the linear alternative, it introduces oscillations in the region close to the border of the known data set that make the interpolation very poor in these regions: this effect is known as Runge's phenomenon. Also, it has to be noted that the evaluation of the polynomial interpolation is computationally demanding with respect to linear interpolation. The polynomial interpolation family also includes another well known approach called trigonometric (or Fourier) interpolation, which results to be especially suitable for the interpolation of periodic functions. In this case, the interpolant is given by the sum of sines and cosines of given periods. An important special case is when the given data points are equally spaced and the exact solution is given by the discrete Fourier transform [4].

The disadvantages of polynomial interpolation can be limited by using spline interpolation. The spline interpolation uses low-degree polynomials to approximate each data set segment, selecting the polynomial pieces such that they fit smoothly together. The resulting function is called spline. For instance, the natural cubic spline is piecewise cubic and twice continuously differentiable. Like polynomial interpolation, spline interpolation incurs a smaller error than linear interpolation and the interpolant is smoother. Additionally, the interpolant is easier to evaluate than the high-degree polynomials used in polynomial interpolation and it is also able to limit the Runge's phenomenon. The interpolation methods briefly described above are detailed in the following, reporting the achievable interpolation performance when processing actual ephemeris data. In particular, the analysis reported in the following has been conducted using GPS ephemeredes that are given at 900 sec (15 min) in accordance with *.sp3 file format defined in 1991 by Remondi [2, 3].

2.1 Lagrange Polynomial Interpolation

Given the $n + 1$ ephemeris values $f(t_0), \ldots, f(t_n)$ at the distinct times $t_0, \ldots, t_n$, it exists an unique interpolating polynomial $p_n(t)$ satisfying the condition

$$p_n(t_i) = f(t_i) \text{ for } i = 0, \ldots, n \tag{1}$$

The polynomial $p_n(t)$ is usually identified as Lagrange interpolation polynomial and can be written in the form

$$p_n(t) = \sum_{i=0}^{n} f(t_i) l_i(t) \tag{2}$$

where the polynomial coefficients are given by

$$l_i(t) = \frac{\prod_{k=0}^{i-1}(t - t_k)}{\prod_{k=0}^{i-1}(t_i - t_k)} \cdot \frac{\prod_{k=i+1}^{n}(t - t_k)}{\prod_{k=i+1}^{n}(t_i - t_k)} \tag{3}$$

Several evaluations of the accuracy of this method has been made, and it is generally found that an 8-th order Lagrange polynomial interpolation is able to extrapolate data in the center of 8 points spaced by 900 sec with a precision of 1 cm [6]. Unfortunately, this high level accuracy is not reached when the entire set of ephemeris data, covering 24 hours, is interpolated. For example, in Fig. 2 it is reported the Euclidean distance between the interpolated points and the precise ephemeris of the 7th and the 23rd order Lagrange polynomial obtained by using a subset of 8 and 24 values, respectively, extracted by down-sampling the set of 96 precise ephemeris values (one every 900s) describing the orbit for the entire day. It can be seen that the distance between the interpolated values and the exact ones is very large for the 7th order polynomial because the spacing between the input data is too large to be compensated. Differently, the interpolating accuracy is more acceptable for the 23rd order polynomial, at least out of the Runge's phenomenon region. Obviously this higher precision comes at the price of increased complexity.

To reduce the Runge's phenomenon, the k-th order Lagrange method is usually adopted with successive intervals that overlap in time [6].

2.2 Trigonometric (Fourier) Polynomial Interpolation

Differently from Lagrange method that is a standard interpolation technique typically used for continuous differentiable functions defined on compact intervals, the trigonometric (or Fourier) polynomial interpolation is particularly suited when the

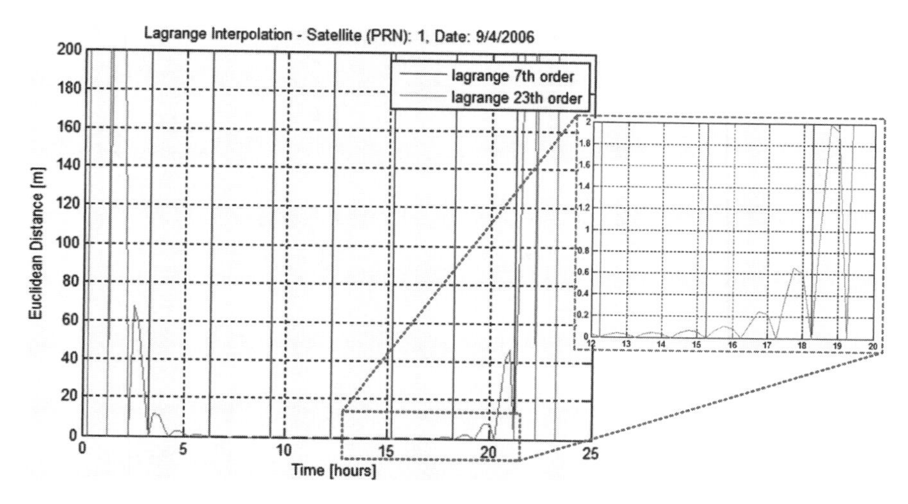

Fig. 2. Euclidean distance between the interpolated pointsand the precise ephemeris of the 7th and the 23rd order Lagrange polynomial.

data set to be interpolated has a clear periodic trend. This is exactly the case of ephemeris data, which results to be almost periodic, with a period of 24 hours if the satellite coordinates are referred to a Earth-centre, Earth-fixed Cartesian coordinate system, as reported in Fig. 3 [5] [6]. The Fourier interpolation approach is based on the idea to assume an interpolating function defined over the interval [0, 2π], defined as

$$p_n(t) = a_0 + \sum_{k=1}^{n} (a_k \cos(\omega t) + b_k \sin(\omega t)) \tag{4}$$

Considering the ephemeris periodicity, we can restrict our attention to a single 24 hour period and generate a trigonometric polynomial using all data available over that period. In fact, the complexity of the method is given by the order of the polynomial function and not by the number of points to be interpolated. In particular, because the satellite orbit is not truly periodic, the polynomial coefficients, including ω, are iteratively obtained by minimizing the error between the interpolated and the exact satellite positions. For this reason, the interpolation accuracy increase by considering a larger known data set. Since the error incurred by assuming the data to be periodic over a k day period would be almost k times greater than the error incurred from assuming the orbit to be periodic over a single day, we have to adopt this approach over intervals that do not exceed the fundamental period (24 hours) of the ephemeris data [6]. The precision that can be achieved by the trigonometric interpolation is reported in Fig. 4 in terms of Euclidean distance between the interpolated and the precise ephemeris for the 5th and 8th order on a 24 hour period. It can be seen that the 8th order polynomial provides a very good approximation, by providing an error lower than 20 m for more than 18-20 hours per day, thanks to a limited Runge's effect, which makes the approximation poor only at the border of the data set.

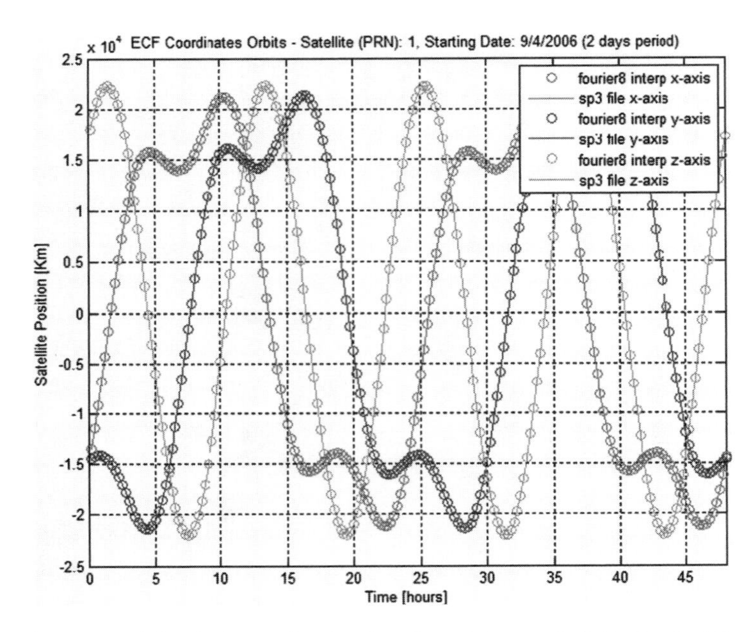

Fig. 3. Ephemeris and fourier interpolated values for a 2 days period. Notice that the function is quasi-periodic.

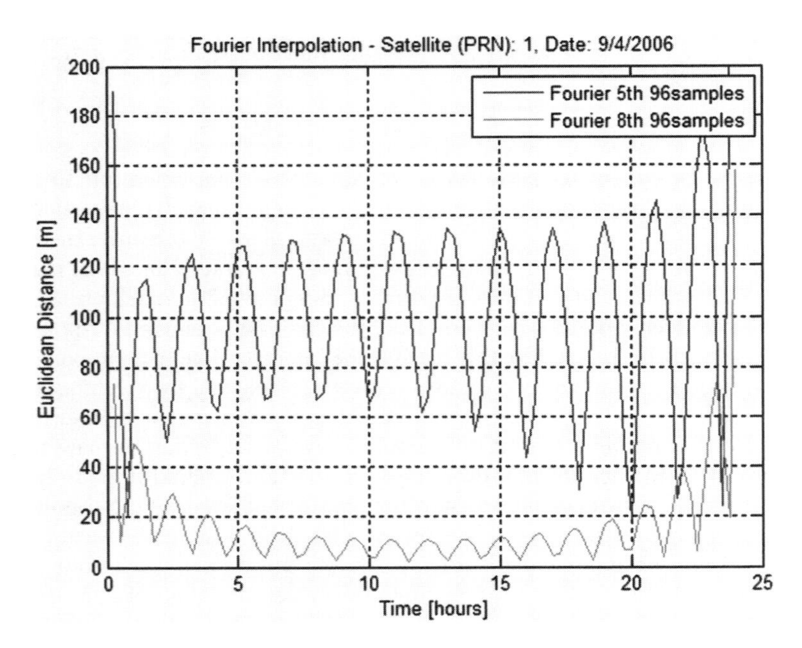

Fig. 4. Fourier 5th and 8th order polynomial interpolation precision in terms of euclidean distance between exact and interpolated values for 1 day period.

2.3 Spline Interpolation

The spline interpolation is an approach that interpolates using a function piecewise defined by polynomials. The spline interpolation is often preferred to polynomial interpolation because it yields similar results, even when using lower degree polynomials, so reducing the Runge's phenomenon characteristic of higher degrees. Additionally, spline is computationally efficient and has an advantage with respect to Lagrange interpolation because it allows to calculate the interpolating polynomials over the entire interval only a single time, at the beginning of the interpolation process. This calculation involves solving a system of equations each of degree k [6].

In its most general form, a polynomial spline $S(t)$: $[a,b] \to \mathbb{R}$ consists of polynomial pieces P_i: $[t_i, t_{i+1}] \to \mathbb{R}$, where $a = t_0 < t_1 < \cdots < t_{k-1} < t_{k-2} = b$. The given k points t_i are called knots. If the knots are equidistantly distributed in the interval $[a, b]$ we say the spline is uniform, otherwise we say it is non-uniform.

Given $n + 1$ distinct knots t_i with $n + 1$ knot values y_i the spline interpolation finds an interpolant of degree n as

$$S(t) = \begin{cases} S_0(t) \to t \in [t_0, t_1] \\ S_1(t) \to t \in [t_1, t_2] \\ \quad \ldots \\ S_{n-1}(t) \to t \in [t_{n-1}, t_n] \end{cases} \tag{5}$$

where each $S_i(t)$ is a polynomial of degree k, which identifies the spline order (for example quadratic spline adopts $k = 2$, cubic spline $k = 3$, etc.). The unique interpolant $S(t)$ is obtained applying boundary conditions between different intervals calculating derivates in the discontinuity points. For our study we have adopted two different spline interpolation functions embedded in MatLab 7.0 called *csape* and *spapi.* The former, *csape* implements a cubic spline interpolation with the possibility to insert a condition to process the end points of the data set in order to reduce border effects (in this deliverable we have used default as ending condition). Differently, *spapi* is the traditional spline interpolation of kth order. Note that, the fact that each subinterval is represented by a kth order polynomial (where $k < n$, in general) means that the evaluation on each interval is much quicker than the Lagrange n-th order counterpart. Obviously, if we process a great set of data, the dimension of the system of equations increase at the price of larger computational complexity. In Fig. 5, the Euclidean distance between the interpolated and precise ephemeris is reported for spline interpolation with 3, 8 and 24 knots, respectively. It is possible to note that only using 24-th order polynomial we can reach a precision in order of few centimeters for one day validity, although the Rounge effect limits the validity of the interpolation accuracy.

3 Ephemeris Data Definition for Assistance Service

The different interpolation techniques discussed in the previous section have positive and negative aspects. In particular, Lagrange and spline of high order are able to ensure a very large accuracy, at least out of the Rounge region. This results is very

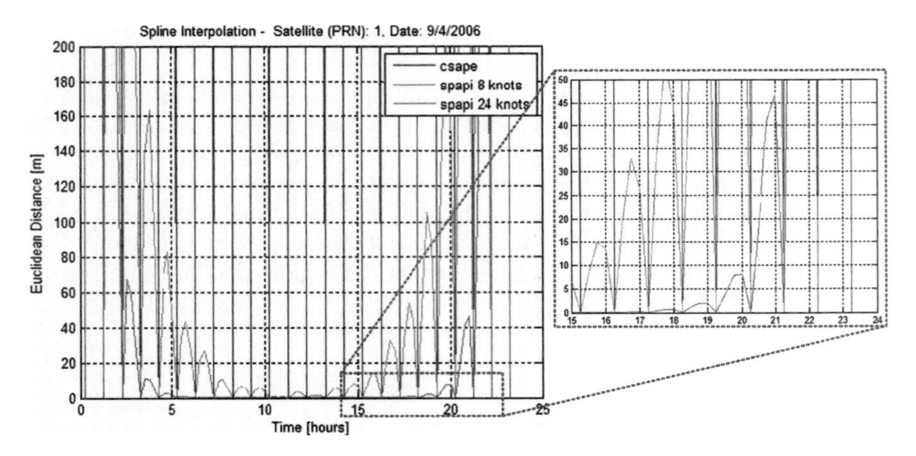

Fig. 5. Cubic, 8th and 24th order spline interpolation precision in terms of Euclidean distance between real data and interpolated results for 24 hours validity.

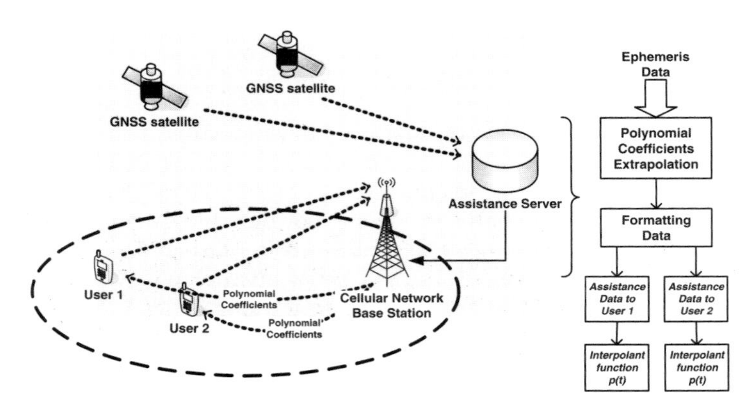

Fig. 6. Assistance GNSS network architecture for the proposed ICT (Interpolation Coefficients Transmission) method.

interesting even though a large accuracy requires a large number of coefficients to describe the interpolation function. Conversely, the trigonometric polynomial interpolation is characterized by a more limited accuracy, but the interpolation function can be described by a more limited number of coefficients. The different characteristics of the proposed interpolation methods can be exploited to provide two alternative solutions for the ephemeris data definition, which are identified as Interpolation Coefficient Transmission (ICT) and Satellite Coordinates Transmission (SCT), respectively. In the ICT solution, precise ephemerides are processed by the assistance server, which then broadcasts the interpolation function coefficients opportunely formatted, as depicted Fig. 6. Because the Fourier interpolation provides, a good trade-off between precision (within 20 m for more than

Table 1. Fourier 8th order polynomial coefficients for 4 GPS satellite orbits on 09/04/2006.

PRN 1	a0	a1	b1	a2	b2	a3	b3	a4	b4	a5	b5	a6	b6	a7	b7	a8	b8	w
x	0.018595	−10391	17641	−0.19771	0.084445	−5000.6	3137.2	0.10498	−0.00846	18.433	1.3311	0.023298	0.003296	−0.02002	−0.01322	0.011524	0.003335	0.065635
y	0.63236	−17918	−10391	−0.96537	−1.1133	3186.2	5119	0.20918	0.30531	1.3964	−18.553	0.040973	0.047285	0.011532	0.061515	0.01783	0.015918	0.065638
z	204.86	−0.14608	0.010998	15908	15441	0.07049	0.000688	−11.664	−67.982	0.021555	0.005715	−0.01101	0.14478	0.008354	0.001515	0.004906	0.001466	0.065636
PRN 2	**a0**	**a1**	**b1**	**a2**	**b2**	**a3**	**b3**	**a4**	**b4**	**a5**	**b5**	**a6**	**b6**	**a7**	**b7**	**a8**	**b8**	**w**
x	0.14776	12823	−16714	−0.08337	0.36025	21.191	5656.4	−0.04442	0.026425	−18.731	−18.113	−0.02298	0.034367	0.13347	0.062799	−0.0176	0.028182	0.065637
y	−0.13186	16893	12361	−0.13135	−0.05226	5466.2	9.7661	0.15503	0.005604	−17.236	17.926	0.037545	0.003117	0.052779	−0.14424	0.020019	0.001771	0.065638
z	−267.61	0.066031	−0.04781	−6820.2	20508	−0.02451	0.034587	−45.706	−87.852	−0.00653	−0.0063	0.51226	0.28713	−0.00322	−0.00193	−0.00655	−0.00272	0.065637
PRN 3	**a0**	**a1**	**b1**	**a2**	**b2**	**a3**	**b3**	**a4**	**b4**	**a5**	**b5**	**a6**	**b6**	**a7**	**b7**	**a8**	**b8**	**w**
x	0.024956	−15898	14221	−0.14739	−0.02449	−4551	2563.9	0.049048	0.005681	18.399	−9.345	0.010413	0.001602	−0.09515	0.09312	0.004917	0.001827	0.065636
y	−0.65332	−14513	−15424	1.0555	1.7146	2676.6	4678.4	−0.24613	−0.53046	−10.39	−19.191	−0.0422	−0.09217	0.076747	0.046322	−0.0171	−0.03974	0.065633
z	−155.73	0.14136	−0.0018	−12499	−17172	−0.07433	−0.00992	47.445	69.896	−0.01378	−0.00684	−0.44138	−0.37007	−0.00766	−0.00224	−0.00049	−0.00023	0.065636
PRN 4	**a0**	**a1**	**b1**	**a2**	**b2**	**a3**	**b3**	**a4**	**b4**	**a5**	**b5**	**a6**	**b6**	**a7**	**b7**	**a8**	**b8**	**w**
x	−1.0986	3322.6	−21015	−1.7921	−7.5443	2384.2	4946.9	1.567	2.4124	9.3026	18.944	0.42526	0.50447	0.3583	0.44697	0.20468	0.22337	0.065656
y	2.2072	20447	3430.7	−1.1419	−0.60776	5095	−2370.7	−0.45789	−0.06617	18.918	−8.645	−0.16115	−0.07136	0.030598	−0.13895	−0.07969	−0.05707	0.065639
z	−35.729	0.068564	0.00279	3034.2	21391	−0.0138	0.001118	10.451	78.965	−0.00404	−0.00245	0.11857	0.58655	−0.00274	−0.00073	−3.88E−05	0.005408	0.065641

18 hours per day) and bit requirements (18 coefficients for the 8th order solution), it represents the best solution to make the ICT method effective.

In Table 1, an exemplary coefficient set is provided for 4 GPS satellites (PRN1 to PRN4) orbits on 9/04/2006.

Each interpolation function coefficient can be inserted in the navigation message with a 32-bit occupation and this implies a total length of 1728 bits for the ephemeris assistance package, as reported in Table 2. The assistance message length required by Fourier interpolation approach to describe the satellite orbits is greater than the conventional technique but its larger time validity makes this solution interesting for assistance purpose.

Note that, the coefficients of Table 2 have been by rounding the original double format numbers processed by MatLab, in order to have a limited number of decimals. The effect of this truncation is reported in Fig. 7, where it is reported the Euclidean distance between the exact interpolation function and the approximated version, obtained with a reduced number of bit for the coefficients representation. It can be seen that the use of a reduced accuracy coefficients (with only a few decimal digits) do not introduce a significant precision loss in comparison with the case of the exact 32 floating points coefficient description. This is a very interesting starting point in order to reduce the total assistance message length. In particular, it can be advantageous to employ a fixed point representation for each coefficient, optimizing the number of digits separately. Even though this activity is left for future investigations, it is possible to foresee that the assistance message length can be consistently shortened with respect to the value indicated in Table 2, which can be seen as an upper bound.

Table 2. Bit occupation and time validity for traditional and ICT assistance methods.

Ephemeris assistance approach	Required fields in the assistance message	Bit occupation in the assistance message for 4 hours validity	Bit occupation in the assistance message for 24 hours validity
Traditional Approach (GPS)	15 (for the 3D orbit)	362 (only spatial coordinates)	–
	$15 \times 6 = 90$ (6 retransmission for 24h)	–	2172 (only spatial coordinates)
	$15 \times 12 = 180$ (12 retransmission per 24h, high reliability)	–	4344 (only spatial coordinates)
Traditional Approach (Galileo)	15 (for the 3D orbit)	362 (only spatial coordinates)	–
	$15 \times 8 = 120$ (8 retransmission for 24h)	–	2896 (only spatial coordinates)
Interpolation Coefficients Transmission (ICT)	$18 \times 3 = 54$ (for the 3D orbit)	–	**1728** (only spatial coordinates)

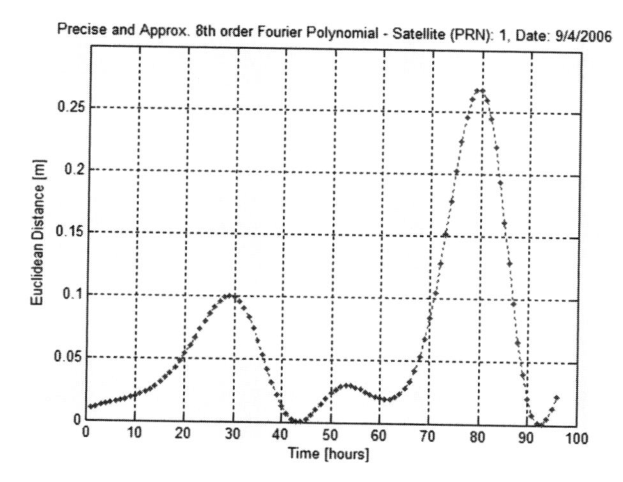

Fig. 7. Coefficient approximation impact for the Fourier polynomial.

Differently from ICT, SCT is based on the transmission of the satellite coordinates, sampled at regular intervals. The number of points to be transmitted to cover 24 hours depends on the desired accuracy and on the interpolation function adopted at the receiver side, as reported in Fig. 8.

In this case, the assistance message size is directly determined by the number of points composing the data set to be transmitted, which, however, strongly depends on the interpolation techniques adopted at the receiver side. To this purpose, Lagrange and spline interpolation can be fruitful employed, as explained in previous sections, although different terminal classes can employ different interpolation functions. To roughly quantify the amount of data to be transmitted, in Table 3 the minimal number of bit to be transmitted is reported for a 26 bit coordinates representation. This representation considers precise ephemeris rounded within 1 meter precision. The impairment of this approximation is quantified in Fig. 9, where the

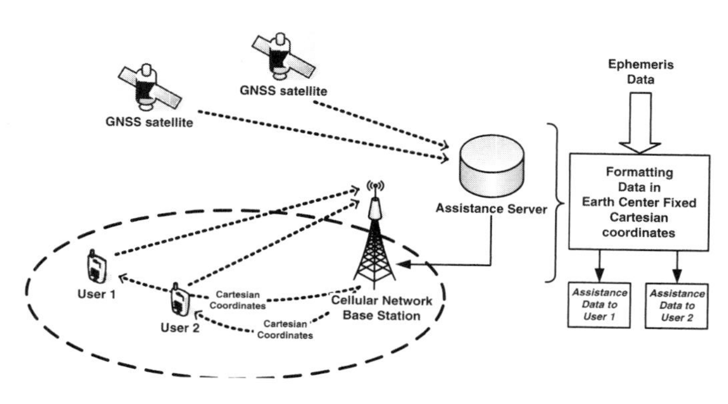

Fig. 8. SCT (Satellite Coordinate Transmission) system architecture for ephemeris assistance data.

Table 3. Bit occupation and time validity for traditional and SCT assistance methods.

Ephemeris assistance approach	Required fields in the assistance message	Bit occupation in the assistance message for 4 hours validity	Bit occupation in the assistance message for 24 hours validity
Traditional Approach (GPS)	15 (for the 3D orbit)	362 (only spatial coordinates)	–
	15 × 6 = 90 (6 retransmission for 24h)	–	2172 (only spatial coordinates)
	15 × 12 = 180 (12 retransmission for 24h, high reliability)	–	4344 (only spatial coordinates)
Traditional Approach (Galileo)	15 (for the 3D orbit)	362 (only spatial coordinates)	–
	15 × 8 = 120 (8 retransmission for 24h)	–	2896 (only spatial coordinates)
Satellite Coordinate Transmission (SCT)	24 × 3 = 72 (for the 3D orbit)	–	**1872** (only spatial coordinates)

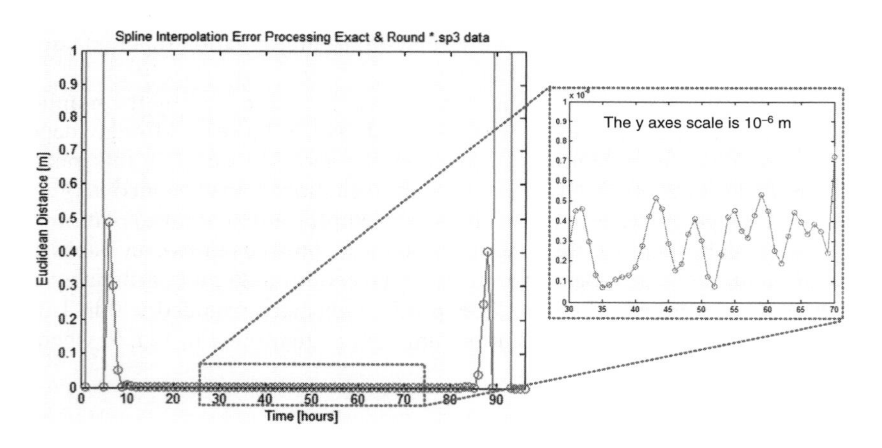

Fig. 9. Euclidean distance between exact ephemeris and rounded ephemeris spline interpolation.

Euclidean distance between precise ephemeris interpolation and approximate ephemeris interpolation is reported.

4 Conclusions

In this paper we have proposed two possible approaches to define novel data sets to describe satellite ephemeris for one day validity. The objective of this study has been the reduction of the number of bits required to describe the satellite orbit with respect to the solution adopted in satellite GNSS systems, under the constraint that the introduced error has to be lower than 20 m. This is in fact a reasonable approximation for most of commercial applications. The first solution, identified as Interpolation Coefficients Transmission (ICT), the A-GNSS server transmits the coefficients of a trigonometric interpolation polynomial that is employed by terminals to evaluate the satellite orbit with one day validity. In this case, the 8th order Fourier interpolation approach, which requires 18 coefficients to completely define the interpolation function, allows achieving the 20 m desired precision. The second approach, identified as Satellite Coordinates Transmission (SCT), foresees the transmission of a sampled version of precise ephemeris. It has been shown that terminals using a 24 knots spline interpolation method are able to largely meet the 20 m precision requirement with only 24 satellite position points with 26 bit-precision in one day. The main result is that both techniques allow an effective reduction of the number of bit to be transmitted to describe the satellite orbit, in a complementary way. In fact, ICT reduces terminal complexity because the interpolation processing is performed by the assistance server. Differently, SCT, which is more computationally demanding for terminals, introduces flexibility in the service provision, being the achievable precision depending on the terminal computational capability.

Acknowledgment

This work has been supported by the GJU GAMMA project (GJU 05/2413 CTR GAMMA).

References

[1] European Space Agency / Galileo Joint Undertaking, "Galileo Open Service Signal In Space Interface Control Document (OS SIS ICD) Draft 0", Draft 0, 23/05/2006.

[2] B. W. Remondi, "NGS Second Generation ASCII and Binary Orbit Formats and Associated Interpolation Studies", Proc. of the Twentieth General Assembly of the IUGG, 1991.

[3] M. Horemuz, and J. V. Andersson, "Polynomial interpolation of GPS satellite coordinates", GPS Solutions, vol. 10, n. 1, 2006, pp. 67–72.

[4] M. T. Heideman, D. H. Johnson, and C. S. Burrus, "Gauss and the history of the fast Fourier transform", IEEE ASSP Magazine, vol. 4, n. 1, 1984, pp. 14–21.

[5] M. Schenewerk, "A brief review of basic GPS orbit interpolation strategies", GPS Solutions, vol. 6, n. 4, 2003, pp. 265–267.

[6] B. Neta, C. P. Sagovac, D. A. Danielson, and J. R. Clynch, "Fast Interpolation for Global Positioning System (GPS) Satellite Orbits", Proc. of the AIAA/AAS Astrodynamics Conference, June 1996.

[7] P. Korvenoja, and R. Piché, "Efficient Satellite Orbit Approximation", Proc. of the ION GPS 2000, September 2000.

[8] D. J. R. Van Nee, and A. J. R. M. Coenen, "New Fast GPS Code-Acquisition Technique Using FFT", Electronics Letters, vol. 27, n. 2, 1991.

GNSS Based Attitude Determination Systems for High Altitude Platforms

Luigi Boccia, Giandomenico Amendola, Giuseppe Di Massa

Università della Calabria
Dipartimento di Elettronica, Informatica e Sistemistica
Via Bucci, 42-C8
87036 Rende CS, Italy
Phone: +39 0984 494700, Fax: +39 0984 494743, e-mail: dimassa@deis.unical.it

Abstract. High Altitude Platforms (HAPs) are a new, promising means of providing innovative wireless services. HAPs can be successfully applied for mobile or broadband communications and for disaster monitoring or response. One of the open issues is whether HAP stations can provide reliable services without temporal outages due to stratospheric winds that can cause an inclination of the platforms. As a possible solution in this paper it is proposed the use of a GNSS based attitude determination system. This technique, which has been successfully applied for both aircrafts and spacecrafts can provide real time three axis attitude data using the GNSS receiver present onboard the platform. In particular, it will be shown how the usage of a particular class of low multipath and lightweight antennas can provide high accuracy without altering the avionic ballast.

1 Introduction

High Altitude Platforms (HAPs) are an innovative technology developed to provide new means to implement innovative wireless services. HAP [1–2] may be either airplanes or airships autonomously operating during long time (up to several years) at altitudes between 15 and 25km and covering a service area up to 1,000 km. The key advantage of HAPs systems with respect to satellite technologies is that these platforms can be deployed in a relatively short time offering a wide range of new opportunities and enabling services that take advantage of the best features of both terrestrial and satellite communications. Using HAP stations can be beneficial to develop and implement regional wireless communication networks providing users with high rate and quality access to internet or with Third Generation (3-G) mobile services. In this context, a single aerial platform can replace a large number of terrestrial masts, along with their associated costs, environmental impact and backhaul constrains. HAPs can be also extremely beneficial to supplement existing services in the event of a disaster (e.g. earthquake, flood, volcano eruption). In a disaster scenario HAP can provide immediate coverage of the disaster area for both communications and monitoring applications.

HAPs are generally large some 200 meters and, even if their operating altitudes can be chosen to limit wind speeds and atmospheric turbulences, sudden gusts of atmospheric currents can alter the inclination and the positional stability of the platform.

In general, the horizontal displacements can be evaluated using an on-board Global Navigation Satellite System (GNSS) such as GPS or Galileo and counterbalanced employing a propulsion mechanism. However, for certain applications, the main HAP limitations derive from the attitude uncertainty. In fact, even if small inclinations can be in some cases compensated by affixing the antennas to a gimbaling system, greater variations require a realignment of the entire platform. In this latter case it is essential to have onboard the aircraft a real-time three-axis attitude determination system.

In this paper it is illustrated the possibility to use of a GNSS for real-time attitude determination of HAP stations. In fact, GNSS based attitude determination systems appear to be very well suited for HAP platforms provided a small, light weight antenna with low multipath capabilities is adopted. In the following, an introduction to GNSS based attitude determination systems will be presented focusing the attention on the receiving antennas, which characteristics strongly affect the system accuracy. Then, an innovative class of low-multipath low profile GNSS antennas will be presented, namely the Shorted Annular Patch (SAP) antennas. It will be shown how these radiators have performances comparable with other high precision GNSS antennas while having smaller size and light weight.

2 GNSS-Based Attitude Determinatio Systems

The possibility to use GNSS for real-time attitude determination of both spacecrafts and aircrafts has been recently introduced in [3]. A GNSS-based attitude determination system is a hybrid sensor that gives a continuous pointing knowledge, completely immune to drift phenomena and therefore without the necessity to be calibrated by a reference sensor thus being well suited for application onboard HAPs which are supposed to operate independently for long time. Furthermore, it gives a reduction in size, power and cost of the sensor hardware as it uses a GNSS receiver already present onboard the platform.

As it is shown in Fig. 1, the attitude information is derived measuring the spatial orientation of three baselines defined on the HAP body as the connection lines from a master antenna to three slave antennas. The basic measurable in GNSS-based attitude determination is a differential measurement across two antennas, the master and one of the three slaves, of the phase of the GNSS signal (Fig. 2). Two GNSS antennas are placed at the two ends of a baseline **b.** They are connected to two different channels of the same GNSS receiver. The channels are made to track the same GNSS satellite. The difference in the measurement of the GPS carrier between the two antennas is proportional to the projection of the baseline vector **b** onto the direction of arrival of the GNSS signal. If **s** designates the unitary vector along this direction of arrival, then one can write:

$$\underline{b} \cdot \underline{s} = m\lambda + \Delta\varphi + \Delta\varphi_{error} \quad (1)$$

The phase difference between the two antennas is shown as an integer number m of carrier wavelengths λ. plus the fractional part $\Delta\varphi$.. plus measurement errors $\Delta\varphi_{error}$. Most of the errors in these measurements, such as atmospheric delays or orbital and clock inaccuracies are spatially correlated are generally cancelled through the

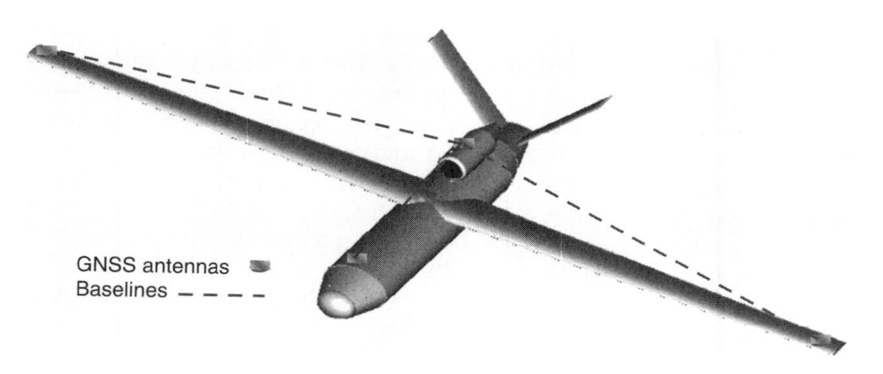

Fig. 1. A schematic of the GNSS-based attitude determination system applied to a HAP platform.

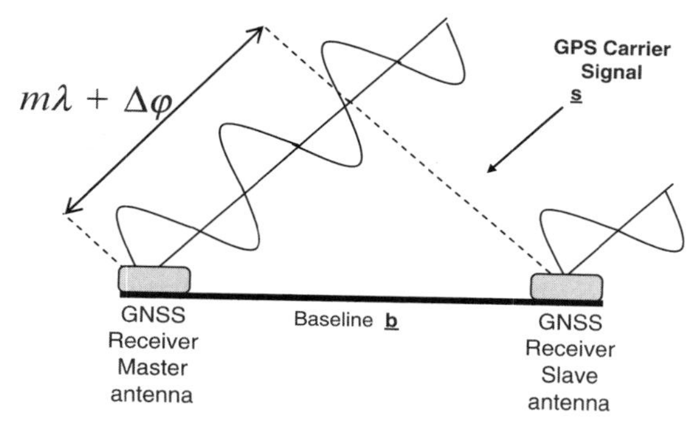

Fig. 2. Principle of GPS-based attitude determination.

differencing process. The major limitation that reduces the accuracy of these systems is owing to the multipath reflections of the GPS signal from surfaces around the antenna. When the length of a reflection path exceeds that of the direct path by more than 10–20 m then multipath errors can be reduced by signal processing techniques at the receiver. Unfortunately, in the most common case the strongest reflected signal component has an excess path length of less than 10 m which makes the receiver unable to detect and remove the multipath contamination. As a further consideration, it should be noticed that, in general, the accuracy of GNSS based attitude determination systems increases with the baseline length $\mathbf{\underline{b}}$ thus being very promising for HAP stations where a large spacing can be used between the antennas.

More effectively, the accuracy of a GNSS-based attitude determination system can be increased by using an antenna capable of rejecting multipath interferences. Antennas can be optimised for GNSS-based attitude determination in two ways. Firstly, they can be designed with high rejection to left-hand circularly polarised (LHCP) signals. This reduces the impact of multipath because the GPS, Glonass and Galileo signal are right-hand circularly polarised (RHCP) while odd reflections are LHCP. Hence, using antennas with a good rejection of LHCP signals, multipath effects arising from direct reflections can be potentially eliminated. Effects due to double reflections remain but they are normally much weaker. Secondly, attitude measurement performance can be improved by shaping the antenna gain pattern to reject low-elevation signals. This is beneficial because reflected signals often impinge on the antenna at low elevations. Hence, a narrow beam minimises their impact. Notice, however, that in order to perform attitude determination, at least two GNSS satellites must be tracked at all times. For this reason, the antenna field of view must be large enough to ensure that two or more satellites are visible throughout the satellite orbit. In practice, this means that the antenna pattern aperture must be greater than 120 degrees (supposing that only GPS is used).

Besides, a new antenna will be suitable for HAP applications only if its size is small enough and lightweight to permit an easy installation on the aircraft. Indeed one essential advantage of using GNSS for attitude determination is that different sensors and their interfaces can be eliminated and, in turn, costs, power requirements, weight and complexity reduced. Attempts to simultaneously meet these antenna requirements have been made in several ways and various types of GPS and Glonass antenna design have been proposed including spiral or helix antennas, patch elements placed on choke rings or "stealth" ground planes and several array configurations. Even if these solutions can be designed to reduce the multipath error they result in large and heavy structures that are not well suited for HAP applications especially if it considered that these antennas should be mounted at the opposite corners of a platform.

As a possible solution, in this paper it is proposed the use of an innovative class of patch antennas, namely the Shorted Annular Patch (SAP) antennas that demonstrated accuracy comparable to other larger and heavier solutions [4]. The key advantage of SAP antennas with respect to other concurrent radiators is that their weight is usually less than 500 gr while their overall size do not exceed 15 cm of radius. In practice, this implies that such antennas can be easily installed on HAP stations without affecting the overall avionic ballast and providing high accuracy attitude information.

3 Shorted Annular Patch Antennas

The SAP antenna geometry is shown in Fig. 3. An annular patch with external and internal radii a and b respectively is printed onto a dielectric grounded slab having relative dielectric constant ε_r and height h. At variance of conventional annular patches the inner SAP border is shorted to the ground plane thus making the dominant cavity mode a TM_{11} field variation. The SAP radiation pattern is therefore similar to that of the circular disk. However, once the dielectric characteristics and the operating frequency are fixed, the disk radius is uniquely determined and its radiation pattern cannot be modified. Conversely, the radiation characteristics of the shorted ring can be easily controlled varying the antenna geometry so that larger patches have higher amplitude roll-off near the horizon. Thanks to this feature, SAP antennas can be in fact optimized for high precision GNSS applications choosing the outer radius to minimize the multipath interference and adjusting the inner radius to make the patch resonate at the desired frequency.

As a proof of the SAP peculiarity, three shorted annular patch antennas resonating at the nominal GPS L1 frequency, 1.57542 GHz, with an external radius of 35,

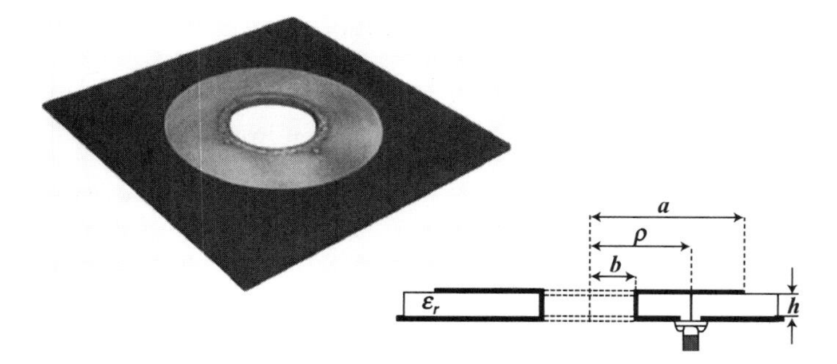

Fig. 3. Shorted annular patch geometry.

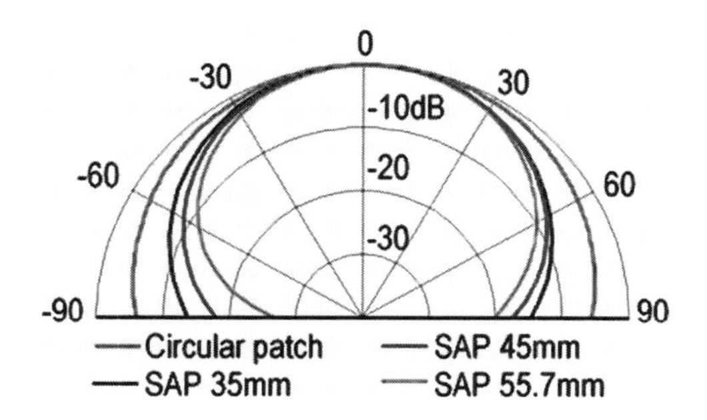

Fig. 4. SAP radiation pattern flexibility.

45, and 55.7 mm, have been designed considering a substrate with dielectric constant $\varepsilon_r = 2.55$ and thickness of 3.2 mm. Adequate circular polarization purity is attained by feeding the antenna by means of two 50 Ω coaxial probes located 90 deg. apart and having 90 deg. of phase difference.

The effect of a larger external radius is shown in Fig. 4 where the co-polar radiation patterns of the three SAP antennas have been compared with the one of a conventional circular patch resonating at the same frequency and designed using the same substrate. As expected, a larger outer radius of the antenna results in a narrower beam. Obviously similar characteristics can be obtained considering the Galileo contellation.

3.1 Radiation Characteristics

Prototypes of the three shorted annular patch antennas were then fabricated machining the inner hole in the dielectric substrate and shorting the internal boundary by means of a soldered copper foil. The geometrical characteristics of the three prototypes are shown in Table 1. In order to keep the circular polarization characteristics of each antenna as much as possible independent from the feed network

Table 1. Inner and outer radii, feed positions and dielectric size.

Antenna	a	b	d	ρ
SAP-G	55.7 mm	30.08 mm	140 mm	36.5 mm
SAP-M	45.0 mm	18.83 mm	150 mm	25 mm
SAP-P	35.0 mm	6.0 mm	160 mm	12 mm

design, each prototype was driven by means of an external quadrature hybrid (Pasternack PE2051) providing 90° ± 0.2° of phase difference within the GPS-L1 bandwidth and having a maximum VSWR equal to 1.07.

3.1.1 Radiation Patterns. In a first assessment the radiation characteristics of the three shorted rings were evaluated considering different figures of merit like the amplitude roll-off from broadside to the horizon, the polarization purity over all the hemispherical coverage and the phase response uniformity. All the measurements presented in this section have been taken at 1.575 GHz. Circular polarization characteristics have been obtained using a linearly polarized probe with a co-polar to cross-polar ratio higher than 40 dB in the broadside direction.

The radiation pattern of the SAP-M antenna was also measured and results are provided in Fig. 5. As expected, with respect to the SAP-G radiator, the SAP-M shows a reduced amplitude roll-off from broadside to the horizon that is around 20 dB. The on-axis gain is 8.97 dB while the RHCP to LHCP isolation is 23 dB. Coherently, the SAP-P antenna provides a more uniform hemispherical coverage with a gain at the horizon 15 dB lower than the one on axis (Fig. 6). In the broadside direction, the gain and the RHCP to LHCP ratio are 8.06 dB and 26 dB respectively.

It should be noticed that the amplitude response of the three prototypes is not uniform. However, this amplitude inhomogeneity is not so critical for a GPS system [5] as the only requirement is to have a signal level sufficient for all the coverage angles so that the receiver electronics can maintain lock with adequate signal to noise ratio.

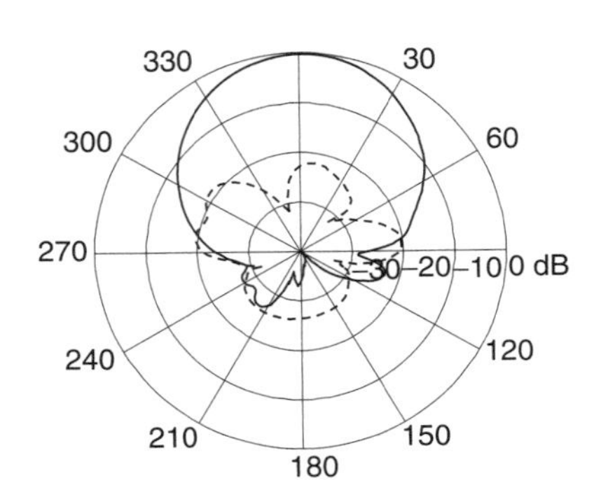

Fig. 5. Measured radiation pattern of the SAP-G antenna in the cut plane Φ = 45°: solid line RHCP, dashed line LHCP.

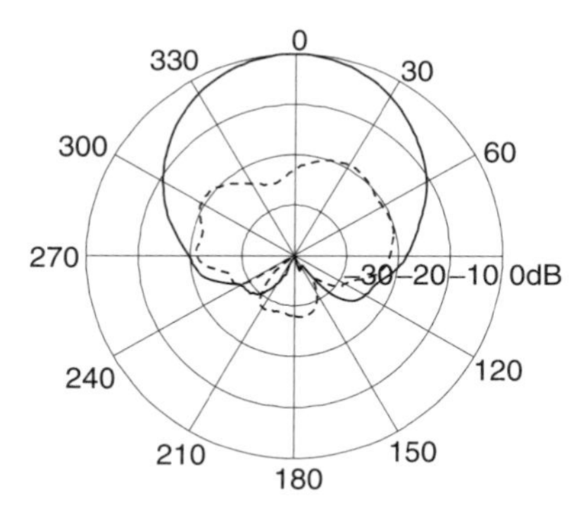

Fig. 6. Measured radiation pattern of the SAP-M antenna in the cut plane $\Phi = 45°$: solid line RHCP, dashed line LHCP.

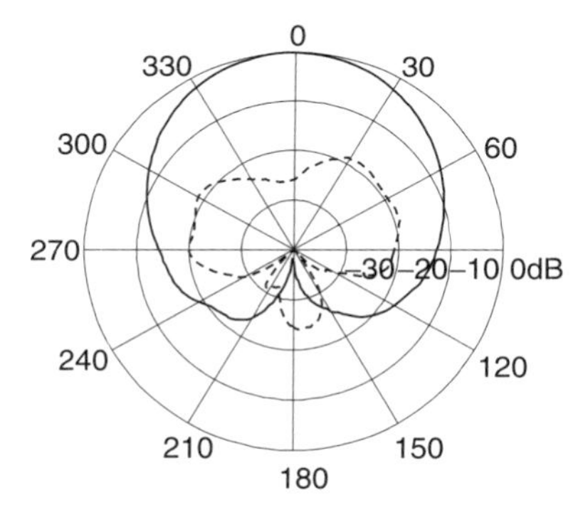

Fig. 7. Measured radiation pattern of the SAP-P antenna in the cut plane $\Phi = 45°$: solid line RHCP, dashed line LHCP.

3.1.2 Phase Centre. The phase response uniformity was estimated considering the phase centre variations versus the observation angle. For each prototype, the phase centre location was determined positioning the antenna to be coaxial with the positioner axes of rotation and elaborating the RHCP phase measurements by means of the code proposed in [6]. The horizontal and vertical phase centre offsets with respect to the mechanical centres in the cut planes $\Phi = 0°$, $45°$, $90°$ and $135°$ are shown in Figs. 6–10. As can be seen, for all the antennas both the horizontal and vertical phase centre locations diverge when the observation angle is taken near the horizon. However, a fair evaluation can be obtained taking into account only the variations achieved for observation angles comprised between +/–80°. Under this condition, the maxima of the horizontal and vertical phase offsets calculated for the three prototypes are reported in Table 2. These results indicate that the larger the antenna radius the more distributed is the phase centre. However, this effect is not exclusively related to shorted rings as a similar behaviour is typical of many other antennas such as horns or helixes [5]. In fact, the pattern cut-off near the horizon increases in antennas with a wide radiating surface due to a process of phase interference which in turn deteriorates the antenna phase front and polarization [5].

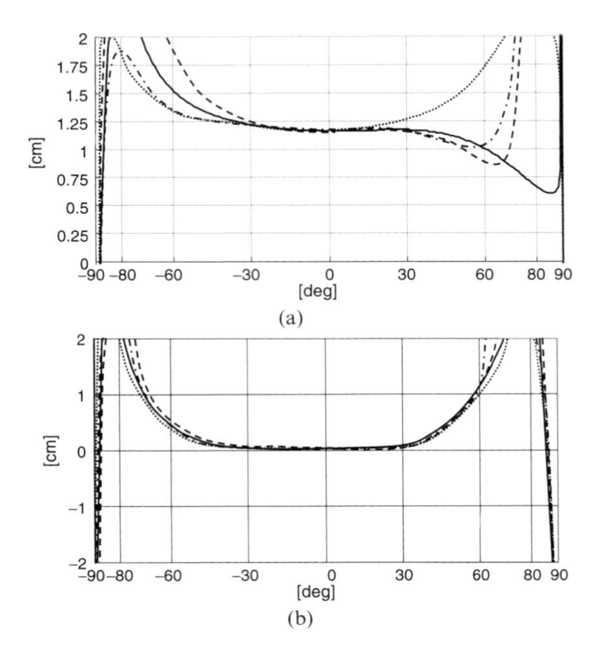

Fig. 8. SAP-G phase centre variations vs. observation angle θ with respect to the mechanical centre of the antenna. a) Vertical offset, b) Horizontal offset.
—— $\Phi = 0°$, --- $\Phi = 45°$; --- $\Phi = 90°$; ...... $\Phi = 135°$.

Table 2. Maximum horizontal and vertical phase centre variations for the SAP-G, -M and -P.

Antenna	Vertical [cm]	Horizontal (cm)
SAP-G	±2.24	±2.53
SAP-M	±0.543	±0.812
SAP-P	±0.29	±0.40

3.2 On Field Assessment

The results presented in the previous section indicate that the axial ratio and the phase stability of SAPs deteriorate as the external radius increases. Thus, while larger SAP antennas have higher amplitude roll-off near the horizon and therefore better immunity to grazing signals, they lack in terms of phase uniformity and polarization purity. As has been noticed before, this consideration is not limited to shorted rings, but it can be extended to many other radiators. In fact, it is very difficult to have a single GPS antenna that simultaneously satisfies all the high precision radiation requirements, especially when physical constraints such as limited size

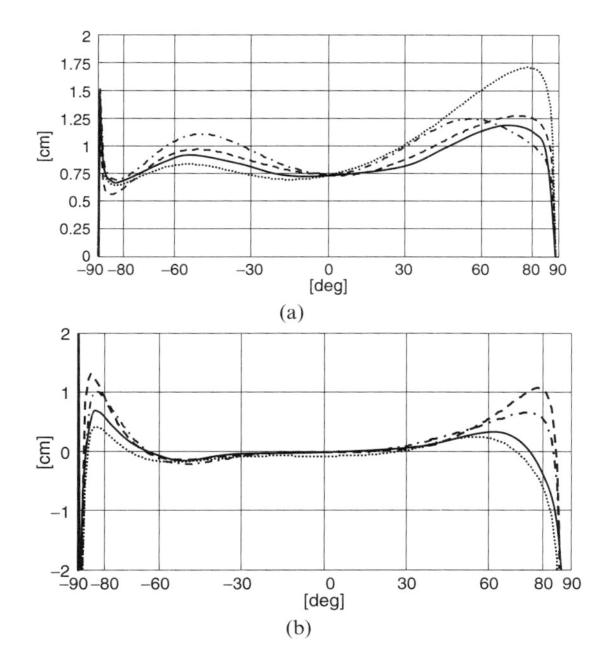

Fig. 9. SAP-M phase centre variations vs. observation angle θ with respect to the mechanical centre of the antenna. a) Vertical offset, b) Horizontal offset.
—— $\Phi = 0°$, ---$\Phi = 45°$; -·- $\Phi = 90°$; ...... $\Phi = 135°$.

are also considered. As a consequence, the performance evaluation uniquely based on the radiation characteristics can be ambiguous and an experimental on-field assessment is necessary to find the optimal design.

In this paper, in order to clearly identify the SAP design with best immunity to the multipath error, a comparative test campaign was conducted at the GEO-GPS facility, an on-ground GPS test range of the National Research Centre (Centro Nazionale delle Ricerche, CNR) in Rende (Cs), Italy. This facility, normally used to test DGPS-based geodetic systems, is constituted of two identical GPS receivers with 10 Hz reporting capability connected to a workstation. Each receiver is paired with a 30 dB amplifier and with an antenna mounted atop a rigid support and aligned with True North. The system is fixed on the rooftop of a 15 m tall building located in a dense urban zone and with an unobstructed view for elevations above 7°. The basic measurable quantity of this differential GPS configuration is the baseline separation between the two antennas which essentially depends upon the differential path delay of the received GPS signal. Therefore, this kind of measurement provides a valid means to assess the performances of a GPS antenna, since it is the major error source owing to multipath interferences. In fact, other inaccuracies such as differential line bias can be easily cancelled correcting the baseline reference vector through a calibration process. However, it should be noticed that the test set-up used in these experiments is not intended to be representative of the best multipath performances attainable with the antennas under test. Indeed, a precise

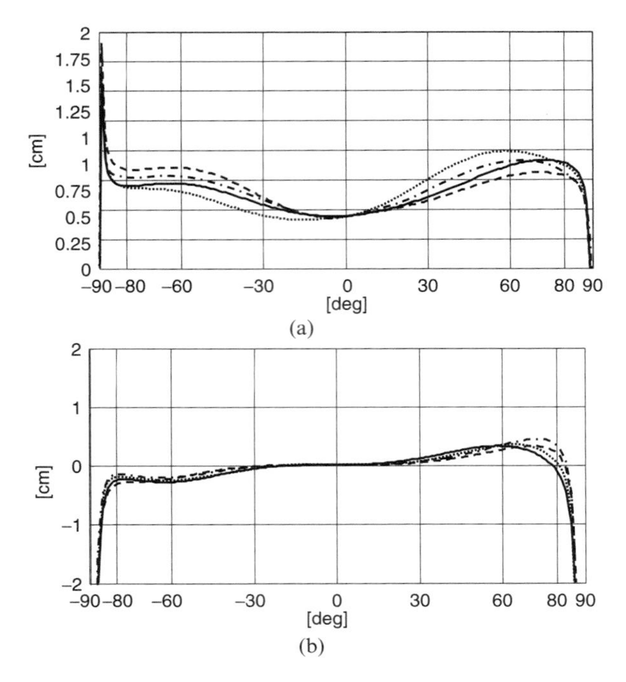

Fig. 10. SAP-P phase centre variations vs. observation angle θ with respect to the mechanical centre of the antenna. a) Vertical offset, b) Horizontal offset.
—— $\Phi = 0°$, ---$\Phi = 45°$; -·- $\Phi = 90°$; ...... $\Phi = 135°$.

assessment would be strongly influenced by the configuration of the environment surrounding the receiving antennas and by the baseline distance. As a consequence, a fair evaluation can be only inferred on the basis of a comparative test campaign.

The low multipath performances of the three SAP antennas were evaluated fabricating pairs of identical prototypes and performing 24-h tests to collect differential baseline displacements. This experiment duration is optimal because it evens out daily temperature oscillations and because it is equal to the repeatability period of the GPS constellation as seen from the ground. Thanks to the high gain of the SAP antennas and to the amplifier present in the receiving chain, it was possible to lock the GPS signals coming from all the satellites with elevations above 10°. Furthermore, in order to provide an additional reference for the evaluation of the SAP performances, two pairs of commercial GPS antennas were also tested in the same facility; namely, a single feed patch (Ma-Com 1141 [8]) and a dual-band multi-feed GPS antenna (AT2775-42A W from AeroAntenna Technology, Inc [24]). These radiators are referred to as Ref-1 and Ref-2 respectively. The Ref-2 element is a high precision antenna [9] specifically designed for GPS-based geodetic applications and it was tested in the GEO-GPS facility with a ground plane extension having 20 cm of external radius. Both the Ref-1 and the Ref-2 antennas having an internal amplifier, it was possible to test these two radiators without any additional amplifier.

Table 3. Experimental results RMS of the differential baseline displacement.

Antenna under test	RMS [mm]
Ref-1	2.021
SAP-G	1.697
SAP-P	1.215
Ref-2	0.979
SAP-M	0.759

For each antenna under test, the measured data were statistically evaluated considering the RMS of the differential baseline displacement whereas the nominal baseline length was 5m in all the experiments. For uniformity, the data collected from satellites with elevation lower than 10 degrees were filtered out in all the test cases and the Ref-2 data for the L2 band were discarded. The experimental results are presented in Table 3 showing that the three SAP elements have very different performances. As expected, the accuracy achieved with all the shorted rings is better than the one of the Ref-1 patch. In particular, the minimum baseline displacement is obtained with the SAP-M antenna whose RMS is 0.759 mm whereas the SAP-G and SAP-P errors are 1.697 mm and 1.215 mm respectively. Hence, the SAP-M performances are 55% better than the ones of the SAP-G antenna and this result might be even optimized designing and testing other SAP prototypes and using a multi-feed arrangement. However, this is beyond the scope of this work which shows that the shorted annular patch accuracy can be significantly improved when the RSW criterion is abandoned and a trade-off between all the antenna radiation parameters is considered. As an additional achievement, it should be noticed that under this condition the SAP performances are competitive even when compared with other high accuracy GPS antennas such as the Ref-2 antenna whose measured error is 0.979 mm.

A similar on field assessment has been performed at the European Space Agency GPS Test Facility at Estec and obtaining similar results thus confirming the correctness of the presented performance assessment.

4 Conclusions

GNSS based attitude determination systems are a promising candidate for HAP stations. In general, this class of sensors provides real time attitude determination with an accuracy that is strongly influenced by the antenna immunity to multipath signals. In this paper it has been presented a class of low multipath antennas very promising for usage in HAP stations. The proposed solution, namely the Shorted Annular Patch antenna, has been selected for its appealing characteristics in terms radiation pattern flexibility. SAP radiators, in fact, are low profile antennas which can be designed to have high precision performances while keeping their size and weight low. A comparative analysis of the performances of different SAP antennas has been proposed showing that such antennas can provide an accuracy comparable to that of other high precision antennas but with a reduced size and weight. This feature makes these antennas very attractive for usage in space applications or for HAP stations. It should be noticed that a GNSS-based attitude determination sensor could be employed onboard a HAP station provided the antennas are sufficiently immune to multipath signals and only when the radiators fulfil the physical constrains. For this reasons, it appears extremely important the individuation of a class of GNSS receiving antennas satisfying both this important features.

As a further possibility to improve the accuracy of SAP antennas, the simultaneous employment of both GPS and Galileo systems could be considered.

References

[1] DJUKNIC, G. M., FREIDENFELDS, J., and OKUNEV, Y.: "Establishing wireless communications services via highaltitude aeronautical platforms: a concept whose time has come?", IEEE Commun. Mag., September 1997, pp. 128–135

[2] SkyStation, see http://skystation.com

[3] C.E. Choen, "Attitude Determination Using GPS", PhD Thesis, Stanford University – 1992

[4] L. Boccia, G. Amendola, G. Di Massa, L. Giulicchi, "Shorted Annular patch antennas for multipath rejection in GPS-based attitude determination", Microwave and Optical Technology Letters, January 2001

[5] Ray, J. K. and M. E. Cannon (1999), Characterization of GPS Carrier Phase Multipath, Proceedings of ION National Technical Meeting, San Diego, January 25–27, pp. 243–252

[6] Schrank, H.; Hakkak, M.; "Antenna designer's notebook-design of pyramidal horns for fixed phase center as well as optimum gain", IEEE Antennas and Propagation Magazine, Volume 33, Issue 3, June 1991 pp. 53–55

[7] S. Best, Tranquilla, "A numerical technique to determine antenna phase centre location from computed or measured phase data", ACES Journal, Vol. 10, No. 2, 1995

[8] http://www.macom.com/

[9] http://www.aeroantenna.com/

Galileo IOV System Initialization and LCVTT Technique Exploitation

M. Gotta[+], F. Gottifredi[+], S. Piazza[+], D. Cretoni*, P.F. Lombardo*, E. Detoma°

[+] Alcatel Alenia Space Italia S.p.A.,
e-mail: monica.gotta@aleniaspazio.it, Phone: +39–06–4151.3134
* Univ. La Sapienza (Rome),
e-mail: lombardo@infocom.uniromal.it, Phone: +39–06–44585.472
° Sepa S.p.A., e-mail: edoardo.detoma@sepatorino.it, Phone: +39–011–0519220

Abstract. Satellite-based navigation systems uses one-way ranging measurements for system orbit estimation and time-keeping, due to its operational advantage when compared with two-way ranging technique, in terms of complexity of ground monitoring stations (completely passive and requiring a simple omnidirectional antenna to track all the satellites in view). However, a sufficient number of simultaneous independent measurements is required to solve the system unknowns: in particular simultaneous visibility of multiple stations by an individual satellite (allowing to separate the ground stations clocks contributions since the SVs clocks disappear), as well as simultaneity of observations from the same monitor stations of a large number of satellites (allowing to recover the SVs clocks parameters, since the GSS clocks drop out) is the key to an effective separation in the solution of the clock contributions from the pseudo-ranges.

In the Galileo IOV phase (consisting of 4 satellites on two orbital planes and a ground network of 20 Sensor Stations), the first condition is clearly fulfilled, however the second condition is not met for a considerable part of the time. If two GSSs do not see simultaneously a single Galileo satellite, they will not be able to estimate their clocks time and frequency drifts, i.e. they will not be synchronized. The free running clocks will essentially enter a holdover mode, were the relative time between the two stations will be slowly drifting as a function of the initial conditions and the stability of the clocks. The ground stations synchronization will gradually degrade with time and when a satellite will rise on the horizon they will be essentially not synchronized to the extent required to carry on a one-way-based Orbit Determination & Time Synchronization (OD&TS).

During the IOV phase, the limited number of satellites available and the peculiar characteristics of the Galileo orbits will make difficult for the Orbit Determination and Time Synchronization to start producing meaningful data, therefore some form of intermediate operational configuration must be sought to help in the OD&TS process initialization.

The paper will address the proposed solution to overcome the problem of Galileo system initialization starting from the intermediate configuration with first 2 satellites (first IOV Launch) up to final IOV Configuration after second IOV launch. The proposed solution will be based on a limited use of GPS to insure the synchronization of the Galileo Sensor Stations, relying on the exploitation of the Linked Common View Time Transfer (LCVTT) Technique, while the Galileo Orbit Determination and SVs clocks characterization will be carried on autonomously and independently by GPS, in a two step process, up to the achievement of the IOV Configuration with 4 satellites, when the nominal Orbit Determination and Time Synchronization process will be operated.

Moreover the paper will address the development of the LCVTT Algorithm, carried on as part of development of the infrastructure aiming to support the Galileo Verification Phase currently under definition as part of Galileo Phase C/D/E1 contract. The algorithm design and implementation will be presented together with the validation carried out (both for LCVTT and MLCVTT) to verify that the synchronization accuracy is adequate to support the Galileo System Initialization.

1 Introduction

Galileo, as the European-controlled world-wide satellite navigation system, is conceived to be the contribution to the next GNSS (Global Navigation Satellite System) system, namely the global infrastructure for the integrated management of the multimodal mobility on world scale (see [1]). Galileo will be an autonomous system but contemporarily compatible and, possibly, interoperable at the maximum extent with other navigation systems, particularly with the GPS system. Galileo will be under the control of a civil authority and will provide basic services at global coverage level for a wide range of applications in different transport domains, like road, railway, air, maritime and personal mobility, and suitable to fulfil various user needs spread over a wide professional areas.

The Galileo Programme is jointly supported by the European Commission and by the European Space Agency and is currently facing its C/D/E1 Phase, focused to the development, deployment and validation of an initial part of the System, known as IOV Configuration, composed of a reduced Space Segment and a reduced Ground Segment compared with the Final Operation Capability (FOC).

The Galileo IOV constellation shall be a sub-set of the Galileo FOC constellation of 30 satellites, comprising four satellites, in two different planes. The Ground Segment will be also a sub-set of the final one and is reported in Table 1.

The IOV reduced architecture has a high impact on Navigation Performance in IOV, leading of course to a degradation on accuracy and availability of navigation solution.

Several are the problems encountered when trying to work with an IOV constellation:

- The lack of measurements associated to each GSS station make difficult to compute a snapshot bias per epoch as currently envisaged for FOC.
- Furthermore, for long intervals of time, the satellites are not in view of the master clock station, and hence all the measurements at those epochs are rejected (more than half the total number of observations), hence degrading the orbit and clock estimations.
- Finally, the Orbit Determination and Time Synchronisation algorithm often fails because the normal matrix cannot be inverted, due to the bad conditioning of the system from the mathematical perspective. This can be sometimes overcome by restricting the a-priori covariance of the clock estimation (that implies constraining the normal matrix).

Moreover, in order to reach its on-line IOV Operation, the System needs to be initialized, as to reach convergence (especially in its Ground Processing, namely Orbit

Table 1. Galileo IOV configuration.

	IOV system configuration
Satellites	4
S-band TTC stations	2
C-band up link stations	5
L-band sensor stations	20
Galileo Control Centre	1

Determination & Time Synchronization) and allow starting the nominal IOV Test Campaign. In the following section the problem of system initialization, especially in the reduced IOV Configuration, will be treated and solution will be presented to support this activity.

2 IOV System Inizialization

Satellite-based navigation systems use one-way ranging measurements for system orbit estimation and time-keeping. The operational advantage of one-way ranging versus two-way ranging is obvious when one considers the complexity of the ground monitoring stations in the two approaches. One way requires a simple omni-directional antenna to track all the satellites in view, is completely passive (non-transmitting) and the station can be deployed or redeployed with minimum effort, requiring only a surveyed location, making it ideal for a military system. On the contrary, a two way ranging station requires a complex transmitting equipment, a large directional antenna and, as a consequence, will not be able to simultaneously track multiple satellites and it is expensive to deploy.

One-way ranging measurements are termed "pseudo-ranges" since they contain the system clocks contributions, namely the space vehicle (SV) clock and the monitor station (GSS in Galileo) clock in addition to the propagation delay caused by finite propagation velocity v_p over the range:

$$\tilde{\rho} = v_p \cdot (\Delta t_{prop\,(SV-GSS)} + \Delta t_{GSS-SV}) = v_p \cdot \lfloor \Delta t_{prop\,(SV-GSS)} + (\Delta t_{SV} + \Delta t_{GSS}) \rfloor \quad (1)$$

where in the last term I have expressed the contribution to the pseudo-range measurement as the sum of the offsets of the individual clocks with respect to the system time which, for a composite clock solution, must satisfy, in principle, the relationship:

$$\sum_i w_i \cdot \Delta t_{SV,i} + \sum_j w_j \cdot \Delta t_{GSS,j} = 0 \quad (2)$$

and the sums are carried on over all SVs and GSSs clocks in the system, each properly weighted with weights w_i and w_j.

The fundamental assumption of the one-way ranging technique is the capability to separate the three contributions to the pseudorange measurements, given a sufficient number of measurements. This will yield range observables, used to update the estimate of the orbit, and time offsets observables, to estimate the clock offsets and derive from the latter the clock parameters (phase and frequency offsets and drift) subject to the condition (2).

However, a sufficient number of independent measurements is required to solve the system unknowns, but additional constraints apply for the term separation to be effective, namely that a sufficient number of simultaneous independent measurements is available. Simultaneity of observations from the same monitor stations of a large number of satellites, as well as simultaneous visibility of multiple stations by an individual satellite, is the key to an effective separation in the solution of the clock contributions from the pseudo-ranges.

Single and double-differencing techniques, widely used for data reduction in the geodetic community, will be of help in understanding the underlying physical rationale. When two satellites are simultaneously observed by a single monitor station (Fig. 1), the first difference of the two pseudorange drops the common MS clock term from the observable.

Notice that we assume that the ionospheric propagation effects are completely removed by the use of the two-frequency technique and tropospheric effects are common to the two measurements, so they cancel out too. This yields a nice observable for the SV clocks. When a single satellite is simultaneous in view of two ground stations, the situation depicted in Fig. 2 applies.

Again, the first difference drops the SV clock and yields an observable which contains only the MS clocks contribution. However, while the same considerations for the ionospheric propagation still applies as before, now the tropospheric delay is not "common mode" and may (will) affect the final estimation of the range and GSS clock estimation.

Monitor stations clocks estimation (prediction) is affected by "local" (mainly due to the wet component of the troposphere) propagation delays and by the stability in

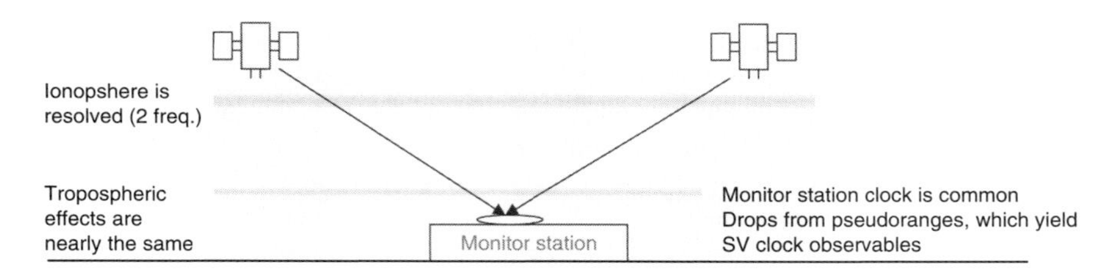

Fig. 1. Simultaneous observations from a single monitor station.

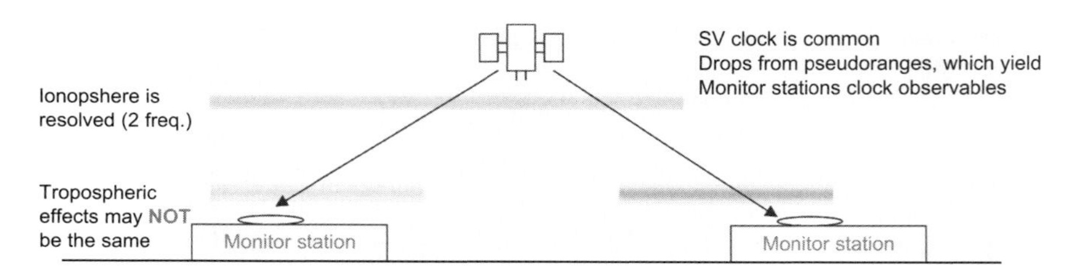

Fig. 2. Single satellite in simultaneous view of two monitor stations.

the equipment delays which are not correlated. Hence, errors propagate to orbit determination and indirectly affect the final user positioning/timing accuracy.

Since a second difference of a number of simultaneous observations having in common the SVs and the MSs will yield the orbit estimation free of clock terms, it is intuitive that the tropospheric effects are the major source of error left in the GSS clocks state estimate and, as a consequence, of the GSS clock states error projection on the orbit estimate.

Therefore, use and improvement of meteo data and tropospheric propagation models is of importance in the overall system error budget. In parallel, an independent monitoring capability of the MS clock behaviour (by two-way time transfer, for instance) may help in highlighting possible mismodeling effects in the troposphere as well as improving the capability to verify the MS clocks state estimate and their final contribution toward the orbit estimation errors.

From the previous considerations, it is clear that for the Orbit Determination and Time Synchronization Process to produce an optimum solution for the SVs orbits and system clocks two conditions must be fulfilled: that each satellite be continuously and simultaneously in view of more ground stations; this allows to separate the ground stations clocks contributions since the SVs clocks disappear; that each station be continuously and simultaneously in view of more than one satellite; this allows to recover the SVs clocks parameters, since the GSS clocks drop out.

In the Galileo IOV phase, the first condition is clearly fulfilled, however the second condition is not met for a considerable part of the time (as explained in Section 1). If two GSSs do not see simultaneously a single Galileo satellite, they will not be able to estimate their clocks time and frequency drifts, i.e.: they will not be synchronized. The free running clocks will essentially enter a holdover mode, were the relative time between the two stations will be slowly drifting as a function of the initial conditions and the stability of the clocks. The ground stations synchronization will gradually degrade with time and when a satellite will rise on the horizon they will be essentially not synchronized to the extent required to carry on a one-way-based OD&TS.

During the IOV phase, the limited number of satellites available and the peculiar characteristics of the Galileo orbits will make difficult for the OD&TS to start producing meaningful data, therefore some form of intermediate operational configuration must be sought to help in the OD&TS process initialization. The proposed solution to overcome this problem is based on a limited use of GPS to insure the synchronization of the GSSs, while the orbit determination and SVs clocks characterization will be carried on autonomously and independently by GPS.

The IOV phase will be characterized by three distinct temporal situations:

- prior to the availability of the first two satellites in orbit, the Ground Mission Segment will be the only component of the system supporting the navigation function that will be fully operative (phase I);
- the second phase (phase II) starts with the availability of at least two Galileo satellites in orbit, on the same orbital plane;
- phase III allows four Galileo satellites in orbit, on two orbital planes.

Due to the nature of the Galileo orbits, in full deployment the constellation repeats the same geometrical visibility with respect to a ground user every 8 hours, but with different satellites[1]. The same satellites will be visible with the same geometry by a fixed point on the Earth surface only every 10 days. Therefore, when a limited number of satellites are available, as in the IOV phase, it is understandable that the visibility conditions will occur at relatively sparse intervals.

The decision to implement a Master Clock configuration for the Galileo System Time (GST) turns into a distinct advantage under these rather restrictive conditions, since:

- GST becomes independent from the number of deployed system clocks, and is only based on the clocks ensemble at the PTF; therefore no discontinuity arises as new SVs or GSSs are added due to to the new clocks or a redistribution of weights;
- the previous consideration implies that GST can be maintained at nominal performances well ahead of the deployment of the Space Segment and even before th full GMS is deployed, as long as the PTF is operative.

Therefore, during phase I we may safely assume that GST is running and available at nominal performances and that the deployed GSSs can be referred to GTS via a GPS-based Linked Common View Technique, independently by the OD&TS but with a synchronization technique, the Common View, which is based on pseudorange measurements and therefore with resulting biases correlated to the OD&TS solution.

This allows to solve the problem discussed previously, that th GSSs needs some form of external T&F synchronization due to the lack of continuous and simultaneous visibility of orbiting Galileo satellites.

Having a sufficient[2] knowledge of the GSSs relative time offset, independent of the availability of Galileo satellites, the OD&TS process can be started by exploiting the condition (i) above, i.e., that at least two stations are simultaneously in view of each SVs[3] in orbit.

By basing, at this stage, the OD&TS process on single differences of the observables, the SVs clocks cancel and all the observations contribute only to the orbit determination, i.e.: the Keplerian parameters plus the modelled (solar pressure) and unmodelled accelerations[4].

Since:

- all the observables will contribute to the orbit determination only and
- a sufficient number of GSSs exist at this stage to provide an overdetermined solution and continuity of observations along the full orbit of the SVs, and moreover the observations can be time-correlated with a small degradation due to the

---

[1] For the GPS, the constellation repeats every (sidereal) day with the same satellites.

[2] It is assumed that the LCVTT will yield a relative synchronization between the GSSs (including the one located at the PTF) with an accuracy in the order of 5 ns, which yields an upper bound on one-way ranging of ≈ 1.5 m.

[3] In this phase only two satellites will be available.

[4] We assume that the prior knowledge of the gravitational field and perturbations from other bodies of the solar system or from the Earth (liquid and solid tides) is available, as it is, with the required accuracy to support an orbit determination with no degradation with respect to the system requirements.

parallel LCVTT process (due to the age of data in Crosslink Navigation Update Mode, see [2]), and
- because of the intrinsic high stability of the orbit with respect to the clocks

there is a high degree of confidence that the orbit determination process will converge quickly to an accurate solution.

Once the orbit is determined, the on-board clocks can be characterized "a posteriori" using the same orbital arc on which the orbit has been computed and the original observations. This is equivalent to an absolute one-way time transfer, where the ground clocks are known and the ranges are derived from the computed orbit. Subtracting these two quantities by the pseudoranges (observables) provided by all the stations in visibility of each satellite, yields the SVs clocks offsets in the form of a time series, from which the time and frequency offset (and frequency drift) can be estimated.

This two-steps process differs substantially by the final OD&TS operation by the fact that the solution is not provided in real-time but only in post-processing, no prediction is possible until the post-processing has produced a workable and stable solution, which in turn has to wait for the orbit determination to achieve a degree of stability and accuracy sufficient to support the SVs clocks characterization. The other major difference, at the algorithmic level, is in that two observables are used instead than one as in the operational OD&TS: the first difference of pseudoranges for orbit determination and the pseudorange for the SVs clocks characterization, both corrected "a priori" for the GSSs clocks offsets.

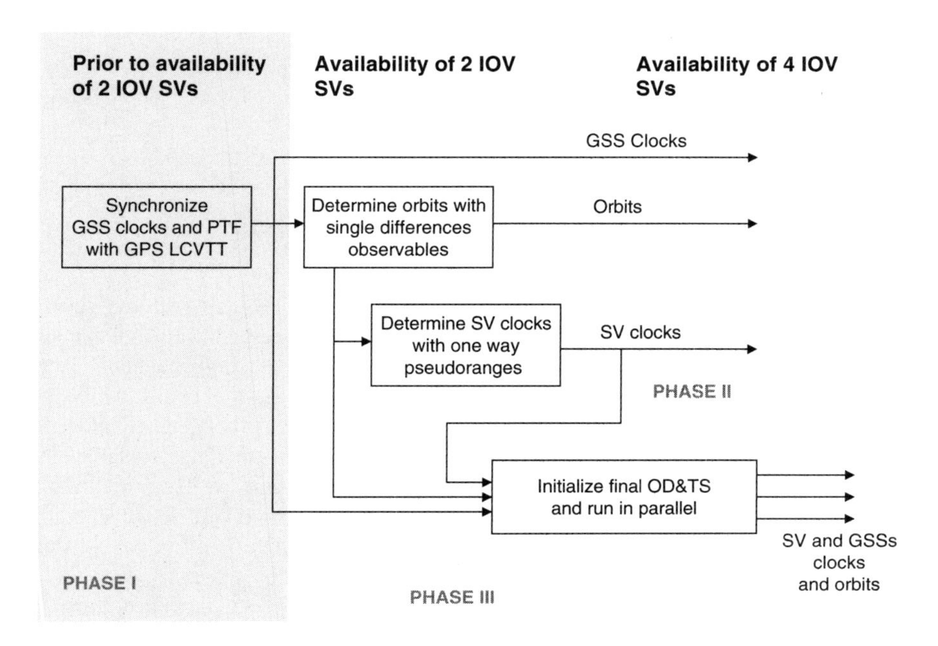

Fig. 3. Evolution of the OD&TS process prior & during IOV.

Accepting these limitations, the process should provide a good solution for the system parameters since what we have termed above as the phase II, i.e.: with only two satellites in orbit.

Once the orbit and clocks parameters are known with a sufficient degree of stability and accuracy, the final OD&TS process (Fig. 3) can be started in parallel, with initial conditions as provided by the previous process.

The rationale of the initialization is based on the "a priori" synchronization of the GSSs clocks by independent and external means. The use of Cs clocks under these conditions would provide benefits in insuring a better synchronization (small prediction error, see [3]) in the holdover mode, i.e. when the GSSs do not "see" any Galileo satellite that can be used for synchronization and they must rely on the clock stability and external measurements to keep a relative synchronization and a synchronization with GST.

In the next paragraph the development of a synchronization algorithm able to support System Inizialization by processing of GPS measurements collected at Galileo Sensor Stations is presented. Both the theoretical background and the implementation aspects will be treated, both with reference to Linked Common View and Multiple Path Linked Common View techniques (see [4], [5]).

3 LCVTT and MP-LCVTT

Using the Common-View Technique, provided there are enough satellites in common-view visibility between pair of stations, a number of sensor stations can be linked by implementation of LCVTT technique, to provide:

- the time offset between individual pairs of station clocks;
- the time offset between remote sites not in common view.

The situation is shown in Fig. 4 below, where only a few links are shown not to unnecessarily clutter the picture. The LCVTT allows not only to recover the time offset between adjacent stations, but by taking multiple differences also to measure the time offset between non adjacent stations, for instance, between Papeete and Kransoyarsk for the links shown.

This technique, although simple and computationally efficient, suffered several disadvantages. One of the most important is that using linked common view a single noisy site can decrease the precision of synchronization. An improvement of this technique can be obtained by synchronizing two remote station using a multiple common view path approach. In fact many possible links exist between two far stations, as depicted in Fig. 5. In order to increase the amount of data available, and therefore increase the precision of the synchronization one can use as many links as possible. By providing multiple independent measurements that can be averaged the noise measurement can be reduced. This approach is statistically more robust than the single linked common-view.

The error contributions that affect both LCVTT and MP-LCVTT synchronization are the same of Common View. In the next section the description of the most important errors is provided.

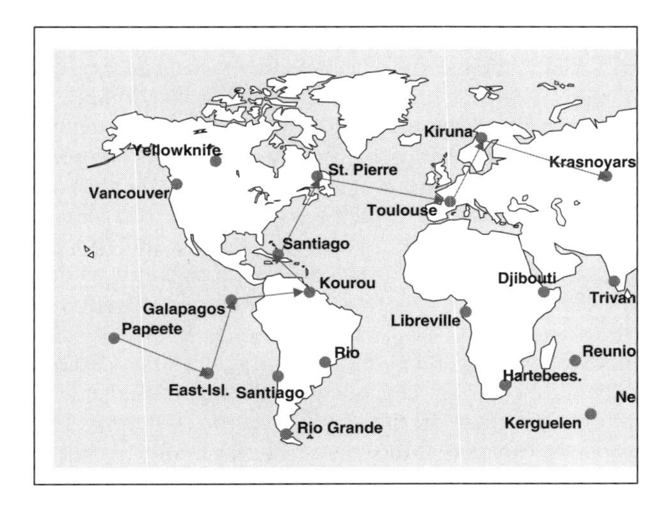

Fig. 4. Example of linked common-view time transfer.

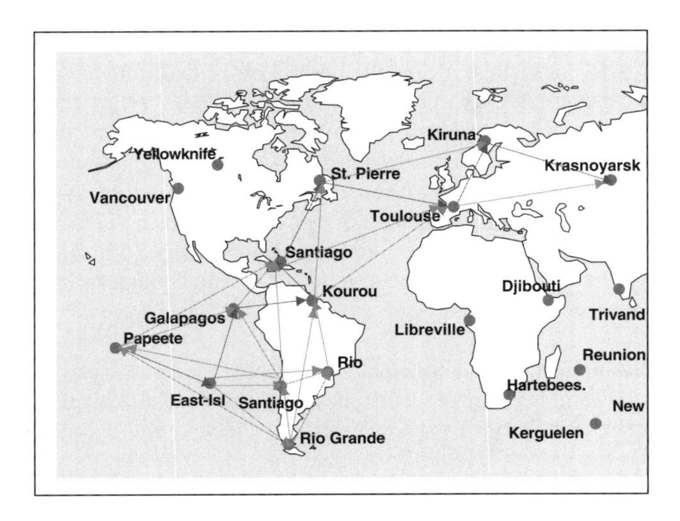

Fig. 5. Example of multiple-path linked common-view time transfer.

3.1 Error Contributions

3.1.1 Errors Resulting from Satellite Ephemeris. The time transfer error is dependent upon the ephemeris or position error of a satellite.

Common-view time transfer yields a great reduction in the effect of these errors between two stations, A and B, as compared to transfer of time from the satellite to the ground.

Common-view time transfer is accomplished as follows: stations A and B receive a common signal from a satellite and each records the local time of arrival, t_A and t_B respectively. From a knowledge of station and satellite position in a common coordinate system, the range between the satellite and each of the stations is computed, ρ_A and ρ_B. The time of transmission of the common signal according to each station is computed by subtracting from the times of arrival, the times of propagation from the satellite to each station, i.e., the time to travel the distances, ρ_A and ρ_B, are τ_A and τ_B (the range delays) and are given by $\tau_A = \rho_A / c$ and $\tau_B = \rho_B / c$ where c is the speed of light. This speed is subject to other corrections as are treated later. Finally, the time difference, τ_{AB}, of station A's clock minus station B's clock a the times the signals arrived is: $\tau_{AB} = (t_A - \tau_A) - (t_B - \tau_B)$.

If the ephemeris of the satellite is off, the computed ranges from the stations to the satellite will be off an amount dependent on the way the ephemeris is wrong and the geometrical configuration of the satellite-station systems. The advantage of common-view time transfer is that the computed bias is affected not by range errors to individual stations, but by the difference of the two range errors. Thus, much of the ephemeris error cancels out.

To see how this works in detail, suppose the ephemeris data implies range delays of τ^1_A and τ^1_B, but the actual position of the satellite, if known correctly, would give range delays of $\tau_A = \tau^1_A - \Delta\tau_A$ and $\tau_B = \tau^1_B - \Delta\tau_B$. Then the error in time transfer would be $\Delta\tau_{AB} = \Delta\tau_A - \Delta\tau_B$, where $\tau_{AB} = \tau^1_{AB} - \Delta\tau_{AB}$ is the true time difference (clock A - clock B) and where τ^1_{AB} is the computed time difference from the actual time of arrival measurements and ephemeris data. Thus, $\Delta\tau_{AB}$, the time transfer error due to ephemeris error, depends not on the magnitude of the range errors, but on how much they differ. The error in time transfer, $\Delta\tau_{AB}$, as mentioned above, depends on the locations of the two stations and of the satellite, as well as the orientation of the actual position error of the satellite. Since the GPS satellites are so far out, 4.2 earth radii approximately, the direction vectors pointing to the satellite tend to be close to parallel, thus cancelling most of the ephemeris error in all cases where common-view is available.

3.1.2 Errors Resulting from Ionosphere. The ionospheric time delay is given by (3) where TEC is the total number of electrons, called the Total Electron Content, along the path from the transmitter to the receiver, c is the velocity of light in meters per second, and f is the carrier frequency in Hz.

$$\Delta t = 40.3/cf^2\, TEC(\mathrm{s}) \tag{3}$$

TEC is usually expressed as the number of electrons in a unit cross-section column of 1 square meter area along the path and ranges from 10^{16} electrons per meter squared to 10^{19} electrons per meter squared.

For low latitudes and solar exposed regions of the world, time delays exceeding 100 ns are possible specially during periods of solar maximum. It's possible to show that the total delay at night time and/or high latitude is much smaller than at day time, and that the correlation in absolute delay time covers much larger distances when one moves away from the equator and the vicinity of noon; the conclusion being that a significant amount of common-mode cancellation will occur through the ionosphere at large distances if all observations are made at either high latitudes and/or at night time. These cancellation effects, over several thousand km, will cause errors of less than 5 ns. For short baselines less than 1000 km, this common-mode cancellation will cause errors of the order of or less than about 2 ns.

Since the ionosphere is a dispersive medium, when pseudo-range (code) measurements are available both at L1 and L2, an ionospheric delay-free pseudo-range $\rho_{iono-free}$ can be constructed with the following relationship:

$$\rho_{iono-free} = \frac{f_{L1}^2}{f_{L1}^2 - f_{L2}^2}\rho_{L1} - \frac{f_{L2}^2}{f_{L1}^2 - f_{L2}^2}\rho_{L2} = \frac{\rho_{L2} - \gamma\rho_{L1}}{1-\gamma}\ (\text{m}) \tag{4}$$

where $\gamma = (f_{L1}/f_{L2})$.

It's also possible computed ionospheric delay (seconds) with the following equation:

$$\Delta t_{iono} = \frac{1}{c}\frac{f_{L2}}{f_{L1} - f_{L2}}(\rho_{L1} - \rho_{L2})(\text{s}) \tag{5}$$

If double frequency measurements aren't available, the ionospheric delay can be estimate using Klubachar model (possibly corrected by the difference in height of the satellite with respect to an Earth observer). This model use the eight ionospheric coefficient that each satellite transmit with the Navigation Message. This model can be used to determine the delay in the vertical direction relative to a certain position from four amplitude components and four periodic components; this method is said to be capable of correcting about 50% of ionospheric delay.

3.1.3 Errors Resulting From Troposphere. In transferring time between ground stations via common-view satellite, one records the time arrival of the signal and computes the time of transmission by subtracting the propagation time. The propagation time is found by dividing the range to the satel lite by the velocity of light. However, moisture and oxygen in the troposphere have an effect on the velocity of propagation of the signal, thus affecting the computed time transmission and therefore, the time transfer. This effect is dependent on the geometry, the latitude, the pressure, and the temperature, and may vary in magnitude from 3 ns to 300 ns. However, by employing reasonable models and using high elevation angles, the uncertainties in the differential delay between two sites should be well below 10 ns. Later on, if needed, the magnitude of the troposphere delay can be calculated with uncertainties which will approach a nanosecond.

For the implementation of the Synchronization algorithm same tropospheric model are used. If for a station measurements of pressure, temperature and relative humidity are available it's possible use a mathematical model to estimate and remove the tropospheric delay. In the following Hopfild Model, with Seeber Mapping function or trough Series Expansion of Integrand, and Saastamoinen model are used.

3.1.4 Errors Consideration in Receiver Design. A common concern for all modes of extracting time from GPS/GALILEO is the calibration. The precise calibration bias of a GPS/GALILEO system (including receiver, antenna and cabling) typically is one of largest errors in providing time offset between two stations.

Absolute calibration can be achieved by using a GPS/GALILEO signal simulator to calibrate the group delay trough the GPS/GALILEO antenna, receiver and cables. Calibration is more commonly achieved by using a GPS/GALILEO receiver whose calibration has been previously determined. Using great care, the calibration bias can be reduced to less than 5 ns.

Even in the best-designed system, GPS/GALILEO receivers can vary by several nanoseconds in their calibration over time (months to years). It's important to frequently check the calibration of the GPS/GALILEO System.

Multipath is a well know error source for all forms of GPS/GALILEO observations. However, timekeeping has an added multipath concern due to reflections in the cabling. Care should be taken in impedance matching between the elements of the user's GPS/GALILEO System. Failure to do so can cause large temperature and time-dependent variations in the measurements (up to 10 ns).

It's not possible decrease the impact of the receiver delay error averaging many independent time offset value because this error component affect alike all measurements. For this reason it's important know precise receiver time delay. If the calibration delay is known, after removing, the pseudo-range observables can be considered free from this error contribution. In the following section the LCVTT and Multiple Path LCVTT algorithm implementation is discussed.

3.2 Synchronization Algorithm Implementation

The planning of the multiple path LCVTT requires an evaluation process of average number of satellite in common view for the path linking each station. Through SVs constellation simulator it's possible to obtain the common view visibility data that can be used to evaluate each possible link path. This, in turn, allows to optimize the path selection (Fig. 6) used to synchronize all network stations.

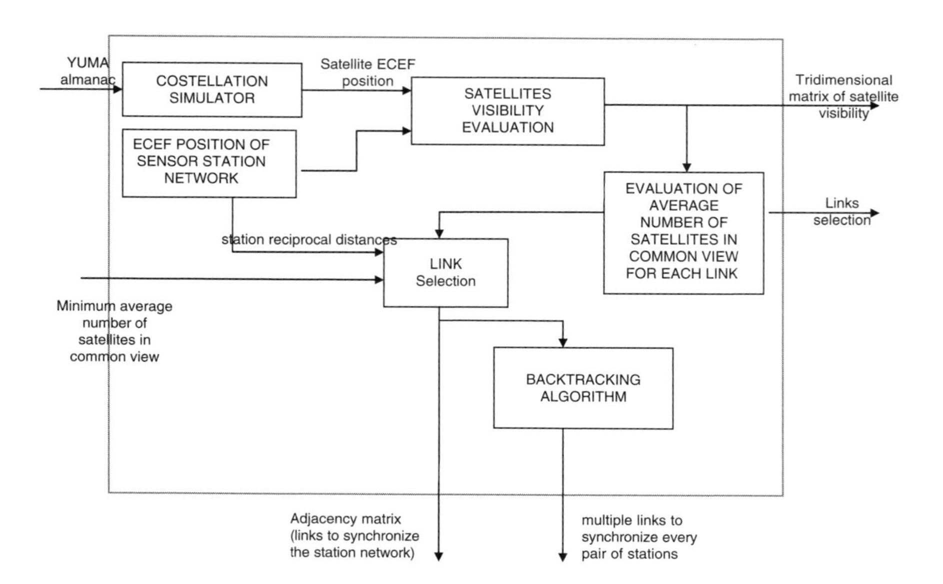

Fig. 6. Path selection block diagram.

For GPS, the metric used to select a contributing link is based on common observation times for each station connected to adjacent stations via LCVTT. It is possible to precompute the number of common observations from the reference site to each of the adjacent sites, and the higher this value, the better the link will be in providing useful data for the LCVTT. A second metric is based on the average number of SVs in common view over each of the paths; a large number of SVs implies again that more time transfer will be available and that the measurement noise will be smaller as a consequence of averaging a large number of raw time transfers, each of which will results from one of the SVs in common-view.

Common-view visibility data, provided by the SVs constellation simulator, is used to fill the metrics used for each path link evaluation. This in turn allows to optimize the path selection, leading to a ground network topology for the LCVTT. In operations, the raw data from the Sensor Stations is sent to the LCVTT computations process, for elaboration and estimation of time offset between individual pairs of station clocks (synchronization value).

For each link identified at the output of Link Selection Block, the CV Synchronization Algorithm is applied as depicted in Fig. 4. By receiving in input the pseudorange measurements of the identified satellites, k values of CV synchronization can be computed for that link (where k is the number of satellites in CV of station I and j), by applying the CV Synchronization Block (detailed in Fig. 5). The final synchronization value at instant t for station i and j can be obtained by averaging all synchronization values obtained for this pair of stations from the pseudorange related to all the k satellites in CV.

By means of the "Adjacency Matrix" (output of the Link Selection Block of Path Selection algorithm), this can be iteratively applied in the MLCVTT to all the possible links that cannot be directly synchronized via single CV, by applying a backtracking algorithm to identify the possible multiple paths. The synchronization values for each link is obtained by applying the LCVTT Block (Figs. 7 and 8) for those paths identified at the output of the Backtracking Block (average value for each pair of station is again obtained for each instant by averaging over all the satellites in CV for each individual link).

The Stand Alone Synchronization Block returns the pseudorange of each Sensor Station to be synchronized, after removal of ionosphere, troposphere, equipment delays and true slant range, in order to obtain the Space Vehicle (SV) and Sensor Station (rx) clock contribution, considering that measured pseudorange (as obtained at output of Sensor Station receiver) can be defined as:

$$\rho_{measured} = \rho_{true-range} + c\Delta t_{iono} + c\Delta t_{tropo} + c\Delta t_{sv} + c\Delta t_{rx} + c\Delta t_{equip} + \varepsilon \tag{6}$$

Please refer to Fig. 9 for the Stand-Alone Synchronization algorithm Block Diagram.

Validation of the developed algorithms in terms of the achievable synchronization accuracy is performed by running the developed algorithm with input pseudorange measurements as collected at IGS worldwide Ground Stations and comparing the Ground Station Synchronization results obtained as output of LCVTT and MPLCVTT with IGS clock Products as provided by International Geodetic Service (IGS). In the following section the experimentation results are presented.

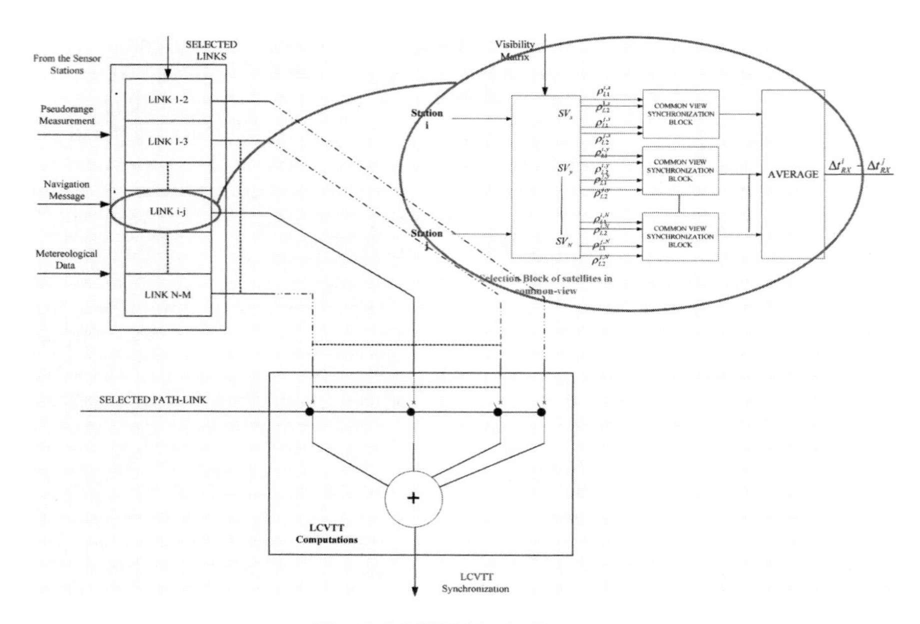

Fig. 7. LCVTT block diagram.

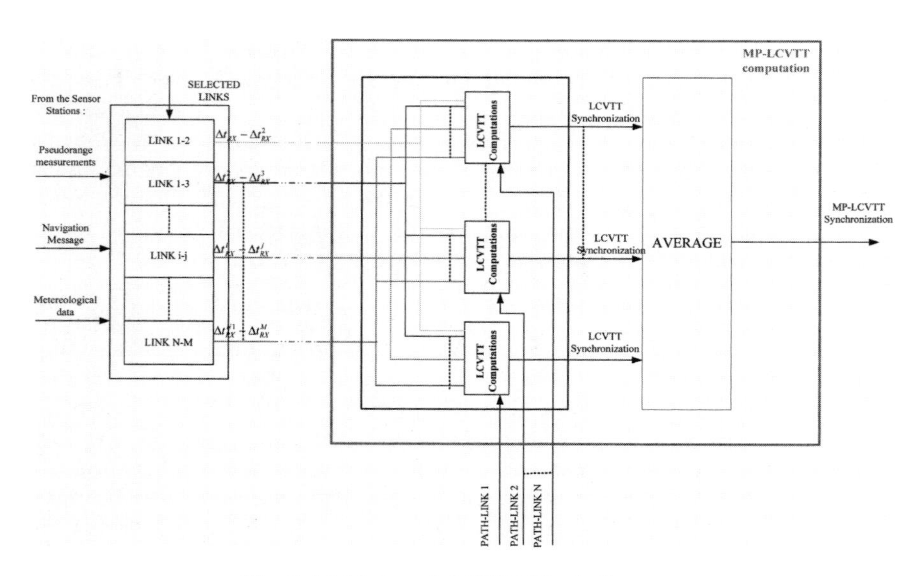

Fig. 8. Multiple path LCVTT block diagram.

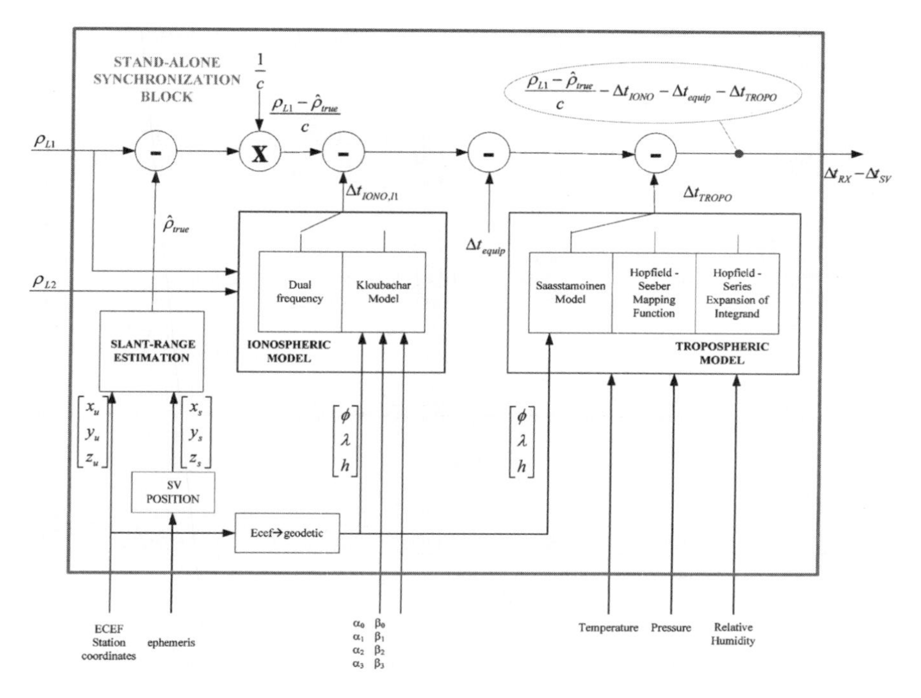

Fig. 9. Stand-alone synchronization block diagram.

4 Experimentation Results

Preliminary experimentation results are obtained running the prototype algorithm (either LCVTT and MLCVTT) by processing in input code measurements acquired at the IGS Sensor Stations as indicated in Fig. 10 (and retrieved at http://igscb.jpl.nasa.gov) and by comparing the obtained outputs with IGS clock products (available at same web site). Only data from Sensor Station that are found complete of all needed informations to efficiently remove pseudorange error contribution (meteo data, equipment calibration data etc.) are used to obtain the results shown in Figs. 11–13. The figures shows the comparison of the synchronization between pair of stations, obtained with single-path and multiple-path linking, with the synchronization computed by IGS for the same pair of station (considering IGS as the "true" reference).

As expected, the synchronization error decreases with the distance of the two stations because pseudorange measurements are affected by error that are not common and cancel out only at first order; the advantage is that using LCVTT it's possible recover the time offset between two stations not in common visibility. Moreover, by averaging multiple independent estimates of the time transfer from the multiple path, the LCV timing stability (at 1 day averaging) can be increased with Multiple Linked Common View.

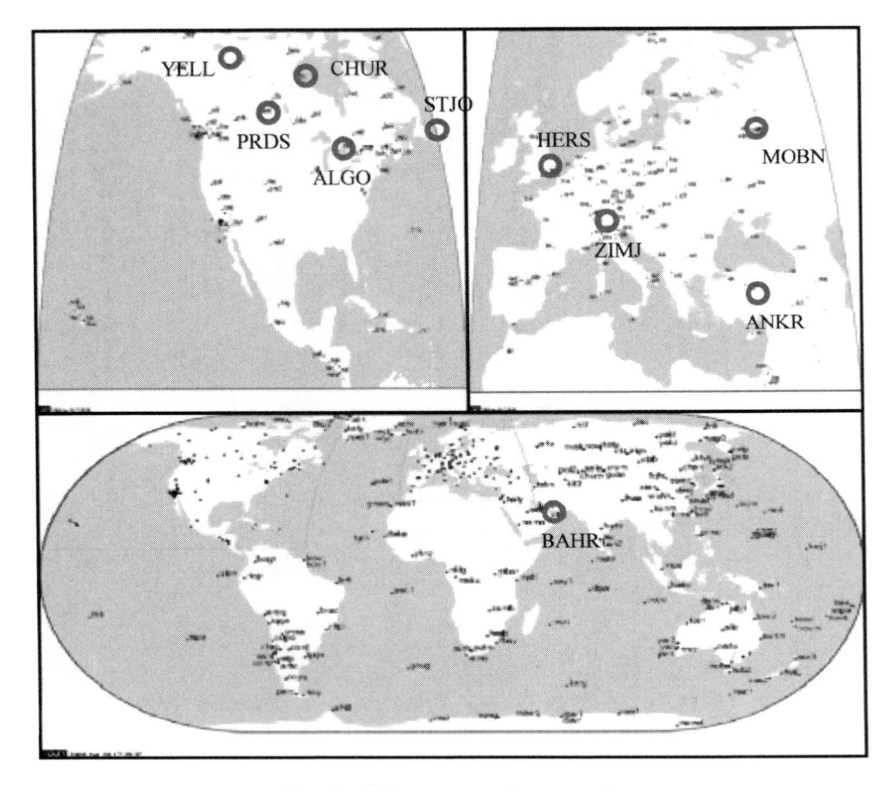

Fig. 10. IGS sensor station network.

To be able to obtain the same performance of single Common View (stability around 5 ns rms at 1 day) for all the Sensor Station Synchronization (including Stations not in view) using Multiple LCVTT we need many stations to increase the number of independent measure of the clock offset. If it's not possible to use an enough number of stations, the synchronization suffers the dominance on the linking process of single noisy link.

5 Conclusions

Both Linked Common View and Multiple Path Linked Common View have been implemented in a combined algorithm. The algorithm is run by processing the measurements as collected by the International GPS Service (IGS) and output synchronization products are compared with IGS products for algorithm validation. In particular the improvement obtained with Multiple Path LCVTT when compared with Linked Common View is shown based on experimentation results. MP-LCVTT can be advantageously implemented to guarantee the synchronization of those stations that are located at intercontinental baseline and for which a continuous

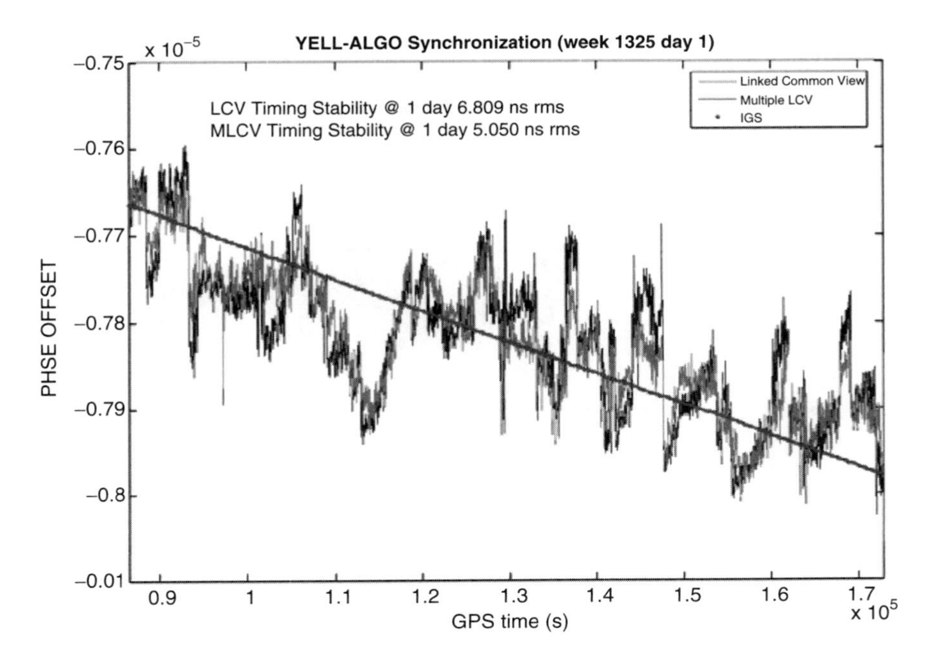

Fig. 11. YELL-ALGO synchronization.

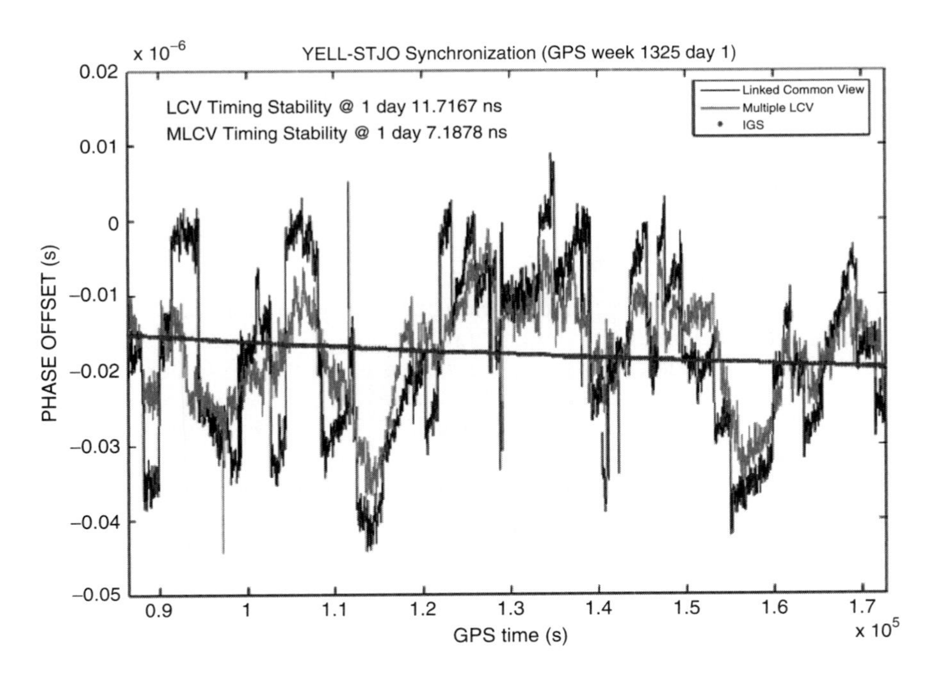

Fig. 12. YELL-STJO synchronization.

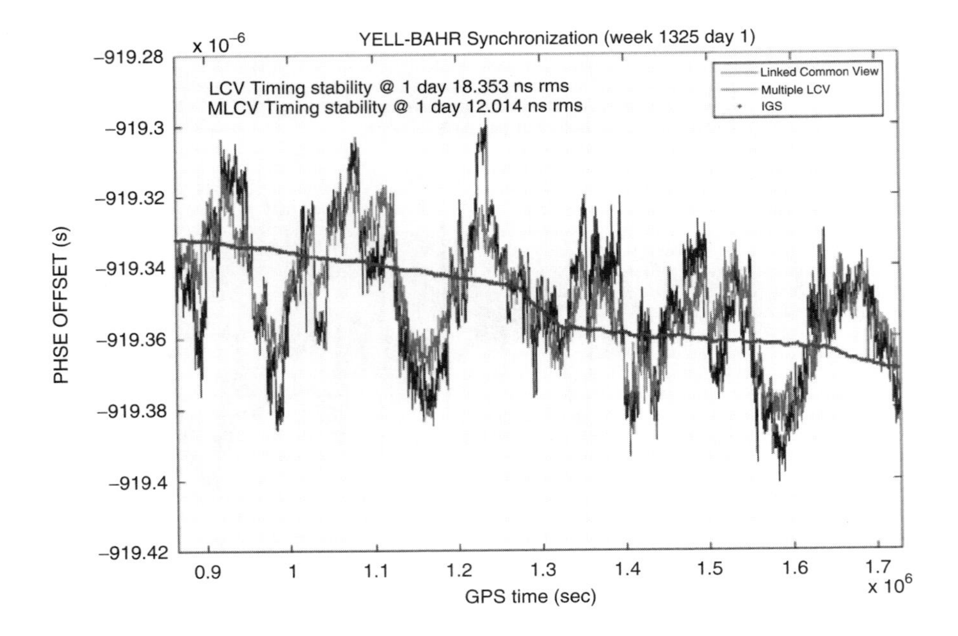

Fig. 13. YELL-BAHR synchronization.

synchronization would not be differently possible. As expected, the performance of synchronization achievable by combining adjacent links is worst that single CV (CV time stability around 5 ns rms at 1 day) as the CV errors sums as the square root. However this can be compensating, in MP-LCVTT, by the reduction in the measurement noise resulting from the averaging of multiple, independent time transfers from multiple links to a given site. The experimentation campaign allowed to confirm this expectation, considering that:

- All the considered sensor station (world-wide distributed) are synchronized with an error between 2 and 12 ns rms at 1 day
- This results is considered a worst case, as the experimentation results are limited and constrained by the sensor station measurement quality and choice that has been used for the experimentation campaign.

The sensor station network is in fact non-optimal in terms of number of stations and quality of measurements acquired by them. This is due to the fact that the selection of the sensor station network has been driven by the availability, for each of them, of all the information necessary to efficiently reduce the error contribution (e.g. availability of dual-frequency receiver to remove the ionospheric delay and of meteo data to remove the tropospheric delay) and, at the same time, availability of the IGS products (for those Sensor Stations) to be used as reference for the algorithm validation. These constraints led to the selection of number of stations that is limited and characterized by an highly variable and not optimized measurement quality. It has to be noted that the noise affecting the pseudorange measurements is directly impacted by the quality of the oscillator feeding the receiver (that in the selected network is not always represented by an high stable atomic clock, sometimes being an internal quartz) and by the calibration of the station equipment delays (that are often not available for the selected sensor stations and thus was not possible to remove it).

These limitations would not apply when the same algorithm would be used by processing Galileo measurements as acquired at Galileo Sensor Stations (GSS); in this case calibration data will be regularly available and all the Sensor Stations will be equipped with high stable atomic clocks, whose performance will be aided by use of temperature-stabilization systems.

Therefore it is expected that the performance achievable by the developed algorithm, when applied to Galileo Sensor Station Synchronization, is sensibly better and adequate to the initialization of System in its IOV Configuration.

References

[1] O. Galimberti, M. Gotta, F. Gottifredi, S. Greco, M. Leonardi, F. Lo Zito, F. Martinino, S. Piazza, M. Sanna, "Galileo: The European Satellite Navigation System", Proceedings of ATTI dell'Istituto Italiano di Navigazione, March 2005, p. 50–96

[2] K. Ghassemi, S.C. Fisher, "Performance Projections of GPS IIF", Proceedings of the 10th International Technical Meeting of the Satellite Division of the Institute of Navigation, ION GPS 1997 Meeting (Kansas City, Mo. – September 16–19, 1997), pp. 407–415

[3] P. Tavella, M. Gotta, "Uncertainty and prediction of clock errors in space and ground applications", in Proc. 14th European Time and Frequency Forum, Turin, March 2000.

[4] W. G. Reid, 1997, "Continuous Observation of Navstar Clock Offset from the DoD Master Clock Using Linked Common View-Time Transfer," in Proceedings of the 28th Annual Precise Time and Time Interval (PTTI) Systems and Applications Meeting, 3–5 December 1996, Dana Point, California, USA (U.S. Naval Observatory, Washington, D.C.), pp. 397–408.

[5] W. G. Reid, 2000, "Multiple-Path Linked Common-View Time Transfer," in Proceedings of the 31st Annual Precise Time and Time Interval (PTTI) Systems and Applications Meeting, 7–9 December 1999, Dana Point, California, USA (U.S. Naval Observatory, Washington, D.C.), pp. 43–53.

[6] T. E. Parker, D. Mstsakis, "Time and Frequency Dissemination – Advances in GPS Transfer Techniques", GPS world, November 2004.

Impact of Atmosphere Turbulence on Satellite Navigation Signals

Per Høeg, Ramjee Prasad, Kai Borre

Aalborg University
Institute of Electronic Systems
Niels Jernes Vej 14, 9220 Aalborg, Denmark
Phone: +45 9635 9828, Fax: +45 9815 1583, e-mail: hoeg@kom.aau.dk

Abstract. Atmosphere turbulence for low elevation angle reception is a noise source that is not well defined in existing systems using satellite navigation signals. For high precision aviation purposes atmosphere turbulence needs to be assessed to meet the future stringent requirements.

High precision receivers using open-loop mode data sampling at high sampling rate enables investigations of the characteristics of the noise and the multi-path signal errors through the determination of the refractive index structure constant C_n^2. The main modulation of GPS signals in low-elevation measurements is attenuation and frequency shift due to ray bending. Whereas the presence of turbulence results in a spectral broadening of the signal. Analysis of the trends of the spectral mean slope for different frequency domains will be discussed in relation to the characteristics of atmosphere turbulence. Additionally we present results from phase-lock receivers loosing lock during strong perturbations.

1 Introduction

High altitude field tests have established experimental knowledge on the influence of atmosphere turbulence on receiver performance in tropical regions [1]. Moist air turbulence measurements from Haleakala, Hawaii, are studied and presented here for spectral signal structure characteristics.

The performed spectral properties of the received signals by a high precision GPS instrument in both phase-locked mode (PL) and open-loop mode (OL) are compared to theoretical results. PL or closed loop tracking is the standard technique used by most GPS receivers. It reduces the needed bandwidth of the measurements by having the carrier phase locked to the received signal. Open loop tracking (OL) or raw mode sampling requires the full measurements relatively to an on-board Doppler model. The method requires more advanced receivers with high sampling rates. The advantage is the full information on phase and amplitude, which reduces relative errors in the signal-to-noise ratio.

The OL mode of the applied GPS instrument provides sampling rates up to 1000 Hz, which enables investigation of spectral signatures that are normally not seen in GPS data [2, 3]. The use of directive antennas pointed towards the horizon give signal recordings down to the lowest layers of the atmosphere.

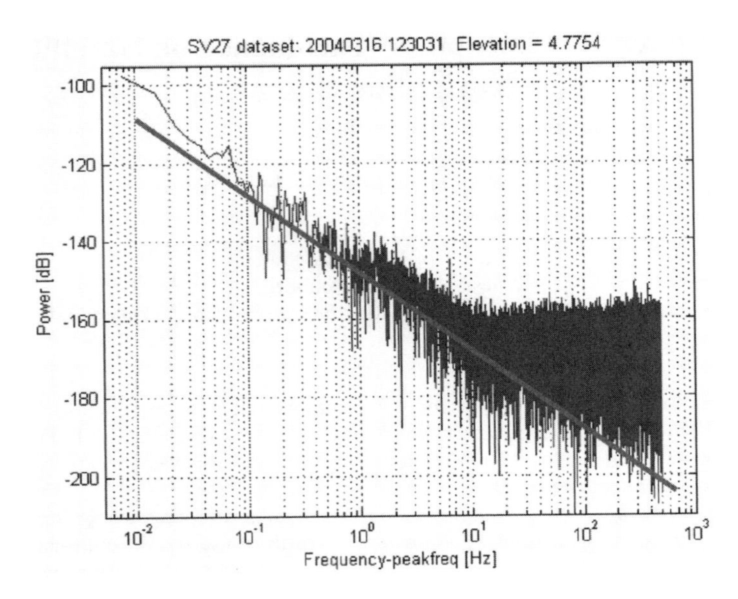

Fig. 1. Spectral characteristics for turbulence conditions.

Figure 1 shows a spectrum for a low-elevation measurement in raw mode sampling (OL) when turbulence is present in the direction towards the transmitting GPS satellite.

The high-frequency part of the signal, for frequencies larger than 100 Hz, is dominated by thermal noise. While the lower frequency part is dominated by clock-noise, which for the receiver rubidium clock falls off as the inverse of the frequency squared. So in order to study atmospheric low-elevation turbulence by spectral analysis, we investigated spectral fluctuations above the noise characteristics of the clock caused by the turbulent atmosphere.

The experiments are from the top of Haleakala at an altitude of more than 2500 meters (Fig. 2). The analysis focuses on observations from the south-west, since most measurements in this geometry of ascending and descending GPS trajectories are close to a vertical plane surface. This geometry leads to faster scanning of the troposphere, which makes it possible to compare results from different altitude regions of the troposphere. Additionally, this sector also turned out to have the lowest multipaths reflections from the nearby islands and objects on the islands.

2 Instrumental Setup

The instrument setup consists of separate L1 and L2 antennas placed right next to each other and oriented with the main gain lobe toward the horizon. The signals are fed into a prototype version of a satellite high precision GPS receiver (Fig. 3), where the instrument software is modified for ground-based signal Doppler conditions. An ultra-stable rubidium frequency reference is used to control the receiver clock for

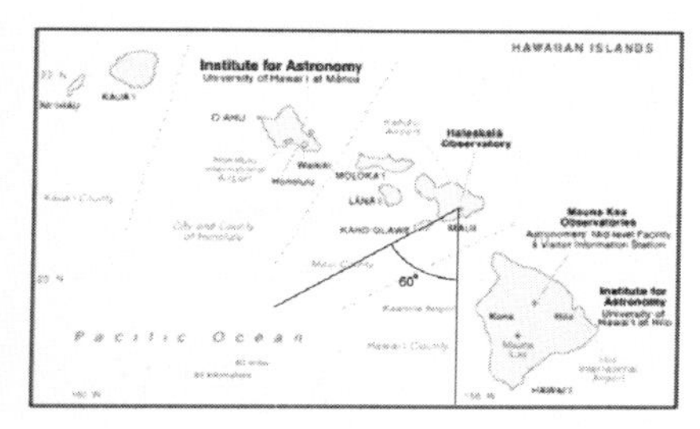

Fig. 2. Left panel is a picture of the field of view for the observations, which for most of the time in the horizontal direction is above the clouds. The adjacent panel gives the directional cone for the observation presented here.

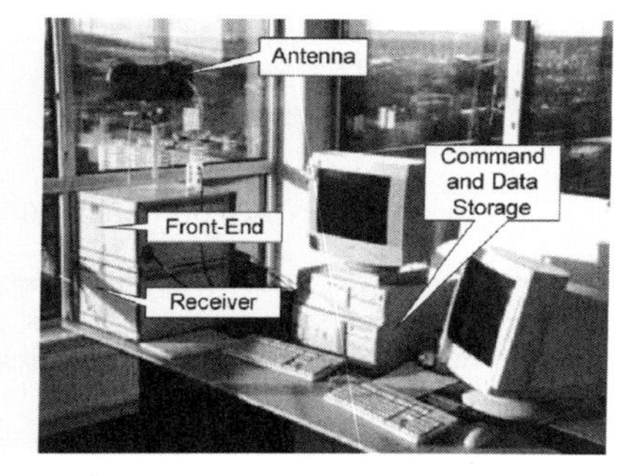

Fig. 3. Left panel shows the applied antenna type. While the picture on the right shows the prototype high precision GPS receiver used for the observations.

precise timing of the measurements. Signals are tracked in both PL and OL mode at the same time in separate receiver channels.

3 Experimental Results

During multi-path conditions, traditional PL tracking receivers may loose signal lock and hence fail to track the signal. During multi-path situations, where rapid phase and amplitude variations occur, signal conditions cause the PL tracking to fail.

The complex signal strength at the antenna is given as,

$$V_{Ant}(t(n,m)) = \sqrt{G_k^{-1}(ch)} \cdot (I(n,m) + jQ(n,m)) \cdot e^{j\varphi(t)}$$

where, t represents the time (at data packet number n and sample number m), $G_k(ch)$ the receiver gain for channel ch at wave number k. I and Q are the quadrature amplitudes, while φ is the phase. For coherent signals (Fig. 4) the I and Q terms give a unique solution. While for non-coherent signals, as is the case for the data presented in this paper, the spectral spread gives a multitude of solutions. PL-based receivers loose lock under these conditions since no unique solution is possible to identify.

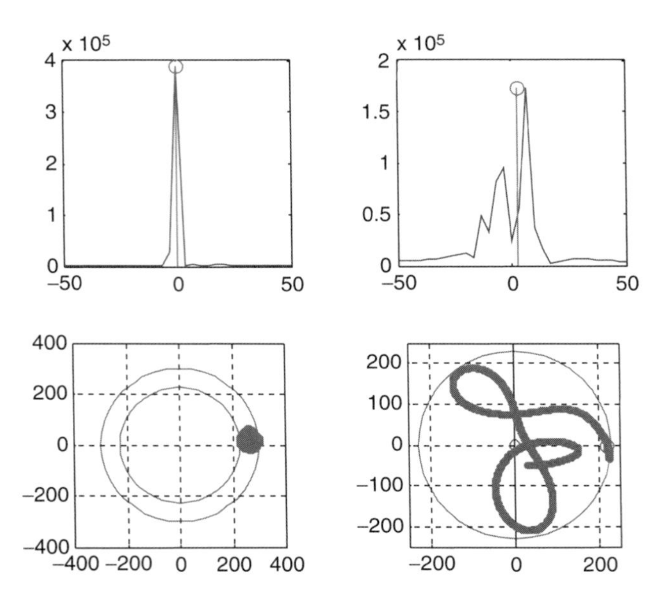

Fig. 4. Power spectra (upper panels) and their respective *I* and *Q* measurements (lower panels) for coherent (left graphs) and non-coherent atmosphere conditions (right graphs). The observations to the right represent the general situation when atmosphere turbulence is present.

The OL data sampling rate of 1000 Hz enables detection and investigation of the characteristics of the noise and the multi-path signal error sources through the determination of the refractive index structure constant C_n^2 of the atmosphere turbulence [4, 5]. The main atmospheric modulation of GPS signals in low-elevation measurements is attenuation and frequency shift due to ray bending, whereas the presence of turbulence is causing a spectral broadening of the signal. Displaying the power spectrum as function of frequency difference from the main signal peak reveals the characteristic domains of the spectrum. Up to 10 Hz, the spectrum is approximately sloping as the inverse of the frequency squared (Fig. 1). While for higher frequencies, in the range 10–500 Hz, the spectrum flattens. The latter part of the spectrum originates from thermal noise, while the first sloping part is characteristic for the rubidium frequency reference used in both the GPS transmitter and the receiver [6]. Analysis of the trend of the mean slope in the spectra for different frequency domains showed an increased slope as function of the elevation of the received signal above the horizon, indicating turbulence and eddies in the beam direction [1].

Here we shall focus on the observations, where meteorological conditions have an important impact on the generation of atmosphere turbulence. Most conditions are driven by either dynamical phenomena (as large vertical winds, horizontal wind shear, mountain lee waves or gravity waves) or ther-modynamically unstable air mass conditions (strong high/low pressure systems, some types of clouds, and lightning). The five time-series chosen here depict the generation of enhanced turbulence and larger inner scale lengths for the dissipative processes of the scintillations during such conditions. The horizontal wavelengths are estimated to range from 0.5 m to 10 km, with vertical extents of 100–2000 m.

The *orange* dataset in Fig. 5 are from a meteorological situation, where the magnitude of the wind is decreasing with low vertical wind components. The gradients in the pressure surfaces are at the same time increasing. During the *red* time series the winds are very variable with high vertical wind speeds and large gradients in the temperature and the pressure. The *blue* and the *red* datasets are during similar atmosphere conditions. The difference is that in the *blue* set lower vertical winds speeds are observed.

The spectral changes as function of elevation angle show a broadening once the turbulent layers are in the field-of-view of the direct signal. The right panel in Fig. 6 shows the enhanced variances from the turbulent region. The turbulent layers show a lot of vertical fine structure with vertical extents smaller than 100 m.

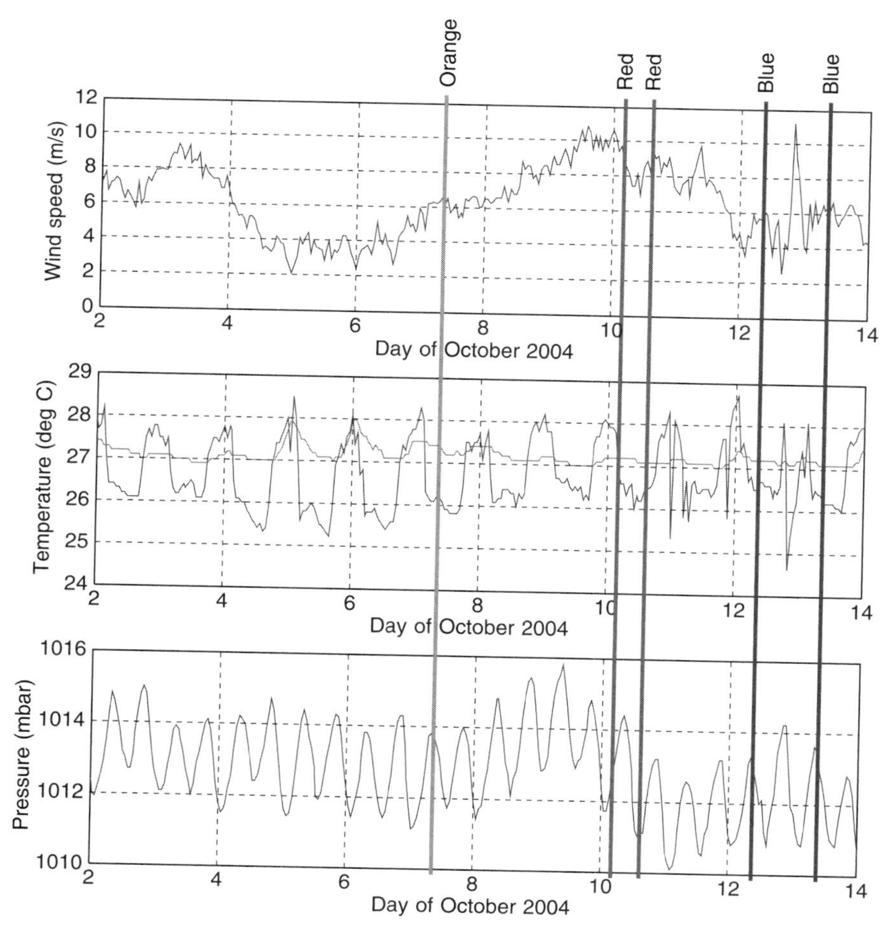

Fig. 5. Wind, temperature and pressure conditions during situations of spectral turbulence for the monitored region of troposphere. The orange, red and blue datasets are marked since they describe meteorological conditions leading to weak and strong turbulence.

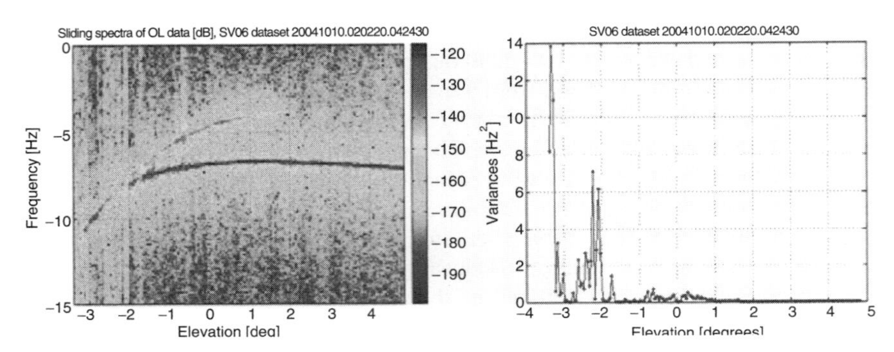

Fig. 6. Stack plot of power spectra in the left panel. Both the direct and the ocean-reflected signal are present. The right panel gives average variances for the turbulent troposphere in the first time series of the *red* datasets.

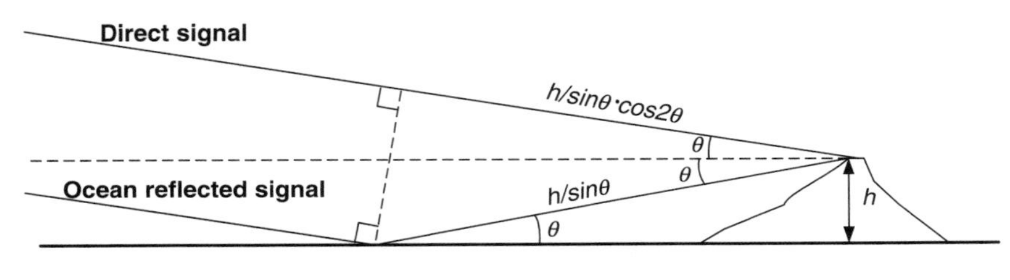

Fig. 7. Schematic presentation of the received signals for low elevation angle measurements.

Apart from the direct signal, having the highest power, the spectra also show the ocean-reflected signal, which can be seen in Fig. 6 for elevation angles lower than 3 degrees. Fig. 7 gives a schematic presentation of the geometry leading to the two main spectral signals.

The prominent characteristic scales of the turbulence region are confined between the outer scale L_0 and the inner scale l_i of the turbulence [4, 5]. For scale lengths larger than the outer scale, eddies introduce kinetic energy into the turbulence region. While eddies smaller than the inner scale remove kinetic energy through dissipative processes. Applying these assumptions leads to the definition of the amplitude variance and the turbulence structure constant [7, 8].

$$\sigma_\chi^2 = \int_0^L C_n^2(x) \left[\frac{\chi(L-\chi)}{L} \right]^{5/6} dx$$

$$C_n^2 = C_{n,0}^2 e^{-\left[\frac{h}{H} + \frac{(x-L_1)^2}{2(R_E+h)H} \right]}$$

$C_{n,0}^2$ is the turbulence structure constant for the given region. L is a characteristic scale length, R_E is the radius of the Earth, h the altitude of the region in the atmosphere, and H the atmospheric scale height for the turbulence region. Assuming that the turbulent layers has a scale size relation as,

$$a^2 = b(R + z_0)$$

leads to the definition of five different regions of turbulence for the presented observations [8]. Here, a is the horizontal extend of the turbulent layer, b the scale height of the structure constant, and z_0 the tangent altitude of the line-of-sight. The next chapter contains an estimation of the regions that specifically defines the observed turbulence.

Figures 8 and 9 show observational time series from the *orange* and the *blue* meteorological conditions. The vertical extent of the turbulence region is quite different from the situations in the *red* datasets. For the *orange* situation (Fig. 8) the weak turbulent layers cover altitudes from above the boundary layer up to altitudes of 5 km with a lot of vertical fine structure. The situation in the Fig. 9 (the *blue* data sets) is mostly related to the troposphere processes in the last two kilometers above the surface of the ocean, leading to less fine vertical structures.

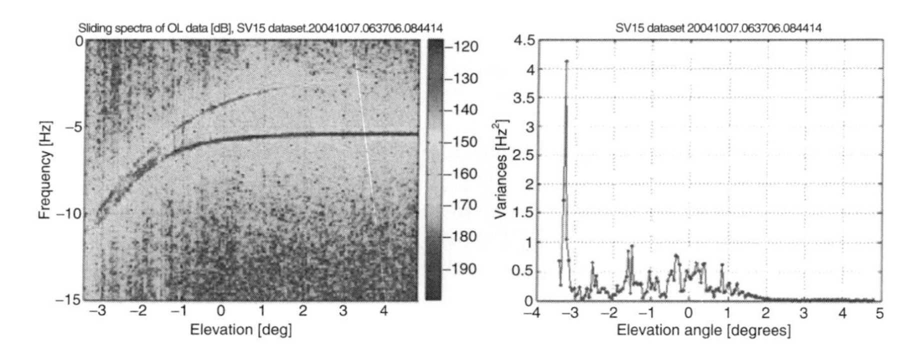

Fig. 8. Power spectra stack plot (left panel) and variances for the turbulent troposphere in the *orange* dataset.

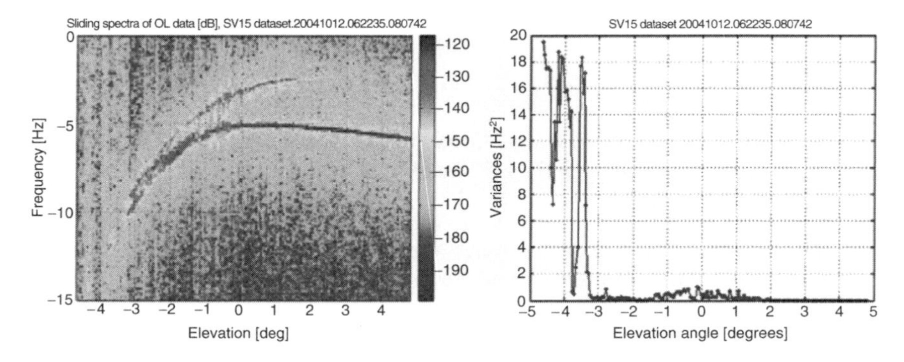

Fig. 9. Power spectra stack plot (left panel) and variances for the turbulent troposphere in the *blue* datasets.

4 Discussions

All the spectral information of turbulence for the presented datasets shows a strong resemblance to the region 2 and 3 variances, defined in [8]. Region 2 and 3 turbulence variances becomes respectively,

$$\sigma_{\chi}^{2} = C_{2} k^{2} L_{0}^{5/3} a C_{n}^{2}$$
$$\sigma_{\chi}^{2} = C_{3} k^{3} L_{0}^{8/3} (ba/L_{1}) C_{n}^{2}$$

for situations when the product of the wavelength and the characteristic turbulence scale height is much larger than the outer scale length squared and much smaller than the product of the scale height of the structure constant and the outer scale length [8, 7, 5]. The conclusions from our measurements follow nicely other similar observational results [9]. They found for 9.6 and 34.5 GHz frequency measurements a power law relation in the power spectra with a mean slope of –8/3, which is similar to the region 3 variances. Our analysis of the statistical estimates of the mean slopes (for the frequency interval 0.1–10 Hz) for our observations gives slopes ranging from –2.8 to –1.5.

5 Conclusions

We have shown that high precision open-loop GPS receivers are capable of determining the troposphere turbulence in tropical regions. The OL observations revealed characteristics of the vertical layering of turbulence. For the strongest cases the vertical fine structure becomes less than 100 m. During such cases it was also identified that standard PL receivers loose signal lock. The dominant cause for the presence of troposphere turbulence is a combination of larger vertical wind components combined with thermodynamically unstable air driven by temperature and pressure gradients. This is also reflected in the spectral variances for the layers, which directly link to the turbulence structure function constant. The statistical analyses of all the measurements lead to region 2 and 3 turbulence with spectral slopes varying from –2.8 to –1.5.

References

[1] L. Olsen, A. Carlström, and P. Høeg, "Ground Based Radio Occultation Measurements Using the GRAS Receiver", ION 17th Sat. Div. Techn. Meeting, Proceedings, ION, pp. 2370–2377, 2004.
[2] P. Høeg, M.S. Lohmann, L. Olsen, H.H. Benzon, and A.S. Nielsen, "Simulations of Scintillation Impacts on the ACE+ Water Vapour Retrieval Using Satellite-to-Satellite Measurements", ESA Atmos. Remote Sensing Symposium, Proceedings, ESA, pp. 148–161, 2003.
[3] P. Høeg, and F. Cuccoli, Measuring Atmosphere Turbulence, Humidity, and Atmospheric Water Content (MATH-AWC). ESA Science Report, ESA, EOP-SM–1297, 2005.
[4] V. I. Tatarskii, The Effects of the Turbulent Atmosphere on Wave Propagation. U.S. Dept. of Commerce, Springfield, USA, 1971.

[5] A. D. Wheelon, Electromagnetic Scintillations. I. Geometrical Optics. Cambridge Univ. Press, Cambridge, 2001.

[6] A. S. Nielsen, M. S. Lohmann, P. Høeg, H.-H. Benzon, A. S. Jensen, T. Kuhn, C. Melsheimer, S. A. Buehler, P. Eriksson, L. Gradinarsky, C. Jiménez, G. Elgered, Characterization of ACE+ LEO-LEO Radio Occultation Measurements. ESA Science Report, ESA, 16743–2, 2003.

[7] A. Ishimaru, Wave Propagation and Scattering in Random Media, Vol. 2. Academic Press, New York, USA, 1978.

[8] R. Woo and A. Ishimaru, "Effects of Turbulence in a Planetary Atmosphere on Radio Occultation", IEEE Transactions on Antennas and Propagation, AP-22, pp. 566–573, 1974.

[9] H. B. Janes, M. C. Thompson and D. Smith, "Tropospheric Noise in Microwave Range-Difference Measurements", IEEE Transactions on Antennas and Propagation, AP-21, No. 2, pp.566–573, 1973.

GIOVE-A SIS Experimentation and Receiver Validation: Laboratory Activities at ESTEC

Massimiliano Spelat[1], Massimo Crisci[2], Martin Hollreiser[2], Marco Falcone[2]

[1]Politecnico di Torino/Electronics Department
C.so Duca degli Abruzzi 24, 10129 Torino, ITALY
Phone: +39(011)2276436, Fax: +39(011)2276299
e-mail: massimiliano.spelat@polito.it
[2]European Space Agency (ESTEC)
Keplerlaan 1, P.O. Box 299, 2200 AG Noordwijk,The Netherlands
e-mail: Marco.Falcone@esa.int, Martin.Hollreiser@esa.int, Massimo.Crisci@esa.int

Abstract. The European Space Agency (ESA) and the Surrey Satellite Technology LTD (SSTL) have completed the on-orbit preparation and activated the payload of GIOVE-A, the first Galileo satellite launched last December, the 28th. After successful launch and platform commissioning achievement, GIOVE-A started signals transmission on 12 January 2006. For the time being the quality of the signal broadcast by GIOVE-A is under examination by mean of sophisticated equipments and facilities, including the ESA ground station in Redu (Belgium) and the Rutherford Appleton Laboratory (RAL) Chilbolton Observatory in the United Kingdom. It is clear that the European Galileo satellite navigation system is moving into a crucial phase concerning the development process; therefore the possibility of testing and validating hardware/software tools (e.g. user receivers) will play a key role from the manufacturers point of view. In this context the navigation laboratory at ESA's European Space Research and Technology Centre (ESTEC), in the Netherlands, could be considered relevant in the receivers validation procedures, as well as in the Signal-In-Space (SIS) experimentation activity, where the GSTB-v2 Experimental Test Receiver (GETR) plays a key role.

The paper will provide the overview of the set-up available in the navigation laboratory at ESTEC, describing the equipments composing the test bench. The Galileo Signal Validation Facility (GSVF-v2) will be presented pointing out the capabilities in the Galileo-like signal generation. In particular, the Galileo L1 Open Service (OS) signal will be analyzed, and the corresponding GETR tracking performance will be presented in terms of code tracking noise curves, autocorrelation function and multipath envelope. Tracking performance for the Galileo L1 OS signal in multipath environments will be evaluated in terms of static and dynamic contributions.

Finally, some screenshots of the GETR graphical user interface (while tracking GIOVE-A signals) will also be included in the paper, as the prove that the entire set-up has been fully integrated with the Space Engineering's Galileo antenna for the reception and process of live GIOVE-A signals.

1 Introduction

The European navigation satellite system Galileo is now becoming a reality entering the so called *Galileo System Test Bed – phase2* (GSTB-v2) development phase, where live radio-navigation signals are broadcast worldwide by the first of the two test satellites, GIOVE-A [1]. This name stands for *Galileo In-Orbit Validation Element* being part of the future IOV constellation, which will be constituted by 4 satellites. Formerly called GSTB-v2, the GIOVE mission has to secure the frequencies allocated for the Galileo system, characterise the radiation environment of the orbits (the *Medium Earth Orbit* – MEO environment), confirm technologies for the navigation payloads architecture of future operational Galileo satellites and perform the *Signal-In-Space* (SIS) experimentation, where the *GSTB-v2 Experimental Test Receiver* (GETR) developed by Septentrio plays a key role [2]. GIOVE-A has been placed in orbit by a Soyuz-Fregat rocket operated by Starsem on 28 December 2005 from the Baikonur Cosmodrome, with the aim of representing the starting point of the Galileo In-Orbit Validation phase. GIOVE-A has been transmitting Galileo-like signals from the beginning of January 2006, carrying a payload is able to generate and transmit the nominal GALILEO L1, E6 and E5 modulations and multiplexing schemes including the *Binary Offset Carrier* (BOC) modulated signals (e.g. BOC(15,2.5) and BOC(1,1)) as well as the wideband *AlternativeBOC* (AltBOC) modulation [3]. This 600 kg satellite, built by *Surrey Satellite Technology Ltd* (SSTL) of Guildford in the United Kingdom, carries two redundant, small-size rubidium atomic clocks, each with a stability of 10 nanoseconds per day, and two signal generation units, one able to generate a simple Galileo signal and the nominal one, more representative in terms of Galileo signals. These two signals are broadcast through an L-band phased-array antenna designed to cover all of the visible Earth under the satellite. Finally, two on-board instruments are monitoring the types of radiation to which the satellite is exposed during its two year mission.

For the time being the quality of the signals broadcast by GIOVE-A is monitored (continuosly or during specific measurement campaign) by means of several facilities, including the *Rutherford Appleton Laboratory* (RAL) Chilbolton Observatory in the United Kingdom, the *European Space Agency* (ESA) ground station at Redu, in Belgium, and the Navigation Laboratory at ESA's *European Space Research and Technology Centre* (ESTEC), in the Netherlands. As far as the SIS experimentation activity is concerned, it is meant to provide supporting data for the frequency filing, characterise the performance of the Galileo SIS and confirm the GETR performance and consolidate the receiver design. Given the nature of the measurements required by such an activity, the SIS experimentation phase has been designed to be carried out using both the Chilbolton 25m-diameter antenna with high gain, and the L1-E6-E5 Galileo Reference Antenna developed by Space Engineering. In order to prepare and fully support the activities, a dedicated test bench has been set-up in the Navigation Laboratory at ESTEC. The paper gives an overview of this test bench, considering that the procurement of hardware and software has been driven by the necessity to reproduce the SIS experimentation environments for the GETR before the real signal in space was available. The content of the paper is focused on the analysis of results provided by the receiver processing the L1 Opens Service (OS) signal. In particular, results are presented identifying two different phases:

- **Phase1,** functional verification and validation of the equipments employed in the test bench
- **Phase2,** results on the GETR performance validation

The verification and validation of the laboratory setup (**Phase1**) consists in testing the equipments with the aim of getting results that have to match the expected ones (theoretical or simulated). For such a reason both the *Galileo Signal Validation Facility* (GSVF-v2) [4] and the GETR have been carefully tested, and some results in terms of spectrum of the signals, correlation functions, multipath envelopes and code tracking noise curves are presented. The GETR performance validation (**Phase2**) consists in testing the receiver performance under GSVF-2 simulated environment conditions (multipath and interference contributions). The paper analyses the GETR in terms of code tracking error (standard deviation and bias) describing as an example the following user environments: *Rural Vehicle* (RV), *Rural Pedestrian* (RP) and fixed scenarios.

The paper presents also the reception of 'live' GIOVE-A signals. Therefore screenshots of the GETR are included in the paper proving the functionality.

2 Test Bench Description

Considerable amount of data will be collected during this phase in order to confirm and steer the design development of the Galileo program segments. It is clear that a reliable understanding of the results analysis can be achieved only if consolidated reference performance results, obtained in realistic and controlled environments, are available. In this way the isolation of each error contribution becomes possible, and it is useful to prevent possible malfunctioning being able to test the receiver in exhaustive number of environment conditions. That is the idea behind the realisation of the test bench shown in Fig. 1, where both the "Real-Time" and "Post-Processing" branches allow for the acquisition and analysis of GIOVE-A signals, whether they are coming from the satellite or the Galileo RF signal generator.

The Galileo Signal Validation Facility together with the Spirent GPS constellation simulator [5] have been integrated with the early prototype of the Galileo receiver, the GETR, being able to acquire and track the ad-hoc generated SIS and the "live" signal coming from the Galileo User Antenna capable of operating also in the GPS bands (L1, L2 and L5). The quality of these signals can be checked in real-time by means of sophisticated real-time analysis tools such as the spectrum analyser and the digital oscilloscope, or in post-processing storing samples of the signal using the bitgrabber (a flexible digitaliser in terms of sampling frequency and quantisation) jointly with specific software tools.

Considering the equipment shown in Fig. 1, a Matlab®-based tool has been developed in order to process and analyse raw output data of the GETR, being able to characterise the list parameters defined for the SIS experimentation activity. The tool has been called *GETRdat* and was used whether during the functional verification and validation of the equipments or in the GETR performance analysis and validation. An introduction on the architecture of the tool is given in 3, pointing out the capabilities concerning in terms of data analysis.

2.1 Galileo Signal Validation Facility

The Galileo Signal Validation Facility, the GSVF-v2, has been developed by *Thales Research and Technology UK* (TRT-UK) for ESA representing the reference in the

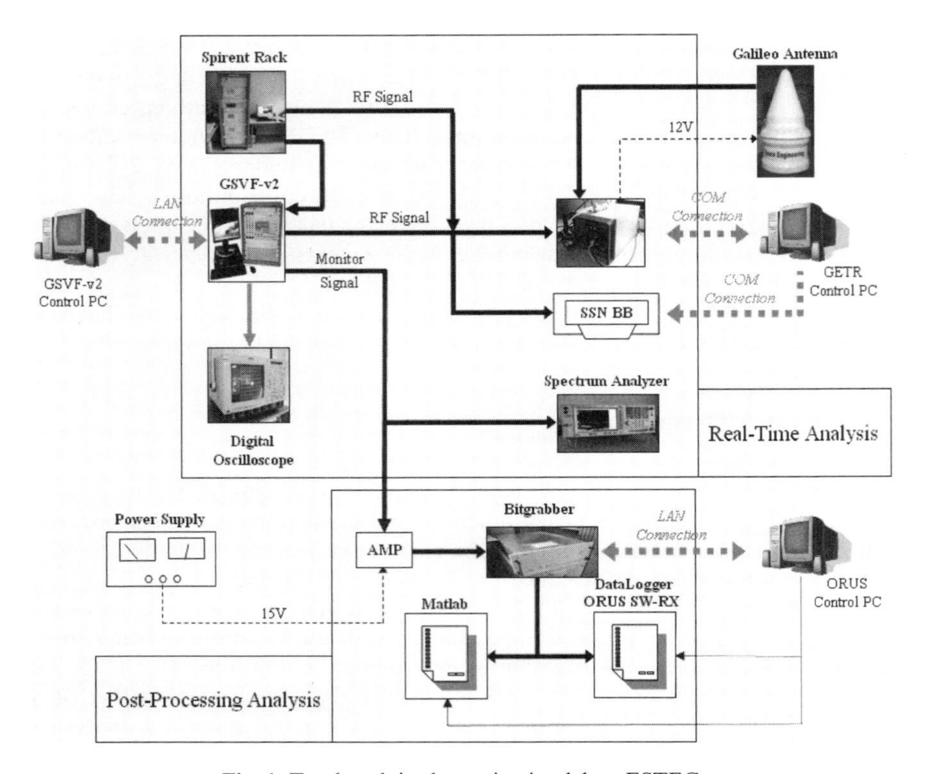

Fig. 1. Test bench in the navigation lab at ESTEC.

field of signal generators for Galileo. The GSVF-v2 constellation RF simulator shown in Fig. 2 is capable of generating a single composite L-band signal fully representative for all three Galileo frequencies from all satellites in view of the user. As it is possible to see in Fig. 2, the simulator consists of the *Control PC* (CPC) and a 19" rack, where three embedded PCs are installed for the 50 Hz pseudorange computations, antenna, clock and high-fidelity multipath modelling, and navigation data elaboration and formatting. Each PC drives a dedicated FPGA-based baseband board responsible for generation of code, phase, gain and navigation data updates at 50 Hz. Each baseband board can generate a single baseband signal for up to 16 satellites (channels) [4].

A generic channel implements models for code and chip modulations generation, Doppler and amplitude variations as well as for the *High Power Amplifier* (HPA) distortion. Moreover, each channel has the possibility of generating fading, shadowing and multipath delay on top of the *Line-Of-Sight* (LOS) signal. Finally, the RF signal is generated by the RF up-conversion module, which combines the three baseband signal to L-band with the possibility of mixing external antenna or interferer signals.

Figure 3 shows the *Graphical User Interface* (GUI) of the simulator, which provides the user with an high level of flexibility in configuring the models for the Galileo signal generation. The simulator is fully compliant against Galileo and

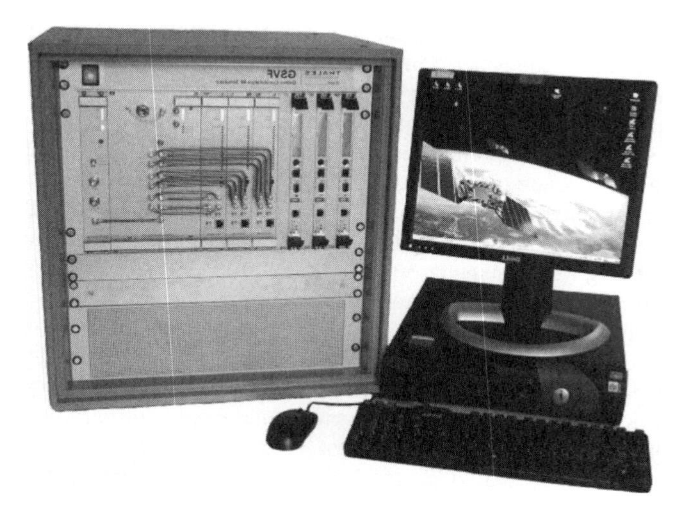

Fig. 2. The Galileo signal validation facility.

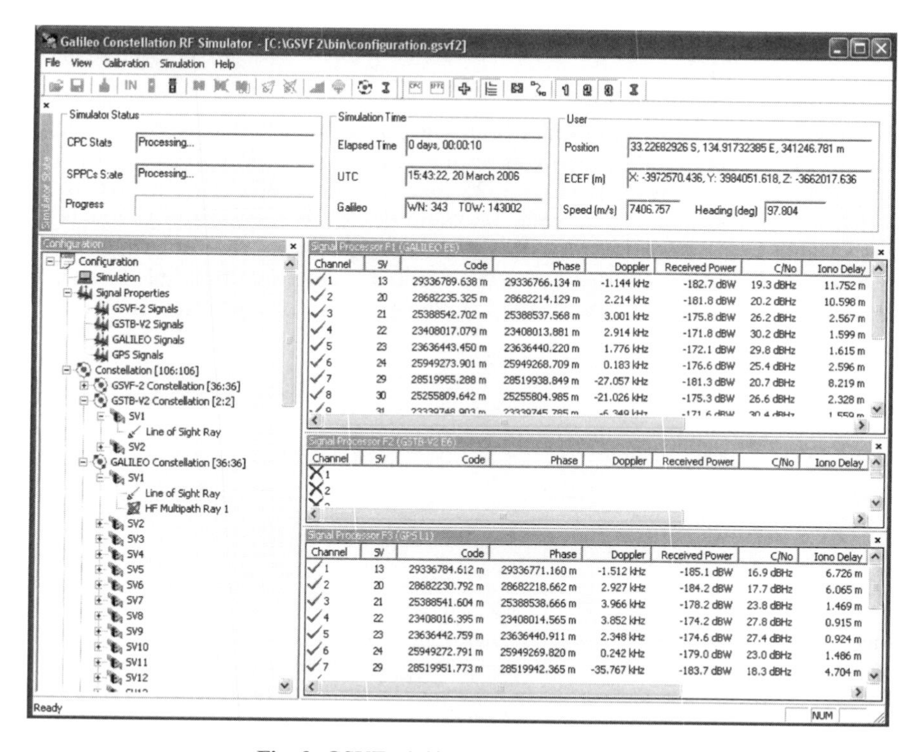

Fig. 3. GSVF-v2 Simulator user interface.

GSTB-v2 SIS *Interface Control Documents* (ICD) for those parts of the specifications which would provide benefits in terms of receiver testing and validation, waiting for the consolidation of the Galileo C/NAV and I/NAV navigation message definition (see Table 1). Anyway, although the implementation of both the C/NAV and I/NAV data streams are not finalised yet, the simulator message contains random data with the appropriate CRC, FEC and interleaving implemented. The GSVF-v2 does not support generation of nominal *Public Regulated Services* (PRS) signals in terms of spreading codes and navigation messages. Since these signals are classified, the simulator emulates them using very long random codes with the correct chip rate and random bits representing the modulated data message [4].

It is important to remark that the GSVF-v2 environment models are fully inline with the Galileo reference specification documents, including those for the troposphere, ionosphere and multipath. As far as the multipath is concerned, elevation and azimuth-dependent models are supported generating multipath ray delay, phase shift and relative amplitude on the basis of actual site survey data. In addition to these geometrical models, the GSVF-v2 also implements linearly varying periodic multipath ray characteristics, as well as the possibility of reading these characteristics from a file as a time-based series. Since the flexibility in modelling the multipath environment is one of the most important feature, the simulator provides two classes of multipath ray generation:

- High-fidelity, quantising the generated multipath delay in 11.1 ns steps
- Low-fidelity ray, using dedicated hardware resources without restriction in terms of ray modelling

Four low-fidelity rays for each space vehicle are always supported by the simulator, while the maximum configuration using high-fidelity rays is represented by a 47 high-fidelity rays on top of a single LOS (because the number of channels in the simulator is limited to 48, 16 for each baseband board). Rayleigh fading models can be applied to the multipath ray specifying the fading bandwidth in the range 1mHz – 2.4 kHz. Finally, the LOS supports Rician fading settings (configurable mean state duration) for both the "good" or "bad" states controlled by a Markov model.

Table 1. GSVF-v2 signal compliance.

System specifications	E5			
	E5a	E5b	E6	L1
GSVF-v2 Iss.4 Galileo Iss.11 Rev.2	Full signal And F/NAV compliance	Full compliance Full signal compliance. Compliant navigation message framing and encoding. E6-A and L1-A PRS codes emulated with random codes.		
GSTB-v2 Iss.2 Rev.2	Full signal and OS compliance	Full signal compliance. Compliant navigation message framing and encoding		

Fig. 4. GSTB-v2 experimental test receiver.

2.2 GSTB-v2 Experimentation Test Receiver

The GSTB-v2 Experimental Test Receiver is an all-in-view dual-frequency GPS receiver, which can simultaneously track up to 7 GSTB-v2 and/or Galileo signals. Fig. 4 shows the GETR, where tracking of Galileo signals as well as of GPS C/A Code is implemented in baseband modules using FPGA technology. The internal architecture of the GETR foresees the presence of the GPS dual-frequency Polarx2 receiver that working in parallel with the GSTB-v2/GPS C/A chipset, allows the synchronisation with the GPS signal. Anyway, the GPS L1 signal is processed by the same front-end and digital logic as the Galileo/GSTB-v2 signal in order to avoid any inter-system bias. As far as the compliance with the GSTB-v2 and Galileo signals is concerned, the GETR is fully representative in terms of modulations, chip length, chip rate and *Binary Offset Carrier* (BOC) sub-modulation, also for the BOC(15,2.5) and BOC(10,5) PRS signals. For each BOC modulation, the GETR supports both the sine and cosine type, with default setting as specified in the SIS ICDs. Six independent channels can be allocated to acquire and track any Galileo/GPS signal, apart form the *Alternative BOC* (AltBOC) modulation; since the AltBOC carries two different data streams, a dedicated channel has been implemented with some shrewdness for the internal architecture point of view [2].

The GETR is capable of producing and storing different type of data. In particular:

- Raw data (code phase, carrier phase, Doppler, C/No, etc. . .)
- Navigation data (message, CRC, Interleaving, etc. . .)
- IF samples of the received signal
- Samples of the autocorrelation function

The *GETRdat* tool has been developed on the basis of these output, being able of testing the receiver functionalities and performance.

2.3 Other Equipments

Apart from the Galileo Signal Validation Facility and the GETR, the test bench shown in Fig. 1 presents other hardware tools playing a key role in the real/simulated GIOVE-A signal, even if they are not used to perform the tests described in this paper. First of all, the set-up includes the Space Engineering Galileo/GPS antenna, see Fig. 5. This is a *Right Hand Circular Polarised* (RHCP) antenna able to receive L1, L2, L5, E6 and E5 signals and to amplify them by means of the internal LNA with 27 dB gain (considering connectors and cables). Another important tool to be cited is the Spirent GPS/Glonass Constellation simulator. It is capable of generating L1, L2 and L5 signals fully in-line with the SIS ICDs [5]. The role of this tool is mainly crucial for the evaluation of intersystem interference, as well as for providing GPS time synchronisation to the GETR.

Considering the diagram of Fig. 1, the section called "Real-Time" analysis is completed by the spectrum analyser and the digital oscilloscope, providing the user with high flexibility for SIS analysis (e.g. examining the spectrum of the GSVF-simulated signal in terms of shape and relative power sharing between different channels in the same band, for instance L1 A,B and C). In particular, the test bench has been equipped with an Agilent E4448A PSA spectrum analyser with 3 Hz – 50 GHz bandwidth, and a LeCroy LC334AM digital oscilloscope with a variable sample rate in the range 2Gsamples/500 Msamples per second depending on the number of channels used.

Finally, the "Post-Processing" section of the test bench is represented by the bitgrabber with some ad-hoc-developed software tools, for instance the *GETRdat* tool. The bitgrabber is a quantiser, being able to sample the RF signal at the input. It guarantees flexibility in the sampling procedure, with the possibility of setting the sampling frequency up to 250 MHz and the quantisation up to 10 bit. It produces a binary output file containing the stream of bits representing the samples of the quantised signal.

It is important to remark that, the software tools designed for the integration with the test bench equipment have been implemented in Matlab®, with the aim of communicating with both the GETR and the bitgrabber output files.

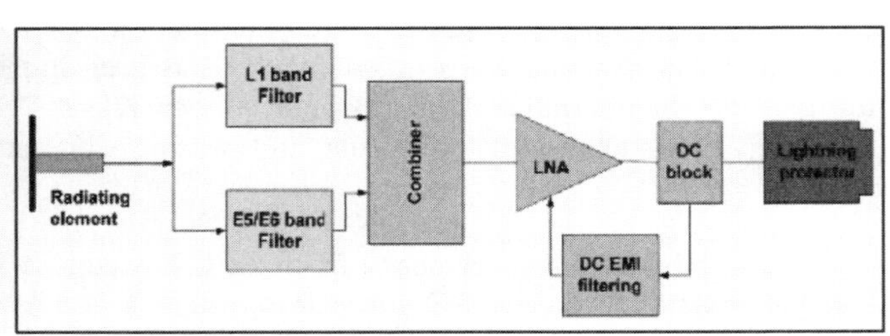

Fig. 5. Space engineering Galileo/GPS antenna.

3 Getrdat Tool

In order to be able to accomplish both Phase 1 and Phase2 described in section 1, which consist in the functional verification and validation of the equipments employed in the test bench as well as the analysis of results on the GETR performance validation, a really flexible software tool has been developed using Matlab® platform. It is also important to remark that the final intent in developing such a tool is to provide a generic user with the possibility of analysing the real GIOVE-A signals in the context of the SIS experimentation activity.

As shown in Fig. 6, the *GETRdat* tool contains 4 different sub-tools:

- *Dual Channel Tool*, to analyse the raw data files generated by the GETR
- *IF Samples Tool*, to process the IF samples of the signal whether they are stored using the GETR front-end or the bitgrabber
- *Correlation Data Tool*, to reconstruct the correlation function by means of processing correlation samples generated by the GETR
- *Navigation Data Tool*, to analyze the navigation data demodulated and stored in the GETR output file

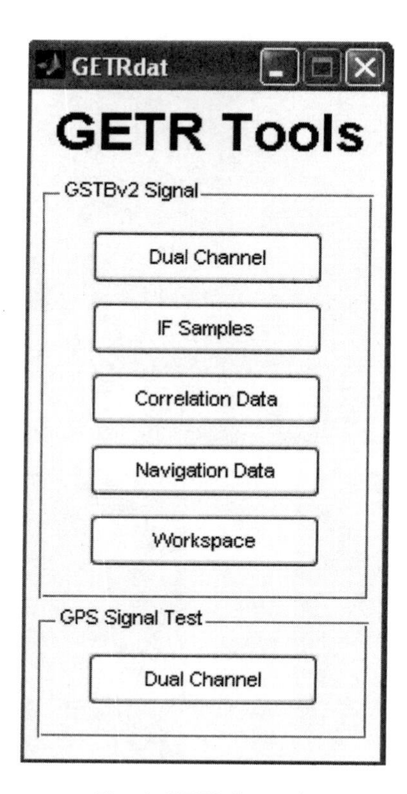

Fig. 6. GETRdat tool.

These sub-tools have been grouped in two sections, on the basis of the signal type (GPS or GSTBV-2) they are suppose to be used with.

4 Functional Verification and Validation of the Equipments

As it has been identified in Section 1, the first step in the validation of the test bench is the functional verification of the equipments involved in the integration. This section provides an overview of the set-up validation, mainly focusing the attention on the OS signals with particular reference to the L1 BOC(1,1). The set of results presented are related to GSVF-generated signals, being able to create a controlled environment surrounding the user and the signal path. Fig. 7 shows the spectrum of the signal generated by means of the GSVF-v2, where the components on the Galileo L1 band are clearly visible. The upper part of Fig. 7 shows the L1 A, B and C channels composition, implementing the CASM modulation between the nominal BOC(15,2.5)c and BOC(1,1) for the PRS and the OS signals respectively. The bottom part of the figure presents these nominal sub-modulation independently, using BPSK or QPSK scheme for the generation.

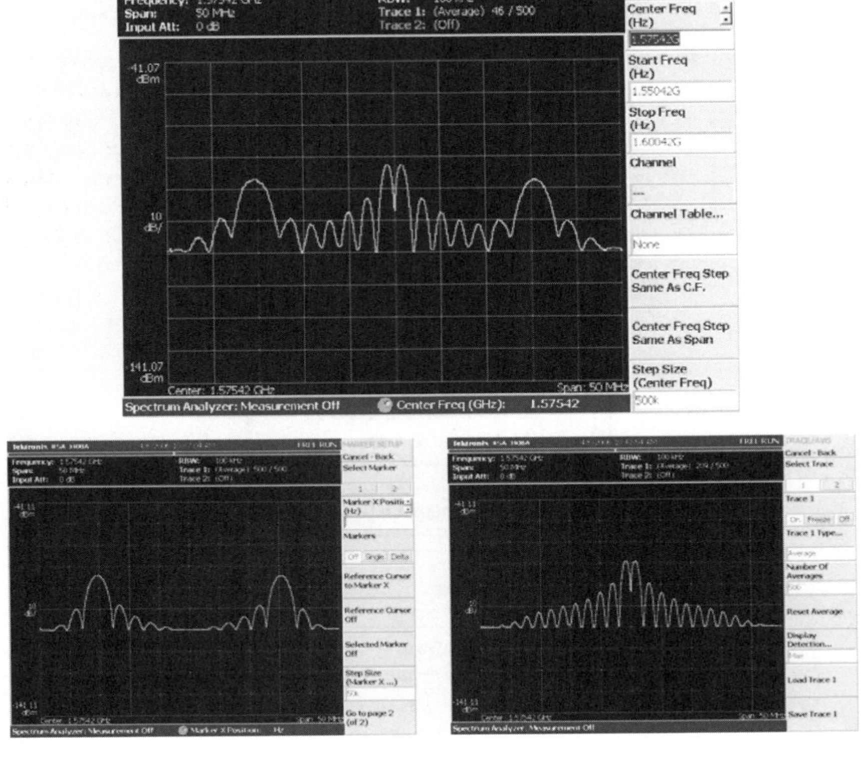

Fig. 7. GSVF-v2 spectrum of simulated GIOVE-A L1 signal.

The GSVF-generated L1 signal has been directly sent to the RF input connector of the GETR, being able of testing and validating some aspects of the receiver. First of all, the code noise tracking error curves for all the nominal GIOVE-A signals has been evaluated analysing the code-carrier phase measurement provided by the GETR. Nevertheless, the paper presents results only in the case of BOC(1,1) signal since it represents the nominal modulation foreseen for L1 OS signals. The signal has been generated sweeping the signal-to-noise ratio at the input of the GETR in the range 29–50 dBHz, considering a clean environment from both the user and signal path points of view. In such a way, it is possible to isolate the error on the code tracking due to the noise. Fig. 8 shows the results for the L1 pilot and data channels, comparing the obtained curves with the theoretical one. In this figure the theoretical curves are derived form the following formula, which represents the variance of the code noise error (expressed in m^2) [6][7]:

$$\sigma_{DLL}^2 = T_C^2 \frac{B_L \int_{-\beta_r/2}^{\beta_r/2} G(f)\sin^2(\pi f \Delta)\,df}{\frac{C}{N_0}\left(2\pi \int_{-\beta_r/2}^{\beta_r/2} fG(f)\sin(\pi f \Delta)\,df\right)^2} \times \left[1 + \frac{\int_{-\beta_r/2}^{\beta_r/2} G(f)\cos^2(\pi f \Delta)\,df}{T\frac{C}{N_0}\left(\int_{-\beta_r/2}^{\beta_r/2} G(f)\cos(\pi f \Delta)\,df\right)^2}\right]$$

where T_c is chip duration, B_L is the DLL noise bandwidth, $G(f)$ is the spectrum of the signal, C/N_0 is the carrier-to-noise ratio and T is the DLL predetection time.

The second part of the functionalities verification and validation of the equipments has been devoted to the analysis of the autocorrelation function at the output of the GETR's correlators. The stream of correlation samples have been collected and processed by means of the *GETRdat* tool, with the aim of reconstructing the shape of the function checking for possible asymmetries. The left side of Figs. 9 and 10 point out the shape of the correlation function for both the BOC(15,2.5)c and BOC(1,1) modulations considering high C/No scenario (around 50 dBHz).

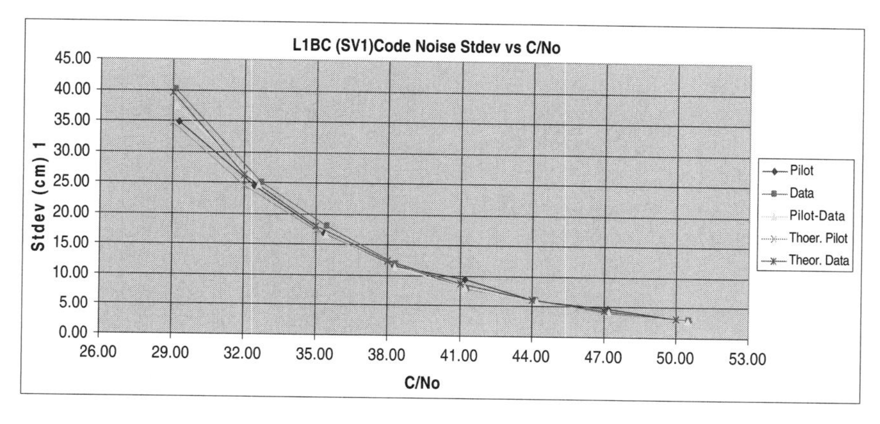

Fig. 8. L1BC code tracking noise error (GETR and theoretical).

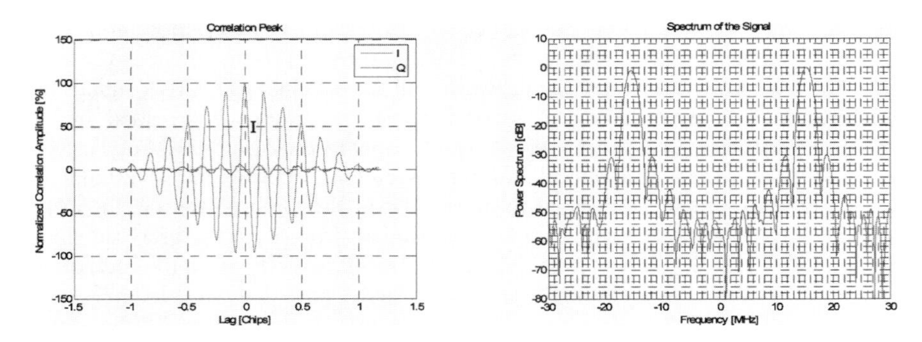

Fig. 9. BOC(15,2.5)c Correlation function and spectrum of the signal.

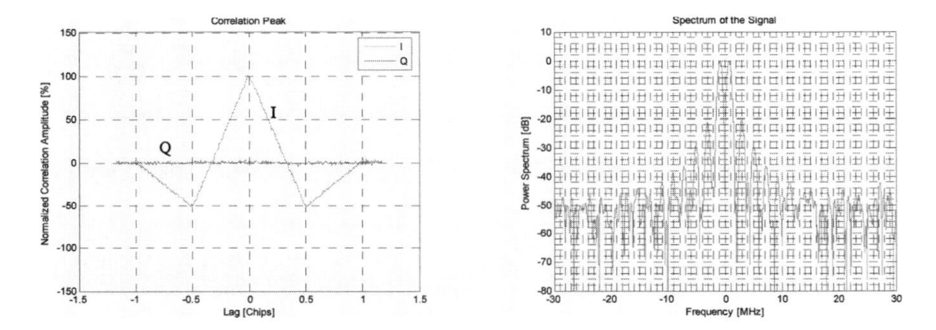

Fig. 10. BOC(1,1) Correlation function and spectrum of the signal.

The plots do not present any relevant asymmetry in the lobes amplitude, which could results in fixed bias on the pseudorange computation. Figures 9 and 10 show also the spectrum of both the modulations, derived through post processing operations on the correlation functions. The main lobes of the spectrum are clearly recognisable, with the correct distance from the central frequency (identified with 0 MHz) and nominal bandwidth.

Finally, the GETR functionalities under multipath conditions have been analysed and validated in terms of multipath envelope. The shape of the envelope derived directly from the GETR for the nominal signals (except the PRS ones) are shown in Fig. 11, where the zoom on the first 80 meters of ray delay with respect to the LOS is also presented in the case of multipath-to-signal ratio of 6 dB. Code phase error amplitudes versus multipath ray delay are in line with the expectations.

5 GETR Performance Results

This section describes results about the characterization of the receiver performance in presence of multipath. Only the tracking performance has been analyzed since it is considered the first and most important step towards the assessment of the

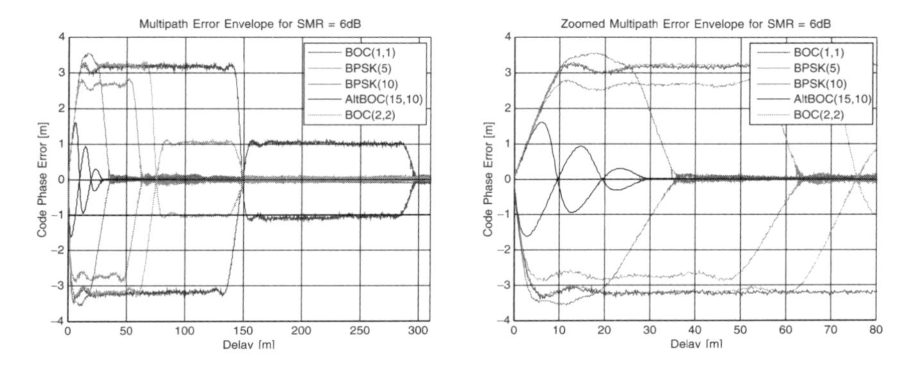

Fig. 11. GETR multipath envelope, SMR = 6 dB.

positioning performance. As a first approximation, the position performance can be simply derived including the satellite to user geometry, and projecting in the position domain the standard deviation of the error in the range domain. Thus the starting point becomes the ranging accuracy. Only multipath and noise contributions are considered for this analysis. No other errors (ionosphere, troposphere, interference, etc.) have been included. The measurement are performed in a high (around 50 dBHz) and low (around 34 dBHz) C/N0 conditions. Only L1 BOC(1,1) signal performances are shown.

To better understand and evaluate the obtained results in terms of static and dynamic errors, the added multipath is generated according to the model of Fig. 12. The GSVF-v2 multipath model that has been used to perform the test includes a Direct Path (Shadowing and Fading) and multipath (Fading), where the

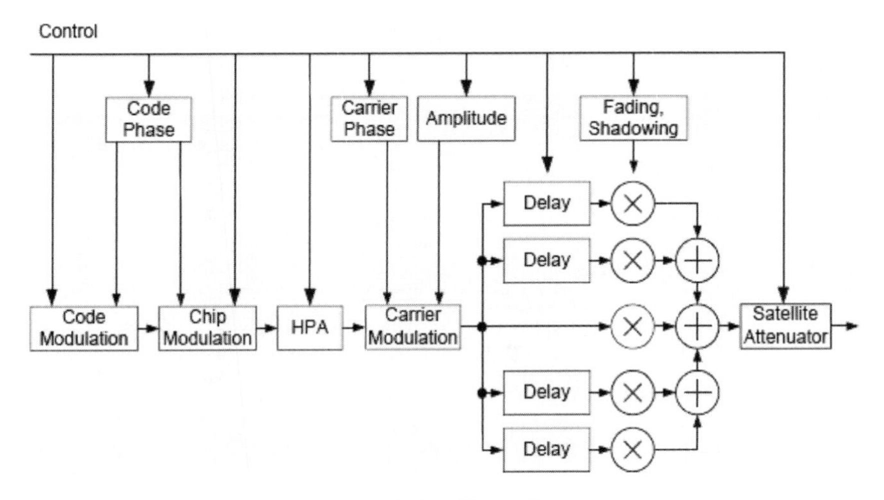

Fig. 12. Satellite path.

computation of the complex path scale factor to implement the fading and shadowing is according to

$$A = G \cdot \frac{K + \frac{X + jY}{\sqrt{2}}}{\sqrt{1 + K^2}} \cdot e^{\frac{S}{20 \cdot \log_{10}(2)} \cdot Z} \cdot e^{j\Phi}$$

where A is the complex scale factor, $X/Y/Z$ are band-limited unit variance Gaussian noise values, G represents the required path gain (in linear units), K is the Rice factor (in linear units), S is the standard deviation of the shadowing in dB and Φ is an additive phase for static multipath.

The model consists of 4 multipath rays with fixed delays and randomly varying phase added to the line of sight. The mean power relative to the line of sight of the Rayleigh fading is specified in Table 2. Rician fading is also applied to the line of sight. The delay, the Rician fading factor and the Bandwidth of the fading process are also specified in Table 2.

The parameters of the model have been selected according to Table 2, in line with Galileo Reference multipath model. The values selected constitute only one of the possible infinite choices, and they are considered representative of an average multipath scenario for the indicated type of environment (rural) and user category (pedestrian, fixed and vehicle).

Table 2. GSVF-v2 Multipath model configuration.

Rural Pedestrian	LOS ray
Enable LOS Ray	YES
Mean Time in Good State (s)	1.0
Enable Good State Fading	YES
Rician Fading Factor (dB)	13.71
Enable Bad State Fading	NO
Fading Bandwidth (Hz)	4

	Relative power (dB)	Path delay (ns)	Dynamic bandwidth (Hz)
Ray 1	−13.91	20	4
Ray 2	−14.11	40	4
Ray 3	−14.46	75	4
Ray 4	−14.86	115	4

Rural Vehicle	LOS ray
Enable LOS Ray	YES
Mean Time in Good State (s)	1.0
Enable Good State Fading	YES
Rician Fading Factor (dB)	13.71
Enable Bad State Fading	NO
Fading Bandwidth (Hz)	140

	Relative power (dB)	Path delay (ns)	Dynamic bandwidth (Hz)
Ray 1	−13.91	20	140
Ray 2	−14.11	40	140
Ray 3	−14.46	75	140
Ray 4	−14.86	115	140

Fixed	LOS ray
Enable LOS Ray	YES
Mean Time in Good State (s)	1.0
Enable Good State Fading	YES
Rician Fading Factor (dB)	13.71
Enable Bad State Fading	NO
Fading Bandwidth (Hz)	0.0025 (0.1)

	Relative power (dB)	Path delay (ns)	Dynamic bandwidth (Hz)
Ray 1	−13.91	20	0.0025 (0.1)
Ray 2	−14.11	40	0.0025 (0.1)
Ray 3	−14.46	75	0.0025 (0.1)
Ray 4	−14.86	115	0.0025 (0.1)

The GETR was configured with a PPL and DLL bandwidth of 10 Hz and 0.25 Hz respectively, while the predetection time was 10 ms for the PLL and 100 ms for the DLL. A non-coherent dot-product power discriminator for the code phase tracking:

$$\Delta\tau = I_{E-L} I_P + Q_{E-L} Q_P$$

where I_E, I_P, I_L represent the early, punctual and late replicas of the in-phase correlators output, while Q_E, Q_P, Q_L are the output of the correlators on the quadrature branch. Table 3 summarises the characterisation of the multipath contribution on L1 BOC(1,1) signal. The static contribution on the tracking error is calculated as the difference between the code measurements of two cloned satellites; one affected by multipath the other one in a multipath-free scenario. In fact in order to accurately isolate the MP contribution (bias plus standard deviation) from the code phase, a cloned satellite is generated (same orbit, same clock etc) but with a different PRN

Table 3. Multipath contribution on L1 BOC(1,1) signal.

Rural pedestrian			C/No ~ 34 dBHz
SV 1	PRN 1	No MP Model	Std = 19.59 cm
SV 2	PRN 2	Yes MP Model	std = 40.20 cm
			bias = 50 cm
Rural vehicle			C/No ~ 34 dBHz
SV 1	PRN 1	No MP Model	std = 25.14 cm
SV 2	PRN 2	Yes MP Model	std = 26.41 cm
			bias = 1 cm
Fixed			C/No ~ 34 dBHz
SV 1	PRN 1	No MP Model	std = 17.5 cm
SV 2	PRN 2	Yes MP Model	std = 166.58 cm
			bias = 41 cm

Rural pedestrian			C/No ~ 50 dBHz
SV 1	PRN 1	No MP Model	std = 3.85 cm
SV 2	PRN 2	Yes MP Model	std = 34.07 cm
			bias = 51 cm
Rural vehicle			C/No ~ 50 dBHz
SV 1	PRN 1	No MP Model	std = 5.98 cm
SV 2	PRN 2	Yes MP Model	std = 6.96 cm
			bias = 2 cm
Fixed			C/No ~ 50 dBHz
SV 1	PRN 1	No MP Model	std = 3.27 cm
SV 2	PRN 2	Yes MP Model	std = 162.5 cm
			bias = 42 cm

and without applying the multipath model; in other words, this space vehicle is used as a reference. Both standard deviation and bias can then be computed.

The results show the impact of multipath for the scenarios considered, RV, RP and fixed user. The dynamic of the user as well as the bandwidth of the fading determine the presence and the amount of the bias in the error, which cannot be filtered or averaged away. These preliminary results are also in line with the expectations, as specified in the requirements for the L1 OS signal in case of both the rural pedestrian, rural vehicle and fixed.

6 GIOVE-A SIS Acquisition and Tracking

As it has been introduced already in Section 2, the test bench set-up was finally integrated with the Galileo user antenna, being able of tracking real GIOVE-A and GPS signals. Fig. 13 shows an example of tracking, while the GETR was receiving and processing real signals from GIOVE A using the Space Engineering Antenna. The final results of such an integration is making comparison between real and simulated data possible, as well as validating the reference models (e.g multipath for fixed users).

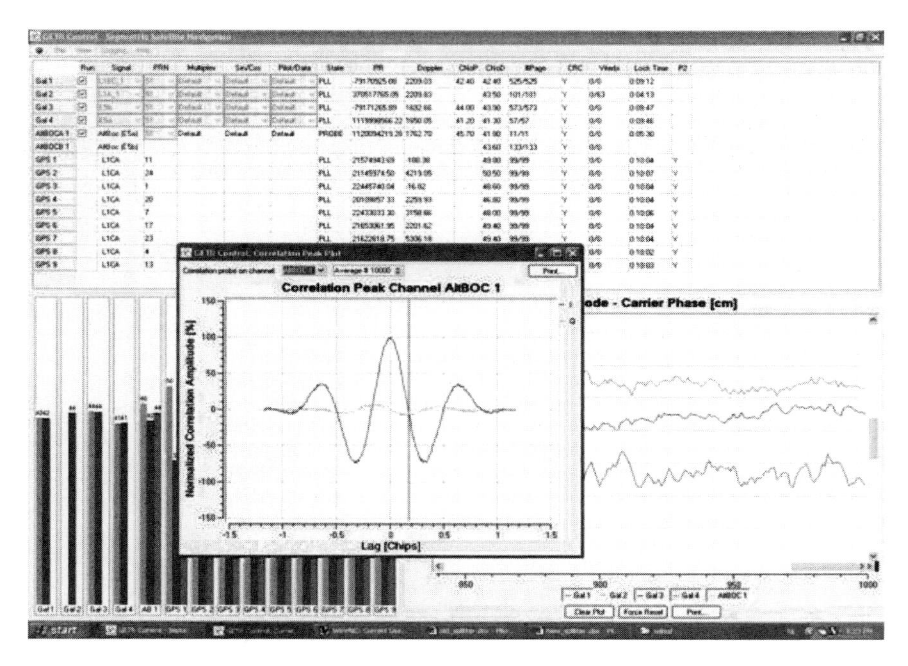

Fig. 13. GETR operating with real GIOVE-A SIS.

References

[1] http://www.esa.int/esaNA/galileo.html
[2] A. Simsky, J. Sleewaegen, W. De Wilde, F. Wilms, "Overview of Septentrio's Galileo Receiver Development Strategy", *ION-GNSS (Long Beach)*, September 2005.
[3] J. W. Betz, "Binary Offset Carrier Modulation for Radionavigation", *Navigation*, Volume 48, pp. 227–246, Winter 2001–2002.
[4] P. J. Harris, M. Spelat, G. J. Burden, M. Crisci, "GSVF: The Galileo Reference Constellation RF Signal Simulator", *ENC-GNSS (Manchester)*, May 2006.
[5] http://www.spirentcom.com
[6] E. D. Kaplan, C. J. Hegarty, "Understanding GPS, Principles and Applications", Second Edition, 2006.
[7] J. W. Betz, "Extended Theory of Early-Late Code Tracking for a Bandlimited GPS Receiver," *Navigation: Journal of the ION*, vol. 41, no. 3, pp. 211–226.

Overview of Galileo Receivers

S. Di Girolamo[1], M. Marinelli[1], F. Palamidessi[1], F. Luongo[1], M. Hollreiser[2]

[1]Galileo Industries SpA, V. G.V. Bona, 85 – 00156 Rome – IT
[2] ESA – ESTEC, Keplerlaan 1 P.O. Box 299-2200-AG Noordwijk – NL

Abstract. After the successful launch of the first experimental Galileo satellite, i.e. GIOVE-A, and after the signature of the Galileo-IOV-CDE1 contract, the constituted European enterprise consortium is working hard in order to achieve the Galileo In-Orbit-Validation by 2009. During this IOV phase, the first four operational satellites will be launched and the ground segment will be set up and validated, including the development of the first test user receivers.

After an introduction on the Galileo overall system, this paper will describe the Galileo receivers under development in the frame of the IOV contract, focusing on the impacts of the Galileo system requirements towards the receivers design.

1 Galileo System High Level Overview

Figure 1 outlines the Galileo Overall Architecture, including the following segments:

- the **Galileo Space Segment** including a constellation of 30 satellites placed at an altitude of in Medium Earth Orbit (MEO) altitude km
- the **Ground Mission Segment** (GMS) providing the determination and uplink of the navigation data messages and integrity data messages needed for the provision of navigation services and UTC time transfer service. The GMS includes a worldwide network of Galileo Sensor Stations (GSS network) providing the collection of the input observable data, which are subsequently processed at Galileo Control Center (GCC) to determination the Galileo navigation data messages and integrity data messages.
- the **Ground Control Segment** (GCS) providing the telemetry, telecommand and control function for the whole Galileo satellite constellation.
- the **Test User Segment** (TUS) including a number of Test User Receiver (TUR) Configurations.

Figure 1 highlights the two different types of receivers are under development in the frame of the Galileo IOV contract, namely:

- The Test User Receiver (TUR): used to perform Galileo IOV system tests and hence demonstrating the capability of the Galileo System to meet the Galileo performance requirements.
- The Galileo Reference Chain (GRC) installed in each GSS. The GRC provides code phase and carrier measurements of the Galileo L-band navigation signals, as

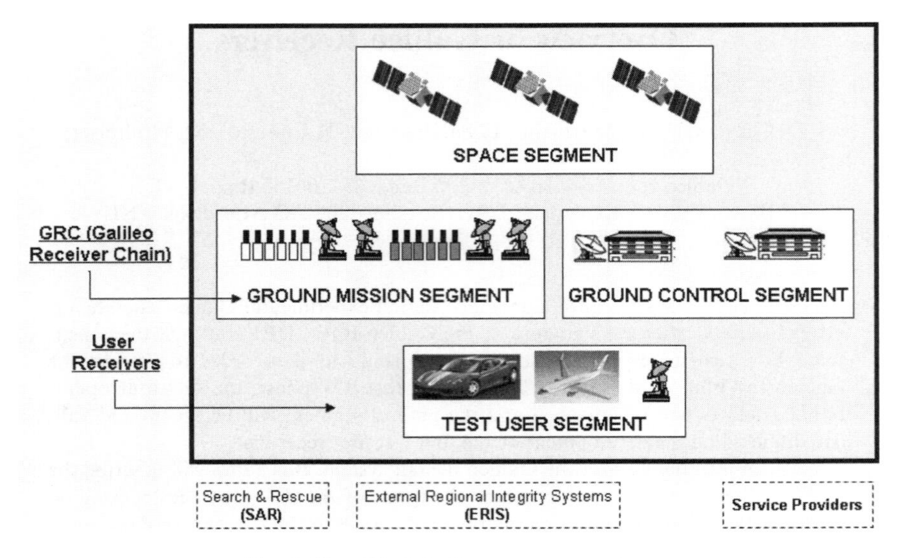

Fig. 1. Simplified Galileo system overview.

required to derive the navigation data and integrity data broadcast to user through the Galileo signals.

2 System Design Drivers

The design of the navigation satellite systems (including Galileo) is usually done taking into account a lot of system requirements. Among them, a subset could heavily impact the design and performance of the system: they are usually called "design drivers". Tables 1–3 summarise the main ones for Galileo. Three main groups can be identified:

- The frequency bands
- The services
- The navigation performance requirements

Table 1. Galileo frequencies.

Navigation signal in space	
Design aspect	Value
Carrier Frequency/Bandwidth	E5: 1191.795MHz / 92.07MHz E6: 1278.750MHz / 40.92MHz L1: 1575.420MHz / 40.92MHz
Min Receiver Power Level	–152.2 dBW (at 10° elev. angle)

Table 2. Galileo services.

Galileo services	
Acronym	Meaning
OS	Open Service
CS	Commercial Service
SoL	Safety of Life
PRS	Public Regulated Service

Table 3. Galileo navigation performance requirements.

Service required navigation performance			
Galileo required navigation performance	Open service	Commercial service	Safety of life service
Coverage	Global	Global	Global
Position Accuracy (95% confidence level)	4 m H - 8 m V (dual frequency)	NA	4 m H - 8 m V (dual frequency)
UTC Time Transfer Accuracy (only for dual frequency)	Accuracy: 30 nsec (95% confidence level)	NA	NA
Availability	99.5%	NA	99.5%
Integrity		NA	Required
Alert Limit		NA	12 m H - 20 m V
Time to Alert	NA	NA	6 seconds
Non-Integrity Risk		NA	2.0×10^{-7} / 150 sec (Excluding user receiver contribution)
Discontinuity Risk		NA	8.0×10^{-6} / 15 sec (Excluding user receiver contribution)
Access Control	Free Open Access	Controlled Access of Ranging Code and Nav Data Message	Controlled Access of Nav Data Message

3 Test User Segment (TUS)

The TUS under development in the frame of the Galileo IOV contract will be used during IOV system test campaign to demonstrate the capability of the Galileo System to meet the performance requirements specified in [1] for the following Galileo Satellite-only services:

- Single-Frequency Open Services (SF-OS)
- Dual-Frequency Open Services (DF-OS), including dual frequency navigation service without integrity and UTC dissemination service

- Safety of Life Services (SOL), including integrity
- Dual-Frequency Public Regulated Services (DF-PRS), including integrity
- Single-Frequency Public Regulated Services (SF-PRS), without integrity
- Commercial Services

The different Galileo Services are to be provided in a number of environments. The mapping og Galileo services on the specified environments is summarized in Table 4.

These environments are characterized by various parameters including satellite masking angle (10°), user dynamics environment, tropospheric environment, ionospheric environment, multipath environment, and external interference (see Table 5).

Table 4. Mapping of Galileo service on environments.

Environment	Environment ID	Service
Rural Pedestrian	RP	OS, SF-PRS
Rural Vehicle	RV	OS, CS, SOL
Aeronautical	AR	OS, CS, SOL, DF-PRS
Fixed	FX	OS, CS

Table 5. Definition of environment conditions applicable to test user receiver.

Env. ID	Dynamics	Troposphere	Ionosphere	Multipath	Interference
RV	Vel: 100m/s Acc: 10m/s^2 Jerk: 20m/s^3	Vertical delay: 2.7 mt	-S4 Scintillation: 0.2 -Vertical TEC: 50 TECU (single freq) 250 TECU (dual freq)	-Average delay 50 ns; -Linear decay slope 10 dB/μs; -Doppler bandwidth 140 Hz; -relative power −7.2 dB	−141.3 dBW/MHz
RP	Vel: 10m/sec	As above	As above	As above, except Doppler bandwidth 4 Hz;	As above
FX	None	As above	As above	As above, except Doppler bandwidth: 2.5 mHz	As above
AR	Vel: 128.6m/s Acc: 20m/s^2 H 15m/s^2 V Jerk: 7.4 m/s^3 -Banking angle: 0°	As above	As above	-Diffuse Component: Delay: 0s; Relative Power: 14.2 dB -Fuselage Reflective Compon.: Delay: 1.5 ns Relative Power: 14.2 dB -Ground Reflective Compon.: Delay: 10 ns Relative Power: 14.2 dB	−141.3 dBW/MHz plus ICAO DME

3.1 TUS Development Concept

In order to cope with the different Galileo Services and environments. the Test User Segment is conceived to provide the emulation of different Test User Receiver (TUR) classes.

Several types (i.e. aeronautical, pedestrian, vehicle, fixed) of TUR are envisaged with different configurations, namely:

1. Single or Dual Frequency Band
2. With or Without Integrity
3. Position/Velocity/Time or Precise Timing/Frequency Calibration
4. Service: PRS or SoL or OS or CS.

The TUS developmental concept can be summarized as follows:

- **Modular design**: the TUS includes certain core functionalities common to all receiver classes plus certain service-specific functionalities:
  Example of service-specific functionalities are:
  - the implementation of Geographic Denial in the Galileo PRS Receivers.
  - The implementation of specific algorithms to determine the integrity and continuity of position solution for TUR with integrity
    The emulation capabilities of different receiver classes is achieved using different antennas (gain and multipath mitigation) and changing the Radiofrequency Front-end (RF FE) performance (e.g. by suitability attenuating the signals, by changing the input bandwith, by applying different quality of reference frequency (quartz, OCXO, . . .))
- **Capability for Gradual Implementation**: The TUS is designed to allow its gradual evolution for re-use during the Galileo Full Operational phase
- **Navigation Signal Flexibility**: The TUS is designed to cope with changes of the characteristics of the navigation signal (any code of the code family, data modulation,. . .).

3.2 TUS Architecture

The Test User Segment is made up with the development of two TUR products:

- Non PRS Test User Receiver (Non PRS TUR) for receiving Open Service, Commercial Service and Safety of Life Service in a reconfigurable receiver;
- PRS Test User Receiver (PRS TUR) for classified signals reception.

Figures 2 and 3 provide both the non-PRS TUS and the PRS TUS architecture overview.

The Test User Receiver non-PRS architecture is based on the following physical elements:

- The antenna and the preamplifier,
- The antenna cable (30m),
- The Core Receiver (To process the Galileo Signals),

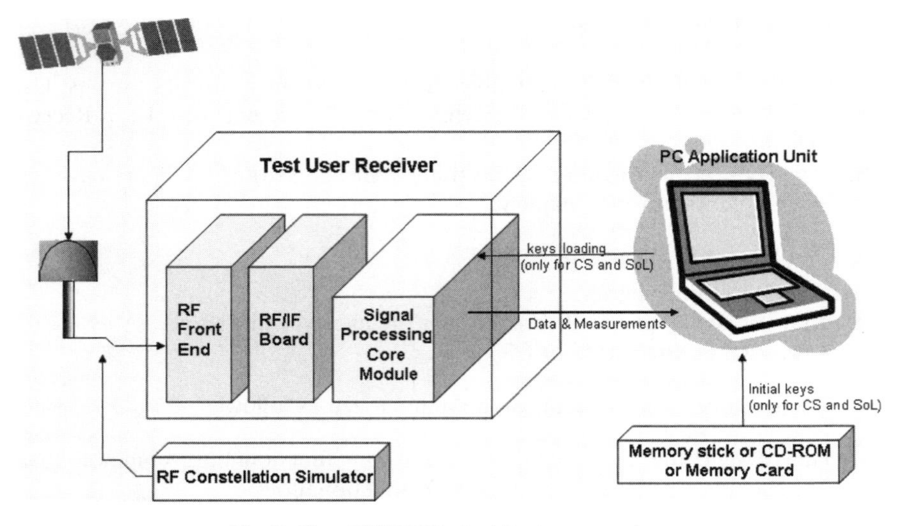

Fig. 2. Non-PRS TUR Architecture overview.

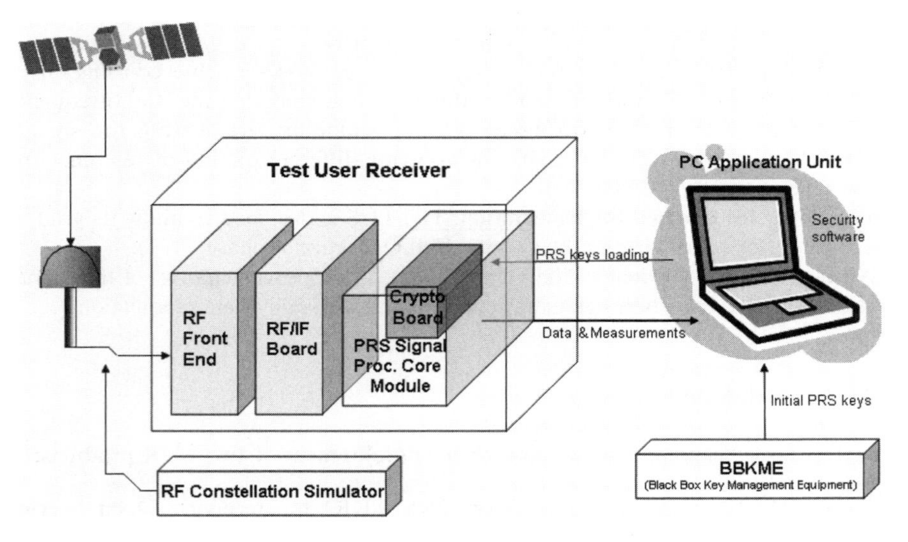

Fig. 3. PRS TUR Architecture overview.

- The PC Platform with the Application Unit Software to compute the navigation solution and perform data processing, display, post-processing and analysis.

The PRS TUR is based on the same architecture with the same physical decomposition. The only differences result in:

- The addition inside the Core receiver of a specific "Crypto Board" to process the PRS keys and restitute the PRS secret code,
- The development of a specific software module (implementing the security functions) on the Application Unit software.

A dedicated link from the PC towards the Crypto Board is foreseen to transmit PRS-specific data.

The PRS version due to security constraints integrates a Crypto Board in the Core receiver part to process the NAVSEC function and a crypto software module in the PC Application Unit to manage the keys and process the COMSEC function.

4 TUS Sub-Systems

4.1 Antenna

Depending on the environment conditions to test, two versions of antenna will be available:

- An "Aero & Reference" antenna, based on microstrips filters and patch antenna for installation purposes, with "reasonable" LNA noise figure and weak group delay bias requirements,
- An "High-End" antenna, based on cavity filters for fix applications and cross-dipoles based antenna, with "low noise figure".

The types of antenna are common for Non PRS and PRS TUR. No specific antennas will be dedicated to the TUR PRS.

4.2 TUS Core Receiver

The Core Receiver built with a modular concept, is broken down into the following sub-systems:

- **One RF sub-system:**
  - One Clock & **RF Front End** board to provide reference clock and to separate and distribute all the RF frequency bands. The RF Front End components is composed by the following parts:
    - The receiver RF Front End which is to recover, filter and suitably distribute the RF signal and clock reference;
    - The clock generation & interface and HW synchronization functions and resources. This specific board RF FE board is dedicated to the purpose of: LO Frequencies synthesis, Clock generation and distribution (in charge to distribute the reference frequency locally generated), 1 PPS output signal synthesis.
- Two **RF/IF boards (E5, L1/E6)** to carry out the frequency down-conversion for each RF frequency band, the filtering and amplification. It is configurable in order to be used with more than one RF frequency (not at the same time). A common RF/IF module is expected for all the environment configurations (Reference, High End, Aero).
- Two **Core Module boards (E5, L1/E6)** to amplify and digitize the IF signal, to acquire and track N satellites per frequency and provide code and carrier raw measurements as well as the navigation data. The core module performs the digital signal processing and is mainly in charge of performing pseudo-range measurements

on the SIS ranging codes and pseudo-phase measurements on the SIS carrier. 16 parallel channels allows to process all the satellite in view (Maximum visible satellite = 11 according to Galileo visibility analysis with a mask elevation angle of 10 degrees). The main functions performed are:

  - Carrier Tracking
  - Carrier phase measurement
  - Ranging code tracking
  - Ranging code measurement
  - Multipath mitigation
  - Interference mitigation
  - Cycle-slip detection and correction
  - Signal monitoring

  and also the recovering of the navigation data transmitted by the Navigation Signals (satellite parameters, integrity flags, SISA, . . .). To carry out this function the following tasks are needed:

  - Bit synchronisation
  - Frame synchronisation
  - Frame Demultiplexing
  - Symbol recovery
  - CRC check
  - Viterbi decoding
  - De-interleaving
  - Bit recovery

- One **Mainframe** sub-system including: Enclosure, Power Supply and Backplane to allow the assembly, the power supply and the interconnection of all the boards.
- One **Crypto Board** in the TUR PRS dedicated to the processing of the security code deciphering and the NAVSEC (i.e. Navigation Security) function in relation with the Core Module.

In order to combine the synergies between each Non PRS services and between Non PRS and PRS services, a common core module architecture (a generic processing board) will be developed. This board can address the Non-PRS services with its generic signal processing resource or the PRS service:

- A provision of HW resources (included in the common core module design),
- A software customization,
- The addition of a mezzanine crypto board having an external direct connexion for transmission of specific security management related to PRS.

4.3 TUS PC Application Unit

The Application Unit is decomposed in:

- A PC platform (COTS laptop PC)
- A PC SW composed of:
- The Navigation/Integrity Module which process the following main functions: Navigation Data Recovery, Raw Measurement preprocessing, Navigation Solution

Determination, SIS integrity determination, Autonomous Integrity Monitoring and HMI related Computations.
- The Performances Analysis Software Module (PAS) which analyses, stores and displays the data provided by:
  - the Core Modules
  - the Navigation PC Software
  - the Integrity PC software
  - The PAS can control the RF signal and constellation simulators, recover the "truth data" from the simulator and compare with the Receiver outputs. The PAS can also send commands and control the operating parameters of the different subsystems in order to enable automated tests (MMI and configuration files).
- The Data Server Module to process the data interface between the different software module and the Core receiver.
- A security SW, to implement the security function (excepted the NAVSEC function) (PRS Version only).

Certain services impose Service-specific constraints and/or mandate particular algorithms. These are defined for each specific service.

Figure 4 shows the relationship between the main algorithms to be implemented by the TUS PC Application Unit:

- Pseudorange and Phase Measurements
- Navigation Message Recovery
- Position, Velocity, and Time Determination
- Integrity Determination (only for TUR versions including integrity), including the following algorithms

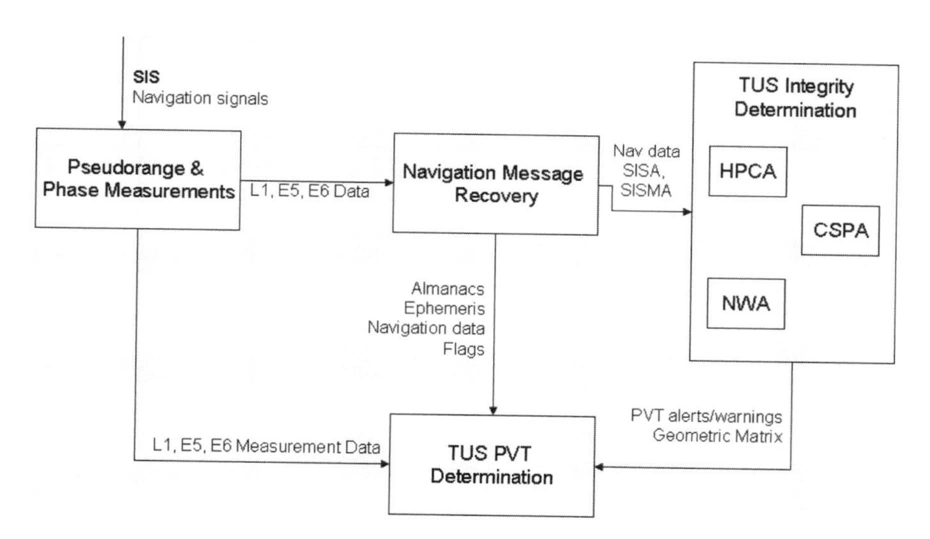

Fig. 4. PC Application unit algorithms overview.

- NWA (Navigation Warning Algorithm)
- HCPA (HMI Probability Computation Algorithm)
- CSPA (Critical Satellite Prediction Algorithm)

4.3.1 Pseudorange and Phase Measurements. This function is in charge of performing pseudo-range measurements on the SIS ranging codes and pseudo-phase measurements on the SIS carrier. The following tasks are executed:

- Generation of reference clock
- Acquisition of SIS
- Carrier tracking
- Carrier phase measurement
- Ranging code tracking
- Ranging code measurement
- Multipath mitigation
- Interference mitigation
- Cycle-slip detection and correction

4.3.2 Navigation Message Recovery. This function is in charge of recovering the navigation data transmitted by the Navigation Signals (satellite parameters, integrity flags, SISA, . . .).

4.3.3 Position, Velocity, and Time Determination. This function is in charge of elaborating the navigation solution based on valid Pseudorange Measurement.

The following tasks are executed:

- Measurement pre-processing including correction of the code phase and carrier phase measurements
- Determination of navigation solution (Position, Velocity and Time)

4.3.4 Integrity Determination. The TUR versions including the integrity service implements a set of algorithms to determine the integrity and continuity of the position solution computed at each 1 second fixing epoch.

To this purpose the following service-specific functionalities are implemented at each epoch:

1. Authentication of the integrity information extracted from the received message. This check is performed to verify that the received integrity data stream is the integrity information generated by the integrity function of the Galileo ground infrastructure;
2. Selection of the redundant and positively checked integrity data-streams the integrity data stream to be used;
3. Determination from the selected and positive checked integrity information and the navigation information which signals are valid;
4. Computation through the HPCA of the integrity risk at the specified Vertical/Horizontal Alert Limits;
5. Computation through the CSPA of the number of critical satellites for the critical operation period;

6. Determination through the NWA of the availability of the service and generation of warnings to the end user.

More details on the integrity determination function are given in the next section.

5 Integrity Determination

5.1 Integrity Information Authentication

The integrity information generated by the integrity function of the ground segment is signed (authenticated) so that it can be validated by the user receiver. This validation has to be performed in the integrity information validation function. The validation will ensure that only integrity information that was not changed at all or that was changed during dissemination with the allocated probability will be positively checked.

The validation information is provided in the integrity information data stream to the user receiver at every epoch, even if no other integrity information is broadcast to the user. This allows the user to determine at any epoch whether all integrity information has been received or not.

The validation will be performed for every integrity data stream that the user receiver will receive during nominal operation. There are at least two independent data streams that the user receiver receives.

5.2 Integrity Information Selection

Out of the positively checked integrity information data streams the user receiver has to select one integrity data stream to be used for further processing. This will normally be the same integrity data stream used at the epoch before.

The integrity information from one of the other positively checked data streams will only be used, if the integrity data stream selected at the epoch before is no longer available or if it is predicted that the integrity data stream selected at the epoch before will be not available for at least one epoch during the integrity exposure time.

If both streams are positively checked at the beginning of the operation one of them has arbitrary to be selected.

5.3 Valid Signal Determination

The valid signals to be included in the user geometry at each position fixing epoch are all the signals that are predicted to be received above the defined masking angle over the continuity exposure time, and have

1. the satellite health status flag not set to "healthy",
2. the integrity flag not set to "OK"
3. the user receiver has not detected internally any anomalous condition
4. Valid navigation data batch and valid integrity data over the continuity exposure time

5.4 HMI Probability Computation Algorithm -HPCA

The HPCA provides the computation of the probability of HMI using the Galileo Integrity Equation given in Appendix A.

The computation is based on the knowledge of the applicable user geometry, the SISA data and SISMA data extracted from the navigation messages, as well as other data stored inside the TUR memory such as satellite failure rate and "a priori" estimate of range error component due to TUR local effects (thermal noise, multipath, interference and troposphere).

The predicted HMI probability rate represents the contribution due to the following events:

- adverse stochastical combination of random pseudorange errors and user geometry leading to anomalous behaviour in the position domain (Fault Free HMI)
- system integrity failures resulting in an undetected degradation of the signal measurement accuracy, so that one and only one faulted satellite is included in the user geometry currently used for computation of the position solution (Single Failure HMI).

5.5 Critical Satellite Prediction Algorithm (CSPA)

The CSPA provides the prediction, at each epoch the position fixing epoch To, of the service continuity performance by counting the number of critical satellites.

Indeed, a critical satellite is defined as a satellite in the current user geometry whose loss or exclusion unconditionally leads the HMI probability to exceed the tolerated value. This means that the service discontinuity risk at a given epoch can be predicted by

- counting the number of critical satellites included in the current geometry, and
- taking into account the discontinuity risk allocated to each satellite in the Galileo system continuity tree

Given the current Galileo system discontinuity allocation to each satellite, the continuity performance at a given epoch To is fulfilled when no more than 6 critical satellites are included in the user geometry applicable over the time interval To + 15 s.

The CSPA computes the number of critical satellites using an istance of the Galilelo integrity equation given in appendix A. The input geometries used for such computation correspond to the reduced geometries achieved excluding one at the time the satellites included in the current user geometry. The number of critical satellites is then achieved by counting the number of satellites whose exclusion leads the HMI probability to exceed a prefixed value.

5.6 Navigation Warning Algorithm (NWA)

The NWA provides the implementation of the set of rules, in order to decide whether the navigation service with integrity is available or not at the current epoch To, as well as to predict its availability for the incoming critical period Tc. To this end, this algorithm shall provide three levels of outputs, namely:

1. "normal operation" or "use" message, which indicates that the navigation service is available at epoch To, and foreseen to be available over the next critical operation period with the required level of end-to-end performance. In this condition the user is enabled to start or continue operations at epoch To.
2. "don't initiate" warning message, which indicates that the system is available at epoch To, but discontinuity risk is not guaranteed to be acceptably low in the next critical operation period. This warning message indicates that a critical operation (e.g. aircraft approach) must not be commenced, but a user shall be permitted to finish his current Critical Operation.
3. "don't use" alert, which indicates that the user must instantly abort its current critical operation because the HMI probability exceeds the specified value or the PVT (position, velocity, and time) solution is lost.

6 Conclusions

In order to cope with the very challenging Galileo system requirements and driver criteria, the Test User Receivers will be developed by using the technology at the state of the art. The design has been conceived by trying to satisfy all Galileo major driver criteria, that are very challenging specially if compared with GPS and EGNOS. The receiver will have very sophisticate functions such as multipath and interference mitigation. Moreover, a sensible effort has been made in order to meet the integrity requirements: dedicated integrity algorithms have been identified and will be tested as part of TUS activities.

References

[1] GALILEO Global Component System Requirements Document, ESA-APPNS-REQ-00011, issue 4.2 27 July 2004, doc. ESA/ESTEC.
[2] Test User Segment Requirement Document, GAL-RQS-GLI-SYST-A0549 issue 4.3, Galileo Industries
[3] GALILEO Signal in Space Interface Control Document, GAL-ICD-GLI-SYST-A/0258 issue 11.2, Galileo Industries
[4] TUS System Design Document & Design Justification File, GAL-DJF-THA-TUS-R-00004 issue 4.1, Thales Avionics
[5] TUS Design, Development & Verification Plan, GAL-PL-THA-TUS-A-00030 issue 4.1, Thales Avionics
[6] TUS UERE/UERRE High Level Performance Justification and Assessment, GAL-RPT-AASIM-TUS-I-0013 issue 2.0, Alcatel Alenia Space Italy (Laben).
[7] TUS PVT High Level Performance Justification and Assessment, GAL-RPT-AASIM-TUS-I-0012 issue 2.0, Alcatel Alenia Space Italy (Laben).
[8] TUS Antenna Design Document And Verification Plan, GAL-DDP-SE-TUS-A-0005 issue 4.0, Space Engineering

Performance Assessment of the TurboDLL for Satellite Navigation Receivers

Fabio Dovis†, Marco Pini†, Paolo Mulassano‡

†Politecnico di Torino – Dipartimento di Elettronica
Corso Duca degli Abruzzi 24, 10129, Torino, Italy
Phone: +39 011 5644175, Fax: +39 011 5644099, e-mail: name.surname@polito.it
‡Istituto Superiore Mario Boella
via Pier Carlo Boggio 61, 10128, Torino, Italy
Phone: +39 011 2276414, Fax: +39 011 2276299, e-mail: paolo.mulassano@ismb.it

Abstract. In this paper a detailed evaluation of the performance of the architecture named "Turbo Delay Lock Loop" (TurboDLL) is presented. Such an architecture has been introduced by the authors in [1], as an innovative solution for improving the performance of satellite navigation receivers in multipath affected scenarios. The relevant innovation resides in the fact that the architecture aims at tracking each multipath component and, after a transient time, use them to wipe the multipath components off the input signal. The iterative procedure allows for a major improvement in the error induced in the code-based pseudorange measurement.

The architecture uses a preliminary estimation of the propagation channel in terms of number of not negligible reflections, and of their relative amplitude. In this paper the robustness of the TurboDLL architecture with respect to imperfect channel estimation is demonstrated.

1 The TurboDLL Architecture

In order to compute the user's location, the GNSS receiver must estimate the distances with respect to, at least, four satellites, through a fine alignment of the incoming and local codes. After the acquisition phase, such an operation, usually named "code tracking process", is carried out using a DLL for each digital channel within the receiver. In case a coherent Early-minus-Late DLL is used, a Phase Lock Loop (PLL) co-operating with the DLL is required [3]. The baseband input signal is correlated with the prompt (P), Early (E) and Late (L) versions of the locally generated code through a multiplication and an integration along a pre-detection integration period. The early and late correlation values are directly used in the code tracking process. In fact, the feedback control signal is calculated on the basis of an odd discriminator function obtained through the difference between the early and the late correlation values. The multipath presence affects such a discriminator function. In fact, if a multipath component with a delay lower than 1.5 chip with respect to the LOS is present at the input, a bias error on the code alignment is experienced.

The architecture at the basis of Turbo DLL is quite different from common rejection techniques, like the narrow correlator, where multipaths are not tracked. The new system employs a set of more than one DLL per each channel and its strategy

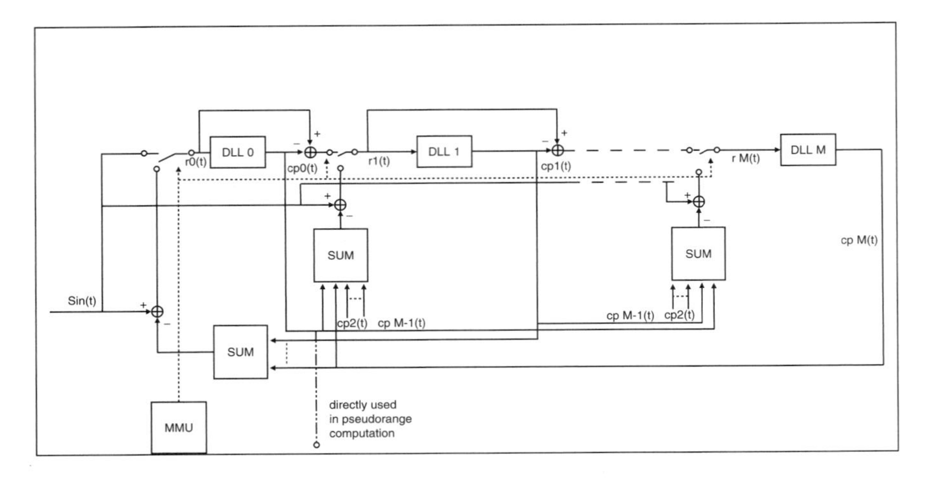

Fig. 1. Turbo DLL functional diagram.

is similar to the RAKE approach used in communication receiver. The complexity of the Turbo DLL is increased, but the proposed scheme is able to track the incoming signal replicas, since the delayed versions of the LOS are treated as additional input signal.

Referring to Fig. 1, once DLL_i is able to follow the evolution of *i-th* replica, its local code is used in order to cancel the *i-th* multipath component from the incoming signal.

In this sense, the tracking of multipath is a sort of information, which is fed back to the first DLL in order to improve the overall performance of the receiver tracking system. The word "Turbo" is thus referred as the capability of the system to boost the overall performance.

As already remarked the TurboDLL architecture is based on the multiple DLL scheme explained in [2] and two different phases can be identified:

- **Transient Time:** in which the tracking algorithm employed in the multiple DLL scheme is applied to the incoming signal and each stage of the system tracks a multipath component;
- **Steady State:** in which the code loop is closed and the DLL_0 exploits other DLLs' information to track a "cleaned" version of LOS component.

In order to allow a better understanding of the overall tracking system a detailed explanation of each phase is provided in the following.

Transient Time. As previously mentioned before in this phase the tracking algorithm developed for the Multiple DLL structure is applied. As in the case of the Multiple DLL the MMU unit is present and plays a key role. In fact it is in charge of determining whether it is necessary or not to activate the turbo architecture in presence of deleterious multipath. In particular it has to estimate the number of replicas and their amplitude with a sufficient accuracy. Furthermore, it has to compare them with a predefined threshold in order to determine how they will distort the S-curve. Otherwise if no replicas are detected or their effect can be assumed as negligible the MMU unit will force the digital channel to use standard DLL (e.g., DLL E-L narrow-correlator).

Since in this stage the aim is to study the behaviour of the Turbo architecture, a MMU being able to perfectly determine the feature (amplitude) of multi-paths that degradate the incoming signal has been assumed, while the effect of non perfect estimation will be discussed in the following sections. With reference to the general case shown in Fig. 1, in which the MMU has been able to detect M replicas of the useful signal, once the generic DLL_i is locked on the *i-th* replica then its local code, $c_{Pi}(t)$, is subtracted from the corresponding input signal $r_i(t)$ and then is fed to the next DLL, DLL_{i+1} in a sort of chain.

Steady State. As soon as the last DLL is locked on its corresponding MP then the system enters this phase and the loop is closed. This means that all the local codes generated by each DLL are subtracted from the overall incoming signal obtaining, in this way, a new input signal $r_0(t)$ for DLL_0 that will work on almost the LOS component only. The more accurate the other DLLs have tracked their corresponding MP component the better the DLL_0 works providing a local code that can be used to compute pseudorange due to the fact that multi-path distortion has been largely reduced.

As far as the other DLLs are concerned they do not have to work anymore on the signal representing the difference between the input signal of DLL_0 and the local code generated by DLL_0 itself because such a signal is just made of the residual tracking error of the LOS. To overcome this fact, the new input signal of the generic DLL_i will be the difference between the overall incoming signal and the sum of all the local codes generated by the other M DLLs; in other words if we refer, for instance, to DLL_1 in Fig. 1 its input will be given by the difference from $S_{in}(t)$ and the sum of $c_{P0}(t)$, $c_{P2}(t)$, . . ., $c_{PM}(t)$. In this way each DLL is forced to work on the corresponding multipath component according to the philosophy of the TurboDLL.

Each DLL's contribution is iteratively employed to better the performance in pseudorange estimation.

A system of switches is then necessary to allow the DLLs to commute to the right input signal as well as one adder for each DLL_i (with $i = 1, \ldots, M$) increasing the overall complexity. The need of additional switches and adders determines a difference with the structure presented in [2], where, anyway a final DLL stage was needed.

2 A Functional Example

To better understand the way the architecture works a complete example, showing the evolution of the tracking error of each DLL, will be given in the following. In particular a situation characterized by the presence of two multi-paths in addition to the LOS will be taken into account. This assumption can be appropriate for those situation in which besides to one dominant reflective surface there is also an obstacle that causes diffraction of the direct ray generating a more delayed and attenuated replica of the LOS itself.

The simulation system taken into account is made up by three DLLs narrow-correlator with a spacing of 0.25 chips between Early and Late versions of the local code. The integration period has been assumed equal to two code periods; the amplitude of the first multipath is 0.5 (half of the LOS amplitude) while the amplitude of the second one is 0.3 and they are delayed of 0.7 chips and 1.1 chips with respect to the LOS, respectively.

Figures 2–4 show the way the system works during the *Transient Time* phase. By observing Fig. 2, which depicts the tracking error of DLL_0, it can be noticed that DLL_0, after tracking the incoming signal for about 790 periods (pointed out by the black dashed line), can be considered locked to a certain value that matches the value of 0.0779 chips that is the zero-crossing point of the distorted S-curve. Figure 3 which depicts the evolution of the tracking error of DLL_1 shows the fact that DLL_1 does not work until DLL_0 is locked; after 1090 periods DLL_1 can be considered locked and its output is used to feed up the DLL_2 whose tracking error is represented in Fig. 4. Similarly to the behaviour previously described, DLL_2 begins to work as soon as DLL_1 locks and starts tracking the second multipath.

When entering the Steady State (i.e. closing the loop), the tracking error is highly reduced as demonstrated by the behavior of DLL_0 whose tracking error is represented in Fig. 5. It is evident that after closing the loop its tracking error falls down to a value that oscillates approximately around zero with small variance, so that a significant enhancement in the pseudorange estimation can be expected.

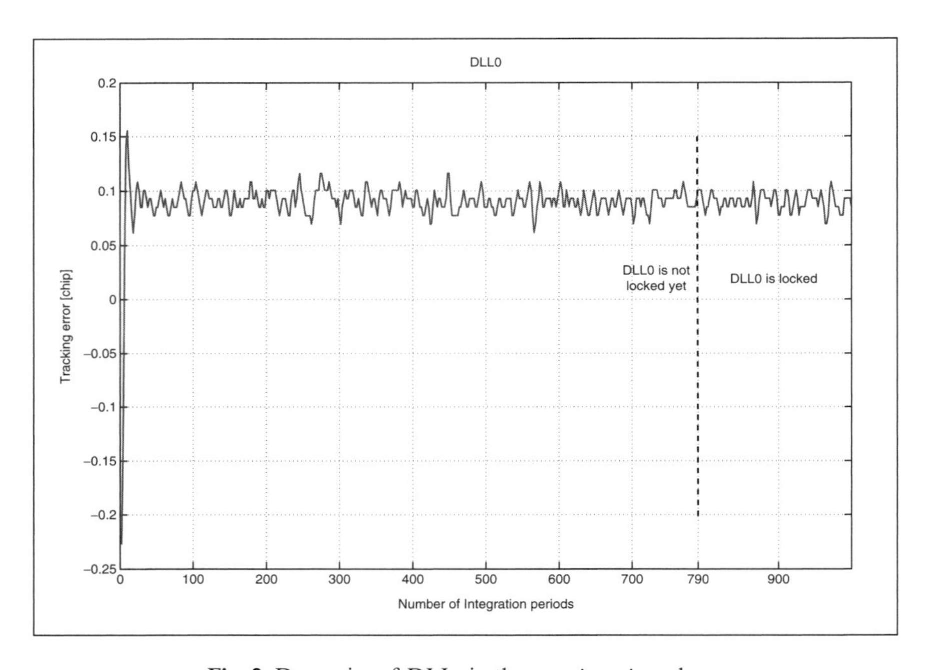

Fig. 2. Dynamics of DLL_0 in the *transient time* phase.

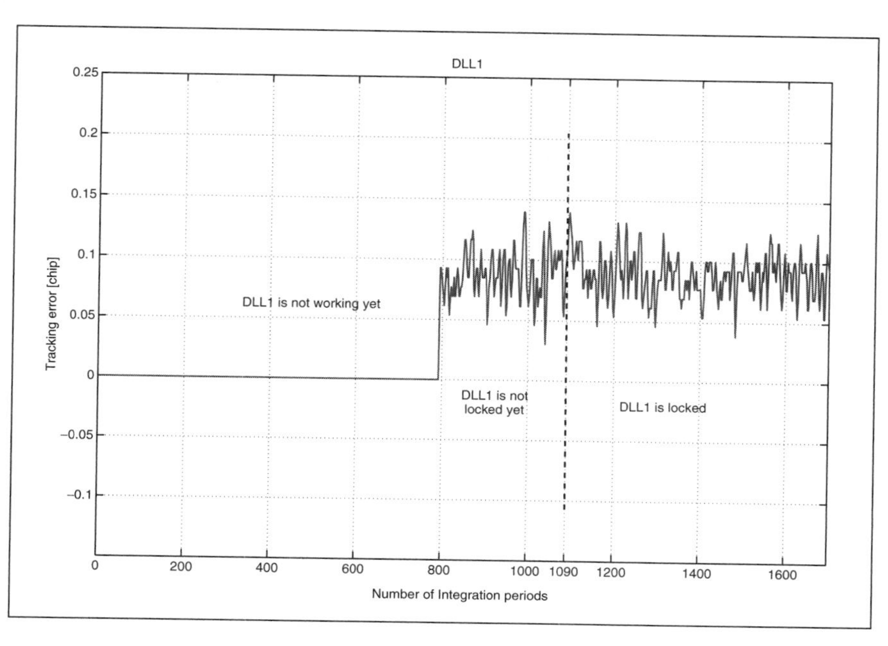

Fig. 3. Dynamics of DLL_1 in the *transient time* phase.

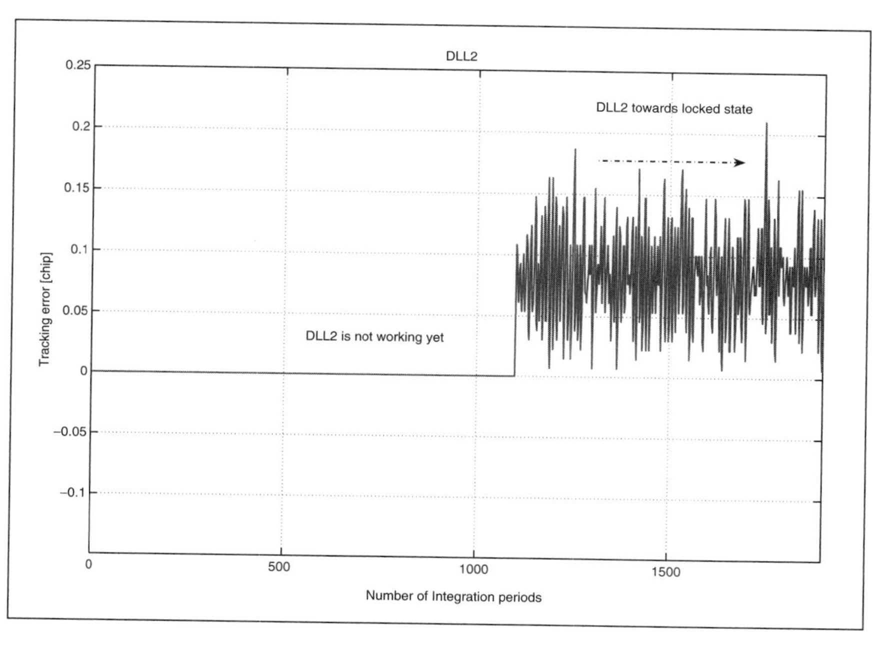

Fig. 4. Dynamics of DLL_2 in the *transient time* phase.

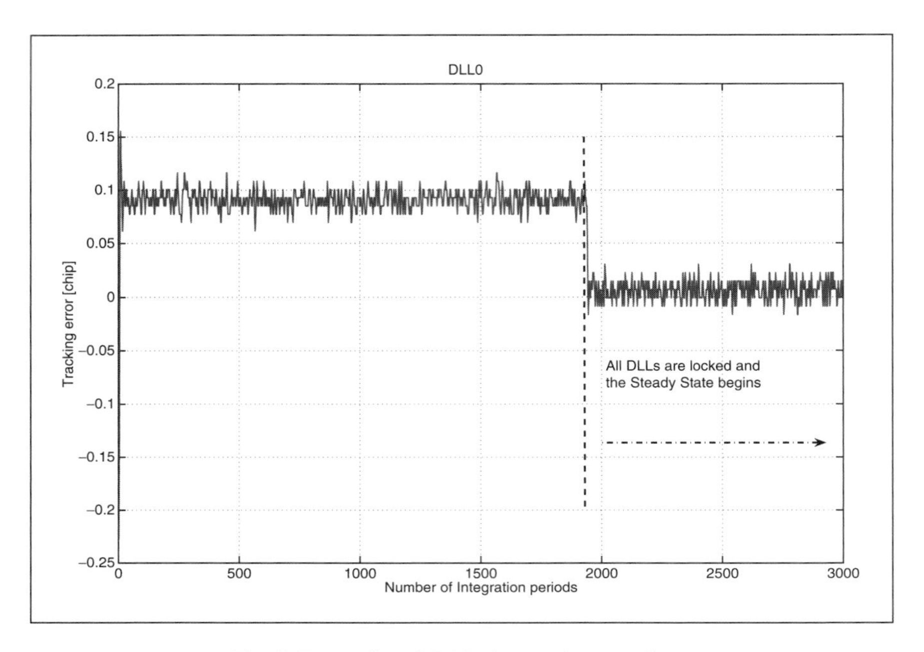

Fig. 5. Dynamics of DLL_0 in *steady state* phase.

2.1 The Propagation Channel Estimation

The complexity of the TurboDLL architecture is larger than conventional DLLs, in order to allow for a separate tracking of the input signal components. For this reason it could be desirable to start it only when multipath is conditions are detected. Furthermore, an estimation of the amplitude of each *i-th* component is required in order to adapt the dynamic of each component to the DLL_i in charge to track it. With this aim, a block in charge of the channel estimation, named Multipath Monitoring Unit (MMU) is required.

The MMU plays an important role for the functioning of the overall tracking system. In fact, it has to estimate the number of replicas and their amplitude. Furthermore, it has to compare them with a predefined threshold in order to determine their distortion effect on the S-curve. It is useful to remark that the MMU actives the Turbo DLL only in case of multipath, while, if the multipath presence is not detected, the digital channel still continue to use a standard DLL (e.g., DLL E-L narrow-correlator) as shown in Fig. 6.

The estimation performed by the MMU is affected by errors; the study presented in the paper demonstrates the ability of the iterative architecture to recover from not perfect estimation of the amplitude of the multipath components.

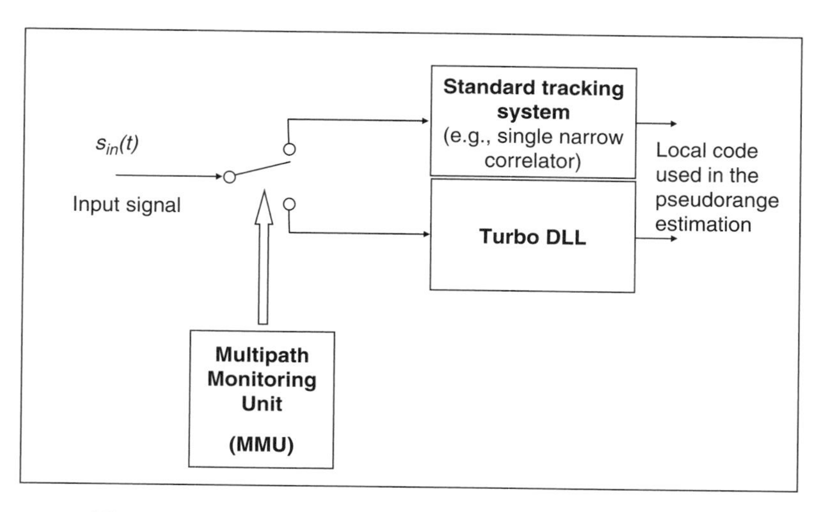

Fig. 6. Multipath monitoring unit as selector of tracking systems.

3 Performance Evaluation

Figure 7 depicts some examples of the results obtained for GPS C/A code with the TurboDLL architecture using the multipath envelope diagram. Such a plot represents the error due to the presence of one reflected ray (in fraction of chip) versus the delay of the multipath with respect to the Line-of-Sight path [3].

3.1 Imperfect Estimation of the Amplitude

The MMU plays a key role in the exploitation of the TurboDLL architecture within a receiver, since it is in charge of detecting the multipath presence and it must be able to extract some features of the received signal that are needed by the TurboDLL architecture.

The results presented for sake of example, have been obtained considering a TurboDLL made of two stages, and the presence of one single reflection with half the amplitude of the direct ray.

In Fig. 7 the theoretical curve (solid bold line) and the simulated values for a classical narrow correlator single DLL are reported. From the comparison with the values simulated for a TurboDLL with perfect estimation of the amplitude of the MP, the gain advantage of the architecture can be appreciated.

If the MMU is not precise in the estimation of the replica amplitude, providing a value which is lower than the real one, the system is still robust providing performance that are better than the single narrow correlator DLL.

Figure 7 reports the results obtained for the architecture dealing with a GPS C/A code. The considered estimation error ranges from 10% to 60% of the real

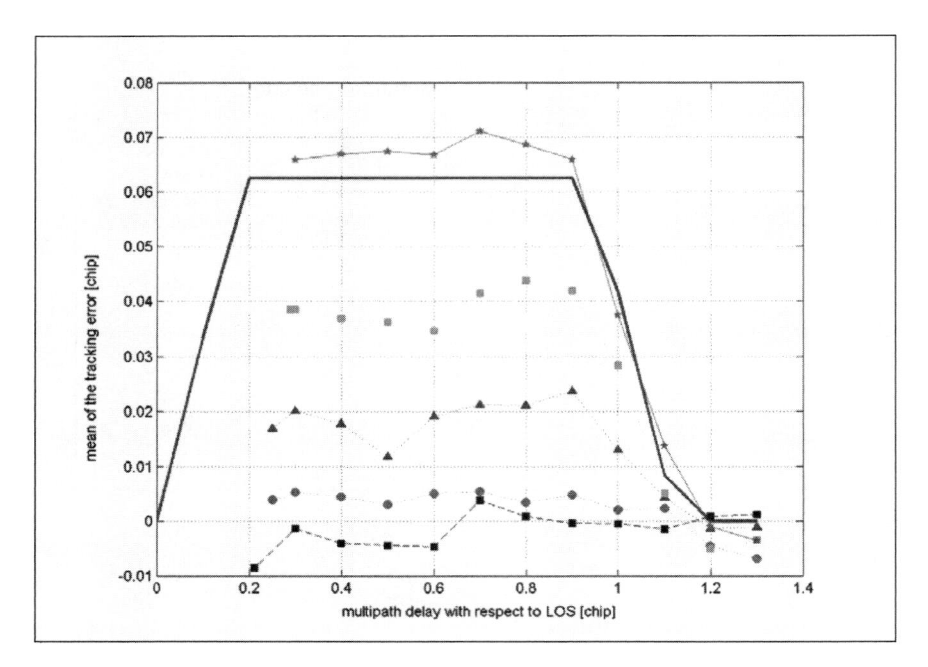

Fig. 7. GPS C/A code: theoretical MP envelope of a single DLL (solid bold line), simulated single DLL (solid line with star markers) and simulated turbo DLL architecture with perfect estimation of the MP (dashed line with square marker) versus simulated turbo DLL scheme with error of MP amplitude estimation equal to: –10% (dotted line with circle markers); –30% (dotted line with triangle markers); –60% (dotted line with square markers).

amplitude value. For an error of the 90% the system is not able to keep the tracking phase. The dotted line with square markers represents the performance in case an underestimation error of 60%. Also in this case the bias error is lower than the one of the narrow correlator single DLL. As already remarked in [1] the system performance is bounded by the ability of "separating" the multipath from the LOS contribution, and a stable state cannot be reached for very short delays of the MP with respect to the LOS (less than 0.2 chip).

A similar behavior is obtained for the Galileo BOC(1,1) signals, as depicted in Fig. 8. Also in this case the system is robust to underestimation of the multipath amplitude up to the 60% of the actual value, and it outperforms in all the cases the single narrow-correlator DLL architecture.

4 Conclusions

In this paper the performance of the TurboDLL have been discussed, showing how the architecture is able to recover the multipath contributions from the received signal, and then wiping them off the signal in an iterative procedure. The system

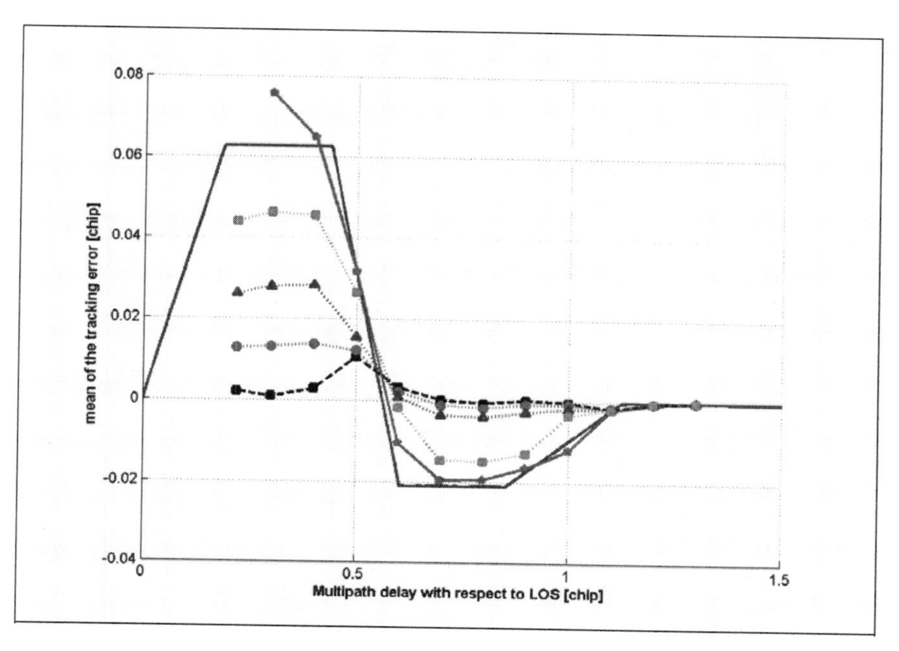

Fig. 8. Galileo BOC(1,1): theoretical MP envelope of a single DLL (solid bold line), simulated single DLL (solid line with star markers) and simulated turbo DLL architecture with perfect estimation of the MP (dashed line with square marker) versus simulated turbo DLL scheme with error of MP amplitude estimation equal to:–10% (dotted line with circle markers); –30% (dotted line with triangle markers); –60% (dotted line with square markers).

provides relevant improvements in the performance of code-based pseudoranges with respect to classical narrow-correlator discriminators, trading off the performance with the overall complexity of the tracking stage. In particular the robustness of the architecture with respect to inaccurate estimation of the multipath amplitude has been demonstrated.

Acknowledgement

Authors would like to thank the "Istituto Superiore Mario Boella" research center for its support trough the Navigation Lab. In addition, the authors would like to thank Luca Greco and Paolo Favaro for their valuable help in conducting the performance analysis.

References

[1] Fabio Dovis, Marco Pini, Paolo Mulassano, "Turbo DLL: an Innovative Architecture for Multipath Mitigation in GNSS Receivers", *ION GNSS 2004*, Long Beach, CA (USA), 21–24 Sept., 2004.

[2] Fabio Dovis, Marco Pini, Paolo Mulassano, "Analysis of the Multiple DLL Architecture: a Novel Solution for Reducing the Multipath Effect in GNSS Receivers", *ION GNSS 2004*, Long Beach, CA (USA), 21–24 Sept., 2004.
[3] P. Misra, P. Enge, *Global Positioning System. Signals, Measurements, and Performance*, Ganga-Jamuna Press, 2004.

Analysis of GNSS Signals using the Robert C. Byrd Green Bank Telescope

Marco Pini[1], Dennis M. Akos[2]

[1]Politecnico di Torino, Electronics Department
C.so Duca degli Abruzzi n.24, 12138, Torino, Italy
Phone: +392276436, e-mail: marco.pini@polito.
[2]University of Colorado, Aerospace Engineering Sciences
429 UCB, Boulder, CO 80309-0429
Phone: +303 7352987, e-mail: dma@colorado.edu

Abstract. Recent experiments have shown that a valuable way to monitor the quality of the signal broadcast by Global Navigation Satellite System (GNSS) satellites is to use a high gain antenna. Signal monitoring experiments are important to check the health of the electronic devices on board of the satellite just after the launch, but also to characterize the signal quality over time. In fact, for high performance applications such as Global Positioning Systems (GPS)-based aircraft navigation and landing systems, even small errors due to signal distortions must be considered in the error budget.

This paper describes the experiment performed in Green Bank, West Virginia (U.S.A.), where a 110 meter high gain antenna has been used to track several GNSS satellites. After the description of the system set up, the paper will present interesting results obtained in post processing through a toolset, specifically developed for this type of analysis.

1 Introduction

Using traditional antennas and receiving hardware, the received power of the GPS signal is below the thermal noise floor. This implies that the received signal broadcast by the satellites is masked by the noise; thus the chips of the Coarse Acquisition (C/A) and Precision (encrypted) P(Y) codes are not discernable. The signal processing relies on the gain obtained from the spread spectrum nature of the signal structure. In fact, the receiver is able to demodulate the navigation data and recover the precise time of transmission by correlating the incoming signal with a local version of the known spreading code transmitted by satellites [1]. GPS satellites operate at an altitude of over 20,000 Km and the transmitted power is not sufficient to observe their complete signal structure directly using conventional demodulation methods.

An alternate approach is to use a high gain antenna to receive the signal transmitted by the satellites and perform measurements directly on the signal at the Radio Frequency (RF) output. The high gain antenna provides a positive Signal-to-Noise Ratio (SNR), such that both the individual chips and the navigation bits can be demodulated without performing the despreading procedure.

This paper presents the analysis on the data sets collected using a high gain antenna. After introducing the motivation of the work, the experimental setup and the data collection approach will be explained. The paper will then focus on the most important results obtained in post processing through an analysis toolset developed for this type of experiment. Such a toolset has been developed on the basis of the previous work carried out by Mitelman [8]. The most significant diagrams will be shown and the differences between the signals broadcast by various satellites will be underlined.

2 Background and Motivation

Using traditional GPS antennas and receiving hardware, the signal at the front end output is masked by noise. The received signal power spectral density is approximately 20 dB below the thermal noise floor and both the navigation data bits and the chips of the spreading code are not discernable. GNSS receivers are based on the "de-spreading" process and assume that the Pseudorandom (PRN) codes are perfect square wave signals. This assumption can not be taken for some high accuracy applications like aircraft navigation, where even small deformations of the signal must be considered to assure the integrity of the positioning procedure. Thus, for the most demanding users of satellite navigation, it is important to characterize the nominal signal structure in order to detect minimal variations resulting from hardware-based errors.

A high gain antenna can improve the Signal-to-Noise Ratio (SNR) at the front end output and drastically increase the signal observability.

As an example, Fig. 1 (a) clearly shows that the signal at the front end output using a commercial GPS antenna looks like noise, but, looking at Fig. 1 (b), one can see that the high gain antenna drastically increases the SNR and the chips of both the C/A and the P(Y) codes become visible.

Recently, high gain antenna measurements have been employed to characterize the signal quality of GPS satellites belonging to different blocks [3] [4]. This type of investigation started on the early 90's when an anomalous behaviour was observed on Satellite Vehicle Number (SVN) 19. After several investigations, it was found out that the problem was due to a misalignment between the Coarse Acquisition (C/A) and the Precision (encrypted) (P(Y)) codes [5].

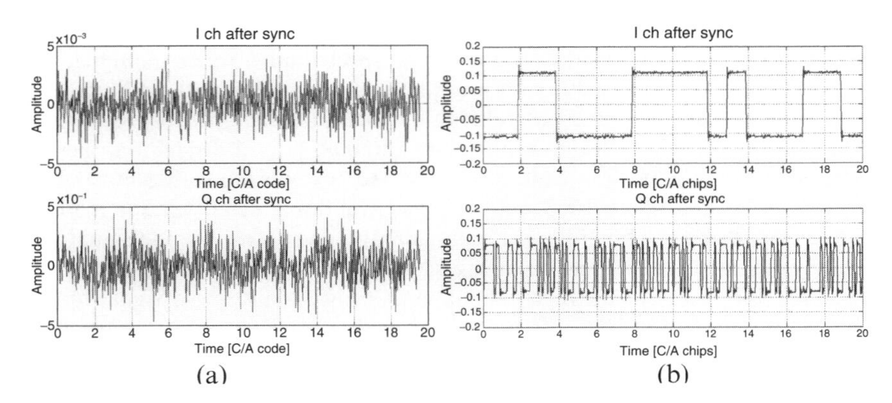

Fig. 1. Inphase and quadrature signal at the front end output (a) using a commercial GPS antenna and (b) a high gain antenna.

It was only possible to fully understand the problem source using a high gain parabolic antenna. In fact, such an antenna guaranteed that the signal power rose above the thermal noise floor. Today, the signal analysis with high gain antennas is widely used by scientists. For example, the first Galileo satellite (GIOVE-A) has been closely monitored just after the launch by radio telescopes in Redu (Belgium), Chilbolton (UK) and Weilhein (Germany) as it is described in [6].

3 System Set Up Description

Figure 2 shows the block diagram of the system used to collect data during the experiment performed on the 24th of July 2005 in Green Bank, West Virginia (USA).

In this particular case, the Robert C. Byrd radio telescope was used. It is the world's largest fully steerable radio telescope equipped with a 110 meter dish diameter antenna (see Fig. 3). The GBT antenna has a gain of approximately 70dB in the L-band and guarantees the received GPS signal power rises above the noise floor. The surface of the telescope can be adjusted and the overall structure is a wheel-and-track design that allows the telescope to view the entire sky above 5 degrees elevation [7].

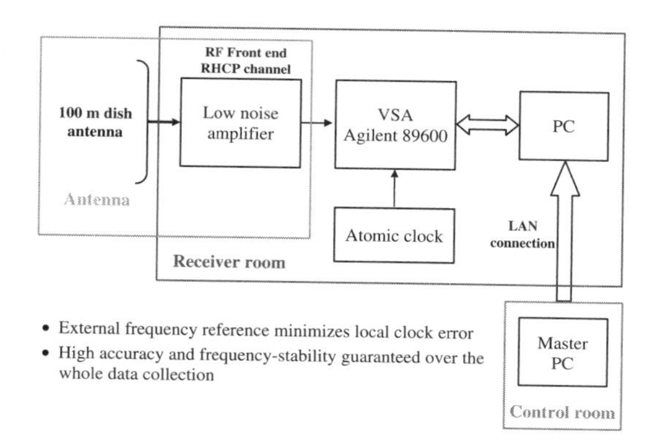

Fig. 2. Block diagram of the system used to collect data.

Fig. 3. Robert C. Byrd radio telescope, Green Bank, West Virginia USA.

The received signal was amplified, filtered and finally sent to a Agilent 89600 Vector Signal Analyzer (VSA) which was used to downconvert the signal to a lower intermediate frequency and then to a digital format. The raw samples of the signal were stored into a disk for post processing analysis. The VSA was slaved to a 10 MHz rubidium clock, which guaranteed high accuracy and frequency stability over the whole duration of the data collection.

The VSA was installed inside the receiving room at the top of the telescopes and was controlled from a remote centre, through a Local Area Network (LAN) connection. It is important to underline that all the instruments installed in the receiving room were previously tested in the lab, in order to avoid breakdowns during the experiment.

The experiment lasted about 12 hours and thanks to a good data collection schedule, 24 satellites were observed.

4 Post Processing Results

Monitoring of the GNSS Signal Quality Via a High Gain Antenna

With a 70 dB gain antenna the structure of the signals transmitted by GNSS satellites is well visible at the front end output. As an example, Fig. 4 shows a zoomed view of the C/A and the P(Y) code, once the unknown phase offset between the incoming carrier and the local oscillator has been recovered in post processing with a Phase Lock Loop (PLL).

The signal is no longer masked with noise and it is possible to use appropriate diagrams (e.g.: IQ diagrams, eye patterns) usually used in the communication field, to better analyze the signal structure.

The block diagram depicted in Fig. 5 summarizes the main features of the software toolset developed to support the GNSS signal quality analysis performed using high gain antennas.

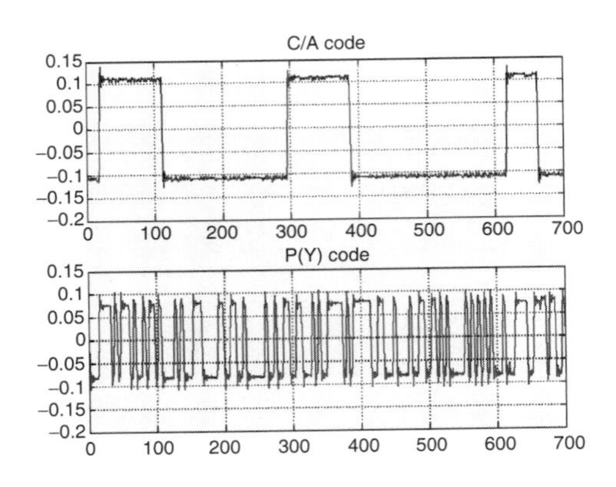

Fig. 4. Inphase (C/A code) and quadrature (P(Y) code) signals broadcast on L1 by satellite SVN 24 (a), IQ diagram (b).

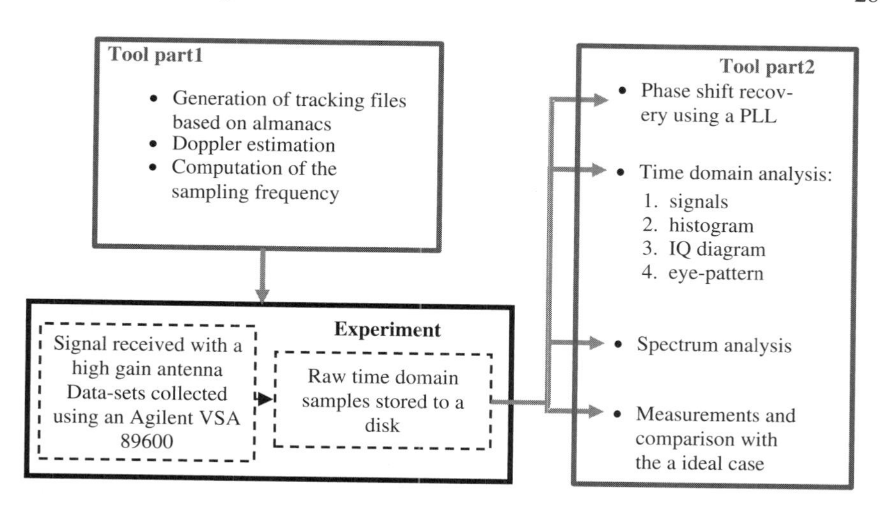

Fig. 5. Structure of the software toolset used in the quality monitoring of GNSS signals.

Such a software toolset is composed by two parts:

- **Tool part 1** generates the tracking files used during the experiment to aim the antenna at the satellite under test. Furthermore, it is useful to predict the Doppler effect on the incoming signal and compute the desired sampling rate. Note that if the dithered sampling frequency [8] [9] is implemented, the computation of the sampling rate require an accurate estimation of the code rate, which is affected by Doppler and changes as the satellite moves.
- **Tool part 2** is used to analyze the collected data sets in post processing. This part of the tool automatically generates important plots of interest (i.e.: IQ diagrams, histograms, eye patterns. . .). These plots are compared to the theoretical ones and used to quantify possible distortions or anomalies on the signal structure.

As an example of the plots that can be obtained with the developed toolset, Fig. 6 compares the IQ diagrams of the signal broadcast by Satellite Vehicle (SV) 24 and SV 59. The IQ diagrams show the Quadrature signal versus the Inphase signal and reveals additional details that are not readily apparent from the time domain data [4].

While the first IQ diagram does not show particular anomalies, the IQ diagram in Fig. 6 (b) does not match the ideal one (grey) when only the C/A code change signs. For this satellite it was found out that the ratio between the amplitudes of the C/A and the P(Y) codes is approximately 1dB lower than 3dB, which is the expected value.

The IQ diagram is also extremely useful to check the synchronization between the C/A and the P(Y) codes. In fact, when both C/A and P(Y) codes change sign at the same time, a transition occurs along one of the diagonals of the diagram [4]. Ideally, the diagonals pass through the origin, but looking at real diagrams, it is evident that this is not the case. SV 59 shows an asymmetry across the origin, while for SV 24 the chip transitions are more symmetrical. It was interesting to verify that all the satellites of Block IIR present the distortion observed on SV 59, while no satellite of

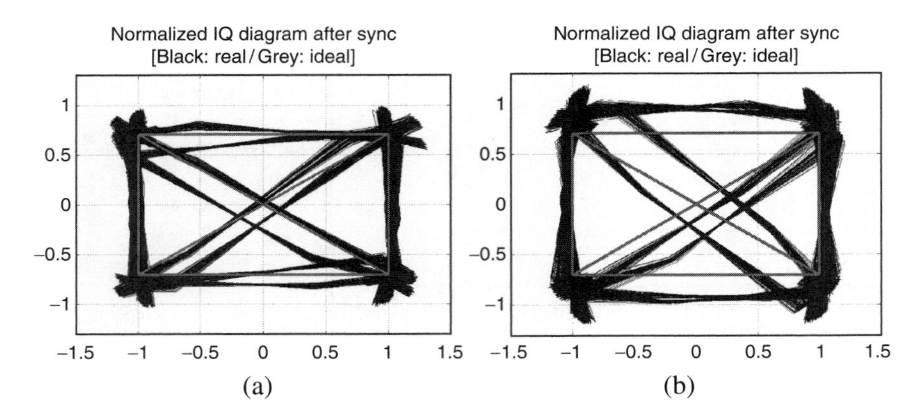

Fig. 6. Normalized IQ diagrams of the signal broadcast on L1 (a) by SV 24 and (b) by SV 59, compared to the theoretical IQ diagrams (grey).

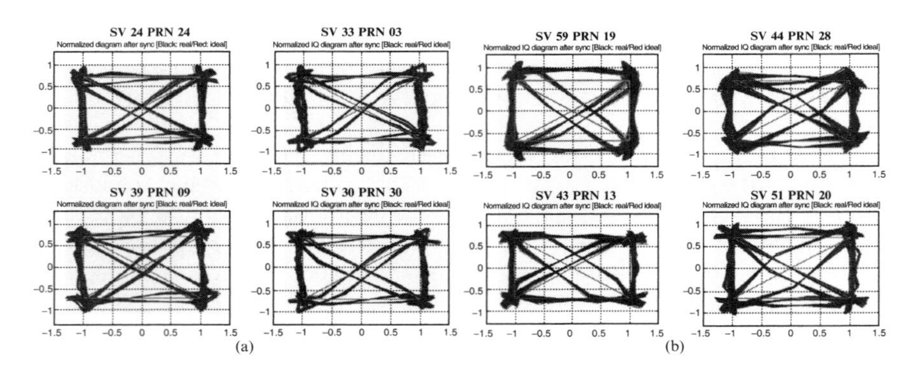

Fig. 7. IQ diagrams of the signal broadcast on L1 by several satellites belonging to (a) Block IIA and (b) Block IIR.

Block IIA has this unusual asymmetry on the IQ diagram of the L1 signal. Figure 7 compares the IQ diagram of 8 different satellites. Four of them (SV 24, SV 33, SV 39, SV 30) belong to the Block IIA and are gathered together in Fig. 7 (a), while those of Block IIR (SV 59, SV 44, SV 43 and SV 51) are shown in Fig. 7 (b).

For all the satellites of Fig. 7 (b) the distortion across the origin is quite evident and is due to a not perfect synchronization between the C/A and the P(Y) codes, which can be quantified with an average delay as it has been shown in [4].

Monitoring of the GNSS Signals Using Innovative Signal Processing Techniques

The described analysis can be further improved using a particular sampling rate, called dithered sampling frequency. Such a sampling frequency follows the same approach adopted for repetitive signals in sampling oscilloscopes and has been proposed for use in GNSS by Mitelman [8].

Considering repetitive signals (as GNSS signals after a proper Doppler removal stage), this technique allows for a high resolution of the time domain representation through a samples superimposition in post processing.

Moreover, the dithered sampling frequency preserves the synchronization. Thus, the samples of the signal can be averaged over time and the SNR can be further increased in post processing. These advanced signal processing techniques have been adopted in GNSS signal monitoring experiments using conventional receiving hardware/antenna and their effectiveness has been shown in [9].

However, both the dithered sampling strategy and the averaging technique can also be used with high gain antennas. In this case, the noise contribution is further reduced in post processing and the resolution is increased, thus the structure of the PRN code broadcast by the satellite becomes extremely clear. Figure 8 shows a zoomed view of one of the C/A code chips broadcast by SV 24, when both the dithered sampling rate and the averaging technique are applied to the signal received with the Green Bank telescope. In this case, the virtual sampling rate achieved in post processing by the use of the dithered strategy is approximately equal to 460 MHz, whereas the SNR increment due to the averaging technique is equal to 11.46 dB.

The chip shape is well defined and it is even possible to note the ringing effect due to the VSA filtering. In conclusion, it is possible to state that these innovative post processing techniques allows for performing accurate measurements on the time domain signal.

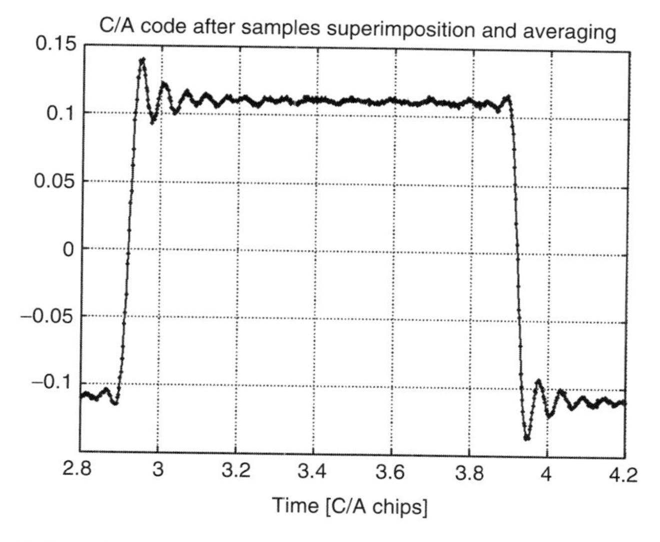

Fig. 8. Zoomed view of a PRN chip, when both the dithered sampling frequency and averaging technique are applied to the signal received with the Green Bank Telescope.

5 Conclusion

This paper has mainly focused on the analysis of GNSS signals. The experiment performed at Green Bank (West Virginia, USA) on the 24th of July 2005, has been described. In the experiment the "Robert C. Byrd" telescope was used to observe the signals broadcasted by GNSS satellites. The high gain antenna drastically increases the SNR and makes the raw code chips directly observable on a vector signal analyzer. This type of investigation is extremely useful to check the quality of the signal broadcast by new satellites as soon as they are in orbit. The same type of investigation has been performed in Europe in January 2006 to quality monitor the first Galileo satellite, called GIOVE-A [6].

The paper has also shown the main features of the software toolset specifically developed to analyze the collected data sets in post processing. It helps to monitor the quality of GNSS signals through appropriate plots and measurements. It has been shown how it is possible to detect differences between satellites as SV 24 (Block IIA) and SV 59 (Block IIR) as well as other possible anomalies.

The paper has also shown that the sampling resolution and the fidelity of the plots can be improved, using the dithered sampling frequency, which achieves high sampling rates preserving synchronization. Furthermore, it has been shown that the noise contribution can be further reduced by averaging several sequences of superimposed samples. It is important to remark that both the dithered sampling strategy and the averaging procedure can be easily applied to noisier signals. This means that the presented experiment can be repeated using smaller directive antennas or a multiple antenna array, reducing the overall cost of the experiment.

References

[1] P. Misra, P. Enge, Global Positioning System. Signal Measurements and Performance, Ganga-Jamuna Press, 2004, ISBN 0970954409
[2] E.D. Kaplan, Understanding GPS: Principle and Application, Artech House Publishers, ISBN 0890067937
[3] R.E. Phelts, D.M. Akos, "Nominal Signal Deformations: Limits on GPS Range Accuracy" GNSS 2004, the 2004 International Symposium on GNSS/GPS, 6–8 Dec 2004, Sydney, Australia
[4] M. Pini, D.M. Akos, S. Esterhuizen, A. Mitelman, "Analysis of GNSS Signals as Observed via a High Gain Parabolic Antenna" *in Proceedings of the 18th International Technical Meeting of the Satellite Division of the Institute of Navigation*, ION-GNSS 2005, 13–16 Sept 2005, Long Beach, California (USA)
[5] C. Edgar et al. "A Co-operative Anomaly — Resolution on PRN-19," *in Proceedings of ION GPS-93, Institute of Navigation*, September 1993
[6] O. Montenbruck et al. "GIOVE-A Initial Signal Analysis," submitted to *GPS solution* on 1st of March 2006
[7] http://www.gb.nrao.edu/GBT/GBT.html
[8] A. M. Mitelman, "Signal Quality Monitoring for GPS Augmentation System", Ph.D. Dissertation, Stanford University, Dec 2004
[9] M. Pini, D.M. Akos, "Exploiting Global Navigation Satellite System (GNSS) Signal Structure to Enhance Observability" to appear on *IEEE Transaction on Aerospace and Electronic Systems*

First Results on Acquisition and Tracking of the GIOVE-A Signal-in-Space

Fabio Dovis, Marco Pini, Andrea Tomatis

Politecnico di Torino, Electronics Department
C.so Duca degli Abruzzi 24 10129 Torino – Italy
Tel: +39 011 2276416, Fax: +39 011 5644099
e-mail: name.surname@polito.it

Abstract. This paper presents the first results obtained processing the Signal-In-Space broadcast by the Galileo In – Orbit Validation Element A (GIOVE-A) satellite, the first experimental Galileo satellite. The paper analyzes both the acquisition and tracking phase of a Galileo software receiver able to process raw samples of the signal collected by means of a commercial front end. Starting from the description of the system set up used to collect the data set the paper introduces the Partial Correlation approach adopted in the acquisition phase and describes the tracking structures used to synchronize the local and the incoming codes. Techniques tailored to the new structure of the Galileo signal have been employed. In fact, the GIOVE-A signal uses novel modulation schemes and it is made of longer spreading codes that required modification to both the acquisition and tracking algorithms usually implemented within a GPS receiver.

1 Introduction

On 28th of December 2005 the first Galileo satellite was launched from Baikonur, Kazakhstan. The satellite named GIOVE-A started broadcasting the signal on 12th of January 2006 and the event was definitely a milestone in the development of the new European Global Navigation Satellite System (GNSS).

After the launch, several experiments have been performed from many different sites to monitor the quality of the signal and check the health of the electronic devices on board of the satellite. Due to the low power of the received signal (i.e. the signal power spectral density is approximately 20 dB lower than the thermal noise floor) in all the experiments, a high gain antenna has been used to track the GIOVE-A satellite.

At the beginning of March, the initial results of the analysis performed in Weilheim (Germany) have been published by Montenbruck et al. [1]. Then, on 31st of March, on the basis of the previous work, the GPS group at the Cornell University has been able to decode and publish the Galileo codes on the web [2].

The availability of the codes represented a new opportunity for all the scientists working in the navigation field to test the first Galileo receivers/algorithms with real data sets.

2 GIOVE-A Signal

As it is well described in [1], for the L1 band, the transmission chain on board of GIOVE-A follows the block diagram depicted in Fig. 1.

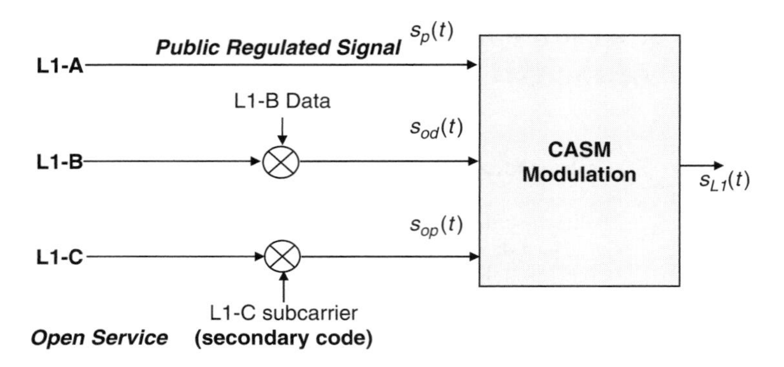

Fig. 1. GIOVE-A signal block diagram.

The Galileo signal modulation has two specific characteristics: filtered Binary Offset Carrier(BOC) pulse form and Coherent Adaptive Subcarrier Modulation (CASM) [3]. The BOC waveform was originally conceived in order to allow a spectral separation between the existing single carrier GPS signals and the new signals. In addition, if properly exploited, the new signals also provide a better performance in multipath environments [4]. To ease the acquisition and tracking under weak signal conditions, Galileo includes a pilot on each carrier. As a consequence, the L1-band accommodate a total of three signals: a pilot $s_{op}(t)$, a data signal $s_{od}(t)$ for the Open Service (OS) and Safety Of Life Service (SOL) and a data signal $s_p(t)$ for the Public Regulated Service (PRS). Priorities and jamming constrains led to the following mapping of signals [5].

$$
\begin{aligned}
s_{L1} = {} & \frac{\sqrt{2}}{3}\left[s_{od}(t) - s_{op}(t)\right]\cos(2\pi f_{L1} t) \\
& + \frac{1}{3}\left[2s_p(t) + s_{od}(t)s_{op}(t)s_p(t)\right]\sin(2\pi f_{L1} t)
\end{aligned}
\tag{1}
$$

with f_{L1} = 1.57542 MHz, and

$$
\begin{aligned}
s_{od}(t) &= d_{od}(t)\,c_{od}(t)\,sign(\sin(2\pi f_s t)) \\
s_{op}(t) &= c_{op}(t)\,sign(\sin(2\pi f_s t)) \\
s_p(t) &= d_p(t)\,c_p(t)\,sign(\cos(30\pi f_s t)
\end{aligned}
\tag{2}
$$

In this expression, f_s denotes the subcarrier frequency of the OS signal which has been selected to match the OS chip rate f_c = 1.023 MHz. The modulated data bits $d(t)$ and code bits $c(t)$ are given by

$$
\begin{aligned}
d_{od}(t) &= d_m^{(od)}\ if\ mT_b \le t < (m+1)\,T_b \\
c_{od}(t) &= c_m^{(od)}\ if\ mT_c \le t < (m+1)\,T_c \\
c_{op}(t) &= d_m^{(op)}\ if\ mT_c \le t < (m+1)\,T_c \\
c_p(t) &= d_m^{(p)}\ if\ mT_c \le 2.5t < (m+1)\,T_c
\end{aligned}
\tag{3}
$$

Here $T_c = \frac{1}{f_c} \approx 1\mu s$ and $T_b = 4.902 \cdot T_c = 4ms$ denote the duration of an individual OS code chip and the duration of the entire OS data code sequence, respectively. As mentioned before Galileo include also a Pilot channel which a code sequence longer than the Data channel one. Its duration is 8ms.

3 System Setup

In this Section the experimental system set up used to collect and to post process the real data of GIOVE-A satellite is described.

Figure 2 shows the block diagram of the system set up: the signal is received by an L1 antenna connected to a commercial front-end and recorded on a storage support.

The front-end is based on a Application Specific Integrated Circuit (ASIC) basic front-end with a bandwidth of about 4MHz, an intermediate frequency f_{IF} = 4.1304MHz and a sampling frequency f_s = 16.3676 sample/s and quantized over 4 bits.

The raw data collected have been post processed by means of a software tool able to perform the signal acquisition and tracking. Such a software platform developed by the authors enables the possibility to test novel techniques tailored to the new features of the Galileo signals (BOC modulation and longer primary and secondary codes).

The software tool is composed by a novel acquisition stage for a coarse evaluation of the satellite parameters and an adapted tracking loop made by a Delay Lock Loop (DLL) and Phase Lock Loop (PLL) loops designed for the GIOVE-A signals. Thus, it allows the post processing of the collected data and to evaluate the main satellite signal parameters.

4 Data Collection

Using the system described in the previous section, several minutes of the signal have been recorded.

The signal samples have been collected on the hard disk drive during the pass of GIOVE-A above Torino (Italy) on 1st March, 2006 starting form 12 AM to 12.30 AM. During that time the satellite reached a maximum elevation angle of 39°.

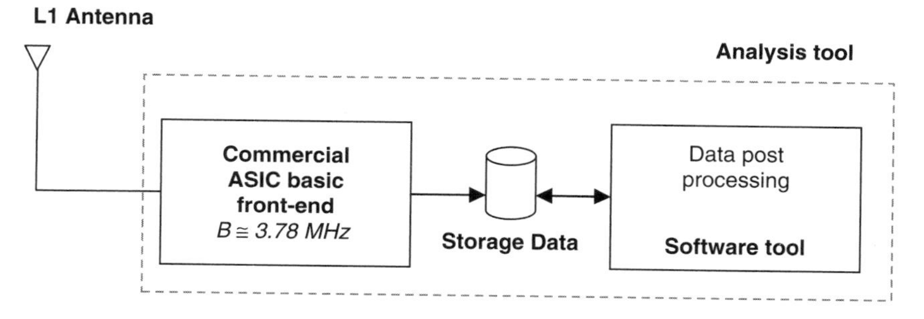

Fig. 2. Block diagram of the experimental setup.

5 Processing Results

The aim of this work is the evaluation of both the acquisition and tracking phases on the signal broadcasted by GIOVE-A. In this paragraph the most significant results obtained post processing the real GIOVE-A signal sample are shown, as far as the acquisition and tracking phase are concerned.

5.1 Acquisition

This section describes the GIOVE-A signal acquisition process deployed in the Software tool.

Acquisition is a coarse synchronization process which produce outputs on estimation of the PRN code offset and of the carrier Doppler shift. This information is then used to initialize the tracking loops.

As already mentioned, Galileo is expected to use longer codes with respect to GPS, as the one currently transmitted by GIOVE-A. Moreover, the code structure foreseen is the so called "tiered codes" where a longer primary code is modulated by a short, lower rate, secondary code.

In order to deal with the presence of bit transitions on the data channel and of the secondary code on the pilot channel, in the acquisition phase a Partial Correlation method has been employed in the search space evaluation. This approach consists in correlating the incoming signal with the local code, as for the case of the standard approach, but using a time window shorter than a code period.

As Fig. 3 shows, considering the code period of the incoming signal 4 ms long for the Data channel or 8 ms long for the Pilot channel, the local code could be created only 2 ms long instead of using the whole code length. The principle behind this method is to perform the correlation between the shorter local code and the incoming signal moving the local code from a particular position as in Fig. 4(a) to the position with the maximum correlation peak as in Fig. 4(b) finding the best matching between the two sub-portion of the PRN codes. Even if the correlations are not made on the whole code period the acquisition is still possible, even if, of course loss in the correlation performance is experienced.

In order to reduce as much as possible the losses due to the coarse Doppler shift estimation, the doppler bin size in the search space must be accurately designed. It can be proved that the doppler resolution is related to the coherent integration time:

$$D = \frac{1}{T} \tag{4}$$

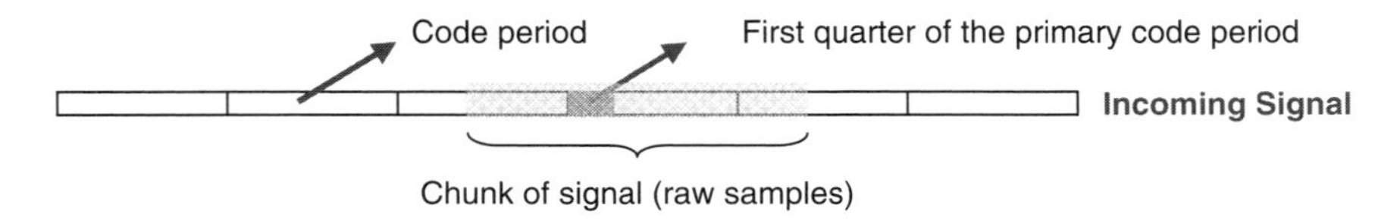

Fig. 3. Partial Correlation example.

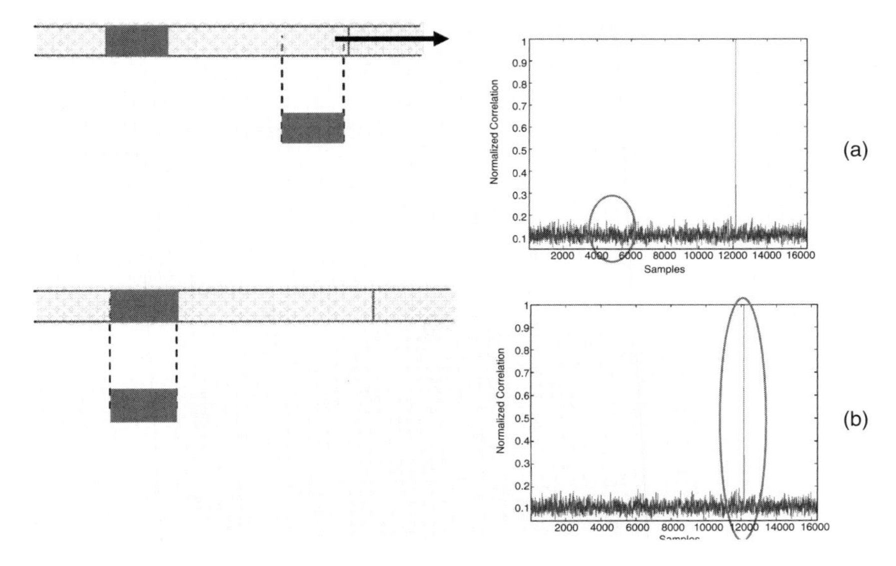

Fig. 4. Partial correlation method.

where D is a frequency bin width in Hz and T is the predetection integration time in seconds [5]. The problem arises with long code periods as in the GIOVE-A signal. In this case the bin width is small (250 Hz for Data channel e 125 Hz for Pilot channel) and this means that in order to cover the full range of uncertainty (usually from –5KHz to 5KHz for a user) the computational burden for the acquisition algorithm becomes unacceptable. For this reason, if the correlation is performed over a shorter time window, the frequency step can be increased maintaining the losses identical to the one with long integration time.

In other words, this means that this method has the advantages of reducing the computational load and the so called receiver time to first fix, but due to the fact that it does not fully exploit the code correlation characteristics the acquisition performed are reduced due to the larger correlation loss [6].

It must be noted that with the new structure of the signal this techniques can be applied to both the Data and the Pilot channels. As an example, Fig. 5 shows the signal acquisition on the Pilot Channel. As it can be observed in the Search Space plot of Fig. 5, the correlation peak pops out form the noise floor resulting in a visible and clean peak. Fig. 6 depicts the code correlation for the doppler row which provides the highest peak, it is possible to see both the main peak and the two side peaks due to the BOC modulation.

5.2 Tracking

As far as the signal tracking is concerned, the paper shows how the adopted double loops, composed by a PLL and DLL track the GIOVE-A signal.

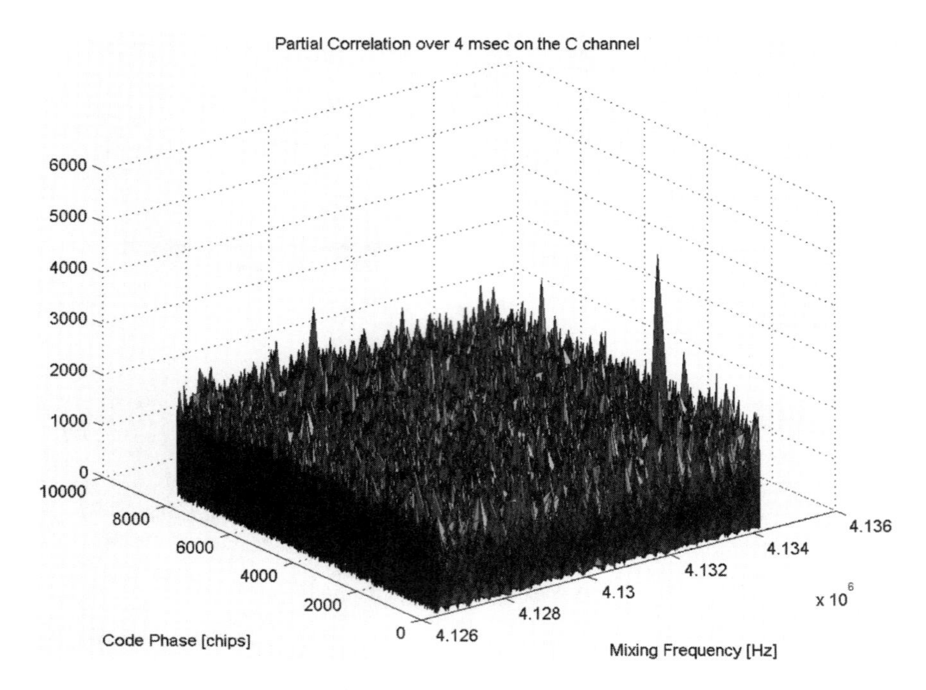

Fig. 5. Acquisition on pilot channel – search space.

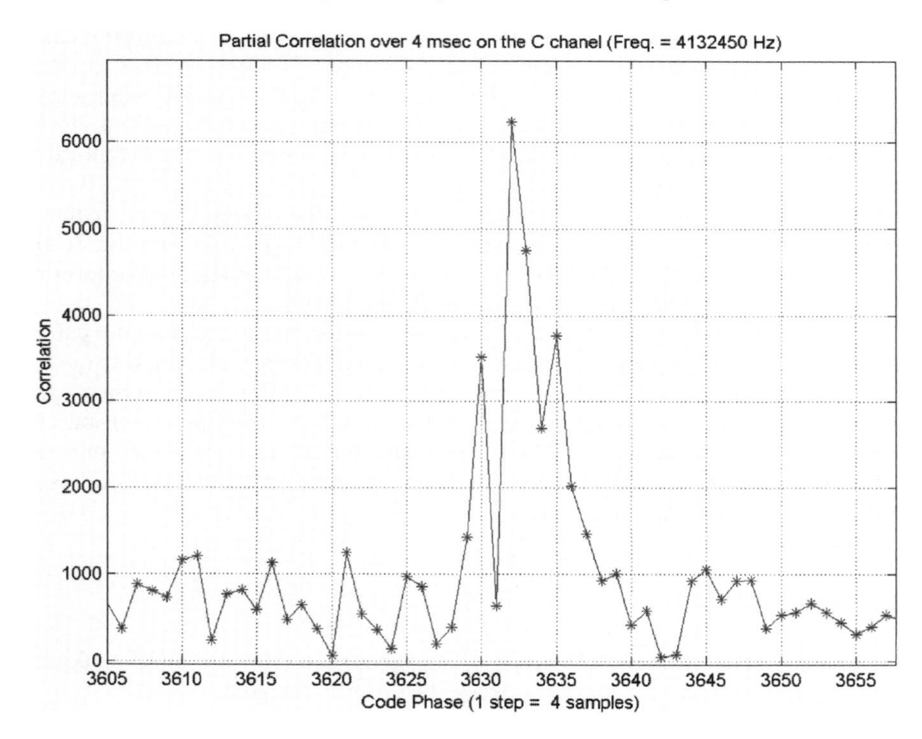

Fig. 6. Acquisition on pilot channel – autocorrelation peak.

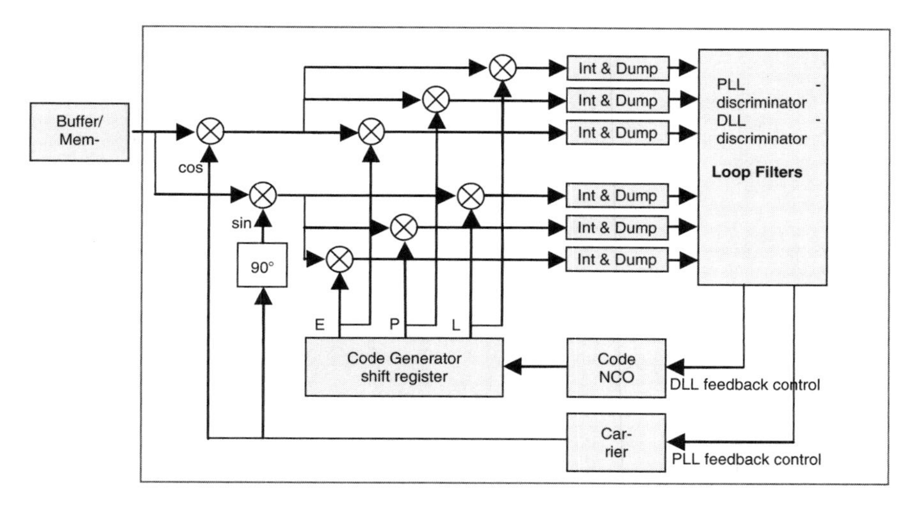

Fig. 7. Tracking block diagram.

After the acquisition process the control is then handed to the tracking loops. This part is able to track the variation in the carrier Doppler and code offset due to the line of sight dynamics between satellite and receiver. Another important function of the tracking loops is to demodulate the navigation data from the incoming signal.

Figure 7 shows the block diagram of the tracking core of the Software tool, where the received PRN code is synchronized with a locally generated code. This operation is at the basis of a GNSS receiver and the overall performance depends on the accuracy of such synchronization.

The tracking system was set up with a second order tracking and code loops with a bandwidth equal to 60 Hz and 5 Hz respectively. The PLL was a Costas PLL and the DLL was configured with a normalized dot product discriminator. Both the loops use a PI filter and the E-L spacing was set equal to 0.5 chips.

Choosing the right local code in the Software tool both the Data channel and the Pilot channel can be tracked.

As it is shown in Fig. 8, once the DLL and PLL are locked, data bits are visible at the output of Prompt in phase channel while the output of the quadrature channel is noisy.

Looking at Fig. 9, which shows the early, prompt and late correlator outputs for the Data channel, it is evident that the DLL is able to keep the local and incoming codes synchronized. In fact the early and the late outputs assume the same values over all the seconds of processed data. The DLL loop become stable after a short transient time, which depends on the loop bandwidth set.

Furthermore, such values are equal to half the prompt output, since the early-minus-late DLL was used (delay between the early and late is equal to 0.5).

Figure 10 shows the increment of the local carrier frequency with respect to the nominal value (4.1304MHz). Note that the mean of the local frequency tends to increase. Since the satellite is moving with respect to the receiver, the incoming

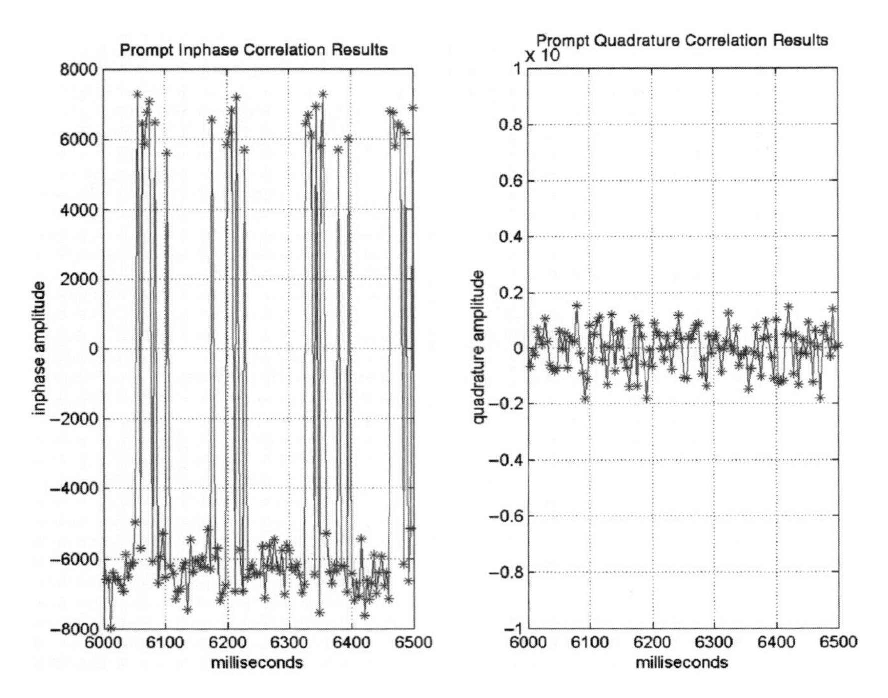

Fig. 8. Prompt quadrature and inphase channels – data channel.

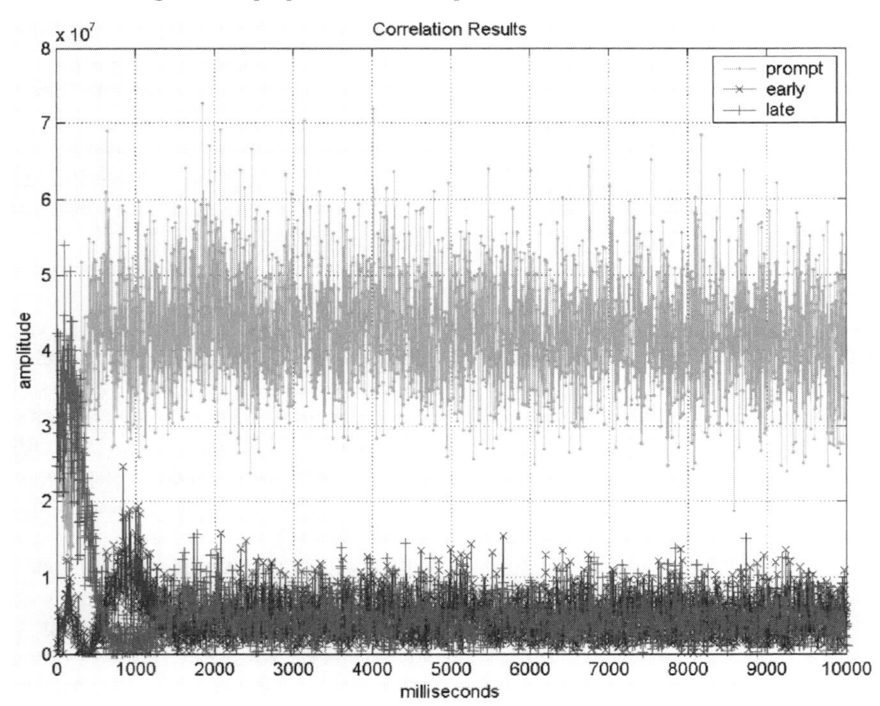

Fig. 9. Code discriminator output of the data channel.

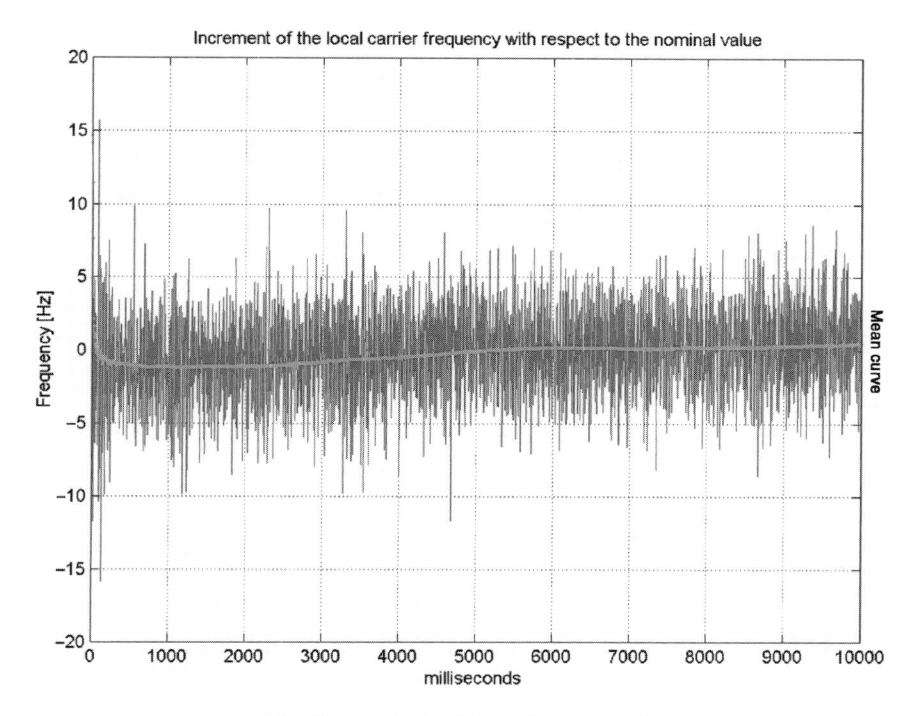

Fig. 10. Carrier tracking – data channel.

signal is frequency modulated due to the Doppler effect. The PLL is locked and the local frequency changes as well.

Thanks to the modularity of the Software tool the GIOVE-A signal has been tracked using also the Pilot channel. The Pilot channel has long code period, and integrating over a longer period, the loop control signals are sent back only once every 8 ms, but the noise impact is reduced as Figs. 12 and 13 show.

At the output of Prompt in phase channel it is visible the secondary code of GIOVE-A Pilot channel as depicted in Fig. 11, while only noise is present at the quadrature channel.

Looking at Fig. 12 it is evident that the DLL is synchronized with the incoming signal in fact the Prompt channel is higher than the early and late, which assume almost the same values.

Also the PLL, after a transient time, follows the carrier frequency and the shape of the output curve shown in Fig. 13 is consistent with the one depicted in Fig. 10.

6 Conclusion

This work has shown the basic signal processing for a GNSS receiver on the signal broadcast by the first Galileo satellite: GIOVE-A.

The new structure of the signal (longer primary codes, BOC modulation, use of a pilot channel with a secondary code, higher data rates) forces to change the common

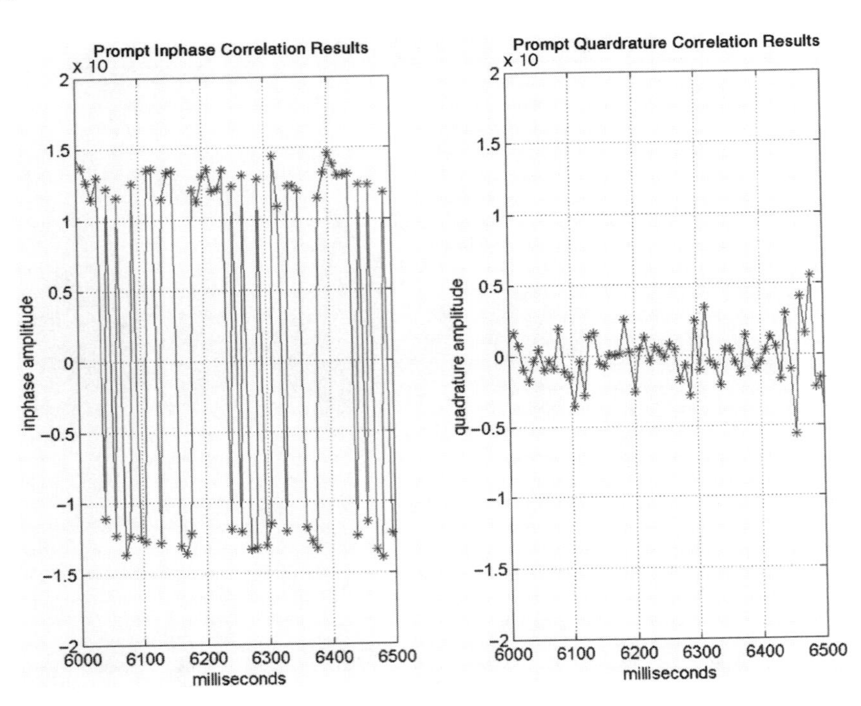

Fig. 11. Prompt quadrature and inphase channels – pilot channel.

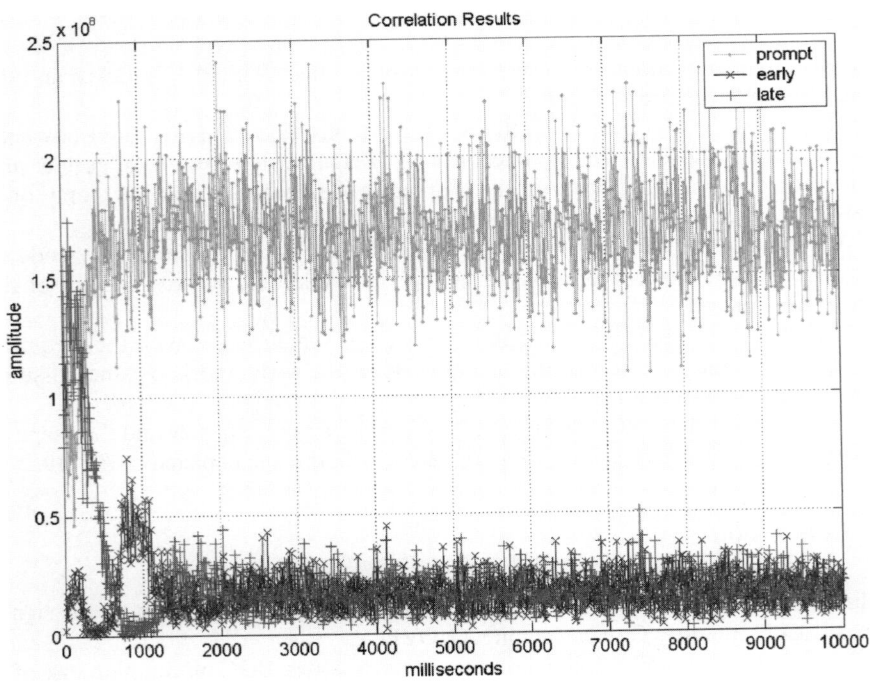

Fig. 12. Code discriminator output of the pilot channel.

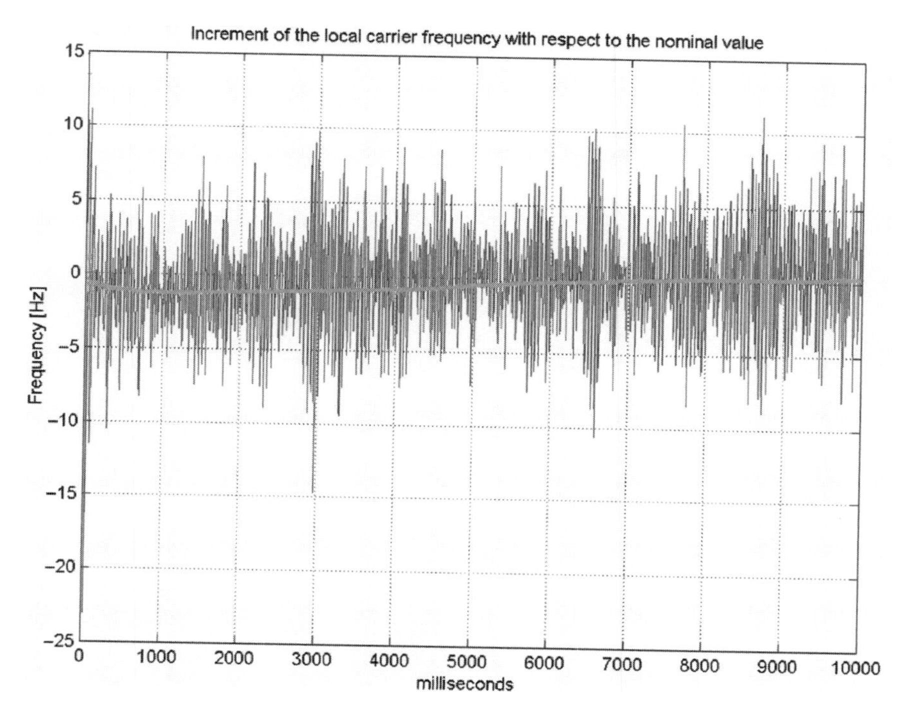

Fig. 13. Carrier tracking – pilot channel.

acquisition and tracking strategies usually adopted for GPS. For this reason the signal broadcast by GIOVE-A was acquired using the Partial Correlation and tracked on both the Data and Pilot channels.

The experiment performed on real data sets allowed to validate and better understand previous works based on simulated Galileo signals, and represent a first step for the test of novel receiver architectures tailored to the new features of the open service Galileo Signal In Space (SIS).

References

[1] Oliver Montenbruck et al., "GIOVE-A Initial Signal Analysis", GPS Solution, Volume 10, Number 2, pp.146–153, May 2006.

[2] http://gps.ece.cornell.edu/galileo/

[3] Dafesh PA, Nguyen TM, Lazar S, "Coherent Adaptive Subcarrier Modulation (CASM) for GPS Modernization.", Institute of navigation, National Technical Meeting, ION-NTM-1999, 1999.

[4] Hein GW. et al., "Performances of Galileo L1 Signal Candidates", GNSS 2004, 16–19 May 2004, Rotterdam, The Netherlands.

[5] Falcone M. et al. in: Kaplan E., Hegarty ChJ (eds), Understanding GPS-Principles and Applications, 2nd edition. Artech House.

[6] Borio D., Camoriano L., Lo Presti L., Fantino M., "Acquisition analysis for Galileo BOC modulated Signals: theory and simulation", European Navigation Conference (ENC) 2006, Manchester, UK.

3 PTF Major Design Drivers

The PTF is based on the experience of the Experimental Precise Timing Station. Particularly important with respect to the new PTF targets are the lessons learnt in the field of GST Generation Algorithms, Time generation & measurement performance, Operations & Logistics.

Key functional Req's are the following:

- Generation of the Galileo System Time / MasterClock [GST(MC)]
- Fully unmanned operations under remote Control/Monitor of the Galileo Asset Control Facility (GACF),
- Cooperation for metrological timekeeping with an entity external to GMS, the Time Service Provider (TSP)
- Capability of fully autonomous generation of GST in lack of the TSP.

The PTF Performance Requirements are very challenging; they are dimensioned in front of the results from the research phase conducted by the European Time & Frequency Laboratories, and establish new technology limits for the industrial products (see section 7).

They cover mainly the following parameters:

- GST Time and Frequency Stability
- GST Time and Frequency Offsets wrt. UTC
- GST to PTF2 Time/Frequency Offset and stability
- GST to GPS Time/Frequency Offset and stability Accuracy

Additional requirements are established to ensure that the PTF is a real industrial product; they include Quality, design and environmental req'ts as the following:

- Reliability, Availability and Maintainability
- SW Quality as per the Galileo SW standards
- Electro-Magnetic Compatibility and Interference
- Thermal and Magnetic environments (due to the sensitivity of the instrumentation)

4 CTT PTF Overall Configuration

The PTF configuration is based on the following S/S's (see Fig. 2):

- Measurement / Control S/S including the Data Processing Platforms that run the SW Components, including the main Algorithms, and the Measurement Instrumentation.
- Time Generation S/S including the pure HW items capable to generate and distribute GST with high dependability, namely the short and medium term stability oscillators
- Time Transfer S/S including the Two Way Station, the GPS CV Receiver, the GSS I/F and the relevant Algorithms.

The Fig. 3 shows the PTF from a control view point:

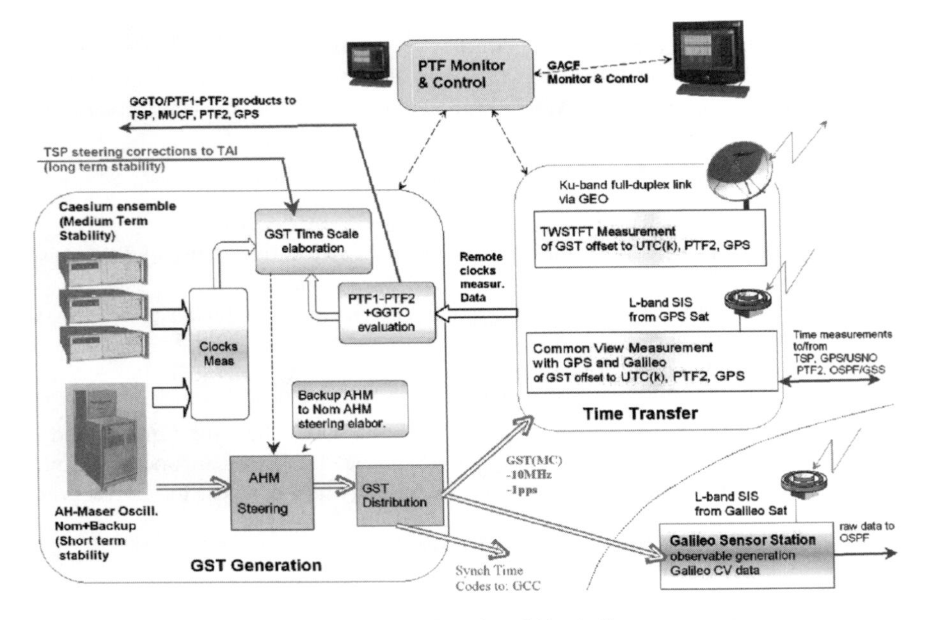

Fig. 2. CTT PTF functional block diagram.

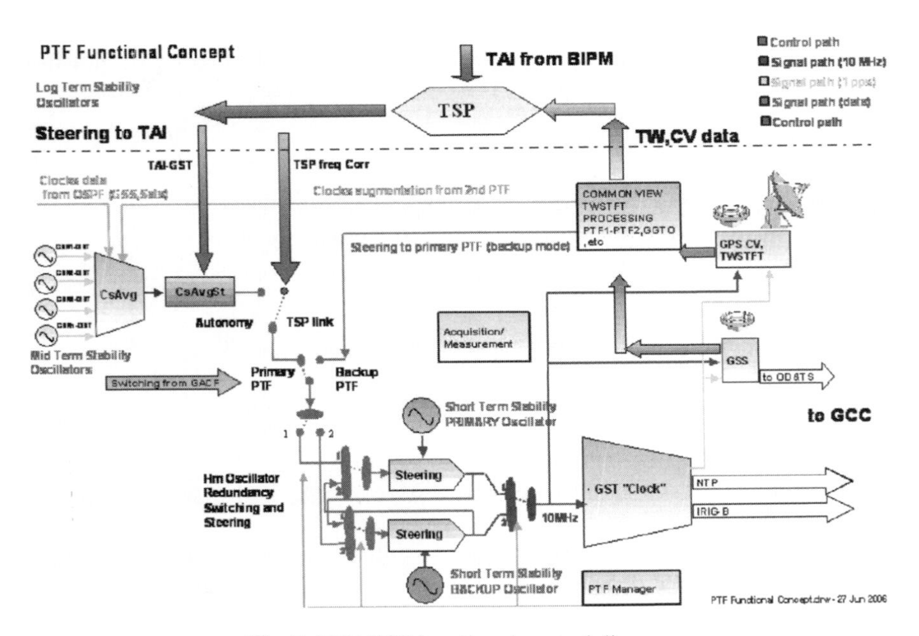

Fig. 3. CTT PTF functional control diagram.

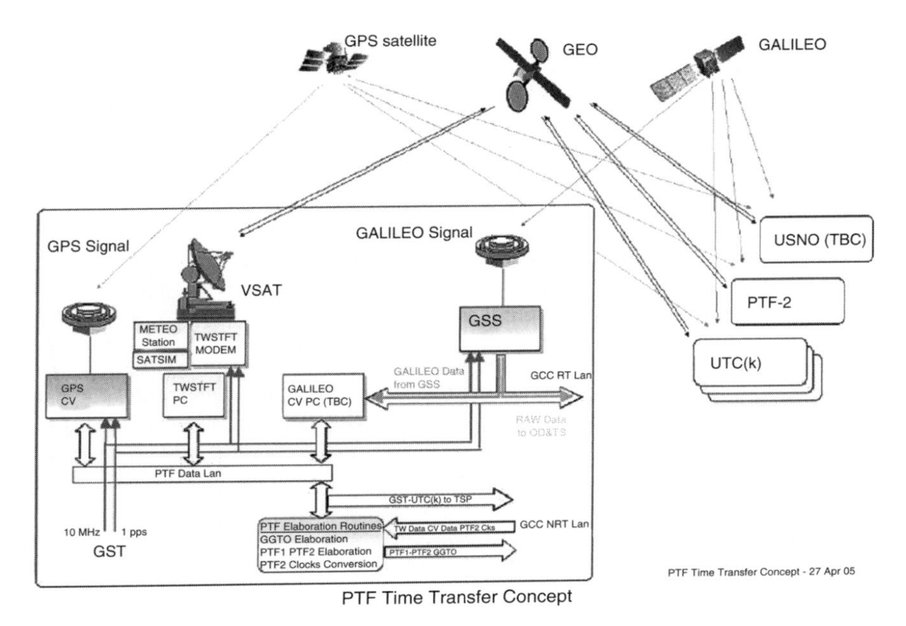

Fig. 4. CTT PTF time transfer concept.

The Time Transfer S/S is designed to actuate the measurement of the Time Offsets wrt the other Time keeping facilities as shown in the concept diagram of Fig. 4.

5 CTT PTF Technological Issues

The PTF technology is aimed to ensure the performance requirements of the Galileo System Time scale (see section 7).
To afford such performance the key technological issues of the PTF are:

- Eccellent short term stability provided by Active Hydrogen MASER (AHM): Allan Deviation in 24 hours $\leq 2.\ 10^{-15}$
- Eccellent medium term stability provided by a Caesium clock ensemble: Allan Deviation in 5 to 30 days $\leq 1.\ 10^{-14}$
- State-of-the-art measurement and data acquisition instrumentation capable of monitoring in real time the clocks performances
- GST generation algorithms used both to check the TSP corrections in nominal operations and to replace them in autonomy. These algorithms are based on the data processing of the local and remote clock measurements.
- Physical realisation of GST(MC) by controlling the output of one of the available Active Hydrogen Maser clocks via a precision Phase Pico-Stepper
- Redundant AHM continuously steered to the primary one by means of dedicated algorithm to ensure a smooth switch in case of failure

- High-accuracy synchronisation links to external laboratories (UTC(k)) to provide data to TSP for GST to TAI/(UTC) steering corrections elaboration. These synchronisation links are performed by state-of-the-art two-way modems and common-view GPS time transfer receivers
- Direct reference of GST to the OD&TS functions (Orbite Determination & Time Synchronization) via a co-located GSS receiver (feeding ranging data to the OD&TS) driven by physical realisation of GST(MC)

6 CTT PTF Algorithms

The PTF is operating on the basis of different algorithms, including the GST Generation Algorithm, that is the core, the Algorithms for the steering of the backup AH MASER to the nominal one, the steering of the backup PTF on the master one and those for the measurement and evaluation of the time offsets between the generated GST scale and the external time references of the UTC(k) European and GPS laboratories.

The GST Generation Algorithm is based on the prototype developed by INRIM (Fig. 5) and Politecnico di Torino since year 2000 and in particular through the experience and lessons learned on the Experimental Timing Station (EPTS) developed during the Galileo GSTB-v1 phase (see Reference [1]).

Fig. 5. Experimental Precise timing station at INRIM (2005, GSTB-v1).

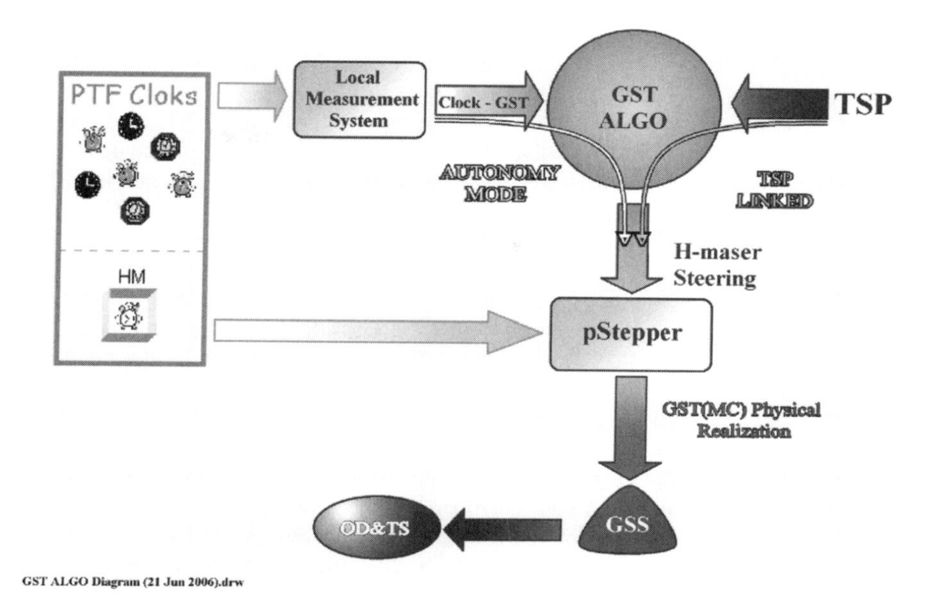

Fig. 6. GST generation algorithm concept.

The current GST Algoritms (Fig. 6) are being developed in front of PTF requirements, including difference wrt. the EPTS as:

- Corrections provided by the TSP (in nominal mode)
- Steering to the other PTF (in slave mode).

7 CTT PTF Performance

The PTF performance parameters are relevant to the goals of the Galileo System Time scale in the domains of Navigation and Time Metrology.
So far no performance results are available as the PFT implementation process is at preliminary design stage. Performance data are here-after given in terms of applicable requirements and design analyses.
In the domain of Navigation, the PTF is required to generate the GST(MC) physical time scale with the following main performance:

- Frequency stability to TAI < 4.3 × 10–15 in terms of Allan standard deviation in 24 hours, including the contribution due to corrections for steering toward TAI

The main goals of the Time Metrology are implemented by the PTF in cooperation with TSP, BIPM and Time Metrology Laboratories and are established with the following requirements applicable at level of Galileo system:

- GST(MC) to UTC *(mod 1 sec)* Time Offset < 50 ns (2σ) over any yearly time interval
- Uncertainty of GST(MC) to TAI Offset < 28 ns (2σ).

Anyway the PTF, thanks to its reliability, shall ensure fully autonomous operations in case of failure of TSP or Time Metrology Laboratories (that are not industrial products) with the following requirement:

- GST(MC) degradation ≤ 20 ns (2σ) over 10 days (still under revision).

In addition, to ensure interoperability with the US GPS, the PTF shall accurately compute the GPS to Galileo Time Offset (GGTO) with following requirement:

- GGTO Accuracy ≤ 5 ns (2σ) over any 24 hours.

The PTF design analyses are on going in parallel to the design definition. The performance analyses have been conducted on the basis of the documentation of the foreseen instrumentation and of the results obtained from the GSTB-v1 experience.

These results are significant as the E-PTS Algorithms and its time instrumentation are basically reutilized for the PTF.

The Fig. 7 shows the results obtained in 10 months of test during the GSTB-v1 phase in terms of Experimental-GST to TAI Offset (Reference [1]).

The results of the current analyses show that the applicable requirements are expected to be met.

Figure 8 shows in particular the estimation of the GST stability.

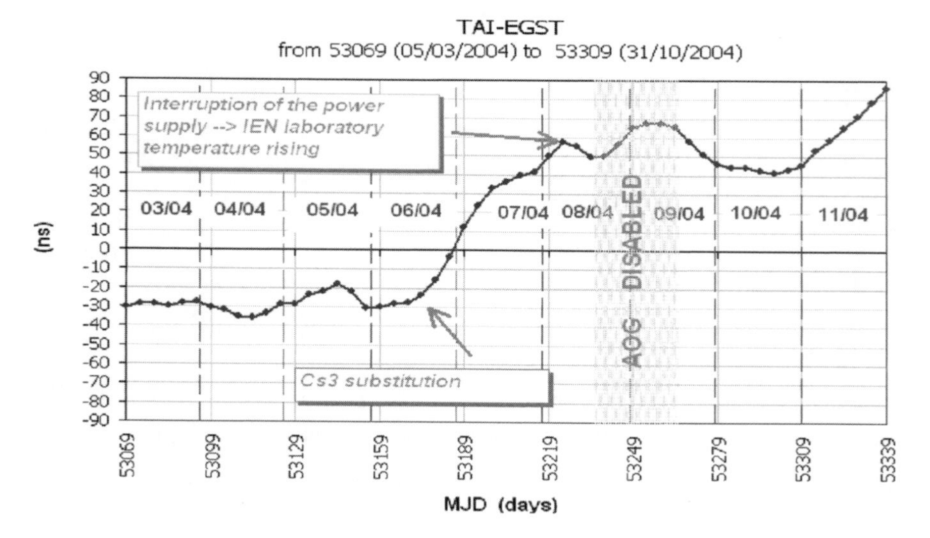

Fig. 7. E-PTS performance (2005, GSTB-v1).

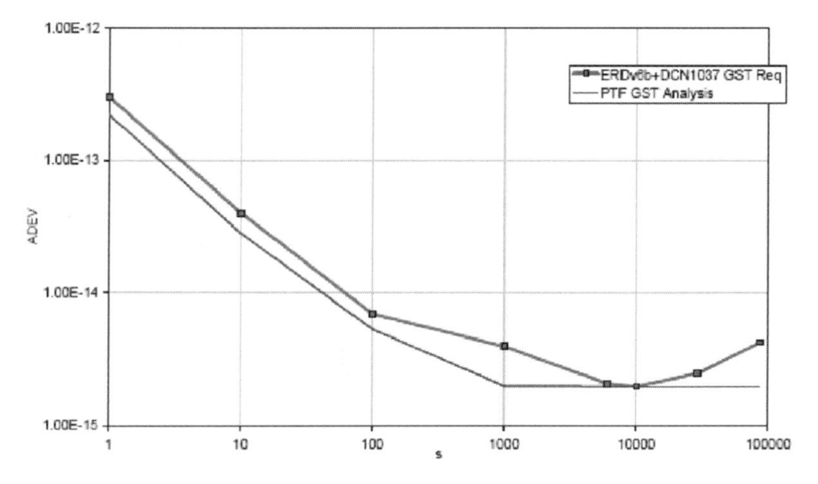

Fig. 8. GST(MC) expected stability vs. requirements.

References

[1] Cordara,Costa,Lorini,Orgiazzi,Pettiti,Sesia, Tavella (IENGF, Torino, Italy), Elia, Mascarello (Alenia Spazio, Torino, Italy), Falcone, Hahn (ESA, Noordwijk, The Netherlands): **E-GST: one year of real time experiment, 36th Annual PTTI Mtg, 2005**

GIRASOLE Receiver Development for Safety of Life Applications

Livio Marradi[1], Lucio Foglia[1], Gianluca Franzoni[1], Antonella Albanese[1], Stella Di Raimondo[1], Vincent Gabaglio[2]

[1]Alcatel Alenia Space Italia S.p.A.
SS. Padana Superiore 290
20090 - Vimodrone (Mi) - Italy
e-mail: livio.marradi@aleniaspazio.it
[2]Galileo Joint Undertaking
Rue du Luxembourg, 3 B-1000 Bruxelles – Belgium
e-mail: vincent.gabaglio@galileoju.com

Abstract. The GIRASOLE Safety of Life Receiver is developed in the frame of the Galileo Joint Undertaking (GJU) Research and Development activities, in the context of the European Commission 6th Framework Programme 2nd call by ALCATEL ALENIA SPACE Italia (AASI), manufacturer leader of advanced, high performance GPS/EGNOS/Galileo receivers for space and safety-critical applications.

Differently from GPS, which relies on an external signal (EGNOS), one of the most appealing features of Galileo is the signal embedded integrity. As this characteristic provides the users with information about the availability and correctness of the Galileo system and signal, Galileo is particularly attractive for all the critical applications that undergo the general name of Safety of Life, or SoL for brevity.

The Safety of Life Services (SoL) are targeted at users who need assurance of service performance in real-time. Typically they are safety critical users, for example Aviation, Maritime and Rail, whose applications or operations require stringent performance levels.

Other applications can be envisaged like emergency and road.

In these types of applications, the receiver plays an important role since is one of the key elements of the safety chain.

The main characteristic of such type of receivers is their capability to detect failures that can come from different sources like the Signal In Space, the environment (ionosphere, troposphere, interference and multipath effects), the constellation etc. and be capable to interpret the integrity information broadcast by the satellites.

In the frame of its 2nd Call, the Galileo Joint Undertaking (GJU) has launched several activities aiming to provide basic technological elements (i.e. receivers) useful for the different services offered by Galileo.

Within this frame, GJU has selected the GIRASOLE project as the one aiming at development of receivers for SoL application. In the framework of the project, Alcatel-Alenia-Space Italia S.p.A. (AAS-I) is leading a Consortium of several companies from eight different countries all over the world.

The GIRASOLE project aims to allow strategic developments of technologies and basic elements of a Safety of life (SoL) Galileo Receiver.

The GIRASOLE system architecture is based on GARDA heritage, a project developed by AASI-Milano under GJU contract (1st call) with the key objective to build an advanced user receiver prototype, configurable to simultaneously track Galileo/GPS/SBAS satellites, supporting all Galileo frequencies and modulations.

The possibility to have Galileo receivers prototypes available at an early stage will provide benefits for standardization and certification of the receivers within each user communities, while facilitating the market penetration of Galileo.

The receiver processes Galileo signals on the L1, E5b bands and GPS/EGNOS signals on the L1 band. It also provides combined Galileo + GPS Navigation solution, Integrity calculations (HMI, critical satellite prediction and Navigation warning) using Galileo and EGNOS Integrity message, interference and multipath mitigation, support to the use of Local Elements and output raw measurements data of each satellite.

GIRASOLE is conceived to be flexible and easily customized to the user needs in the following applications:

- *Aviation*
- *Maritime*
- *Rail*

The receiver architecture is based on a common core and some specific parts related to the application.

The common core includes: main part of the RF to IF down conversion, HW/SW Signal Processing, Standard Navigation SW, Integrity Processing. The application specific parts include: antenna, RF Front End and filtering and Application dependant HW and SW (e.g. Interfaces.)

1 Introduction and Project Overview

One of the main differentiators of Galileo with respect to GPS and GLONASS is the provision of the integrity information and guarantee of service. These characteristics make systems based on Galileo SoL service suitable to be used in safety related applications where guarantees for integrity and continuity of service are essential.

One of the key elements of the ground components that contribute to the overall system integrity is the receiver, since guarantee for integrity and continuity of service can be achieved only if the user embeds in its equipments Galileo receiver of the Safety of Life class.

There is therefore the need to foster the development of technological elements, like the receivers, for the Galileo SoL service. The GIRASOLE projects aims to provide a comprehensive answer to these needs either in terms of technology development or in term of covering the widest area of the safety critical application.

The main objectives of the GIRASOLE project are:

- Investigation and identification of the main core technologies for Galileo SoL receivers;
- Development of breadboards for the three main safety critical applications (i.e. Aviation, Maritime and Rail) in order to support standardization, foster certification process within each User Community and facilitate Galileo market penetration;
- Allow early availability of receiver prototypes;
- Develop tools useful for SoL receivers development.

The project Consortium gathers expertise in different fields from several countries all around Europe and world wide. Three main receiver manufacturers, each involved in the development of one specific SoL receiver, are supported by Research Institute and specialized company for the investigation and development of relevant core technologies suitable for SoL receivers.

The project baseline approach is constituted by three main elements linked together:

- The main Inputs (SiS ICDs, Standards, External co-ordination activities)
- The main Tasks (Technology Investigation, Requirements definition, Common Platform development, Breadboards development and tests, Final architecture and prototypes)
- The Target Results (consolidated specifications, validated key techniques and components)

Several inputs are foreseen for the projects and among them those coming from the GARDA project (a GJU project under the frame of the 1st Call) have a particular importance. They include studies about the development activities of the Galileo receivers, identification and investigation on some core technologies, development tools and preliminary development activities of a Galileo receiver. All these inputs are taken into account and further deepened within the project.

Three issues of the Receiver Requirements and Specification document will be created: first one on the basis of the input to the project, as described above, a second issue will be released taking into account feedbacks coming from the Core Technologies investigation task and the third and final issue is completed at the end of the project with the feedback from the Breadboard development task.

An important part of the project is the identification and investigation of the main Core Technologies that act as common base for all the SoL applications and

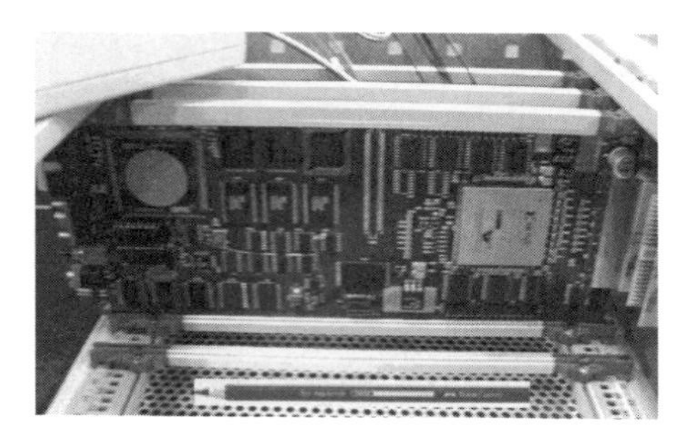

The Garda Prototype.

identification and investigation of technologies specifically related to the targeted SoL applications.

Once identified, they will be analysed by means of the GRANADA SW receiver tool (another heritage of GARDA project) in order to produce solutions and any other indications that will be used as starting point in the Breadboard development task.

Another important concept developed in the frame of GIRASOLE is the Common Platform.

The concept is based on the consideration that the receiver is made of several functions, some of which are common and some others are specific for the three applications.

The Common Platform objective is therefore twofold:

- A way to study and test common core technologies, providing partners with a bench on which main common basic SOL receiver functions can be developed and tested, like the Integrity concept, possibly anticipating problems and solutions to be reflected on final breadboard;
- A way to develop common building blocks shareable among all the three receiver types.

The SoL receivers breadboard development activity derives and is conducted in parallel with the Core Technologies investigation, to which it provides feedback for a better identification, and is aimed to the design and development of a Receiver Breadboard for each of the three identified applications. The breadboards will therefore provide feedback to the Requirements and Specification activity and the Architecture and the Architecture and Building Block definition for the final receiver.

Galileo Simulator test tools are also developed in the frame of the project as elements supporting the receivers development activity.

Based on the above consideration GIRASOLE has several interesting and attractive elements that can be summarized as follows:

- Three GNSS receiver manufacturers are involved in the design and development, thus allowing to exploit different and complementary expertise;

- each GNSS receiver manufacturer is pushed to develop basic technologies for the Galileo system, to be afterwards transferred to other potential users;
- the activities are distributed among different manufacturers and countries, thus allowing a well spread Galileo Safety of Life Service promotion.

1.1 Use of GNSS in Safety of Life Applications

Several applications can be envisaged that require integrity information and continuity of service. The three main areas for safety critical application are identified in the Aviation, Maritime and Rail sectors.

The level of safety provided by EGNOS and Galileo, together with the use of interoperable GPS/Galileo receivers, will grant pilots assistance in all flight phases, from on the ground movement, to take-off, en-route flying, and landing in all weather conditions.

Aircraft separation can be reduced in congested airspace thank to higher accuracy and service integrity, allowing a significant increase in traffic capacity.

Surface movements and guidance control are application of the aviation sector that can benefit of an improved safety service.

Also helicopters guidance in bad weather conditions can benefit of the Galileo SoL service, thus allowing an improvement in the availability of rescue services (like medical helicopter) and generally in all emergencies management.

A wide variety of vessels moves around the world every day and sea and waterways represent one of the most widely used mean for good transportation. The increased accuracy, integrity, high availability and certified services that Galileo can provide, will improve efficiency, safety and optimization of marine transportations.

Usage of GNSS SoL based equipments can be envisaged in every phase of marine navigation: ocean, coastal, port approach and port manoeuvres, under all weather conditions.

The characteristics of a SoL receiver are ideal for navigation in the open sea and the integrity improve adds confidence in the calculated position of a vessel.

Also inland Waterways Navigation can benefit of precise navigation provided by a SoL receiver. Navigation on rivers and canals needs accuracy and integrity of navigation data as fundamental requirement, especially in critical geographical environments or in bad weather conditions.

The main users of SoL receivers for maritime application are conventional sea and river vessels, i.e., vessels whose navigation is fully regulated by the safety requirements adopted by National Maritime Administrations, namely:

- deep-sea and coasting ships, including cargo ships, tankers, passenger ships and fishing ships;
- sea ships and harbor boats, auxiliary, specialty and service ships, including icebreakers, floating workshops and floating cranes;
- river ships, including tankers, passenger ships, transport ships, self-propelled auxiliary and specialty ships and boats.

Railway operations can be significantly improved by the use of global positioning systems in terms of navigation accuracy, safety and assistance to the operations.

By integrating the integrity concepts, Galileo will ensure the accuracy and integrity for multi-modal transport applications, thus making user applications more reliable and more accurate.

The use of the GNSS signals by the railways represents a technical, industrial and operational challenge. The railways have a consolidated experience using other means of navigation, which may not be ideal, but are well known and familiar to the operators. On the other hand, railways, like other modes, have to cope with a number of new challenges and are under economic pressure to improve and optimize significantly their operations in terms of track occupancy, safety, productivity and customer satisfaction.

The primary objective of the integration of GNSS in the train equipment is to demonstrate the improvement of the train self-capability in determining its own position and velocity, with limited or no support from the track side and to show that the equipment can comply with the European Railway Train Management System (ERTMS) requirements enabling a cost-effective modernization and increasing the efficiency.

2 Receiver Architecture Overview

The receiver architecture of the AAS-I breadboard is composed of three main modules:

- Antenna and RF Front-end: this is a single element covering both the lower band (E5, E6) and the higher band (L1). The antenna is designed by Satimo targeting the professional/safety-of-life receiver applications. The active section is based on a Low Noise Amplifier and filters providing required amplification with very good noise characteristics.
- RF/IF section: the RF/IF section is in charge of RF signals amplification, RF to IF down-conversion and Local Oscillator synthesis. The down-conversion is based on a single mixing stage, while the final conversion to base band is accomplished after the signal digitization, by the digital channel. The IF signal sampling is also performed within this section, using a high speed ADC and coding the samples on three bits.
- Digital Section: a new proprietary digital channel was specifically designed and developed by AAS-I to process Galileo and GPS signals. The channel, named GALVANI, was designed and simulated through a Simulink design process

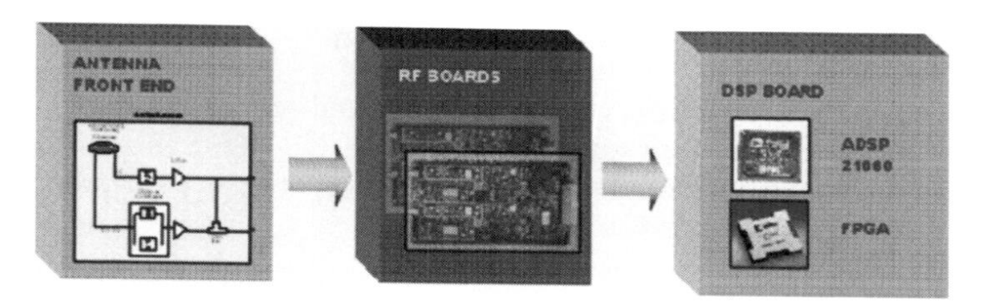

Receiver prototype block diagram.

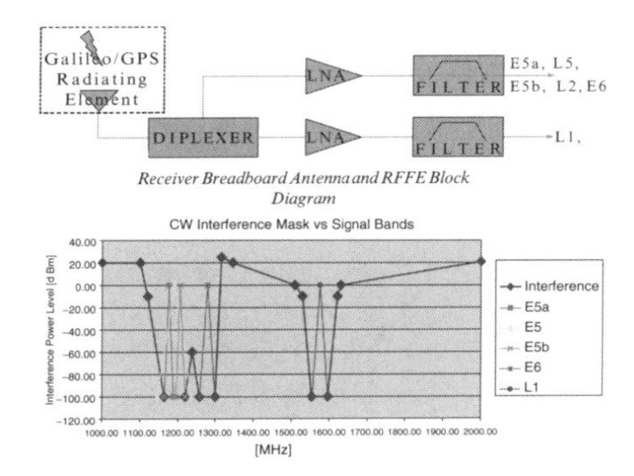

Breadboard interference rejection mask.

and ported to VHDL for implementation on a Xilinx Virtex II FPGA. The digital section includes an Analog Devices ADSP21060 that runs all the acquisition signal processing and tracking loops software. Digital channels can be flexibly configured as Single Frequency Channels (SFC) including data and pilot signals or as complex Multi-Frequency Channels (MFC), each handling multiple Galileo carriers.

2.1 Antenna and RF Front End

The antenna breadboard design has been driven by the following concepts: phase centre stability, minimization of multi-path and overall phase error, minimization of antenna size, suppression of unwanted out-of-band signals and minimization of manufacturing costs.

The antenna is a broadband element covering the L1 and E5 and E6 bands. It is based on a patch element, and includes the input RF filter (interference mitigation), based on a custom made diplexing element. The LNAs are physically located in the antenna envelope and provide two separate outputs feeding the RF/IF board. Band separation is achieved through a diplexer with reasonable rejection slope and low losses to be trade-off with the overall antenna performance. The additional separation of the E5 bands is better accomplished after the LNA.

2.2 RF/IF Section

The design of RF/IF section includes the hardware for two complete (L1, E5b) signal paths with identical architecture. Each path is in charge of generating the Local Oscillator signal, mixing the incoming RF signal to Intermediate Frequency, according to a properly defined frequency plan, providing the signal amplification, filtering and, finally, digitizing the IF signal.

The down-conversion is based on a single mixing stage, while the final conversion to base-band is accomplished after the signal digitization, by the digital channel.

Reconfigurable PLLs, image filters, IF filters and digitally controlled AGCs are located in each signal branch.

The IF signal sampling is also performed within this section, using a high speed ADC. The ADCs used are AD9480 with a maximum resolution of 8 bit at 250 Msps, which also provide LVDS outputs suitable for signals transfer to the Signal Processing. The digital samples are coded using three bits and the sampling frequency of the ADC is 95 Msps: this sampling frequency has been selected adopting standard precautions against aliasing and excess signal loss.

The frequency plan has been studied considering spurious frequency suppression. This means that harmonics or non-harmonics related to the clock generation and mixing process must be kept under control and possibly located in non dangerous frequency regions. The Local Oscillators phase noise is kept under tight control. The frequency synthesizer has been designed with the goal of achieving best phase noise performances.

The master oscillator is a standard 10 MHz clock and the choice of 70 MHz as IF frequency has allowed the use of standard SAW filters.

2.3 Digital Section and Channel Correlators

The Digital Signal Processing board is a re-programmable computing platform for high data-throughput signal processing Galileo/GPS applications. The module is able to process an entire Galileo/GPS digital channel acquired via LVDS Interface. The digital channel (called GALVANI) main functionalities have been implemented on a Virtex-II XC2V8000 FPGA, while signal processing algorithms run on Analog Device DSP 21060 SHARC. The board is also equipped with SPI and RS232 connectors: the former is used to connect an external memory card useful for testing purposes, while the latter to download data from the on board DSP and communicate with the user interface program.

The reference block diagram of the GALVANI chip is sketched in the figure: it is made up by a matrix of processing elements referenced as Digital Channel (DC). Each DC is a flexible processing unit that can be configured and controlled to demodulate and track any channel of the Galileo signal, including both pilot and the data sub-channels. Indeed the DC matrix is controlled and configured by a software routine running on DSP core in order to implement acquisition and tracking algorithms and symbols demodulation. A DC can also be referred to as a Single Frequency Channel (SFC) to highlight that each processing element is dedicated

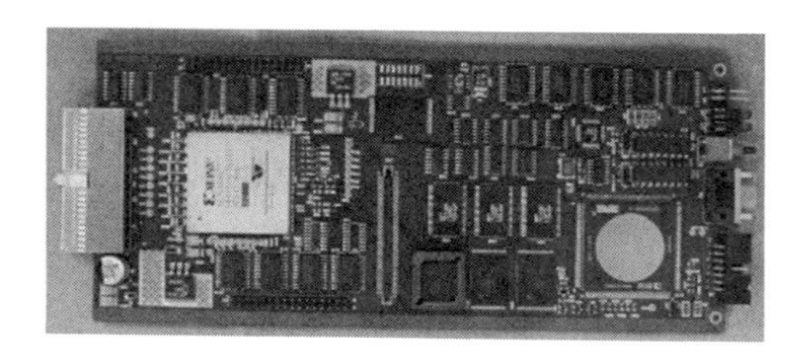

Digital signal processing board.

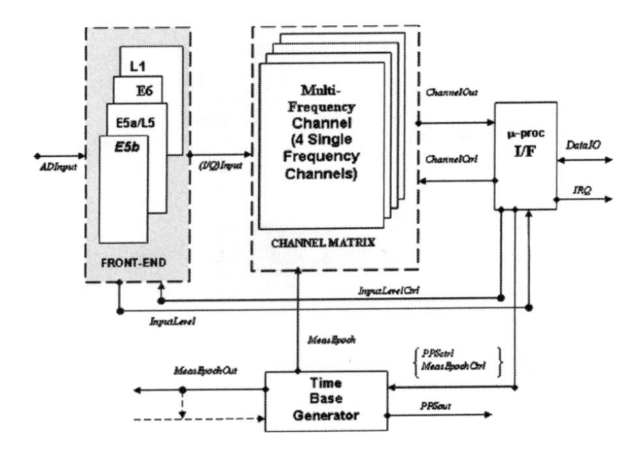

Digital channels architecture.

to a single channel on a particular Galileo carrier. In line with this notation, a collection of 4 DC, one for each carrier, is referred to as Multi Frequency Channel (MFC), and it is the basic unit which can track all the signals from a Galileo satellite.

By properly programming the DSP it is possible to reconfigure each DC realizing a flexible reuse of the FPGA resources. Within the FPGA a dedicated micro-processor interface is also needed to interface the internal signal processing with the DSP.

2.4 Application Breadboards

In the framework of GIRASOLE, AAS-I has developed an integrated Galileo/GPS/EGNOS breadboard receiver for on-board train safety-critical applications.

The receiver processes the Galileo signals on the L1 and E5b bands and the GPS/SBAS signals on the L1 band. The receiver baseline is to use E5b and L1 for Safety of Life Service. In future configurations, to allow also dual-freq GPS L1-L5 measurements, the receiver may be configured with an additional E5a/L5 RF path. The Rail breadboard receiver performs the following main functions:

- Search visible Galileo/GPS/ SBAS SVs and allocate HW channels to SVs on the basis of predefined strategies;
- Acquire and Track Galileo/GPS/ SBAS specified signals;
- Maintain Code Lock and Carrier Lock, demodulate and decode data messages and recover Navigation Data from each received GNSS satellite;
- Implement interference and multipath mitigation techniques at the most appropriate level (HW or signal processing);
- Perform position, time and velocity calculation with GPS, Galileo and/or a combination of GPS + Galileo SVs in view;

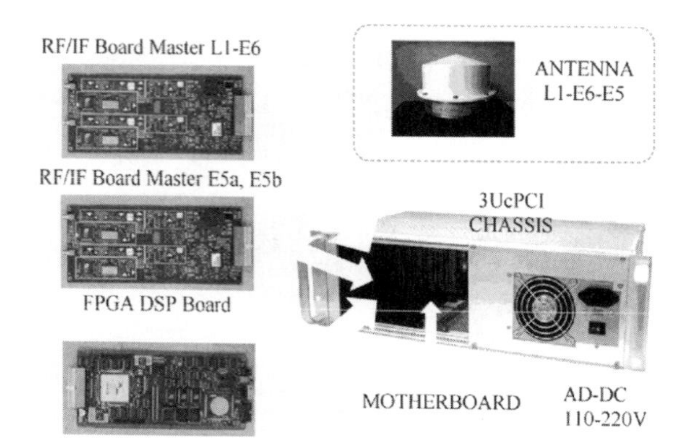

Receiver breadboard functional architecture.

- Use calculated position information to establish geometrical line of sight information of each acquired GNSS satellite with respect to the receiver platform and predict a tracking list of visible satellites;
- Perform integrity related calculations (xPL, HMI, Critical Satellites Prediction and Navigation Warning) to alert the user in case of integrity risks;
- Provide raw measurements data for each GNSS satellite in lock;
- Monitor receiver health status;
- Allow receiver control by the user through its Command & Control interface.

In addition, the following features are under investigation:

1. the Rail SoL Receiver capability to accept, at its serial interface port, pseudorange measurement correction data from a trackside Local Reference Station, in order to perform a code-based local-precision positioning solution,
2. the Rail SoL Receiver capability to accept, at its serial interface port aiding data from an inertial measurement unit (including three axis acceleration and angular rates) in order to propagate position/velocity and satellites predictions in conditions of no visibility (tunnels), to aid satellites acquisition/reacquisition and to perform additional data integrity checking.

3 Core Technologies

3.1 Integrity

Integrity is a key issue in satellite navigation for safety-of-life applications. Integrity is a measure of the trust which can be placed in the correctness of the information. Integrity includes the ability of a system to provide timely and valid warnings to the user (alerts) when the system must not be used for the intended operation. The integrity performance is specified by means of three parameters:

- ***Integrity risk***
  This is the probability during the period of operation that an error, whatever is the source, might result in a computed position error exceeding a maximum allowed value, called **Alert Limit**, and the user be not informed within the specific time to alarm.
- ***Alert limit***
  This is the maximum allowable error in the user position solution before an alarm is to be raised within the specific time to alarm. This alarm limit is dependent on the considered operation, and each user is responsible for determining its own integrity in regard of this limit for a given operation following the information provided by GNSS SIS.
- ***Time-to-alert***
  The time to alert is defined as the time starting when an alarm condition occurs to the time that the alarm is displayed at the user interface. Time to detect the alarm condition is included as a component of this requirement.

The integrity capabilities of the Girasole receiver consist of the Galileo SoL service integrity messages, the RAIM, the SBAS integrity messages and the GBAS integrity messages.

The integrity concept introduced by Galileo is innovative and has the aim to provide the user with a more powerful mean to check the integrity of the system. Integrity concepts have been established and optimized for present Space Based Augmentation Systems (SBAS) like WAAS (Wide Area Augmentation System) or EGNOS (European Geostationary Navigation Overlay System) according to their required performances in terms of availability, integrity, and continuity.

The performance requirements for Galileo are one order of magnitude more demanding compared to these present systems and therefore a new integrity concept, based on the established approaches, has been developed.

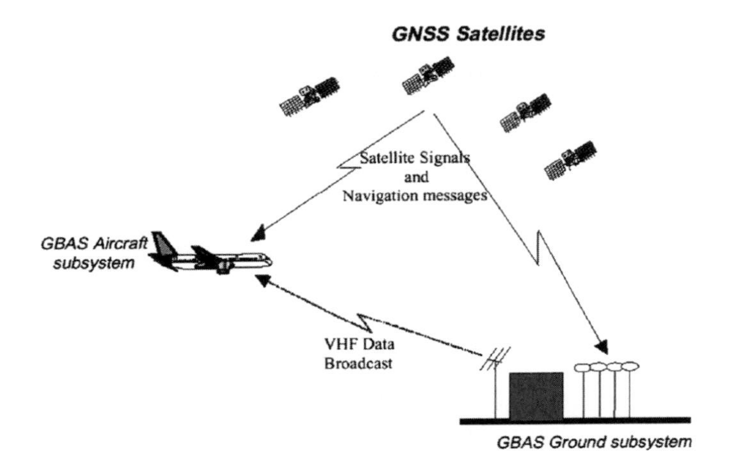

GNSS/GBAS system overview.

In the frame of Girasole project, the concept of multisystem integrity has been addressed with the purpose to establish a link between two generations of GNSS, to find the best way to integrate the Galileo and EGNOS integrity information into a new multi-system integrity algorithm and then to identify the implications in the use of the additional information provided by different RAIM schemes.

The integrity concepts have been investigated in the frame of Girasole specific environments; because of the military prime nature of navigation satellite systems, integrity requirements were first defined only in the avionic environment. As these systems were intended to be used also for civil purposes, it became necessary to extend the requirements also for the other applications. So, at the present, the requirements are standardized only for aviation, while for maritime and rail environment they are recommended values. Furthermore, while in the aviation both vertical and horizontal parameters have to be considered (3D-analysis), in maritime and rail the analysis is focused only in the horizontal plane (2D-analysis). In particular for rail environment the parameter to be controlled is the along-track position, that is the distance of the train along the railway path from a fixed point (1D-analysis).

3.2 Interference

It well known that a GNSS receiver is in principle vulnerable to several types of interference, which can lead to a complete signal disruption. This is an intrinsic feature of this type of receiver, because of the way it extracts the pseudorange information from the SIS (Signal in Space), and to the very low SIS power.

Hence, it becomes of main interest to evaluate the possible impact of potential interferences in bands of interest and in particular in the Galileo frequencies SoL bands, illustrated in Figure.

Among different transportation system scenarios, the railway environment represents a source of new developments incorporating different advanced radio systems that supports Automatic Train Control (ATC) and Automatic Train Protection (ATP). There are different kinds of interferences that could disturb a GNSS receiver in train applications and they are related with the environment in which the receiver operates. A classification of main potential interference sources is presented:

- Unintentional Electromagnetic Interference (EMI) sources from devices strictly related to railway environment.
- Unintentional EMI sources from high-voltage transmission lines.

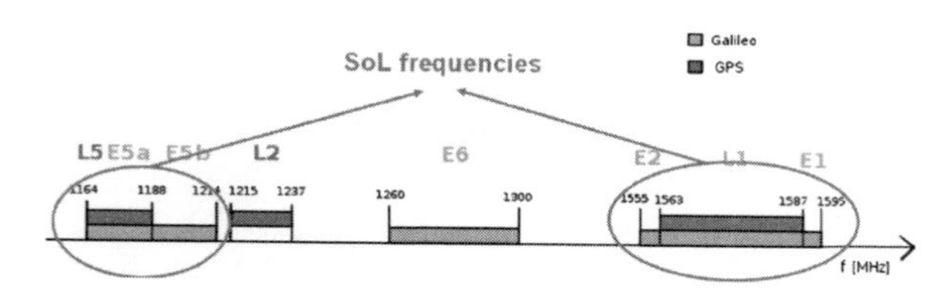

Galileo frequency bands.

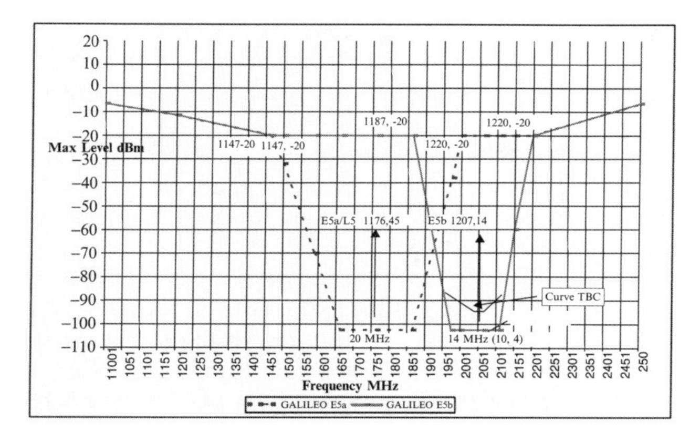

E5 band interference mask for avionic receiver.

- Unintentional RF disturbances that probably could interfere in GNSS receiver but not necessarily related to railway environment. These could be indicated in higher harmonics of out-of-band interferers as Continuous Wave (CW) or Wide Band (WB) interferences, or in-band interferers as Ultra Wide Band (UWB).

For GNSS civil aviation application, the standard received signals and interference environment applicable to the GNSS receivers are defined in the aviation standards issued by ICAO, Eurocae and or RTCA. These are defining the minimum equipment applicable environmental conditions by referring to dedicated sections of ED 14 D (Europe) or D0 160 C (US) documents. Additional constraints are set by the aircraft manufacturer depending upon more detailed installation conditions.

In Maritime environment, there are a lot of electric, electronic and radio equipments on the vessel board, which are the sources of the electromagnetic radiations. Vessel external environment is more propitious for electromagnetic compatibility (EMC) as shipborn not connected to power cable supply and communication, which can perceive interference signals. Even in harbour, where many shipborn systems don't work in fact or theirs work is forbid, unlikely find oneself nearer then 500 m from permanent commercial or industrial interferers, or then 1 km from transmitter.

Out of band and secondary radiation values are regulated in accordance with ITU recommendations. Special department controls level of these radiations. But there is a risk of appearance of such interferences and it can be an object of additional investigations in ship-borne equipment (which is not included in conventional, required for IMO approval) interference chart compiling.

3.3 Multipath Mitigation Techniques

Multipath is the phenomenon whereby a non-direct signal arrives at the receiver antenna. Non-direct signals makes the tracking loops to detect the maximum correlation power in a different instant than that corresponding to the isolated direct signal. Since this error source is the main contribution to the pseudorange measurement error and cannot be removed by local or wide area augmentations, a multipath mitigation strategy at the antenna (HW) or signal processing (SW) level must be included in the receiver.

Taking classical multipath mitigation techniques used in GPS as baseline, new techniques have been investigated in the frame of Girasole project taking profit of the Galileo signal characteristics (E5 wideband, or BOC modulations) to develop new multipath mitigation algorithms.

In order to be able to efficiently mitigate multipath and interference one has to drop classical methods which are based on correlator techniques. This also comprises to partly leave classical synchronization architectures which give astoundingly good performance, while not being designed for severe multipath scenarios like urban environments. New methods and architectures need to be introduced to the field of navigation in order to enhance quality of TOA (time-of-arrival) parameter estimation.

Different methods have been proposed to track a GNSS signal in presence of multipath propagation, mitigating the errors induced by it. These techniques can be classified according to the approach used as reported in the following.

Discriminator Based Techniques. These algorithms rely on DLL to track the signal. However, in presence of multipath the shape of the ACF is distorted, and some error appears in the estimation of the code delay. This kind of techniques tries to modify the discriminator function to reduce the error.

Tracking Error Compensation Algorithms. These techniques are based on the concept of Multipath Invariance (MPI), which states that there exist regions or properties of the ACF that do not vary as a function of the multipath. Knowing the position of these regions, the tracking loop solution can be corrected by the difference between the measured and ideal position.

Multipath Estimation Techniques. In the *A-Posteriori Multipath Estimation* (APME) technique, the tracking is done in a conventional narrow correlator DLL that achieves low noise. The multipath error is estimated in an independent module

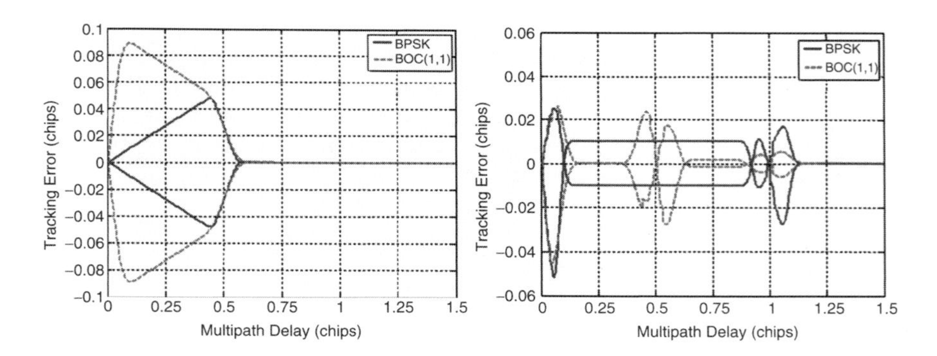

Multipath error profile for E1/E2 slope and Early/Late slope technique.

using the signal amplitude measurements. The use of *Kalman filtering* for tracking the time delays in CDMA systems is an approximation to the minimum variance estimator when the observation sequence is nonlinear in the state variables. In this case, the state variables are the multipath complex coefficients, multipath delays and Doppler shift. Some adaptive filtering techniques have been applied to multipath mitigation: in this approach the adaptive filter (RLS, LMS . . .) is used for estimating the multipath delay profile, which is subtracted from the measured correlation function of the signal with the code. The *Maximum Likelihood (ML) Estimation Techniques* family of algorithms try to cancel the multipath interference subtracting from the correlation function of the estimated contribution of the multipath. Within this class of algorithms lay the MEDLL, Deconvolution Methods, Subspace-Based Algorithms, Quadratic Optimization Methods, Teager-Kaiser (TK) Operator-Based Algorithms. *Wavelet Filtering* is a kind of multi-resolution analysis that gives simultaneously time and frequency information of a signal sequence. It can be applied to non-stationary signals, as the case of satellite navigation systems signals. The signal is passed through a series of high pass and low pass filters to analyze the signal at multiple resolution.

4 Conclusions

One of the main differentiators of Galileo with respect to GPS and GLONASS is the provision of the integrity information and guarantee of service. These characteristics make systems based on Galileo SoL service suitable to be used in safety related applications, where guarantees for integrity and continuity of service are essential. There is therefore the need to foster the development of technologies that will allow the receiver to meet the safety-critical requirements of several applications ranging from aviation to rail, maritime and emergency services. The Girasole project is the step towards this definition and development. The Girasole project is also keeping a tight link with other on-going safety-critical application development programs so that the Galileo SoL receiver could be used as a basic element in future generation safety-critical transport systems.

References

[1] L. Marradi et al., The Garda Galileo Receiver Prototype, ENC GNSS 2006. Manchester, UK May 2006
[2] João S. Silva, Signal Acquisition Techniques for Galileo Safety-of-Life Receivers, ENC GNSS 2006. Manchester, UK May 2006
[3] G. Franzoni et al., The GAlileo Receiver Development Activities (GARDA Project) Receiver Development Approach, Core Technology Overview and Prototype, Navitec 2005, ESA-ESTEC

Galileo Performance Verification in IOV Phase

M. Gotta+, F. Martinino+, S. Piazza+, F. Lo Zito+, E. Breeuwer*

+Alcatel Alenia Space Italia S.p.A.,
e-mail: monica.gotta@aleniaspazio.it, Phone: +39–06–41513134
*European Space Agency ESTEC,
e-mail: Edward.Breeuwer@esa.int, Phone: + 31–715653519

Abstract. The Galileo IOV (In Orbit Validation) Phase is an intermediate step of the Galileo system deployment. The main objective of the IOV Phase is to demonstrate that the Galileo Full Operation Capability (FOC) requirements (as specified in the Galileo System Requirements [2]) can be met, with the support of analyses and simulations, before to complete the deployment of the full system. To this respect, the IOV is a "break" in the deployment to get sufficient confidence that the final system will properly work.
The direct consequence of the above statement is twofold:

a) the IOV system configuration is reduced with respect to the final one, but it has to be designed in order to be easily upgraded to the FOC configuration. Therefore, the design, the development and the deployment of this configuration has to be driven by the FOC requirements
b) the IOV configuration may require additional functions/means to support the verification campaign that are not strictly required for the final system.

The authors of this paper, members of the Galileo System Integration and Verification (SI&V) team are responsible for the definition and execution of the IOV test campaign within the industrial consortium called Galileo Industries (GAIN) that is currently building the Galileo IOV configuration. This paper presents the current state of definition of the IOV Test Campaign.

Taking into account the constraints of the IOV configuration, a specific approach had to be developed to allow the verification in IOV of all GSRD requirements ([2]). In this paper the verification approach is presented, with particular focus on the System Performance Verification.

1 Introduction

The main constraint in defining the Verification activities in IOV Phase is linked to the fact that IOV Configuration is heavily reduced w.r.t. FOC configuration (see [6] for complete description of Galileo System in FOC Configuration), versus which the design of the Galileo System is performed (Fig. 1). In fact IOV Configuration foresees to include 4 satellites out of 30, 20 Sensor Stations out of 40, 5 Up-Link Stations out of 9, 2 Telemetry Tracking & Commands (TT&C) Stations out of 5 (see [1]). This implies that the User Service Performance in terms of positioning/timing accuracy, integrity, availability and continuity cannot be tested and assessed completely by test. This is due to the fact that the processing algorithms need to operate on a reduced number of observables and therefore the navigation message parameters will be

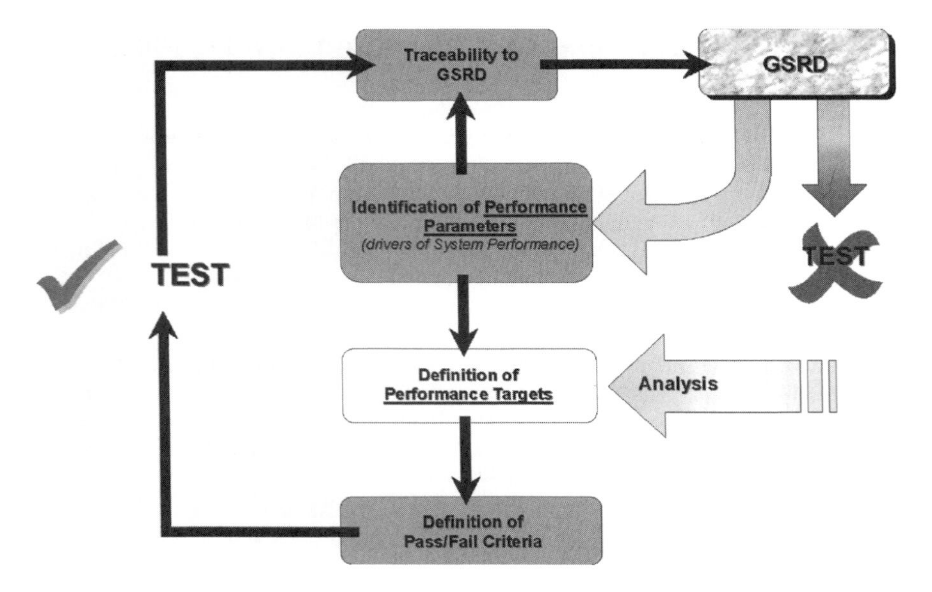

Fig. 1. IOV performance verification approach.

degraded w.r.t. final performance. Moreover the user testing equipment will not be continuously able to solve for its position due to the reduced constellation and the holes of visibilities of the complete set of 4 satellites during the test campaign.
Therefore the approach selected to verify the System Performance is:

- To identify, through the System Performance Allocation, a set of lower-level parameters that are directly linked to the User Service Performance and that can be observed in the IOV configuration. This set of parameters will allow the system to be accepted when the related IOV Performance Target (tailored to IOV reduced configuration) is met according to the correspondent Pass/Fail Criteria.
- The GSRD requirements ([2]) that are linked to the identified parameters are verified according to the verification methods identified in the Requirement Verification Matrix ([3]), by eventually complementing the verification by test with analysis/review of design/simulation.

Performance verification activity makes use of an integrated platform called GALileo System Evaluation Equipments (GALSEE). In the frame of the performance verification activities, GALSEE is deeply exploited for both Galileo algorithm performance targets verification and IOV Test Campaign performance targets verification.

Taking maximum benefit from the reuse of tools already available from Galileo segments, GALSEE will support the performance verification activity by processing real data flowing into a controlled environment compliant with the one under which the GSRD service requirements has been specified, but flexible enough to allow System Sensitivity Analysis on the identified performance targets

The Performance requirements and parameters to be verified during IOV Test Campaign have been grouped into the following Verification Scenarios:

- Sensor Station Data Quality, including tests of Galileo Sensor Station data quality
- Signal In Space Monitoring, including monitoring of Signal RF characteristics
- Navigation, including tests of Navigation Determination Processing Function
- Integrity, including tests of Integrity Determination Processing Function, including verification of the Time-To-Alarm
- Timing, including tests of Galileo System Time (GST) Generation and Steering Function
- Search And Rescue, including one test end-to-end of the Galileo supporting function to Cospar-Sarsat External System
- UERE, including tests on filed to assess the User Equivalent Range Error impacting User Service Performances.

In the following the logic of the System Integration and Verification activities is provided in order to present the overall frame in which the IOV Test Campaign is placed, representing the last 6 month of the overall System Verification, dedicated mainly to System Performance Verification.

2 System Integration & Verification Logic

The purpose of the System Verification activities (concluding with the IOV Test Campaign) is to demonstrate that the System has been designed, implemented and tested to meet the GSRD requirements ([2]), as customized for IOV Configuration. In particular the IOV Test Campaign focuses on the verification of the System Performance, being the verification of Functional and Inter-Segment/External Interface concentrated at the maximum extent during Phase D, prior to IOV Readiness Review 1 (IOV-RR1).

In order to provide the overall picture of the Verification Strategy in which IOV Test Campaign is planned, an overview of the System Verification Phases and content is provided in the following.

The System Integration and Verification activities are infact divided into three main categories:

- The Ground Segment Integration & Verification, including the Integration of Ground Mission Segment (GMS) and the Ground Control Segment (GCS) inside the Galileo Control Centre (GCC), the integration of the Remote Sites with the GCC and the verification of the Interfaces between the GMS and GCS and their compliance versus relevant ICD
- The System External Entities Integration & Verification, including the Integration of all the Galileo External Entities to the Ground Segment and the verification of their interfaces and their compliance versus relevant ICD
- The Overall System Integration & Verification, including Ground-Space Compatibility Testing (with the satellite on-ground), System Functional Test and Overall Integration with Space Segment and Test User Receiver (after Satellite Hand-over to Galileo Control Center)
- The IOV Test Campaign that is the In Orbit Verification of the System Performances against System Requirements as tailored w.r.t. the IOV reduced configuration.

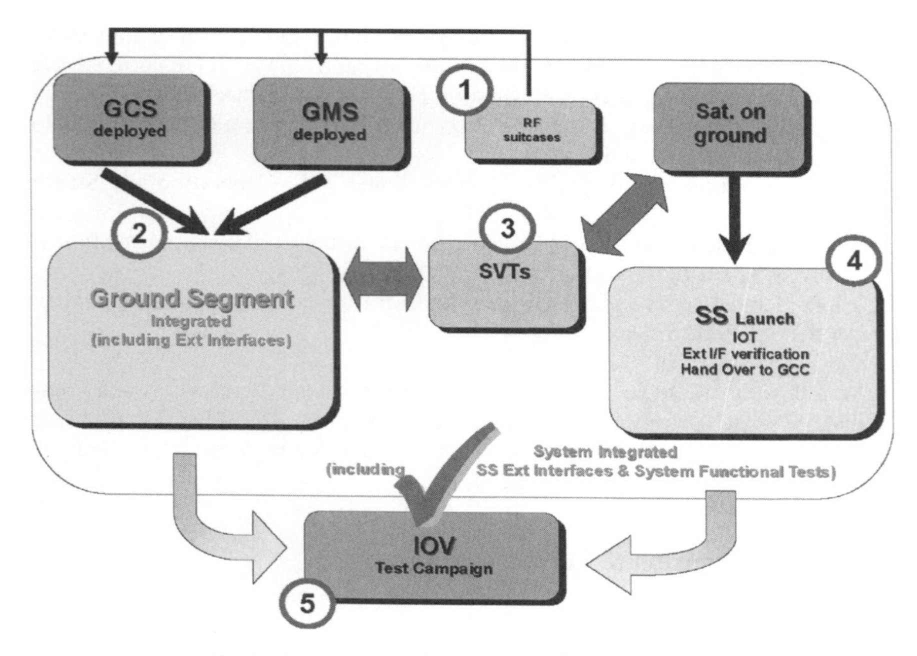

Fig. 2. System integration & verification activity flow.

The SI&V activity flow showing the staggered approach described above is depicted in Fig. 2 (numbering represents activity sequence). For more information on Galileo System Architecture and Design on which System Integration & Verification (SI&V) approach is built (including Segments composition and Interconnections) please refer to [6].

The System Integration will be performed, in an incremental way up to the IOV Readiness Review 1 (IOV-RR1) involving the Overall Galileo Ground Segment, the Galileo System External Interfaces and, following the handover of the Space Segment, the Overall Ground Segment Integration with the in-orbit Satellites, the External Interfaces with Satellite and the integration of the Test User Segment (TUS) with the satellite. The System Integration activities will be conducted as part or in parallel with the completion of verification and qualification activities at segment level.

The completion of the System Integration Activities will ensure the correct integration of all the elements of the Galileo System, including the Satellite in orbit.

The System Verification testing activities will be mainly developed in parallel with System Integration activities, starting from the Qualification Review (QR) of the Segments. The verification activities are performed into two separate steps:

- Verification activities to be performed during the system integration up to the first readiness of the IOV system (IOV-RR1)
- Verification activities to be performed during the IOV Test Campaign on which basis the system is accepted by the Customer.

The scope of the verification activities is to support the verification of the internal interfaces between segments and the external interfaces of Galileo system. In order to minimize the risks, verification activities concerning the compatibility of Space Segment with Ground Segment will be performed first with satellite on-ground and after with the satellite in-orbit, in the frame of In-Orbit Test (IOT), conditioning Satellite Acceptance.

The first In-Orbit Testing (IOT) after first launch before Space Segment Hand Over 1, will be performed in conjunction with completion of integration activities and prior to IOV RR1. The IOT of the last two satellites will be executed in parallel to IOV Test Campaign execution and before Space Segment Hand Over 2 (SS-HVR2).

The completion of the System Verification activities will ensure the correct verification of System Interfaces and System Functional Requirements and will allow starting the IOV Test Campaign, that is the final set of Tests focused on System Performance Verification.

The first task to be performed in preparation to IOV Test Campaign is the Start-Up of the System. This activity aims to inject in the system the necessary data in order to start all the processes and to initialize the System in its operative status. The Start-up of the System is executed by Operations after Space Segment Hand Over 1 (SS-HVR1) and is to be concluded at IOV-RR1. IOV Start up will be considered successfully concluded (allowing readiness to be declared) only once the System has reached its operative status, delivering Services that are functionally working (SIS broadcasted by each Satellite available and no more transmitting dummy frames, but real frames as operationally generated on-ground).

The IOV Test Campaign will start, organized in a certain number of System Acceptance Verification Scenarios, as provided in the following section, aiming to provide testing of main system/segment performance parameters that are considered affecting the overall Galileo Service Performances (Table 1).

The System Verification and Acceptance in the IOV Configuration will be reached by providing evidence of the successful testing, during IOV Test Campaign, of all the identified Parameters/requirements, the acceptance criteria for the GSRD

Table 1. IOV Test campaign test matrix.

Test case	Test objective	Elementary tests	Performance parameters
Sensor StationData Quality	To verify a certain set of parameters related to the quality of Sensor Station data (observables) as it directly impact the performance of Galileo processing algorithm (mainly navigation and integrity Determination functions)	GSS data availability Test GSS Output Quality Test	GSS data availability GSS data continuity GSS Cycle slip GSS Multipath and Receiver Pseudorange and Carrier Phase Noise

(*Continued*)

Table 1. IOV Test campaign test matrix—cont'd

Test case	Test objective	Elementary tests	Performance parameters
Navigation	To test the accuracy of OD&TS products (orbit and clock estimation and prediction accuracy) and SISA computation representativeness of real SISE. It has to be noted that OD&TS orbit and clock prediction accuracy directly impacts UERE budget and, at the end, User Service Performance	OD&TS Computation Accuracy Assessment Test	OD&TS Orbit and clock estimation accuracy OD&TS Orbit and clock prediction accuracy SISA and SISA/SREW ratio
		OD&TS Modelling Quality Test	Iono Model Performance Broadcast Group Delay (IFB) performance
		Satellite Station keeping accuracy Test	Across track orbit keeping (relative RAAN variations) Relative inclination variations Relative along track orbit keeping between any two adjacent operational satellites in the same orbit plane
Integrity	To test the Integrity Processing main actors, namely Integrity Algorithm synchronization error verification, SISMA values, TTA	IPF Sensor Station Synchronization Test	Sensor Station Synchronization Error
		Min/Max SISMA values	SISMA
		TTA Test	Time-To-Alarm
Search And Rescue	To verify the Galileo External Interface to SAR (namely GMS to Return Link Service Provider Interface) main performance (RLM Delivery Time)	SAR Return Link Service Delivery Time Test	SAR Return Link Service Delivery Time
Timing	To verify the main Timing-related Performances, both at System Level and at User Level	GST to TAI Time and Frequency offset Test	GST and TAI. time and frequency offset
		GGTO Performance requirements Test	GGTO computation accuracy
		UTC Time and Frequency accuracy distribution Test	time offset and frequency offset between the physical realization of GST and the GST (and UTC) Time as reproduced at user

Table 1. IOV Test campaign test matrix—cont'd

Test case	Test objective	Elementary tests	Performance parameters
			level by a standard timing/calibration laboratory Galileo receiver receiving Galileo SIS.
UERE	To test the UERE as the main contributor (apart from geometric diluition of precision) to User Service Performance	Open and PRS Service (Single Frequency) UERE Test	Open and PRS Service (Single Frequency) User Equivalent Range Error
		Open Service (Dual Frequency) UERE Test	Open Service (Dual Frequency) User Equivalent Range Error
		SOL Service UERE Test	SOL Service User Equivalent Range Error

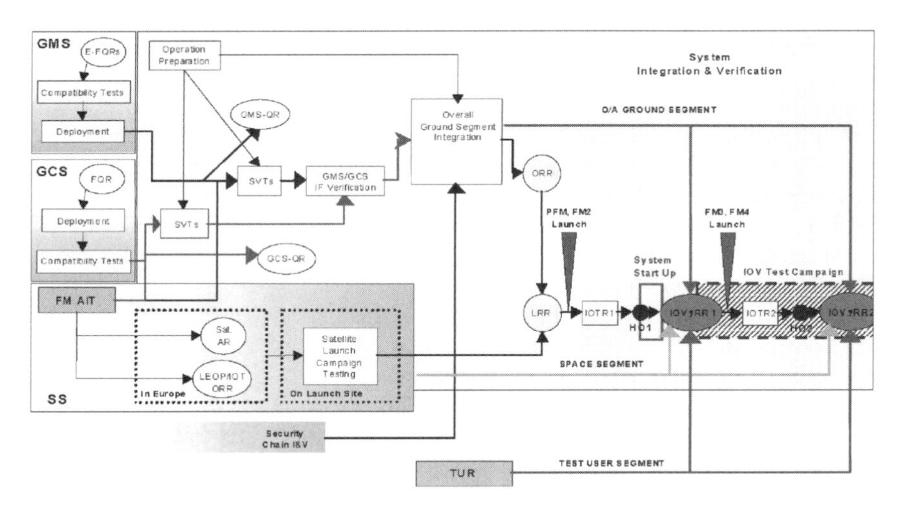

Fig. 3. System integration & verification process overall logic.

requirement ([2]) being defined as the achievement of all criteria relevant to all parameters concurring in that requirement. This should lead to IOV Review (IOV-R), when the IOV Test Results are reviewed to release the System Acceptance and consequently the Segment final acceptance at Segments Final Acceptance Review (FAR). The overall SI&V logic is provided in Fig. 3.

3 IOV Test Campaign Logic

The first main area of activities to be executed during the first months of IOV Test Campaign is represented by the GSS Data Quality Test Case. The main focus of this Test case is to verify a certain set of parameters related to the quality of Sensor Station data (observables) as it directly impact the performance of Galileo processing algorithm (mainly navigation and integrity Determination functions). The Test Case foresees the verification of the Sensor Station UERE (User Equivalent range Error), mainly receiver and multipath noise after the compensation of propagation errors (local components therefore giving an indication of quality of measurements as impacted by Sensor Station itself) and GSS Cycle slip occurrence monitoring. As the correctness of the signal has been verified during IOT, any problems occurred during GSS Output Data Quality should be reasonably due to Sensor Station themselves; SIS ICD ([7]) verification is, in this sense, a pre-condition for the Sensor Station Data Quality Test Case. Moreover the verification of the availability and continuity of the GSS data at Galileo Control Centre (GCC) is verified as part of this Test Case, as the loss of link and exchanged data can affect the System Performance in term of Navigation and Integrity Accuracy and Availability. It comes out that Navigation and Integrity Test Cases, at least, cannot be verified until the assessment of the input quality is carried out.

At the same time, the Test User Receiver can start to acquire the SIS and carried out on-field tests to verify with the real Signal that the Test Receiver is functioning correctly and with the expected performance (focusing on the function and performance strictly characteristic of the receiver itself and leaving out the propagation effects). Therefore the Signal/Receiver Parameters verification is expected to be carried out by TUS in the first months of IOV Test Campaign in order to get the confidence that the receiver (than will be used by System Verification Team during UERE Test Case) is verified and that minor contribution ca be expected by TUS itself in case of UERE Test anomalies; in this sense the on-field verification of TUS is a precondition for UERE Test Case.

Once the GSS Data Quality Test Case is completely carried out, the Navigation Test Case can start by including tests of accuracy of OD&TS products (orbit and clock estimation and prediction accuracy) and Signal-In-Space Accuracy (SISA) computation representativeness of real Signal-In-Space Error (SISE). It has to be noted that OD&TS orbit and clock prediction accuracy directly impacts UERE budget and, at the end, User Service Performance. Also the assessment of the representativeness of some internal OD&TS modeling will be carried out, in particular for what regards the Broadcast Group Delay and the Ionospheric Model Implementation (for more details on SISA, SISE and the others parameters, please refer to [6]). Moreover the verification of the Satellite Station keeping accuracy is carried out within the present Test Case, as its performance impacts OD&TS accuracy itself (the system is designed in such a way that satellite does not need a maneuver for a time span of several years and Ground Control Segment will generate the necessary maneuvers to guarantee that the satellite does not violate the "deadband" during a determined time span, that is, a maneuver-free time span); it will be performed based on offline OD&TS estimation of satellite position and extrapolation over the maneuver-free time to demonstrate that the station keeping thresholds are not violated in the whole maneuver-free time span (extrapolation of IOV measurements will be needed).

When Navigation Test Case is carried out and at least SISA Computation representativeness of SISE is confirmed, the Integrity Test Case can start by including the testing of the Integrity Processing main actors, namely Integrity Algorithm synchronization error verification, Signal-In-Space Monitored Accuracy (SISMA) values. In parallel the TTA Test is carried out with the aim of measuring the delay of several parts of the system and to evaluate the complete TTA (Time-To-Alert), being TTA defined as the time occurred between the beginning of a sampling period, in GSS receiver, during which a satellite SIS Misleading Information (MI) will be received (start event) and the time of reception of the last bit of the navigation message (containing an Alert condition) at the input of a user receiver (end event).

The UERE Test Case is the main IOV Test Case as UERE is the main contributor (apart from geometric dilution of precision) to User Service Performance. The Test should be executed as the last one, as all the parameters tested within the other Test Cases (especially Navigation and Integrity) impact UERE.

However, in order to fit the tight schedule of IOV Test Campaign, the UERE Test Case is considered to start in parallel to Navigation Determination. In this case the OD&TS contribution to UERE can be estimated separately though the Navigation Test Case and can be eventually subtracted by UERE itself in order not to waste UERE assessment in case OD&TS error experiences strange behavior as its verification is not yet complete. A certain number of UERE budgets need to be assessed in order to cover the different kind of services and users specified in GSRD. Actually an assessment is ongoing trying to optimize the UERE Test Set-up and Configuration as to combine more that one UERE test under a common configuration allowing still matching the tight schedule.

Parallel Tests can be conducted that are quite self-standing: Timing Test Case, and SAR Test Case.

Timing Test Scenario is focused to verify the main Timing-related Performances, both at System Level and at User Level. In particular the verification of Galileo System Time Performance in terms of GST Offset and Frequency Stability w.r.t. UTC/TAI will be evaluated, together with the capability of the user to retrieve UTC by receiving the SIS (and broadcasted navigation message), within the specified performances.

Moreover the GGTO computation performance will be evaluated as part of present Test Scenario. Please note that GST will be available also prior to satellite launch and the Timing Test Scenario is planned to start just after the IOV Readiness Review, in order to allow sufficient data collection to be able to verify the performance with a statistically significant sample.

The SAR Test Case is currently foreseen to cover mainly verification of Galileo External Interface to SAR (namely GMS to Return Link Service Provider Interface). In particular the verification of the Delivery Time of SAR Return link Message to the User will be provided as part of SAR Test Case. Furthermore, performances related the External Interface to the Cospas-Sarsat organization will be tested already in advanced, as part of In Orbit Test. The verification of all those performances that are related to the Cospas- Sarsat Capabilities and performances (including MEO Local User Terminal – MEOLUT – ones) is expected to be performed as a part of the European Union's 6FP programme activities.

An activity of Signal-In-Space Monitoring is run throughout the IOV test campaign, in support to System Test Results Analysis and Troubleshooting. This activity will be conducted by means of adequate infrastructure inside GALSEE (GALSEE Signal Monitoring Facility) and will allow to support IOV Test Results analysis and the troubleshooting activities, with the possibility to correlate strange behavior in on-field measurement collection for UERE Tests with the behavior of the monitored SIS in terms of RF characteristics and TUS receiver in terms of tracking performances.

The above mentioned activity consists in to monitoring the RF and Code & Data characteristics of both the Open signals and the PRS signals (TBC) and allow both real-time and post-processing analysis of the Galileo/GPS signals received on ground (nominal power level, spectrum of modulated signals, in-band and out-band interference, etc.) during the IOV Test Campaign.

The SIS Spectrum of one pre-selected satellite in time is monitored in order to allow its tracking with a high-gain steerable antenna.

Moreover some important parameters that impact the Quality of the TUS Receivers output data are monitored through GALSEE, in particular:

- C/NO analysis of all the tracking channels of the receiver and comparison with expected values based on link budget analyses;
- Code and Carrier Phase measurements statistical analysis (e.g., least square fitting and standard deviation of error);
- Doppler measurements statistical analysis (e.g. least square fitting and standard deviation);
- Correlation Function evaluation;
- Multipath Analysis computation;
- Chip rate and modulation evaluation.

A dedicated Spectrum Analyzer and Test User Receiver monitoring Galileo SIS will be operated during the IOV test campaign as part of GALSEE.

Different analyses can be provided during the SIS monitoring activity allowing the characterization of several parameters such as signal bandwidth, minimum received power level, C/NO, signal carrier stability, interference, code phase and carrier phase error, etc.

The provided activities implementation and logic of activities precedence is depicted in Fig. 4. Please note that exact duration of each Test Case is not yet defined, therefore, at the time being, the provided figure should be considered only as qualitative. The exact planning of IOV Test Campaign will be provided as soon as the assessment on each test duration will be carried out and the planning will be updated accordingly.

Please note that the staggered approach as presented in Fig. 4 is still to be confirmed, pending its feasibility to be carried out in the 6 months allocated for IOV

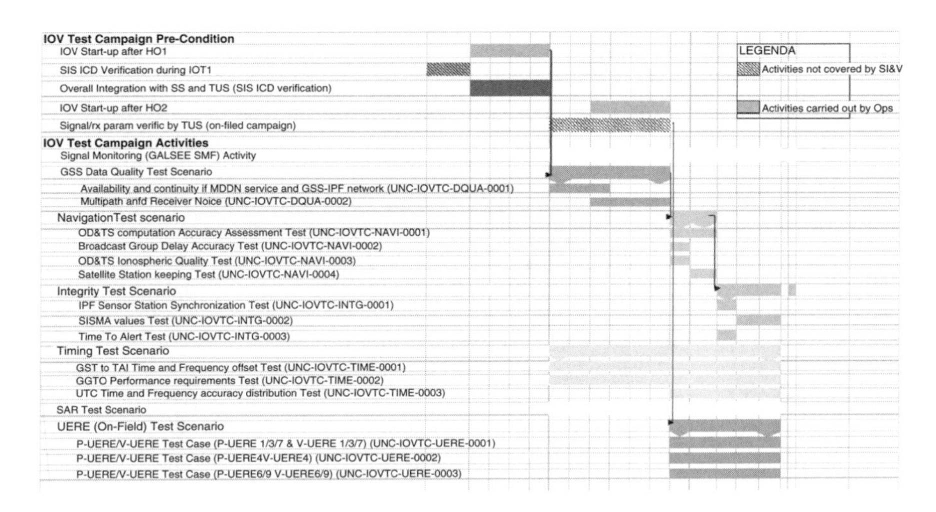

Fig. 4. IOV test campaign planning.

Test Campaign. In case this will be not confirmed (once the exact Tests Duration will be finalized), the approach will be changed (from the staggered one) into the "concurrent" one, that foresees to parallelize – as much as possible – Tests, in order to fit the IOV Test Campaign schedule. It has to be noted that the second approach (the "concurrent" one) implies to accept an increased risk in the IOV Test Campaign activities, as problems occurring with test failures cannot be duly allocated to one area, as everything is tested in parallel. The choice between the 2 approaches is kept open until a detailed assessment on test duration is done that allows to demonstrate which of the approaches is feasible with the time and resources constraint.

Moreover constraints coming from security-related aspects are still under assessment that can lead to re-consideration of provided approach.

4 Conclusions

An overview of the System Verification activities during IOV Phase has been provided in the present paper, with focus on the last 6 months of IOV Test Campaign dedicated to verification of System Performance. The Logic of the activities in terms of identification of Test Cases, preliminary test plan and inter-tests dependency and pre-conditions has been provided, highlighting the main difficulties in terms of schedule and feasibility of tests. In particular a staggered approach for IOV Test Campaign is proposed that is the one mitigating the risks by conducting the tests in the correct sequence as to identify the contribution to test execution in a step-wise approach. It has been highlighted that feasibility of this approach is still to be confirmed pending exact assessment of elementary tests duration that is to be carried out as part of normal work in the following months. In case the feasibility of this approach would not be confirmed, alternative solutions should be envisaged that carries more risks in the overall process of Galileo Verification.

Acknowledgment

The authors would like to acknowledge the contributions received by the Galileo Project Office of European Space Agency for its fruitful and constructive collaboration in defining the verification strategy of Galileo System.

Moreover a particular acknowledge is to be given to the industrial team of the Galileo Industries (GaIn), grouping all the main European companies working on Galileo since Phase A, for the useful discussion held on the subject and the precious suggestions received by GaIn Team about System Integration and Verification activities planning.

References

[1] Galileo IOV Implementation Requirements – ESA-EUING-GALCDE1-So W/01000-A4
[2] Galileo System Requirements Document – ESA –APPNS-REQ – 00011
[3] GSRD Requirements Verification Matrix (RVM) – GAL-DVM-ALS-SYST-A/0347
[4] System Integration and Verification Plan – GAL-PLN-ALS-SYST-A/0349
[5] "Galileo In-Orbit Validation (IOV)", F. Gottifredi, F. Martinino, S. Piazza, R. Dellago, 2005 DASIA Conference, Edinburgh
[6] "Galileo: the European Satellite Navigation System", O. Galimberti, M. Gotta, F. Gottifredi, S. Greco, M. Leonardi, F. Lo Zito, F. Martinino, S. Piazza, M. Sanna, ATTI dell'Istituto Italiano di Navigazione, n°178, March 2005
[7] Signal-In-Space Interface Control Document (SIS-ICD) - GAL-ICD-GLI-SYST-A/0258

Different Acquisition Algorithms for the Galileo L1 Signal with BOC(1,1) Modulation

R. Campana[1], F. Gottifredi[1], V. Valle[2], P.F. Lombardo[2]

[1]Alcatel Alenia Space Italia S.p.A.,
Phone: +39 06 41514189, Fax: +39 06 4191287,
e-mail: Roberto.Campana@alcatelaleniaspace.com
[2]University La Sapienza (Rome)

Abstract. The aim of this work is to analyse, implement and test different acquisition algorithms for the European Global Navigation Satellite System Galileo signal. The focus of the work is on the L1 signal and the target of the acquisition algorithms is mainly relevant to a Software Receiver Application. Both GPS signal and Galileo signal have been taken into account in order to perform a comparison between these two systems, in terms of computational burden and acquisition sensibility.

The motivation of this work comes from the actual tendency of studying algorithms for the realization of a software receiver for the GPS signals (whose principal characteristics are a good flexibility, when compared to the hardware solution, and the possibility of getting "on the fly" updates), and it's aim is to bring advantages of software-based approach to the new Galileo system, in particular for the acquisition schemes of the L1 signal. The principal fact is that acquisition has to be performed on a normal PC, without using a dedicate hardware. In this context, have been developed different acquisition schemes to provide an analysis on their performance, regarding to a software acquisition. All the schemes are finally compared: then, real advantages of employing BOC (Binary Offset Carrier) digital modulation are investigated, provided that this modulation has been developed to transmit GPS and Galileo signals over the same bands to obtain a greater degree of compatibility and interoperability between the two systems (and the L1 signal represents, in this way, the principal example of such interoperability).

1 Introduction

A global navigation software receiver provides several advantages, mainly when involved in studying new type of signals such as Galileo signals.

The principal difference between a software receiver and a standard hardware receiver is that the correlator chip functions are moved to software that runs on a general-purpose microprocessor. This means that the operations of acquisition and tracking are made through software routines structured in a high-level language. A typical example of software receiver architecture is depicted in Fig. 1. It consists of an antenna and a RF front-end and an Analog-to-Digital Converter (ADC). The RF front-end device is necessary to down convert the GPS signal to an intermediate frequency (IF). The IF signal is then sampled and digitize through the ADC.

In a conventional GPS/Galileo receiver, the acquisition and tracking of the signals are all processed by the hardware. However, in a software GPS receiver, the signal is

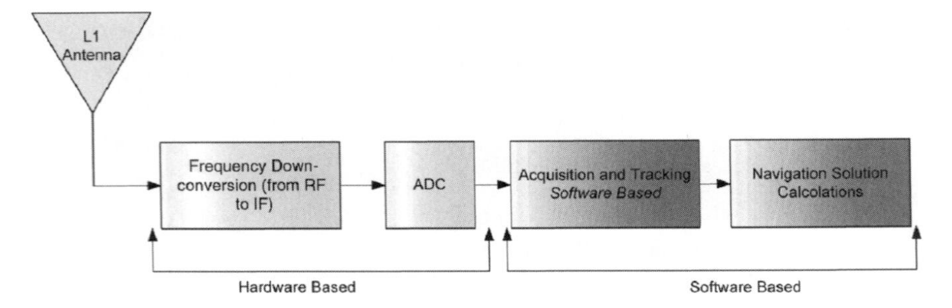

Fig. 1. Software receiver architecture.

digitized using an Analog-to-Digital Converter (ADC). The digitized input signal is then processed using the software receiver. The acquisition process searches for the presence of a signal from a particular satellite. Then, the tracking function refines the analysis finding the phase transition of the navigation data from that satellite. Ephemeris data and pseudoranges can be recovered from the navigation data bits. Finally the user position can be calculated using the measured pseudoranges and the so obtained ephemeris.

A receiver that uses a hardware digital correlator will require hardware modifications in order to use Galileo new signals. A software receiver is able to elaborate new possible signals without changing the correlator chip. Given a suitable RF front end, new frequencies and new pseudo-random number (PRN) codes can be used simply by making software changes. Thus, there are good reasons to develop practical software Galileo/GPS receivers.

In this paper, different acquisition schemes are proposed and implemented considering a software receiver solution.

2 Analysis Approach and Implementation

To carry out the analysis on the acquisition schemes, developed under Simulink®, the entire transmission and reception environment, including a realistic simulation of in-band white Gaussian noise has been initially simulated. In order to simplify the discussion, are presented here three sections that are, in natural order, the GPS/Galileo signal simulator, the RF Front End and Down-Conversion, and finally the acquisition block, as illustrated in Fig. 2.

2.1 GNSS Signal Simulator

2.1.1 GPS Signal Simulator. The GPS L1 signal has been simulated with the P(Y) code (Quadrature component) and the C/A code (In-phase component) modulating the navigation message, as illustrated in Fig. 3.

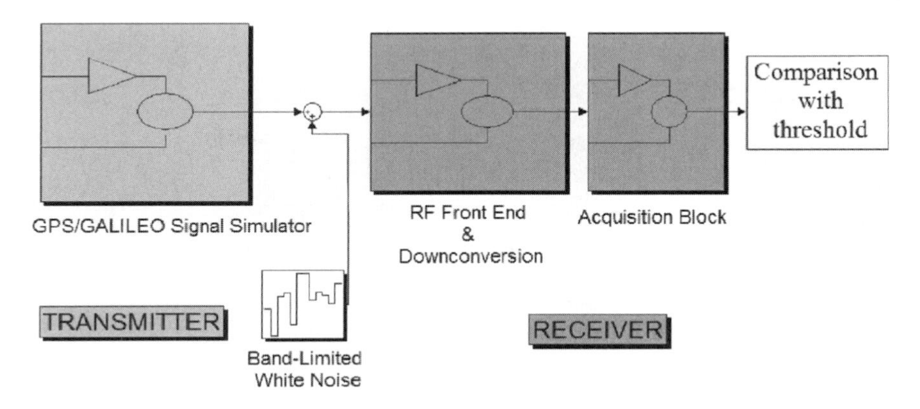

Fig. 2. Overview of the overall simulink® scheme.

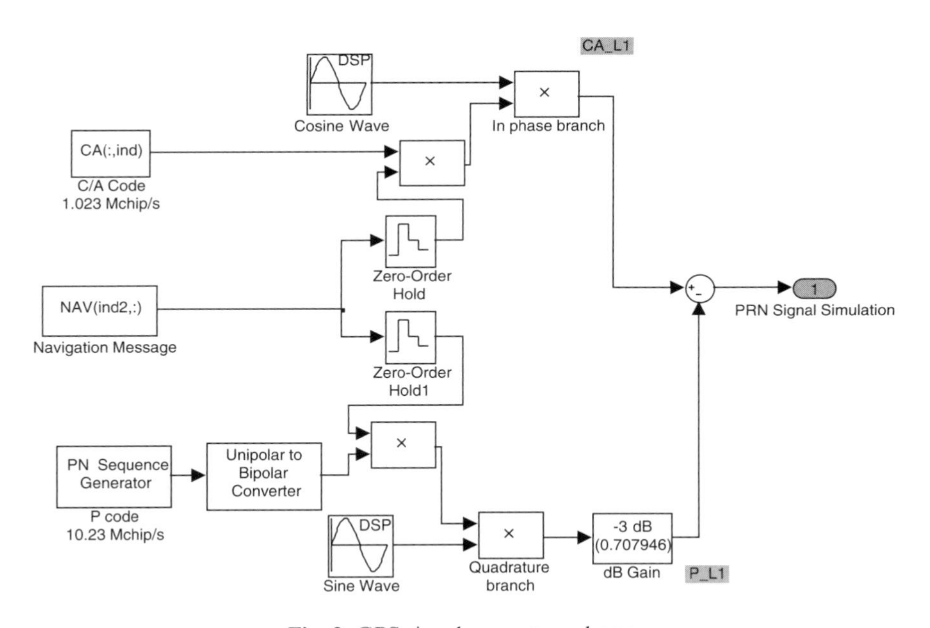

Fig. 3. GPS signal generator scheme.

2.1.2 Galileo Signal Simulator. The Galileo L1 signal is composed by Pilot Channel, Data Channel (I/NAV) and Encrypted Channel (G/NAV). Pilot and Data Channel are transmitted with a BOC(1,1) modulation scheme, while Encrypted Channel is modulated with a BOCc(15,2.5). On channels modulated by BOC(1,1) are simulated the two Gold(13) codes, each 4092 bits long, and also the Secondary code for Pilot channel. These three channels are modulated following CASM (Coherent Adaptive Sub-Carrier Modulation) modulation technique, as can be seen in Fig. 4.

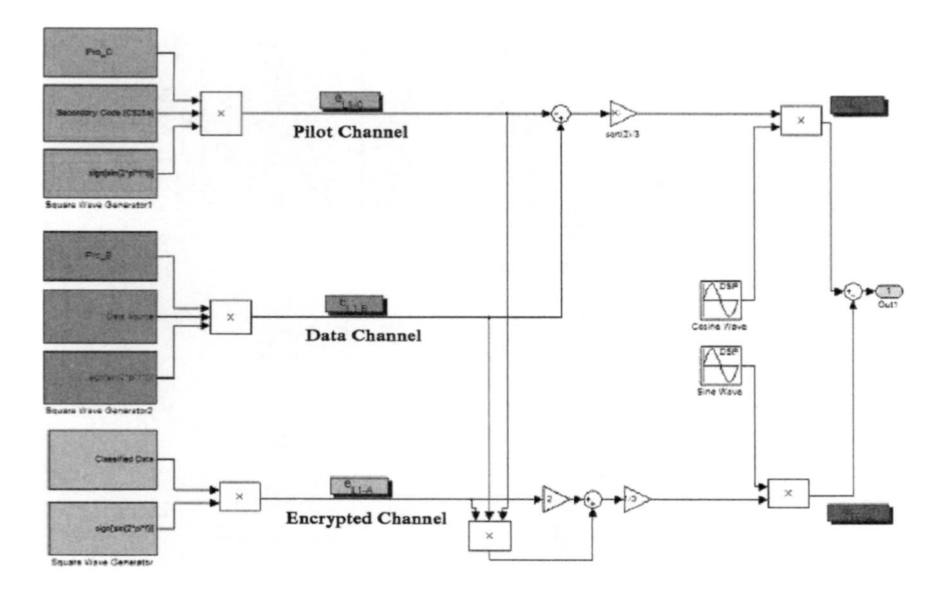

Fig. 4. Galileo signal generator scheme.

To obtain an enough realistic simulation, signals are transmitted in a noisy environment in which power levels are adjusted to reach effective C/N_0 values (as reported in respective ICDs of both systems).

2.2 RF Front-End

Simulated signals became then the input for the second section, represented by RF Front-End together with Down-conversion, followed by an ADC. Front End system consist of a pre-selection filter (to cut off the out of band noise), while Down-conversion is done adopting a coherent superheterodyne scheme with DICO (DIrect Conversion) approach. Finally, input signal is digitalized using a 3-bit Analog-to-Digital Converter, as depicted in Fig. 5.

2.3 Acquisition Algorithms

The section of acquisition has been developed to be suitable for a software approach. In fact, recent works for GPS signals have clearly pointed out that FFT-based algorithms appear very cheap under computational cost and execution time aspects. Those algorithms allow parallel search in Code Phase domain or in Doppler Frequency domain, meaning that the bins are all searched contemporaneously.

Attention has been focused mainly on a Parallel Code Phase Search Acquisition, depicted in Fig. 6, in which a serial search over all Doppler bins and a parallel search over all Code Phase bins are realized. In particular the input signal, on which has

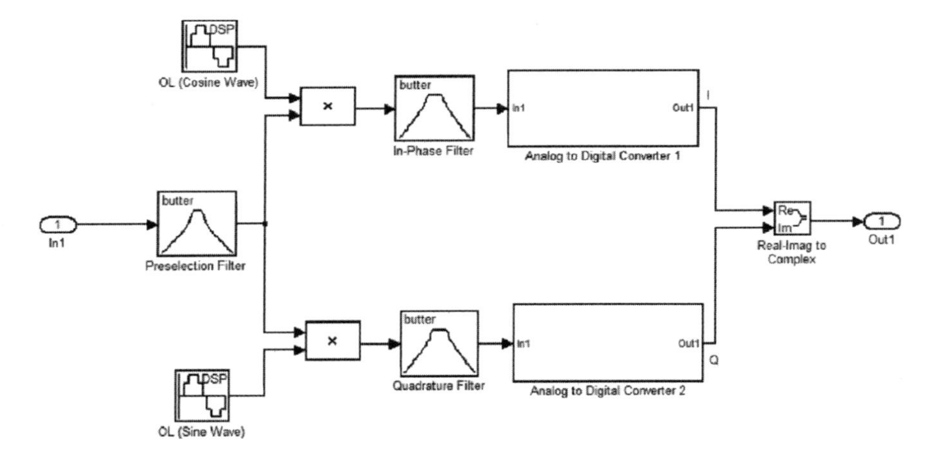

Fig. 5. Front-End scheme.

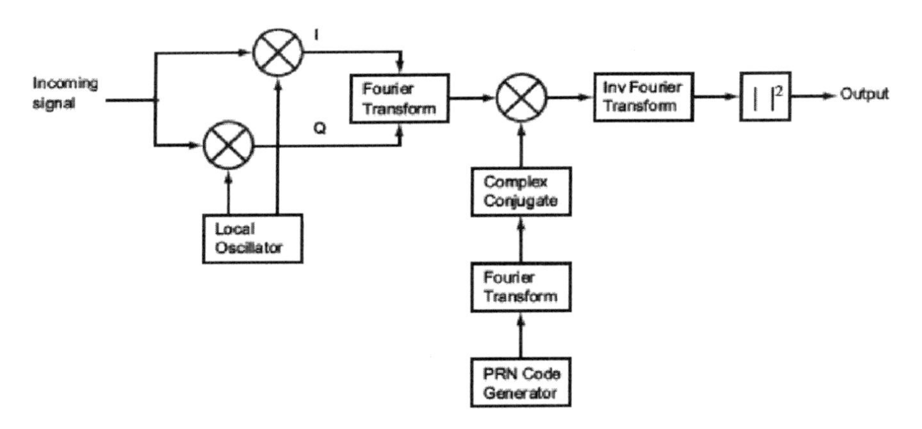

Fig. 6. Parallel code phase search acquisition scheme.

already been performed the Doppler frequency search, is correlated with stored codes using circular correlation performed in Frequency domain.

The advantage of the FFT version is that it calculates the correlation for an entire range dimension (selected Doppler) in a single step. In this technique a FFT is applied to the incoming GPS signal and multiplied by the conjugate FFT of the reference signal. Taking the inverse FFT of the product gives the correlation result in the time domain for all the 1023/4092 code phase offsets. Multiplying the FFTs of two signals $x(n)$ and $y(n)$ and taking the inverse FFT of the product corresponds to convolution in the time domain. However since we correlate the incoming GPS/Galileo signal with the reference signal in the time domain, this corresponds to multiplying the conjugate FFT of $x(n)$ with FFT of $y(n)$, and then taking inverse FFT of the product. Simulink implementation of above mentioned algorithm is shown in Fig. 7.

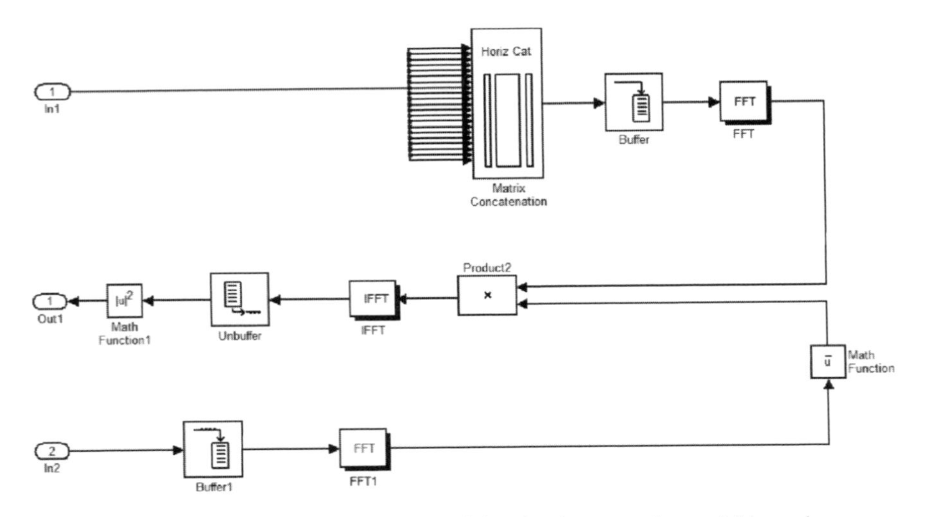

Fig. 7. Simulink implementation of parallel code phase search acquisition scheme.

This algorithm has been already realized and tested for GPS signals and allows avoiding serial search over 1023 code phase bins of GPS C/A code.

About Doppler Frequency, we search over 21 bins, 1 kHz spaced, on a range – 10 kHz: 10 kHz: this range is used for a receiver typically used on a highspeed vehicle. In fact, in designing a GNSS receiver, if the receiver is used for a low-speed vehicle, the Doppler shift can be considered as ±5 kHz. If the receiver is used in a high-speed vehicle, it is reasonable to assume that the maximum Doppler shift is ± 10 kHz. These values determine the searching frequency range in the acquisition program. The output of circular correlation, shown in Fig. 8, is finally compared with an acquisition threshold, calculated taking into account noise power to obtain a decided false alarm (FA) probability.

The same Parallel Code Phase Search Acquisition of Fig. 7 has been used also for acquisition of Galileo L1 signals (for Pilot Channel): a better advantage in parallelizing code phase search is obtained, because Galileo L1 code phase has to be searched over 4092 bins, instead of 1023. Moreover, it must be emphasized the difference from GPS and Galileo system regards the searching frequency step choice. Since the code length for the GPS L1 signal is 1 ms, to avoid a phase reverse in the integration period, the maximum frequency error allowed is 500 Hz; so a frequency step of 1 KHz can be chosen. Differently, since the code length for the Galileo L1 signal is 4ms, it's necessary to choose a frequency step of, at least, 250 Hz and that increases the processing time. The frequency step has been set just to 250 Hz: so, still considering a ± 10 kHz range, there are 81 Doppler bins to search for in Galileo system.

Figure 9 illustrates the results for L1 Galileo signals acquisition in our noisy environment. Figures 8 and 9 can be used for a comparison between BPSK conventional modulation and BOC innovative modulation.

Besides Parallel Code Phase Search algorithm, have been implemented also a Parallel Frequency Search and a Serial Search, whose theoretical scheme are presented in Figs. 10 and 11.

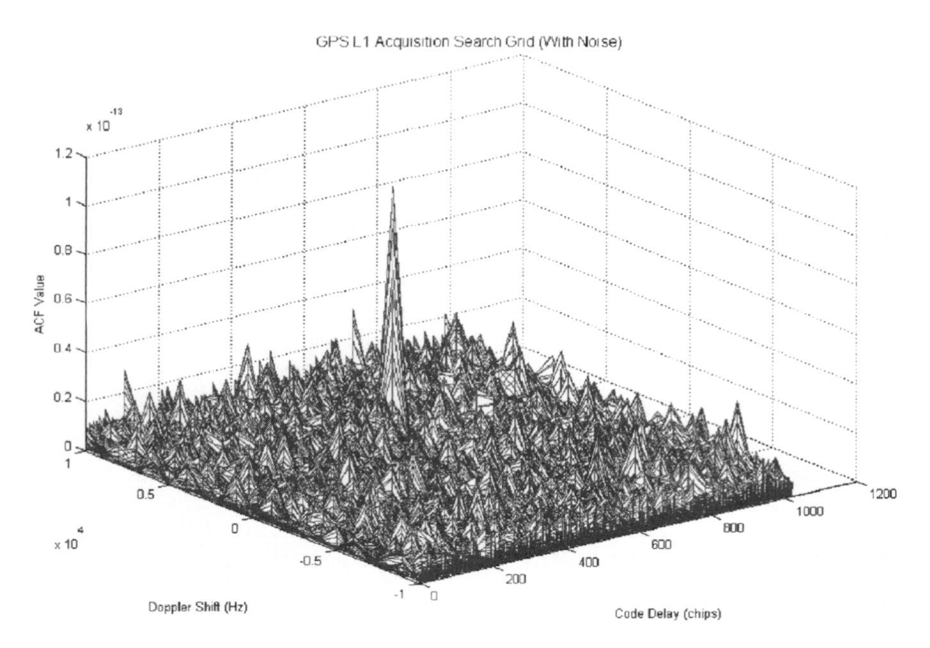

Fig. 8. Parallel code phase search acquisition grid for GPS L1.

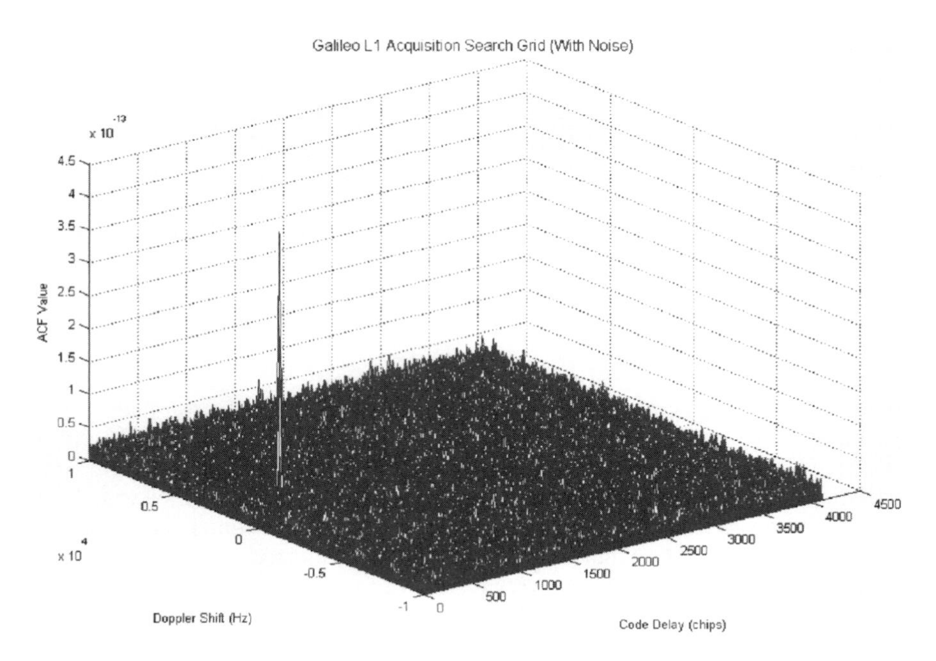

Fig. 9. Parallel code phase search acquisition grid for Galileo L1 (pilot channel).

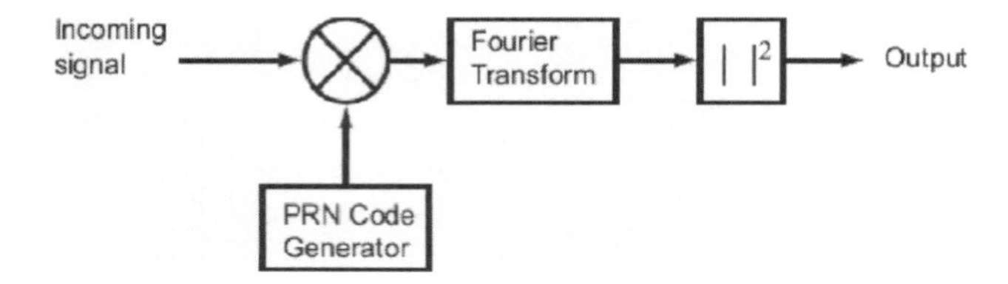

Fig. 10. Parallel frequency search acquisition scheme.

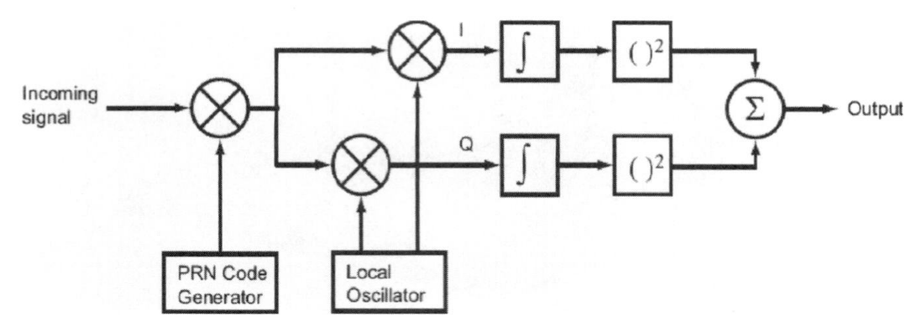

Fig. 11. Serial search acquisition scheme.

Acquisition search grids for Serial Search are the same of Figs. 8 and 9. For Parallel Frequency Search are presented only plots representing the Doppler axis acquisition peak (Figs. 12 and 13).

3 Analysis Results

All algorithms described above have been tested under computational burden and acquisition sensitivity aspects. The results are examined in the followings paragraphs.

3.1 Computational Burden

A key feature of an acquisition algorithm is the computational load, mainly when regarding a software implementation. The schemes are been tested over a Pentium® IV PC, 3.4 GHz clock with 1024 MB of RAM. Number of Floating Point Operations (flops) has been estimate for every algorithm, based on the calculation of both PC speed (in terms of flops/sec) and time used to acquire the signal. The results are summarized in Table 1 for both systems and for every algorithm.

First of all, it can be see that is very expensive for both system, under computational load aspect, to switch from Parallel Code to Parallel Frequency acquisition, with flops ratios of 68.166 (GPS) and 84.788 (Galileo). That is, IFFT computational cost added for first algorithm is widely justified by parallelized search advantage.

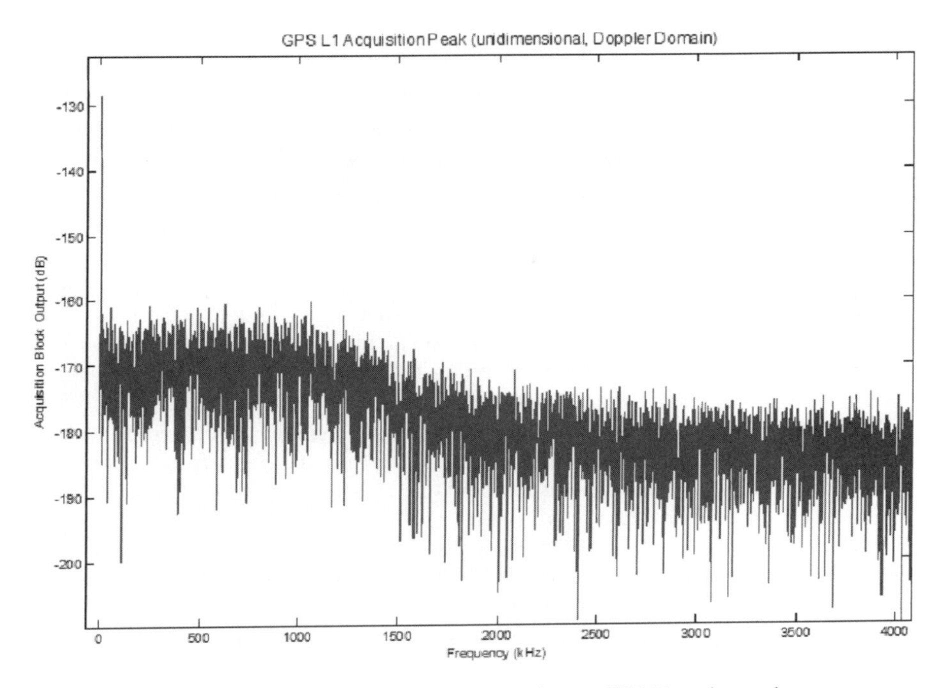

Fig. 12. Parallel frequency acquisition scheme: GPS Doppler peak.

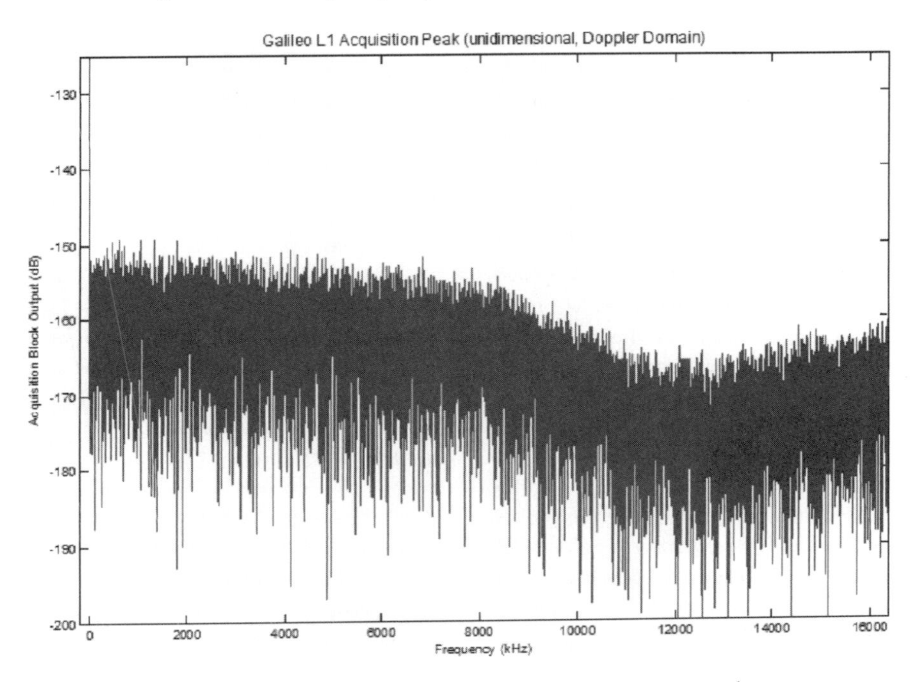

Fig. 13. Parallel frequency acquisition scheme: Doppler peak.

Table 1. Acquisition algorithms flops estimation.

	Parallel code	Parallel frequency	Serial
GPS	52.435 Mflops	3574.3 Mflops	7562.7 Mflops
Galileo	808.7 Mflops	68568 Mflops	480730 Mflops

On the other hand, it is not so expensive for GPS changing from Parallel Frequency to Serial acquisition, with a flops ratio of 2.11: thus, frequency parallelization is not so convenient for GPS system when compared with Serial search, due to low number of Doppler bins. For Galileo acquisition, flops ratio become 7.01 because of Doppler bins increasing, meaning that the Parallel Frequency search is still opportune with respect to serial search.

The principal feature to point out is that Parallel Code search represent best scheme for both system when considering computational load aspect, with flops ratios calculated on serial search equal to 144.23 (GPS) and 594.45 (Galileo).
The comparison between the systems shows that Galileo acquisition has, regardless of what algorithm is used, a higher computational cost than GPS acquisition, due to the use of longer spreading codes (4092 chips vs. 1023).

3.2 Acquisition Sensitivity

Another important element for a receiver in general is the acquisition sensitivity. Here, sensitivity is defined as the ratio between the main (acquisition) peak and the higher disturbance peak, in case of right acquisition. Noise level is specified in term of C/N_0, which is considered the same for both systems, regardless of the difference in transmitted power.

First plot (Fig. 14) shows a comparison between systems, in which sensitivity is referred to a Parallel Code Phase Acquisition.

From above plot it is clear that, using Parallel Code Phase acquisition, the GPS Signal acquisition is lost before than for Galileo, confirming better performance of the latter (concerning acquisition process). This happens despite has been ignored the fact that Galileo satellites will transmit signal with higher power than GPS ones. The difference between the systems is due to the longer spreading codes used by Galileo satellites, which allow obtaining a higher acquisition peak value

Another plot is proposed, which represents a comparison between algorithms too. In fact, Fig. 15 shows acquisition sensitivity performance for both Parallel Code Phase and Parallel Frequency Search, considering GPS and Galileo signals. This comparison points out that better performances are obtained when using Parallel Code acquisition instead of Parallel Frequency scheme. Besides this, it is clear also that Galileo lost acquisition before GPS when considering Parallel Frequency acquisition, overturning results observed using Parallel Code Phase acquisition. Both those characteristics can be explained considering that Parallel Frequency acquisition peak is not a correlation peak, and so correlation properties (and advantage for Galileo) are no longer relevant. Thus the only difference between the two systems is in frequency domain: Galileo signals occupy twice the bandwidth occupied by GPS C/A signal, so in Galileo there is a higher noise contribution.

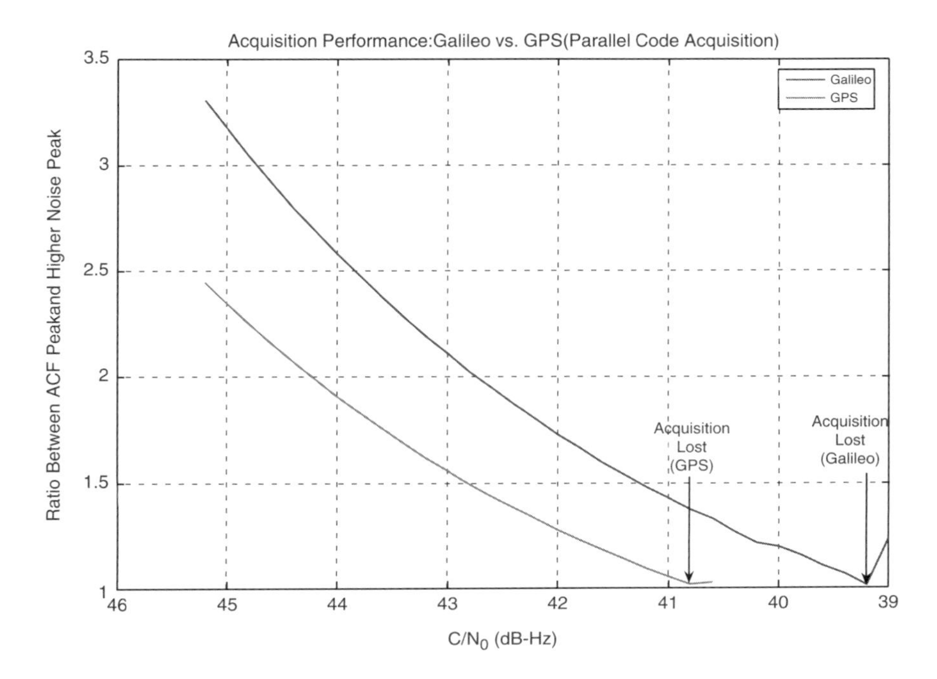

Fig. 14. Comparison between GPS and Galileo acquisition sensitivity using parallel code phase algorithm.

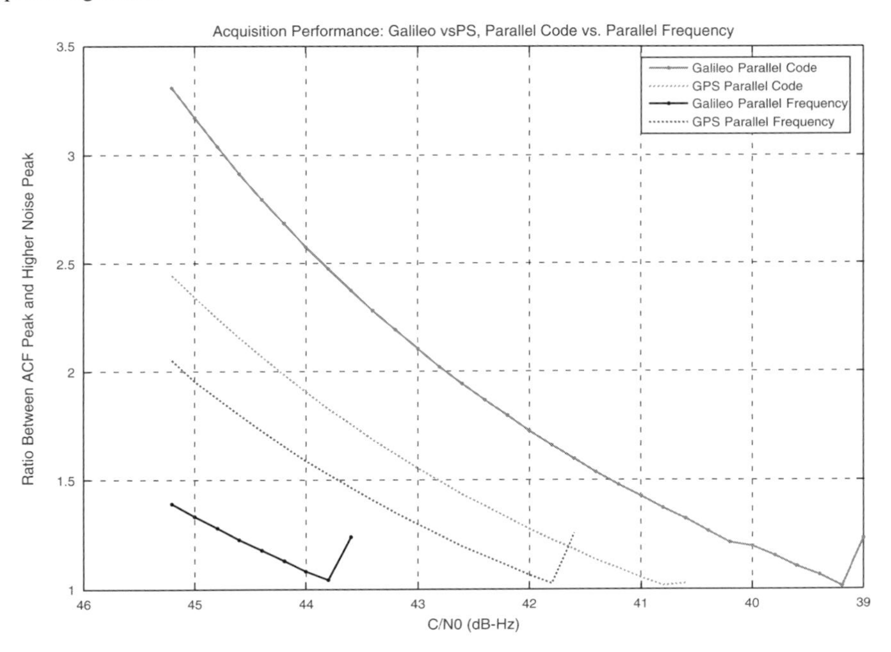

Fig. 15. Comparison between GPS and Galileo acquisition sensitivity using parallel code phase and parallel frequency algorithms.

4 Conclusions

In this work have been analyzed several acquisition algorithms for L1 Galileo and GPS signal, considering a software receiver approach. Three schemes (Parallel Code Phase Search, Parallel Frequency Search and Serial Search) have been developed and compared each other, under computational burden and acquisition sensitivity aspects.

From an algorithmic point of view, Parallel Code Phase Acquisition has shown the best performances in terms of both computational load (evaluated in number of flops) and acquisition sensitivity.

From a system point of view, Galileo has shown better performance in terms of acquisition sensitivity when employing Parallel Code Phase search (satellites can be acquired with lover C/N_0 values), but has been underlined that new European system employs longer spreading codes (i.e. a higher number of samples), so increasing computational burden regardless of what acquisition algorithm is applied.

References

[1] B. Parkinson, J. Spilker, P. Axelrad, P. Enge (Eds.), "Global Positioning System: Theory and Applications Volume I", American Institute of Aeronautics and Astronautics, Inc., 1996.

[2] J.B.Y. Tsui, "Fundamentals of Global Positioning System Receivers a Software Approach. 2nd Edition", *Wiley & Sons*, 2005.

[3] D.M. Lin, J.B.Y. Tsui, "Acquisition Schemes for Software GPS Receiver", Proceedings of 11th ION-GPS, sess. D2, pp. 317–324, 1998.

[4] D.J.R. Van Nee and A.J.R.M. Coenen, "New Fast GPS Acquisition Technique Using FFT", Electronic Letters 17 Vol. 27 No. 2, pp. 158–160, January 1991.

[5] C. Hegarty, M. Tran and A.J. Van Dierendonck, "Acquisition Algorithms for the GPS L5 Signal", Proceeding of ION GPS/GNSS 2003, Portland, OR, pp. 165–177, September 2003.

Chapter III

Satellite Navigation: Perspectives and Applications

Galileo: Current Status, Prospects and Applications

Vidal Ashkenazi

Nottingham Scientific Ltd
Loxley House, Riverside Business Park, Nottingham NG2 1RT, UK
Phone: +44 115 9682960, e-mail: vidal.ashkenazi@nsl.eu.com

Abstract. This article is the written version of an introductory presentation made at the Opening of the Workshop session entitled "Satellite Navigation: Prospects and Applications". The paper describes the present GNSS scene, which has changed significantly with the successful launch of GIOVE A, the first Galileo test satellite. A summary description of the planned modernisation and governance of the two existing global systems, GPS and GLONASS, is followed by brief details of the proposed Galileo Public-Private-Partnership (PPP) scheme, and a brief summary of the various regional Space Based Augmentation Systems (SBAS). The paper then concentrates on the proposed control of operations and governance of Galileo, a summary of the proposed Galileo services, and the applications that are likely to emerge in the near future.

1 Current GNSS Scene

With Galileo in its Development and Validation Phase, the future developments in GPS IIF and GPS III, the renewed interest in GLONASS, and the satellite navigation initiatives in Japan, China, India, Australia and several other countries, GNSS or Global Navigation Satellite System, is moving from being a concept, largely based on GPS alone, to a full global reality. To describe Galileo in its proper context, it is important to start with an overview of the current status of GNSS.

At present, GPS is the only fully operational global satellite navigation system. Despite its advancing age and unparalleled success of GPS, until recently the basic system had changed very little, since its inception in the 1970's, in terms of orbits and signals. In 1998, the White House announced the addition of a second civil GPS signal to improve the accuracy and reliability of GPS for civilian users. This was followed in 2000 by the removal of SA (Selective Availability).

These were the first steps in a comprehensive programme of GPS modernisation, which is currently underway, and aims to deliver significant improvements to both military and civil users. The programme covers modernisation of the space and control segments, including improved signals and stricter monitoring, which together will deliver a significantly improved performance to all users.

The first batch of Block II-F (Follow on) satellites will be in orbit in 2007 (or as recently announced in 2008), still one or more years before the European Galileo satellite system becomes operational. This space segment upgrade, coupled with several ground segment improvements, leading to increased accuracies of satellite orbit positions and clock data, will result in a much enhanced user positioning performance. For dual frequency civil users, the resulting navigation accuracy could be as good as one metre.

The US Department of Defense is also planning to incrementally upgrade and improve the system, through a process which is known as GPS III, which will address the future needs of military and civil users over the next 30 years. As a result of extensive consultations with industry, academia and users, there is a substantial amount of information in the public domain on the proposed and planned features of GPS III. However, one would expect that the final configuration and full specifications of GPS III will not be defined until 2008, or soon thereafter, when Galileo-1 is due to be fully deployed and declared operational. Only then will one be able to assess fully the capabilities and the resulting commercial advantages of Galileo, from the point of view of service providers, government, receiver manufacturers, safety critical transportation users, and ordinary citizens. By then one would also have a clearer idea on the uptake of PRS (the so-called Public Regulated Service) of Galileo, and especially the penetration of the Commercial Service into the mass-market commercial applications, on a global basis. This is when the US will show its full hand, or most of it, and define GPS III.

The Russian GLONASS system is also undergoing an extensive Programme of Modernisation, with several recent successful launches of GLONASS-M satellites, to supplement the current constellation. This will be followed by the GLONASS-K satellites, which will be launched over the period 2008-2015 and are expected to be operational until 2025. GLONASS is now controlled by ROSKOSMOS, with an Interagency Coordinating Board, which involves several ministries, including Transport, Defence, Industry and Energy. This is not unlike GPS, which is now directed by an Executive Committee of Positioning, Navigation and Timing (PNT), co-chaired by the DoD and the DoT, and including representatives from the Departments of Commerce, State, Homeland Security, the Joint Chiefs of Staff, NASA and other government departments and agencies, as required. This is in sharp contrast to Galileo which, for the time being at any rate, is controlled by the Galileo Joint Undertaking (GJU) on behalf of the European Commission (EC) and the European Space Agency (ESA). However, this will change soon.

2 Galileo Prospects

Much has already been written about Galileo which, at present, is in its Development and In-Orbit-Validation phase. This will be followed by Full Deployment and start of Operations sometime around 2010 or soon thereafter. Galileo is Europe's initiative to develop a civil global navigation satellite system, which will provide highly accurate and reliable positioning, navigation and timing services. Galileo will be compatible and interoperable with GPS and GLONASS, and will offer multiple civil frequencies. Galileo will also provide instantaneous positioning services at the one-metre level as a result of improved orbits, better clocks, and dual frequency enhanced navigation algorithms.

Through its different services, Galileo will also offer a level of guarantees of service availability, and will inform users within 6 seconds of a failure of any satellite.

This will allow the system to be used for several safety-critical, mission-critical and business-dependent applications. The combined use of Galileo, GPS and GLONASS will offer a very high level of performance for a large variety of user communities and businesses. In its present configuration, the Galileo design is comparable to GPS Block IIF which was defined in 2000. The European Space Agency has now tabled a Proposal for the Evolution of the European GNSS Programme, call it if you wish Galileo-II, for consideration by the Member States, but these are early days. Unlike GPS, which is fully funded by the US Department of Defense, that is the tax payer, Galileo is expected to be funded and evolved through a PPP (Private-Public-Partnership) scheme, which is still to be fully fleshed out.

3 Regional Augmentations

In parallel to the development of the 3 global satellite navigation systems, there are also several initiatives to develop satellite augmentation systems. Space Based Augmentation Systems (SBAS) have been designed to provide the necessary levels of accuracy, integrity, availability and continuity from GPS (and GLONASS), in order to facilitate the migration towards a satellite-based, global navigation infrastructure. In recent years, there has been a significant interest in the development of SBAS by an increasing number of regions in the world. The US, Europe and Japan were the first countries to commit themselves to the development of a regional SBAS, and led the way with WAAS, EGNOS and MSAS. More recently they have been joined by India, and now there is also a variety of other SBAS trials which are being conducted elsewhere in Asia, Australia, Latin America, and Africa.

Additionally, some countries are designing or developing independent regional satellite positioning systems, based on geostationary (GEO) and/or inclined geostationary satellite orbits (IGSO). Most notable among these are the QZSS in Japan, based on 3 satellites with highly elliptical orbits (HEO), BEIDOU in China which is based on 3 GEOs, and IRNSS in India which is planned to include 3 GEOs and 4 IGSOs. Lastly, China is proposing to have its own global satellite positioning system COMPASS, which will be based on 24 MEO's, just like GPS and Galileo, and include a security signal which will operate on the same frequencies as Galileo's PRS.

4 Galileo Operation and Governance

The European Galileo Project is moving ahead, and preparing to face the many current and future challenges. Over the last 18 months, the Galileo Joint Undertaking (GJU) has carried out an extensive process of inviting competitive tenders for the Galileo Concession, selecting and merging of the two "preferred bidders", and the submission of a single proposal by the merged Consortium followed by negotiations. Meanwhile, a new licensing authority, which will manage the interests of the public sector vis-à-vis the Concessionnaire, in relation to Galileo and EGNOS, has been set up. The Galileo Supervisory Authority (GSA), which will take over the activities of the GJU by the end of 2006, will not only supervise the private concession holder, which will be known as the Galileo Operating Company

(GOC), but will also contribute to the development of equipment and applications through the licensing of intellectual property rights (IPRs) vested in it. The day-to-day control and operations of Galileo will be carried out by the Galileo Operations Company (OpCo), which will manage the system and services on behalf of the Galileo Concession Holder (GOC), under contract. Like all the other global, regional and local satellite navigation systems, Galileo is all about applications. This is where the real challenges lie ahead. With GPS the civilian community was offered a free signal, with no guarantees whatsoever, whether on performance, integrity, coverage or continuity. Nevertheless, the all pervasive human ingenuity came along and provided a variety of tools to overcome the drawbacks of standalone GPS. These ranged from the development of differential GPS (DGPS) and carrier phase or kinematic GPS, which offered users centimetric and millimetric accuracies, to wide area augmentation systems, like WAAS and EGNOS, which improved integrity. These developments led not only to mass market, scientific and professional applications, but also to the development of safety-critical transportation applications, such the landing of civilian aircraft and the docking of ships entering harbour.

5 Proposed Services and Applications

With the impending arrival and start of operations of Galileo, there will a whole range of new applications based on the proposed 5 Galileo services, namely the Open Service (OS), the Commercial Service (CS), the Safety-of-Life (SoL), Search-and-Rescue (SaR), and the Public Regulated Service (PRS). These new applications will include business-critical, environmental-critical, financial-critical, legal-critical and government-policy-critical applications. Like GPS, the OS will cater for Mass Market applications, such as Location Based Services ("where am I?" or "where is the nearest . . . ?"), Telematics (fleet management, asset tracking, etc), and leisure (sport and recreation). Safety critical transportation, like the landing of airplanes and the docking of ships will be catered for by the SoL service. As the name indicates, the Galileo SaR service, which will also involve communications, will be used in emergency situations.

The precise role of the remaining two Galileo services, namely PRS and CS, is not yet fully defined. The declared intention is to use PRS for the police, security services, firefighting and ambulance services, where there is a need of additional integrity, coverage and continuity of the satellite signals. The CS will also provide the same degrees of integrity and coverage, but not necessarily of continuity, especially in emergency situations at certain geographical locations. However, the main difficulty of the CS is due to the fact that this will be the main Galileo service which is expected to generate the necessary financial returns for the Concessionnaire, so that the latter can fulfil its obligations under the Galileo Public Private Partnership (PPP) principle.

Among the potential candidate applications which could use the Commercial Service (CS), one could list Location Based Security (LBS), which would be used with portable PC's or laptops containing business sensitive data, and Galileo Time Synchronisation for time stamping of financially critical transactions and the synchronisation of data communication networks, including the world-wide-web.

Other potential CS user communities could include the offshore oil and gas industry, civil engineering and land surveying, and possibly the science and engineering community. All this would be subject, of course, to developing appropriate commercial services targeting the specific needs of these professional communities.

At present, it is not yet clear to what extent critical applications, such as Road User Charging (RUC) which is being seriously contemplated by the British government and has financially-critical aspects, Train Signalling which is safety-critical, and offender-tracking which is legal-critical will be catered for by the CS or the PRS. Furthermore, offender tracking, which will inevitably involve indoor as well as outdoor positioning, will require a suitable combination of High Sensitivity Galileo (HSG) with WiFi Local Area Networks (LAN). We are heading for interesting times, with ample opportunities for human ingenuity and enterprise.

The Galileo Test Range

G. Lancia, M. Manca, F. Rodriguez, F. Gottifredi

Atena Centre, Filas S.p.A.
piazzale della Libertà 20, 00192 Rome, Italy
Phone: +39 06 326959218, Fax: +39 06 36006804, e-mail: lancia@filas.it

Abstract. The GTR (Galileo Test Range) project is an initiative of Regione Lazio in the frame of its support to technical research and innovation in satellite navigation. Its main target is the development of a laboratory for the test and analysis of the Galileo signals, the support for development, test and certification of user terminals (GPS, EGNOS, Galileo) and of applications in different user domains.

The Phase A of the project started on July 2005.

The project is coordinated by a Consortium Agreement (C.A) composed by Telespazio, Alcatel Alenia Space Italia and Finmeccanica under the supervisory of FILAS S.p.A. the regional Financial Investment Agency dedicated to the support of innovation.

In the phase A, the C.A. has the responsibility of developing an infrastructure able to acquire and process signals coming from a constellation of pseudolites, from GPS, from EGNOS and from GSTB V2 satellites and able to support the testing of applications demanding high accuracy positioning providing augmentation to the users in the covered area.

Actually the proposal of constituting a center of excellence for satellite navigation follows the strategy of growth of industrial and technological capabilities in the frame of an industrial development policy pursued by Regione Lazio.

1 The GTR Center – Overview

The Galileo Test Range (GTR) project is an initiative of Regione Lazio in the frame of its support to technical research and innovation in satellite navigation. It is born with the scope of supporting the following high level missions:

- Characterization of the Galileo signal: the GTR aims to support the activity of analysing the performance of the Galileo navigation system, through the analysis of signal measurements in an environment suitably characterized and controlled. For this purpose, the GTR shall be able to gather raw navigation data within an experimental area and to process them in its analysis laboratories. Such objective is limited, as far as Phase A is concerned, to the characterization of the signal GTSB-V2, as it is propedeutical to the final scope of characterizing the Galileo signal, this latter assigned to the Final Phase of the Programme. This Centre will support the certification of the Galileo receivers the GTR must represent a suitable Test Bed for Galileo terminals, besides the GPS and EGNOS, placing the own navigation infrastructures and all the necessary hardware and software support instruments at disposal.

- Realization and Distribution of Services: The GTR, with the know-how and the experience gathered in the first operative phase of the project, must be set as a base for the realization of the Services to be distributed to the end users both public bodies, companies and privates. The support for the certification of applications, at a system level and at user terminal level, the possibility of providing universities and research centres with laboratories and testing areas, the continuous monitoring of the Galileo constellation are the basis of the GTR offer. Moreover the GTR aims to support the definition and the development of new High-Tech applications in the various user domains, for the utilization of the services offered by Galileo once operative.
  The development of the GTR is foreseen in three phases, in order to match the capabilities of the system with the development plan of Galileo:
  - Phase A = Definition and Start up: implementation of the initial system, based on the generation on ground of navigation signals (GPS-like) using pseudolite technology and based on the analysis and use of signals in space coming from GPS and EGNOS.
  - Phase B1 = Preparation to the development and deployment of Galileo system: implementation of the GTR in a configuration able to generate Galileo-like signals with ground equipment and to receive real signals coming from GSTB V2.
  - Phase B2 = Full deployment and initialisation of the GTR: implementation of the GTR final configuration, not only able to generate Galileo–like signals, but also to receive and process real signals coming from Galileo IOV satellites.

The Phase A of the project started on July 2005.

The project is coordinated by a Consortium Agreement (C.A) composed by Telespazio, Alcatel Alenia Space Italia and Finmeccanica with the FILAS S.p.A., the regional Financial Investment Agency dedicated to the support of innovation, as customer.

2 The Objectives of Phase A

In the phase A, the C.A. has the responsibility of developing an infrastructure able to acquire and process signals coming from 4 pseudolites, from GPS, from EGNOS and from the GSTB V2 experimental satellite and able to support the testing of applications demanding high accuracy positioning providing augmentation to the users in the covered area.

During Phase A, the GTR will achieve a basic configuration, called "start-up" and its main objectives are:

To provide a preliminary validation of the whole set of pseudolites implemented from this first stage, eventually in combination with the reception of the satellite GSTB-V2 that is already in orbit from the beginning of 2006, waiting for the Galileo signal.

To carry out analysis and experimentation on the GPS, EGNOS and GSTB-V2 signals (the latter at least for the frequency in the L1 band) that bring to the evaluation of their navigation performance, through the suitable acquisition, filing and processing of the gathered data.

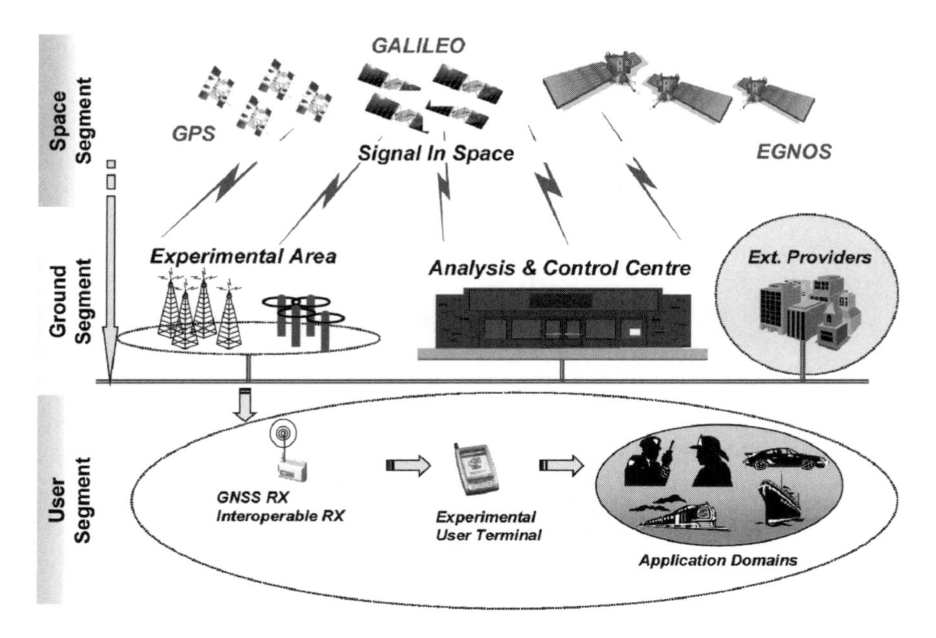

Fig. 1. GTR final high level architecture.

The characterization of the environment, by means of the use of an appropriate meteo station foreseen in the GTR architecture.

The physical realization of a local time reference, through the development and implementation of a time laboratory within the GTR.

The implementation of local "augmentation" of the navigation performance for the development of applications prototypes based on the use of navigation GPS + EGNOS signals.

The Phase A architecture has been designed taking into account the mentioned technical requirements and it is composed by the following macro segments (Fig. 1):

- The Space Segment: it is not part of the GTR system. Nevertheless the GTR is structured in order to allow the reception and the evaluation of the signals coming from the GPS, GSTB-V2, EGNOS and Galileo satellites.
- The Analysis & Control Centre
- The Experimental Area (that contains the Test Area)
- The User Segment.

3 The Analysis & Control Center

The characterization of the Galileo signal, with the consequent evaluation of the navigation performance in relation to its more prominent design aspects, such as the modulation and coding scheme it adopts, cannot leave out of consideration special calculation infrastructures. From the above follows the identification of a certain

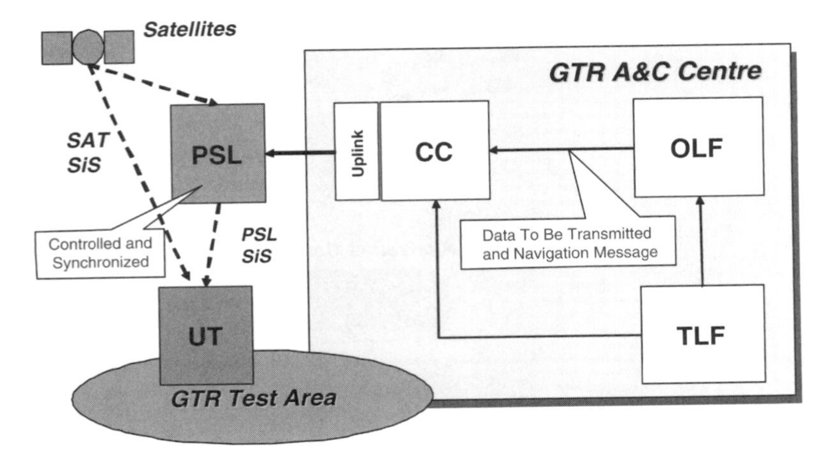

Fig. 2. Synchronization concept.

number of laboratories (for the generation of the time, the computation and analysis of orbitography and integrity) that the GTR requires in its configuration to be able to process and file the navigation data gathered with suitable Galileo receivers (available on site).

The Analysis & Control Center, the heart of the GTR, is composed by a Control Centre, Processing Facilities and Specialised Laboratories, and it is sited in the Tecnopolo Area (in the East of Rome).

It includes the structures dedicated to the processing of the navigation data produced in the experimental area. It includes the Signal Generator (SGF), that provides a controlled environment for User Terminal Qualification Tests (Fig.2).

The Analysis & Control Center includes moreover the Time Laboratory Facility (TLF), used to provide a reference time scale to the GTR system. This laboratory is composed by a group of high stability atomic clocks (one H-Maser and four Caesium atomic clocks) operating in a controlled environment. This guarantees a high stability both at short term and at medium/long term; the GTR-ST time scale is moreover steered to the TAI by means of GPS System Time.

The Integrity (ILF) and Orbitography (OLF) Laboratory facilities provide the support for the implementation of the integrity and navigation algorithms both on the GPS and Galileo constellation and on the pseudolite constellation.

Two types of processing chains are implemented in the GTR:

- Real time Processing for orbit determination and time synchronization, with the aim to provide the synchronization to all the elements composing the testing area and provided by part of the OLF.
- Real time integrity determination, with the aim to monitor the quality of the signals generated in the testing area and to rise alarm flags if system errors exceed certain thresholds, provided by part of the ILF.

Finally the Control Center contains the infrastructures for the monitoring and control of all the GTR elements besides the filing center of all the data produced by it.

Moreover the general purpose GTR-Laboratory has all the instruments necessary to support more innovative studies and activities, such as analysis, modelling and compensation of errors affecting GPS and Galileo measurements, development of Galileo Receivers technology, development of prototype applications and certification (with focus on the applications certification).

4 The Analysis & Control Center

Two main areas (See Fig. 3) can be identified in the ground segment:

- The ***Experimental Area***: for the support to the Galileo receivers certification, an experimental area, in east of Rome, has been identified, where it is possible to perform the relevant tests, taking into account the most typical environmental conditions that cannot be reproduced in an extremely good way in a laboratory.
- The ***Test Area***: that is the area covered by the Pseudolite Signals and where it is then possible to conduct Tests in a fully controlled (also in terms of Signals) environment. The Test Area is a subset of the Experimental Area. At the same time special test campaigns are foreseen for the testing of the prototypes of applications and relevant added value services in portions of the experimental area that

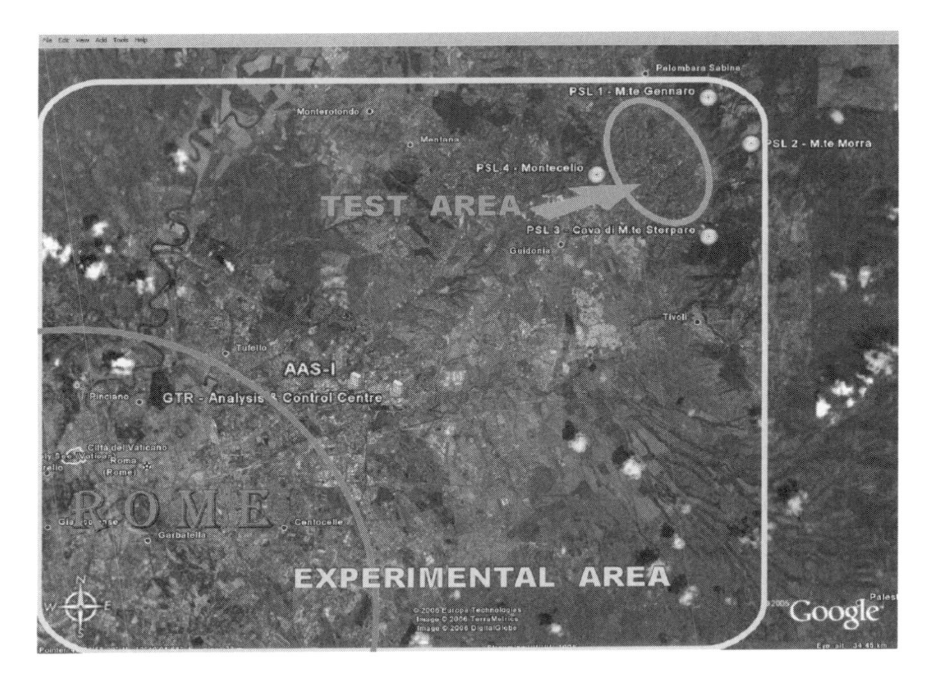

Fig. 3. GTR experimental & test areas.

contains the infrastructures representative of the concerned users domain, such as parts of roads, railways, sections of urban environment and others.

The Experimental Area is mainly characterized by the following elements:

- A constellation of pseudolites (PSL) that broadcast navigation signals representative of the GPS and/or Galileo System (only GPS in phase A) to the users present in the area,. They are synchronized by the GTR-OLF with respect to the reference time scale of the GTR (GTR-ST), generated by the Time Laboratory (TLF), also part of the Analysis & Control Center.
  The Sensor Station (SS): the navigation signals generated both from the ground network of pseudolites and from the satellites in orbit, are also processed by two monitoring stations one sited in the test area and the other co-located into the TLF. The two SS gather the observables useful for the estimation of the synchronization parameters of the pseudolites to be sent to the users. The synchronization algorithm is based on the well known Common View technique, processing the measurements acquired by the two sensor stations and computing the clock biases of the different elements.
- The Differential Reference Stations (DRS): the experimental area will include infrastructures of augmentation for the implementation of applications prototypes: besides the pseudolites, two Differential Reference Stations will be installed. The two DRS are collecting raw data that are then archived in the Control Centre as RINEX files, but they are also distributing corrections in real time through the NTRIP protocol (message types 1, 3, 16, 20 and 21), so that users can have access to high accuracy positioning service directly through internet.

5 The User Segment

The user segment includes the following elements:

- GPS/EGNOS Receivers: necessary for the evaluation of the user position calculated with the GPS system, as a reference for the tests carried out with other receivers.
- GSTB-V2 Receivers: used for the reception of the GSTB-V2 signal, that will be representative of the final Galileo signal. In combination with the processing of the signals transmitted by the pseudolites (transmitting GPS Signals), it allows a first evaluation of performance at user level, obtainable with the Galileo signals.

6 Conclusions

The Regione Lazio represents an area of absolute international relevance in the aerospace sector. With 5 billion € turnover, more than 30,000 employees and 250 prominent sized companies the region is characterised by strong technical capabilities, high quality productivity and broad diversification in national and international projects.

Actually the proposal of constituting an international center of excellence for satellite navigation follows the strategy of growth of industrial and technological capabilities in the frame of an industrial development policy pursued by Regione Lazio.

In particular, SMEs that wish to step into the new business of satellite navigation, have been particularly addressed during the development of the project, supporting the C.A. in the research and development activities concerning the set up of the GTR.

Moreover the role of 10 public research centers, 5 Universities (La Sapienza, Tor Vergata and La Terza, Cassino, Viterbo) and 4 aerospace engineering faculties in the Lazio region is of paramount importance in the entire framework since they participate proactively both to the definition and to the development of the Galileo Test Range.

Perspective of Galileo in Geophysical Monitoring: The Geolocalnet Project

M. Chersich*, M. Fermi*, M. C. de Lacy**, A. J. Gil**, M. Osmo*, R. Sabadini***, B. Stopar****

*Galileian Plus S.r.l.
Via Tiburtina 755, 00159 Roma, Italy
Tel.: +39 6 40696500, Fax: +39 6 4069627, e-mail: mchersich@galileianplus.it
**University of Jaen
Campus Las Lagunillas s/n 23071 Jaén, Spain
Tel.: +34 953212467, Fax: +34 953212854, e-mail: ajgil@ujean.es
***University of Milan
Via Mangiagalli 34 20133 Milano, Italy
Tel.: +39 2 503154936, Fax: +39 2 50315494, e-mail: roberto.sabadini@unimi.it
****University of Lubljana
Jamova 2, 1000 Ljubljana, Slovenia
Tel.: +386 1 476 85 00, Fax: +386 1 425 07 04, e-mail: bstopar@fgg.uni-lj.si

Abstract. Earth crust deformation continuous monitoring using Global Navigation Satellite System (GNSS) local geodetic networks demands suitable computational and informatics tools based on solid scientific background. This is particularly true if geophysical applications are concerned. Applications like seismic hazard mitigation, subsidence and landslides monitoring requires the highest accuracy in positioning with as short as possible measurements sessions length. The GEOLOCALNET project, co-funded by the Galileo Joint Undertaking (GJU) under the 6th Framework Program and managed by a consortium of Research Unit (RU) and Small and Medium Enterprises (SME), investigates the three carriers based Galileo Satellite System positioning capability developing and validating innovative algorithms, models and estimation procedures.

The project addresses many and critical issues affecting precision and stability in GNSS processing strategies and promotes the usage of local geodetic networks for deformation evolution monitoring.

1 Introduction

GNSS has been used for many years for the deformation monitoring of manmade structures such as bridges, dams and buildings, as well as geophysical applications, including the measurement of crustal motion, and the monitoring of the ground subsidence and volcanic activity.

Modernized GNSS systems, such as GALILEO, based on three-carriers signal, offer the opportunity to address the NRT high precision positioning issue, and the GEOLOCALNET project objective is the utilization of GALILEO multiple frequencies to improve the accuracy in differential carrier-phase based positioning techniques and to promote the use of the local geodetic networks for Earth crust deformation monitoring.

The present work is organized as follow. In section 2 the whole project is briefly described, emphasizing the application background and the monitoring requirements in particular. Translation of requirements into algorithms is discussed in section 3, where the software prototype characteristics are presented and the implemented processing strategy discussed. After giving an insight into the established test plan for the GEOLOCALNET prototype validation in section 4, section 5 illustrates some preliminary numerical results. Conclusions are drawn in section 6.

2 Geolocalnet Project

2.1 Consortium Partners

GEOLOCALNET is a one-year project co-funded by the GJU aiming to fully exploit the new expected performances of GALILEO system.

Both RU and SME, belonging to the European Union, compose the Consortium involved in the GEOLOCALNET project. The Prime Contractor of the project is Galileian Plus S.r.l. (Italian SME). Other contractors are Space Engineering (Italian SME), Harpha Sea (Slovenian SME), University Ljubljana (Slovenian RU), University of Milan (Italian RU) and University of Jaen (Spanish RU).

Partners' background guarantees solid scientific knowledge, computational and informatics capability to properly address the GALILEO data processing issue in the frame of high accuracy positioning.

2.2 Application Background

The project main purpose is to develop and to validate innovative algorithms, models and procedures to improve the accuracy in differential positioning and to promote the use of local geodetic networks for Earth crust deformation monitoring. A local geodetic network can be considered a local element dedicated to high accuracy relative positioning measurements, through a set of GNSS receivers located in a limited area (typically 20*20 km^2). These networks are established to monitor the Earth crust deformations due to seismic motion, landslides and subsidence. The local deformation is measured by repeated estimations of baseline vectors between couples of fixed markers.

The accuracy in deformation determination strictly depends on the type of application, ranging from 1 cm order of magnitude for fast landslide monitoring [3] to millimetre level for active faults monitoring, which is the main application scenario of GEOLOCALNET project.

Among many geophysical events requiring high precise and fast update monitoring, seism is the most demanding one. Hereafter, discussion concerning accuracy will be focused on seismic monitoring, being aware that other deformation monitoring applications caused by other geophysical events (landslide, subsidence, etc.) will be by-product derived from this reference application.

Actually the currently available GNSS (GPS, GLONASS) applied to seismic monitoring allow high precise measurements indeed. Nevertheless the data amount required to obtain millimetre level precision is very large, thus preventing the possibility of NRT monitoring, as required, for example, during pre and post seismic events.

Deformation monitoring through GPS measurements, integrated with seismological studies and geophysical forward modelling, is becoming of paramount importance to discriminate areas prone to earthquake events with a given magnitude [1]. These results are obtained applying the so called intermediate-term middle-range earthquake prediction algorithms and the geophysical forward modelling, which translate surface strain fields obtained by GNSS data analysis in deep stress field at the level of the seismogenetic fault. The objective of geophysical forward modelling is to derive seismic hazard maps. A conceptual picture of the described approach is provided in Fig. 1. However, seismic hazard maps are associated to time and space uncertainties that, at present, are respectively of few years and of a few hundred kilometres. The order of magnitude of these uncertainties does not allow an effective early warning service to protect population.

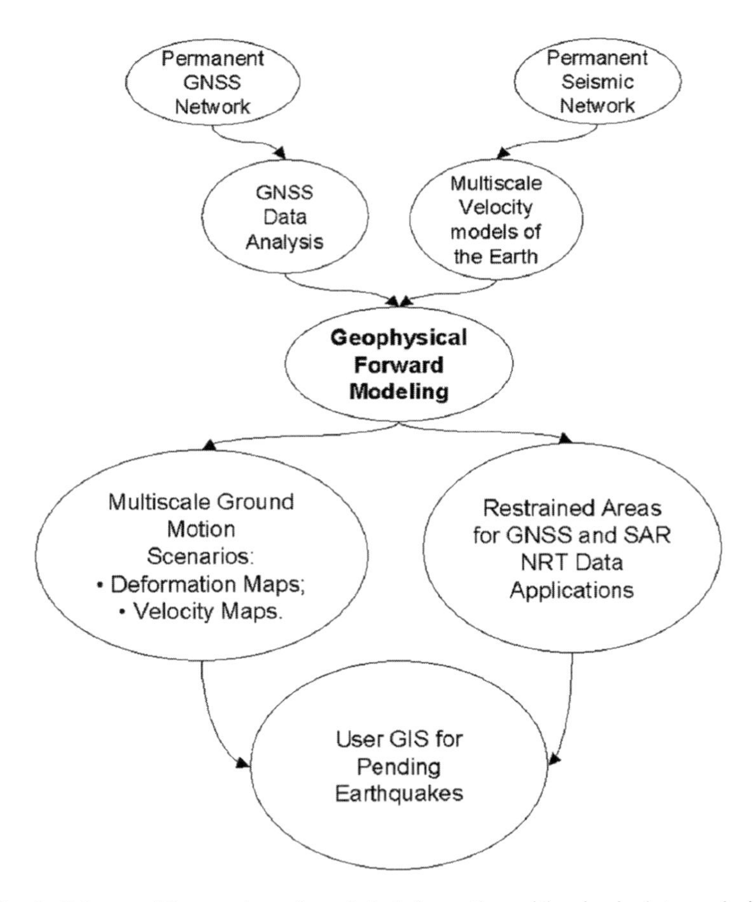

Fig. 1. Scheme of integration of geodetic information with seismic data analysis.

More refined monitoring of active faults, based on a NRT processing of data collected by local geodetic networks and on local seismicity analysis bridged together through geophysical forward modelling in the areas prone to earthquake events, is the current goal to further reduce time and space uncertainties in the prediction of the seismic events. Note that GNSS networks of permanent receivers play an important role also in the identification of the restrained areas where more refined monitoring is needed. A scheme of the NRT applications is depicted in Fig. 2.

The proposed scheme involves permanent GNSS network and analysis of seismicity to identify restrained areas for GNSS and SAR NRT monitoring (as better shown in Fig. 1), then, on a local basis, space geodetic data are analysed to provide NRT more refined information to geophysical forward modelling.

Concerning the space and time accuracy requirements, it is worth noticing that, in geodynamic monitoring of local networks, GPS accuracy is of the order of few millimetres using several hours of observations (at least three hours, with 30 seconds or higher sampling rate), while, as stated hereafter, the required accuracy is of the order of one millimetre in NRT.

What is really challenging is to explore the GALILEO capability for NRT high accuracy applications to provide a first step, in the overall approach depicted in Fig. 2, toward the provision of the essential contribution to reduce space and time uncertainties in the prevision of Earthquake events. This application is the frontier for a real support to Civil Protection in emergencies management.

2.3 Requirements for Seismic Hazard Assessment and Monitoring

Synergic use of GNSS and geophysical forward modelling complement the information gained from purely statistical analyses of earthquake historical records. In such a way the rules of seismic hazard estimate in terms of observational data and of sound physical methodologies are established. GNSS techniques, at the spatial

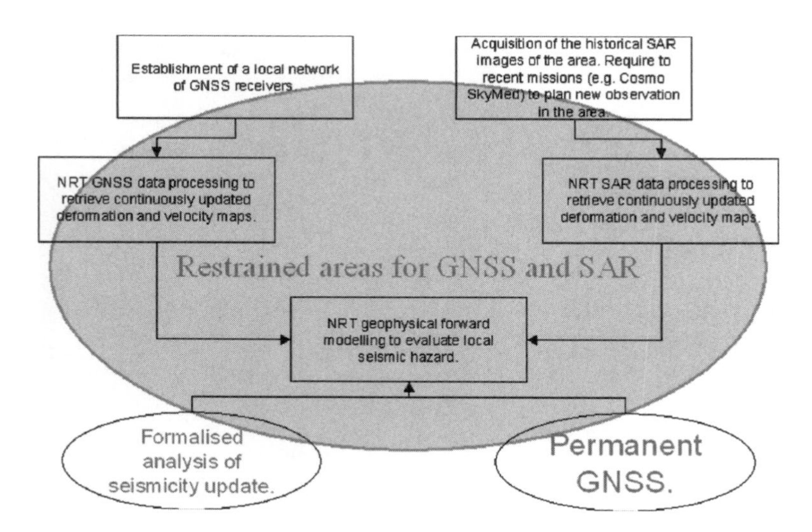

Fig. 2. Scheme of NRT applications.

scale of the seismogenic zones, coupled with expressly developed models for post-seismic, inter-seismic and pre-seismic phases within proper inversion and assimilation schemes based on GNSS data, can be used to retrieve the deformation style and stress evolution within the seismogenic zones, thus providing the tools for establishing earthquake warning criteria based on deterministic grounds.

In general, co-seismic deformation is well understood, also in terms of GNSS data, as shown for example in [5] which makes it possible to provide a static description of the event, before and after the earthquake; post-seismic deformation also started to be understood and detected in the Mediterranean region, as first shown in [1]. Thanks to the expected performance of GALILEO, it is now possible to make a step ahead in the mathematical simulation of the fault behaviour during the pre-seismic phase, by inversion of the stress field within the fault gouge from accurate, high resolution GNSS data, collected at the Earth's surface in the seismogenic zone.

An essential requirement that the new earthquake warning scheme based on GALILEO data comes directly from the observation that during the post-seismic and inter-seismic phases, relative motions across typical Mediterranean faults is of the order of millimetres per year rather than centimetres per year, as for California's faults, for example, from which it is immediately possible to establish that in order to catch the expected pre-seismic signals at the Earth's surface of Mediterranean seismogenic faults it is necessary to go beyond the performances of actual GPS receivers, which already reached the following resolution during the most advanced studies of Mediterranean post-seismic phases, over baselines of few kilometres, namely:

- 1 mm/yr, in the horizontal component;
- 2–3 mm/yr, in the vertical component;

based on yearly sampled data.

In order to detect expected pre-seismic phase Earth's crust deformation and in particular possible acceleration in deformation rates at the Earth's surface over the fault zones during the final stage of the pre-seismic phase, as expected for strongly non-linear systems, it is thus necessary to detect Earth's surface deformation, over baselines ranging from few kilometres to tens of kilometres, with the following accuracy:

- 0.3 – 0.5 mm/yr, in the horizontal component;
- 0.5 – 1 mm/yr, in the vertical component.

Besides these "static" requirements of GNSS accuracy, it is necessary to establish time interval criteria over which that accuracy should be assured, in order to catch the expected acceleration of deformation rates during the pre-seismic phase: due to the strongly non-linear dynamics of the fault, it is required that such an accuracy could be reached at the weekly rate at the most.

2.4 Prototype Overview

The proposed research is focused on GALILEO NRT data analysis SW prototype development, aiming to reduce space and time uncertainties in the frame of Earth crust monitoring having in mind the accuracy requirements for seismic hazard mitigation and therefore to realize a preliminary step forward a faster and more refined updating of seismic hazard maps.

The goal shall not be only high accuracy relative positioning, but as fast as possible reaching of a precise and reliable solution, keeping unchanged the above accuracy requirement. This is the reason for having addressed what NRT means for the GEOLOCALNET processing technique. During requirement definition analysis it has been evidenced how GPS measurements campaign, in a short baselines scenario (few hundreds meters between receivers), allow to reach millimetre level accuracy with at least three hours of data, sampled at 15 sec. Similar experiments confirms this datum [2].

This project is focused on exploiting the three-carriers capability of GALILEO system and in particular for improving at least:

- ionosphere delay modeling/estimation;
- convergence time of phase ambiguity fixing.

The second feature essential for the application proposed, as indicated in Fig. 2, is the NRT response of the algorithm. To reach this goal a dedicated ambiguity fixing algorithm have been implemented and adapted to the three-carriers capabilities of the GALILEO system.

The algorithm approach is based on double differences (DD) building technique as usually applied in local networks data analysis. This technique does not completely eliminate the ionosphere and troposphere errors, which appear to be significant for baselines greater than few kilometres. Therefore, the development of ionosphere and troposphere models is essential to maintain the precision required for geodetic networks deformation monitoring and seismic risk mitigation.

The project does not include the procurement of new hardware or tools, but the reuse of existing facilities. Data processing will thus use the upgrade of a product developed by Galileian Plus, called Network Deformation Analysis (NDA) [4]. NDA has been developed from scratch and it is based on standard geodetic processing technique [13] designed for local network of GPS receivers. NDA is able to perform single baseline adjustment using L1 and L3 (ionosphere-free phases combination) double differenced data. The resulting prototype processing strategy is outlined in the next section.

As we do not have GALILEO data available yet, one of the main tasks of this project will be the generation of simulated data. This will be achieved by means of the Galileo System Simulation Facility (GSSF), an existing GALILEO simulator tool of ESA.

2.5 Progress Status

GEOLOCALNET activity started in the end of 2005 and it will last 12 months. At the time of writing this article the project entered the second half of its duration. The first six-months activity was mainly devoted to requirement definitions, GSSF simulator analysis and some preliminary implementation activities such as new RINEX 3.0 format data reader implementation, the upgrade of pre-existing cycle slips and outliers detection and removal algorithms and to focus on the processing strategy for baseline adjustment using the three-carriers GALILEO observations data.

After the mid-term review meeting occurred on late June 2006, the project entered into the validation phase, where both simulated GALILEO data and real and simulated GPS data will be generated and used to examine the performance of the developed prototype.

3 Algorithms and Processing Strategy

Upgrading NDA version from the dual frequencies GPS data processing to the prototype that allows baseline adjustment based on three-carriers signal of modernized GNSS has been the most time consuming part of GEOLOCALNET project. Algorithms implementation ended with the midterm review milestone when debugging and preliminary test activities started. In the following, phase linear combinations used in the processing strategy are introduced before outlining the whole processing strategy, from GALILEO data acquisition to baseline solution.

3.1 Galileo Phase Linear Combinations

Wide-lane combinations, due to their large wavelength, play a prominent role in many of the GNSS ambiguity fixing procedures that have been proposed and published (see, for instance, [8]).

In this section we introduce the GALILEO wide-lanes, medium-lane and extra narrow-lane combinations and their corresponding ambiguities in the Open Service (OS) and Commercial Service (CS) frequencies scheme. In Table 1 the wavelengths of these combinations are shown.

Extra Wide-lane (EWL) linear combination is given by:

$$L_{EWL}=\lambda_{EWL}\left(\frac{L_2}{\lambda_2}-\frac{L_3}{\lambda_3}\right)\text{ with }\lambda_{EWL}=\frac{\lambda_2\lambda_3}{\lambda_3-\lambda_2}. \tag{1}$$

The observation equation for L_{EWL} reads:

$$\begin{aligned}L_{EWL}(t)=&\,R(t)+dR(t)+T(t)+c\,(dt(t)-d\tilde{t}(t-\tau))\\&-\lambda_{EWL}\left(\frac{I_2(t)}{\lambda_2}-\frac{I_3(t)}{\lambda_3}\right)+\lambda_{EWL}(N_2-N_3)+\varepsilon_{EWL}.\end{aligned}$$

where,
L_k is the phase observation on the k-th carrier frequency, in metres (k = 1,2,3);
λ_k is the wavelength of the k-th carrier frequency;

Table 1. Wavelength of some useful phase linear combination.

	EWL (m)	WL (m)	ML (m)	NL (m)
OS	9.768	0.814	0.751	0.125
CS	3.455	0.789	1.011	0.121

R is the geometric range between satellite and receiver (metres);
dR is the orbital error (metres);
I_i is the ionospheric delay (metres) on the i-th carrier;
T is the tropospheric delay (metres);
N_k is the phase ambiguity on the k-th carrier frequency;
ε_k is the noise of the observations;
$dt(t)$ and $d\tilde{t}(t-\tau)$ represent receiver and satellite clock offset at receiving and emission epoch (τ is the signal travelling time).

In the same way, the Wide Lane (WL) phase combination is

$$L_{WL} = \lambda_{WL}\left(\frac{L_1}{\lambda_1} - \frac{L_2}{\lambda_2}\right) \text{with } \lambda_{WL} = \frac{\lambda_1\lambda_2}{\lambda_2 - \lambda_1}. \tag{2}$$

and the corresponding WL observation equation is given by:

$$L_{WL}(t) = R(t) + dR(t) + T(t) + c\left(dt(t) - d\tilde{t}(t-\tau)\right) - \lambda_{WL}\left(\frac{I_1(t)}{\lambda_1} - \frac{I_2(t)}{\lambda_2}\right) + \lambda_{WL}(N_1 - N_2) + \varepsilon_{EWL}.$$

It is important to mention another important linear combination used in the ambiguity fixing procedure: the Extra Narrow-Lane (ENL):

$$L_{ENL} = \lambda_{ENL}\left(\frac{L_2}{\lambda_2} + \frac{L_3}{\lambda_3}\right) \text{with } \lambda_{ENL} = \frac{\lambda_2\lambda_3}{\lambda_3 + \lambda_2},$$

L_{ENL} observation equation is:

$$L_{ENL}(t) = R(t) + dR(t) + T(t) + c\left(dt(t) - d\tilde{t}(t-\tau)\right) - \lambda_{EWL}\left(\frac{I_2(t)}{\lambda_2} + \frac{I_3(t)}{\lambda_3}\right) + \lambda_{ENL}(N_2 + N_3) + \varepsilon_{EWL}.$$

Taking into account these linear combinations, the initial ambiguity N_1, N_2 and N_3 can be expressed in term of N_{EWL}, N_{WL} and N_{ENL} using the following relations:

$$\begin{cases} N_{EWL} = N_2 - N_3 \\ N_{WL} = N_1 - N_2 \\ N_{ENL} = N_2 + N_3 \end{cases} \Rightarrow \begin{cases} N_1 = N_{WL} + \frac{1}{2} N_{EWL} + \frac{1}{2} N_{ENL} \\ N_2 = \frac{1}{2}(N_{ENL} + N_{EWL}) \\ N_3 = \frac{1}{2}(N_{ENL} - N_{EWL}). \end{cases} \tag{3}$$

Relations in equation (3) will be used in baseline adjustment performed with the ionospheric-free (IF) three-carriers phases combination.

Since ionospheric effect is frequency dependent it is possible to exploit the full three-carrier capability of modernized GNSS for the first order ionospheric path error elimination. Among the many IF linear combinations, in this work it has been considered the triple frequency minimum noise combination.

A general form for the triple frequency linear combination is given by the following expression:

$$L_{IF} = \alpha L_1 + \beta L_2 + \gamma L_3, \tag{4}$$

and the correspondent associated noise can be written as:

$$\sigma_{IF} = \sqrt{\alpha^2 + \beta^2 + \gamma^2}\,\sigma_0.$$

If parameters fulfil the following two relations:

$$\begin{cases} \alpha + \beta + \gamma = 1 \\ \alpha + \dfrac{\lambda_2^2}{\lambda_1^2}\beta + \dfrac{\lambda_3^2}{\lambda_1^2}\gamma = 0, \end{cases}$$

L_{IF} is a IF phase combination, and it is possible to obtain a single parameter depending expression for the noise σ_{IF}. Minimizing it, the three values for the parameters of the minimum noise IF three carriers combination results for CS and OS frequencies respectively:

$$\begin{cases} \overline{\alpha} \cong 2.380 \\ \overline{\beta} \cong -0.134 \\ \overline{\gamma} \cong -1.246 \end{cases} \text{and} \quad \begin{cases} \overline{\alpha} \cong 2.315 \\ \overline{\beta} \cong -0.479 \\ \overline{\gamma} \cong -0.836. \end{cases}$$

Supposing phase measurements noise $\sigma_0 \cong 0.0030m$, double differenced L_{IF} combination of equation (4) associated noise is: $\sigma_{IF} \cong 0.0161m$ (CS frequencies) and $\sigma_{IF} \cong 0.0150m$ (OS frequencies).

ML is for the Medium Lane phase combination, given by:

$$L_{ML} = \lambda_{ML}\left(\frac{L_1}{\lambda_1} - \frac{L_3}{\lambda_3}\right) \text{with } \lambda_{ML} = \frac{\lambda_1 \lambda_3}{\lambda_3 - \lambda_1}.$$

3.2 Baseline Estimation Strategy

The upgraded version of NDA is able to manage and process GALILEO raw data to compute the estimated baseline coordinates. The baseline estimation procedure is carried out in two parts: pre-processing and processing. Both data flowcharts are introduced in order to give insight into the overall baseline estimation strategy and to give evidence of the mitigation models used in the computation.

NDA allows the user to choose between processing GPS or Galileo (OS or CS frequencies) and to set up some processing inputs, i.e.:

- To define the measurement session length to be processed;
- To import observation and ephemeris files;
- To import antenna phase centre file and phase centre variation table (if available);
- To choose the elevation mask (cut-off angle) for each station of the network;
- To choose the baselines to be processed, fixing the station to be considered as reference;
- To choose the appropriate model to handle the atmosphere (troposphere and ionosphere) effects;
- To activate the residual tropospheric zenith path delay estimation in baseline adjustment (i.e. using double differenced observations);
- To choose the observable to be processed in baseline adjustment, i.e. E1/L1 or IF.

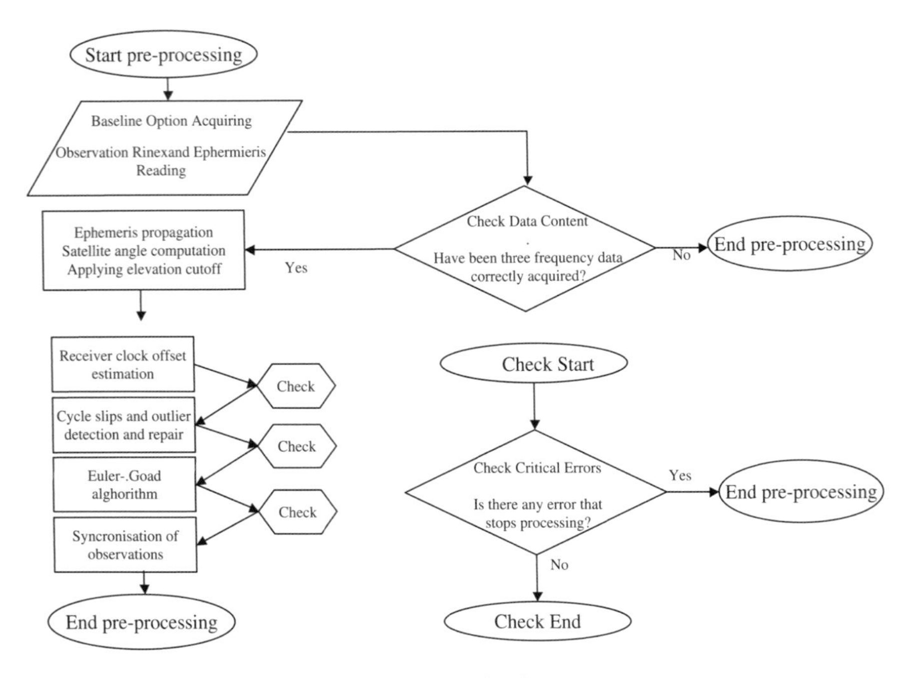

Fig. 3. Pre-processing flowchart.

After choosing operational set-up, single station data pre-processing, for every receiver that constitutes the baseline to be adjusted, can be started.

Pre-processing flowchart is shown in Fig. 3.

Operations performed in this phase include single station data acquisition, managing cycle slips and outliers, the noisy data and data under the cut-off mask editing, computing quantities such as geometrical variables, satellite coordinates at emission epoch, etc., and estimating the receiver clock offset, the dispersive and non-dispersive epoch-by-epoch path delay and the initial float ambiguity.

During each pre-processing step, controls are active to check every computational operation, verifying their reliability, and preventing the prototype collapse in presence of critical errors. Euler-Goad algorithm refers to the originally proposed algorithm [7] for single station epoch-by-epoch estimation of the dispersive, non-dispersive path delay together with the initial (float) ambiguity. The original algorithm has been upgraded for managing three-carriers GALILEO signal, as well as modernised GPS.

The purpose of this computational block is to prepare the observation data batch to processing block, i.e. to the baseline adjustment of synchronised data.

Processing allows to obtain the final baseline coordinates by keeping fixed the reference receiver coordinates to their a-priori values and applying the estimated corrections to the rover receiver coordinates only. Data processing flowchart is shown in Fig. 4.

NDA approach to be chosen (E1/L1 or IF) depends on the baseline length. If the baseline length is less than 5 km, it is suggested the E1/L1 processing, since ionospheric

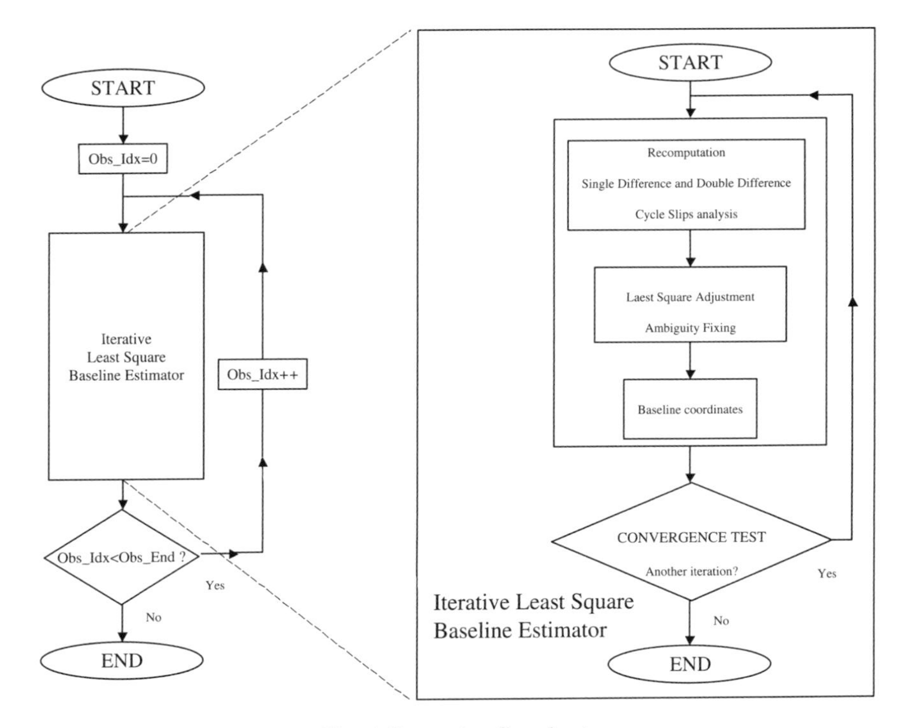

Fig. 4. Processing flowchart.

effect is common to both stations and it is mainly removed in DD procedure. When baseline length is greater than 5 km it is recommended to use the IF observable.

Data processing module mainly performs an iterative batch least squares (LS) adjustment following these steps:

- Recomputation: once coordinates corrections have been estimated, they are applied to compute satellite coordinates, geometric distance between satellite and receiver and tropospheric path delay. This step is obviously omitted in the first iteration;
- Single difference construction;
- DD construction;
- Cycle slips and outliers analysis on DD;
- LS approach: a weighted least square batch estimator is performed to estimate the corrections to (float) baseline coordinates, float ambiguities and, eventually, residual tropospheric zenith path delay, based on Saastamoinen slant model [11] or projecting the dry and wet zenith delay using the corresponding Niell mapping function [10];
- Ambiguity fixing by using LAMBDA method [12];
- Integer Least Square approach to obtain the fixed solution;
- Convergence test, if it fails another iteration starts;

- Storing the solution: baseline coordinates, ambiguities, and the estimated correction to the tropospheric zenith path delay are stored in the observation matrix for post-processing purposes.

This iteration procedure is used if single frequency (E1 / L1) DD observables are processed. If ionospheric-free DD are considered, a loop (that in Fig. 4 is represented as a loop over ObsIdx variable) over three different phase combinations starts. Phase combinations are considered in this order: EWL, WL (due to their decreasing wavelength) and IF DD. In case of IF approach, the iteration is carried out as following:

1. DD of EWL and WL combinations of equation (1) and equation (2) respectively, are considered to estimate float EWL and WL ambiguities in an LS approach. In this case the ionospheric effect is reduced by Klobuchar model [9] (with broadcast coefficients or coefficients estimated by the CODE centre - http://www.aiub.unibe.ch/ionosphere - improving the mitigation of the ionospheric effect), or the technique explained in [6]. Alternatively, float EWL ambiguities, due to their long wavelength and being the mismodeled residuals well below this value, could be obtained from the Euler-Goad algorithm extension applied to the DD of observations.
2. LAMBDA method is used to estimate integer EWL and WL ambiguities $DD\check{N}_{EWL}$, $DD\check{N}_{WL}$.
3. $DD\check{N}_{ENL}$, $DD\check{N}_{WL}$ enter as known parameters in the minimum noise ionospheric combination written in terms of DDN_{EWL}, DDN_{WL}, DDN_{ENL}, as equation (3) states. The LS approach is used to estimate a float solution (coordinates, tropospheric residual on DD, and ENL ambiguity) and LAMBDA method is used again to solve the ENL ambiguities. It is worth noticing that ENL ambiguity has an associated wavelength of 12 cm order, while the intrinsic noise of the combination is at one-tenth level. Furthermore, parity bound link between EWL fixed ambiguities and the ENL ones is used, with the effect of doubling the associated ENL wavelength.
4. $DD\check{N}_{EWL}$, $DD\check{N}_{WL}$, $DD\check{N}_{ENL}$ enter as known parameters in the minimum noise ionospheric combination. Fixed solution is obtained from an LS approach.

The iteration loop on EWL or WL observables stops when the corresponding ambiguities are all fixed, or ambiguity validation fails. In case of IF observations are processed, the iteration stops if the LS estimator reaches convergence, i.e. if the new coordinate estimated corrections do not produce any relevant change on residual variance.

4 Test Plan and Validation

The validation of the developed prototype starts at time of writing this article and it is performed through the usage of pre-existing GPS networks data (real) and the corresponding GPS and GALILEO simulated data with GSFF. Tests will make use of real and simulated GPS data to verify the consistency of the simulation environment. Simulated GALILEO data will be used to verify the capability of the developed prototype to produce baselines estimates with the millimetre level accuracy using progressively shorter time span. Furthermore NRT data analysis on GPS and GALILEO simulated data shall provide a test bed for evaluating expected improvement of GALILEO with respect to GPS system.

This approach will also have an added value, being test cases built on the basis of real existing GPS networks, thus demonstrating the capability of the developed prototype to properly address the target monitoring needs. Two test methodologies will be adopted.

Repeatability Test: using four different baselines lengths (zero, up to a Km, less than 10 Km, ≈ 30 Km) and different data session length (from one solution with 24 hours data, to 24 daily solution with one hour batch data), three different data batch will be collected and processed, namely: simulated GALILEO data, simulated GPS data and real GPS data extending over 10 consecutive days at 15 s sampling rate. With the estimated coordinates, the standard deviation of the samplings will be computed for addressing NRT performance issue, since the estimated time series repeatability shall assume the meaning of nominal accuracy for different measurement sessions. In this sense, repeatability datum represents the smallest detectable receiver displacement that the prototype will be able to detect in the considered time span.

NRT Capability Test: it shall start from the repeatability results obtained from the above-depicted analysis. Simulating GALILEO data needs to know the a-priori coordinates of the two receivers. As a consequence the only way to compare a GPS measurement with a GALILEO measurements without using a-priori coordinates estimated with the GPS itself is a system allowing knowing the baseline components of the two receivers system at a given epoch with an independent measurement equipment.

For this reason NRT capability test shall use a GPS antenna mover mechanism. It will be powered by a stepper motor and steered by software to generate linear

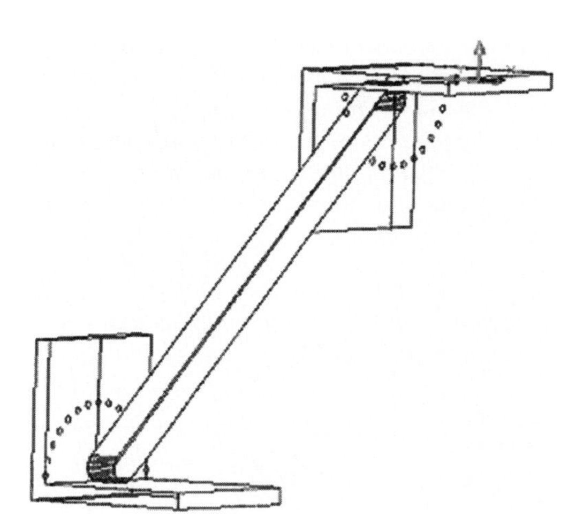

Fig. 5. Antenna mover mechanism schema.

velocities in the range of interest (a few mm/h). The accuracy of the position of the mobile equipment is less than one millimetre. The concept is shown in the schematic diagram in Fig. 5.

Different test will be performed with the aim of putting in evidence the time resolution of the a-priori known moving position of the mobile antenna, and comparing the obtained results with monitoring requirements.

5 Preliminary Numerical Results

For prototype debugging purpose two hours observations data for the GALILEO constellation and for two receivers PRO1 and PRO2 have been simulated using the GSSF. Observation are sampled at 15 s rate and they have not been corrupted by receiver noise or multipath, i.e. observations have been simulated in as best environmental condition as possible.

In Table 2, receivers coordinates, used for generating simulated RINEX 3.0 observables files, and "true" baseline components are shown.

Ephemerides have been generated in .sp3 'c' data format (the new International GNSS Service format for precise ephemeris).

Processing options used are:

- IF observable is used;
- Estimation of tropospheric zenith residual delay during baseline adjustment over both receivers;
- Integer ambiguity determination required;
- Ionospheric delay modelled with Klobuchar model with CODE estimated coefficients;
- Tropospheric delay modelled with Saastamoinen zenith model and Niell mapping function.

Figure 6 shows the NDA interface messaging during processing.

Estimated coordinates have an associated errors below the 10^{-4} and this is mainly due to the optimal simulated environmental conditions.

Results are summarized in Table 3, while in Table 4 differences between "true" and estimated baseline components and module are shown.

Table 2. GALILEO receivers' coordinates and PRO2–PRO1 baseline components and module (ITRF 2000) used to simulate the RINEX 3.0 data.

	X (m)	Y (m)	Z (m)	Module (m)
PRO1	4562839.3221	1040349.2812	4320428.0735	–
PRO2	4556822.0410	1070672.8784	4320758.4559	–
Baseline	−6017.2811	30323.5972	330.3824	30916.6196

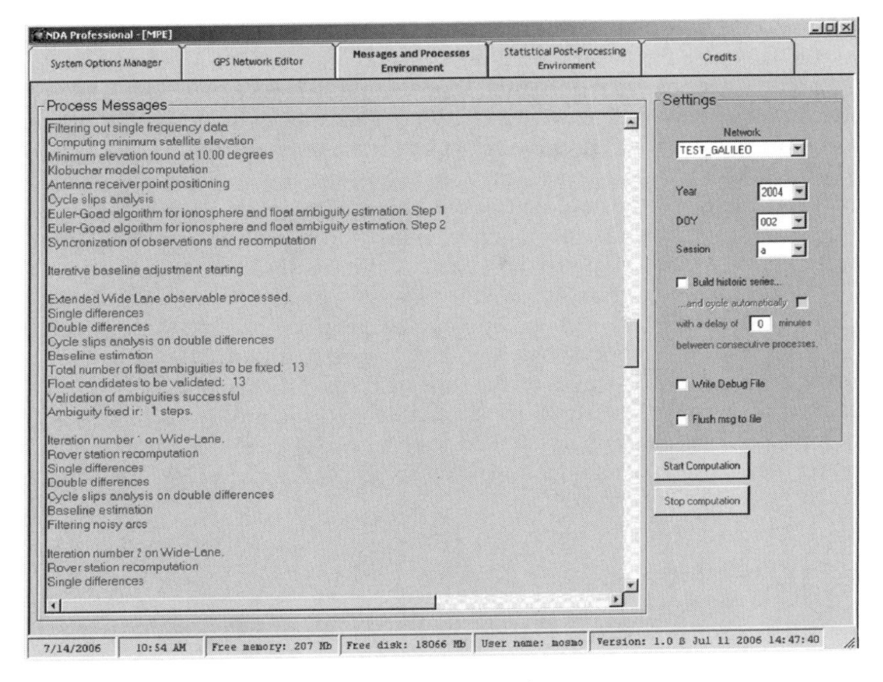

Fig. 6. NDA processes enviroment.

Table 3. Baseline adjustment results. In last column amount of data batch used in processing is shown.

Mod. (m)	X (m)	Y (m)	Z (m)	Data amount
30916.62085	−6017.2798	30323.5986	330.3841	1 hour
30916.62068	−6017.2800	30323.5984	330.3838	2 hours

Table 4. Differences between "true" values and estimated values.

Δ Mod. (m)	Δ X (m)	Δ Y (m)	Δ Z (m)	Data amount
0.0011784	0.0012440	0.0014350	0.0016793	1 hour
0.0010074	0.0010882	0.0012324	0.0014434	2 hours

6 Conclusions

GELOCALNET is a one-year long project co-funded by the GJU under the 6th Framework Program realised by a consortium of European RUs and SMEs aiming to address the high accuracy NRT monitoring issue using modernised GNSS such as GALILEO.

The project is based on upgrading an existing software module for dual frequencies GPS baseline adjustment, called NDA, to explore the GALILEO capability to provide a first step toward the essential contribution to reduce space and time uncertainties in the prevision of Earthquake events. This application is the frontier for a real support to Civil Protection in emergencies management.

GEOLOCALNET thus represents a step forward the synergic use of a solid underlying computational tool and interfacing capability for establishing warning criteria based on deterministic grounds, since future perspectives shall give the opportunity, with an integrated approach of GNSS and geophysical modelling results and methodologies, to achieve the two-fold objective of cross-validating the GALILEO performances and calibrating the geophysical models proposed for the spatial scale of the seismogenic faults, targeted towards the detection and comprehension of the dynamics of the earthquake pre-seismic phase.

At time of writing this article, the project entered the prototype validation and testing phase. This phase is devoted to test the performance of the prototype in reaching the objective of millimetre level accuracy in baseline estimation using few hours of GALILEO observations.

References

[1] K. Aoudia, A. Borghi, R. Riva, R. Barzaghi, B.A.C. Ambrosius, R. Sabadini, L.L.A. Vermeersen and G.F. Panza, Postseismic deformation following the 1997 Umbria-Marche (Italy) moderate normal faulting earthquakes, Geophys. Res. Lett., 30(7), 1390, doi:10.1029/2002GL016339, 2003.

[2] J. Bond, D. Kim, R.B. Langley, and A. Chrzanowski, An investigation on the use of GPS for deformation monitoring in open pit mines, UNB, 2003.

[3] M. Chersich, G.B. Crosta, “La frana di Cortenova (LC)”, 8th ASITA National Conference Acts, Roma, December, 2004.

[4] M. Chersich, A. De Giovanni, M. Osmo, NDA: un tool italiano per il processamento automatico di dati da reti GPS permanenti, Atti della 6a Conferenza Nazionale ASITA, Perugia, Novembre 2002, pagg. 769–774. P. Briano, M. Chersich, “NDA-DQE: Un Software Italiano per l'Analisi di Qualità in Automatico del Dato GPS”, Atti della 7a Conferenza Nazionale ASITA, Verona, Ottobre 2003, pp. 499–504.

[5] G. Dalla Via, R. Sabadini, G. De Natale and F. Pingue, Lithospheric rheology in southern Id from postseismic viscoelastic relaxation following the 1980 Irpinia eartquake, Journal of Geophysical Research, Vol. 110, doi:10.1029/2004JB003539, 2005.

[6] M. C. de Lacy, F. Sansò, A. J. Gil, G. Rodríguez Caderot, A method for the ionospheric delay estimation and interpolation in a local GPS network, Studia Geophysica & Geodaetica, Vol. 49, pp. 63–84, 2005.

[7] H. J. Euler, C. C. Goad, On optimal filtering of GPS dual frequency observations without using orbit information. Bulletin Geodesique, 65: 130–143, 1991.

[8] B. Forssell, M. Martin-Neira, and R. A. Harris, Carrier Phase Ambiguity Resolution in GNSS-2. Proceedings of ION GPS-97, Kansas City, USA, 1997, pp. 1727–1736, 1997. U. Vollath, S. Birnbach, H. Landau, Analysis of Three-Carrier Ambiguity Resolution (TCAR) Technique for Precise Relative Positioning in GNSS-2. Proceedings of ION GPS-98, Nashville, USA, 1998, pp. 417–426, 1998. W. Zhang, E. Cannon, O. Julien, P. Alves, Investigation of Combined GPS/Galileo Cascading Ambiguity Resolution Schemes. ION-GPS 2003, 9–12 September 2003, Portland OR.

[9] Klobuchar J.A., 1987. Ionospheric time-delay algorithm for single frequency GPS users, IEEE Transactions on Aerospace and Electronic Systems, AES-23 (3), pp. 325–331.

[10] A.E. Niell, Global mapping functions for the atmospheric delay at radio wavelengths, J. Geophys. Res., 101(B2), 3227–3246, 1996.

[11] J. Saastamoinen, Atmospheric correction for the troposphere and stratosphere in radio ranging of satellites, Geophys. Monogr. Ser., vol. 15, pp. 247–251, American Geophysical Union, Washington D.C., 1972. J. Saastamoinen, Contribution to theory of atmospheric refractions, Bulletin Geodesique, n. 105–107, pp. 279–298; 383–397; 13–34, 1973.

[12] P.J.G. Teunissen, Least square estimation of the integer GPS ambiguities, Invited lecture, Section IV Theory and Methodology, IAG General Meeting, Beijing, China, August 8–13, also in Delft Geodetic Computing Centre LGR series, No. 6, 16 pp, 1993.

[13] P.J.G. Teunissen, A. Kleusberg, GPS for Geodesy, 2nd Edition, ed. Springer; 1998. A. Leick, GPS Satellite Surveying, John Wiley & Sons, 2nd edition, 1995.

Common–View Technique Application: An Italian Use Case

E. Varriale, M. Gotta, F. Gottifredi, F. Lo Zito

Alcatel Alenia Space Italia S.p.A.,
Franco.Gottifredi@alcatelaleniaspace.com, Phone: +390641512082

Abstract. Common view (CV) is a well known synchronization technique usually adopted to determine synchronization values to be analyzed in post processing to monitor clock stability. In the Italian GALILEO Test Range (GTR) project, a near real-time, CV based, synchronization technique has been implemented to ensure performance and reliability of test services offered to the GTR users.

1 Introduction

The GALILEO Test Range (GTR) represents the Italian initiative for the realization of a National and International centre of excellence in Satellite Navigation field.

The GTR project main objective is to provide a feature rich, technologically advanced test area where a number of services are offered to users for improvement of already available satellite navigation technology as well as for development of new technological and methodological solutions and applications.

The main objectives of the GTR project can be summarized as follow:

- Characterization & Verification of the new Navigation Signals: GPS L2C, GPS L5, EGNOS L1, EGNOS L5, GSTB-V2 (GIOVE-A&B) & GALILEO
- Support the certification process of Galileo, GPS & EGNOS Receivers
- Definition & Experimentation of Application Prototypes
- Definition and Provision of services

Setting up such an advanced test environment is a challenging engineering task since the whole system will be made of many complex parts which have to work together simultaneously and with a non linear input/output data flow. In order to achieve this challenging objective, a uniform and highly efficient resources meta-management must be implemented. In this context, the synchronization plays a central role in keeping data semantically coherent, implying that the selected synchronization technique will affect the whole system reliability and efficiency. In this paper, an application of Common-View, a well known synchronization technique, to a near real-time and complex system architecture is discussed. Main aspects of CV method implementation are taken into account, in particular:

- information exchange architecture,
- data acquisition policies,

- data flow,
- algorithm implementation details,
- testing and post processing analysis,
- validation techniques and tools.

2 GTR Elements Layout

The GTR can be viewed as an ensemble of logically independent components: the facilities.

GTR components can be grouped in 'remote' and 'local' facilities, where 'remote' means that there are no direct connections to other GTR elements as opposed to 'local'.

Local facility are the following:

- the *Control Center* (CC) which acts as controller of the data flow and supervisor of the efficiency of all elements,
- the *Time Laboratory Facility* (TLF) which runs the System Time reference by mean of an Active Hydrogen Maser atomic clock,
- the *Orbitography Laboratory Facility* (OLF) whose task is to provide clock synchronization and orbit determination and propagation services,
- the *Integrity Laboratory Facility* (ILF) which has to implement algorithms for System Integrity control,

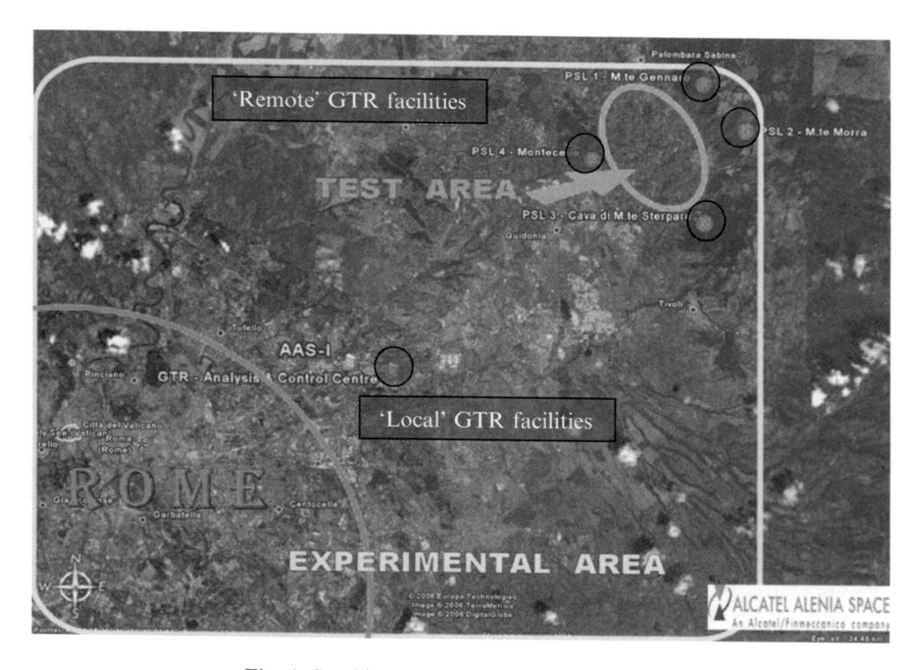

Fig. 1. Satellite image of Rome & GTR areas.

- the *GTR Lab* that has the scope to analyze navigation messages, from test area or from satellite constellations (Fig. 1), to validate algorithms and HW/SW equipment and to support receiver and application prototyping.

All 'local' facilities are physically arranged in the same building and interconnected by high speed network infrastructures and appliances.

'Remote' facilities are the *Pseudolites* (PSL), whose main task is to generate GPS-like and GALILEO-like signals on ground. Such pseudolites are equipped with an atomic clock system composed by a Rubidium that drives an OCXO and are able to broadcast navigation messages and to receive clock correction parameters. Furthermore all pseudolites are equipped with GPS receivers making them able to act in turn as observing stations (i.e. recording and delivering observables data).

3 The Choice of Synchronization Technique

Synchronization accuracy has a great impact on navigation messages and range measurements, affecting the reliability and performance of the services made available by the GTR.

Particularly critical is the synchronization between remote and local facilities both for the impossibility of direct communication, and hence for the reliability of the data exchange medium, and for the relevance of the role played by pseudolites in the GTR setup.

It's clear that the choice of the synchronization method must be carefully assessed against performance (accuracy and stability), cost effectiveness, scalability (both in term of number of actors and in performance), availability, computation time and clock corrections data rate.

A number of methods are studied in literature, each with its pros and cons; a brief summary is given Table 1 summarizing the performance of some well-known techniques.

Table 1. Synchronization techniques and their performances. [1]

Type	Time accuracy	Time stability	Freq. accuracy 24 h
Internet	20 ms-1s	20 ms	2×10^{-7}
Dial-Up ACTS[1]	5 ms-1s	5 ms	6×10^{-8}
HF Radio	2–200 ms	2 ms	2×10^{-8}
LF Radio WWVB[2]	0.5–20 ms	1 µs	1×10^{-11}
LF Radio LORAN	1 µs	100 ns	1×10^{-12}
GPS One way (no SA)	10–40 ns	2–7 ns	2×10^{-14}
GPS Common View	1–10 ns	1–2 ns	1×10^{-14}
TWSTFT[3]	1–5 ns	0.1–2 ns	$0.1\text{–}1 \times 10^{-14}$

[1] ACTS = Automated Computer Time Service
[2] WWVB = World-Wide Voice Broadcast
[3] TWSTFT = Two-Way Satellite Time & Frequency Transfer

The synchronization technique selected for the GTR System is represented by GPS Common-View because it is the only method, among the ones listed above, fulfilling all the requirements imposed by GTR design constraints. In particular the following aspects has driven this technical solution choice:

- performance guaranteed by GPS CV is adequate for meter level precision,
- GPS CV is cost effective (could be COTS based, only some minor equipment configuration is required),
- scalability in number of actors is assured as no change has to be made for new stations to synchronize them,
- scalability in performance is not easy but possible using phase measurements instead of code ones,
- availability of Navigation Signals is granted by high availability of GPS constellation and it will double when Galileo will be available,
- computation time and clock corrections data rate could be very flexible provided an adequate data sharing infrastructure is available as will be explained later.

4 The GPS Common View Technique

The GPS Common-View technique (Fig. 2) is used to double compare ground based time references by mean of GPS signal as received on ground.

The idea behind this technique is to compute the bias between two clocks by differencing against a third, common clock; in the case of GPS based common view the common time scale is the GPS Time as broadcasted by the satellites on board atomic clock. The whole process and concept can be sketched as follows:

Each station, A and B, receives the GPS signal broadcasted from a satellite in common view at time t_0: A receives signal at time t_1 and B at time t_2. Knowing the

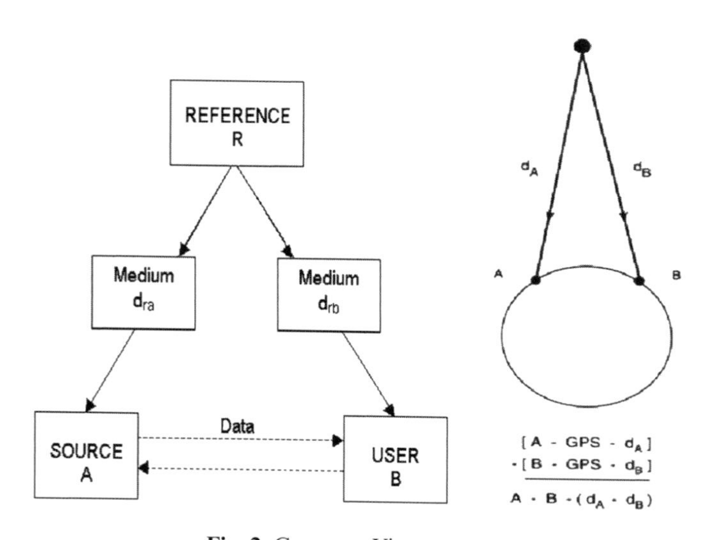

Fig. 2. Common-View concept.

time of transmission t_0, A and B can compute a pseudorange with respect to the satellite (SV) expressed as:

$$PseudoRange_{A,SV_i}(t_0) = SlantRange_{A,SV_i}(t_0) + c \cdot \left((t_1 - t_0) - \Delta T_{SV_i} + E_A + K_A\right) \quad (1)$$

$$PseudoRange_{B,SV_i}(t_0) = SlantRange_{B,SV_i}(t_0) + c \cdot \left((t_2 - t_0) - \Delta T_{SV_i} + E_B + K_B\right) \quad (2)$$

where:

c .... is light speed in vacuum,
ΔT_{SV} .... is the time offset of transmitting satellite's clock,
$E_{A\&B}$ .... are ionospheric and tropospheric delays of stations A & B
K .... represents time contributes from multi-path errors, thermal noise and calibration factors.

Differencing the equations (1) and (2), the following is obtained:

$$\Delta T^{SV_i}_{A,B}(t_0) = \frac{(\Delta \text{Pseudorange} - \Delta \text{Slantrange})}{c} + \Delta K_{A,B} + \Delta E_{A,B} \quad (3)$$

which is the time offset between the two clocks running at A and B for epoch t_0 and a given satellite SV_i in common view.

If multiple satellites SV_i, $i = 1 \ldots n$, are being observed simultaneously, an averaged clock offset can be computed:

$$\frac{1}{n}\sum_{i=1}^{n} \Delta T^{SV_i}_{A,B}(t_0) \quad (4)$$

computing the average of multiple observations, relatives to the same epoch, contributes to mitigate/smooth the errors (like orbit determination, tropospheric and ionospheric errors).

5 Implementation

5.1 General Considerations

While describing the Common-View techniques some assumptions has been made:

- it was assumed that time of transmission t_0 could be used as a matching criterion between data collected at ground stations but this is generally not feasible because GPS receivers are able to dump recorded data only at discrete time intervals
- it was assumed the capability to determine the slant range between a ground station and a given satellite using satellite ephemerides broadcasted by the Navigation Satellites.

Ionospheric and tropospheric delays may play a major role in keeping away from theoretical performance of the common view technique, especially over long baselines between the two ground stations, then some technique should be adopted to mitigate the impact of such delays.

In general, while the conceptual ideas behind common view technique are very simple, when achieving the implementation/system design phase, all side conditions must be accurately assessed.

While designing the synchronization algorithm the main assumption made was to consider ionospheric and tropospheric delays to be equal at every observing facility i.e.:

$$E_A \cong E_B \tag{5}$$

this is justified by the short baseline between the GTR receiving stations (i.e. less than 20 km along).

This assumption gives twofold advantages: handling of errors coming from satellite signal propagation medium can be skipped and, with those short baselines, the synchronization process can get really close to theoretical CV performances.

Other error sources has been assessed during system architecture design phase, in particular thermal noise and calibration bound errors are kept at minimum by the ability to constantly monitor and operate all installed equipments; moreover multipath errors, albeit not thoroughly modeled, are reduced by using choke ring antennas. Finally prediction error during update message time is kept small due to the stability of time reference clock (running at Time Laboratory Facility – TLF) that is ensured by a clock ensemble composed by an Active Hydrogen Maser (AHM) and 4 Caesium atomic clocks kept in a thermally controlled environment.

Some analyses on pseudolites clock stability have been conducted during the GTR development phase. Such studies pointed out the possibility to keep time errors within 3 ns (1 -sigma) if a clock correction model is provided every 8 minutes. Such an error in synchronization will translate in 90 cm pseudorange estimate error at user level (i.e. in the UERE).

One of the main performance limitations for this technique remains then the orbit determination and propagation of the satellites. In this aspect the GTR synchronization system is bound to ephemeris data broadcasted from satellites by mean of navigation messages; more advanced, and hence accurate, determination and propagation algorithms presently cannot be used due to performance issues and the near real time design of synchronization process that is required to compute clock biases every 8 minutes. Pseudorange accuracy of P code measurements is another synchronization performance limiting condition. It can be found in literature that picoseconds accuracy can be exploited by Common-View technique by using phase measures and thus resolving integer cycles indetermination. Practical applications of these improvements are under development (experimentation phase).

Summarizing, a high level list of requirements/side conditions to be taken into account for Common View near-real time application follows:

- Baseline between stations (affects both error budget and data exchange system design):
- Error budget affecting Common View performances
- Stability of clocks running at ground stations (affects the choice of synchronization cycles scheduling)
- GPS receivers recording rate
- Type of satellites orbital determination & propagation method
- File format of recorded GPS observables
- Data exchange infrastructure and architecture

In the following sections the GTR facility interconnection, data flow and software implementation of synchronization process will be emphasized.

5.2 The GTR Information Management Architecture

There are a number of reasons to ask for an optimized data distribution infrastructure and control policies other than the obvious necessity of reliable data exchange among facilities:

- facilities require and produce a great amount of data for their own tasks
- the near real time design of some processes introduces the need to acquire input and export output data as soon as available
- facilities run independently but are related by input/output dependency chains and this requires that some level of fault detection must be implemented to avoid deadlocks and starvation
- heterogeneity of throughput and protocols of communication channels (between local and remote facilities)
- all experiments involving recording of observables must agree over common acquisition timelines in order to produce time overlapping data chunks for coherence of further elaborations.

The GTR Control Center facility copes with all those issues implementing a centralized control and data distribution architecture. Back to synchronization, the following collaboration diagram (Fig. 3) depicts data flow between facilities involved in the process.

Data flow in the GTR can be bound to a producer/consumer paradigm where the Control Center (CC) has the role to capture and hand out every data resource conveyed through the GTR network to proper subscribers (consumers). Three resource types are required by the synchronization task:

- observation files produced by pseudolites and sensor stations (one remote and one local running in the TLF playing time reference role) and consumed by the synchronization algorithm running at OLF processing facility
- navigation files produced by the two sensor stations and consumed by the clock synchronizer
- clock correction files produced by the clock synchronization software and subscribed by pseudolites

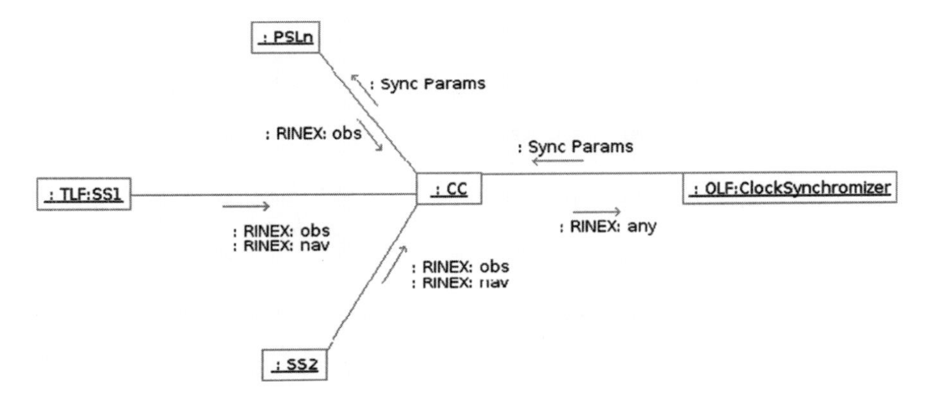

Fig. 3. Synchronization UML collaboration diagram.

All data coming from GPS receivers are wrapped into RINEX file format, which is a common choice to other similar applications.

Networking interconnection between local (i.e. CC, TLF, OLF, etc) and remote facilities (i.e. pseudolites) is provided by Control Center by mean of the GSM (PSTN as backup) network and a callback protocol in order to make both actors to request a connection.

The drawback of this choice is the slow data transfer rate provided by the GSM channel (9.6Kbit/s), which imposes restrictions over the size of observables data files if clock correction model have to be computed and delivered to pseudolites every 8 minutes (as established in the system architecture document).

5.3 Software Synchronization Service

What is required for the GTR project is a software tool that implements CV technique taking into account all the above considerations in order to offer a synchronization service to all GTR facilities requiring it from within OLF facility.

A client/server, multitasking, software architecture was chosen to implement the synchronization service: this is a common choice when talking about asynchronous services offered to a variable number of clients for this approach both maximizing throughput and minimizing resource required on the machine running the server side.

The serving cycle can be sketched as in Fig. 4:

When the server is started it begins polling a specific directory (the inbox) looking for RINEX files, when one appears, different actions are taken:

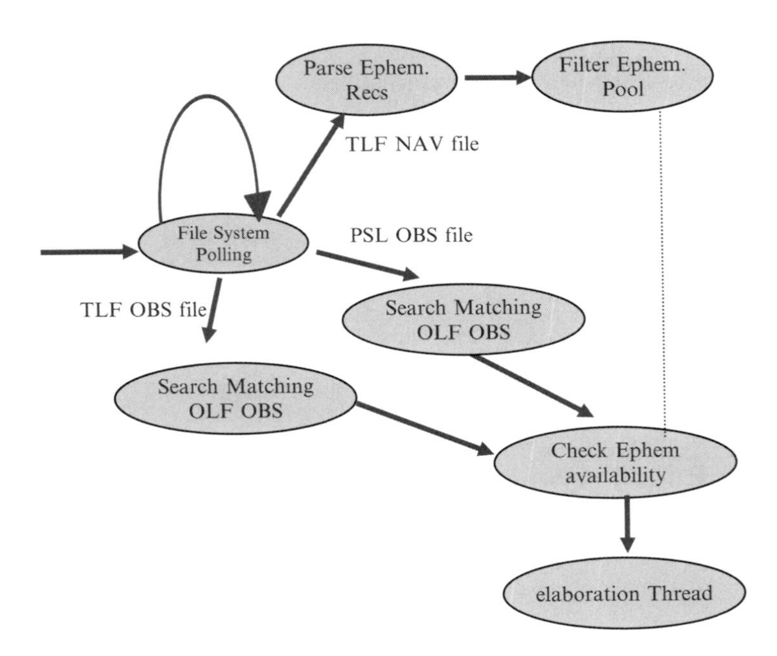

Fig. 4. Serving cycle diagram.

- if the incoming file is a RINEX Navigation file then the file is parsed and the obtained ephemerides are stored into an ephemeris pool. This pool is periodically filtered against older than 2 hours ephemerides.
- if file is a RINEX observation file coming from the TLF (the facility running the time reference) then a search procedure is started to find all observables files coming from pseudolites (PSL) referring to the same time interval. If some matching is found, ephemeris pool is checked to ensure that there are enough ephemerides to successfully compute the common view algorithm, if this is true, for each matching PSL observation file a new elaboration thread is started.
- if file is a RINEX observation file from a PSL unit, then the server search for a timestamp matching file coming from the time reference facility: if found the software will check the ephemeris pool and awake a synchronization thread as in the above case.

When a new elaboration thread is spawned it has all the elements to compute the synchronization independently from the main, serving, thread. Each thread implements the CV algorithm merging both the theoretical technique formulation and the side requisites imposed by GTR design.

The resulting algorithm can be expressed in pseudo code as:

INPUT: two RINEX observation files, REF and COMP, where REF is coming from time reference station and COMP is the facility running clock to synchronize, an ephemeris pool.

OUTPUT: an ASCII file containing the two coefficients of the linear regression computed from a vector of synchronization samples and a timestamp.

***For REF and COMP*:**

- build a dictionary CVDICT having key defined as
  - ✓ (epoch of observation, observer SV ID) couples and value defined as (pseudorange, slant range) couples
- filter out records with elevation angle below 5 degrees

***Initialize a dictionary, SYNC_POOL, of clock biases indexed by epoch For every matching key of CVDICT(REF) against CVDICT(COMP)*:**

- use common view formula to compute clock bias, B
- add B to SYNC_POOL

***For every epoch, E, in SYNC_POOL*:**

- compute the mean of clock biases, BM, contributed by every satellite in sight
- set SYN_POOL[E] = BM

Compute a linear regression fitting model of SYN_POOL as $Y = a_1*X + a_0$ and save a_0, a_1 to output file.

The computational bottlenecks of such algorithm are:

1) the slant-range evaluation
2) the matching over CVDICT keys.

Point 1 requires some floating point arithmetic evaluation mainly to:

- retrieve GPS signal broadcast time (time of transmission from SV) from the pseudorange measure and reception time.
- compute satellite position in ECEF coordinates at time of signal transmission. This involves:
  - best ephemeris search
  - computing Cartesian components from navigation messages
  - apply clock corrections model to SV clock
  - apply corrections for Earth rotation
  - apply correction for relativistic effect

Hence complexity depends on the cardinality of Ephemeris Pool set |E|, on the number of observations in RINEX files and on the cost of computing a single slant-range. As stated before, Ephemeris Pool is periodically purged from 'expired' ephemeris so its max cardinality is equal to the maximum number of satellites. The cost of a single slant-range computation is floating point intensive but can be regarded as constant in time (given the above limitation on |E|). The total cost of slant-range computation can be expressed as:

$$T_{slant} = n \bullet F_{fp}(|E|) = O(n) \tag{6}$$

where n is the number of observations in each RINEX file and F_{fp} is the floating point evaluation cost depending on |E|.

So ***Point 1*** complexity is at most linear in input size.

Point 2 complexity is completely determined by the choice of data structures. Red-black trees were chosen which guarantee logarithmic access and insertion time. Matching over CVDICT keys means to exhaustively scan one container, say $CVDICT_{REF}$, and use each key, in turn, as search key in $CVDICT_{COMP}$. Hence the matching requires, at most:

$$T_{match} = O(m \cdot \log_2(m)) \tag{7}$$

where

$$\mathrm{m} = \max\left(|\mathrm{CVDICTREF}|, |\mathrm{CVDICTCOMP}|\right) \tag{8}$$

5.4 Development

The synchronization software was entirely coded in the C++ language using Object Oriented design paradigm following the SW Development Standards of [7]. The OLF workstations have been installed with the GNU/Linux operating system.

The choice of GNU/Linux as operating system comes from a number of considerations:

- the software must run both on high end server computers and on canonical workstations (where development took place) with no changes.
- the services offered by OLF must have the highest possible reliability and experience the lowest possible downtimes

- OLF operating system must provide a multi-user/multitasking environment which can be scaled as easy and effectively as possible against number of processors
- OLF algorithms can get very CPU demanding, hence all software must be able to run in distributed/parallelized computing environments with few or no changes
- many OpenSource scientific libraries are available (GNU Scientific Library – GSL)
- a well designed, easily extensible OpenSource GPS library (gpstk) offering, among others features, RINEX parsing and ephemerides management, is available.
- 2.6 Linux kernel offers event driven, low grained filesystem polling facility (inotify)

The choice of Object Oriented paradigm and C++ language arises mainly from the necessity to achieve high performance and still have the highest flexibility in software design, in particular the use of the Unified Modelling Language was very useful, especially because the development of the synchronization software went in parallel with the development the other facility/services among which the Control Centre and the data delivering system interfaces and policies. By sketching the synchronization system functionalities and interfaces with the GTR system it was possible to proceed to further development phases in a top-down fashion seamlessly integrating software objects/components as they became available without the need to evaluate every time the relationships between parts and the whole.

5.5 Testing

Design and development phases of software synchronizer took place separately from GTR context, that implied that preliminary testing and validation should have been conducted 'on factory'. To achieve this task IGS observables and navigation RINEX files was used.

IGS is "*. . . the International GNSS Service (IGS), formerly the International GPS Service, and is a voluntary federation of more than 200 worldwide agencies that pool resources and permanent GPS & GLONASS station . . .*" (from http://igscb.jpl.nasa.gov/).

From IGS stations network only mutually co-located receivers, or along short baselines, was taken into account in order to reproduce the same conditions holding for GTR.

IGS observables and navigation files are produced on daily basis and observations are recorded every 30 seconds (GTR specifications impose a 1 second acquisition schedule), so files were splitted in elements of 300 samples each and fed by facility emulators to the synchronization software server which, in turn, returned clock correction parameters files according to GTR interfaces and data flow specifications.

5.6 Post-Processing and Statistical Data Analysis

Data fed in as input to the software synchronization system come from GPS receivers which are affected both by systematic and stochastic errors; while systematic errors, like circuitry delays, can be compensated by calibration procedures, stochastic errors are more difficult to eliminate and to estimate on the short period. This is the common problem of outliers identification.

This problem could be relevant to the synchronization system in particular for the fast rate at which synchronization parameters are computed and for the short time span of data used in computing clock corrections coefficients.

The policy adopted involves both runtime data pre-processing filtering and post processing data analysis.

Pre-processing is done via a 3-sigma cut-off outliers filtering which is a good compromise given the low number of samples and the computational restrictions imposed by the fast pace at which synchronizations are computed.

Post-processing analysis is relative to linearity of fitted synchronization samples and clock stability.

Linearity of fitted data is tested by evaluating the variance of least square residuals and by comparison with a locally weighted least square smoothing function (LOWESS). In the following picture (Fig. 5) a sub sample of a daily synchronization between IGS stations USNO and USN3 (for day 94, year 2006) is shown together with linear regression fit and LOWESS fit, for the considered biases sample the variance of the linear least squares regression was of 2.11e-18.

Clock stability check is achieved by computing the Allan deviation (Fig. 6) from computed synchronization biases (i.e. obtained by common view technique) over a period of 24 hours and comparing the resulting values with the stability models of clocks involved. Such analysis gives useful hints about poor performing clocks

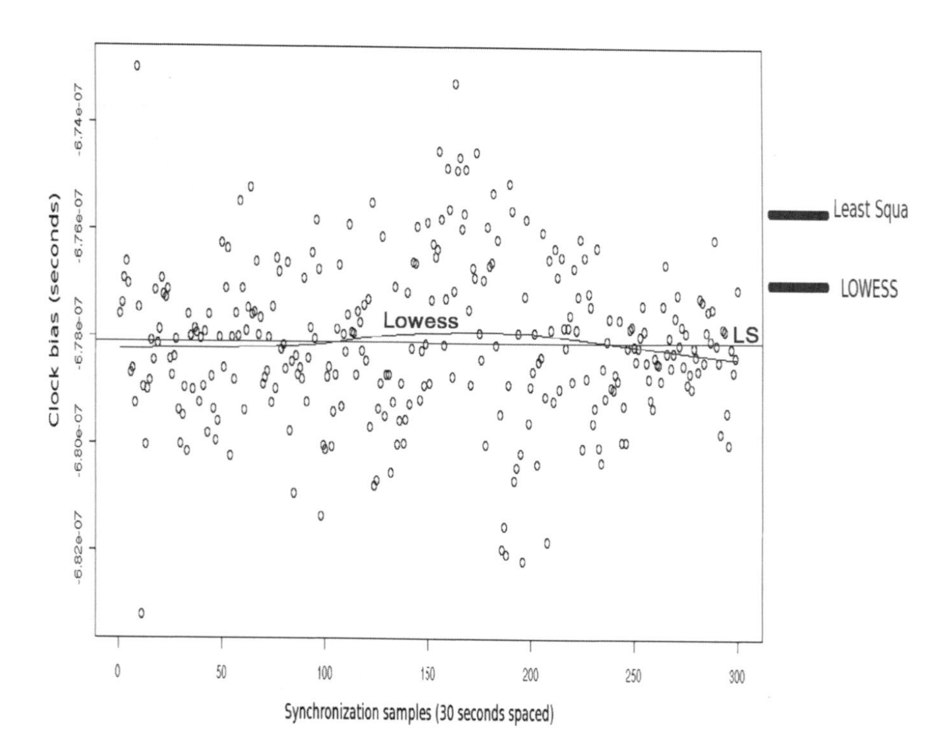

Fig. 5. Clock bias residuals with LS and LOWESS.

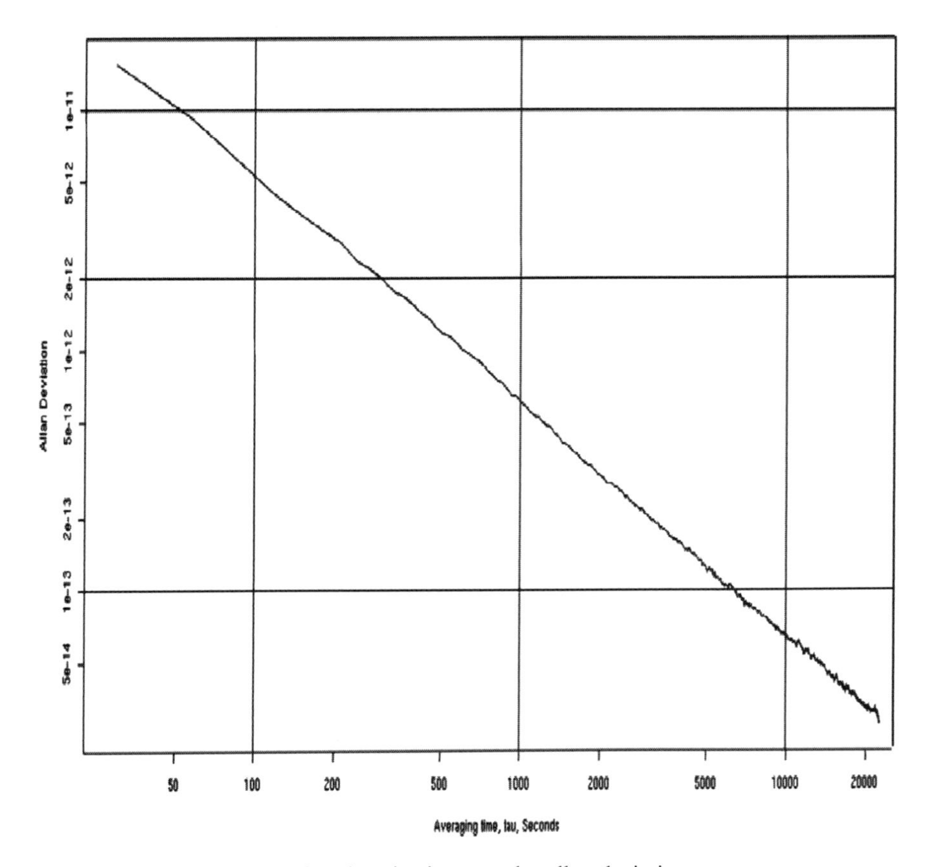

Fig. 6. Synchronization samples allan deviation.

and/or synchronization process and is a common starting point for further stability concerned analysis. An Allan deviation for a daily set of synchronization samples relative to IGS stations USNO and USN (day 94, year 2006) computed by the GTR ClockSynchronizer follows:

6 Conclusions & Way Forward

The work presented in this paper shows how the requirements and needs coming from the GTR project have been met using the Common View Technique in a real-time application. In order to achieve the required performance, the identified key-parameters have been:

- the availability of high-stable clocks (i.e. in the GTR ad-hoc atomic clocks ensemble for the PSL have been developed),

- the availability of GNSS observables (i.e. in the GTR GPS Receivers collect RINEX obs and nav data continuously),
- the duration of the observation data sample (i.e. in the GTR 5 minutes have been selected also due to the liability and the limited data rate of the "remote" site connections),
- the optimization of the SW in order to allow real-time computation and linear/quadratic regressions.

The ClockSynchronizer SW will be upgraded in the next phase of the GTR when a more reliable and fast connection will be implemented among the remote sites and the GTR-CC. This will allow the possibility to use a longer observation data set so as to improve the clock synchronization performance of the GTR. Furthermore, the OLF will have to work with and increased number of remote facilities (up to 9 PSL and 2 SS) and this will maybe require an upgrade of the HW in terms of processing power.

Finally it is important to underline that the GTR is a Local Application of this type of technique that has been developed and it is foreseen to be also used with non-local architectures (i.e. regional and world-wide) as, for example, Galileo (as shown in [8]) during its In-Orbit Validation (IOV) Phase.

References

[1] D. W. Allan, M. A. Weiss, "*Accurate Time and Frequency Transfer During Common-View of a GPS Satellite*", 34th Annual Frequency Control Symposium, pp. 334–346, May 1980.

[2] V. S. Zhang, T. E. Parker, M. A. Weiss, F. M. Vannicola, "*Multi-Channel GPS/GLONASS Common-View between NIST and USNO*", IEEE International Frequency Control Symposium, pp. 598–606, June 2000.

[3] M. A. Lombardi, L. M. Nelson, A. N. Novick, V. S. Zhang, "*Time and Frequency Measurements Using the Global Positioning System*", Cal. Lab. Int. J. Metrology, pp. 26–33, (July-September 2001).

[4] D. B. Sullivan, D. W. Allan, D. A. Howe, F. L. Walls, "*Characterization of Clocks and Oscillators*", NIST Technical Note 1337, Mar. 1990.

[5] K. M. Larson J. Levine, "*Carrier-Phase Time Transfer*", IEEE Transactions on Ultrasonics, Ferroelectrics and Frequency Control vol. 46, pp. 1001–1012, Jul. 1999.

[6] F. Gottifredi, G. Lancia, M. Manca, F. Rodriguez, "*The GALILEO Test Range*", Location conference, 2006.

[7] ECSS Standards, "*Space Engineering – Software*", ECSS-E-40 Part 1B, 28 November 2003

[8] M. Gotta, F. Gottifredi, S. Piazza, D. Cretoni, P. F. Lombardo, E. Detoma, "*GALILEO IOV System Initialization and LCVTT Technique Exploitation*", TIWDC'06 Conference, Ponza (Italy).

MARKAB: A Toolset to Analyze EGNOS SBAS Signal in Space for Civil Aviation

N. Caccioppoli, A. Pacifico, V. Nastro

Dipartimento per le Tecnologie, Università degli Studi di Napoli
"PARTHENOPE" Via Medina 40, I-80133, Naples, Italy
Phone: +390815474747, Fax: +390815474777,
e-mail: caccioppoli@uniparthenope.it, a.pacifico@email.it,
vincenzo.nastro@uniparthenope.it

Abstract. The aim of this paper is to describe a toolset named MARKAB, recently developed to analyse the EGNOS Signal in Space. This toolset is intended to analyse the performance that can be obtained using the augmentation navigation system in civil aviation user community. We demonstrate the utility of MARKAB by processing SBAS Signal in Space logged with an EGNOS/WAAS dual frequency receiver installed at our monitoring station. Latest results show that MARKAB establishes an important improvement in the analysis of Required Navigation Performance to the augmentation system, contributing with a new technique to assess the performance in term of Continuity of Service.

1 Introduction

The validation of a Satellite Based Augmentation System (SBAS) requires a careful analysis of performance that can be experienced by the user before the system can be declared operative. The European Geostationary Navigation Overlay Service (EGNOS) is the European SBAS equivalent to the U.S. Wide Area Augmentation System (WAAS). With three geostationary satellites and a network of ground reference stations, this system transmits differential corrections and integrity data to enhance the positioning signals sent out by satellite positioning systems (GPS, GLONASS), and make them suitable for safety-critical applications such as commercial aviation. At the beginning of June 2006 the EGNOS system is ready to broadcast a continuous signal, including the so-called "Message type 0/2" (MT0/2) allowing to offer a graceful transition from EGNOS System Test Bed (ESTB) to EGNOS for Global Navigation Satellite System user communities. The broadcasted signal use the MT0/2 and the Band 9 of the ionospheric grid with the aim of improving the performance in the Northern European latitudes. The addition of MTO/2 is a significant milestone in the development of the navigation system for users of non-safety of life services. In the frame of performance validation activities, the European Organisation for the Safety of Air Navigation (Eurocontrol) has established a standardized data collection environment to evaluate the performance that can be achieved during flight operations for which the navigation system is intended. Thanks to the daily data collection and evaluation a wide expertise has been built up on the tools that are currently being developed to understand how the augmentation system works and how its performance can be evaluated. To be able to tell whether the system is available and can be used during a given period it will therefore be necessary to analyse its performance and compare it to the requirements. The requirements are expressed by the International Civil Aviation Organization (ICAO) under the form of Standards and Recommended Practices [2] in terms of Required Navigation Performance (RNP). The performance objectives for aeronautical applications are usually characterised by four main parameters: Accuracy, Integrity, Availability and Continuity. Among the principal user community the requirements for civil aviation are very strict in terms of Integrity and Continuity and hence the EGNOS performance is driven by those needs. In this contest it is important to establish the performance of the system that would be experienced by a potential user, and to verify the stability of the performance in certain time period. This paper describes a toolset named MARKAB recently developed and tested in MATLAB The Mathworks Inc. environment to analyse the EGNOS Signal In Space (SIS). Our objective is to demonstrate the utility of our toolset processing the SIS logged in a series of static measurements over extended periods with an EGNOS/WAAS dual frequency receiver, installed at our monitoring station. Results show that MARKAB represents a significant improvement in the frame of performance evaluation contributing to define a new technique to assess the Continuity of Service (CoS). Particular attention will be paid on the currently achieved system performance and how the augmentation system is able to fulfil civil aviation mission requirements.

2 Background

Each of the four RNP parameters corresponds to the risk that a certain event occurs that has the potential to lead to an excessive position error. The Accuracy covers the risk that an excessive system error causes a position failure. Unfortunately, real life navigation systems will always suffer from rare failures that cause its performance to degrade beyond the alert limit, which would make the system effectively unusable. However, when the user is made aware of failure, he can revert to backup navigation systems to still enable him to stay within the alert limit. In other words, failure detection can be exploited to mitigate the risk of position failures. The risk associated with latent system failures is covered by the Integrity requirement. The necessary level of integrity for each operation is established with respect to specific alert limits. When the integrity estimates exceed these limits, the pilot is to be alerted within the prescribed time period. Although reversion to a backup system mitigates the risk of using an erroneous system, such a reversion is itself without risk. This is particularly true in landing, the most demanding phase of flight. An unscheduled loss of the ability to determine and display a valid position, due to the detection of a failure condition, is specified by the Continuity requirement. For this concept, the Continuity of Service (CoS) relates to the capability of the navigation system to provide navigation Accuracy and Integrity during a given period for an intended operation. Finally, Availability covers the risk of a lack of guidance at the initiation of an intended operation. The RNP concept assumes that some kind of failure detection mechanism is used to notify the user in case of dangerous malfunctions. Signal availability is the percentage of time that navigational signals transmitted from external sources are available for use. Availability is a function of both the physical characteristics of the transmitter facilities [4]. The SBAS system provides the user with integrity information to compute the protection levels (Horizontal and Vertical Protection levels, HPL and VPL), which represent an upper bound on the position error. For each operational mode, the system is declared as unavailable when the protection level is greater than the Alert Limit (AL) defined in [2]. If the system is available and the position error is not bounded by the protection level, thence the event is considerer as a HPL and VPL failure, since the protection level is always supposed to be an upper bound on the position error (PE). In such a case, the event is declared as Hazardously Misleading Information (HMI), if the position error exceeds the AL or as Misleading Information (MI) if the AL isn't exceeded.

3 Markab

There are several approaches to the validation of satellite navigation systems. These include analysis, modelling and evaluation of collected data. In general, data must be collected and evaluated in order to demonstrate that the implemented system is compliant with the requirements [2]. After this, it is possible define the operational rules and procedures for aircraft to use the system for particular applications. Even if some studies have already been initiated on the operational side still various issues have to be resolved and will have to be adapted as more is learned about the actual operations that EGNOS can be used to support. The objective of MARKAB is to provide an efficient and fast statistical evidence about the performance of EGNOS in the airborne environment of commercial airlines and to determine to what extent it could be safely approved for operational use. It doesn't want substitute other tools developed during the Operational Validation phase, but simply it wants to be a toolset component that allow us to analyse and compare the performance of the augmentation system in a simple way and giving the possibility of easy access to the graphical presentation of the results. As mentioned above, a standardized data collection environment has been established to evaluate the performance by means a prototype tool named PEGASUS. It aims to be a step forward the development of a standard processing and analysing tool to be used for the EGNOS validation. MARKAB consists of several software components called modules. Each module is designed for a specific task like data concatenation, position solution analysis and processing. The toolset is composed of two main standalone components of the PEGASUS:

- **Convertor module**. It converts the binary raw data file producing several ASCII readable files which contain GNSS and SBAS related data.
- **GNSS_Solution module**. Its aim is to deliver a position solution compliant with the MOPS [3] for GNSS receivers used in avionics (GPS, SBAS).

These modules are completely integrated with those of MARKAB. This particular design of the MARKAB architecture allows the access to all the data, even at intermediate stages of data evaluation, and its display and visualisation. A complete data processing sequence can be summarized in Fig. 1.

The core of MARKAB is the developed module named Gen_report. Its aim is to manage, analyse and combine easily data from different input files or processed with

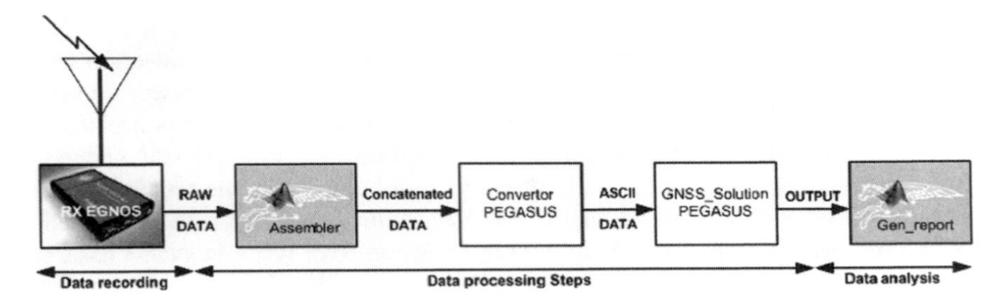

Fig. 1. MARKAB structure.

different modules. This module is composed of several MATLAB functions that implement the First Glance Algorithm [5], in addition to our algorithms to compute new parameters and indexes to evaluate the performance of the augmentation system. The RNP parameters definitions used are those defined in [3]. Gen_report reads from a directory, defined by the user, a series of files, (e.g. *.pos; *.smt; *.rng) generated by the standalone PEGASUS modules. These file are related to days belonging to the time period that the user is interested to analyse. The input data, results and parameters are stored in a dedicated directory on the computer file system, providing an easy access to the results, either by visualising the results by means of plots, or by producing a standard daily report containing the results.

4 Continuity of Service

The key process to verify the EGNOS system performance in terms of Continuity is the relationship between the CoS and the continuity characteristics of the broadcasted corrections [10]. The EGNOS system performance requirements, and in particular the CoS, are specified in the position domain. To assess CoS it is necessary verify that the system is able to provide navigation Accuracy and Integrity during any required time interval for an intended operation; such requirements are indicated in Table 1.

In the position domain, all parameters as HPL and VPL are time discrete series of values or samples with a time period of 1 sec. In according to [5], all valid samples for a given operation are computed by filtering all samples in order to take only those that are valid. With valid samples are identified all samples that have a navigation mode different from "*no position solution available*". At this point, it is important remark that are defined available samples all valid samples that have the corresponding navigation mode set on "*SBAS Precision Approach position solution*".

Table 1. CoS requirements.

Phase of flight	Departure	En-route	Terminal	Initial approach	APV
CoS	150 sec	300 sec	150 sec	150 sec	15 sec

For a given operation, available samples are computed considering all available samples for which the corresponding Protection Levels (*HPL* and *VPL*) are less or equal to the related Alert Limit threshold (*HAL* and *VAL*). For Precision Approach with Vertical guidance procedures (APV) the Alert Limits [4] are reported in Table 2.

We observe that considering all available samples and assigning the value 1 to the samples that are available for a given operation and the value 0 to the remaining samples, we obtain for each operation a time discrete binary series of the same length represented by the total number of all available samples. Therefore, each series represents the instants in which the augmentation system is available (samples of value 1) and when it is unavailable (samples of value 0). We define they as Total Availability for a given operation. Until today the PEGASUS tool analyzes the Continuity criteria counting, for a given operation, only all transition events from unavailable to available. This events are defined as Continuity events [5]. For greater clarity, we define Discontinuity states all instants in which the series is zero (Fig. 2). From the Total Availability, we can assess the CoS during any required time interval for an intended operation [2] performing that for each time interval there aren't discontinuity states.

In this way, for a given operation, to assess the concept of CoS we can filter the binary time discrete series of Total Availability by a Discrete Time Sliding Window Filter (DTSWF), (Fig. 3).

Considering a step size of one second (one sample), for each time sliding Window Size (*WS*), corresponding to the required time interval, it assigns at the current sample belonging to the lower limit of the current sliding window (current output sample) the value 1 if in the related window there aren't discontinuity states otherwise

Table 2. Alert limits in APV.

Facility performance	Horizontal alert limit	Vertical alert limit
APV-I	40 m	50 m
APV-II	40 m	20 m

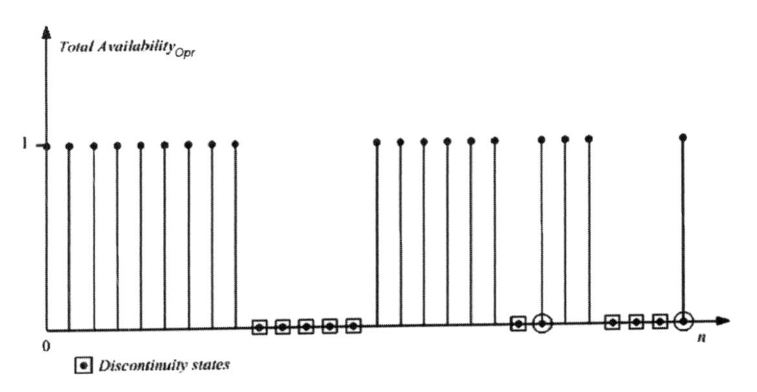

Fig. 2. Total availability for a given operation.

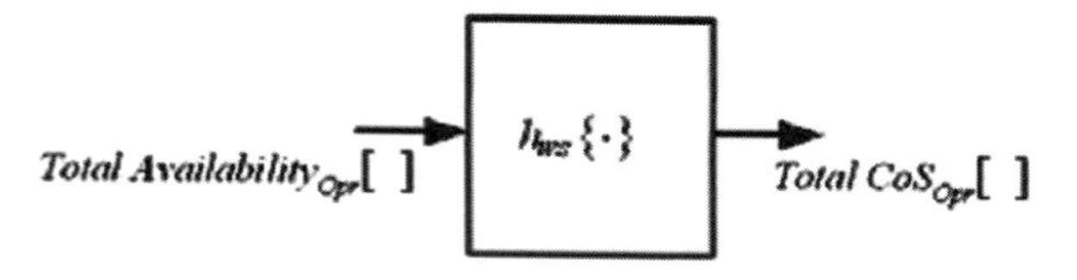

Fig. 3. DTSWF.

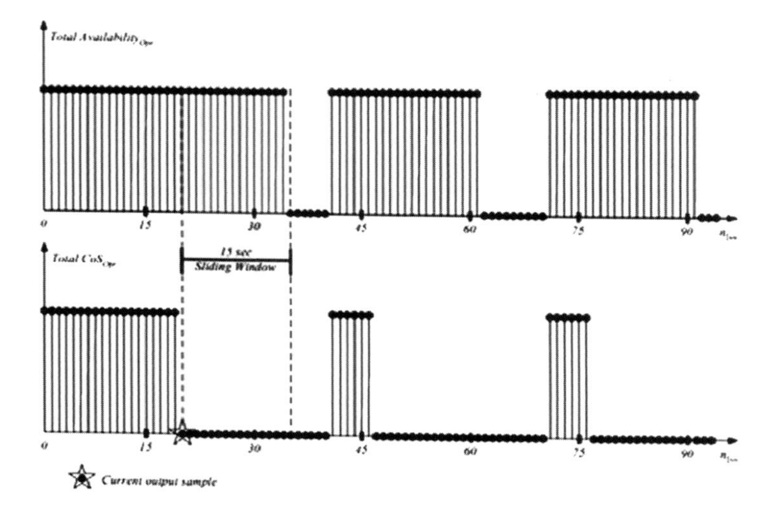

Fig. 4. CoS filtering in APV procedures.

it assigns the value 0 if in the related sliding window there is at least a discontinuity state (Fig. 4).

Thus we have another binary time discrete series, that we declare Total CoS for a given operation, in which any sample that has the value 1 represents the instant from which the navigation system will be available for the required time interval. Then indicating with $N_{available}$ the total number of available samples (duration of the discrete series), we can define the probability that the augmentation system will be continuous in any required time interval for a given operation as follow:

$$CoS_{WS\_Opr} = \frac{\sum_{N_{available}} Total\ CoS_{Opr}}{(N_{available} - WS - 1)} \qquad (1)$$

5 Signal in Space Processing

In this section we show the most features of MARKAB processing the SIS logged in a series of static trials performed during May 2006. A series of results will be presented to assess the performance of our toolset and particular attention will be paid to the achieved system performance and how far the EGNOS system is able to fulfil civil aviation mission requirements.

5.1 Monitoring Station

Our monitoring station is located in a site of Sorrento country provided by our Department. The antenna is located on the roof of a building with good sky visibility. Its precise position has been determined by means GPS phase measurements in the WGS84 coordinate reference system:

- Latitude: 40.62678163 North
- Longitude: 14.38733587 East
- Altitude: 146.132 m (above WGS84 ellipsoid)

The equipment of monitoring station is composed of following elements:

- Septentrio PolaNT dual frequency GPS antenna
- Septentrio-PolaRx2.4 GPS/SBAS receiver
- Logging PC connected to the receiver

5.2 MARKAB Performance

The main objectives of MARKAB are the automation of data processing within any time period (e.g. one day or more days, generally one month), to keep traceability of the process and to find data and results storage solution allowing the combination and the easy access to all data. To assess MARKAB accuracy performance in Fig. 5 is reported a comparison between a daily report generated by PEGASUS and the one by MARKAB. It is evident that the statistical parameters computed with MARKAB are the same with ones computed by PEGASUS. This result encourages us to show data in a series of plots to summarize the performance during any time period. Then we show a series of original plots to analyse the latest EGNOS performance.

Site	[NPU1] NAPLES PARTHENOPE UNIVERSITY						Date	15/05/2006
Location	Lat	40.627		Lon	14.387		Alt	146.13
Receiver	Septentrio PolaRx 2			Software	Pegasus 4.1		PRN	124
Data set	Duration	Start	Stop	Expected	Total	SBAS Msg	Valid	Valid(%)
1 Hz	24h00	00:00	23:59	86400	86400	86400	86021	99.56%

Results per operation

	all valid samples	APV-I	APV-II	CAT-I
	HAL/VAL	40 / 50	40 / 20	40 / 12

Accuracy (m)

		Meas.	Scaled	Req.	Meas.	Scaled	Req.	Meas.	Scaled	Req.
HNSE(95%)	1.12	1.05	4.65	16	0.99	4.82	16	0.89	5.10	16
VNSE(95%)	1.99	1.77	5.66	20	1.55	2.28	8	1.31	1.45	4

Availability (%)

		APV-I	APV-II	CAT-I
samples	86014	82754	66690	20698
Minimum Required		99%	99%	na
Availability		96.202%	77.528%	24.062%

Continuity

	APV-I	APV-II	CAT-I
Events	39	754	1447

Integrity

	MI	HMI APV-I	HMI APV-II	HMI CAT-I
Total	0	0	0	0
Horizontal	0	0	0	0
Vertical	0	0	0	0

Protection level statistics

	99%	95%	50%	mean	std deviation
HPL	54.30	19.35	8.93	11.15	8.34
VPL	68.53	36.49	14.37	17.80	12.69

Position error statistics

	99%	95%	50%	mean	std deviation
HPE	1.63	1.12	0.42	0.50	0.36
VPE	4.62	1.99	0.60	-0.16	1.10

```
--------------------------------REPORT--------------------------------
                          DAY: 15 MAY 2006
----------------------------------------------------------------------
   APV-I                         APV-II                       CAT_I
----------------------------------------------------------------------
                            ACCURACY (m)
  HNSE      Meas.      Scaled.    Meas.     Scaled.    Meas.      Scaled.
  1.12      1.05       4.65       0.99      4.82       0.89       5.10
  VNSE      Meas.      Scaled.    Meas.     Scaled.    Meas.      Scaled.
  1.99      1.77       5.66       1.55      2.28       1.31       1.45
                            AVAILABILITY
Samples :          82754          66690                       20698
----------------------------------------------------------------------
  99 percent                    99 percent                  na percent
    96.202                        77.528                      24.063
                             CONTINUITY
Events :           39             754                         1447
                             INTEGRITY
          MI         HMI APV-I          HMI APV-II            HMI CAT-I
          0             0                   0                     0
          0             0                   0                     0
          0             0                   0                     0
----------------------------------------------------------------------
                     PROTECTION LEVEL STATISTICS
----------------------------------------------------------------------
          99%           95%          50%         mean            std
HPL:      54.30         19.35        8.93        11.15           8.34
VPL:      68.53         36.49        14.37       17.80           12.69
----------------------------------------------------------------------
                      POSITION ERROR STATISTICS
----------------------------------------------------------------------
          99%           95%          50%         mean            std
HPE:      1.63          1.12         0.42        0.50            0.36
VPE:      4.62          1.99         0.60        -0.16           1.10
----------------------------------------------------------------------
                    CONTINUITY of SERVICE 15(sec)
----------------------------------------------------------------------
    95.964                  74.476                       19.074
----------------------------------------------------------------------
                    CONTINUITY of SERVICE 150(sec)
----------------------------------------------------------------------
    94.537                  69.240                       15.765
----------------------------------------------------------------------
                    CONTINUITY of SERVICE 300(sec)
----------------------------------------------------------------------
    92.953                  64.844                       13.336
```

Fig. 5. PEGASUS and MARKAB daily report.

5.3 Monthly Analysis

To have a summary representation of performance we present an analysis performed on raw data collected during May 2006. Briefly we report only the most important daily parameter plots.

5.3.1 Accuracy. Daily Accuracy is computed as a 95 percentile of the error distribution of all valid samples within the assessed period [5], (Fig. 6).

5.3.2 Integrity. Daily Integrity is evaluated computing the minimum observed safety index defined as the ratio of protection level to the related position error (Fig. 7). The dashed line represents the threshold to MI detection.

5.3.3 Availability. Daily Availability is computed as the ratio of the number of available samples for a given operation to the total number of valid samples [5] (Fig. 8).

5.3.4 Continuity of Service. Daily CoS is computed in according to the relation (1) defined above (Fig. 9).

5.4 Daily Analysis

Daily analysis is performed processing raw data collected on 15 May 2006.

5.4.1 Performance in APV Procedures. The main features of MARKAB are a new graphical representation of performance parameters. We show a summary plot of performance provided in APV procedures (Figs. 10 and 11). In the upper and lower plot respectively it is represented the Total Availability and the Total CoS discrete

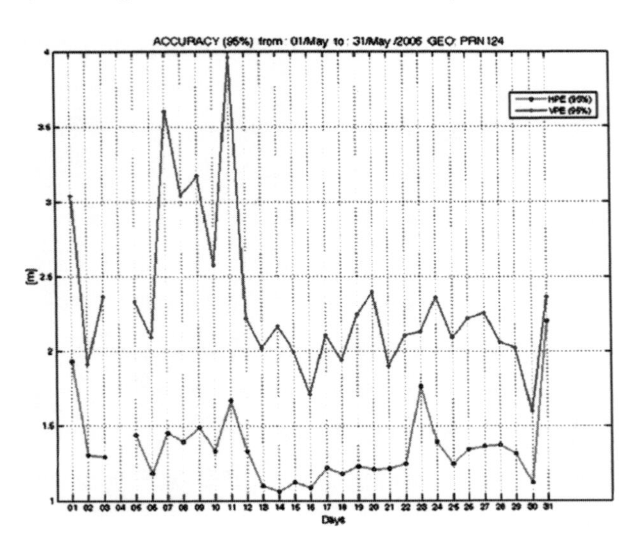

Fig. 6. Accuracy.

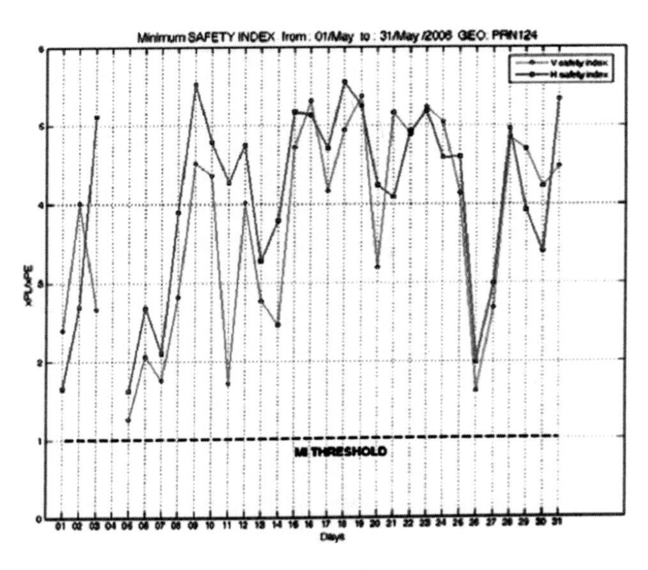

Fig. 7. Safety index.

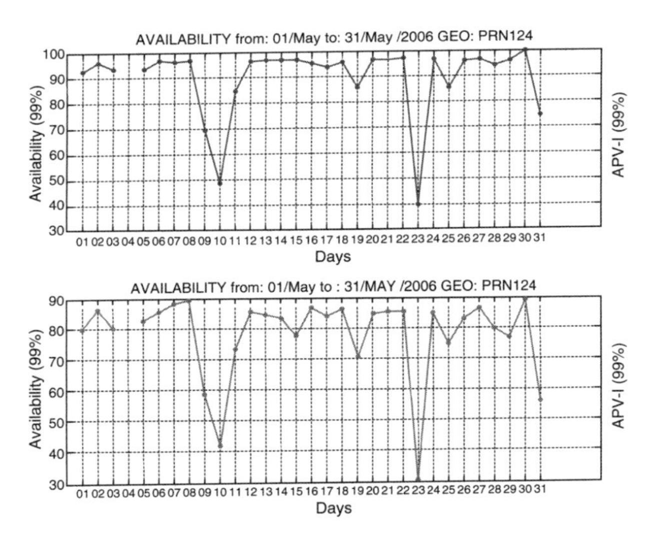

Fig. 8. Availability.

series for an interval of 15 sec, [4]. In the middle plot it is shown the continuity events position and duration referred to the upper plot (grey lines indicate only the transition events).

5.4.2 Integrity. Currently to assess performance in terms of Integrity the most used plot is the so called Stanford plot that summarizes the relations between the protection level (PL) and the position error (PE), Fig. 12.

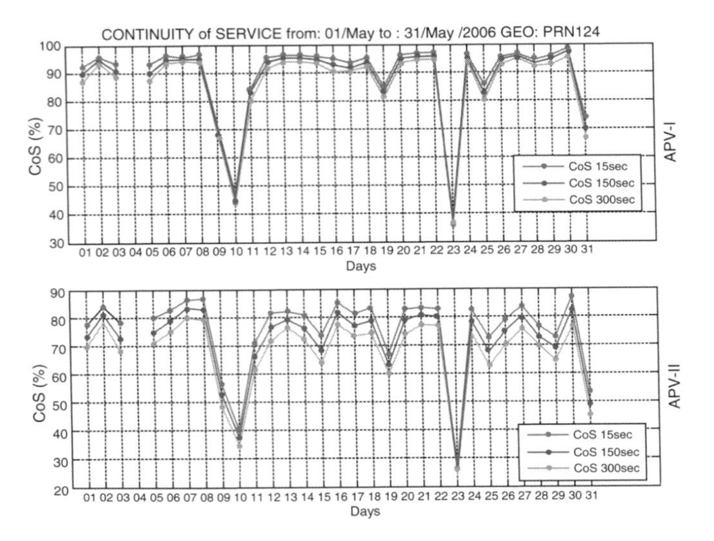

Fig. 9. Accuracy.

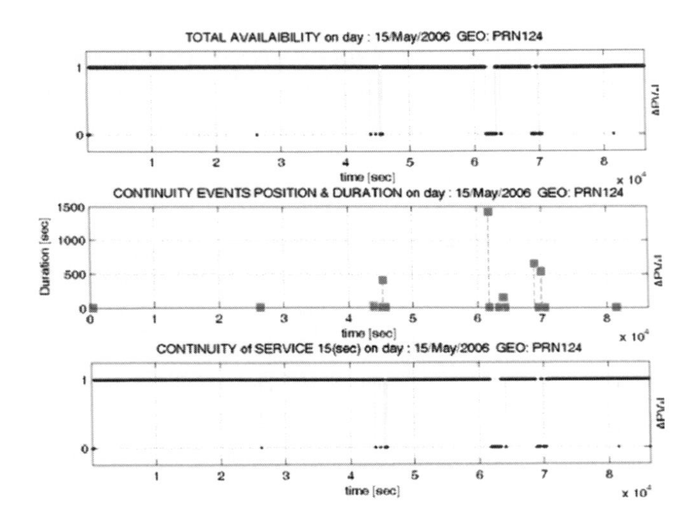

Fig. 10. APV-I performance.

To enhance the information provided with the Stanford plot we have performed a three-dimensional plot (Figs. 13 and 14) in which on x and y axises are represented the east/west and the north/south distances to a reference point of position solution.

On z axis we represents the protection level (horizontal or vertical). In this way we have a summarized representation of Integrity performance and position error. It is simple to notice that a conventional Stanford plot is the half-vertical plane containing the vertical line of the reference position and the corresponding position solution.

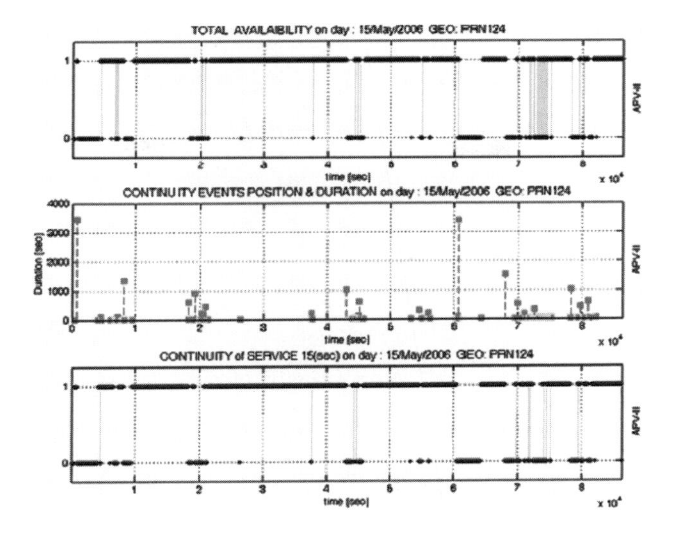

Fig. 11. APV-II performance.

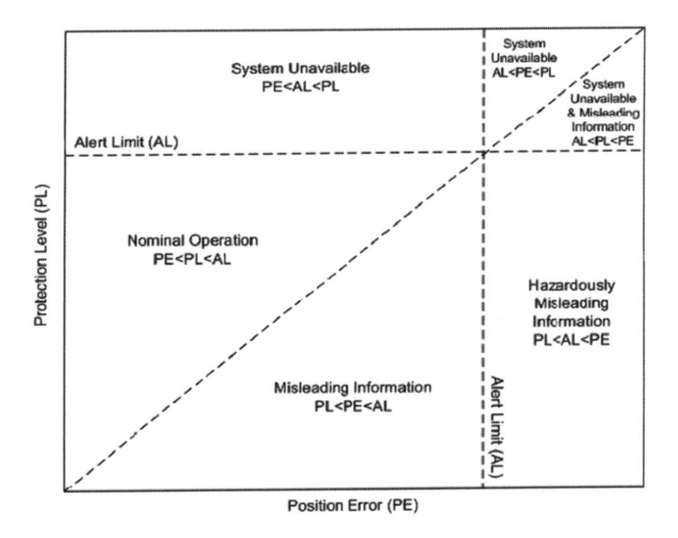

Fig. 12. Stanford plot.

5.4.5 Range Domain Analysis. The fast correction must be utilised in all navigation modes from En Route to Precision Approach. Fast corrections, provided as range correction values, are applied directly to the range measurements. Integrity indicators in the form of User Differential Range Error (UDRE) estimates are provided in the range domain.

This UDRE is an upper bound on the residual error of the pseudorange after the application of fast corrections; it is used to compute protection levels and

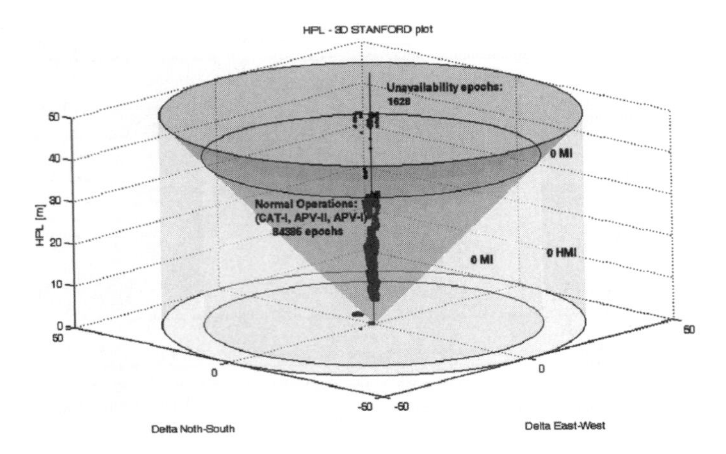

Fig. 13. Three-dimensional horizontal stanford plot.

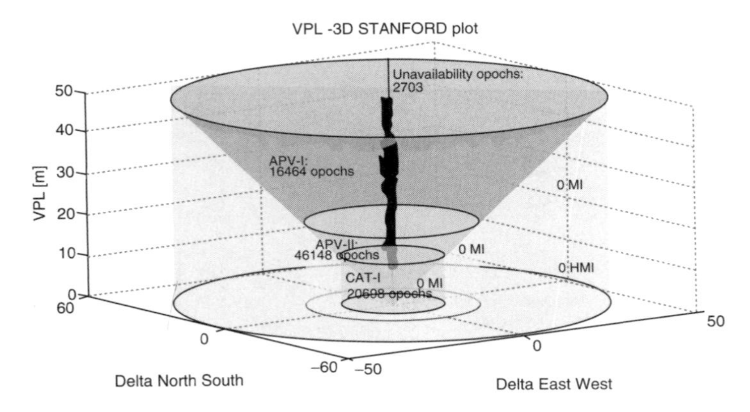

Fig. 14. Three-dimensional vertical stanford plot.

warning flags indicating that an individual satellite should not be used in the position solution.

To analyse these Integrity parameters we perform an innovative SKY plot on a Mercator projection in which for any Pierce point of the direct line-of-sight from the looked satellites to the our position through the ionosphere, we plot the parameter values by means the marker colours. In the Figs. 15 and 16 for all looked satellites, we show respectively the UDRE and the Ionospheric Vertical Delay (UIVD).

The contribution of receiver noise to the residual range error shall represent the accuracy performance of the airborne receiver, including receiver noise and multipath [1], [11]. The tracking accuracy of the receiver is evaluated as part of the accuracy requirements in [3]. The tracking accuracy for either a GPS or a SBAS satellite is depending on the current signal to noise ratio and the time since initialisation of the smoothing filtering. To asses the tracking accuracy we perform in the same way a SKY plot of the signal to noise ratio, Fig. 17.

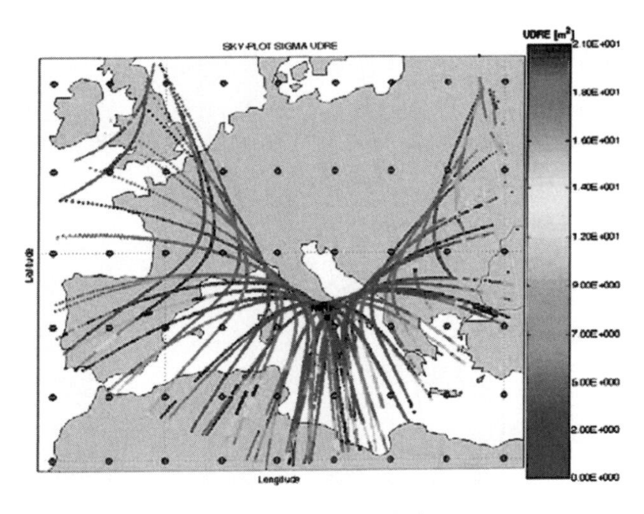

Fig. 15. UDRE SKY plot.

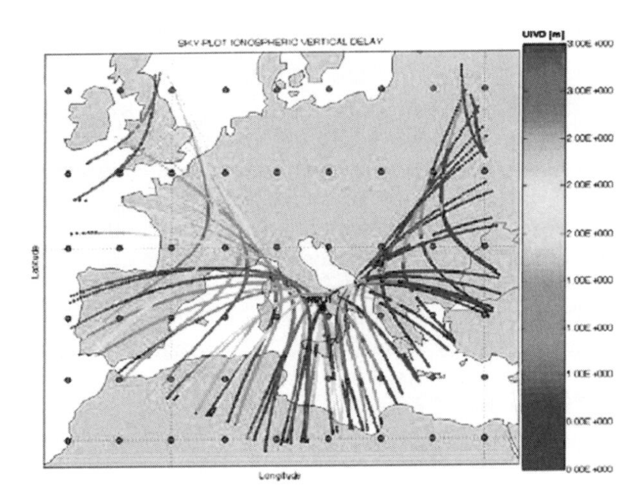

Fig. 16. UIVD SKY plot.

6 Conclusion

Our paper, placed in the frame of the EGNOS performance validation activities, has described a toolset named MARKAB recently developed to analyse the performance that can be experienced by a user, and to verify the stability of the performance for a fixed time interval. The objective of our toolset is to provide an efficient and fast statistical analysis about the performance in the airborne environment and to determine to what extent it could be safely approved for operational use. In order to

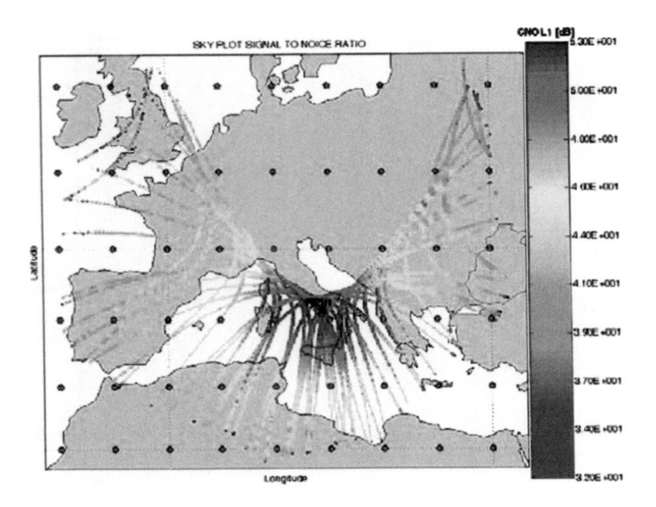

Fig. 17. Signal to noise ratio SKY plot.

gain experience with the European implementation of an SBAS system, confidence in the performance of that system has to be established. The architecture of MARKAB is designed in such a way as to provide easy and immediate access to the data, at all stages of data processing, and easy access to the graphical presentation of the results. Moreover, we have presented a new technique to assess the performance in terms of Continuity of Service which represents an important improvement in the context of performance evaluation. It allows us to define a new parameter to determine the probability that the augmentation system will be continuous in any required time interval for a given operation. The current performance of the augmentation system have been analysed in a series of static trials. These results raise a lot of hope to the validation of EGNOS system.

References

[1] M. Hernández-Pajares, J. M. Juan Zornoza, J. Sanz Subirana, R. Farnaworth and S. Soley, "EGNOS Test Bed Ionospheric Corrections Under the October and November 2003 Storms", IEEE Trans. on Geoscience and Remote Sensing, Vol. 43, NO. 10, pp: 2283–2293, Oct. 2005.

[2] ICAO "International Standards and Recommended Practices" – Annex 10, Volume 1, 5th Edition with amendments up to Amendment 80, November 2005.

[3] RTCA "Minimum Operational Performance Standards for GPS/WAAS Airborne Equipment" Radio Tech. Commiss. Aeronautics, Washington, DC, Doc229C. Nov. 2001.

[4] EUROCONTROL: "Civil Aviation Performance Requirements for EGNOS", Doc. No.: RNAV FG/WP4 Vol. 3, Dec. 2005.

[5] EUROCONTROL: "First Glance Algorithm Description", Doc. No.: APV/ESV/2, Second Meeting, Oct. 2005.

[6] N. Penna, A. Dodson, W. Chen, "Assessment of EGNOS Tropospheric Correction Model", The Journal of Navigation, Vol. 54, pp: 37–55, Jan. 2001.
[7] C. Butzmühlen, R. Stolz, R. Farnworth, E. Breeuwer, "PEGASUS–Prototype Development for EGNOS Data Evaluation–First User Experiences with the EGNOS System Testbed", Proceeding of the ION National Technical Meeting, 2001.
[8] George V. Kinal Fintan Ryan, "Satellite-Based Augmentation Systems: The Need for International Standards", The Journal of Navigation, Vol. 52, pp: 70–79, Jan. 1999.
[9] V. Ashkenazi, W. Chen, C. J. Hill, T. Moore, "Wide Area and Local Area Augmetations: Design Tools and Error Modelling", The Journal of Navigation, Vol. 51, pp: 58–66, Jan. 1998.
[10] M. Sams, A. J. Van Dierendock and Quyen Hua, "Availability and Continuity Performance Modelling", Proceeding of the ION annual Meeting, Cambridge, Jun. 1996.
[11] J. A. Klobuchar, "Ionospheric time-delay algorithm for single-frequency GPS users, IEEE Trans. Aerospace and Electronic Sys, AES-23, pp: 325–331, 1987.

Hybridization of GNSS Receivers with INS Systems for Terrestrial Applications in Airport Environment

Guglielmo Casale*, Patrizio De Marco*, Romano Fantacci**, Simone Menci**

* Selex - Sistemi Integrati - ATM & Airport Systems Division
Via Tiburtina Km 12.4 Rome - 00131 Italy
Phone: +39-06-4150.3976, Fax: +39-06-4150.3728
e-mail: gcasale@selex-si.com, pdemarco@selex-si.com
** University of Florence – Department of Electronics and Telecommunications
Via di S. Marta 3 Florence – 50139 Italy
Phone: +39-0554796467, Fax: +39-0554796485
e-mail: fantacci@lart.det.unifi.it, menci@lart.det.unifi.it

Abstract. This paper deals with the study of low-cost efficient techniques of GNSS-INS integration for the realization of a surveillance system for the terrestrial vehicular traffic in an airport. This system will have to enable accurate, continuous and reliable tracking of each ground vehicle operating in airport area in support of maintenance and management of aircrafts. High accuracies are required to give the ground control a precise and dynamic view of the airport situation, in order to optimize the ATM (Air Traffic Management) activity; this is particularly important in large hubs. In order to integrate satellites and sensors measurements we relied on Kalman filtering, a powerful signal processing tool for optimal blending of heterogeneous data sources. Specifically, we used an integration architecture named tightly coupled, where only one Kalman filter is used to integrate pseudorange, Doppler and sensor measurements. We will show that our hybrid receiver is a rather simple but extremely efficient solution to this problem.

1 Introduction

One of the main drawbacks of using GNSS (*Global Navigation Satellite Systems*) systems like GPS, GLONASS or Galileo for land surveying, navigation (mainly terrestrial) and in general route tracking, is the temporary loss of service we get when the receiver is moving under bridges or other types of obstacles. For example, a terrestrial vehicle equipped with a GPS receiver would experiment a loss of positioning when passing under trees surrounding the street; in the same way an aircraft flying at low altitude above that vehicle or standing still in its proximity, not so rare events in an airport environment, could produce the same effect. In these cases the obstacles are responsible of partial or total obstruction of the satellites navigation signals, and, during these service outages, navigation performance degrades rapidly. So one of the main problems of stand-alone satellite navigation is the service continuity, which is not guaranteed in all application scenarios. Various techniques may be used to compensate this drawback: one of the most important classes goes under name of *data fusion* (or data integration), that is the combination of the information coming from GNSS measurements with that coming from other sensors (vehicular or not) which are not affected by the above mentioned vulnerabilities.

A simple system of this type consists of using the technique, well known to navigators, named *"dead reckoning"* which, in its most simple realization, takes advantage of an odometer and a magnetic compass (or a system of gyroscopes for a three-dimensional system) which measurements are integrated with GNSS measurements in order to assist the receiver in the interpolation of vehicle position during the loss of tracking of satellites. But, in a more general view, the information may also come from accelerometers, tachometers, altimeters, and from any other instrument suited to the application and mounted on the vehicle considered. The most sophisticated approaches involve combining GNSS with INS (*Inertial Navigation Systems*): these systems are usually made of gyroscopes for the determination of the angular motion of vehicle axes with respect to a local reference system and of accelerometers oriented along these axes to measure the accelerations. Starting from a known position and twice integrating in time the accelerations we obtain differences of position that determine the vehicle trajectory, along with its velocity, acceleration and attitude. The simple sensor group that realizes dead reckoning for a terrestrial vehicle can be viewed like a simplified form of INS platform [1].

Inertial systems have some advantages with respect to GNSS: they are autonomous and independent from external sources, they have not visibility problems, and often provide accuracies similar to GNSS when used in short time intervals. *Hence, INS can be used as an interpolator for gaps of service in GNSS.* As a drawback, the estimation error tends to grow with a quadratic trend in time because of sensor errors, hence periodic resets with known and accurate positions are required in order to prevent excessive degradations of the position estimate. Indeed sensors are not ideal, they are affected by noise and distortions, so inertial navigation performance depends strongly from their quality (and hence their cost) [2].

So, GNSS and INS systems have complementary characteristics that render them ideal candidates for the integration. In fact GNSS periodically reinitializes INS, and this last one integrates GNSS during outage periods and assists it in satellites tracking.

This paper deals with the study of low-cost efficient techniques of GNSS-INS integration for the realization of a surveillance system for the terrestrial vehicular traffic in an airport. This system will have to enable accurate, continuous and reliable tracking of each ground vehicle operating in airport area in support of maintenance and management of aircrafts (refuelling, maintenance, baggage transport, safety, etc.). Each vehicle equipped with hybrid receiver will then transmit its position to the airport control room, where the operator will be able to monitor the overall situation and dispatch orders. For these transmissions we don't assume any particular technology; good candidates are the Extended Squitter channel on 1090 MHz frequency or a wireless link realized with Wi-Fi or WiMAX technologies.

High accuracies are required to give the ground control a precise and dynamic view of the airport situation, in order to optimize the ATM (*Air Traffic Management*) activity; this is particularly important in large hubs. It's clear that this objective can be reached only through full continuity of positioning service.

In the continuation of this paper, we will assume a GPS / EGNOS receiver for reasons of rapid implementation, but it's important to point out that these results may be easily extended in a straightforward way to future multisystem GNSS receivers.

2 System Architecture

We can count at least three different architectures for GPS / INS integration [1,2]: they are depicted in Fig. 1. Modes a) (*uncoupled mode*) and b) (*loosely coupled*) initially consider GPS receiver and INS like separate systems and then integrate "optimally" the measures of position, velocity and attitude keeping count of the respective estimation error variances. Mode c) (*tightly coupled*) instead considers GPS and INS like "virtual sensors", that is systems which outputs measurements of pseudorange, pseudorange rate (or, equivalently, Doppler shifts) and variations of velocity and heading. In this case there is only one navigation processor responsible of fusion of all data, obviously more complex than the other cases. The tightly coupled approach is the one we chose, because while it's the more complex to study and implement, it's also potentially the more performance effective.

In order to integrate satellite and sensors measurements we relied on *Extended Kalman Filter (EKF)* [2,3], a powerful signal processing tool for optimal blending of heterogeneous data sources. With tightly coupled architecture only one EKF is used to integrate pseudorange and Doppler satellite and sensor measurements. In general, Kalman filtering allows to optimally estimate the state x_k of a dynamic system based on measurements z_k (at discrete-time step t_k) and on uncertain system dynamic model:

$$x_{k+1} = \Phi_k x_k + G_k u_k + \Gamma v_k$$
$$z_k = H_k x_k + w_k$$

The first equation is the *dynamic equation*, while the second is the *measurement equation*. v_k is the *process noise*, here an acceleration noise, which models uncertainty on

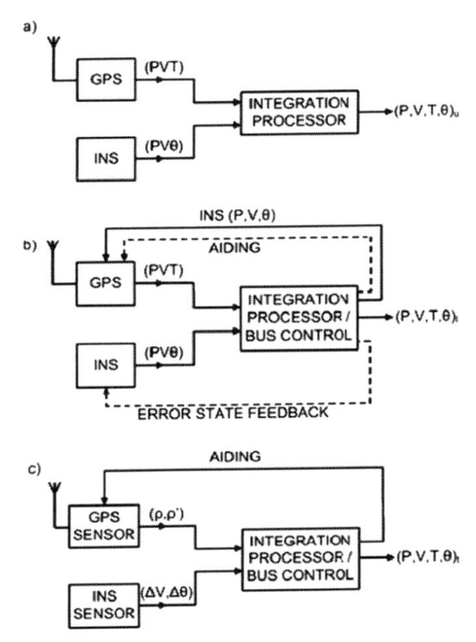

Fig. 1. Generic architecture GPS/INS: a) uncoupled mode; b) loosely coupled mode; c) tightly coupled mode.

Table 1. Essential equations of extended kalman filter.

Predictor (time updates)
Predicted State Vector
$\hat{x}_k^- = \Phi_k \hat{x}_{k-1}^+$
Predicted Covariance Matrix
$P_k^- = \Phi_k P_{k-1}^- \Phi_k^T + Q_{k-1}$

Corrector (measurements updates)
Kalman Gain
$\bar{K}_k = P_k^- H_k^T (H_k P_k^- H_k^T + R_k)^{-1}$
Corrected State Estimation
$\hat{x}_k^+ = \hat{x}_k^- + \bar{K}_K (z_k - H_k \hat{x}_k^-)$
Corrected Covariance Matrix
$P_k^+ = P_k^- - \bar{K}_K H_k P_k^-$

dynamic system model, w_k is the *measurement noise*, which models pseudorange and Doppler measurement errors and sensors errors. Γ is a *noise distribution matrix*, which keeps count of impact of acceleration noise on all state variables. In our system we have no control input u_k, so we will neglect it. The equations that compose EKF are reported in Table 1 with following explanation of quantities and symbols.
where:

Φ_k is the *state transition matrix*.
H_k is the *measurement sensitivity matrix* or *observation matrix*.
$H_k \hat{x}_k^-$ is the *predicted measurement*.
$z_k - H_k \hat{x}_k^-$, the difference between the measurement vector and the predicted measurement, is the *innovations vector*.
$\bar{K}_K$ is the *Kalman gain*.
P_k^- is the *predicted* value or *a priori* of estimation covariance.
P_k^+ is the *corrected* value or *a posteriori* of estimation covariance.
Q_k is the covariance of dynamic disturbance noise.
R_k is the covariance of *sensor noise* or *measurement uncertainty*.
$\hat{x}_k^-$ is the *predicted* or *a priori* value of the estimated state vector.
$\hat{x}_k^+$ is the *corrected* or *a posteriori* value of the estimated state vector.
z_k is the *measurement vector* or *observation vector*.

We chose the following state vector, taking into account two components (along East and North) of horizontal position, velocity and acceleration, and the errors typically associated with receiver clock, bias and drift.

$$x_k = \left[\Delta P_E\ \Delta P_N\ \Delta V_E\ \Delta V_N\ \Delta A_E\ \Delta A_N\ b\ d\right]^T$$

We assumed that in a airport environment (with terrain almost flat) a terrestrial vehicle has only two significant degree of freedom: longitudinal translation (about roll axis) and azimutal rotation (about yaw axis), that is, as a good approximation, we neglected other types of movements. So, we used only two sensors: an accelerometer and a gyroscope.

Table 2. GPS/EGNOS receiver, sensor models and indicative costs.

Reference GPS receiver	u-blox TIM-LR	Chipset 70–80 €
Accelerometer	Analog Devices ADXL 103	7–8 $
Gyroscope	Analog Devices ADXRS 150	30 $

The specific GPS receiver and sensors models assumed in this work are shown in Table 2 [4,5].

It's important to remark that all these components were chosen to have low costs in order to make feasible the implementation of the entire system even in a large airport with many vehicles. Moreover, the inertial sensors are realized with MEMS (*MicroElectroMechanical Systems*) technology, hence they have very interesting characteristics. Next, we summarize the main of them:

- Good performance
- Low cost
- Very low weight, dimensions, power consumption and dissipated heat, so they are well suited for installation on various types of mobile platforms.

The GPS receiver is conceived for integrated applications, because is predisposed to accept sensor inputs and is equipped with EKF software for dead reckoning applications; it supports fully automatic calibration of sensor inputs with temperature compensation and it has a 40 Hz dead reckoning calculation rate for high accuracy calculations (we will see later that this is a very important parameter). The position update rate is 1 Hz.

This receiver can also receive EGNOS signals for the application of differential corrections in order to have integrity and accuracy improvements [6]. We will exploit this capability in order to maximize performance.

After some filter tuning [3], we managed to find the optimal choice for the above mentioned vectors and matrices, here reported:

Initial state estimate and associated covariance matrix:

$$x_0=\vec{0} \quad P_0^+=diag\left(8.46^2, 8.46^2, 10^{-3}, 10^{-3}, 10^{-3}, 10^{-3}, 10^4, (c\cdot 10^{-6})^2\right)$$

(all lengths are expressed in km)

Matrices related to dynamic equation:

$$\Phi_k=\left[\begin{array}{cccccc|cc} 1 & 0 & T & 0 & T^2/2 & 0 & 0 & 0 \\ 0 & 1 & 0 & T & 0 & T^2/2 & 0 & 0 \\ 0 & 0 & 1 & 0 & T & 0 & 0 & 0 \\ 0 & 0 & 0 & 1 & 0 & T & 0 & 0 \\ 0 & 0 & 0 & 0 & 1 & 0 & 0 & 0 \\ 0 & 0 & 0 & 0 & 0 & 1 & 0 & 0 \\ \hline 0 & 0 & 0 & 0 & 0 & 0 & 1 & T \\ 0 & 0 & 0 & 0 & 0 & 0 & 0 & 1 \end{array}\right] \qquad \Gamma=\begin{bmatrix} T^2/2 \\ T \\ 1 \end{bmatrix}$$

$$Q_{k-1}=\Gamma\sigma_v^2\Gamma^T=\left[\begin{array}{cccccc|cc} T^4/4 & 0 & T^3/2 & 0 & T^2/2 & 0 & 0 & 0 \\ 0 & T^4/4 & 0 & T^3/2 & 0 & T^2/2 & 0 & 0 \\ T^3/2 & 0 & T^2 & 0 & T & 0 & 0 & 0 \\ 0 & T^3/2 & 0 & T^2 & 0 & T & 0 & 0 \\ T^2/2 & 0 & T & 0 & 1 & 0 & 0 & 0 \\ 0 & T^2/2 & 0 & T & 0 & 1 & 0 & 0 \\ \hline 0 & 0 & 0 & 0 & 0 & 0 & Q_b & Q_{bf} \\ 0 & 0 & 0 & 0 & 0 & 0 & Q_{bf} & Q_f \end{array}\right]\sigma_v^2$$

The process noise is a DWPA (Discrete-time Wiener Process Acceleration), hence with white noise acceleration increments (*jerk*).

Matrices related to measurement equation:

$$z_k=\left[\begin{array}{c} PR_{s1} \\ \vdots \\ PR_{sM} \\ \hline D_{s1} \\ \vdots \\ D_{sM} \\ \hline a_r \\ \phi_y \end{array}\right] \quad H_k=\left[\begin{array}{cc|cc|cc|cc} -e_{E1} & -e_{N1} & 0 & 0 & 0 & 0 & 1 & T \\ \vdots & \vdots & \vdots & \vdots & \vdots & \vdots & \vdots & \vdots \\ -e_{EM} & -e_{NM} & 0 & 0 & 0 & 0 & 1 & T \\ \hline 0 & 0 & -e_{E1} & -e_{N1} & 0 & 0 & 0 & 1 \\ \vdots & \vdots & \vdots & \vdots & \vdots & \vdots & \vdots & \vdots \\ 0 & 0 & -e_{EM} & -e_{NM} & 0 & 0 & 0 & 1 \\ \hline 0 & 0 & 0 & 0 & a_{Er} & a_{Nr} & 0 & 0 \\ 0 & 0 & 0 & 0 & a_{Ey} & a_{Ny} & 0 & 0 \end{array}\right]$$

$$R_k=diag\left(\sigma_{PR1}^2,\cdots,\sigma_{PRM}^2,\sigma_{D1}^2,\cdots,\sigma_{DM}^2,\sigma_a^2,\sigma_\phi^2\right)$$

where e_{Ei} and e_{Ni} are the components along East and North axis (of local tangent plane (LTP)) of unit vectors pointing i-th satellite from receiver position; PR_{si} and D_{si} are respectively pseudorange and pseudorange rate (derived from Doppler) measurements, a_r and φ_y are the sensors measurements, respectively roll axis acceleration and yaw axis rotation.

In order to better understand how sensors sensibilities terms a_{Er}, a_{Nr}, a_{Ey} and a_{Ny} are calculated, we recall here the basis of dead reckoning navigation. In Fig. 2 the basic analytical formulation of dead reckoning navigation is depicted. We will initially refer to the classical formulation where vehicle is equipped with an odometer for measurement of distance travelled Δl_i and a magnetic compass / gyroscope for measurement of variation of heading θ_i in the i-th time interval ΔT (fixed or variable sampling time).

Coordinates X and Y define the horizontal plane where the vehicle moves in a trajectory from initial point (X_0, Y_0) to destination (X_n, Y_n) (X maybe East coordinate,

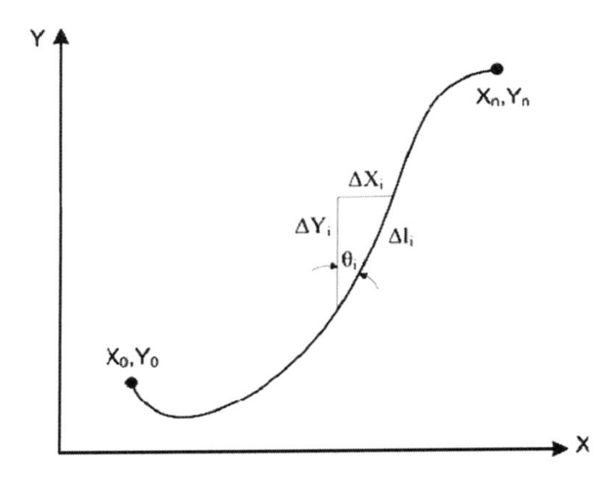

Fig. 2. Dead reckoning formulation.

Y maybe North). For each sampling interval, the vehicle's autonomous navigation unit updates its position with the following first-order approximation:

$$X_n = X_0 + \sum_1^n \Delta X_i = X_0 + \sum_1^n \Delta l_i \sin\theta_i$$

$$Y_n = Y_0 + \sum_1^n \Delta Y_i = Y_0 + \sum_1^n \Delta l_i \cos\theta_i$$

In this way, the reconstructed trajectory is a first-order approximation of the real trajectory, that is a piecewise linear curve. Clearly, the quality of position estimation depends from sampling time and sensors measurement errors, since position errors tend to grow in time because of sensor errors. Odometers are today present on vehicles equipped with ABS, while heading sensors may be implemented with a magnetic compass. The problem is that odometers and compasses yield measurements subjected to serious error effects, like noise, bias, drift and distortion.

We use this formulation for dead reckoning implementation, but in a slightly modified way in order to use more precise sensors like the accelerometers and gyroscopes we chose.

So, starting with geometric relations:

$$a_E = a_r \sin(\phi_y) \quad a_N = a_r \cos(\phi_y)$$

inverting these and taking partial derivatives w. r. t. a_E and a_N, we obtain:

$$a_{Er} = a_E / a_r \quad a_{Nr} = a_N / a_r$$

$$a_{Ey} = a_N / a_r^2 \quad a_{Ny} = -\, a_E / a_r^2$$

which must be calculated using the filter actual better state estimate.

Before Kalman integration we used a particular smoothing procedure of GPS measurements, the *Hatch filter*, which combines pseudorange and Doppler measurements to reduce the effect of multipath and receiver noise.

2.1 Pseudorange Smoothing Filter

In steady state, each L1 pseudorange measurement from GPS receiver is smoothed using the filter:

$$PR_s(k) = \left(\frac{1}{N}\right) PR_r(k) + \left(\frac{N-1}{N}\right)\left[PR_s(k-1) + \phi(k) - \phi(k-1)\right]$$
$$N = S/T$$

where

- PR_r is the raw pseudorange,
- PR_s is the smoothed pseudorange,
- N is the number of samples,
- S is the time filter constant, equal to 100 seconds,
- T is the filter sample interval, nominally equal to 0.5 seconds and not to exceed 1 second,
- ϕ is the accumulated phase measurement,
- k is the current measurement, and
- $k-1$ is the previous measurement.

In principle the more epochs of data are used in the smoothing process the more precise the smoothed pseudorange should become, and should approach the precision of the carrier range (mm-level). In practice there are facts which destroy this ideal situation:

- Since the ionosphere delays the pseudorange and advances the carrier range the change in pseudorange does not equal exactly the change in carrier range (this effect is called the ionospheric divergence).
- If the receiver channel looses lock on the SV momentarily, or if the range rate of change is too high, the carrier phase integration process is disrupted, resulting in a "cycle slip", and an incorrect change in carrier range.

To overcome the above drawbacks the number of observations used to smooth the pseudoranges is limited. At one observation per second a maximum of 100 is a good value. Moreover, large cycle slips can be detected: if the carrier rate of change is larger by a certain margin than the pseudorange rate of change, a cycle slip is declared and the smoothing algorithm is reset ($n = 1$). The margin depends very much on the noise and multipath figures of the receiver and the antenna location. For high quality receivers with optimally located antennas the margin could be as low as 1 m, a value of 15 m is more realistic, which implies that slips of more than 100 cycles (1 cycle is about 0.2 m) remain undetected. The limit value for N limits the error in the smoothed pseudorange, and lets it fade away after one to two minutes.

Code smoothing reduces multipath and receiver noise on the pseudoranges. Although theoretically a reduction of a factor 10 can be reached, we can count on a practical reduction of at least a factor 2, and up to 5. An example of benefit of using this filter is depicted in Fig. 5, where we can see that really the accuracy improvement we get is described by a factor from 2 to 5. We chose this particular smoothing filter in order to further improve estimation accuracy because of its simplicity and cheapness that makes it very attractive to implement.

3 Simulation Scenario

Our activity was carried out through computer simulations with a flexible satellite navigation software simulator.

The simulations involved a vehicle moving in the Roma-Fiumicino airport (Fig. 3); the service vehicle's route (Fig. 4) describes a movement from head of 16R runway to the terminal area. The trajectory is a piecewise linear curve with distributed accelerations and decelerations and changes of heading concentrated in specific waypoints. These particular trajectory, along with accelerations and speed values, is studied to be slighter "extreme" than a realistic case, in order to test satellite tracking and system performance in a sort of "worst case". This route has a length of 4.89 km and, with assumed speeds, the receiver moves from waypoint 1 to waypoint 15 in about 4 minutes; the first part (till waypoint 7) is fairly straight, while second part is much more articulated with more frequent changes of directions, accelerations and decelerations.

Monte Carlo simulations were performed in order to reduce the variability of stochastic factors on position and velocity estimates.

Simulations were made with 11 GPS satellites and the 3 EGNOS satellites all visible from the receiver, a fairly good condition for receiver performance.

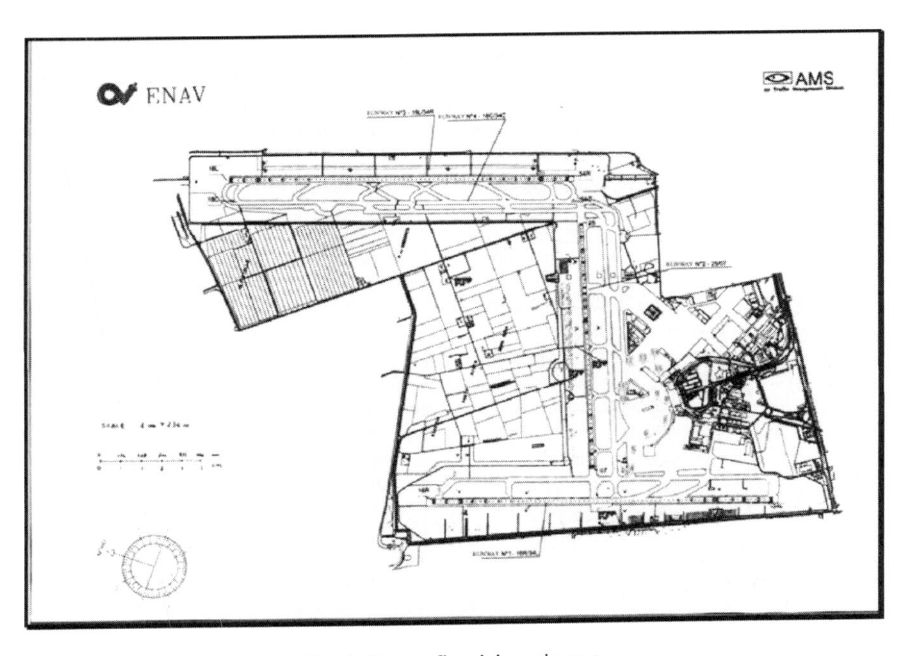

Fig. 3. Roma-fiumicino airport.

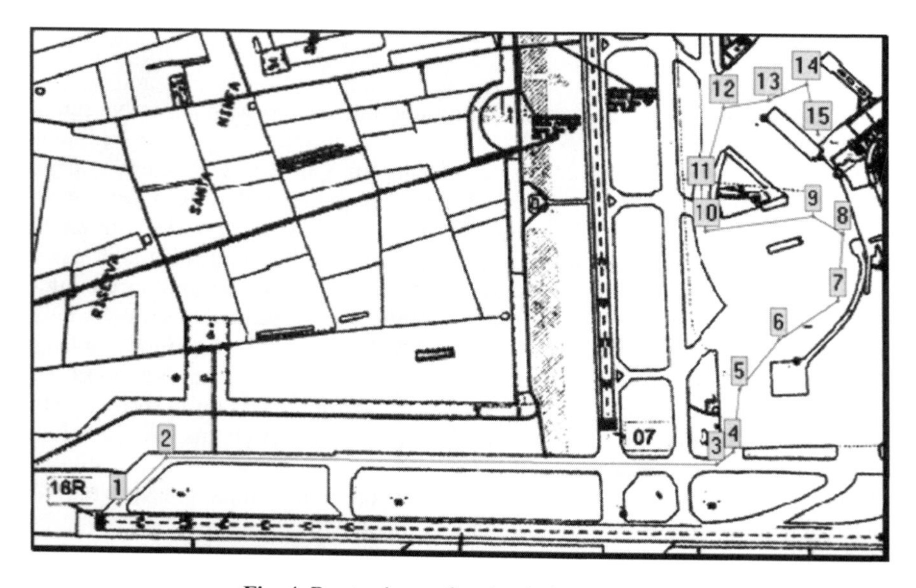

Fig. 4. Route chosen for simulation purposes.

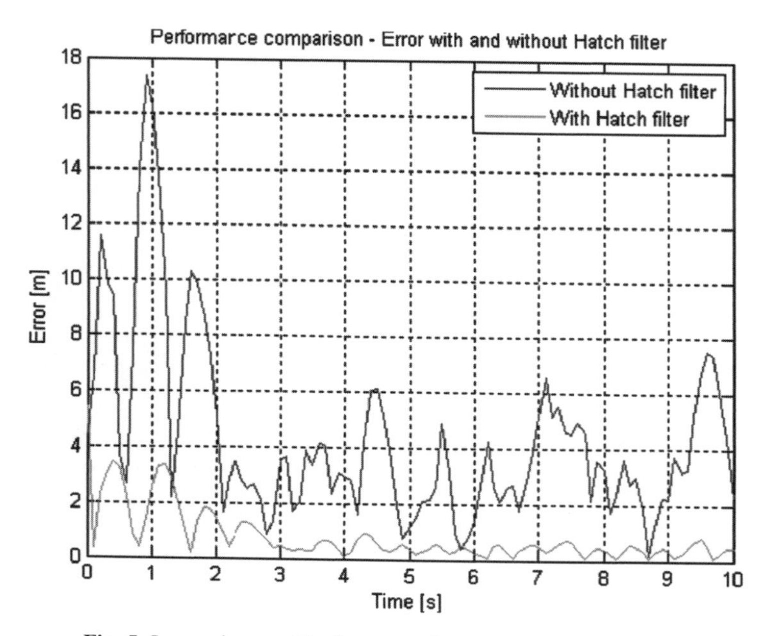

Fig. 5. Improving positioning error through use of Hatch filter.

4 Experimental Results

Next, we see some results obtained from simulations. We set, as the main performance parameter, a target HPL (*Horizontal Protection Level*) of 5 m; *the instantaneous horizontal position estimation error (and its RMS value) must be lower than this threshold in order to declare satisfying navigation performance*. Moreover, we evaluated the *service availability (SA)*, that is *the percentage of time that estimation error is in this acceptable error range*; this parameter must be ≥ 99.5% in order to be satisfying.

Figure 6 shows a plot of estimation error for vehicle movement along the entire chosen route: we can see that GPS stand-alone is not capable of guarantee the target performance in every time, even with many visible satellites and the presence of EGNOS: this is also due to the vehicle dynamic, in fact the error spikes correspond to the turns showed in Fig. 4, more frequent in the second half of route. Though the RMS error is only 2.751 m, the SA is only 81.2%, an unacceptable performance. Instead, hybrid receiver is always capable of maintaining error lower than 3 m, with an RMS error of 0.960 m and a 100% SA, even in cases of rapid unexpected turns, and this is the effect of the sensors presence and of the highest frequency of computing carried out on their measurements with respect to GPS observations.

Figure 7 depicts first 30 s of a situation in which we have a partial loss of tracking (only 6 satellites are visible, with a bad geometry) from 10 s to 20 s. This situation is related to the vehicle passing under an extended obstacle or near a building, and has

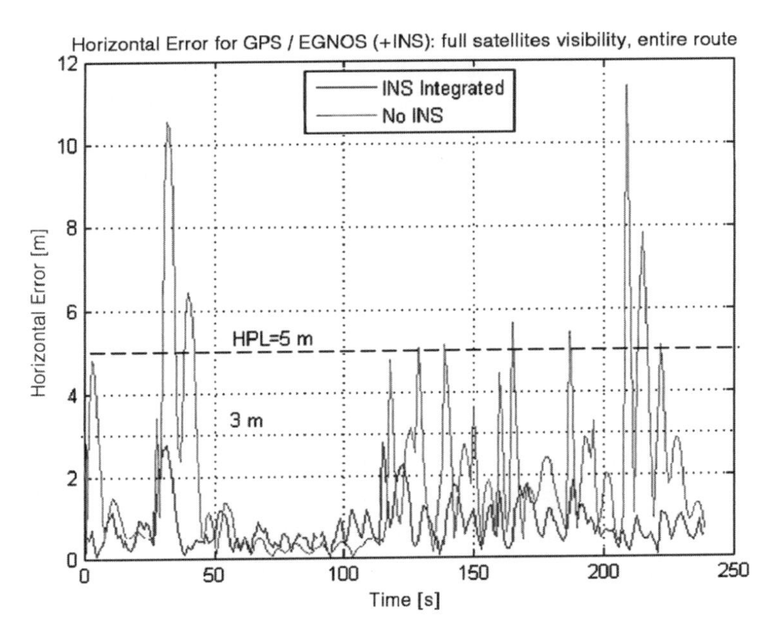

Fig. 6. Performance comparison between GPS / EGNOS receiver and INS-integrated receiver during movement along entire chosen route in case of full satellites visibility.

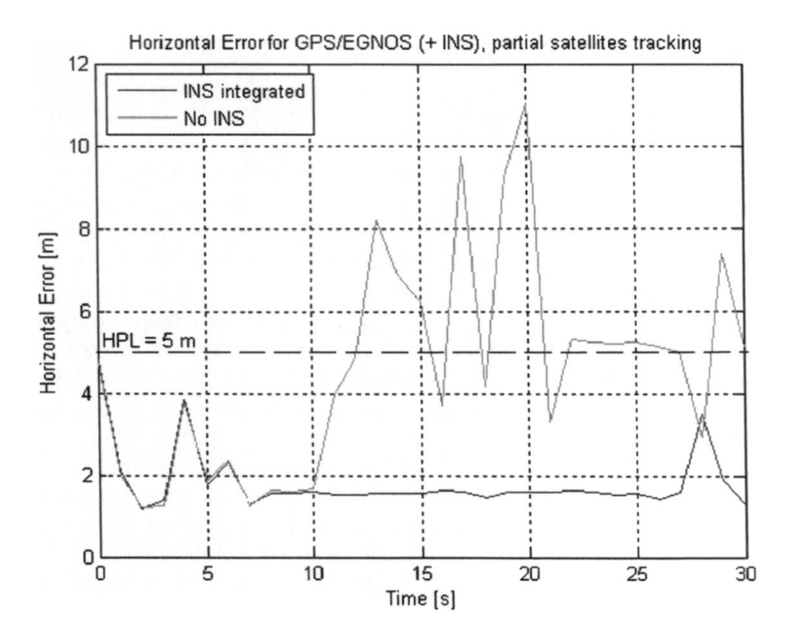

Fig. 7. Performance comparison between GPS / EGNOS receiver and INS-integrated receiver in case of tracking gap of some satellites.

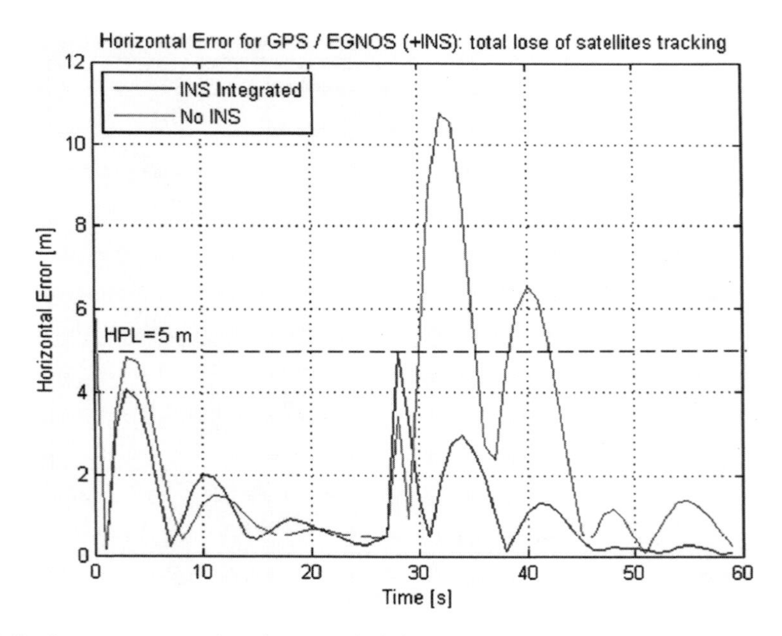

Fig. 8. Performance comparison between GPS / EGNOS receiver and INS-integrated receiver in case of total lose of satellites tracking.

a strong detrimental effect on positioning accuracy. With no integration, the SA is only at 58.1%, and the RMS error is 5.243 m. Our integrated solution is more stable and reliable and presents no discernible degradation, maintaining 100% of service availability and RMS error 1.978 m.

Figure 8 represents an extreme scenario of total loss of tracking of satellites for 20 seconds (20–40): in correspondence of approximately second 28, the vehicle arrives at waypoint 2, and the consequent turn degrades heavily the performance of GPS stand-alone receiver, with error that remains above 5 m for many seconds. Indeed, this is due to the total absence of measurements for correction of the predicted state, which evolves freely in time depending on a mispredicted vehicle dynamic. With support of sensors, the receiver manages to maintain always required performance because of the limited duration of satellites gap, when the sensor measurement errors are not sufficient to degrade position estimation significantly.

5 Conclusions

In this paper we showed that it's possible to augment typical capabilities of GNSS stand-alone receivers with a tightly coupled Kalman filtered integrated solution which uses low cost MEMS sensors of accelerations and rotations. With an expense of a few tens of $, we can equip a terrestrial maintenance vehicle for airport activity with this new hybrid advanced receiver, making feasible an accurate and reliable service of surveillance of overall airport activity for a more efficient ATM. Our hybrid receiver is capable of guarantee an horizontal position estimation error lower than 5 m even in the worst scenarios (high dynamics and bad tracking of satellites) and lower than 3 m (about 1 m) in more common and realistic situations.

References

[1] B. W. Parkinson, J. J. Jr. Spilker, *Global Positioning System: Theory and Applications*, vols. 1 and 2, American Institute of Aeronautics and Astronautics, 370 L'Enfant Promenade, SW, Washington, DC, 1996.

[2] M. S. Grewal, L. R. Weill, A. P. Andrews, *GPS Global Positioning Systems – Inertial Navigation and Integration*, Wiley 2001.

[3] Y. Bar-Shalom, X. Rong Li, T. Kirubarajan, *Estimation with Applications To Tracking and Navigation*, Wiley 2001.

[4] *TIM-LR Sensor-Based GPS Module Data Sheet*, u-blox, 2005.

[5] *Analog Devices ADXL103 and ADXRS150 Data Sheets*, Analog Devices, 2004.

[6] RTCA/DO-229C, *Minimum Operational Performance Standards For Global Positioning System/Wide Area Augmentation System Airborne Equipment*, RTCA Inc., November 28 2001.

[7] J. B.-Y. Tsui, *Fundamentals of Global Positioning System Receivers – A Software Approach*, Wiley, 2005.

[8] J. B. Y. Tsui, M. H. Stockmaster, D. M. Akos, *Block Adjustment of Synchronizing Signal (BASS) for Global Positioning System (GPS) Receiver Signal Processing*, GPS97.

[9] C. T. Brumbaugh, A. W. Love, G. M. Randall, D. K. Waineo, S. H. Wong, *Shaped Beam Antenna for the Global Positioning Satellite System, Rockwell International Corporation*, Antennas and Propagation Society International Symposium, 1976.

[10] F. M. Czopek, S. Shollenberger, *Description and Performance of the GPS Block I and II L-Band Antenna and Link Budget*, GPS93.

[11] *Navstar GPS Space Segment/Navigation User Interfaces, Interface Specification IS-GPS-200, Revision D, November 7 2004*, ARINC Engineering Services, LLC.

[12] *WGS84 Implementation Manual*, EUROCONTROL and Institute of Geodesy and Navigation (IfEN), Version 2.4, February 12 1998.

[13] W. Gurtner, *RINEX: The Receiver Independent Exchange Format Version 2.10*, Astronomical Institute, University of Berne, June 8 2001.

W Band Multi Application Payload for Space and Multiplanetary Missions

Vittorio Dainelli, Giordano Giannantoni, Marcello Muscinelli

Oerlikon Contraves SpA
Via Affile 102, 00131, Roma, Italy
Phone: +390643611, Fax: +390643612877,
e-mail: v.dainelli@oci.it, g.giannantoni@oci.it, m.muscinelli@oci.it

Abstract. The increasing demand for frequency locations with high bandwidth to satisfy the growing satellite communications applications made it necessary to explore higher and higher frequency ranges. W-band (75–110 GHz) can be considered as the new frontier of satellite communications.

Based on the experience gained by Oerlikon Contraves in the development and in the operation of a W band radar sensors operating at 95 GHz for Airport Surface Movement Control and Guidance and on W band point to point ground communication link modular "core" integrated modules based on MMIC technology are described on this paper. The main targets to reach for each "core" module are the minimisation of mass, volume and the modularity in order to have the possibility of use it in payloads that can be embarked on different platforms (nanosatellites, microsatellites, satellites, rovers on Moon and Mars. . . .) or used on ground for different applications including security.

The personalisation of the payload will be done by the antennas, the HPA and the baseband processing depending if the application will be a high rate, high speed communication link or a beacon for propagation measurement or a radar (traditional, SAR, interferometer. . . .)

1 General Information

The increasing demand for frequency locations with high bandwidth to satisfy the growing satellite communications applications made it necessary to explore higher and higher frequency ranges. Frequencies around 60 GHz cannot be utilized effectively due to atmospheric absorption, hence, the W-band (75–110 GHz) can be considered as the new frontier of satellite communications. The large bandwidth availability in W-band allows conceiving and proposing many advanced and innovative services that need high-volume data transfers without tight interactivity constraints for future scenarios.

The reduced wavelength will permit antennas and payloads with reduced volumes and weight doing the possibility of placing them on micro or nanosatellites platforms or on a knob of International Space Station (ISS). These are key payload features since the cost of a mission is closely related with the weight of the payload.

Pencil beam capability at reasonable antenna size will assure improved data security.

Considering that the main disadvantage of the use of W-band frequencies is the atmospheric attenuation, all its benefits could be fully exploited in space out of atmosphere or in interplanetary missions.

For which concern the communications, W band can be used for intersatellite links or for communications between a Lander or a station on the surface of the Moon (or Mars) and an orbiter, that could act as a data-relay node towards an Earth station (for example, the Deep Space Network).

Radars for small debris identification or for rendez-vous and docking operations could be envisaged

Due to the lack of atmosphere on the Moon, the use of millimetre (W band) or sub-millimetre bands of frequency could be the winning strategy for the development of high resolution interferometer radars with reduced mass and volume for the self guidance (instead of optics) of rovers operating on Moon surface or for the development of moon based radiometers for the analysis of atmosphere or for the completion of previous scientific missions (Planck, Herschel..).

W-band is recently considered as a "technological frontier"; thus representing the true challenge towards which the research community and the industry should concentrate their efforts.

2 "Core" Concept

For which concern possible applications relative to ground-satellite, satellite-ground communications, the characterisation of the channel together with some experimentation of the channel itself together with and in-orbit validation of W-band technology and space qualification processes must be considered mandatory before proposing a "commercial" mission operating in this band.

These applications are expected to provide the necessary elements towards the realization pre-operative multi application payload for space and interplanetary mission.

Concerning the hardware, it is important to mention here that since no existing space-qualified hardware has been reported yet, to fulfil the high power requirements and low response to cosmic radiations and space effects as required for GEO and LEO payloads, the research and development of such components is of a vital importance. This is actually the most important technological challenge on W-band frequencies. Actually, previous experience related to the development of platforms for scientific satellites, implied the necessity to anticipate a certain level of flexibility in defining the required characteristics, in which the details of the mission become defined more precisely during the project development (e.g. the detailed allocation of the band, permitted BER, transmission and reception power, the permitted redundancy, etc. . .). Since the characteristics of the on-board processing depend on such details in order to reduce the uncertainties in time schedule, it is necessary to follow a flexible approach that can be adapted according to the variation in requirements. This will offer a number of solutions that can be integrated together on one hardware platform although not defining the series of processing in a unique way. Obviously, the choice of a specific solution depends on the performance to be obtained using the system.

For this reason a multi application W band payload can be conceived composed by "core" integrated modules common to all and personalised in terms of antenna, power amplifier and signal processing depending on the application. The main targets to reach for the "core" front-ends are the minimisation of mass, volume and the modularity in order to have the possibility of use it in payloads that can be embarked on different platforms (nanosatellites, microsatellites, satellites, rovers on Moon and Mars. . . .)

Such constraints will imply a large use of Multifunctional Integrated MMIC on GaAs (MHEMT, PHEMT) reaching for these devices state of the art performances in terms of noise and power(100–200m W).

Other particular devices like MEMS switch could be used in case on such "core" module is required the control of the amplitude and the phase as in case of 3D Imaging radar for self guidance of small rovers on the moon also if, due to difficulties given by the small wavelength, sometimes is preferred move the beam mechanically or electromagnetically (PGB or Metamaterials) at antenna level.

If higher power are required as in GEO mission, HPA transmitters based on vacuum tubes shall be added to such core module. The kind of operation of such SSPA or HPA (CW, pulsed or digitally modulated) will be defined on the basis of the applications.

Other element that will personalize the payload are the baseband generation, the signal processing and the antenna depending if the application will be a high rate, high speed communication link or a beacon for propagation measurement or a radar.

Scope of this paper will be the identification of such "core" integrated modules derived from previous experience of Oerlikon Contraves to be used in different applications and missions space or ground based including security, proposing services, applications, integrated business opportunity able to merge two worlds – communications and navigation – that have been purposely apart for years.

3 Oerlikon Contraves Heritage

Oerlikon Contraves gained a lot of experience in the development and in the operation of a W band radar sensors operating at 95 GHz for Airport Surface Movement Control and Guidance.

The first prototype was installed in 2001 at the Frankfurt Airport and improved on 2003; a second one is installed at the Venice "Marco Polo" airport; presently they are operative and used as gap-filler of the main control radar. Other installations are foreseen within 2006.

The last configuration named SMART High Resolution Radar MK2 is shown in Fig. 1.

The main features of SMART HRR MK2 with respect to the previous one are:

- Klystron Based Transmitter
- Full Redundant TX/RX Chain
- No 'Active' Hardware Rotating
- Frequency agility
- Improved Maintenability Capability

Due to the frequency of operation, the main advantages of such radar are: reduced dimensions, simple installation, interference immunity, high resolution.

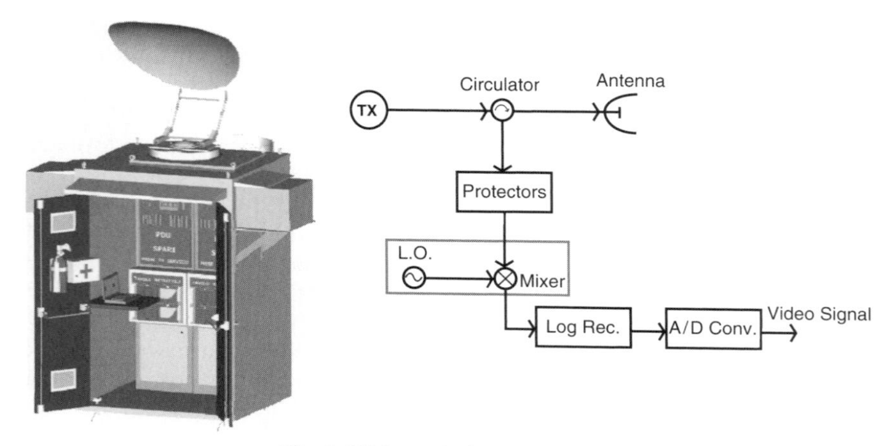

Fig. 1. High resolution rader MK2.

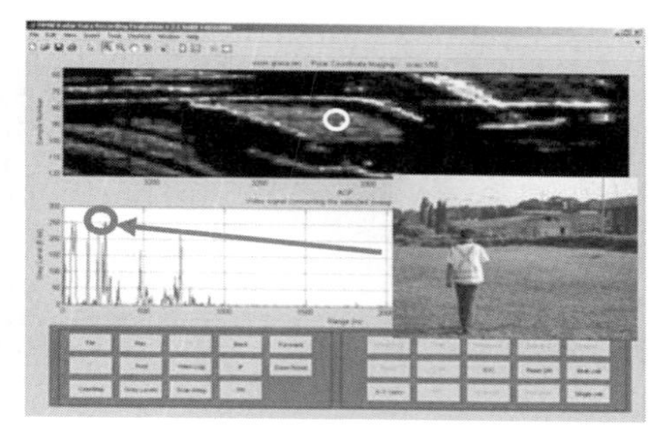

Fig. 2. High resolution rader MK2 test campaign.

This means that the radar can be used other than for **S**urface **M**ovement **S**urveillance (SMS) also for typical homeland security applications like:

Foreign **O**bjects Debris **D**etection (FOD)
Human **M**ovement **D**etection (HMD)

Typical returns from the field are shown in Fig. 2 for SMS, FOD and for HMD

Another field on which Oerlikon Contraves gained a lot of experience was in the development and test of a ground based W band point to point communication link. With this activity, some results relative to the atmospheric attenuation on ground have been found together with the demonstration of a FSK modulated communication link operating at 95 GHz.

Figure 3 shows the results of such experimentation.

Recently Oerlikon Contraves is active in the development and experimentation of a Limited Area W band FOD Warning Radar. The principle of working is shown in Fig. 4.

The radar is FMCW type and the adopted configuration for the prototype development is shown in Fig. 5.

Experimental views of the first prototype in operation are shown in Fig. 6.

In the space contest OCI participated to a lot of programs financed by ASI and ESA and is member of the "Moon base" activities.

In the contest of DAVID and WAVE programs OCI proposed for the respectively LEO and GEO high performance W band communication payloads a W/Ku Frequency conversion unit which scheme is shown in Fig. 7.

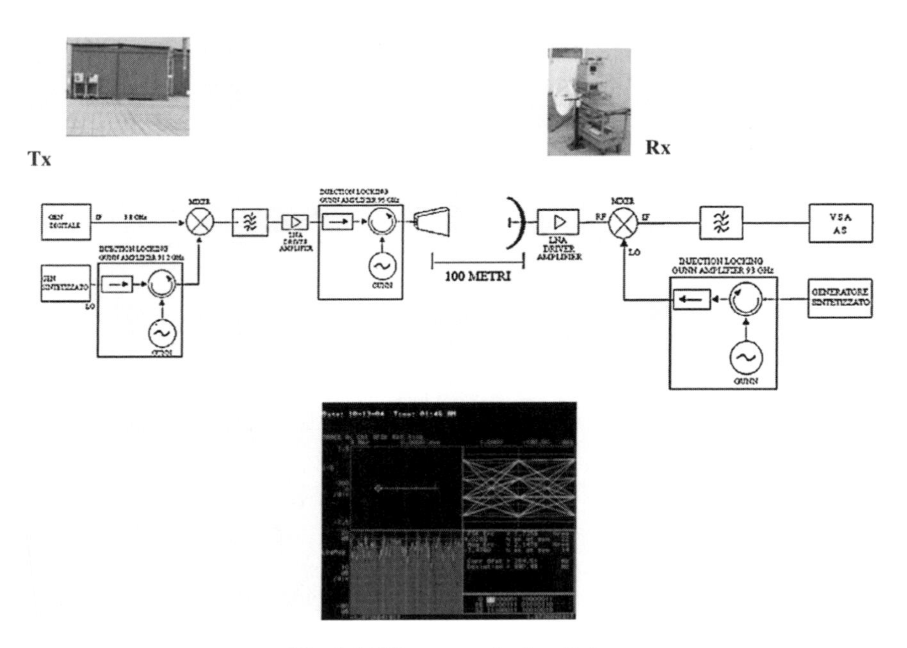

Fig. 3. FSK communication link.

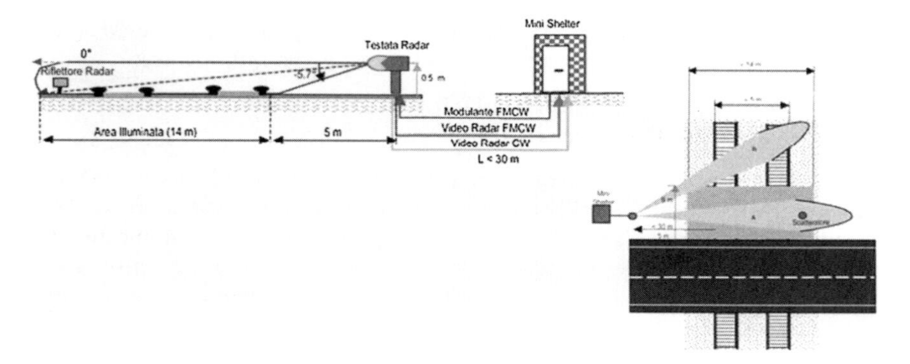

Fig. 4. Limited area W band FOD warning rader principle scheme.

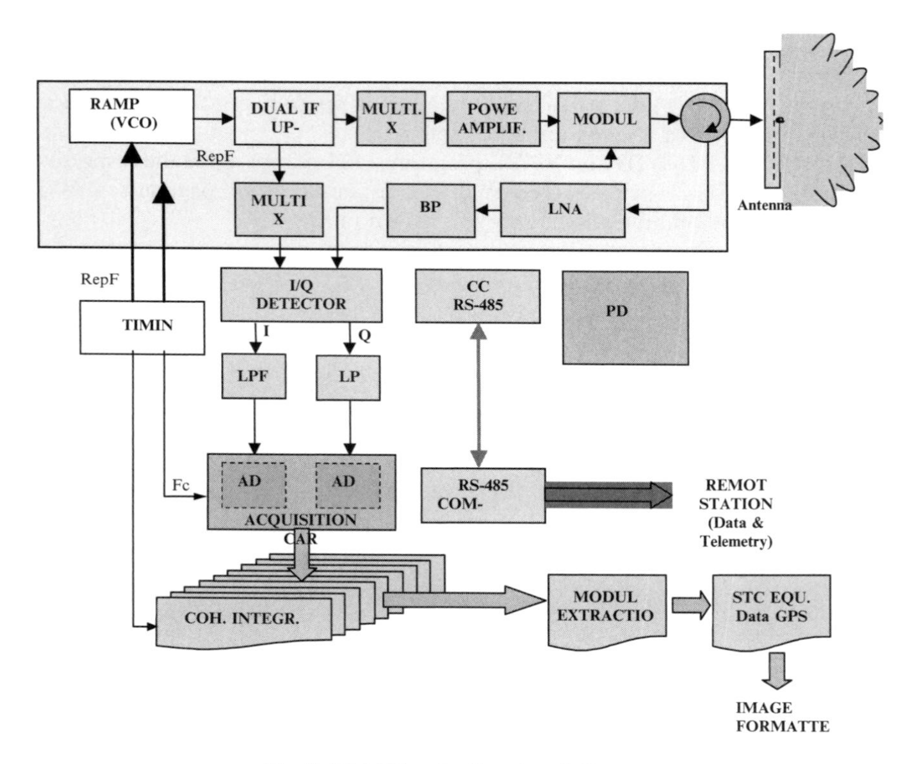

Fig. 5. FMCW rader functional diagram.

The system is considered fully redundant and an up/down conversion is foreseen for each channel. A traditional hybrid technology including low noise and medium power GaAs HEMT MMICs has been considered.

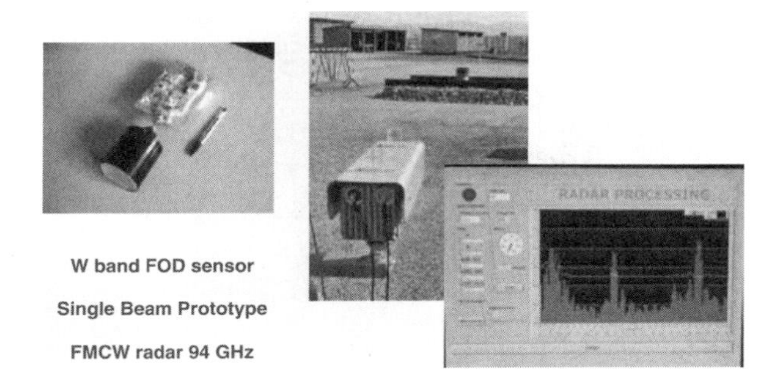

W band FOD sensor
Single Beam Prototype
FMCW radar 94 GHz

Fig. 6. FMCW rader testing results.

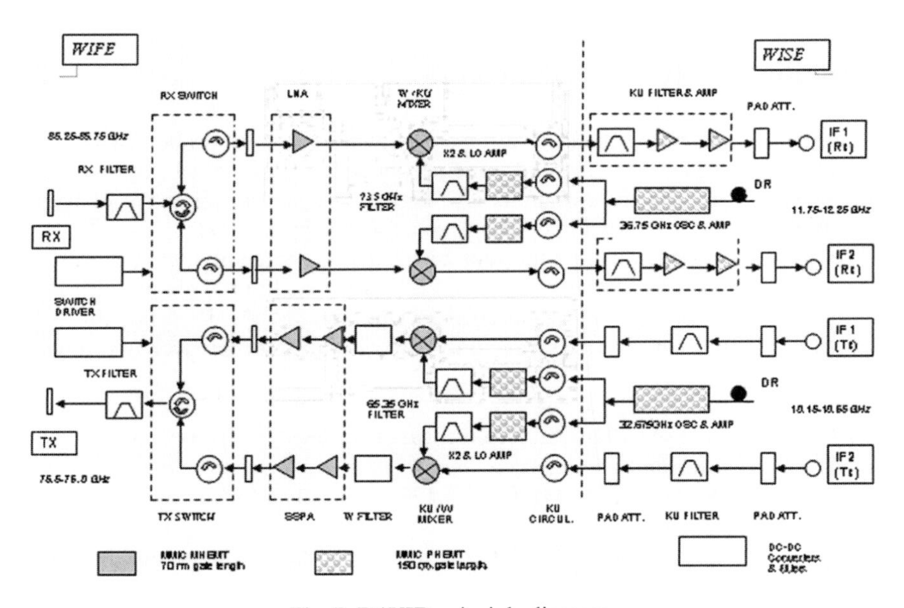

Fig. 7. DAVID principle diagram.

In order to get some additional information regarding the satellite to ground W band channel both for communication and for propagation(beacon), a mission with essential hardware mounted on a stratospheric vehicle (AEROWAVE) has been conceived.

The configuration of the AEROWAVE experiment is shown in Fig. 8.

Also if vacuum tubes, EIO, EIKA or TWT will be used for budget link reasons, monolithic medium power amplifies (MPA) have to be considered as drivers of the tubes itself.

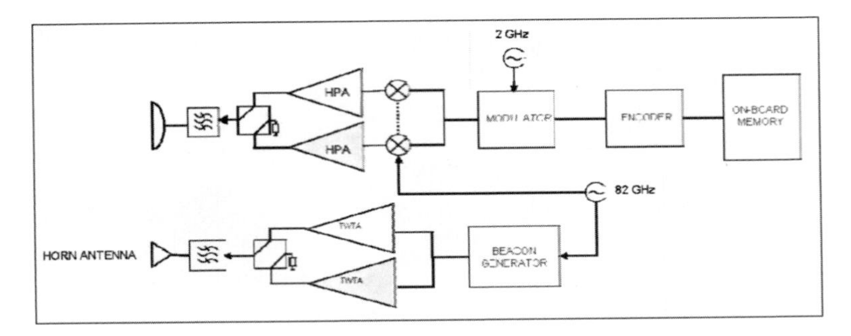

Fig. 8. Aerowave functional diagram.

4 Identification of "Core" Modules

In the block diagrams of the different systems experienced or studied by OCI (see Fig. 1, 5, 7, 8) have been highlighted some particular functional areas that have been studied in order to find the commonalities between them with the scope of finding a minimum number of "core" modules usable for any application.

The first step has been to find the state of the art of the technology in Europe for which concern the development of Low noise and Medium/High power MMIC amplifiers and Low noise MMIC oscillators operating in the W band. From this analysis has been found that in Europe only three foundries or research Institutes are active in this area:

- UMS(F): PH15 GaAs PHEMT 0,15 μm gate length for low noise and low power up to 77 GHz (automotive chipsets)
- OMMIC(F): D01MH GaAs MHEMT 0,130 μm gate length for low noise and low power up to 100 GHz
  D0071H E/D GaAs MHEMT 0,070 μm gate length for low noise and low power up to 160 GHz
- IAF(D): Fraunhofer Research Institute with public foundry GaAs MHEMT 0,070 μm gate length for low noise and low power up to 160 GHz
  GaAs PHEMT 0,150 μm gate length for medium power up to 110 GHz

Between the three IAF, seems to be the more advanced and completed offering both low noise and medium power amplifiers.

The best IAF available results are shown in Fig. 9 and Fig. 10 respectively for Low noise and for medium power MMICs

IAF is active also in the FMCW radar sensor and developed a complete chip shown in Fig. 11

Starting from this reality and working on the commonalities, four core modules (Figs. 12–15 and Tables 1–5) have been extracted in order to cover any need.

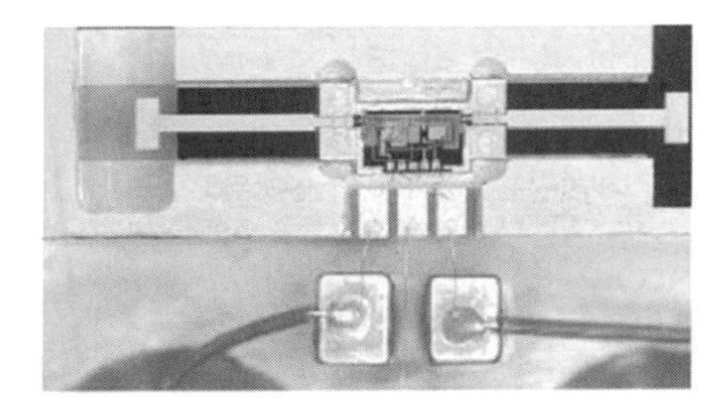

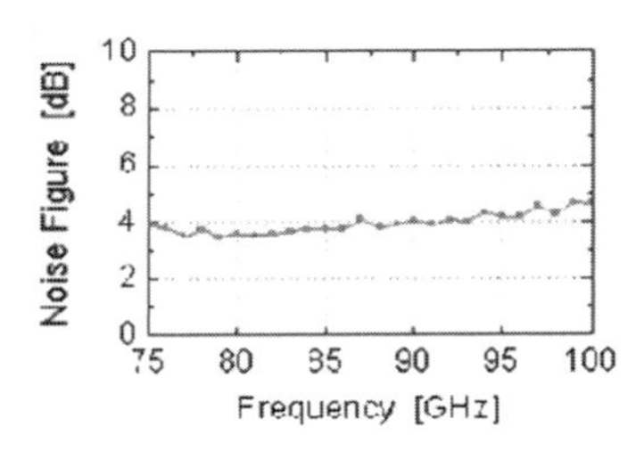

Fig. 9. IAF 94 GHz LNA.

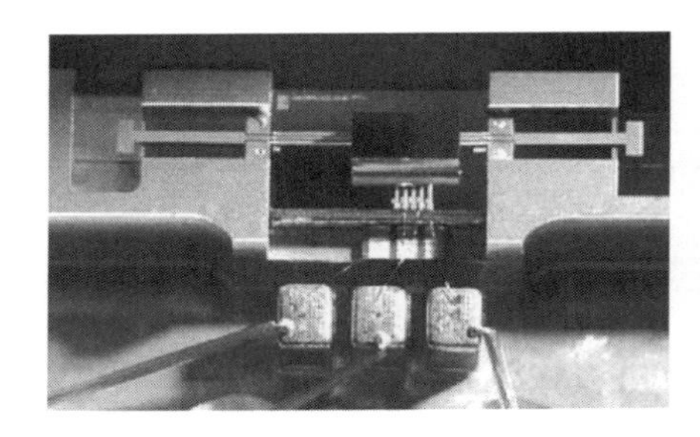

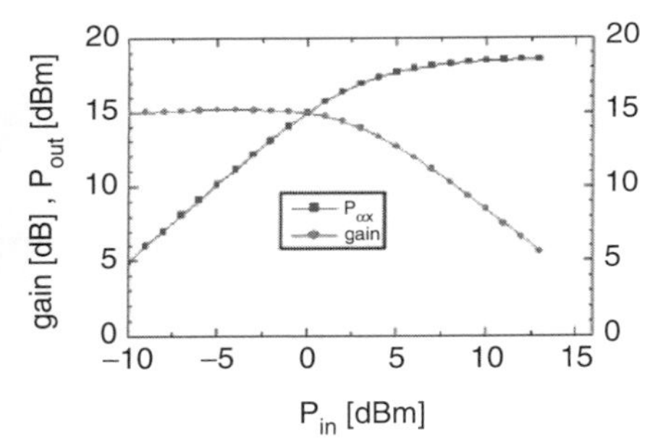

Fig. 10. IAF 94 GHz MPA.

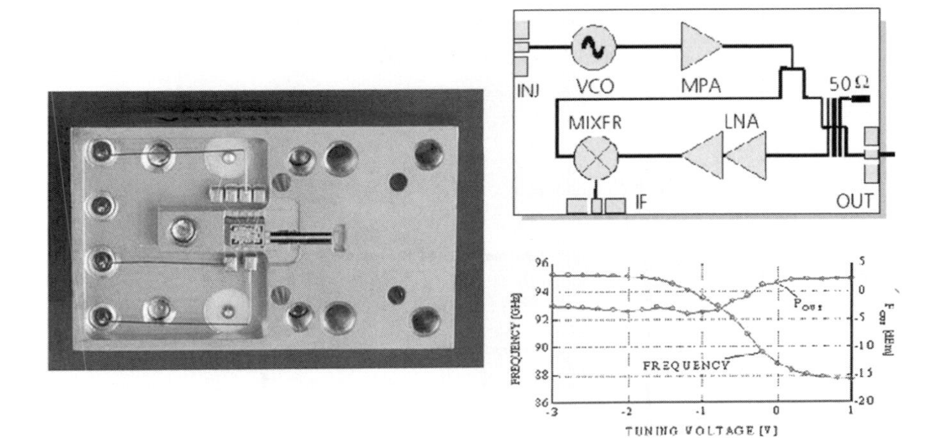

Fig. 11. IAF integrated FMCW.

UP CONVERTER CORE MODULE (UPC)

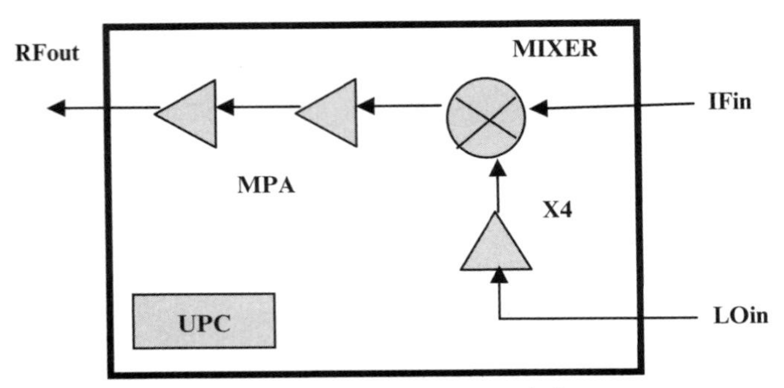

Fig. 12. UPC main characteristics.

DOWN CONVERTER CORE MODULE (DNC)

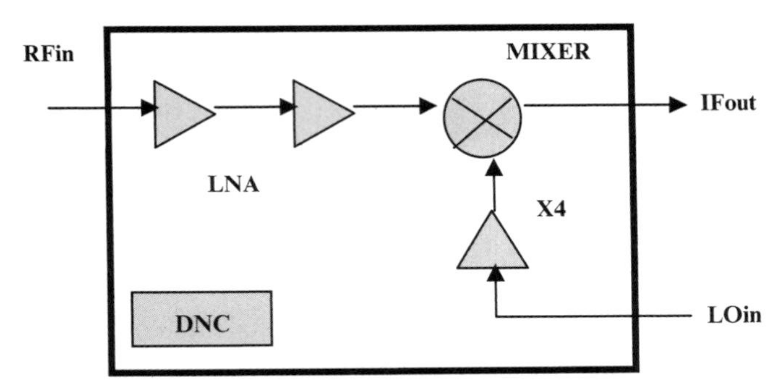

Fig. 13. UPC functional diagram.

MULTIPLIER CORE MODULE (MULT)

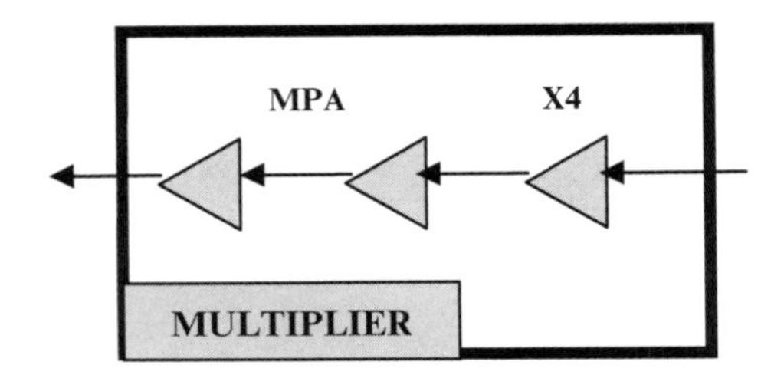

Fig. 14. MULT functional diagram.

FMCW CORE MODULE

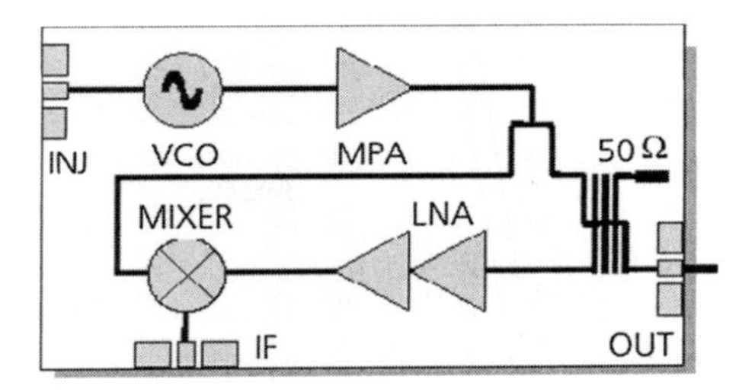

Fig. 15. FMCW functional scheme.

Table 1. UPC main characteristics.

	UP Converter (UPC)	
IF in	2÷10,5 GHz	
RF out	75÷96 GHz	
LO in	18÷24 GHz	
RF out Power	> + 15 dBm	Integrated and with transition to WR10
IF-RF Gain	20 dB	

Table 2. DNC main characteristics.

	Down Converter (DNC)	
RF in	75 ÷ 96 GHz	
LO in	18 ÷ 24 GHz	
IF out	1,5 ÷ 12,5 GHz	
NF	<4 dB	Integrated and with transition to WR10
RF-IF Gain	16 dB	

Table 3. MULT main characteristics.

	Multiplier	
Fin	18÷24 GHz	
F out	73,5÷96 GHz	
Mult. factor	4	
Pout	>15 dBm	Integrated and with transition to WR10

Table 4. FMCW main characteristics.

		FMCW
Fout	94÷96 GHz	
Pout	> +6 dBm	Integrated and with transition to WR10
Tx/Rx isolation	>25 dB	
NF rx	< 10 dB	Integrated and with transition to WR10
Sweeping bandwidth	≤ 200 MHz	
L(f)	−70 dBc /Hz at 100 KHz	

Table 5. Frequency synthesizer main characteristics.

	Frequency Synthesizer
Fout	18000÷24000 MHz
F out/step	40 MHz or less
POUT	>27 dBm
L(f)	< −95 dBc /Hz at 1 KHz −105 dBc/Hz at 100 KHz
Switching time	< 100 nanosec

5 Use of "Core" Modules

Based on these modules an exercise has been done to demonstrate that replacing with these the highlighted parts of the schemes in Figs. 1, 5, 7, 8 is possible to rearrange all the applications just implemented by OCI and to cover all the future demands coming from space and or from homeland security on ground.

The LO in (TX) and LO in (RX) are generated by the same synthesiser switching between two frequencies (Figs. 16–19 and Tables 6–9).

Another application that can be implemented with the same modules is a radiometer operating in W band (see Fig. 20)with the characteristics shown in Table10.

6 Conclusions

Based on the Oerlikon Contraves heritage in the W band gained in about twenty years of activity in different fields a certain number of low mass, small dimensions, low consumption "core" modules have been identified and specified in order to cover all the known possible demands coming from radar sensors, communication links and radiometer.

A demonstration of this is given.

This approach will reduce the development costs of the hardware because all the efforts can be concentrated only in few items that can be used as part of a more complex puzzle.

The system architecture shall be focused on the use of such blocks and will be personalised depending on the application.

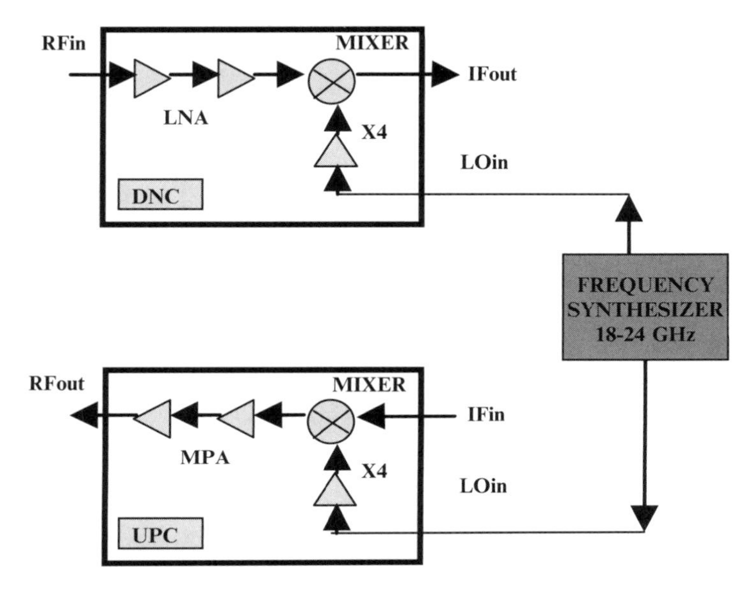

Fig. 16. Wave with core modules functional scheme.

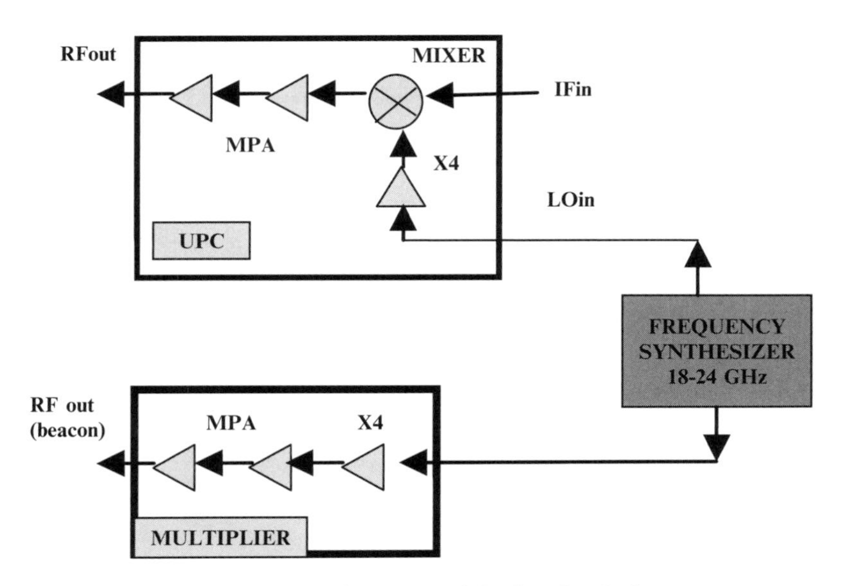

Fig. 17. Aerowave with core modules functional scheme.

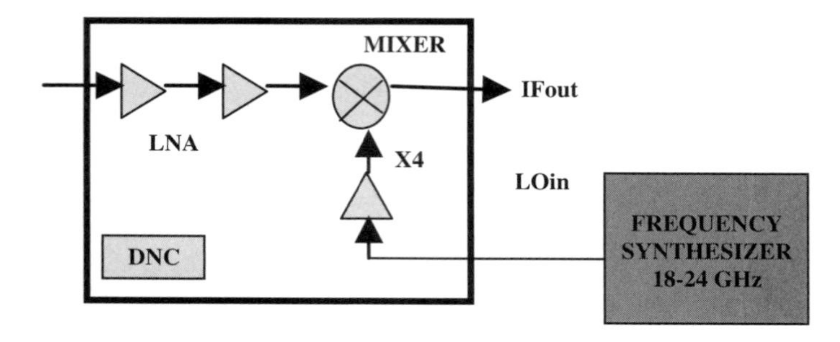

Fig. 18. Smart HRR MK2 with core modules functional scheme.

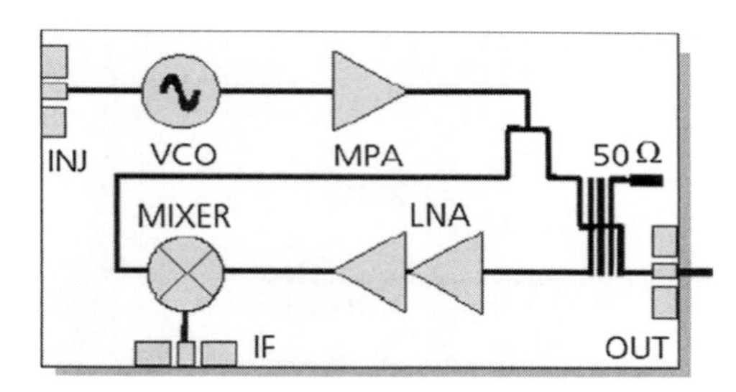

Fig. 19. FOD warning rader with core modulees functional scheme.

Table 6. Wave with core modules main characteristics.

	Wave
RFin (RX)	85500 ± 250 MHz
IF out (Rx)	12000 ± 250 MHz
IF in (TX)	10400 ± 250 MHz
RF out (TX)	75750 ± 250 MHz
POUT (TX)	>15 dBm
NF(Rx)	< 4 dB
LO in (RX)	18380 MHz
Loin (TX)	21540 MHz

Table 7. Wave with core modules main characteristics.

Aerowave		
	IF in (TX)	2000 MHz
	RF out (TX)	84000 MHz
	POUT (TX)	>15 dBm
	LO in (TX)	20500 MHz
	Loin (beacon)	20500 MHz
	RFout (beacon)	82000 MHz
	POUT(beacon)	>+15 dBm

Table 8. Smart HRR MK2 with core modules main characteristics.

	Smart MK2
RF in (RX)	94000–96000 MHz
IF out (RX)	1500 MHz
NF (RX)	<4 dB
LO in	23130–23630 MHz
Lo in step	40 MHz

Table 9. FOD warning rader with core modules functional characteristics.

	W band FOD warning radar
RF out (TX)	94000–96000 MHz
POUT (TX)	>+6 dBm
NF (RX)	<10 dB
VCO sweep BW	200 MHz

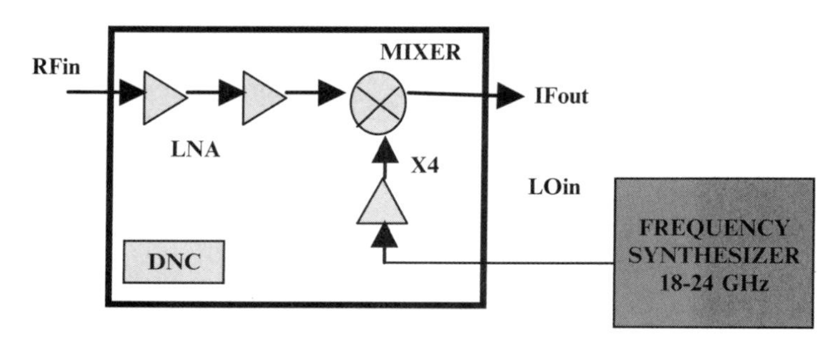

Fig. 20. W band radiometer with core modules functional scheme.

Table 10. W band radiometer with core modules main characteristics.

Radiometer		
	RF in (RX)	75000–96000 MHz
	IF out (RX)	0–250 MHz
	NF (RX)	<4 dB
	LO in	18750–24000 MHz

The use of such miniaturised modules will be fundamental for applications outside the atmosphere for the exploration of Moon and /or Mars.

In particular it can be possible to think to very miniaturised multifunctional payloads to be placed for example on a small rover moving on the Moon surface.

A shared compact hardware can be thought for the guidance of the rover and for communicating big amounts of data to a lunar base station ot to an or-biter rounding around the moon itself.

GNSS Bit-True Signal Simulator
A Test Bed for Receivers and Applications

C. Cosenza, Q. Morante, S. Corvo, F. Gottifredi

Alcatel Alenia Space Italia S.p.A./B.U. Navigation & Integrated Communications
Via Saccomuro 24, 00131, Roma, Italia
Phone: +39 06 4151.2082, Fax: +39 06 4191287,
e-mail: Franco.Gottifredi@alcatelaleniaspace.com

Abstract. This work aims to provide the status and the outcomes of the GNSS Bit-True Signal Simulator (GBTSS) software that is developed, and now it is under completion and optimization, in the Navigation & Integrated Communications Business Unit of Alcatel Alenia Space Italia. This simulator provides the I & Q samples of any GNSS signal as seen at the correlators input of a GNSS receiver.

1 Needs and Applications

Currently the Galileo Navigation System development is on-going under an ESA/GJU contract but the Galileo Signal-In-Space (SIS), although the launch of the experimental GIOVE Satellites, is not yet completely frozen (see also the studies on a Modified BOC modulation proposed by the US/EU SIS working group 0). In parallel to these system engineering activities, several applications are under study and prototyping involving different types of receivers. These two aspects, system engineering activities and receivers design, are strictly correlated by their common physical link: the Navigation Signal.

From the system engineering point of view, the interest on the navigation signal is focused on its performance analysis, on its possible evolutions, on the interoperability with GPS and on its co-existence with other signals in the same bands (i.e. Radars, DME, etc. . .).

From the receivers design point of view, the interest is on the applications in their different domains that drive the receiver architecture also in terms of received signals and bands.

In order to overcome this needs coming from both system engineering and receiver design, it is necessary to have a flexible and configurable facility able to generate any type of navigation signal.

In this paper it is presented a possible solution to meet these needs.

In particular in the following section will be presented a GNSS signal simulator named GBTSS (GNSS Bit-True Signal Simulator) under development at the B.U. Navigation and Integrated Comms of the Alcatel Alenia Space Italia S.p.A. The GBTSS is a facility conceived to generate the I & Q samples of any kind of Navigation Signal; in fact the tool guarantees the flexibility and the configurability so as to give at any user the possibility to reproduce the required GNSS signal.

The architectural design and the current GBTSS status will be described; in the last section, simulation results and the way forward of this project will be provided.

2 System Architecture

Starting from the needs, it is clear that the main functionality of the GBTSS is to be able to generate any type of Navigation taking into account all the effects impacting the signal path from the satellite to the ADC output of the receiver. In order to do that, the architectural choice is of fundamental importance. The solution chosen, after several analyses and trade-off, is to digitally reconstruct the signal path starting from the A/D Converter output (e.g. taking into account a configurable sampling rate) and introducing all the effects from the local errors (i.e. receiver noise, multipath and interference) up to the satellite-user relative (e.g. Doppler).

Figure 1 provides a flow description of the GBTSS Simulated Signal.

How it can be noted from the figure, the simulator is composed of three main blocks:

- the Signal Generator block
- the Environment block
- the Receiver and Local Error block

The **Signal Generator block** is composed by:

- *Satellite* module, that contains all the information about the satellite. In particular, using an Orbit propagation module, the satellite position and velocity are computed.
- *User Receiver* module, that computes the receiver position and velocity, using an Orbit propagator.
- *Kinematics* module, that provides the Doppler components taking into account the satellite and the receiver position and velocity.
- *Signal in Space* module. Inputs to this block are the Doppler components and all the signal characteristics (like the transmitted bandwidth or the carrier frequency) and the output is the generation of the selected signal in time domain.

The **Environment block** is composed by:

- *Elevation and Azimuth* module, that provides the satellite elevation angle and azimuth taking into account the satellite and the final user position and velocity information.
- *Ionosphere* module, that provides the signal delay and the phase variation on the satellite elevation angle and azimuth information.
- *Troposphere* module. This block provides the signal attenuation and delay on the satellite elevation angle and azimuth information.
- *Free Space Propagation* module. This block computes the signal propagation time between the satellite and the receiver.

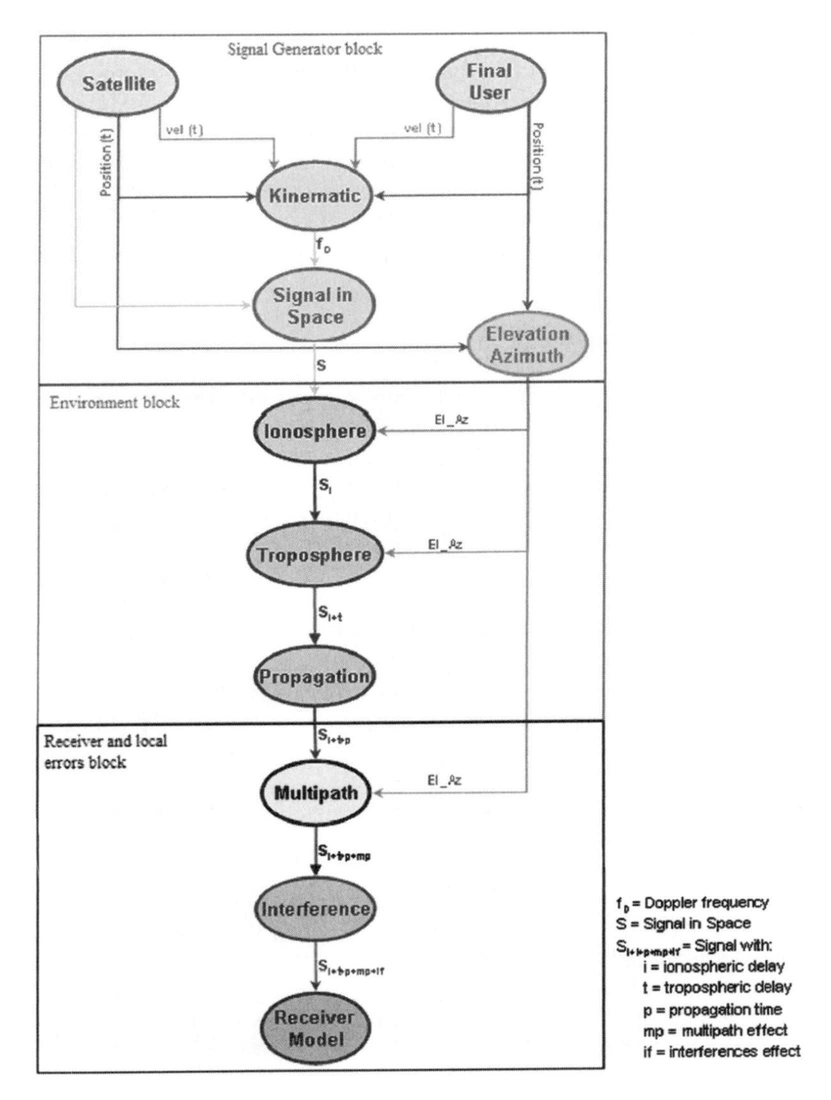

Fig. 1. GNSS signal state flow.

The **Receiver and Local Error block** is composed by:

- *Multipath* module, starting from the signal affected by the environmental effects previously described and the satellite elevation angle and azimuth information, adds to the "real" signal the multipath signal.
- *Interference* module, it represents the possible impact of the external signal to the receiver, like the L-band primary radar signals.

- *User Receiver* module, it adds to the signal, the Gaussian noise created as a random vector of numbers with variance equal to σ. The vector composed by two rows, one for the in-phase component and another one for the quadrature. Since the two components are generated separated to guarantee that both I and Q channels will be statistically independent. Noise is thus added separately for the in-phase and quadrature channels of the received signal. Plus, the *User Receiver* module is able to implement any possible impact of the receiver model, from the antenna input to the A/D Converter level.

2.1 GBTSS Implementation

The GBTSS software aims to simulate any signal type and its navigation path from the satellite to the user receiver and provides the I & Q samples of any Navigation signal at the correlators input of a GNSS receiver.

In order to simulate any signal type, the GBTSS will implement all the modulation schemes and in particular the following Galileo modulations, in addition to the GPS BPSK one, are foreseen:

- The CEM modulation (Coherent Envelope Modulation) modulates the central frequency so to create 3 channels: A, B, and C that operate at the same signal central frequency.
- The BOC(f_S,f_C) modulation (Binary Offset Carrier) (f_S is the subcarrier frequency and f_C is the code frequency) modulates the signal with a rectangular sub-carrier [2]:

$$S(t) = s(t) \cdot sign(\cos(2\pi \cdot f_s \cdot t)) \tag{0}$$

- The AltBOC(f_S,f_C) modulation modulates the signal with a complex rectangular sub-carrier [2]:

$$S(t) = s(t) \cdot sign(\cos(2\pi \cdot f_S \cdot t) + j \cdot \sin(2\pi \cdot f_S \cdot t)) \tag{1}$$

- The BPSK modulation (Binary Phase Shift Keying), that is a particular case of M-PSK modulation with two level (M = 2).

As previous described, the ideal signal is generated by the Signal Generator block in the time domain and passed throw to the Environment block that provides to add the environmental errors.

The *Free Space module* simulates the delay due to free space of a signal transmitted from a satellite to a receiver, that include the corrections for the relativistic Sagnac and Orbit Eccentricity effects.

The *Ionospheric module*, implements the delay due to the free electrons and positively charged ions produced by the solar radiation, Δt_{iono} also called Ionospheric Slant Delay Δs that is a function of the Total Electron Content (TEC).

The GBTSS implements the NeQuick model for the TEC determination that is the model used by the Galileo system.

The NeQuick model is an ionospheric electron density model able to give the electron concentration distribution on both the bottom side and topside of the ionosphere and it is a quick-run model particularly tailored for trans-ionospheric applications.

To guarantee a major flexibility and usability of the GBTSS for both the Galileo and the GPS system, the simulator also implements the Klobuchar model, that is used for the GPS ionospheric effect calculation.

The Klobuchar ionospheric model uses a half cosine to represent the diurnal variation of TEC in the single-frequency user algorithm.

Another deviation from the vacuum speed of light is caused by the *Troposphere*. Variations in temperature, pressure and humidity contribute to the variations of the speed of light and consequently of the radio waves. The total troposheric delay is given by:

$$\Delta t_{totalTropo} = \Delta t_{dryTropo}\, m_{dry}(E) + \Delta t_{wetTropo}\, m_{wet}(E) + (\Delta t_{noise} + \sigma_{noise} \bullet RAN) \quad (2)$$

where $\Delta t_{dryTropo}$, $\Delta t_{wetTropo}$ are the Dry and Wet Tropospheric Zenith Delays [s] and

$$m_{dry}(E) = \frac{1}{\sin\sqrt{E^2 + 6.25}} \quad \text{and} \quad m_{wet}(E) = \frac{1}{\sin\sqrt{E^2 + 2.25}} \quad (3)$$

are the dry and wet delay Mapping Functions for an Elevation Angle E [rad]. The Noise Mean Deviation Δt_{noise} and the Noise Standard Deviation σ_{noise} represent the nominal tropospheric noise parameters that are meant to represent errors due to uncertainty in meteorological situations. These parameters may be overwritten by user-settable Feared-Event values when the tropospheric Feared Event is activated. The random number RAN is a time-correlated Gaussian function [3].

In literature, there are several model to calculate the tropospheric effects:

- Saastamoinen Total Delay Model
- Hopfield Two Quartic Model
- Black and Eisner (B&E) Model
- Water Vapor Zenith Delay Model – Berman
- Davis, Chao, and Marini Mapping Functions
- Altshuler and Kalaghan Delay Model

Both GPS and Galileo system use the Saastamoinen model for the Tropospheric correction.

In the Saastamoinen model, the dry pressure is modelled using the constant lapse rate model for the troposphere and an isothermal model above the tropopause. The vertical gradient of temperature is $T = T_0 + \beta(r - r_0)$, and the resulting pressure profile is $P = P_0\left(T/T_0\right)^{\frac{-Mg}{R\beta}}$ where r is the radius from the Earth centre $(r = (R_e + h))$ and r_0 is the user radius (usually $r_0 = R_e$, the Earth radius), and T_0 is the user temperature. The radius r ranges in value from r_0 to r_T which represents the radius to the tropopause. The corresponding dry refractivity is then $n - 1 = (n_0 - 1)\left(T/T_0\right)^{\mu}$ where $\mu = -M/R\beta - 1$ is a constant exponent. Using an isothermal model above the tropopause, the pressure drops exponentially from its initial value at the tropopause P_T:

$$P = P_T \exp\left[-\frac{gM}{RT_T}(h - h_T)\right] \quad (4)$$

where the subscript T refers to the values at the tropopause.

The wet refraction is dependent on the partial pressure e, which decreases in somewhat the same way as total pressure in the troposphere although much more rapidly.

The Saastamoinen model provide, for elevation angles $E \geq 10$ deg, a delay correction:

$$\Delta = 0.002277\,(1+D)\sec\left(\psi_0\left[P_0 + \left(\frac{1255}{T_0} + 0.005\right)e_0 - B\tan^2\psi_0\right]\right) + \delta_R \qquad (5)$$

where Δ is the delay correction in meters; P_0, e_0 are in millibars and T_0 is in °K. The correction terms B and δ_R are given in Table 1 for various user heights h. The apparent zenith angle $\psi_0 = 90$ deg $- E$. The value of D is $D = 0.0026 \cos 2\phi + 0.00028h$, where ϕ is the local latitude and h is the station height in km [4].

The *Total Environment block* takes in input:

- The signal in space;
- The satellite elevation angle and azimuth information;
- The final user elevation angle and azimuth information;

and calculates the Total Environment Delay of the signal, from its time of emission to time of reception, by summing the Free Space Delay, the Ionospheric Delay and the Tropospheric Delay for each broadcast frequency.

The Total Environment Delay is defined as:

$$\Delta t_{totalEnv} = \Delta t_{freeSpace} + \Delta t_{iono} + \Delta t_{totalTropo} \qquad (6)$$

After this phase, the software passes the delayed signal to the *Receiver and Local Error Block* that adds to the signal the local and the user receiver chain errors (multipath, interferences, etc.).

Table 1. Correction terms for saastamoinen model.

	Apparent Zenith	Station height above sea level							
	Angle	0 km	0.5 km	1 km	1.5 km	2 km	3 km	4 km	5 km
	60 deg 00 min	0.003	0.003	0.002	0.002	0.002	0.002	0.001	0.001
	66 deg 00 min	0.006	0.006	0.005	0.005	0.004	0.003	0.003	0.002
	70 deg 00 min	0.012	0.011	0.010	0.009	0.008	0.006	0.005	0.004
	73 deg 00 min	0.020	0.018	0.017	0.015	0.013	0.011	0.009	0.007
	75 deg 00 min	0.031	0.028	0.025	0.023	0.021	0.017	0.014	0.011
	76 deg 00 min	0.039	0.035	0.032	0.029	0.026	0.021	0.017	0.014
δ_R	77 deg 00 min	0.050	0.045	0.041	0.037	0.033	0.027	0.022	0.018
	78 deg 00 min	0.065	0.059	0.054	0.049	0.044	0.036	0.030	0.024
	78 deg 30 min	0.075	0.068	0.062	0.056	0.051	0.042	0.034	0.028
	79 deg 00 min	0.087	0.079	0.072	0.065	0.059	0.049	0.040	0.033
	79 deg 30 min	0.102	0.093	0.085	0.077	0.070	0.058	0.047	0.039
	79 deg 45 min	0.111	0.101	0.092	0.083	0.076	0.063	0.052	0.043
	80 deg 00 min	0.121	0.110	0.100	0.091	0.083	0.068	0.056	0.047
	B [mbar]	*1.156*	*1.079*	*1.006*	*0.938*	*0.874*	*0.757*	*0.654*	*0.563*

Multipath is the phenomenon whereby a signal arrives at a receiver via multiple paths attributable to reflection and diffraction. Multipath represents the dominant error source in satellite-based precision guidance system.

Multipath distorts the signal modulation and degrades accuracy in conventional and differential systems. Multipath also distorts the phase of the carrier, and hence degrades the accuracy of the interferometric systems. For standard code-based differential system, signal degradation attributable to multipath can be severe. This stems from the fact that multipath is a highly localized phenomenon.

A possible parameterization to simulate the effect of multipath reflections on the measured range, phase and Doppler shift, is follow described. The main input to the algorithm is the transmitter elevation and the main outputs are the multipath range, phase and Doppler errors.

The range error E_{mr} (in meters), phase error E_{mp} (in meters) and Doppler frequency error E_{mf} (in Hz) due to multipath can be computed as follows [3]:

$$E_{mr} = A_{br} + A_r K_{env} K_{rec} K_{ran} \cos(\omega Elev) \tag{7}$$

$$E_{mp} = A_{bp} + A_p K_{env} K_{rec} K_{ran} \cos(\omega Elev) \tag{8}$$

$$E_{mf} = A_{bf} + A_f K_{env} K_{rec} K_{ran} \cos(\omega Elev) \tag{9}$$

where A_r, A_p and A_f are the amplitude of the multipath effect, A_{br}, A_{bp} and A_{bf} are residual multipath bias terms, K_{rec} is the receiver sensitivity factor (0 to 1), ω is the multipath frequency and *Elev* is the elevation angle of the transmitter relative to the receiver.

K_{env} depends on the receiver environment characteristics. Table 2 some of the values that can be used:

K_{ran} is a time-correlated Gaussian distribution noise of mean standard deviation of σ, Correlation time τ and Correlation Enable state.

Through most the above parameters can be user-settable, some representative default values are $A_r = 2\ m$, $A_p = 0.2\ m$, $A_f = 1 Hz$, $K_{rec} = 1$, $\omega = 0.8 cycle / \deg ree$. The residual biases is zero by default.

The GBTSS implements this phenomena as described in the previous equations.

The navigation systems, GPS/EGNOS and Galileo, uses the same frequency band of the primary surveillance radar for Air Traffic Control and DME systems so it is possible to have interference effects coming from this radar (Fig. 2).

Table 2. Relations between the kenv values and the environment characteristics.

$K_{env} = 0.0$	None
$K_{env} = 0.1$	Rural
$K_{env} = 0.2$	Fly
$K_{env} = 0.3$	Airport
$K_{env} = 0.4$	Harbour
$K_{env} = 0.5$	Sail
$K_{env} = 0.6$	SubUrban
$K_{env} = 0.7$	Urban
$K_{env} = 0.8$	Excessive

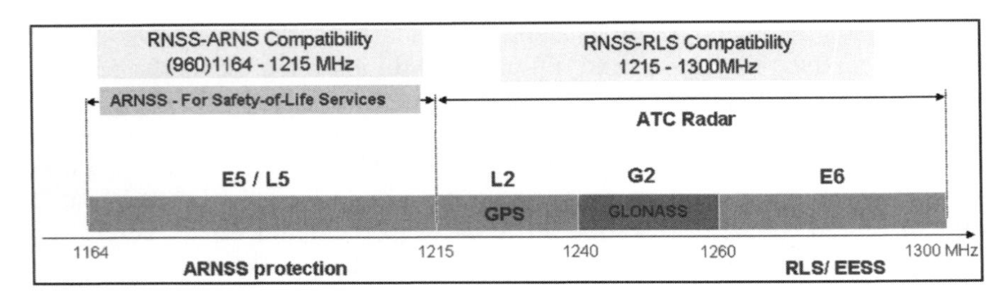

Fig. 2. Allocation frequency.

For example, in Italy, the primary radars that work in L band are the ATCR-44K and the ATCR-44S.

The ATCR 44S is a radar with a solid state power amplifier that employs two coded and different waveforms for the short and long distance coverage.

This radar has two impulse types (Fig. 3):

- short impulse of 32 μs for short distance coverage
- long impulse of 150 μs for long distance coverage

Both impulses are compress at 2.8 μs in order to employ the same receiver chain.

Another possible interference effect can be provided by the DME/TACAN systems.

DME and TACAN are pulse-ranging navigation systems that operate in the 960-1215 MHz frequency band. DME systems provide distance measurement for aircraft; TACAN, a military navigation system, provides both azimuth and distance information. DMEs and TACAN operate in four modes (X, Y, W, Z) as shown in Fig. 4. The DME/TACAN navigation system comprises airborne interrogators and ground-based transponders. The interrogators on board aircraft transmits pulse pairs on one of 126 frequencies with 1 MHz spacing. The ground transponders are usually able to transmit up to 2700 pulses pairs per second (ppps) for DME and 3600 ppps for TACAN, though some DME transponders are capable of transmitting 5400 ppps. However, in normal operation, these rates are only encountered in peak traffic situations.

The DME and TACAN systems operate by transmission from the aircraft of a pair of pulses on the nominal frequency of a visible ground-based transponder, with nominal duration 3.5 microseconds separated by 12 microseconds, as shown in Fig. 5.

The ground-based transponder receives, frequency converts and re-transmits these pulse pairs towards the aircraft receiver, which is able to measurement the delay between transmission and reception and hence calculate the distance to the DME transponder [5].

In the GBTSS the *Interference module* these and other similar effects are modelled.

At the Receiver Level, the signal is down-converted to an intermediate frequency and, after the A/D conversion and the filtering process, it is possible to compute the I & Q samples.

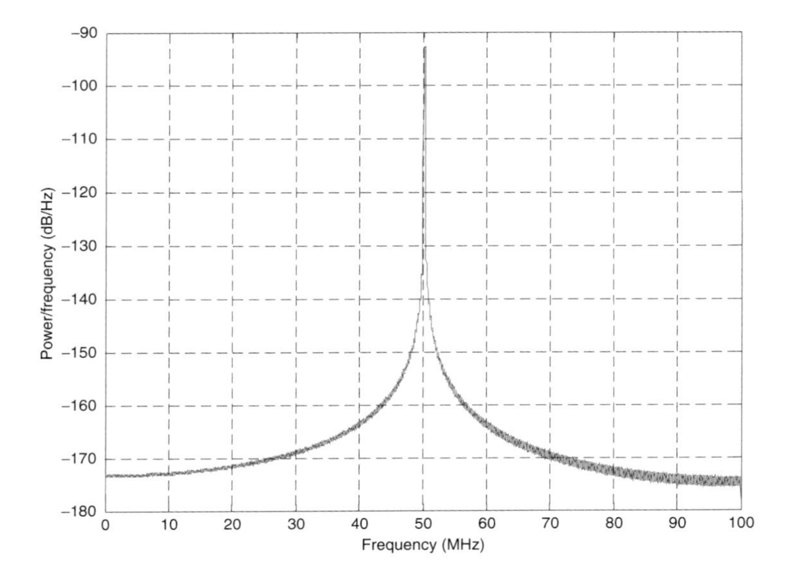

Fig. 3. Long pulse PSD.

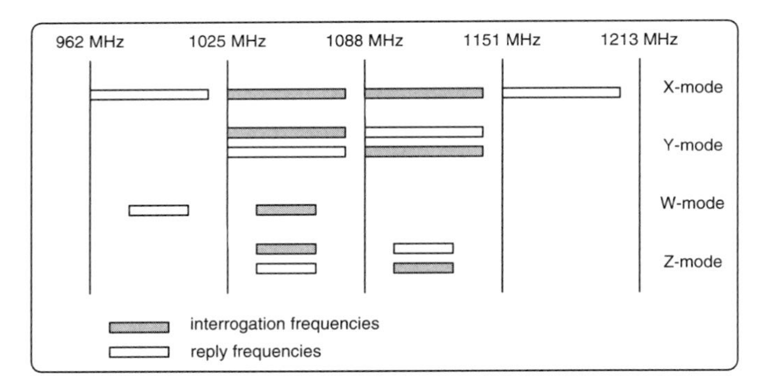

Fig. 4. Standard DME/TACAN channel plain.

The *User Receiver module* contains a Generic Receiver front-end model which receives the GNSS signals from the visible satellites and computes the observables.

Range measurements are simulated by adding measurement errors to the range information provided by the Environment model. These measurement errors are function of noise and multipath effects.

Phase measurements, corrupted by error terms like the clock drift rate and noise, are simulated by integrating the Doppler frequency.

The clocks are modelled as a specific offset relative to the reference time of the segment containing the clock. The reference times are: International Atomic Time, Universal Time Coordinated, Galileo System Time or GPS Time.

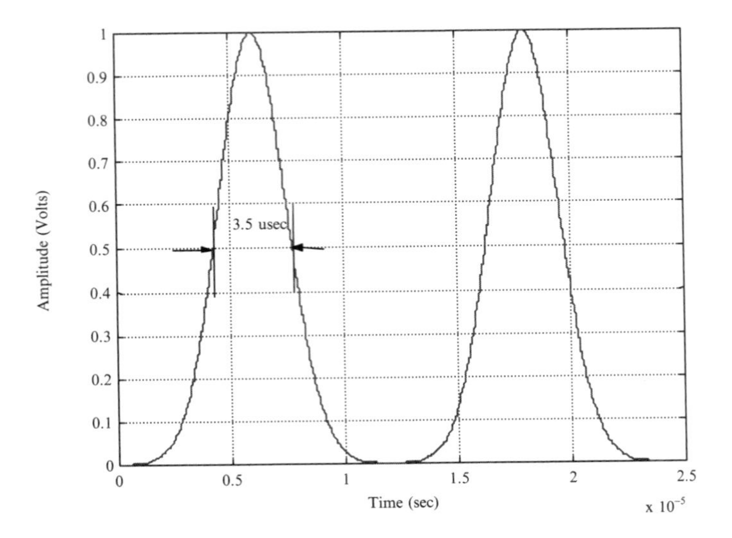

Fig. 5. Ideal gaussian DME pulse-pair.

The clock offset includes:

- Systematic clock error (bias, drift and acceleration)
- Clock drift correction commands
- Thermal noise errors
- Severe drift (feared event)

There are generally five sources of clock noises that can be presented:

- Random walk on frequency
- White noise on frequency
- Flicker noise on frequency
- White noise on phase
- Flicker noise on phase

Random walk is the integration of white noise. It is time-correlated and generates a decreasing power spectral density proportional to 1/f (f is the frequency).

White noise is typically a time independent random sequence with Gaussian distribution. It generates a constant power spectral density (same energy at all frequency, therefore "white").

Flicker noise is a time correlated random sequence also generating a decreasing power spectral density proportional to 1/f (less so than for random walk). More energy at the low frequencies gives more prominent slow varying effects and less prominent fast changing effects [3].

The clock offset error is defined as:

$$\Delta t_{clockDrift} = \Delta t_{clockDrift,\,rx} - \Delta t_{clockDrift,\,sv} \tag{10}$$

where

$\Delta t_{clockDrift,\,rx}$ is the Receiver Clock Drift Error [s]

$\Delta t_{clockDrift,\,sv}$ is the Spacecraft Vehicle Clock Drift Error [s].

3 GBTSS Vallidation Status

Each block, previously described, it will be separately validated.

At the moment it has been implemented the *Signal in Space* block, without considering the Doppler components, and the *Environment* block. This blocks have been implemented in Matlab language and then translated in C language. Particularly it has been simulated the three bands of the Galileo signals: E5 (1191.795 MHz), E6 (1278.750 MHz), L1 (1575.420 MHz).

The Galileo E5 signal is modulated in accordance to the AltBOC(10,15) modulation scheme as reported in Fig. 6:

The AltBOC signal constellation and the baseband power spectral density are showed in Fig. 6.

Other simulation results are showed in the Fig. 7. In particular, the outcomes of Constant Envelope Modulation (CEM) scheme, used in the Galileo E6 signal, is plotted at the constellation level, in Fig. 7a, and at the Power Spectral Level, in Fig. 7b.

The CEM scheme uses a suitable input channel combination in order to have an constant output power level, in particular for the chosen multiplexing scheme (CEM) the constant envelope is maintained by adding to the desired channels A, B and C an additional signal, which is the product of all desired binary signals.

The CEM scheme is also employed in the Galileo L1 signal, the only difference, between L1 and E6, lies in the channel input combination. An BOC(1,1) modulation scheme is used in the L1 Galileo signal for both the A, B CEM channel inputs. The GBTSS output is reported in Fig. 8.

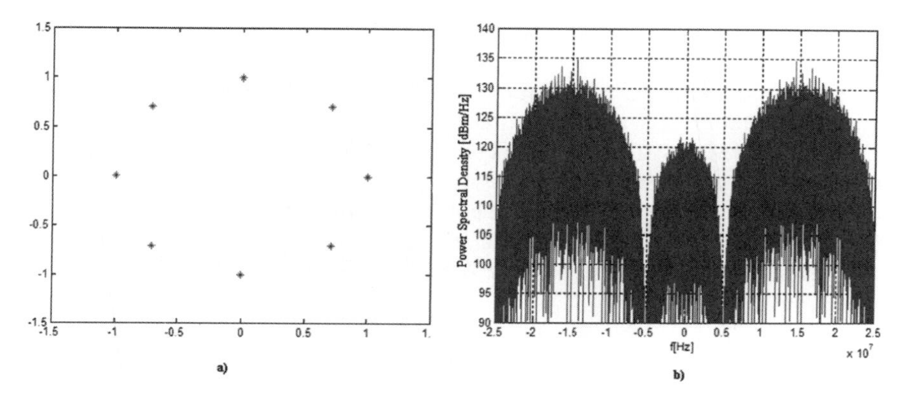

Fig. 6. a) Galileo E5 AltBOC signal constellation; b) Galileo E5 power spectral density.

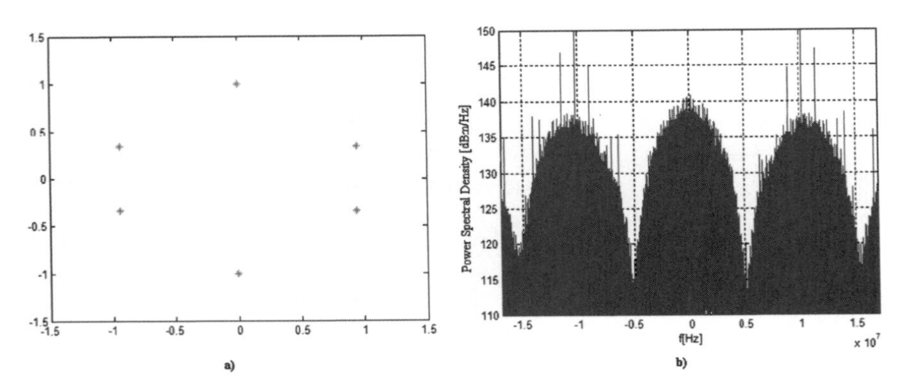

Fig. 7. a) Galileo E6 signal constellation; b) Galileo E6 Power spectral density.

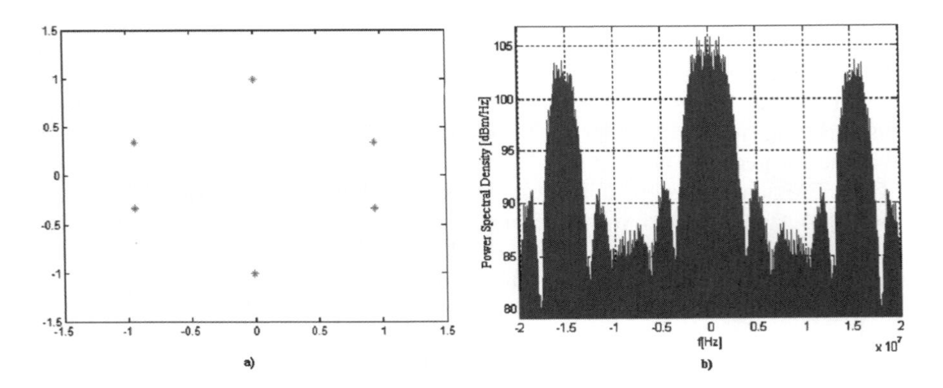

Fig. 8. a) Galileo L1 signal constellation; b) Galileo L1 power spectral density.

The described signals are sent from the satellite to the user receiver through the atmospheric that induces environmental effects (i.e.: delay) on the signals; in particular in this phase is utilized the Klobuchar model for the ionospheric effects and the Saastamoinen for the tropospheric ones.

In the Table 3 are reported the generation parameters utilized to compute the delay introduced in the satellite channel; in particular in Fig. 9 can be noted the delay effects on the sub-carrier of the BOC Galileo E6 signal. The corresponding delay values are:

- Ionospheric delay $\cong$ 0.05 μs (corresponding to an error of about 15 m);
- Tropospheric delay $\cong$ 0.02 μs (corresponding to an error of about 7 m);
- Propagation delay $\cong$ 0.09 s.

Table 3. Inputs value for atmosphere effects determination.

Receiver geodetic latitude [deg]	41
Receiver geodetic longitude [deg]	12
Satellite geodetic azimuth [deg]	210
Year	2006
Month	2
Day	1
Hour	9
Minuts	35
Seconds	0
Satellite elevation angle [deg]	20

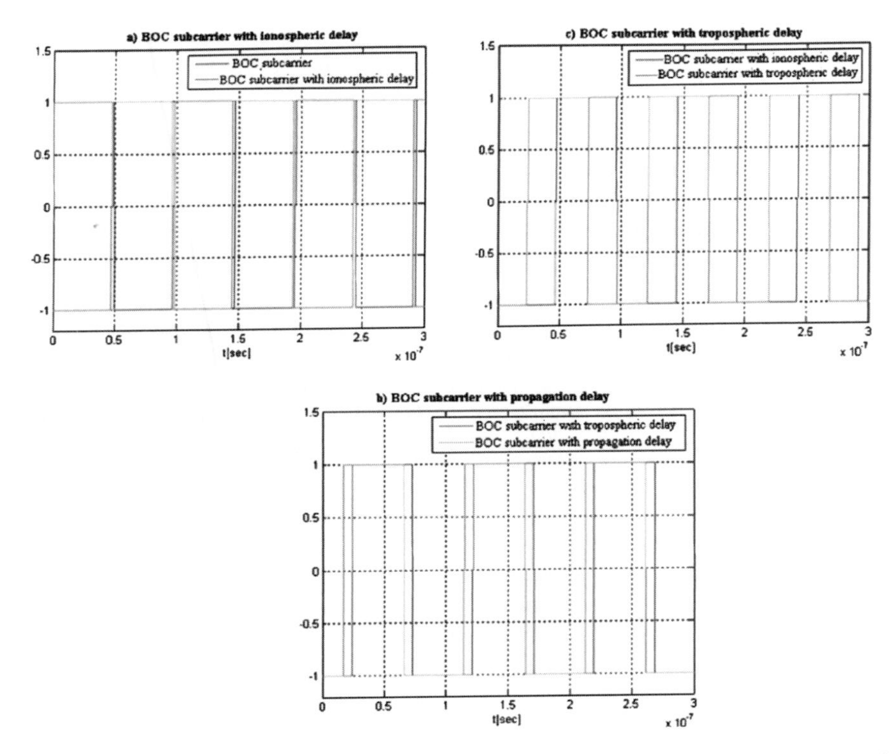

Fig. 9. a) E6 BOC subcarrier with ionospheric delay; b) E6 BOC subcarrier with tropospheric delay; c) E6 BOC subcarrier with propagation delay.

In this section of the paper has been reported the main simulation results; the authors would inform that the whole signals simulator parameters was been implemented in accordance to the Galileo SIS-ICD [6].

4 Conclusion and Outlooks

In the previous sections, an Alcatel Alenia Space Italia GNSS signal simulator architectural design was presented. How it can be noted, the GBTSS is not yet fully developed, for this reason in the paper it was be presented only the main results related to the developed parts.

Currently multipath, interference and receiver error modules are under development, besides new modules that generates the atmospheric effect are in the testing and verification phase.

As previously described the main idea is to realize a GNSS signal simulator able to generate any kind of GNSS signal under any transmission scenario. To reach this goal the GBTSS will be realized like a framework that can contain and drive different modules, an opportune combination of these modules will cover any kind of scenarios.

The future ideas are to fully develop the simulator so to apply it to the B.U. R&D and programmatic activities. In particular, The GBTSS will be employed in the frame of Galileo System Integration and Verification activities, Receivers development, Signal characterization and for all the R&D activities that involve the navigation signals.

References

[1] Guenter W. Hein, John W. Betz, et al., "MBOC: The New Optimized Spreading Modulation", InsideGNSS, vol. 1 number 4 May 1/June 2006

[2] John W. Betz, "Binary Offset Carrier Modulations for Radionavigation", The MITRE Corporation, Bedford, Massachusetts Received September 2001; Revised March 2002.

[3] "Galileo System Simulation Facility - Algorithms and Models", Issue 5, ESTEC 06 June 2005

[4] Elliott D. Kaplan and Christopher Hegarty Editor, Understanding GPS: Principles and Applications - 2nd ed. Artech House, Inc., 685 Canton Street - Norwood (MA), 2006.

[5] "Annex 10 - Aeronautical Telecommunications", ICAO Annexes to the Convention on International Civil Aviation

[6] "Galileo Signal-In-Space Interface Control Document", Issue 11.2, Galileo Industries, February 2006

RUNE (Railway User Navigation Equipment): Architecture & Tests

Livio Marradi, Antonella Albanese, Stella Di Raimondo

Alcatel Alenia Space Italia S.p.A.
SS. Padana Superiore 290
20090 - Vimodrone (Mi) - Italy
e-mail: livio.marradi@aleniaspazio.it

Abstract. Due to European on-going developments (EGNOS, GALILEO), satellite navigation is now an interesting innovation for all fields of transport. One of them is the railway domain, which could considerably profit from the implementation of autonomous on-board positioning systems. For railway transportation, satellite navigation offers new opportunities to increase accuracy of positioning and to implement safety standards everywhere, from high speed trains to local and regional railway lines, enabling a cost-effective modernization and increase of efficiency. Train control poses high demands on positioning with respect to availability, reliability and integrity. Meeting these requirements with a GNSS-based navigation system is the objective of several projects and still needs to be proved.

RUNE is a project developed for the European Space Agency by Alcatel Alenia Space Italia (Laben Directorate), leading a team comprised of VIA Rail Canada Inc. (VIA), Ansaldo Segnalamento Ferroviario (ASF), and INTECS HRT. In this project, the European Geostationary Overlay Service (EGNOS) has been used as part of an integrated solution to improve the train driver's situational awareness.

RUNE integrates positioning sensors with signalling and speed constraint information from a control centre. This can improve safety as a result of a better situational awareness and can also speed up the deployment of drivers on new routes. The primary objective is to demonstrate the improvement of the train self-capability in determining its own position and velocity with limited or no support from the track side, still complying with European Railway Train Management System (ERTMS) requirements. The achievement of such objective could lead to the reduction of physical balises distributed along the track line and needed to reset the train odometer error. Substituting physical balises with GNSS-based virtual balises can lead to reduction of infrastructure costs.

The RUNE project development has gone through a HW-In-the-Loop laboratory set-up up to field-testing. Field tests of the equipment were performed in Italy on the Torino-Chivasso line in April 2005, on board a Trenitalia ALE-601 experimental train. The unit included an LI GPS/EGNOS receiver, Profibus and CTODL interfaces to the train odometer and BTM, an IMU sensor for dead-reckoning positioning, virtual balise and velocity profile databases. This paper presents an overview of RUNE and provides results of the analysis performed on the data collected from the experimental train test campaign. Although testing duration was limited, collecting and analysing real data is important for building expertise in system behaviour and for algorithms evaluation and tuning. Those initial results show achieved performance in the real environment and the capability of the system to provide In-Cab Signalling and Virtual Balise functionalities. It is recognized that only through extensive field testing and validation of the system architecture and algorithms it will be possible to build a robust certifiable GNSS-based train navigation system.

1 Introduction

The railways have a consolidated experience using other means of navigation, which may not always be ideal, but are well known and familiar to the operators. On the other hand, railways, like other modes, have to cope with a number of new challenges. Alternative and more flexible train and freight tracking systems, in combination with existing techniques, are expected to prove attractive, if not indispensable, to be able to meet the new challenges (increased capacity and productivity, higher operating speeds, increased safety requirements, environmental protection), and to cope with the pressure to improve and optimize the economy and profitability of their operations.

Traditional rail operation is based on defined and fixed blocks (length of track of defined limits), the use of which is governed by block signals, cab signals or both. Train movement is constrained by the fixed nature of the block, which impacts the speed and track occupancy. The advantages of GNSS satellite-based systems over traditional rail navigation systems are mainly attributable to the ability to compute real-time accurate and autonomous on-board position and velocity data and provide this information to the on-board Train Control equipment, to the Operations Control Center and to the locomotive engineer. With all of the location and trajectory of each train in the system, a RUNE based operation complemented by conventional practices will provide the flexibility needed to have a "block" move with the train, as opposed to the train moving within a block, thus increasing track capacity while preserving or enhancing safety.

The RUNE system is designed to take advantage of the current EGNOS integrity and wide area differential correction service and extend its availability through an hybrid navigation system based on a Navigation Kalman Filter that integrates data from three main on-board sensors:

- GNSS Receiver: provides a GPS/EGNOS-based PVT solution in addition to EGNOS integrity data;
- Inertial Measurement Unit (IMU): provides three-axis accelerometer and three-axis gyro data for propagation of the solution, specially in case of unavailability of GNSS signals;
- Train Odometer Unit: provides continuous along-track velocity information from two toothed wheels.

Moreover the RUNE equipments uses data contained in a Virtual Balise Map database (VB Map) containing balise identifications with their 3D geographical position and the associated along track distance.

The objective of a such equipment is to introduce an autonomous positioning system into the ERTMS/ETCS architecture in order to:

- Integrate the odometer function by enhancing its accuracy, robustness and integrity estimation function;
- Reduce the need for trackside signalling by substituting as much as possible the physical balises with the virtual balise concept (see section 1.5);

– Provide real time accurate track information to the Train Control equipment and to the locomotive engineer for speed profile supervision, breaking profile computation and alerts to approaching trackside signals.

2 RUNE Architecture Overview

The developed prototype implements the functionalities of the final application together with facilities and functionalities for a real time performance evaluation. The prototype SW allows also to perform off-train tests using the data collected on-board. This capability helps in simulating visibility gaps or slip and slide phenomena or other environmental conditions that cannot be easily to reproduce during a test in a real railway environment.

The RUNE demonstrator is composed of the On Board ATC, the Navigation Sensors and the Recording and Evaluation System (RES) that includes the Navigation Data Fusion Filter (see the figure).

The On Board ATC includes the following modules:

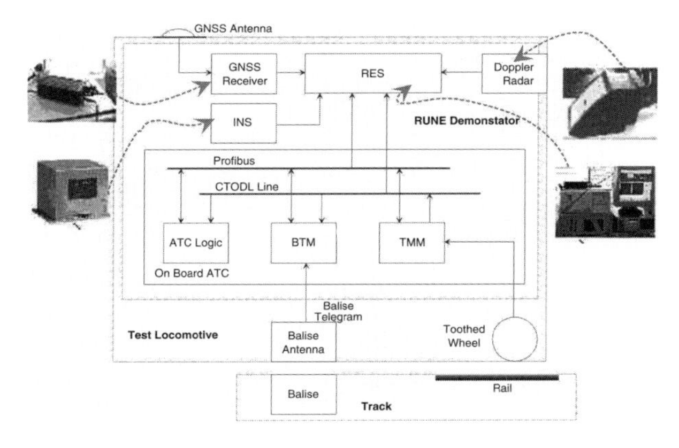

RUNE architecture overview.

- TMM: Train Management Module, which is devoted to the management of the locomotive devices such as the braking system, the juridical recording unit and the toothed wheels for odometry
- ATC Logic: Automatic Train Control Logic which is devoted to the computing functions to control the train speed
- BTM: Balise Transmission Module, which is devoted to read the wayside balises and telegrams
- On Board communication network: based on the field bus PROFIBUS, it is used to exchange data between the on board modules. The use of a field bus minimizes wiring.

In the RUNE demonstrator the On Board ATC is in charge of acquiring and processing the signals generated by the Toothed Wheels and Balises and provide the relevant data on the CTODL Line and PROFIBUS respectively.

The RES software is the heart of the demonstrator: it acquires the Navigation Data generated by the navigation sensors, performs real-time filtering and data fusion to compute the best solution (position, along-track velocity, along-track distance and attitude), performs run-time verification of computed solution accuracy against reference sensors data (eg. Doppler Radar and/or physical balises), records all raw and processed data to allow off-train post-processing, monitors overall RUNE functionality, maintains status logs and provides the Man Machine Interface towards the Test Engineer and the Locomotive Engineer.

The following figure shows the Data Flow Diagram as implemented in the RUNE Architecture.

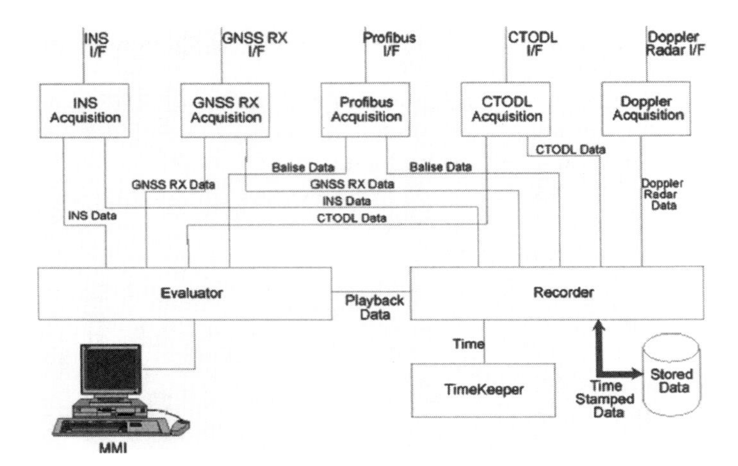

RUNE architecture data flow diagram.

2.1 The RUNE Navigation Data Fusion Filter

Train navigation is somewhat different from all the other type of navigation (aerial, naval, car): train position has only one degree of freedom: the traffic control centres need of the position along the track and not the geographical position.

The odometer is the best suited sensor for such type of measurement; a GNSS sensor, providing absolute position, needs to be assisted by some external aiding information such as, for example, a railway map or inertial sensor in order to correctly translate absolute position into along-track distance.

Odometers performance are typically compromised by slip, slide and creep phenomena; this causes the position error to increase with time and the difficulty to have an accurate velocity profile during brakes (in case of slips). ERTMS foresees the balise as navigation aiding for odometer error reset, and track ambiguity resolution. RUNE inspects the concept of virtual balises. The virtual balise is powered by the GNSS absolute positioning capability. Accuracy and integrity is enhanced by the use of EGNOS and by sensor measurements integration and redundancy. The virtual balise could substitute the physical balise most of the time, offering an unparalleled cost effective solution and adding flexibility to the train management system.

The use of a navigation filter, integrating multiple sensors, solves the problems of odometer error drift, slip/slide and availability and allows to obtain an accurate position and velocity profile estimation.

A Kalman filter is the best method of integrating GPS, IMU and ODO measurements for a number of reasons, not the least its ability to provide a real-time solution. It also has the favourable attributes of being computationally efficient, relatively flexible in that it is able to accept a variety of different measurement types and rates, and has the ability to estimate error sources.

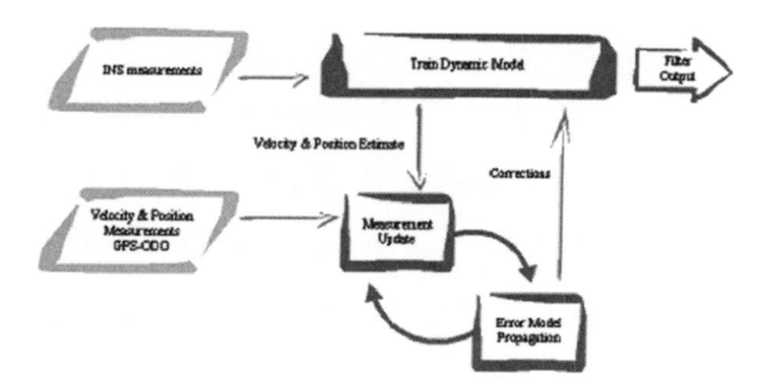

Filter kernel.

The Kalman filter implemented in the RES SW is a time-variant filtering process that deduces the minimum error estimate of the state vector (the unknown parameters) of a linear system, while taking into account knowledge of the system dynamics, measurement model, and the statistics of the system noise and measurement errors.

The set of parameters (8 in RUNE application), called the Kalman state vector, $\vec{x}$, which is allowed to change along the track's path. is defined as:

- [δPx,δPy,δPz]: error on the ECEF position
- [δPS,δVs]: error on the along track position/velocity
- [δA1,δA2,δA3]: error on attitude wrt the navigation frame

The Computation Algorithm of the Filter is based on the following states:

- $\hat{x}_k^- \in \Re^n$: a priori state at step k
- $\hat{x}_k \in \Re^n$: a posteriori state estimate at step k
- $e_k^- = x_k - \hat{x}_k^-$,: a priori estimate errors
- $z_k - \hat{z}_k^-$: the measurement innovation, or the residual.

From these derives a posteriori state estimate $\hat{x}_k = \hat{x}_k^- + K(z_k - \hat{z}_k^-)$ and a posteriori estimate errors $e_k = x_k - \hat{x}_k$.

The matrix K is chosen to be the gain or blending factor that minimizes the a posteriori error covariance

$$K_k = C_k^- H_k^T (H_k C_k^- H_k^T - R_k)^{-1}$$

Here below has been graphically represented the Kalman filter loop.

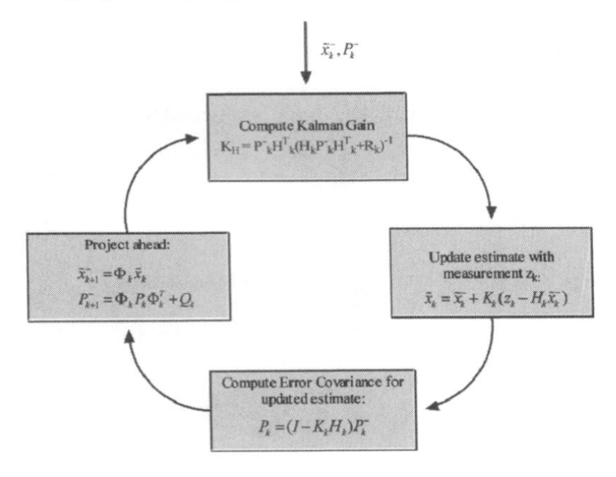

Kalman filter loop.

2.2 The RUNE Performance and Objectives

The RUNE main objective is to satisfy, if not improve, ERTMS levels 2 and 3 performance and integrity requirements.

The main issue of this project is to demonstrate the train capability to self-determine position with an accuracy of 3m and speed with an accuracy of 2Km/h, and through the use of EGNOS, to provide a protected distance of 50m and raise timely alerts when this cannot be guaranteed.

	ERTMS	RUNE objectives
Position Accuracy	5m + 5 % of travelled space	3 m, 95% (GNSS + WAAS/EGNOS)
Velocity Accuracy	2 km/h v < 30km/h 12km/h v < 500 km/h	2 km/h 95% (GNSS + WAAS/EGNOS)
Position Confidence	> 99.9%.	> 99.9%. for a 50 m Protected distance
Availability	better than 99% (Level 3 applications)	better than 99%
Time to Alert	< 5 s	better than 5 sec (Safety Level 3 applications)

The table here reported summarizes ERTMS main requirements and RUNE objectives.

2.3 The Validation Phases

The RUNE equipment validation has been carried out through different incremental demonstration phases. These can be rearranged in three incremental Verification steps as described hereunder:

Step 1. Verify the software performance in a reduced simulated laboratory environment. This step has been performed to mitigate the risk of involving hardware without having clearly verified that the software meets its requirements specification.

Step 2. Build a fully representative RUNE Demonstrator unit and perform a complete validation of the hardware/software equipment in a fully simulated laboratory environment.

Step 3. Perform the RUNE Demonstration on selected railway equipment to verify the performance of the equipment by confirming the data obtained during the above laboratory tests, and to demonstrate the selected user applications.

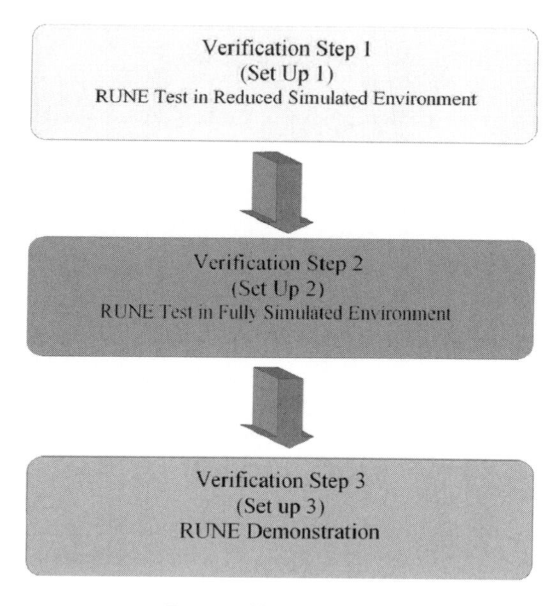

Rune verification steps.

The figure here reported shows the Verification steps as described before: In the following, we will focus our attention on the demonstration phase (step3) providing also the main results of the performance evaluation.

3 RUNE Demonstration Objectives

The RUNE demonstration was performed on the Torino-Chivasso line, on a Trenitaila ALE 601 locomotive located in the "Torino Smistamento" Rail Station (Fig. 1).

The purpose of the test was to collect data to verify the functionality of the RUNE equipment in a real environment, after having performed extensive HW-in-the loop laboratory tests. Demonstration in a real environment is needed also to facilitate the technology acceptance by the railway user communities with respect to ERTMS railway applications.

Fig. 1. The "Ale 601" test train in the "Torino Smistamento" rail station.

The primary objectives of this demonstration were:

- to validate the design of the demonstrator prototype hardware/software;
- to verify sensors behaviour and performance on field to assess typical noise, calibration needs, availability;
- to verify positioning accuracy with respect to trackside references and to assess continuity and availability in a typical railway environment;
- to demonstrate equipment functionality in two user applications including in-cab signalling to assist the locomotive engineer and virtual balises detection for train positioning.

Extensive HW-in-the-loop laboratory testing was carried-out prior to the train demonstration. Summary results of these tests are reported in [1], [2], [3]. Alcatel Alenia Space (AAS-I) laboratory facility includes all necessary simulation tools and equipment: GPS/EGNOS multi-channel simulator, EGNOS receiver, IMU simulator, ASF Balise and Odometer HW equipment and trajectory simulators. This facility has allowed to carry-out RUNE performance verification and sensitivity analysis in representative railway scenarios. In particular those tests included sensitivity of the positioning solution to acceleration, curvilinear trajectories, slip-slide odometer effects and GPS/EGNOS obscuration in tunnels of different length.

3.1 The Torino-Chivasso Route Demonstration

However, it is believed that only through real field testing it is possible to assess the system behaviour, and therefore this paper focuses the attention on the data collected from the train tests. The RUNE equipment installed in the ALE-601 test car is shown in Figs. 2 and 3.

Figure 4 shows the map of the test route from Torino to Chivasso and return. The distance from departure in Torino Smistamento Rail station to Chivasso Rail station is about 20 km. The track offers different scenario conditions, including a tunnel when crossing the city of Torino, curves in the tunnel and straight lines with the train travelling at a max velocity of about 150 Km/hr. Those different conditions allowed to extract interesting information on the equipment behaviour and performance. Figure 4 shows the continuous positioning solution obtained by RUNE during the whole track (bold line) despite the fact that GPS visibility was obscured in the

Fig. 2. Rune installation on ALE-601.

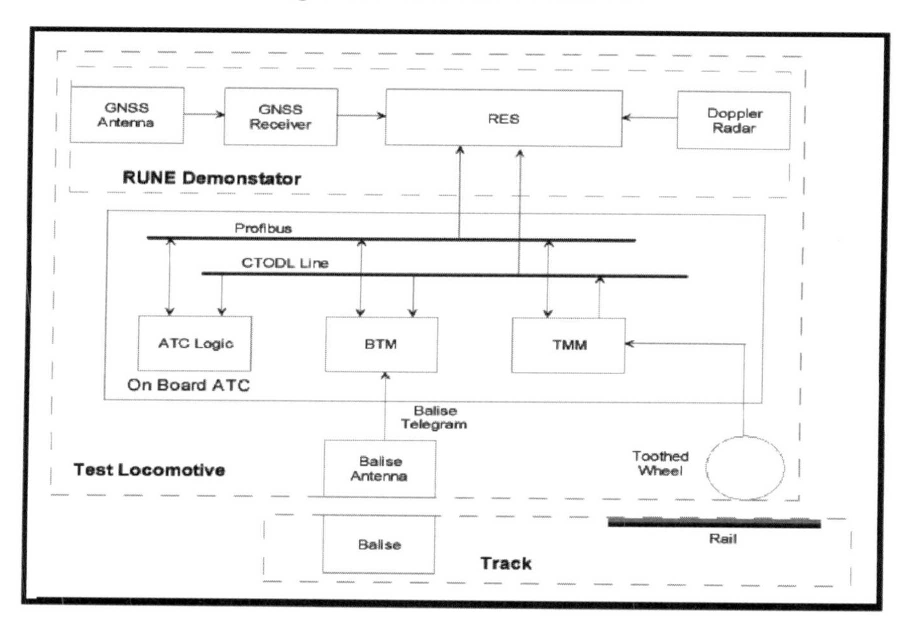

Fig. 3. On train demonstrator set up.

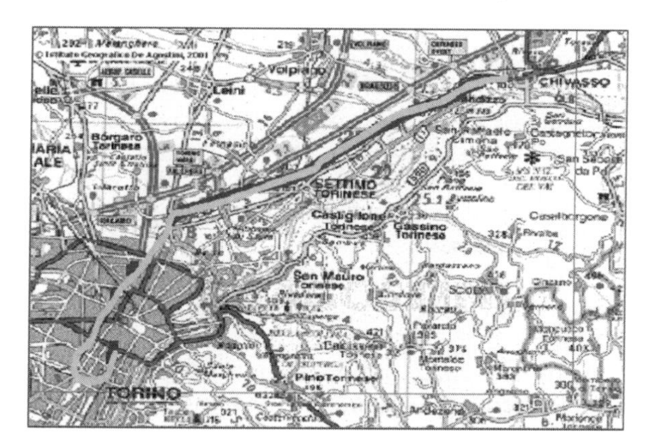

Fig. 4. RUNE solution on the To-Ch route.

Torino tunnel. EGNOS tracking was very limited during the To-Ch route, while during the Ch-To return route PRN 120 was instead in tracking to the receiver for a long period of time.

This is partly explained by the fact that, due to installation constraints, the GPS antenna was applied on the lateral external wall of the train car with only about half-sky visibility.

The RUNE prototype equipment SW allowed to record all received sensors raw measurements during the test run. A fundamental feature of the RUNE SW is to allow off-line playback of collected raw data into the RUNE Data Fusion Filter SW (DFF). This feature was extensively used to analyse RUNE behaviour, applying different calibrations and pre-processing to the raw measurements and, also, allowed to fine-tune the Kalman filter for optimal performance. The only independent position reference that was available during the tests for accuracy performance evaluation was the travelled distance and time information recorded by the on-board train computer (SCMT) when passing over the installed track physical balises. Each balise provides the exact along-track distance from the preceding one.

The following key issues are investigated:

IMU sensors: main issue is related to the misalignment of the IMU axis with the train reference frame, and to the analysis of typical train car vibration noise to apply suitable low-pass filtering;

GNSS receiver: along-track velocity and absolute position accuracy is compared to the balise markers and to the combined-sensors RUNE DFF solution. Integrity information availability in terms of Protection Levels is assessed;

In-Cab Signalling: verify the capability of displaying to the Locomotive Engineer MMI the information on approaching signals and velocity profile warnings based on a pre-stored route database;

Virtual Balise: capability to detect the passage on a Virtual Balise from the VB database (absolute ECEF position and travelled distance) and to generate a balise message on the RUNE Equipment Interface.

3.2 IMU Calibration

IMU calibration in terms of correction of misalignment errors (Fig. 5) and vibration noise filtering is needed before use of the data in the solution DFF. Such calibration was performed in post-processing, having identified suitable phases during the train test, such as train stopped condition, acceleration phase, etc. In an operative system, such calibration could typically be performed on-train, sporadically, during dedicated self-test phases.

During operational phases, the RUNE DFF design disables estimation of IMU biases and misalignment, to reduce processing. Appropriate filtering on acceleration (a_x, a_y, a_z) and angular rate (w_x, w_y, w_z) measurements is determined measuring the noise floor in train stopped, acceleration and turn conditions. Figure 6 shows an example of a_x noise frequency spectrum when train is stopped. A low-pass filter with 0.5 - 1 Hz bandwidth was selected to filter the IMU data for the ALE-601 test train. Gross misalignment errors were calibrated-out by comparing IMU measurements

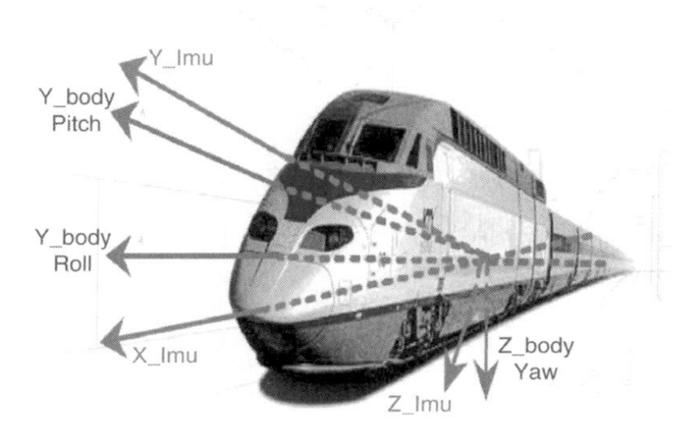

Fig. 5. IMU axes and train axes.

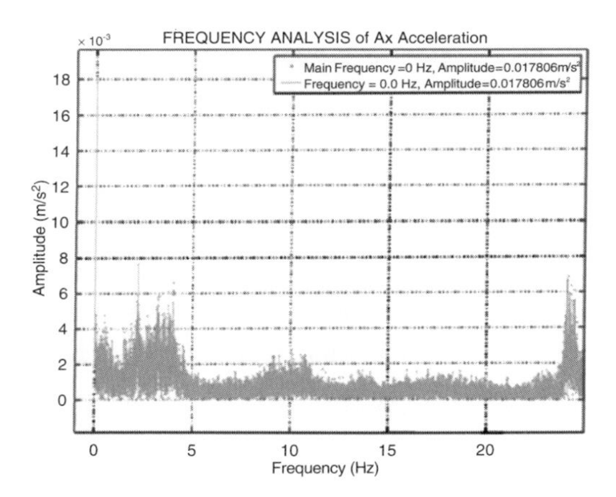

Fig. 6. a_x noise floor.

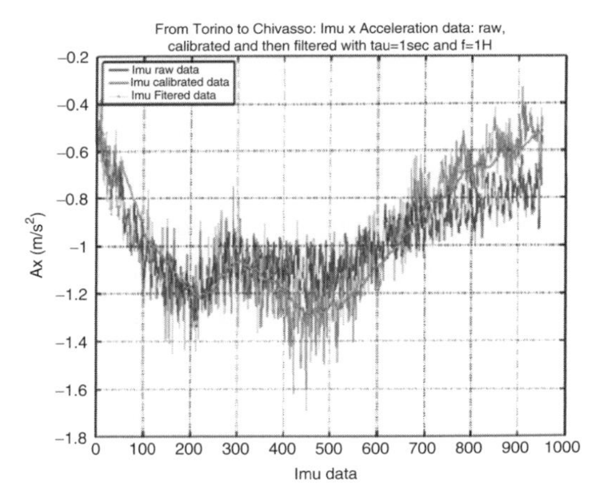

Fig. 7. a_x and w_z calibrated and filtered at 1 Hz.

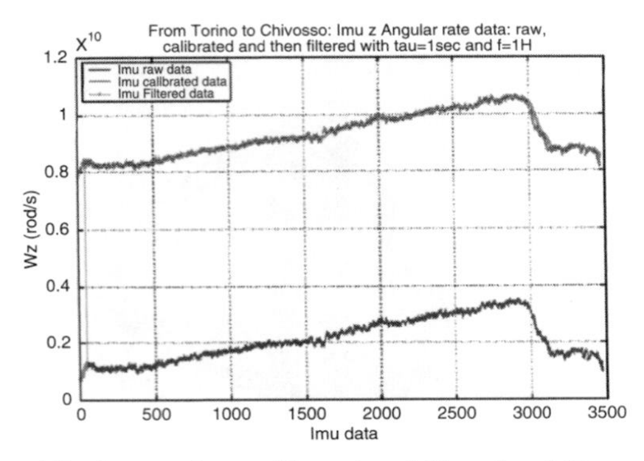

Fig. 8. a_x and w_z calibrated and filtered at 1 Hz.

against expected gravitational and centripetal accelerations with train stopped in horizontal position. Those errors are removed by applying an angular rotation calibration matrix to the IMU data (Figs. 7 and 8).

3.3 RUNE Behaviour and Performance

The accuracy of the RUNE odometry-like function is estimated in terms of accuracy of travelled distance and along track velocity by the RUNE DFF against the balise fixed markers detected and logged by the SCMT along the track. To compare data generated by different sources a time re-alignment of the balise time crossing was necessary.

Figures 9 and 10 show the comparison of velocity profile and along-track estimated distance from RUNE DFF and SCMT/balises after re-alignment. Physical balises are installed in couples on the To-Ch line on a segment of approximately 15 Km with a mean distance of 150m between each balise couple. Figure 11 shows the error in RUNE estimated along-track distance during the runs from To-Ch and return. The error is computed as difference between the distance provided by each physical balise and the along-track distance provided by the RUNE DFF when crossing each balise.

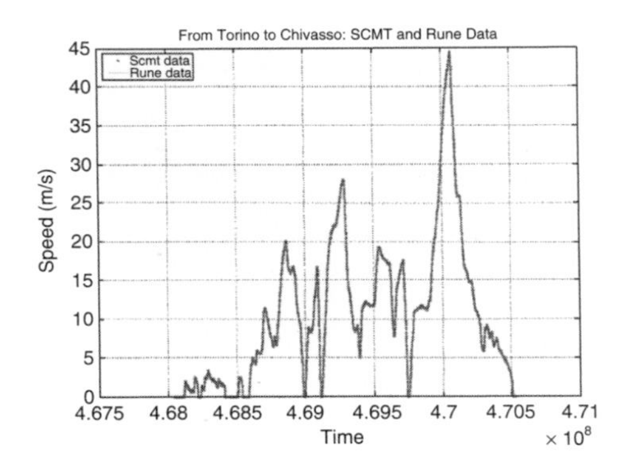

Fig. 9. RUNE along-track velocity vs. SCMT logged velocity.

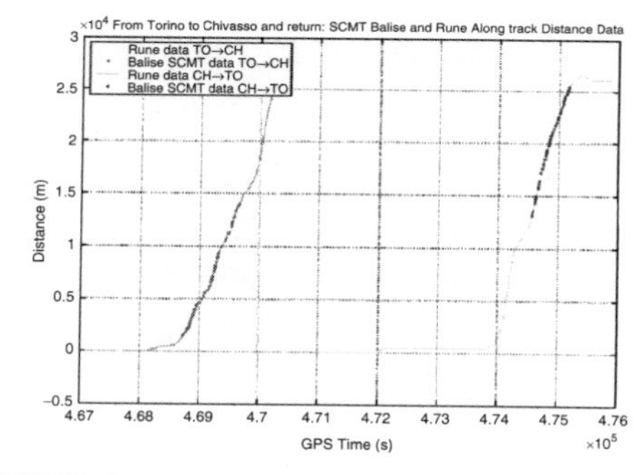

Fig. 10. RUNE along-track estimated distance vs. balise positions.

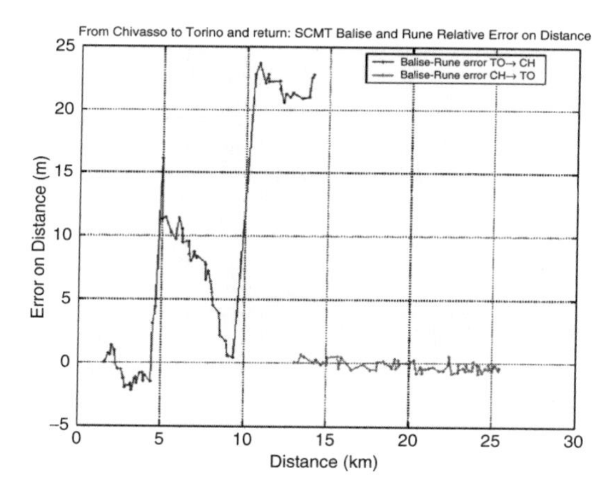

Fig. 11. RUNE along-track distance error vs. physical balise.

No virtual balise was used in this test, therefore this represents the total error accumulated by RUNE over 15 Km. It can be seen that RUNE along-track performance is always well below 2m in the Ch-To return run, while there are two large error spikes in the To-Ch run.

Those were identified as being caused by an incorrect time stamping of sensor raw measurement data samples by Rune. Such an event would have been detected by the raw data integrity checks prior to entering the DFF. Those checks were, however, disabled during the train test data collection. Figure 12 provides the GPS absolute position error on the two runs, showing the accuracy of the solution with and without EGNOS availability. Large error spikes are attributed to multipath effects and to poor SVs geometry also due to the lateral pointing installation of the GPS antenna. The GPS estimation has provided, on average during the runs, an error of 10m in position and 0.5m/s in velocity.

The position in ECEF (in terms of latitude and longitude of the track) estimated by RUNE and recorded from GPS are shown in Figs. 13 and 14 both for To-Ch and return.

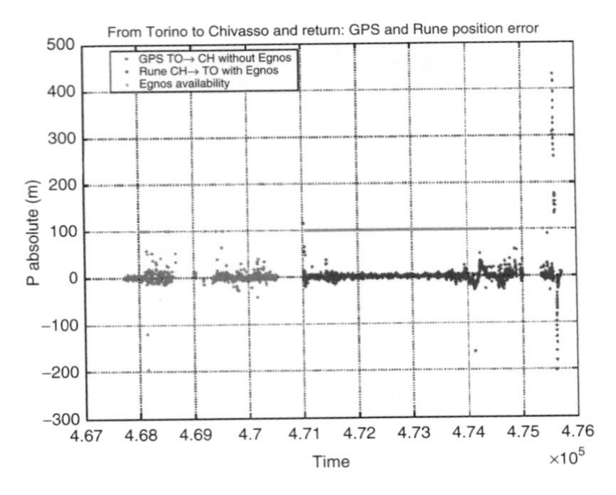

Fig. 12. GPS RMS absolute position error with & without EGNOS (EGNOS solution flag in upper position).

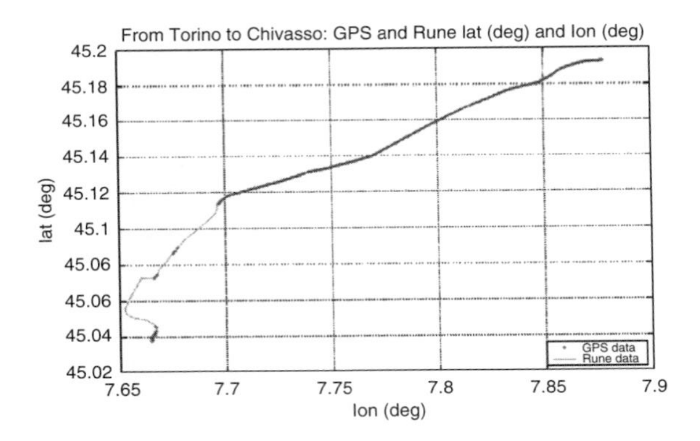

Fig. 13. GPS and RUNE latitude and longitude.

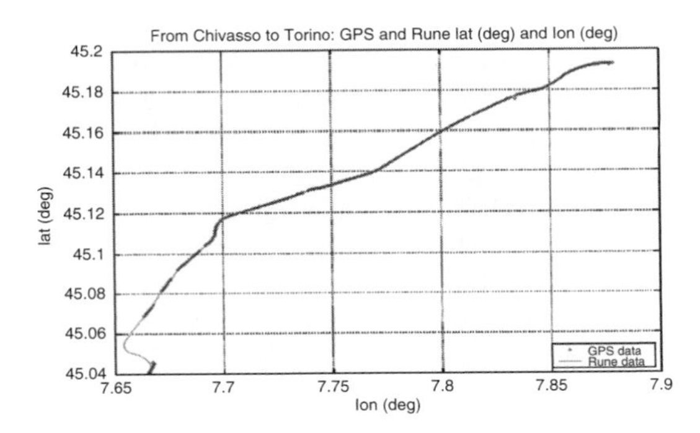

Fig. 14. GPS and RUNE latitude and longitude.

The light line is referred to RUNE DFF solution, and is continuous, while the dark line, represents the GPS solution which is discontinuous when signal is unavailable.

3.4 GNSS Solution Statistics

From the analysis of the GPS files recorded during the test, Fig. 15 shows a histogram of the number of SVs used in the solution and Fig. 16 provides the percentage of time for the three GPS solution cases: no solution, GPS-only, GPS + EGNOS. Figure 17 summarizes the GPS PVT solution availability, subdivided in the different types of environment crossed by the train.

3.5 EGNOS Status During Tests

The EGNOS SIS is broadcast by 3 GEO SVs:

PRN 120 – Inmarsat AOR-E
PRN 124 – ESA's Artemis
PRN 126 – Inmarsat 3F5 IOR-W

The EGNOS system is at present in its Initial Operation Phase (IOP), managed by ESSP, and its performances are increasing and stabilising; it will be soon qualified in order to provide, from the next 2007, a certified service for Safety of Life applications.

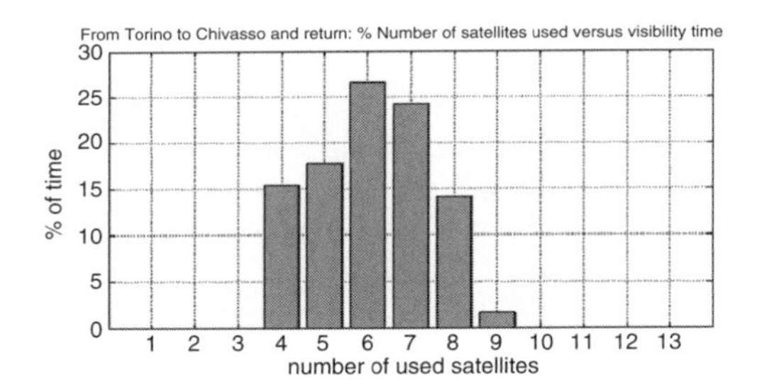

Fig. 15. % Number of SVs used.

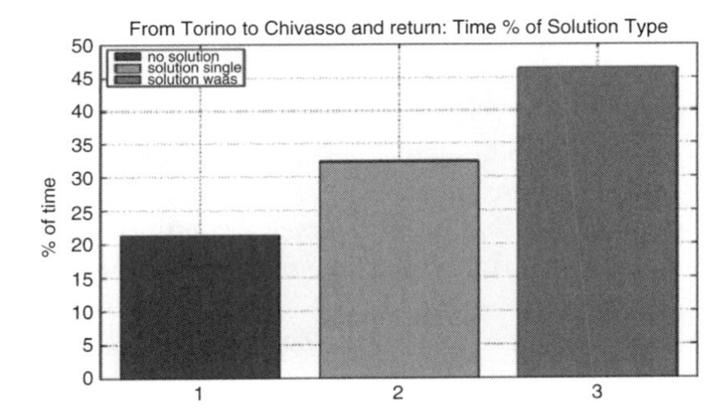

Fig. 16. Global % of solution type.

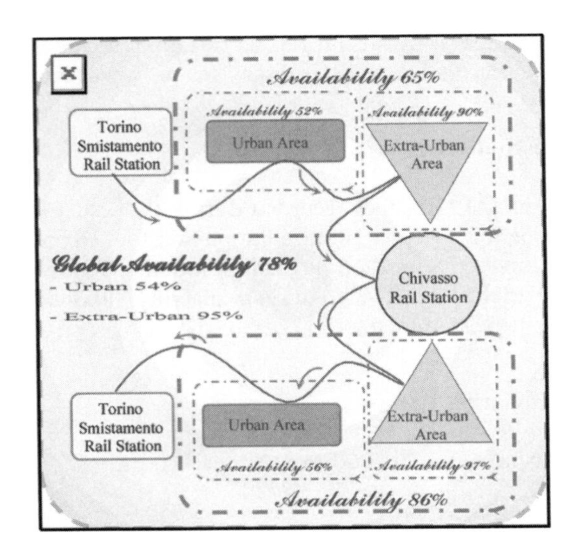

Fig. 17. GPS availability in urban and extra-urban areas.

At the time when the trials have been performed (April 2005), the EGNOS system was under the preparation of the Operation Readiness Review, and as such was subjected to many tests from the Industry and ESA: as a consequence, it was characterised by discontinuity events.

According to the technical reports provided by the IOP consortium, at the date of the trials the satellite better performing was the PRN 124, unfortunately not monitored during the demo. The GNSS receiver used in the demo was able to track only one GEO at a time. The GEO SV tracked in the tests was PRN 120, dedicated to the ESTB activities.

The PRN 124 is going to broadcast the EGNOS message until Jan 2006. PRN 120 will continue to be dedicated to the ESTB, and the PRN 126 reserved to Industry activities.

3.6 GNSS Protection Levels

Using the EGNOS PRN 120 information downloaded by the GPS receiver, the HPL/VPL values have been computed. Figure 18 shows the HPL/VPL during the route Ch-To. A strong increase in HPL/VPL can be observed in the middle of the graph, which seems to be due to bad geometry conditions as confirmed by the PDOP value. If we restrict the analysis to geometry conditions with PDOP <6, we obtain the HPL/VPL values shown in Fig. 19. The HPL mean value is in the order of 10m. Limiting the PDOP to 6 guarantees the availability of the HPL in the 86% of time (Fig. 20). It is important to notice that the HPL concept is defined for a 2D position, while for the railway application a new 1D along-track position protection level indicator needs to be defined (Fig. 21).

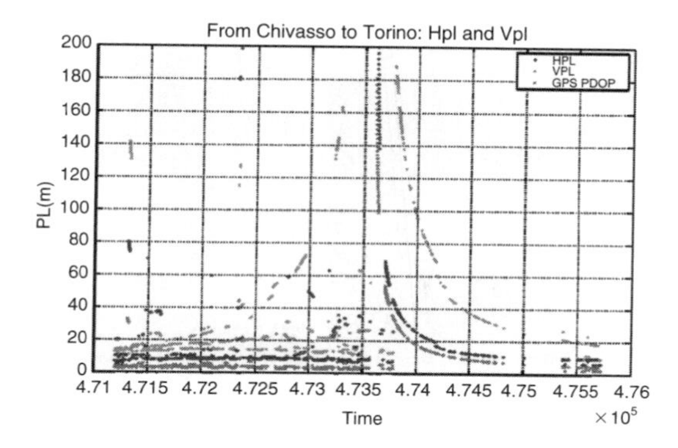

Fig. 18. HPL, VPL and PDOP.

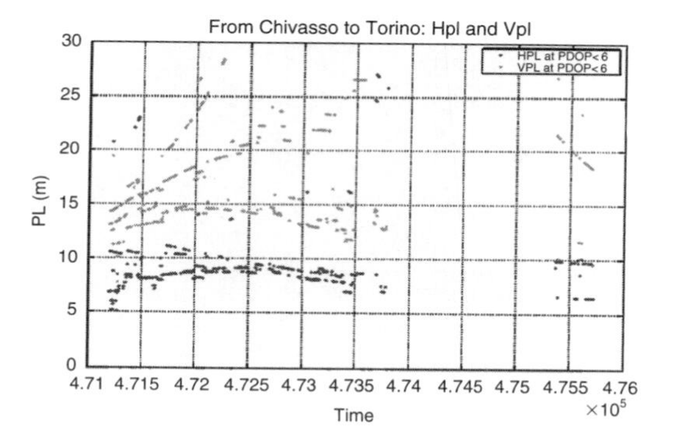

Fig. 19. HPL and VPL at PDOP < 6.

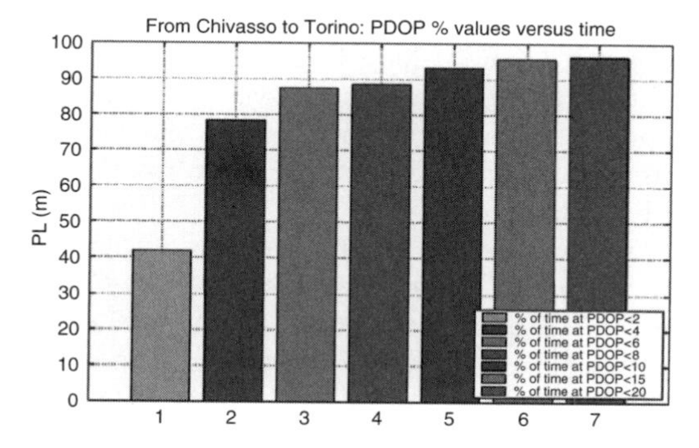

Fig. 20. PDOP % of time.

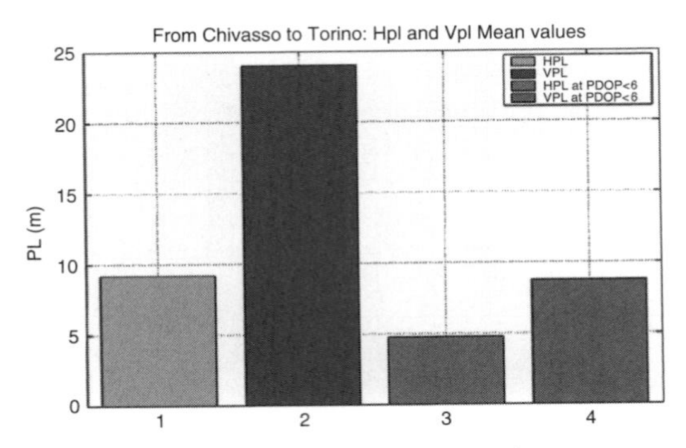

Fig. 21. HPL and VPL mean values.

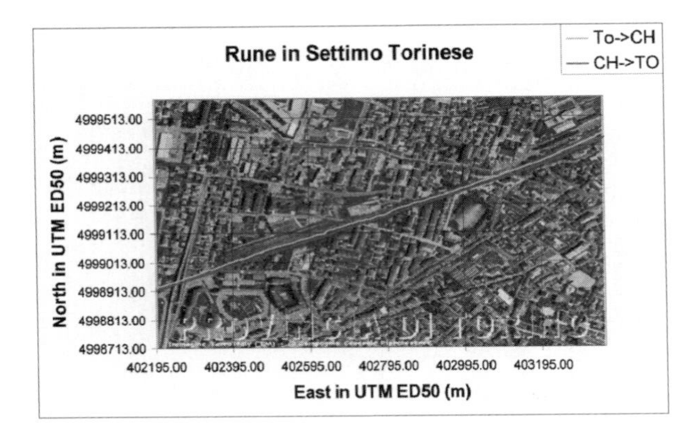

Fig. 22. RUNE in settimo Torinese.

3.7 Track Visualization on Cartographic Maps

Another way of verifying the position estimation obtained by RUNE is to visualize the computed train track on referenced geographical maps.

This data provides the CGR IT2000 Flight in UTM ED50 coordinates. The RUNE position solution during the train test was converted into this coordinate frame and superimposed to the maps. Starting from the Torino Smistamento Rail station, the train passed through the Settimo-Torinese Town and through Brandizzo up to the Chivasso Rail Station.

Figures 22 and 23 show a zoom of the track in Settimo Torinese town (one square = 200m) and a zoom of the train track estimated by RUNE trip when passing close to the Torino-Milano Highway (one square = 50 m). Note that the light line represents the To- Ch route were a GPS-only solution was available, while the dark line represents the return track in which a GPS + EGNOS solution was available. The dark line estimation is clearly more accurate if compared with the underlying picture of the real railway track.

Fig. 23. Rune and the torino-milano highway.

3.8 Future Issues

In the frame of the on-going Galileo Joint Undertaking programmes, the RUNE architecture concept is going to be extended to Galileo within a consortium that includes the major railway signalling companies. This, together with a deep study of the safety issues involved in ERTMS/ETCS standards, will help in the future acceptance of a GNSS-based train on-board navigation system.

References

[1] Genghi A. et al., The RUNE project: Design and Demonstration of a GPS/EGNOS-based Railway User Navigation Equipment, *ION 2003*, Sept. 2003, Portland Oregon USA

[2] Albanese A., Marradi L. et al., The RUNE project: navigation performance of GNSS-based railway user navigation equipment, *Navitec 2004*, Dec. 04, Nordwijck ESA/ESTEC

[3] Albanese A., Marradi L. et al., The RUNE project: Navigation and Integrity Performance of GNSS-Based Railway User Navigation Equipment, *JRC2005*, 2005 Joint Rail Conference, March 16–18, 2005 Pueblo, Colorado.

GPS, Galileo and the Future of High Precision Services: An Interoperability Point of View

R. Capua

Sogei S.p.A./Land Department
99 Via M. Carucci, 00143 Rome, Italy
Phone: +3950253428, Fax: +3950254135, e-mail: rcapua@sogei.it

Abstract. Modernised GPS and Galileo are going to provide, standalone or through code augmentation systems, meter level accuracy. Such accuracy is commonly considered sufficient for the most of the applications. Nevertheless, other higher accuracy applications like Geodesy, Cadastral Surveying, Mapping, Marine surveying, etc. are still stable niche markets. New applications, as High Precision Cartography and ADAS, are emerging. As in any business, the more the offered service precision will be, the more the precision requirements improve. Centimeters level high accuracy is currently achieved through GPS Reference Stations Networks and the integration of mobile terrestrial or satellite communication systems for corrections broadcasting. The weakness of such systems, in RTK mode, are very well known: low density Reference Stations networks coverage for large areas, not homogeneous performances, not reliable communication systems coverage, not guaranteed reliability of the positioning, complex positioning and communication integrated devices, absence of a comprehensive Business Model.

A regional Network backbone design in terms of Reference Stations distribution and separation is needed, integrating existing subnetworks developed for Geodesy, Land Surveying, University research, Marine applications, etc . . . Network spacing is highly dependent on spatial-correlated errors, but MRS/VRS and innovative WARTK approaches, together with three-frequencies availability and TCAR algorithms, will allow to get real instantaneous and homogeneous solutions at large scale. Tight integration with Communication systems is essential for delivering a reliable Real-Time service. Availability of higher mobile communications (e.g. WIMAX) is opening new doors for that. Furthermore, the future availability of a standardised backbone of Galileo Local Elements will provide a great impulse to such integration. Galileo is also studying Business Models for such Services, starting from the availability of innovative services (e.g. Service Guarantee and SoL).

An analysis about GNSS High precision services perspective and interoperability has been performed from different applications points of view, including Galileo future services.

1 State of the Art of High Precision Systems and Services

Satellite geodesy and surveying applications were the first developed since the beginning of GPS operability. The use of carrier phase measurements instead of pure C/A code pseudorange suddenly showed a great potential for getting baselines among receivers with centimetric level accuracy. The cost of using phase measurements was anyway high. The necessity to solve initial phase ambiguities for using phase measurements implies: 1) the need of a fixed Base GPS Station located in the neighbours of the rover; 2) the need for rover receiver phase measurements and ambiguity resolution processing; 3) the need of a robust, relatively low latency communication link between the rover and the Base Station for RTK corrections/ raw data transmission. RTK is by far the main objective of this paper. The first problem has been solved initially trough the use of a couple of receivers on the field, while networks of GPS Reference Stations have been developed by Geodetic institutions, Land Surveying Authorities, Mapping Authorities, Coast Guards, etc. The basic limitation of GPS Networks development was (and it still now is) due to the Reference Stations spacing. RTK solutions are still constrained for ambiguity resolution by distance correlated errors (e.g. ionosphere errors), limiting the rover to Reference Stations separation to 15–30 Km for a good percentage of correct fixing. This implies the development of very dense Networks for a uniform High Precision Service over an entire Country. Concerning ambiguity resolution techniques, many core algorithms have been developed and are today integrated within all code/phase receivers (e.g. LAMBDA and FASF).

Communication link availability has been for years another limiting factor for RTK positioning. VHF/UHF transmitters and receivers at Base Station and rover side were firstly adopted by surveyors, but several limitations are present due to RF licensing and link robustness. Since '90s, the use of mass market mobile communication technology favoured an expansion of GSM and GPRS based RTK systems. Also in this case the corrections transmission costs and links robustness are limiting a full exploitation of such services.

In last ten years, some of those problems have been tackled through the development of Network-RTK technologies (i.e. MRS/VRS approaches). Instead of using a Single-Reference Station approach, a set of Reference Station in the neighbours of the rover is used for modelling and estimating distance-dependent errors. Such an approach allows significantly increasing the Reference Stations separation in the order of 100 Km. In such a way, the number of Reference Stations and relevant cost for developing a nation-wide High precision system can be significantly reduced. Those approaches also lead to an improvement in correction transmission modes, from the generation of a VRS (Virtual Reference Station) close to the rover and relevant RTCM messages sending, to a pure broadcasting of interpolation parameters for errors estimation valid in an entire area surrounded by Reference Stations. The advent of TCP/IP and packet communication leaded recently to the development on IP-based corrections broadcasting protocols, paving the way vs. a real exploitation of this market. Precise Stations Coordinates updates and robust Communication links are also here the requirement.

Another limiting factor is the cost of receivers. It is anyway expected that the future availability of High Precision Services (HPS) and new satellite systems will generate an economy of scale for high precision receivers markets.

On the other hand, satellite augmentation systems (WAAS, EGNOS and MSAS) are currently able to provide meter level accuracy and relevant integrity services for code measurements.

The availability of future GNSS third frequencies leaded to the development of new algorithms for Ambiguity Resolution (TCAR). Anyway, future GNSS constellations and relevant third frequencies cannot improve at great extent ionospheric errors estimation. Reference Stations Networks will always be needed ([5]) for RTK.

TCAR, integrated with innovative WARTK (Wide Area RTK) algorithms, will allow real instantaneous centimetric solutions using long baselines Reference Stations networks.

The future challenge of phase measurement is therefore the development of uniform coverage, sub-centimetric accuracy services in real-time with a high level of reliability and reduced costs for infrastructures. Galileo Local Elements design and relevant development of a Regional Network is an opportunity for Europe for having a real full coverage high precision service to be used for traditional (Mapping, Surveying) and innovative applications (ADAS, Road Lane Keeping, High Precision mapping). The development of dedicated Business models for that is one of the most relevant tasks.

2 Current GNSS Networks Coverage and Integration Perspectives

Current development status of Reference Stations Networks and High Precision Services in the world is very variegated.

Apart from basic regional Geodetic Reference Stations networks, like IGS and EUREF, developed for Reference Frames determination/update, the major part of each Country developed their own Reference Stations for National datum realisation. Such Reference Stations, due to their specific objectives, are commonly wide spaced and not equipped for real-time applications.

In some countries, Mapping or Cadastral Authorities developed national or local networks for RTK applications. In some cases such networks are providing a Post-Processing or Real-Time service for all the Country based on MRS/VRS. This is the case of GPSnet.dk, composed by 26 Stations, providing a VRS service all over Denmark, or of the SAPOS Network, composed by about 300 Stations, providing MRS/VRS services for the whole Germany.

In other Countries the situation is much more fragmented. Local sub-networks based on a single-station RTK or MRS/VRS approach have been developed by Local authorities, Surveyors Organisations, Universities, etc., and are providing Local High Precision Services. In most of the cases such initiatives are in prototyping or test phases and cannot guarantee the required service levels.

An example of such situation is Italy. In the following left picture, Reference Stations currently operating or starting to be operating are reported. As can be seen, while North-West and Centre of Italy are quite well covered, other Italian areas (e.g. coastal regions) are not. In that scenario, it is not cost effective to re-build ex-novo an entire optimum shaped network (less than 100 Reference Stations should be necessary to complete the Country coverage). The re-use as much as possible of existing infrastructures (e.g. Communication networks and available Reference Stations) and the eventual densification with new Reference Stations should be opportune. Taking into account the fact that not all the areas of a Country need the same level of coverage (e.g. mountains regions could be less covered by a Real-Time service), a densification of the network has been studied. In the following right picture, a densification using existing Public Administration sites (triangles) and new Coastal Reference Stations (stars) installation is showed. The total number of stations in this scenario is expected to be in the order of 150.

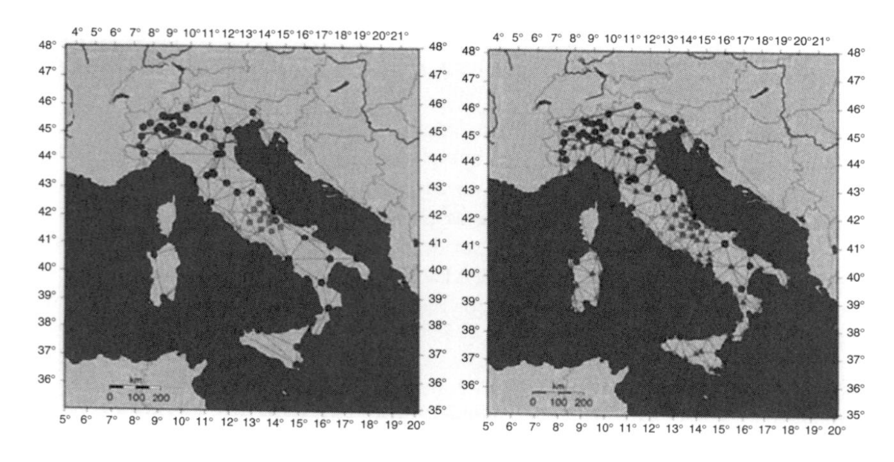

Summarizing, three basic Coverage Scenarios can be found:

Scenario 1: Countries poorly covered by network of Reference Stations or needing a reliable structure for institutional purposes (e.g. Cadastre development in European Eastern Countries); this is the case for the development of a new network of Reference Stations based on Network-RTK

Scenario 2: Countries already partially covered by Reference Stations Networks developed by several entities; integration of different networks and reuse of existing infrastructures under an Institutional umbrella is here the preferred solution

Scenario 3: Countries fully covered by an advanced Reference Stations Network; in this case, the only requirement is the *Open approach* versus future Regional systems, through the use of standard interfaces versus a Regional Control Centre for real-time raw data and ephemeris transmission.

3 Network RTK Technology

Network-RTK concept was developed for overcoming limitations of single-station technologies to provide High Precision Services over a wide area. Such limitations are due to distance-dependent biases (ionosphere and troposphere refraction, orbit errors). Those errors can be accurately modelled using a set of Reference Stations surrounding the rover instead of a single station. This is the basic idea of Network-RTK approach. Basic Network-RTK steps are the following:

1) *Network ambiguity fixing*: ambiguities among Reference Stations receivers are calculated. Only fixed–ambiguities carrier phase measurements can be used for precise errors modelling. This ambiguity fixing method differs from the rover ones due to the fact that precise Stations coordinates are known.
2) *Corrections estimation*: errors corrections are here estimated. Different methods have been developed for modelling errors corrections parameters. Methods working on differenced or undifferenced observables can be distinguished. Differenced

methods estimate errors through statistical modelling of differenced observations, while the undifferenced ones operate on direct observables, through estimation of observations corrections in the State Space domain through dynamic modelling and Kalman filtering.

One example of Differenced method is provided by the analysis of the errors through the collocation method ([1]). In such case the measurements Covariance Matrix is modelled in order to estimate the correlation between errors. Distance dependent errors (orbit errors, ionosphere errors and troposphere errors) are distinguished by uncorrelated errors (receivers clocks, Phase Center Variations-PCV, multipath). A relevant effort has to be put in modelling Covariance and Cross-Covariance matrixes through mapping functions depending on satellite elevations and distance.

Concerning the Undifferenced approach, one of the most common ones is reported in [2]. In this case, all single measurements errors are dynamically estimated though a Kalman filter working in the State Space.

In Network-RTK, reliable Ambiguities solutions require very precise Reference Stations coordinates. PCV derived by antennas calibration are therefore needed.

3) *Corrections transmission*: transmission of corrections can be performed in two basic ways.
   - VRS: the rover sends its approximate position to the Control Centre through an NMEA message; the Control Centre calculates corrections and, starting from a selected Master Reference Station, relevant observations are shifted close to the user in a VRS; raw data or corrections for the new station are sent to the user through RTCM messages (#18 and #19 or #20 and #21). The limitation of such mode is the need for a bi-directional communication link between the rover and the Control Centre. Alternatively, a grid of VRS corrections can be sent to the user, leaving to the rover the VRS selection. A Cluster for providing such service is composed of a minimum 3 Reference Stations (up to 8, depending on computation load on a single server)

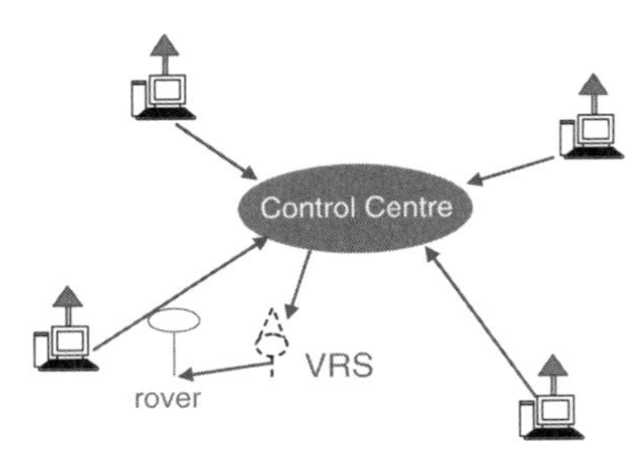

   - Surface corrections parameters: errors are described as a polynomial function (e.g. FKP, [2]) all over the area covered by the Reference Stations. The user is in charge of interpolating such function for obtaining corrections relevant to its position. FKP can be performed through a 1-way broadcasting of corrections functions parameters to the user. The user has to be equipped with relevant processing capabilities (e.g. RTCM 2.3 message #59 or RTCM 3.0 corrections processing). A Cluster is usually composed at minimum by 5 Reference Stations.

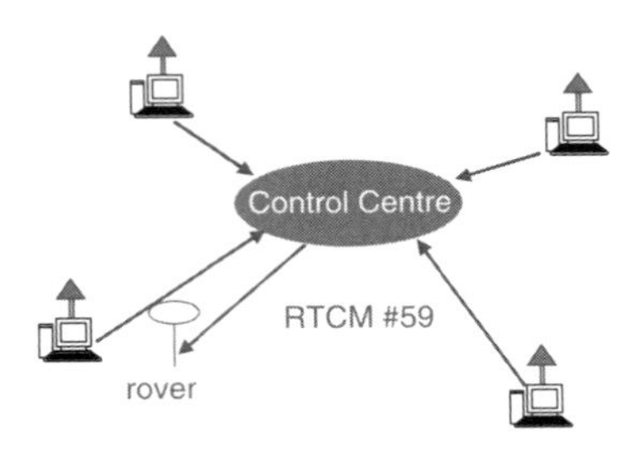

4 Next Generation RTK: WARTK and WARTK-3

If a very accurate ionosphere errors estimation can be provided to the user, e.g. in the form of corrections grid, the distance separation among Reference Stations can be improved to hundreds of kilometres. Several algorithms have been studied during last years on this subject, with the involvement of ESA ([3]). They are referred as WARTK and concentrate their effort on the development of a real-time accurate ionospheric delays estimator through a Kalman Filter. The relevant extension to the future GNSS three frequencies, based on the application of TCAR ambiguity resolutions techniques, is named WARTK-3. TCAR techniques have been extensively studied by some years ([4]). They allow having a real instantaneous correct ambiguity fixing. The integration of WARTK with TCAR will allow reducing inherent limitations of TCAR (e.g. third step Ambiguity Resolution errors introduced by ionosphere biases and multipath errors) and providing a complete RTK European level service through a very low density network of Reference Stations. A high ambiguity fixing success rate has been obtained through simulations. Ionospheric corrections transmission latencies in the order of tenths of seconds and more are accepted.

EGNOS RIMSs, currently used for Wide Area code pseudorange augmentation, can be firstly used at this purpose and eventually densified with external Reference Stations (e.g. EUREF or IGS). Furthermore, EGNOS link (and SISNET) could be used for ionospheric corrections broadcasting to the user. Galileo Local Elements in the future, together with Commercial Services availability, will give the opportunity for Europe for having a first backbone of Reference Stations for the provision of a full coverage real-time European-wide high precision service.

Method	Advantages	Disadvantages
TCAR	Low computational load	Third step of ambiguity fixing limited by ionosperic error
WARTK	Accurate real-time ionospheric error modelling Precise navigation with Reference Stations separation of hundreds of Km	High convergence time
WARTK-3	Use of further widelaning possibilities and accurate ionospheric modelling Single-epoch precise navigation	

The development of accurate standards for WARTK corrections transmission to the user, as well as the integration of WARTK ionospheric corrections with rover generated ones is another step for WARTK solution.

5 The Reliability Problem

Another relevant problem currently slackening a full exploitation of High Precision applications is the evaluation of the RTK solution integrity at user level. Indeed, signal integrity is not sufficient for some applications.

This is particularly relevant for Safety of Life and Professional services. For the first class of applications, like port approach, inland waterways navigation, ADAS, etc., an integrity in the position domain for RTK should be provided. On the other hand, Professional applications (e.g. institutional cadastral surveying or high precision mapping), should guarantee surveying results. Here a wrong correct fix for a point implies repeating an entire survey, with relevant time and money waste. Integrity in the position domain for such applications can therefore improve GNSS High Precision market penetration in a relevant way. Future studies launched by ESA on WARTK will work on such issue.

6 High Precision Services, Networks and Mobile Communication Interoperability

High Precision Services infrastructures development is based on three main functional components:

- Control Centre, in charge of monitoring Reference Stations network status and of providing Real-Time Services or RINEX files to the user
- Communication Network between the Control Centre and Single Reference Stations
- Communication Network between Control Centre and rover

Concerning the Communication Network development between the Reference Station and the Control Centre, it is a critical component for Network-RTK technologies. In order to provide real-time corrections to the user, a continuous stream of raw data from each Reference Station has to be guaranteed. A reliable and high QoS communication network has to be implemented. Various experiences has been performed (e.g. using VPN or WAN/LAN) in many countries for the development of MRS/VRS networks. In any case the problem of the quality and relevant high costs for a reliable communication link is one of the major critical development factors that are slackening the development of such technology.

The solution is the reuse as much as possible of existing networks developed for Nation-wide organisations (e.g. institutional organisations), with favourable service contract with Communication Operators.

On the other hand, the Communication link between the Control Centre and the rover for getting a service is the other face of the same kind of bottleneck.

Starting from the oldest RTK solutions, where a set of modem was available for a GSM user dial-up and a Single Station (or the Control Centre acting as a router), current GPRS and future mobile communication systems (e.g. UMTS) are now offering TCP/IP connections. Major limitations of the GSM dial-up connection are GSM lack or low signal coverage (e.g. in rural or remote areas) and link robustness (every surveyor experienced the need for re-initialisations due to a link failure). Furthermore, GSM dial-up is quite expensive. In the case of 1Hz RTCM messages update, it could imply hundreds €/month for a surveyor daily work. Also in the case of flat RTK services price (as currently provided by some High Precision Service Providers, sometimes coincident with Mobile Operators), ranging in the order of 80–100€/month, a relevant part of the price is due to connection cost.

GPRS, EDGE now or CDMA2000 and UMTS in the future are offering a relevant alternative from the point of view of Internet multicasting capabilities and the service price (based on packets based price figures), but similar problems in terms of link robustness and coverage are expected. A service price in the order of some hundreds of Euros is anyway a possible target without relevant service agreements with Mobile Operators. Other possible broadcasting systems are FM sub-carrier modulation using DARC (Data Radio Channel) or DAB (Digital Audio Broadcasting). WiFi and WIMAX are also emerging as broadcasting means.

Furthermore, other augmentation systems based on a fine real-time modelling of error sources and precise ephemeris are currently emerging, able to provide subdecimeter accuracies through satellite corrections broadcasting (e.g. Star-fire).

Needed bandwidth for RTCM 2.x messages transmission is in the order of 4.8kbps for 12 satellites for single-station RTK and in the order of 2kbps for RTCM 3.0 standards, while VRS corrections transmission bandwidth is in the same order of single-station RTK.

Within the framework of RTK corrections transmission standards, NTRIP (Network Transport of RTCM via Internet Protocol) is the latest development. It is a protocol based on HTTP/1.1 and is designed to broadcast GNSS corrections over the Internet and Mobile IP Networks. It is currently provided in version 1.0 by RTCM and used by EUREF and Network-RTK implementations in the world. Major Professional GNSS manufacturers are starting to integrate NTRIP processing within receivers and it is expected in a couple of years all the receivers will be equipped with such interface. Such protocol requires a few hundreds to few thousands bps. It has to be considered the preferred broadcasting protocol for Network-RTK.

Suitability of Control Centre to rover communication systems will also depend on the Network-RTK method to be developed. While VRS solutions occupation depends on the number of connected users (bi-directional link for each user), other network RTK solutions depend on the number of Reference Stations data to be transmitted.

In any case, the number of data to be directly transmitted depends on the number of satellites. In the future over 12 GNSS satellites could be visible also in shadowed environments and in the order of 20 in Open Sky conditions.

The solution should be to start carrying out service agreements and systems integration between High Precision Service Providers and Mobile Communication

Operators in order to obtain more suitable service price for the user. Such integration should imply a guaranteed coverage and the improvement of the relevant QoS. Interoperability between Reference Stations and Mobile Communication systems can be performed through the architectural components foreseen for LBS services developments (GSM SMLC and UMTS SRNC).

Also in this case, National Institutional actors could drive such development for relevant professional activities (Port Operations, Land Surveying), leaving mass market to Mobile Operators.

Galileo, through the development of Local Elements, has the opportunity of implementing from scratch such kind of strategy.

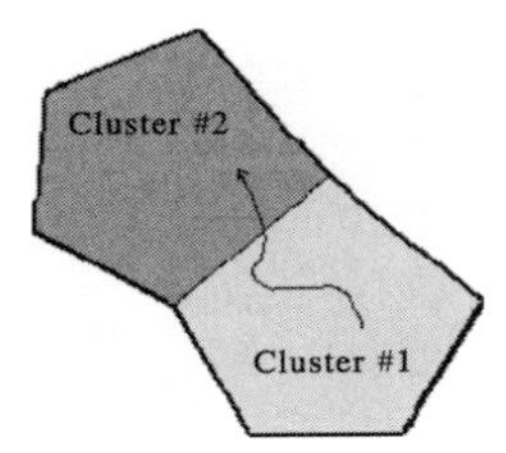

Handover is another relevant problem to be dealt with. Users moving from one Network-RTK Cluster to another should in fact coherently receive a new stream of corrections relevant for the belonging area. While in a VRS approach using bi-directional links, rover position can be updated and a new stream of VRS data transmitted after a Cluster change, a complete broadcast system, not aware of user position, cannot automatically perform such Cluster handover. Such limitation will become more important in the future, where mobile High Precision applications will grow. Such problem can be tackled in the future through WARTK very long-baseline Reference Stations architecture, allowing to provide ionospheric errors corrections for an entire Continental Area.

7 Networks Architectures and Developed Tests

Sogei, within its Institutional tasks, developed extensive R&D activities for maintaining technological leadership for the exploitation of its role of ICT technological partner of the Ministry of Economy and Finance Authorities (Revenues, Land, Customs and State Properties) of Italy.

Within the framework of Geomatics, Sogei participates to Galileo relevant EC Projects on high precision applications development (MARUSE and MONITOR) and it is member of the Consortium Galileo Services.

Concerning R&D activities on high precision surveying (of direct interest for Cadastre), a prototype MRS/VRS Network of five Reference Stations have been developed. The Network covers different environmental conditions (sea, lake, rivers and mountains). A state of the art VRS method based on a Differenced approach has been used for estimating network services performances in such worst case situation.

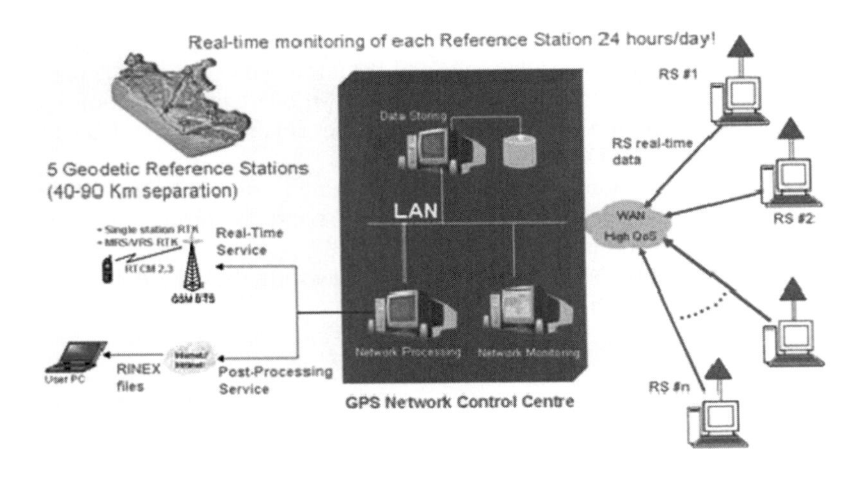

The main objective of such implementation has been to develop a prototype network infrastructure for evaluating Networks architectural constraints from an "Industrial" point of view for the development of a nation-wide MRS/VRS Network and relevant services. Indeed, more than a detailed test on Network performances in terms of accuracy and TTFA (for which an extensive literature exists), Sogei considers essential for a real and successful development of such services a full comprehension of implementation problems concerning Level of Services, Service Guarantee and Reliability. An analysis of constraints for the development of an infrastructure able to provide Professional and Institutional h24 services level has been developed at this purpose.

In the following, main results coming from such analysis developed for about one year of h24 Network monitoring are reported for an infrastructure potentially designed for providing such kind of services.

In order to evaluate interoperability and interfacing constraints, Reference Receivers and processing software from different manufacturers (scientific and commercial) have been integrated within the platform. An MRS/VRS Open software has been adopted for having the possibility of tuning relevant parameters and evaluating impacts on the Service Level. The communication link between the Reference Stations and the Control Centre has been developed using the high quality nationwide WAN network of the Italian Public Administration (RUPA), while GSM supported corrections transmission.

Extensive Post-Processing and VRS RTK tests have been developed within such network. Relevant results are coherent with the ones obtained in the world for similar systems (accuracy between 3 and 5 cm and TTFA in the range of 30s-3 min for most the cases, with millimetric RS repeatability). In order to evaluate land surveying and mapping performances, *Mixed* surveying (integrating traditional topographic and GNSS sensors) have been developed. Main considerations from the analysis are reported hereinafter in terms of requirements for developing such infrastructure:

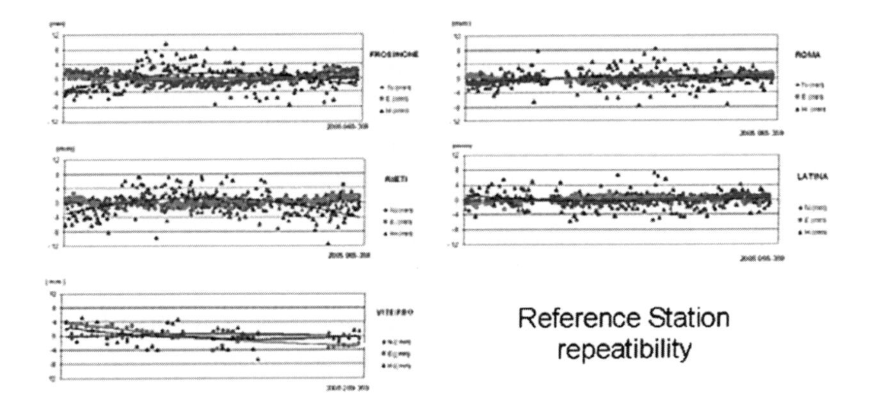

Data Recording:

- About 2.8 GB/month per Reference Station (binary, RINEX files)
- Data recorded hourly at 1 s and daily at 30 s
- Daily automatic data conversion and Quality Check through teqc
- Automatic backup system and monthly storage on DAT
- Batch weekly Reference Stations Coordinates solutions (Bernese)

Communication links between Control Centre and Reference Stations

- High QoS (low latency, robust connection) Communication Networks between Reference Stations and Control Centre and Backup lines
- Real Open standards for raw data streaming from Reference Stations to the Control Centre (RTCM, BINEX, NMEA), with particular reference to ephemeris data
- Development of a distributed architecture, with completely automated Regional Clusters able to monitor and control a set of 5–8 Reference Stations and provide relevant post-processing and real-time services

Reference Stations installation

Availability of logistically equipped sites: power supply with UPS, Network connection all over a Nation (e.g. Public Administrations buildings).

Connection link between the Control Centre and the User

- Post-Processing: Web portal for RINEX data provisioning
- Real-Time: corrections multicasting through NTRIP protocol is the most suitable; it is anyway necessary to guarantee for 2–3 years the transition for users equipped with non-state-of-the-art receivers (e.g. RTCM 2.x interfaces only) through dial-up GSM services

Concerning logistics and engineering of the Control Centre for h24 services, the following critical operation should be performed:

- h24 operations of the Control Centre (provided with semi-automatic control chain at different service levels)
- Integration of different semaphore monitors controlling different architectural components (Communication Network, Reference Stations, Antennas integrity)
- Availability of Automatic Components Controllers (e.g. robots)
- Predictive algorithms for the communication network monitoring
- Customer Care and service assistance
- Availability of a local recovery operators for each Reference Station site and of a Maintenance teams for the whole Nation

Network Design Issues

Concerning the Italian reality, some considerations can be carried out, valid for a general context:

- A National MRS/VRS Network design is necessary for guaranteeing uniform coverage and services: a backbone of Reference Stations distributed at national level and managed by an institution is necessary
- The adoption of a national Reference System based on such network, tied to IGb00, ITRF00 and ETRF
- Development of an *Open System* for the integration of best existing sub-networks able to integrate Reference Stations of different manufacturers and models

A comparative cost analysis for developing a Network-RTK in the Centre-Italy has been performed following two different approaches:

- Use of already existing network of Reference Stations and densification through new developed RS (Scenario 1)
- Development of a new backbone of Reference Stations (e.g. developed by a national institutional actor) and densification with existing networks (Scenario 2): use as much as possible of existing communication/logistic infrastructures owned by the National actors

The cost analysis, including Site monumentations, operations and communication networks costs, leaded to the following cost estimations. It can be seen how, after a few years, the development of RS Network by a unique National actor implies a

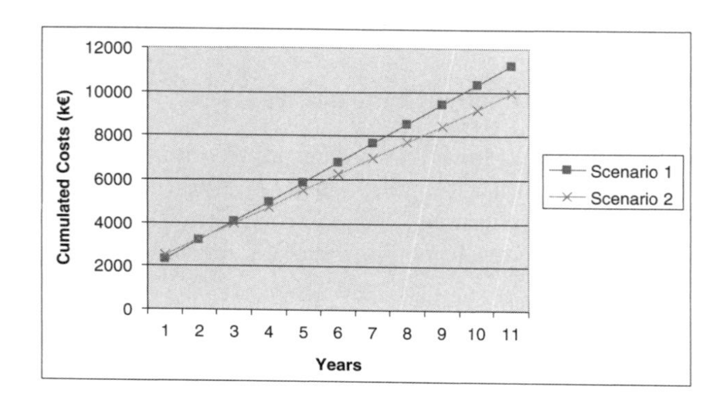

relevant cost saving with respect to a separate integration of existing sub-networks. This is in a great part due to the reduced communication network costs, thanks to previous Service Agreements with the Telecom Operator for a National Customer and reduced sites installation costs.

8 Galileo Local Elements and Business Models

Galileo Local Elements development will allow the definition of a Business Model for High Precision Services implementation at National and Regional Level. The Galileo Service Centre should be in charge of monitoring Commercial Services delivery, assigning Commercial Services encryption keys to National Service Providers and providing Standards for Galileo Local Elements Monumentation, Installations and Operations. National High Precision Services (a unique, possibly Institutional actor) should be responsible of developing at National level a backbone of LE for providing basic HPS. Such national backbone should be developed in MRS/VRS mode (or future WARTK-3). Local Service Providers within each nation can provide HPS and carry out a densification of the backbone. The National Service Provider has to provide a National Communication Network through relevant agreements with National Communication Operators in order to have high QoS at convenient rates.

A Liability Chain should be furthermore established. A *HPS Guarantee mask* of flags, one for each component involved into the chain (e.g. Communication link, Geodetic Frame, LE Raw data, Regulatory Bodies Certification, etc..), should be sent to the rover. A dedicated RTCM messages should be defined in the future for transmitting such Galileo Positioning Guarantee mask.

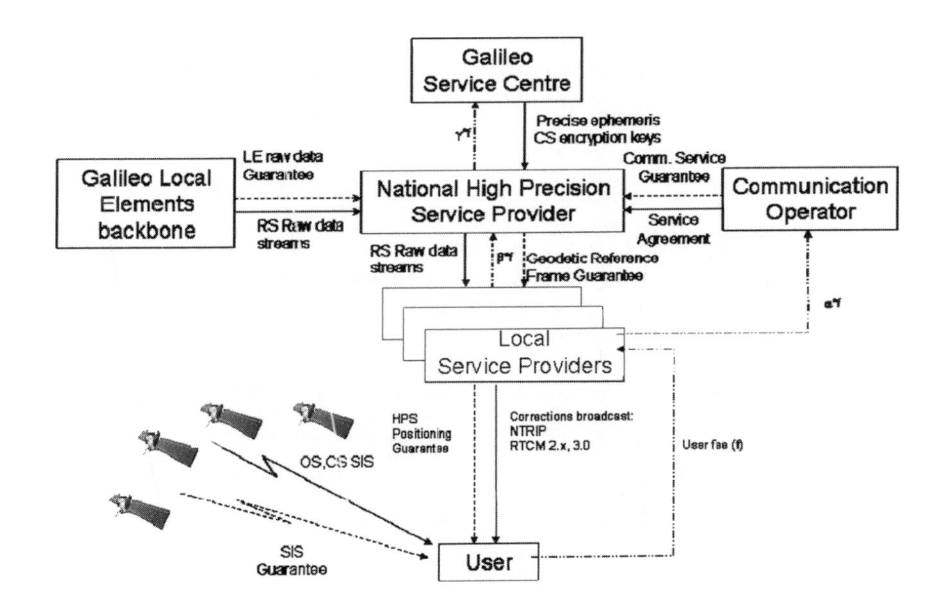

Concerning High Precision Service revenue stream, the National High Precision Service Provider can be the *Central collector of revenues*. Final User can pay Local Service Providers (or directly the National High Precision Service Provider). Part of such revenues should be passed to the Communication Operator and part forwarded to the GSC.

9 Conclusions

The future of High Precision Services is evolving due to Network-RTK and future WARTK technologies. The more HPS will be available, the more innovative applications will require such services (e.g. ADAS, Port approaches and Operations, High Precision mapping, etc.). The implementation of a High Precision Service at National levels implies the following:

- Definition of a National Service Provider (possibly Institutional), providing relevant logistics infrastructures and in charge of Reference Stations/LE developments
- Availability of a high QoS communication network (low latency, high robustness) through relevant agreements and links with GSM SMLC or UMTS SRNC and management of MRS/VRS Clusters handover
- Definition of h24 homogeneous HPS all over the territory through Predictive Systems for Communication performances monitoring
- Adoption of a national Reference System based on such network, tied to IGb00, ITRF00 and ETRF
- Development of real *Open System*, able to integrate best existing sub-networks (different models and manufacturers)
- Definition of Business Models: National Service Providers, to be the revenue stream collectors, and relevant relationships with Local Service Providers, Communication Operators, Institutions and GSC
- Definition of a HPS Service Guarantee mask of flags to be transmitted to the user through an RTCM message to be defined

References

[1] J. Raquet, G. Lachapelle, L. P. S. Fortes, "Use of a Covariance Analysis Technique for Predicting Performances of Regional area Differential Code and Carrier Phase Networks", ION GPS-98

[2] G. Wubbena, A. Badge, M. Schmitz, "Network-Based Techniques for RTK Applications", GPS JIN 2001

[3] M. Hernandez-Pajares, J. Miguel Juan, J. Sanz, R. Orus, A. Garcia-Rodriguez, O. L. Colombo, "Wide Area Real Time Kinematics with Galileo and GPS Signals", ION GNSS 17th Technical Meeting of the Satellite Division, 2004

[4] B. Forssell, M. Martin-Neira, R. A. Harris, "Carrier Phase Ambiguity Resolution in GNSS-2", Proceedings of ION GPS-97

[5] X. Chen, U. Vollath, H. Landau, K. Sauer, "Will GALILEO/Modernized GPS Obsolete Network RTK?", GNSS 2004

GNSS ATC Interface

Giovanni Del Duca°, Claudio Rinaldi°, Carmine Pezzella°,
Alessio Di Salvo°, Stefano Chini*, Massimiliano Crocione*,
Vania Di Francesco*, Luca Pighetti*, Simone Quaglieri*

*Selex-SI, a Finmeccanica company
Via Tiburtina Km 12,300, 00131, Rome, Italy
Phone: +390641503007, e-mail: Rif.vdifrancesco@selex-si.com;
°ENAV S.p.A. –Via Salaria, 716
Phone: +390681661, e-mail: Rif.crinali@enav.it,
gdelduca@enav.it, cpezzella@enav.it

Abstract. The necessity of a navigation system, more flexible and interoperable, has become more and more important and the use of satellite system has been recognized as the main means to obtain this improvement. In the aeronautical field the GNSS has been chosen by the ICAO as fundamental component for the future CNS/ATM systems because of its peculiar characteristics that provide the necessary assistance during all flight phases. The ATC interface developed in the frame of EtoG aims to facilitate the introduction of GNSS services in Italian airspaces. The EtoG programme is a programme for researching and developing of new aeronautic applications to optimize the existing procedures and to find new technologies for the management of critical situations (safety) by using satellite navigation. The introduction of satellite navigation (GPS-EGNOS-GALILEO) allows the management of the aircraft flying phases with remarkable advantage compared to the traditional systems.

1 Introduction

Thanks to the use of the GNSS/EGNOS service and aiming at providing synthetic information to the aircraft controller with regard to the GNSS operating status, ENAV and SELEX–SI have began to develop a new prototype for the monitoring of the GNSS performance on the Italian airspace and presentation to controller. This monitoring tool will be able to provide a great operating support for all people involved in air traffic control and management (ATC/ATM). At the moment a mock-up in term of human machine interface (HMI) has been developed.

The International Civil Aviation Organisation (ICAO) has recognised a need improvements to the existing air navigation system. An ICAO Special Committee of Future Air Navigation Systems (FANS) developed a new concept expressed in terms of communication, navigation, surveillance and air traffic management (CNS/ATM). It is intended to be an evolutionary means of achieving improvements in the global air navigation system. To obtain the benefits of the CNS/ATM concept, aircraft will need to achieve accurate, repeatable and predictable navigation performance. This is referred to as Required Navigation Performance (RNP).

RNP is intended to define the requirement for the navigation performance of each individual aircraft within the airspace.

Table 1. Signal in space performance requirements.

Typical operation	Accuracy horizontal 95% (Notes 1 and 3)	Accuracy vertical 95% (Notes 1 and 3)	Integrity (Note 2)	Time-to-alert (Note 3)	Continuity (Note 4)	Availability (Note 5)
Enroute	3.7 km (2.0 NM) (Note 6)	N/A	$1 - 1 \times 10^{-7}$/h	5 min	$1 - 1 \times 10^{-4}$/h to $1 - 1 \times 10^{-8}$/h	0.99 to 0.99999
Enroute, Terminal	0.74 km (0.4 NM)	N/A	$1 - 1 \times 10^{-7}$/h	15 s	$1 - 1 \times 10^{-4}$/h to $1 - 1 \times 10^{8}$/h	0.99 to 0.99999
Initial approach, Intermediate approach, Nonprecision approach (NPA), Departure	220 m (720 ft)	N/A	$1 - 1 \times 10^{-7}$/h	10 s	$1 - 1 \times 10^{-4}$/h to $1 - 1 \times 10^{-8}$/h	0.99 to 0.99999
Approach operations with vertical guidance (APVI)	16.0 m (52 ft)	20 m (66 ft)	$1 - 2 \times 10^{-7}$ per approach	10 s	$1 - 8 \times 10^{-6}$ in any 15 s	0.99 to 0.99999
Approach operations with vertical guidance (APV-II)	16.0 m (52 ft)	8.0 m (26 ft)	$1 - 2 \times 10^{-7}$ per approach	6 s	$1 - 8 \times 10^{-6}$ in any 15 s	0.99 to 0.99999
Category I precision approach (Note 8)	16.0 m (52 ft)	6.0 m to 4.0 m (20 ft to 13 ft) (Note 7)	$1 - 2 \times 10^{-7}$ per approach	6 s	$1 - 8 \times 10^{-6}$ in any 15 s	0.99 to 0.99999

NOTES:—

1. The 95th percentile values for GNSS position errors are those required for the intended operation at the lowest height above threshold (HAT), if applicable. Detailed requirements are specified in Appendix B and guidance material is given in Attachment D, 3.2.
2. The definition of the integrity requirement includes an alert limit against which the requirement can be assessed. These alert limits are:

A range of vertical limits for Category I precision approach relates to the range of vertical accuracy requirements.

3. The accuracy and time-to-alert requirements include the nominal performance of a fault-free receiver.
4. Ranges of values are given for the continuity requirement for en-route, terminal, initial approach, NPA and departure operations, as this requirement is dependent upon several factors including the intended operation, traffic density, complexity of airspace and availability of alternative navigation aids. The lower value given is the minimum requirement for areas with low traffic density and airspace complexity. The higher value given is appropriate for areas with high traffic density and airspace complexity (see Attachment D, 3.4).
5. A range of values is given for the availability requirements as these requirements are dependent upon the operational need which is based upon several factors including the frequency of operations, weather environments, the size and duration of the outages, availability of alternate navigation aids, radar coverage, traffic density and reversionary operational procedures. The lower values given are the minimum availabilities for which a system is considered to be practical but are not adequate to replace non-GNSS navigation aids. For en-route navigation, the higher values given are adequate for GNSS to be the only navigation aid provided in an area. For approach and departure, the higher values given are based upon the availability requirements at airports with a large amount of traffic assuming that operations to or from multiple runways are affected but reversionary operational procedures ensure the safety of the operation (see Attachment D, 3.5).
6. This requirement is more stringent than the accuracy needed for the associated RNP types but it is well within the accuracy performance achievable by GNSS.
7. A range of values is specified for Category I precision approach. The 4.0 metres (13 feet) requirement is based upon ILS specifications and represents a conservative derivation from these specifications (see Attachment D, 3.2.7).
8. GNSS performance requirements for Category II and III precision approach operations are under review and will be included at a later date.

The new concept of RNP is being applied to develop guidance standards for all phases of aircraft operations, including en route, landing and surface operations. The term RNP is applied as a descriptor for airspace, routes and procedures and can be applied to a unique approach procedure or to a large region of airspace.

This means that RNP is an airspace system function and not a navigation sensor function; the airspace requirements are satisfied independent of the methods by which they are achieved. This is quite different from the method used by regulating agencies at present which requires mandatory carriage of specified equipment for air navigation and thus constraints the optimum application and implementation of modern airborne equipment.

Tables 1 and 2 show the performance requirements of signal in space and alert limit associated to flight phases [1].

With the aim at providing a service to ATC controllers, the system shall provide the following features, with respect to the algorithm defined in [2], [3]:

- A *real time* evaluation of the GNSS availability for any virtual user who flies over the Italian airspace for all phases of flight (from Enroute to Precision approach);
- A *prediction* of the GNSS availability for any virtual user who flies over the Italian airspace for all flight phases (from En-route to Precision approach);
- To display over a particular geographical area or over a specific airways the result of the computation for the GNSS availability;

Moreover the system provides:

- Evaluation of the User Differential Range Error (UDRE) and Grid Ionospheric Vertical Error (GIVE) parameter included within the SBAS augmentation messages provided by EGNOS ATC Server

In order to evaluate the GNSS service in term of the integrity, two parameter, described in the algorithm defined in the RTCA standard document, will be used: the Vertical Protection Level (VPL) and the Horizontal Protection Level (HPL) [2].

Moreover Satellite navigation allow the ADS usage in the CTR and give to the air traffic controller a pseudo radar presentation of the air traffic equipment with ADS.

Table 2. Integrity requirements in terms of alert limit.

Typical operation	Horizontal alert limit	Vertical alert limit
En-route (oceanic/continental low density)	7.4 km (4 NM)	N/A
En-route (continental)	3.7 km (2 NM)	N/A
En-route, Terminal	1.85 km (1 NM)	N/A
NPA	556 m (0.3 NM)	N/A
APVI	40 m (130 ft)	50 m (164 ft)
APV-II	40.0 m (130 ft)	20.0 m (66 ft)
Category I precision approach	40.0 m (130 ft)	15.0 m to 10.0 m (50 ft to 33 ft)

1. The terms APV-I and APV-II refer to two levels of GNSS approach and landing operations with vertical guidance (APV) and these terms are not necessarily intended to be used operationally.

In this way it is possible to increase safety and airport capability.

Data coming from local sensor possibly located in airport field could be taken into account for the evaluation of GNSS availability.

2 Service and Functionalities

The EtoG programme, through the GNSS ATC interface can offer advantages to three main figures:

- *Planner operator* (the person who has to manage and plan the flight within his/her region of interest)
- *Executive operator* (the person who has to directly provide to the airman the guideline for the procedure to be followed)
- *Supervisor operator* (the person who has to monitor the performance of the GNSS system at the moment and for future time)

For each operator the system has been studied to give an appropriate support based on the peculiar characteristic of the operator work. To do this, three main scenarios has been identified and characterized in order to provide a better service for the operator who shall use the system.

In addition for the supervisor operator has been developed, following the ICD provided by ESA [3], the interface for the ATC Client. The ATC Client is the primary mean to acquire data from the EGNOS system. Thanks to the development of this interface it is possible to directly compute data related to the UDRE and GIVE parameters in order to show the result to the supervisor operator. Many other information can be obtain through this connection such as almanacs, satellite status and so on. The information from the ATC Client are particularly useful for performance prediction while, for a real time evaluation this information could be refined with data from a GNSS receiver.

So, the main functionalities provided by the prototype can be summarize as follows:

- Prediction
- Real time evaluation
- Acquisition and elaboration of navigation EGNOS ATC messages

Figure 1 is shown the context diagram related to the main functionalities of the system proposed.

In the following sections the main result in term of sponsored service is presented.

2.1 Planner and Executive Operator

As anticipated before three main scenarios has been identified; in the following each scenario will be described.

This description is driven on a common base, i.e. a full functionality of the satellite navigation systems in all the conditions, and their integration with communication and

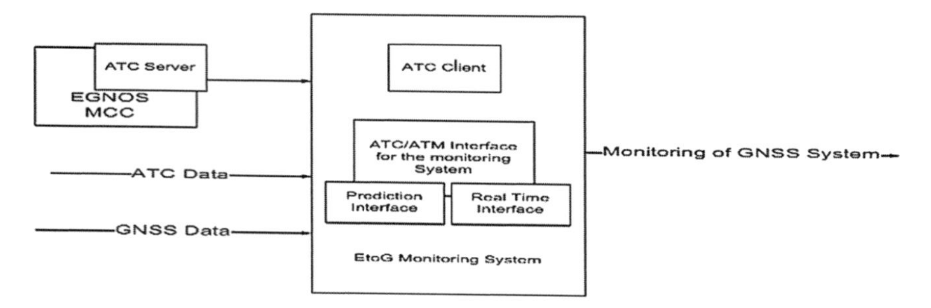

Fig. 1. EtoG tnterface – context diagram.

specific ATC/ATM systems. In particular, it's possible to identify some fundamental topics of scientific research which are strictly correlated to the full and correct definition of the applicative products:

- Interference management
- Augmentation and integrity algorithms

The EtoG project hence foresees to give a fundamental contribution to research in these research fields, beginning to study, develop, and validate at least some of the needed technologies.

2.1.1 En-route Scenario. The en route application product will optimize the air traffic flow from the controller point of view, providing information about the GNSS/SBAS performance within the Italian airspace; in particular it provides a great support for the transition areas (such as north Africa and Middle East areas). Predictive tools, matched with interface towards flow management units will allow to estimate GNSS availability along planned routes.

It is allowed the option to implement in the future the capability to manage Galileo messages too. In the following figures the interface for ACC controller (executive and planner) are presented. The first step is to configure the operating environment as shown in the following. The interfaces show different scenarios that can be set by the operator, for example the interface with the Terminal Manoeuvring Area (TMA) or Airways is presented. The scenario will be set by checking the appropriate checkbox in order to recognize the proper area of competence. Figure 2 is set the checkbox for the TMA. In a similar way it is possible to set different operating environment, such as Control Terminal Regions (CRTs), Flight Information Regions (FIRs) and so on.

In Fig. 3 the airways, within the Italian airspace are presented.

After the setting of the operating environment the ACC operator has to set if the prediction or the real time evaluation will be displayed on a particular airway or on a specific geographic area. In both cases the operator shall select the particular airway or area. The third step is to set the date and time of the evaluation (real time as well as prediction) and then he/she can display the prediction or real time evaluation by click the button *Prediction Data* or *Real Time*, respectively. Figures 4 and 5 show both elaboration.

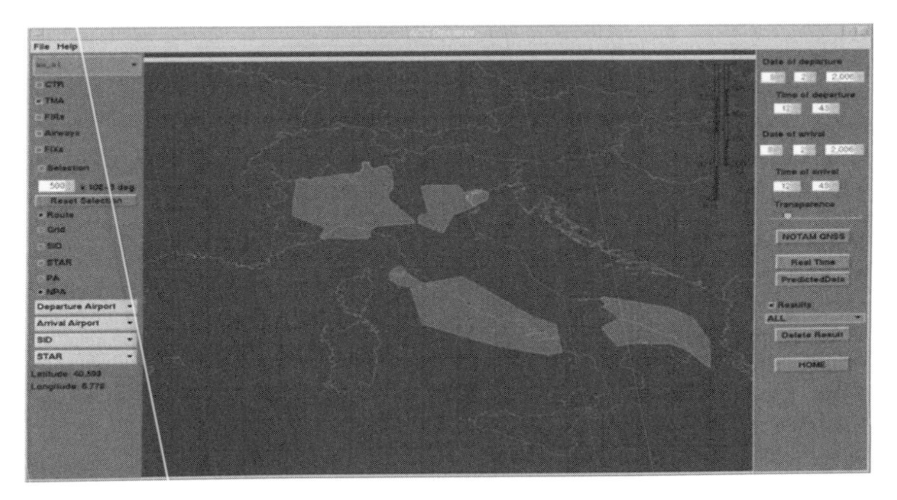

Fig. 2 ACC interface with TMA.

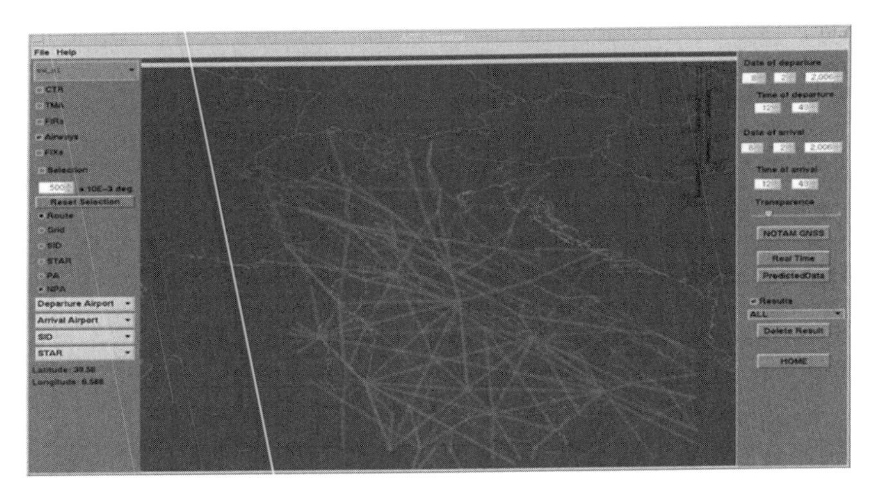

Fig. 3. ACC operator with airways.

For a better and an immediate comprehension of the interface for each type of RNP procedure is associated a given color. In this case the APV-I procedure is represented.

2.1.2 Approach Scenario. For the approaching scenario, the application product is targeted to innovation in the field of satellite navigation, allowing the integration of the satellite functionalities in conjunction with the additional capabilities typical of ADS systems.

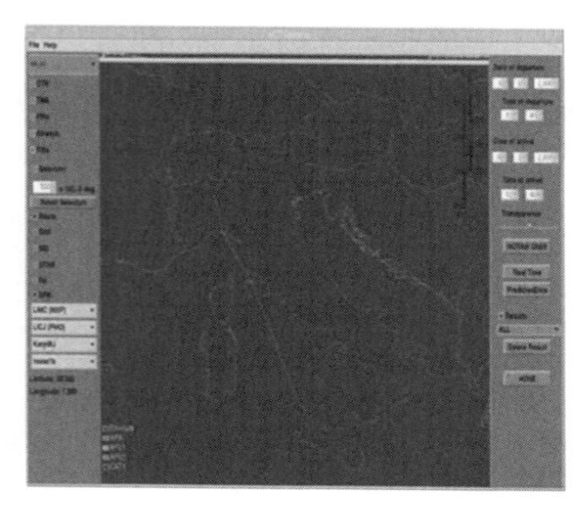

Fig. 4. Real time for a given Route.

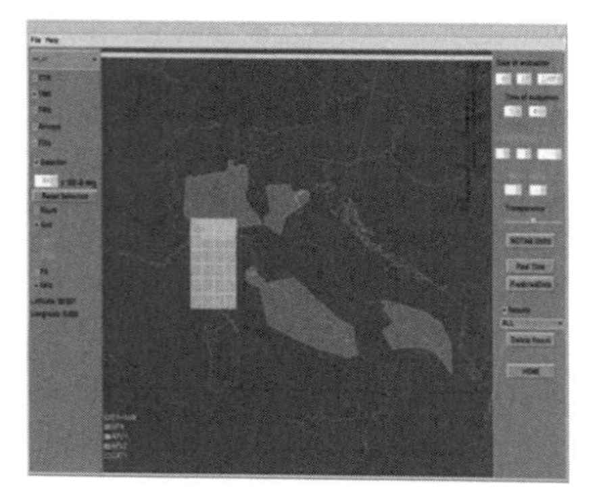

Fig. 5. Prediction in a given area.

As described above, the APP operator has to preliminary set the operating environment, the type of evaluation and the related time and data. The main difference is given by the difference area of interest between the ACC and APP operator.

Figure 6 the prediction on a particular airspace within the Naples CTR and the real time evaluation for one STandard ARrivals (STAR) and two Standard Instrumental Departures (SIDs) for the Rome CTR are presented. In the second case (Fig. 7) the operator will be able to display the distance between the aircraft and the ground (in terms of Flight Level –FL).

2.1.3 Airport Scenario. For the airport operational scenario, the application product is focused on innovation in the field of satellite navigation and its applications. This product focuses its applications to the most critical phases of the flight. Possible benefits are under investigation that local sensor could give to an EM interference analysis potentially impacting approaching procedures.

In Fig. 8 two different SID for the Malpensa Airport are shown. The picture is referred to a prediction.

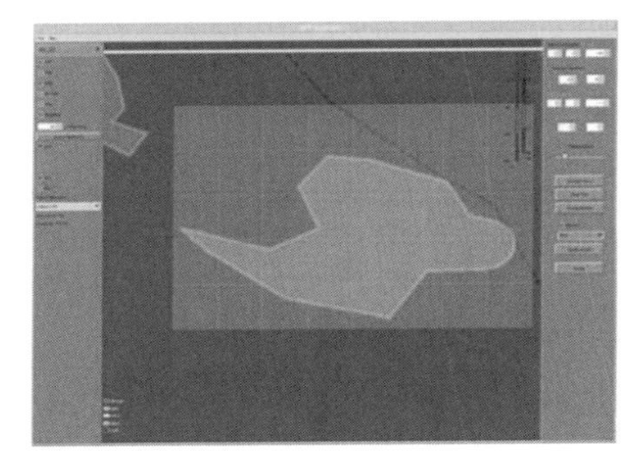

Fig. 6. A prediction in naples CTR.

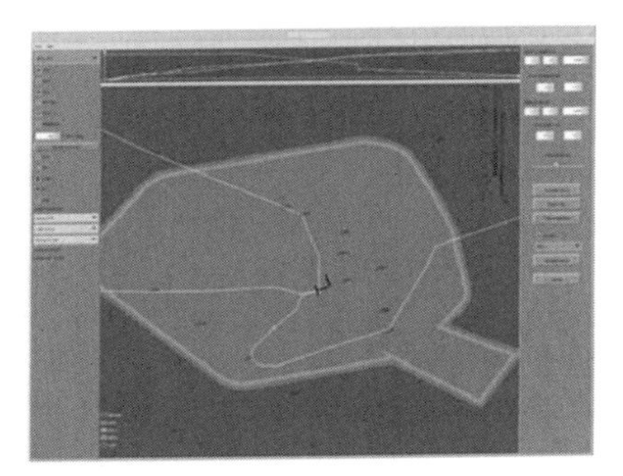

Fig. 7. A real time evaluation in Rome CTR.

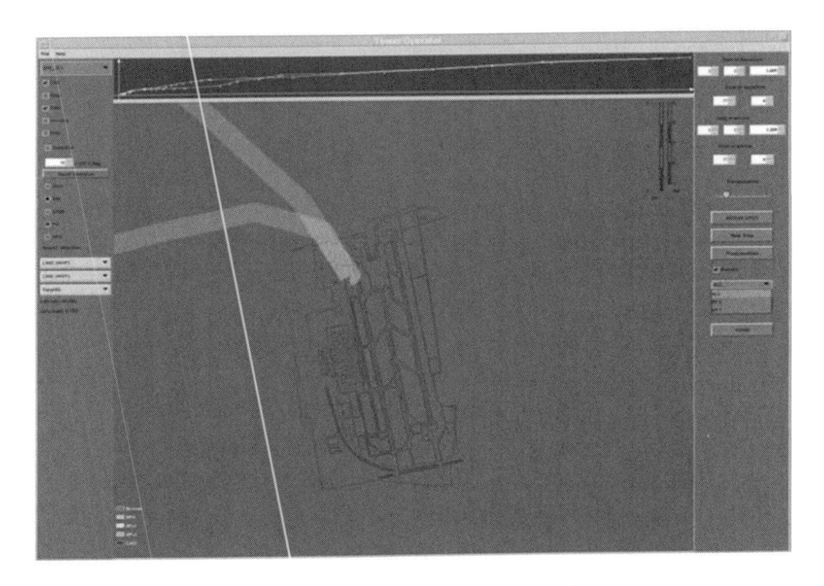

Fig. 8. A real time in airport scenario.

2.2 Supervisor Operator

For the supervisor operator has been developed, following the ICD provided by ESA [3], the interface for the ATC Client which is connected with the ATC Server in the MCC of Ciampino. Thanks to the development of this interface it is possible to predict the performance of the GNSS/SBAS system in terms of integrity parameters. Moreover it is possible to display other elaboration such as the computation of the User Differential Ranging Error and Grid Ionospheric Vertical Error parameter, the average local error, the user fix scattering and so on, as showed in the following pictures. Figure 9 are shown different elaboration in order to verify the GNSS performance of the satellite navigation aid.

The elaboration of the EGNOS-ATC-message type 6 and message type 7 provided by the EGNOS ATC Server will be useful to display the UDRE and GIVE maps in order to evaluate the GNSS behaviour within the Italian airspace.

An example of such elaboration is presented Fig. 10.

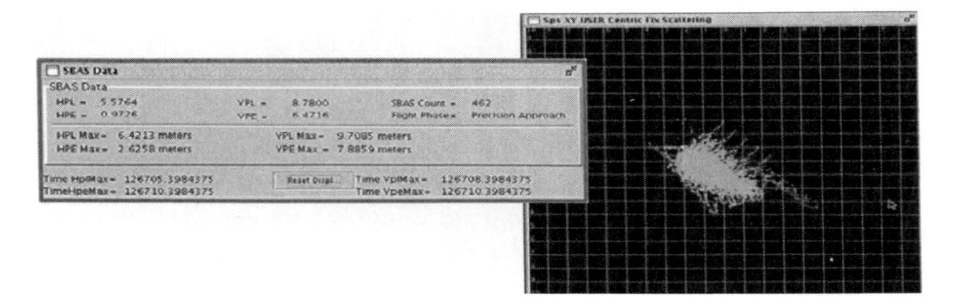

Fig. 9. User fix scattering and local error.

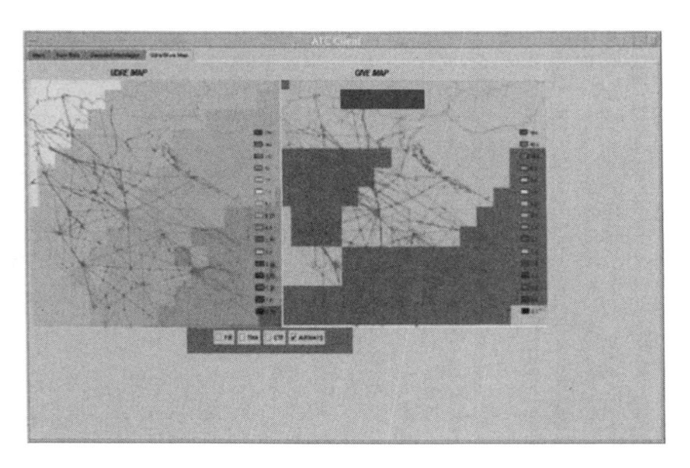

Fig. 10. UDRE and GIVE computing.

The supervisor operator can also know information provided by NANUs messages related to programmed unavailability of GNSS satellites. The unavailability notice are presented in table form or they are displayed, through a simulator, tracking their position as shown in Fig. 11.

2.3 ATC Client Interface

In the scope of EtoG programme, the "ATC Interface" between ATC Server, inside CCF of Ciampino, and the ATC Client (ENAV/SELEX-SI development) that provides data to the ATC users, has been developed following the ICD [3]. Figure 12 the EGNOS ATC block diagram is presented.

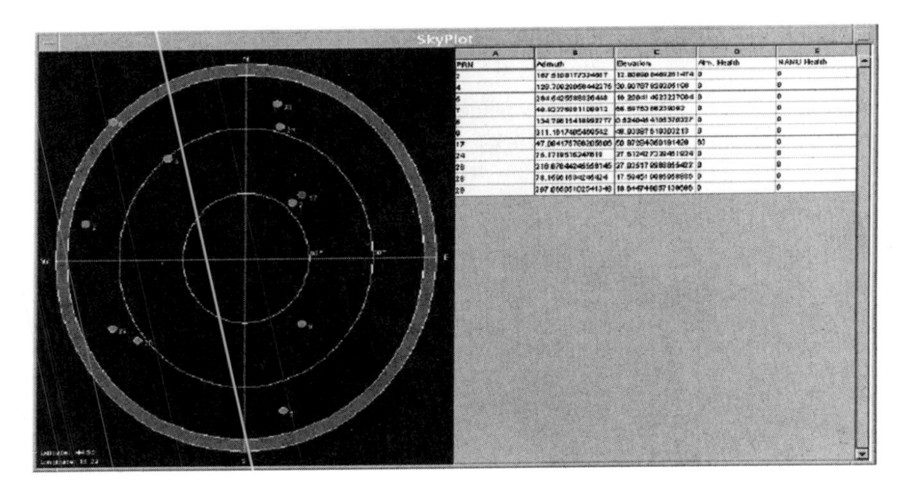

Fig. 11. NANU messages.

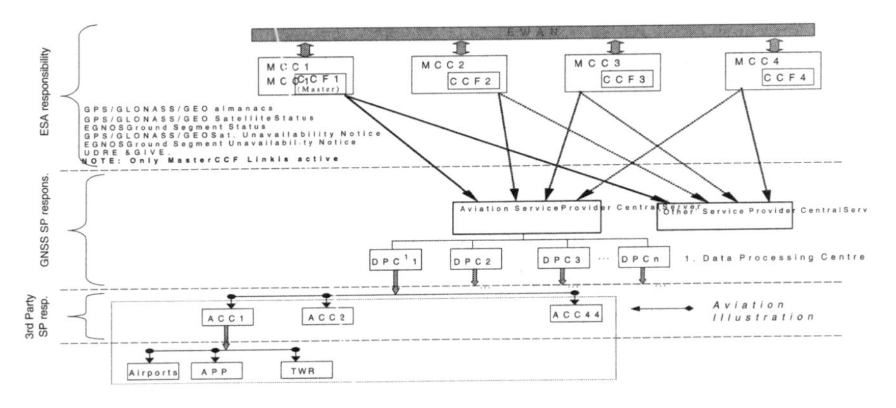

Fig. 12. EGNOS ATC block diagram.

The data sets provided by the CCF are the following:

- 0. ATC Connection Status (MT 0)
- 1. GPS/GLONASS/GEO Almanacs (MT 1)
- 2. GPS/GLONASS/GEO Satellite Status (as monitored by EGNOS) (MT 2)
- 3. EGNOS system Status (MT 3)
- 4. GPS/GLONASS/GEO Satellite Unavailability Notice (MT 4)
- 5. EGNOS Ground Segment Unavailability Notice (planned maintenance) (MT 5)
- 6. CPF processed data: UDRE (MT 6) and GIVE (MT 7)

The data are transmitted to each ATC client connected:

- for the data (1), (4) and (5) repeatedly every 30 minutes and also each time their content are updated
- for the other data (2), (3) and (6) repeatedly every 1 minute

The ATC Server provides data only when the MCC is master; in the future a procedure, described in the ICD, should be developed in order to have the automatic switching from an MCC to another when the first will be not master using a Primary ISDN link. Figure 13 the ATC Client Interface is presented.

It is possible to display the acquired data in two different modes. The *Raw data* window will display the received data without any template, as the ATC server send in broadcast this data. Another way is through the *Decoded Message* window where the received message are formatted in a proper form in order to be read. Figures 14–16 the Raw data window and some example for the Decoded message window.

ATC Client data, possibly merged with data from local sensors (receivers), local environmental conditions (orography, EM environment model) are planned to feed (joined with possible GNSS back-up data from other sources) an ATC Interface for ATC operators.

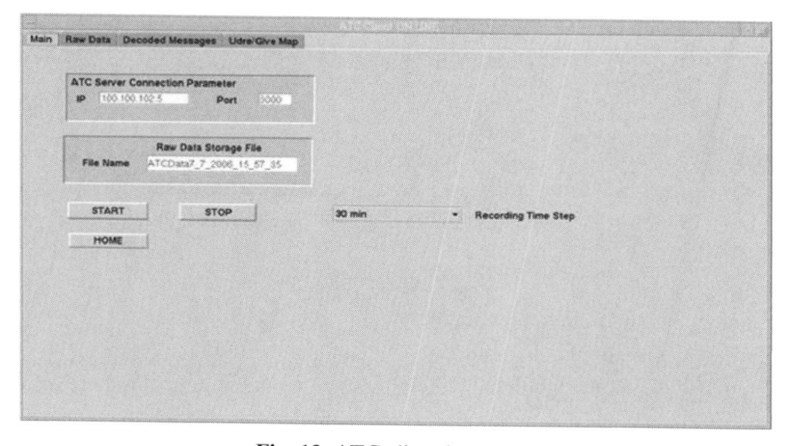

Fig. 13. ATC client interface.

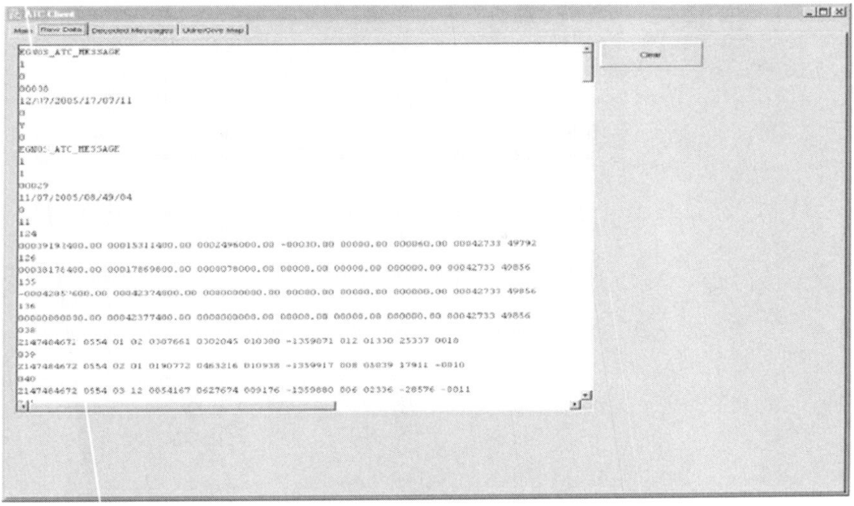

ATC Client

Main | Raw Data | Decoded Messages | Udre/Give Map |

Clear

```
EGNOS_ATC_MESSAGE
1
0
00008
12/07/2005/17/07/11
0
Y
0
EGNOS_ATC_MESSAGE
1
1
00029
11/07/2005/08/49/04
0
11
124
0003919[illegible]400.00 00015311400.00 0002496000.00 -00030.00 00000.00 000060.00 00042733 49792
126
00038176400.00 00017869800.00 0000078000.00 00000.00 00000.00 000000.00 00042733 49856
135
-0004205[illegible]600.00 00042374800.00 0000000000.00 00000.00 00000.00 000000.00 00042733 49856
136
0000000030.00 00042377400.00 0000000000.00 00000.00 00000.00 000000.00 00042733 49856
038
2147484672 0554 01 02 0307661 0302045 010300 -1359871 012 01330 25337 0010
039
2147484672 0554 02 01 0190772 0463216 010938 -1359917 008 03039 17911 -0010
040
2147484672 0554 03 12 0054167 0627674 009176 -1359880 006 02336 -28576 -0011
```

Fig. 14. ATC client - raw data window.

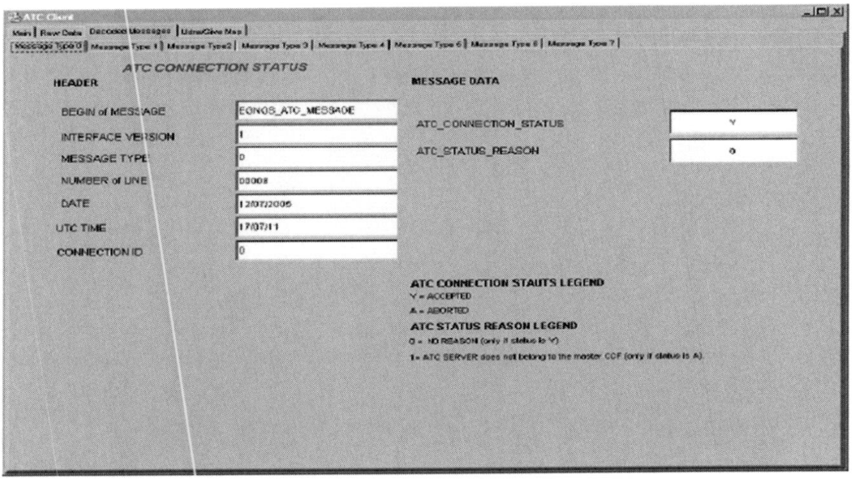

ATC Client

Main | Raw Data | Decoded Messages | Udre/Give Map |

Message Type 0 | Message Type 1 | Message Type2 | Message Type 3 | Message Type 4 | Message Type 5 | Message Type 6 | Message Type 7 |

ATC CONNECTION STATUS

HEADER

BEGIN of MESSAGE	EGNOS_ATC_MESSAGE
INTERFACE VERSION	1
MESSAGE TYPE	0
NUMBER of LINE	00008
DATE	12/07/2005
UTC TIME	17/07/11
CONNECTION ID	0

MESSAGE DATA

ATC_CONNECTION_STATUS	Y
ATC_STATUS_REASON	0

ATC CONNECTION STAUTS LEGEND

Y = ACCEPTED

A = ABORTED

ATC STATUS REASON LEGEND

0 = NO REASON (only if status is Y)

1 = ATC SERVER does not belong to the master CCF (only if status is A)

Fig. 15. ATC Client - decoded message type 0.

Moreover another functionalities is given by the possibility to provide a playback of the recorded EGNOS ATC data. In this way an operator is able to display again, in a separate window, a particular situation while the system runs. Figure 17 the ATC Client in Off Line mode is shown.

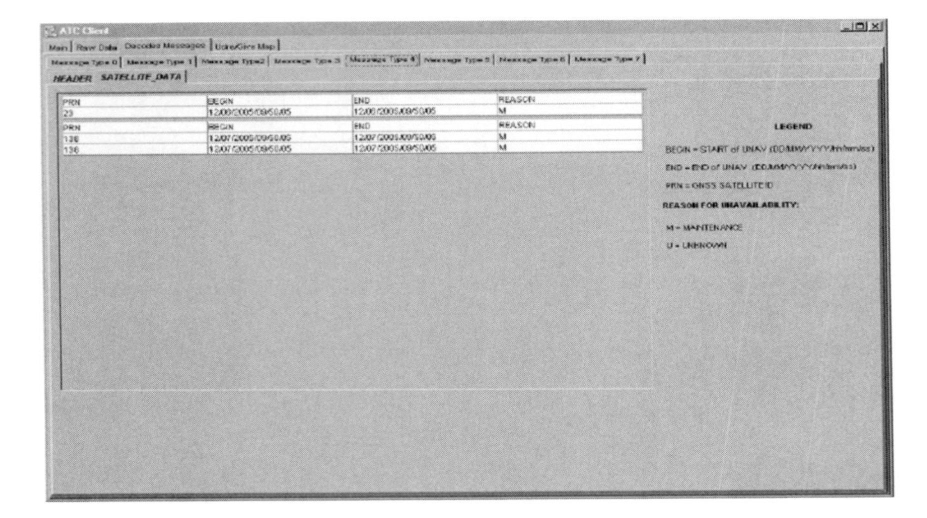

Fig. 16. Decoded message type 4.

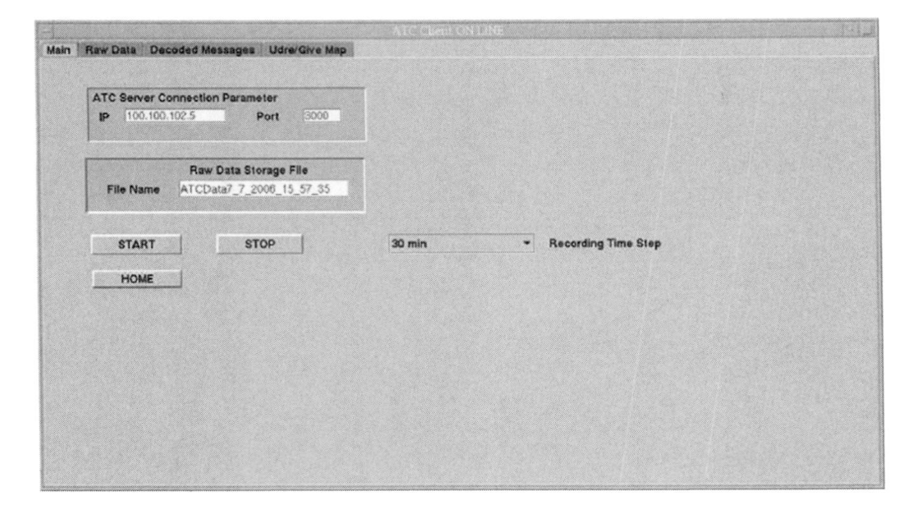

Fig. 17. ATC client for recording and playback mode.

3 Conclusion

Through the EtoG project the ENAV (National Air Navigation Service Provider) and SELEX-SI not only develop satellite application in the important field of air traffic management/control, identifying and planning innovative application products in typical operational scenarios, but also they intend to pursue important targets of research in the field of satellite navigation. More specifically, the project will aim at introducing the GNSS/SBAS applications within Italian airspace.

References

[1] ICAO, International Standards and Recommended Practices – AERONAUTICAL TELECOMMUNICATIONS Annex 10 to the Convention on International Civil Aviation, Volume I (Radio Navigation Aids). Fifth Edition of Volume I, ICAO, July 1996 and Amendments.

[2] RTCA, Minimum Operational Performance Standards for Global Positioning System/Wide Area Augmentation System for the airborne equipment. RTCA/DO 229C release C 28/11/2001

[3] F. FARRE, EGNOS Interfaces Control Document for ATC interface. EGN-ASPI-SYST-DRD 0112/0029 Is: 2 Rev.: A 16/11/2001.

Chapter IV

Advanced Satellite Communications Systems & Services

Advanced Satellite Communication Systems & Services

Dr Satchandi Verma

Session Organizer & Chairman
Universal Satellite Systems
12584 Avocado Way, CA, Riverside, USA
Phone: (951) 688–7600,
e-mail: satchandi@global.net

Since last decade the satellite industry is experiencing growth in advances of satellites in navigation and telecommunications industry. A range of future broadband, navigation, communication and earth observation services are being developed using next generation system architectures transforming the satellite communication and navigation networks for this century and beyond. These various services as shown in Fig. 1 include the following key characteristics:

1. Banking & Retail (Credit authorization of sale, Pricing updates)
2. Transpiration (Inventory control, Fleet management)
3. Entertainment (Video, audio, games, Interactive data)
4. Financial (Brokerage service Electronic payment transactions)
5. Energy (Pipeline monitoring, Power line monitoring)
6. Navigation Services (Navigation, Audio, video, data signals)
7. Telemedicine (Patient evaluation/treatment)
8. Internet (Streaming video, audio & data)
9. Earth Environment (Water and atmospheric Temperatures, weather, ocean changes)
10. Home Security & Information transfer (Audio, video, data)

The next generation system architecture apply advanced techniques in Network operational flexibility (consolidation, diversity, back-up), Global connectivity for symmetrical and asymmetrical applications with low operational cost, System management Flexibility with manageable complexity and Multicast mobility management, Use Networks of Networks, On-demand bandwidth capabilities with Integrated customer infrastructure, Guarantee end-to-end quality of service (premium customers), Multipoint-to-multipoint Internet and broadcasting capability, Backward compatibility/seamless integration into existing protocols and infrastructures (Enhanced TCP, IP routing and ATM switching) and Multicasting using Satellite Multicast Adaptation Protocol (SMAP).

The system architecture is defined using the following key design drivers:

***Satellite Payloads**

Satellite orbital locations (LEO/MEO/HEO/GEO)
High Powered Satellite Buses

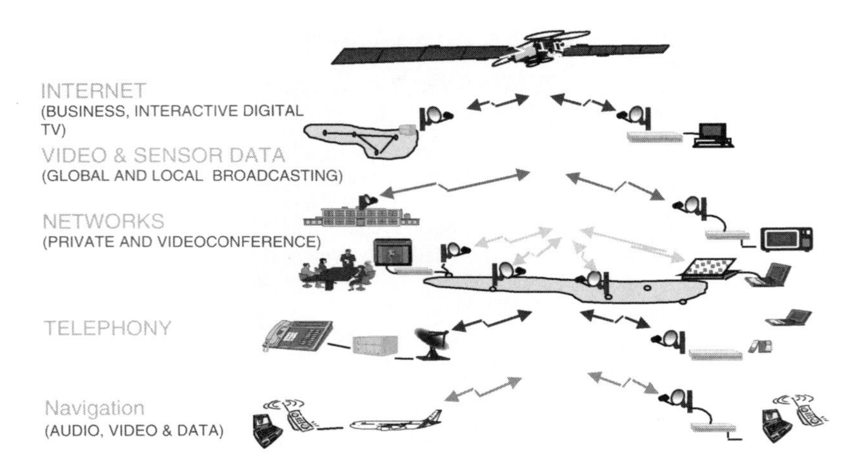

Fig. 1. Advanced satellite systems networks.

Onboard Digital Baseband Processors
Onboard Packet & Beam switching
Progressive Payload
Standard Interfaces

***Satellite Antenna**
Gimbaled Multibeam Antenna
Large Deployable Mesh Antenna
High gain phased arrays Antenna

***Communication Links**
RF & Optical Inter Satellite links
Uplink Power Mitigation techniques
L/S/Ku/Ka Band Up & Down links
Band Efficient Modulation

***Ground Network & Terminals**
Software Defined Radios
Advanced IC/DSP
Intelligent network architectures
Low cost ground terminals
Multi satellite link terminals

***Satellite Platforms**
A new generation of high-speed onboard digital payload processors and high-gain broadband antennas are emerging for either replacing the legacy systems. The major enhancement areas for next generation platforms are:

OBP with multi-satellite accessing, traffic aggregation and routing using inter-satellite links (RF, laser).
On-demand Satellite resource allocation with guarantee end-to-end service quality.
Satellite capability growth for future customer needs.
Progressive Payloads Configurations.

***Satellite transmissions/Antenna**
Network of Networks Architectures using mesh connectivity
Multimedia service enabling networking for dynamic on demand traffic needs
Wide, tailored, land mass coverage with on-orbit flexibility (Regional/local)
Adaptive beam forming and shaped beam antenna (dual polarization)
Efficiency transmissions (coding, packet, cell switching, reduced delays)
Secure Higher Capacity satellite network for open air interfaces/Internet "threats".
Manage core congestion for Internet traffic with flexible bandwidth control.

These advanced systems focus on enhancing the broadband satellite network capacity, billable bits and Quality of Service (QoS) to provide cost effective systems. The system design flexibility is further enhanced by using progressive satellite concepts for enhanced future customer requirement with new technologies. Progressive Supplemental Satellite Provide Cost Effective
System Capability For Future Customer Needs The major features advantages of this concept are:

***Progressive Supplemental satellite features**
Satellite designed to meet changed customer service requirements
Original satellite is designed with the interface to connect with the smaller satellite at a later date
Progressive satellite is connected to original satellite through pre defined Inter satellite link via a pre designed interface
Satellite uses advanced available satellite technologies at launch time
Provides use or future satellite technologies to meet new customer services by sharing the original satellite payloads

***Progressive satellite advantages**
Allows Multimedia satellite System to be upgraded for meeting the new customer service requirements, connectivity and capacity
Advanced new cost effective technologies are used for meeting new system needs after the original satellite launch

This workshop session presentation focuses in the key areas for next generation satellite system and services. In particular papers are presented in the areas of satellite accessing and resources allocation, Integration of Navigation and Communication services, Satellite Terminals Reconfiguration flexibility, Broad Band Mobile terminals, Broadcasting and Telemedicine services

References

[1] Advanced Multimedia Satellite Systems, S Verma, 22th AIAA ISSC Conference, May 2004, Monterey, CA, USA
[2] Next Generation Broadband Satellite Communication Systems, S Verma, E Wiswell, 20th AIAA ISSC Conference, May 2003, Montreal
[3] Broadband Payloads for The Emerging Ka_Band Market, M Bever, S Willoghby, E Wiswell, K Ho, Seventh Ka-Band Utilization Conference, Santa Margherita, Italy, September 26–28, 2001

QOS-Constrained MOP-Based Bandwidth Allocation Over Space Networks

Igor Bisio, Mario Marchese

DIST - Department of Communication, Computer and System Science
University of Genoa, Via Opera Pia 13, 16145, Genoa, Italy
Phone: +39-010-3532806, Fax: +39-010-3532154,
e-mail: Igor.Bisio@unige.it, Mario.Marchese@unige.it

Abstract. The paper formalizes and analyzes the bandwidth allocation process over space communication systems modelled as a Multi – Objective Programming (MOP) in presence of Quality of Service (QoS) constraints. The reference allocation scheme considered is based on GOAL programming and is called "Minimum Distance" algorithm. The work proposes two different versions of the Minimum Distance scheme both aimed at assigning the bandwidth so to approach a non-competitive situation as close as possible and at guaranteeing a fixed performance for each traffic flow traversing the space network. The proposals have been tested over a faded channel by using TCP/IP traffic. The performance evaluation is carried out analytically by varying the degradation level of the channel.

1 Introduction

The advantage of using space communication systems (HAPs, GEO and LEO satellites, possibly integrated with terrestrial links) for TCP/IP applications is clear: to exchange ubiquitous information among geographically remote sites also in hazardous areas with large bandwidth availability. In such environments one of the main cause of degradation is rain attenuation, which generates significant communication detriment, information loss and, consequently, Quality of Service (QoS) degradation [1]. Quality of Service is the ability of a network element (e.g. an application, host or router or an earth station in satellite networks) to have some level of assurance that its traffic and service requirements can be satisfied. Each service has its own set of QoS parameters: *Delay*, which is the time for a packet to be transported from the sender to the receiver, *Jitter*, which is the variation in end-to-end transit delay, *Bandwidth*, that is the maximal data transfer rate that can be sustained between two end points, and *Packet loss*, which is defined as the ratio between the number of undelivered packets and the total number of sent packets.

It is worth noting that the channel capacity (bandwidth) is limited not only by the physical infrastructure of the traffic path within the transit networks, which provides an upper bound to available bandwidth, but is also limited by the number of other flows that share common components of the space network and by the channel error countermeasures typically used by the physical layer of space networks (e.g., Forward Error Correction Codes).

In this work, the proposed bandwidth allocation schemes, conscious of the aforementioned channel capacity limitations, are aimed at guaranteeing a fixed level of quality of service in terms of packet loss probability.

The rationale under this paper is considering bandwidth allocation as a competitive problem where each station is "represented" by a cost function that needs to be minimized at cost of the others. In practice, all the functions must be minimized simultaneously considering the QoS requirements. It is the definition of the Multi-Objective Programming (MOP) class of problems, which is the base of the methods introduced in the paper. The schemes are based on the "Minimum Distance" strategy, which is aimed at approaching the ideal performance obtained when each single station has the availability of all the channel bandwidth, by minimizing the Euclidean distance between the performance vector and the ideal solution of the problem.

The paper is structured as follows: section 2 introduces the network topology. The formalization of the bandwidth allocation is presented in section 3; the TCP packet loss probability model is contained in section 4. The Utopia Minimum Distance (UMD) algorithm, already presented by the authors [2], is revised in section 5. Two alternative mechanisms (CUMD and QDMD) aimed at guaranteeing a QoS requirement are described in section 6. Section 7 reports the performance evaluation and section 8 the conclusions.

2 Network Topology

The network considered (Fig. 1) is composed of Z earth stations connected through a space connection. The choice of the technology does not affect the general behaviour of the schemes and it has been left unspecified here for the sake of generality. It may be applied over GEO/LEO satellites and HAP platforms. The main difference

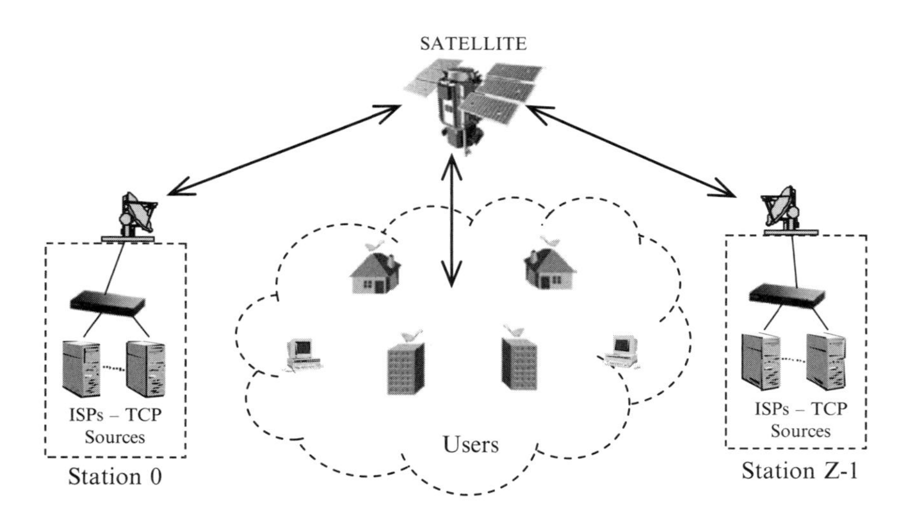

Fig. 1. Network topology.

stands in the round trip time (RTT). The results have been fulfilled by using RTT=520[ms] (GEO environment). The control architecture is centralized: an earth station (or the satellite itself, if switching on board is allowed) represents the master station, which manages the resources and provides the other stations with a portion of the overall bandwidth (e.g., TDMA slots); each station equally shares the assigned portion between its traffic flows (the fairness hypothesis is made).

Each user requests a TCP/IP service (e.g., Web page and File transfer) by using the space channel itself (or also other communication media). The request traffic is supposed negligible. After receiving the request ISPs send traffic through the earth stations and the space link. To carry out the process, each earth station conveys traffic from the directly connected ISPs and accesses the space channel in competition with the other earth stations.

Fading is modelled as bandwidth reduction. From the implementation viewpoint, it means using a FEC code where each earth station may adaptively change the amount of redundancy bits (e.g. the correction power of the code) in dependence on fading, so reducing the real bandwidth availability. Mathematically, it means that the bandwidth $C_z^{real} \in \mathbb{R}$, available for the z-th station, is composed of the nominal bandwidth $C_z \in \mathbb{R}$ and of the factor $\beta_z \in \mathbb{R}$, which is, in this paper, a variable parameter contained in the interval [0,1].

$$C_z^{real} = \beta_z \cdot C_z ; \beta_z \in [0, 1], \beta_z \in \mathbb{R} \tag{1}$$

A specific value β_z corresponds to a fixed attenuation level "seen" by the z-th station. An example of the mapping between Carrier Power to One-Side Noise Spectral Density Ratio (C/N_0) and β_z is contained in Table 1 (from [3]).

The values β_z shown in the table will be used in the performance evaluation section of this paper.

3 Bandwidth Allocation Problem Definition

Each earth station has a single buffer gathering TCP traffic from the sources (ISPs). The practical aim of the allocator is the provision of bandwidth to each buffer server by splitting the overall available capacity among the stations (the competitive entities of the problem). Analytically, the bandwidth allocation defined as a Multi – Objective Programming (MOP) problem may be formalized as:

$$\begin{aligned} \mathbf{C}^{opt} &= \left\{ C_0^{opt}, \ldots, C_z^{opt}, \ldots, C_{Z-1}^{opt} \right\} = \\ &\arg\min \left\{ \mathbf{F}(\mathbf{C}) \right\}; \mathbf{F}(\mathbf{C}) : \mathbf{D} \subset \mathbb{R}^Z \to \mathbb{R}^Z, \mathbf{C} \geq 0 \end{aligned} \tag{2}$$

where: $\mathbf{C} \in \mathbf{D}$, $\mathbf{C} = \{C_0, \ldots, C_z, \ldots, C_{Z-1}\}$ is the vector of the capacities that can be assigned to the earth stations; the element C_z, $\forall z \in [0, Z-1]$, $z \in \mathbb{N}$ is referred to the

Table 1. Signal to noise ratio and related β_z Level.

C/N_0 **[dB]**	66.6–69.6	69.6–72.6	72.6–74.6	74.6–77.1	>77.1
β_z	0.15625	0.3125	0.625	0.8333	1

z-th station; $\mathbf{C}^{opt} \in \mathbf{D}$, is the vector of optimal allocations; and $\mathbf{D} \subset \mathbb{R}^z$ represents the domain of the vector of functions. The solution has to respect the constraint:

$$\sum_{z=0}^{Z-1} C_z = C_{TOT} \tag{3}$$

where C_{tot} is the available overall capacity.

F(C), dependent on the vector **C**, is the *performance vector*

$$\mathbf{F}(\mathbf{C}) = \{ f_0(C_0), \ldots, f_z(C_z), \ldots, f_{Z-1}(C_{Z-1}) \}, \forall z \in [0, Z-1], Z \in \mathbb{N} \tag{4}$$

The single z-th *performance function* is a component of the vector. Each *performance function* $f_z(C_z)$ (or *objective*) of the system is defined here as the average TCP packet loss probability. Actually any other convex and decreasing over bandwidth function may be used. The packet loss probability at TCP layer seems a reasonable choice but it may be regarded also as an operative example for the theory presented. The TCP packet loss probability $P^z_{loss}(\cdot)$ is a function of the bandwidth (C_z) as well as of the number of active sources (N_z) and of the fading level (β_z), for each station z. $P^z_{loss}(\cdot)$ is averaged on the fading level β_z, which is considered a discrete stochastic variable ranging among L possible values β^l_z happening with probability $p_{\beta^l_z}$.

$$\begin{aligned} f_z(C_z) &= \underset{\beta_z}{E}\left[P^z_{loss}(C_z, N_z, \beta_z) \right] \\ &= \sum_{l=0}^{L-1} \left[P^z_{loss}(\beta^l_z \cdot C_z, N_z) \right] \cdot p_{\beta^l_z}; \forall l \in [0, L-1], L \in \mathbb{N} \end{aligned} \tag{5}$$

In general, the problem defined above is a Multi – Object Programming problem where each considered function $f_z(C_z)$ represents a single competitive cost function. In other words, a single *performance function* competes with the others for bandwidth.

The optimal solution for MOP problems is called POP-Pareto Optimal Point [4], coherently with the classical MOP theory. It was adopted in economic environment and may be summarized as follows.

The bandwidth allocation $\mathbf{C}^{opt} \in \mathbf{D}$ is a POP if does not exist a generic allocation $\mathbf{C} \in \mathbf{D}$ so that:

$$\mathbf{F}(\mathbf{C}) \leq_P \mathbf{F}(\mathbf{C}^{opt}), \forall \mathbf{C} \neq \mathbf{C}^{opt} \tag{6}$$

Concerning the operator "$\leq_P$": given two generic performance vectors $\mathbf{F}^1, \mathbf{F}^2 \in \mathbb{R}^z$, $\mathbf{F}^1$ *dominates* $\mathbf{F}^2$ ($\mathbf{F}^1 \leq_p \mathbf{F}^2$) when:

$$\begin{aligned} &f^1_x \leq f^2_x \forall x \in \{0, 1, \ldots, Z-1\} \textit{ and} \\ &f^1_y < f^2_y \text{ for at least one element } y \in \{0, 1, \ldots, Z-1\} \end{aligned} \tag{7}$$

Where f^1_x, f^1_y, f^2_x and f^2_y are the elements of the vector $\mathbf{F}^1$ and $\mathbf{F}^2$, respectively.

In practice, it means that once in a POP, a lower value of one function implies necessarily an increase of at least one of the other functions. In the allocation problem considered, the constraint in (3) defines the set of POP solutions, because, over that constraint, each variation of the allocation, aimed at enhancing the performance of a specific earth station, implies the performance deterioration of at least another station due to the decreasing nature of the considered performance function (as clarified in next section, in formula (8)).

It is worth noting that, in the proposed methodology, no on-line decision method is applied. The system evolution is ruled by stochastic variables. In practice, **F**(**C**) in the optimization problem (2) is considered to be the average value of the performance vector over all the possible realizations of the stochastic processes of the space network. The performance functions are representative of the steady-state behaviour of the system and the allocation is provided with a single infinite-horizon decision.

4 The TCP Packet Loss Probability Model

The TCP model considered is based on previous work of the authors [5]. Considering geostationary space systems, the round trip time *RTT* may be supposed fixed and equal for all the sources. This condition matches the hypothesis of fairness, which is an essential condition to get the used TCP packet loss probability model. Taking TCP Reno as reference and considering only the Congestion Avoidance phase of the TCP, the Packet Loss Probability (used in equation (5)) may be explicitly expressed as a function of the available bandwidth and of the number of TCP active sources as:

$$P_{loss}^{z} = 32N_{z}^{2}\left[3b\,(m+1)^{2}\left(\beta_{z}\tilde{C}_{z}\,RTT + \tilde{Q}_{z}\right)^{2}\right]^{-1} \tag{8}$$

where: N_z is the number of TCP active sources conveyed into the z-th earth station; b is the number of TCP packets covered by one acknowledgment; m is the reduction factor of the TCP transmission window during the Congestion Avoidance phase (typically $m = 1/2$); $\tilde{C}_z$ is the bandwidth "seen" by the TCP aggregate of the z-th earth station expressed in packets/s ($\tilde{C}_z = C_z/d$, where d is the TCP packet size); $\tilde{Q}_z$ is the buffer size, expressed in packets, of the z-th earth station. The model is valid at regime condition of the TCP senders, coherently with the infinite-horizon hypothesis reported in the previous section.

5 Minimum Distance Algorithm

The Minimum Distance method is a flexible methodology that allows the resolution of the allocation problem (2). It is part of the MOP resolution family called GOAL [4]. It bases its decision only on the ideal solution of the problem: the so called *utopia point*. In more detail, the *ideal performance vector* is:

$$\mathbf{F}^{id}(\mathbf{C}^{id}) = \left\{f_{0}^{id}(C_{0}^{id}), \ldots, f_{z}^{id}(C_{z}^{id}), \ldots, f_{Z}^{id}(C_{Z-1}^{id})\right\} \tag{9}$$

where

$$f_{z}^{id}(C_{z}^{id}) = \min_{C_z} \underset{\beta_z}{E}\left[P_{loss}^{z}(C_z, N_z, \beta_z)\right], C_z \in [0, C_{tot}] \tag{10}$$

From equation (10), called *single objective problem*, it is clear that the optimal solution is given by $C_z = C_{TOT}, \forall z \in [0, Z-1]. So, \mathbf{C}^{id} = \{C_{TOT}, C_{TOT}, \ldots, C_{TOT}\}$. In other words: the ideal situation would be when each station has the availability of

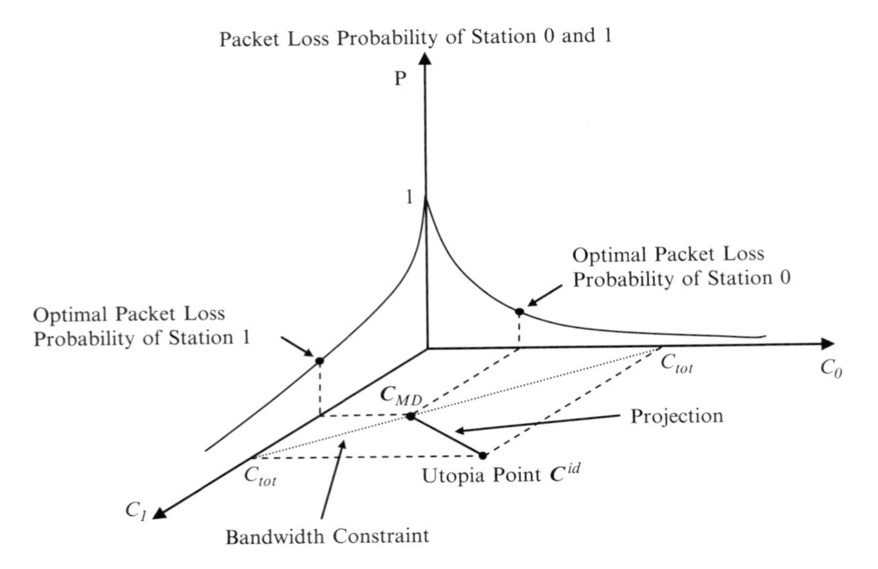

Fig. 2. UMD behaviour.

the overall channel bandwidth. Obviously it is a physically unfeasible condition that can be only approached due to constraint (3).

Starting from the definition of the *ideal performance vector*, the problem in equation (2) can be solved by the following allocation:

$$\mathbf{C}_{MD}^{opt} = \arg\min_{\mathbf{C}} \left(\left\| \mathbf{F}(\mathbf{C}) - \mathbf{F}^{id}(\mathbf{C}^{id}) \right\|_2 \right)^2 \tag{11}$$

where $\|\bullet\|_2$ is the Euclidean norm. The proposed technique allows minimizing the distance between the performance vector and the ideal solution of the problem. It is called Utopia Minimum Distance – UMD scheme, in the reminder of the paper. The Euclidean norm $\left(\| \mathbf{F}(\mathbf{C}) - \mathbf{F}^{id}(\mathbf{C}^{id}) \|_2\right)^2$ is the decisional criterion of the UMD method. The minimization is carried out under the constraint (3).

Figure 2 describes the behaviour of the UMD strategy for 2 earth stations (Station 0 and Station 1) geometrically. On the basis of the fading conditions, UMD computes the projection of the *utopia point* that minimizes the packet loss probabilities of Stations 0 and 1 reported on the axis P simultaneously. C_0 and C_1 are the axes reporting the capacities allocated to Stations 0 and 1, respectively. Station 1 is supposed to be faded in Fig. 2 and its related packet loss probability (shown in the plane (C_1, P)) is higher than the packet loss of Station 0. In practice, the action of the proposed algorithm provides the bandwidth allocation as a solution of the problem (11), representative of the utopia point projection (which is not orthogonal in the capacity domain (C_0, C_1)) over the bandwidth constraint (3). Being a completely competitive environment where each station "makes its own interest", the solution tends to privilege the faded station.

6 QOS Constraints

Additional constraints may be added to match specific QoS performance requirements. The needed capacity may be provided by modifying the UMD method. In more detail, QoS constraints may be fixed for each earth station (i.e. for each *performance function*). Analytically it may be described as:

$$f_z(C_z) \leq \gamma_z, \forall z \in [0, Z-1], Z \in \mathbb{N} \quad (12)$$

where $\gamma_z \in \mathbb{R}$ is the QoS requirement constraint for the z-th station. In terms of packet loss probability typical γ_z values may be set to 10^{-2}, for voice-streaming applications, to 10^{-3}, for more demanding applications. C_z^{thr} needs to be allocated to generic station z to satisfy constraint (12):

$$C_z^{thr} = \{C_z : f_z(C_z) = \gamma_z, \forall z \in [0, Z-1], Z \in \mathbb{N}\} \quad (13)$$

The practical aim is to guarantee that the bandwidth allocated to z-th station is either larger or equal to, C_z^{thr}:

$$C_z \geq C_z^{thr}, \forall z \in [0, Z-1], Z \in \mathbb{N} \quad (14)$$

The main problem is to match the limitation of the overall available capacity (equation (3)). In facts: the sum of the required bandwidths C_z^{thr} may be larger than the overall capacity provided by the space channel. It implies two possible compromises: 1) bandwidth allocation penalizes some stations, whose traffic flows are considered pure best-effort, and guarantees the required bandwidth only for a subset of them, when possible. This approach is called Constrained Utopia Minimum Distance – CUMD; 2) the algorithm provides bandwidth so to approach the requested QoS as close as possible for all the stations. This approach is called QoS Point Minimum Distance - QPMD. This choice may imply that all the Z constraints are not satisfied, even if there is enough bandwidth to satisfy a portion of them.

6.1 Constrained Utopia Minimum Distance (CUMD).

The constraint set reduces the overall possible solutions defined in equation (11) by creating a subset of possible POPs. In more detail, the Euclidean norm $\left(\|\mathbf{F}(\mathbf{C}) - \mathbf{F}^{id}(\mathbf{C}^{id})\|_2\right)^2$ is the decisional criterion also of the CUMD method but the minimization is carried out both under the constraint (3) and under the set of Z constraints defined in (14).

There is no assurance that QoS requirements for each earth station are guaranteed, due to the limited amount of available capacity and to the fading conditions "seen" by the earth stations.

Considering only two earth stations for the sake of simplicity: given the plane (C_0, C_1) of Fig. 2, Figs. 3a and 3b try synthesizing what can happen.

Figure 3a contains the example when both requirements can be satisfied: constraint (14) defines a continuous set of points.

Figure 3b shows the situation when constraints (14) define a discontinuous set of points and cannot be satisfied in the same time.

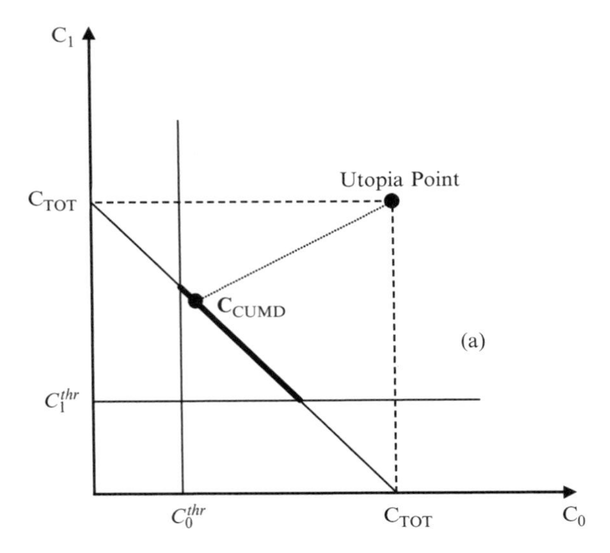

Fig. 3. (a) CUMD behavior – satisfied constraints.

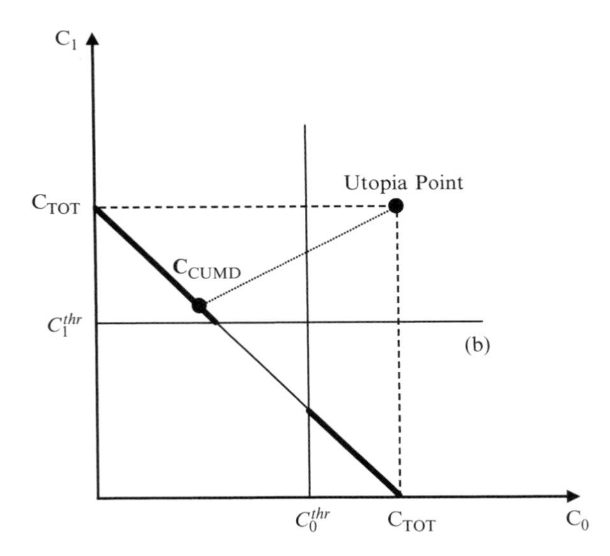

Fig. 3. (b) CUMD behavior – constraints not satisfied in the same time.

It means that just one station can reach its QoS level. The other station needs to provide a full best-effort service to its flows. The choice of the privileged station depends on the distance with the utopia point. The station assuring the minimum distance is chosen: Station 1 in the example. It holds true also for more than two stations. If two or more stations assure the minimum distance, the choice is random (it is true also for the situations described just below). Similar behavior is obtained if at least one of the requirements implies $C_z^{thr} \geq C_{TOT}$: the bandwidth is assigned to the station that requires feasible bandwidth ($C_z^{thr} < C_{TOT}$), so respecting the QoS constraint. The other station gets the residual channel capacity. If there is more than one station where $C_z^{thr} < C_{TOT}$, again the minimum distance choice is taken. The method allows guaranteeing at least one of the performance constraints if the bandwidth needed by one of the stations is lower that the overall capacity available. If $C_z^{thr} \geq C_{TOT}, \forall z$, then one of the station gets all the available channel capacity (through minimum distance choice) and the other(s) is(are) in complete outage condition.

6.2 QoS Point Minimum Distance (QPMD)

QPMD does not minimize the Euclidean distance from the *utopia point* (the situation with no competition) but the distance from the representative point of the desired QoS performance. In practice, it corresponds to use a new definition of the *ideal performance vector*. The new reference point (where QoS is guaranteed) is called *QoS Point* and the related vector is called *QoS performance vector*.

$$\mathbf{F}^{QoS}\left(\mathbf{C}^{QoS}\right)=\left\{f_0^{QoS}\left(C_0^{QoS}\right),\ldots,f_Z^{QoS}\left(C_{Z-1}^{QoS}\right)\right\} \quad (15)$$

where

$$f_z^{QoS}\left(C_z^{thr}\right)=\gamma_z, C_z^{thr}\in\mathbb{R} \quad (16)$$

which is directly derived from (13). From (15) and (16), obviously $\mathbf{C}^{QoS} = \left\{C_0^{thr}, C_1^{thr}, \ldots, C_z^{thr}\right\}$.

The problem in equation (2) can be now solved through:

$$\mathbf{C}_{QPMD}^{opt} = \arg\min_{\mathbf{C}}\left(\|\mathbf{F}(\mathbf{C})-\mathbf{F}^{QoS}(\mathbf{C}^{QoS})\|_2\right)^2 \quad (17)$$

The minimization is obviously carried out under the constraint (3).

Also in this case, the behaviour may be described by using the reference situation where two earth stations are considered. Two cases may happen:

1) the *QoS Point* satisfies the constraint (3) and it is within the set of feasible allocations (Fig. 4a); QoS requirements are matched and the overall performance is surely better than expected because more bandwidth than required is assigned to the stations.
2) the *QoS Point* is outside the feasible allocation region defined by the constraint (3). QPMD provides allocations through equation (17). QoS satisfaction can be only approached (Fig. 4b).

7 Performance Evaluation

The aim of this section is to compare the performance of the allocation techniques (UMD, CUMD and QPMD) in terms of allocated bandwidth and packet loss probability. The action is fulfilled analytically by varying the fading conditions of the earth stations. The considered network scenario is composed of 2 earth stations: Station 0, always in clear sky condition, and Station 1, which varies its fading level according with Table 1. Each station gathers traffic from TCP sources and transmits

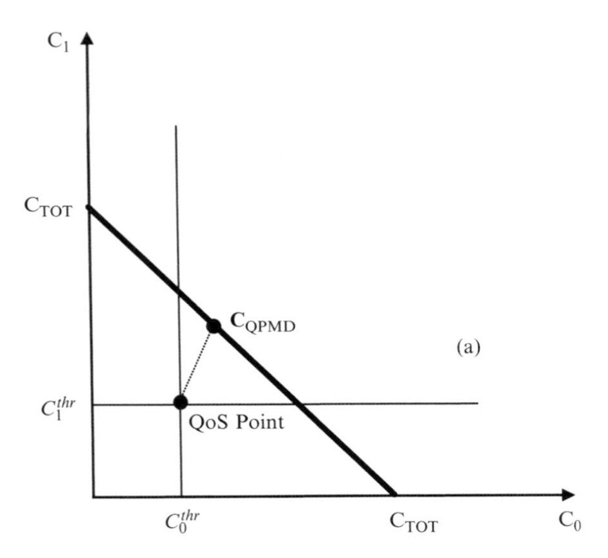

Fig. 4. (a) QPMD behaviour.

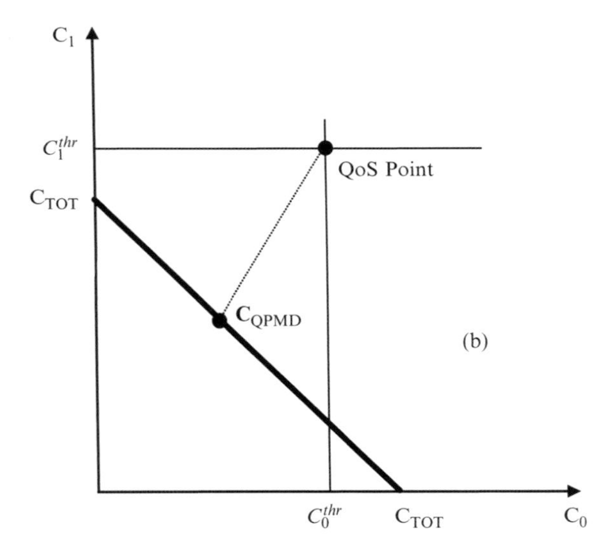

Fig. 4. (b) QPMD behaviour.

it to the terminal users through the space system (GEO satellite in this case). The number of active TCP sources is set to $N_z = 10$, $z = \{0, 1\}$. The fading level is a deterministic quantity ($L = 1$ and $p_{\beta_z^l} = 1$ $\forall z$, $\forall l$) in the tests. The overall bandwidth available C_{TOT} is set to 10.24 [Mbps] (10240 [Kbps]) and the TCP buffer size $\tilde{Q}_z$ is set to 10 packets (of 1500 bytes) for each earth station. The round trip time is supposed fixed and equal to 520 [ms] for all the stations, it is considered comprehensive

Table 2. Minimum bandwidth requirements to match $f_0\ (C_0) \le 0.01, f_1\ (C_1) \le 0.01$, stations 0 and 1.

β_1	C_0^{thr} [Kbps]	C_1^{thr} [Kbps]
0.156	2670	17116
0.312	2670	8558
0.625	2670	4272
0.833	2670	3205
1	2670	2670

of the propagation delay of the GEO channel and of the waiting time spent into the buffers of the earth stations. Bandwidth is considered a discrete quantity and the minimum amount of allocated capacity is set to 128 [kbps].

UMD represents a completely competitive problem with no constraint. Its aim is approaching the ideal point where each station (both, in this case) has the complete availability of $C_{TOT} = 10.24$ [Mbps]. CUMD solves the same problem but adds a set of constraints (in (12)) over each performance function so getting minimum bandwidth requirements (in (14)) for each station. Following the same philosophy QPMD tries respecting (or, if it is not possible, approaching) the constraint set (12) by originating a new ideal point represented by the bandwidth assignations that allow respecting constraints.

The performance analysis is carried out by setting two different sets of performance constraints (12) and showing bandwidth allocations and packet loss probabilities for the two involved stations.

In detail, the first set of tests is obtained by using: $f_0\ (C_0) \le 0.01, f_1\ (C_1) \le 0.01$. Table 2 contains the minimum capacity ($C_0^{thr}\ and C_1^{thr}$) requirements to match performance constraints for Stations 0 and 1. Figures 5 and 6 show the bandwidth assigned to Stations 0 and 1, respectively, versus the fading level of Station 1. Figures 7 and 8 contain the values of the packet loss probability for Stations 0 and 1, respectively, versus the fading level of Station 1.

The second set of tests is performed by setting: $f_0\ (C_0) \le 0.001, f_1\ (C_1) \le 0.001$. Table 3 shows the minimum capacity requirements to match performance constraints in this case. Figures 9 and 10 show the allocated bandwidth, similarly to Figs. 5 and 6, while Figs. 11 and 12 the values of packet loss probability, again for Stations 0 and 1, respectively, versus the fading level of Station 1.

Table 3. Minimum bandwidth requirements to match $f_0\ (C_0) \le 0.001, f_1\ (C_1) \le 0.001$, stations 0 and 1.

β_1	C_0^{thr} [Kb/s]	C_1^{thr} [Kb/s]
0.156	8942	57325
0.312	8942	28663
0.625	8942	14303
0.833	8942	10735
1	8942	8942

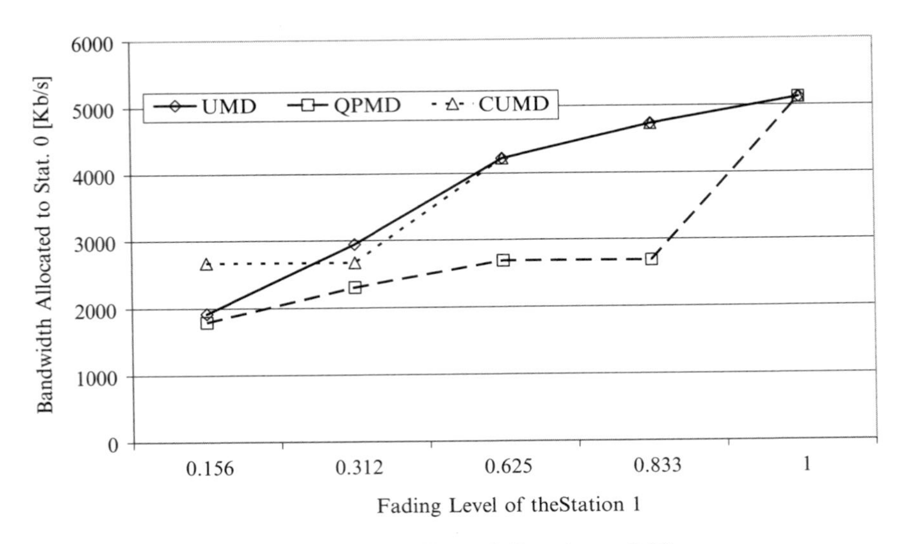

Fig. 5. Bandwidth allocated (Stat. 0, $\gamma_z = 0.01$).

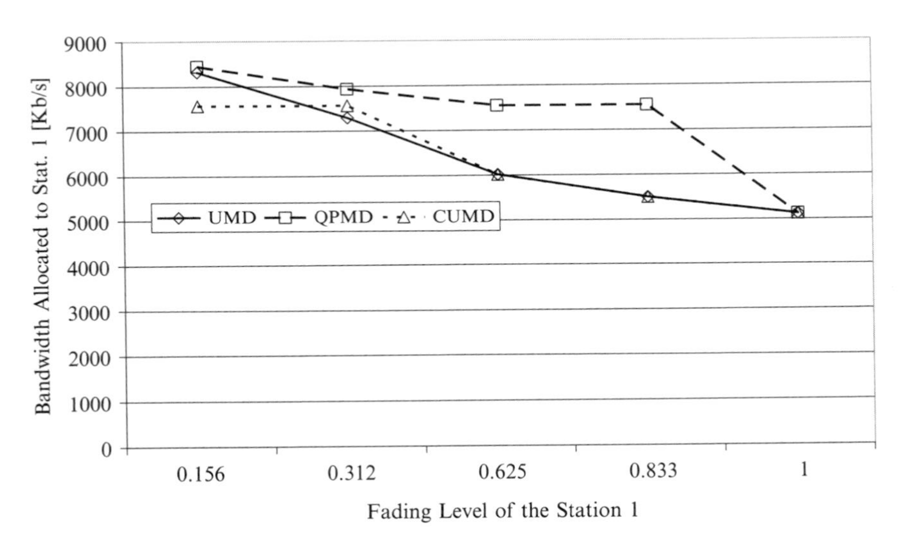

Fig. 6. Bandwidth allocated (Stat. 1, $\gamma_z = 0.01$).

Concerning the first group of tests and CUMD algorithm: values corresponding to $\beta_1 = 0.156$ are the most interesting to check the behaviour of the algorithms. It is the situation where constraints $f_0\,(C_0) \leq 0.01, f_1\,(C_1) \leq 0.01$ cannot be satisfied in the same time because the overall bandwidth is not sufficient ($C_1^{thr} \geq 10.24$ [Mbps], see the first row of Table 2). CUMD chooses to satisfy $f_0\,(C_0) \leq 0.01$ because $C_0^{thr} = 2.67$ [Mbps] ≤ 10.24 [Mbps] and to consider traffic from Station 1 as best effort flows.

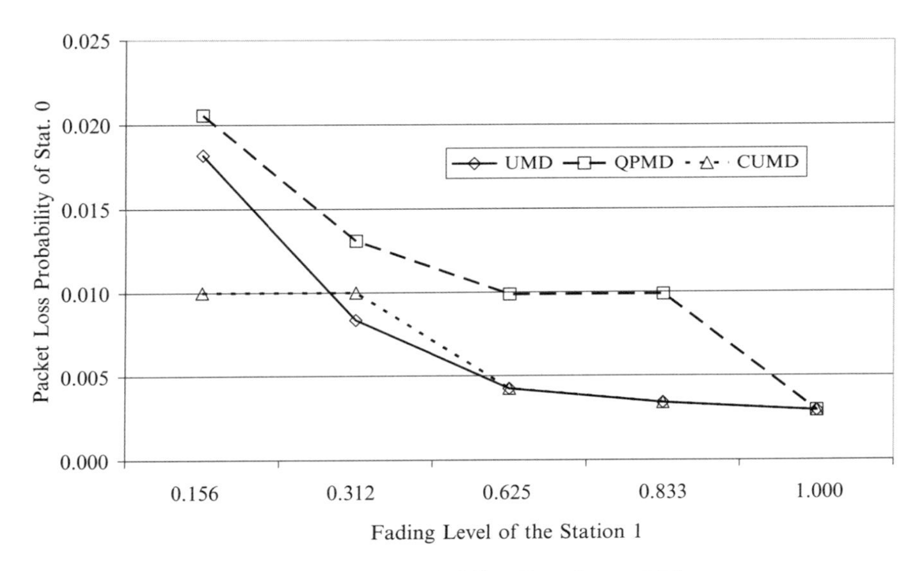

Fig. 7 Packet loss probability (Stat. 0, $\gamma_z = 0.01$).

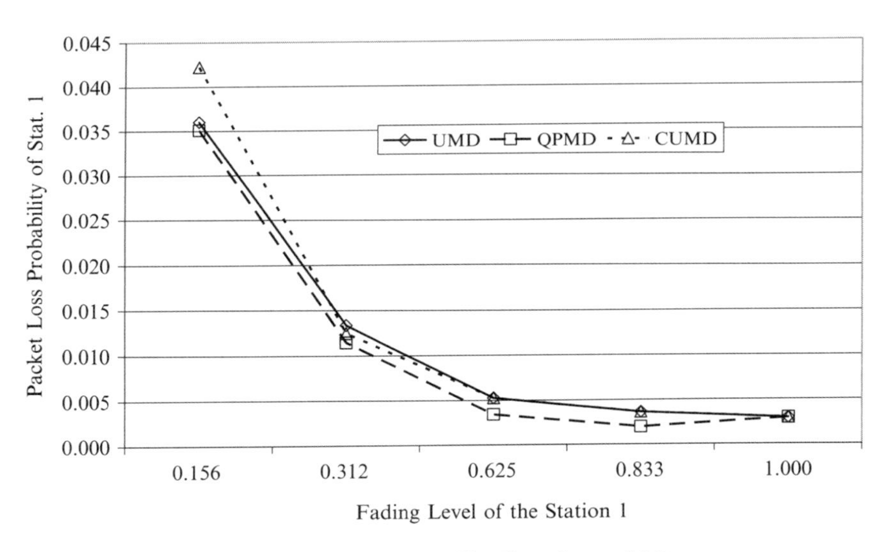

Fig. 8. Packet loss probability (Stat. 1, $\gamma_z = 0.01$).

The minimum bandwidth (2.67 [Mbps]), is assigned to Station 0 and Station 1 gets the residual capacity (7.57 [Mbps]), as clear in Figs. 5 and 6, respectively. The impact on the packet loss may be seen in Figs. 7 and 8. Station 0 value is below the threshold. Station 1 is obviously penalized: the packet loss probability value is far from the threshold. Also values corresponding to $\beta_1 = 0.312$ shows an interesting situation for CUMD because, again, constraints $f_0(C_0) \leq 0.01, f_1(C_1) \leq 0.01$ cannot

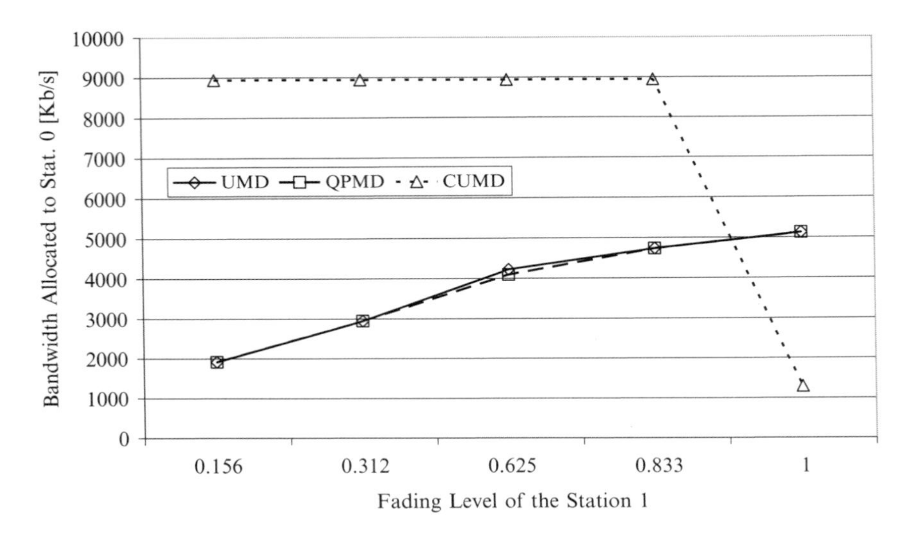

Fig. 9. Bandwidth allocated (Stat. 0, $\gamma_z = 0.001$).

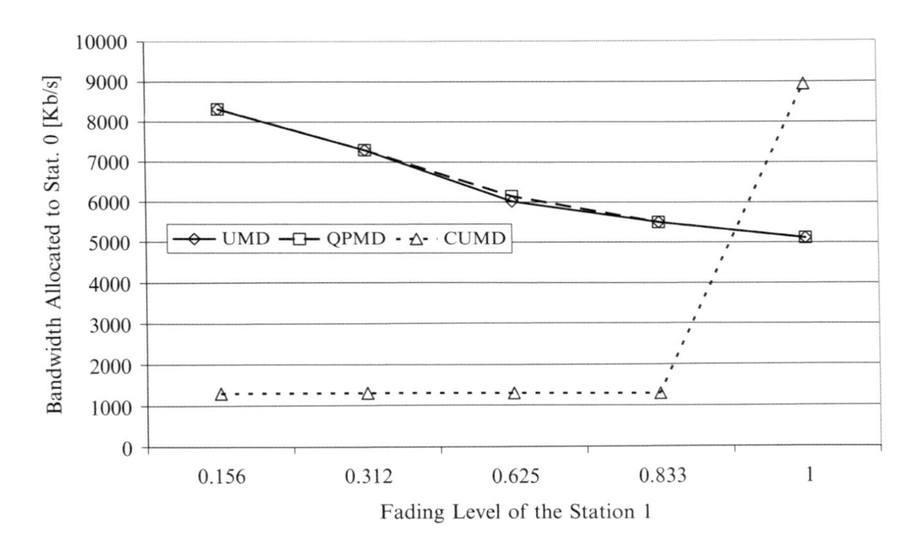

Fig. 10. Bandwidth allocated (Stat. 1, $\gamma_z = 0.001$).

be satisfied in the same time, but both could be satisfied separately. It is exactly the situation shown in Fig. 3.b. CUMD chooses to privilege Station 0 because the choice assures minimum distance with the utopia point. Bandwidth allocations are the same of the previous case (Figs. 5 and 6). The effect on the performance is reported in Figs. 7 and 8. Concerning the behavior of CUMD for $\beta_1 \geq 0.625$, being bandwidth requirements to get performance constraints within the set of feasible allocations, as reported in Table 2, it totally overlaps UMD. Actually, the two schemes have the allocation

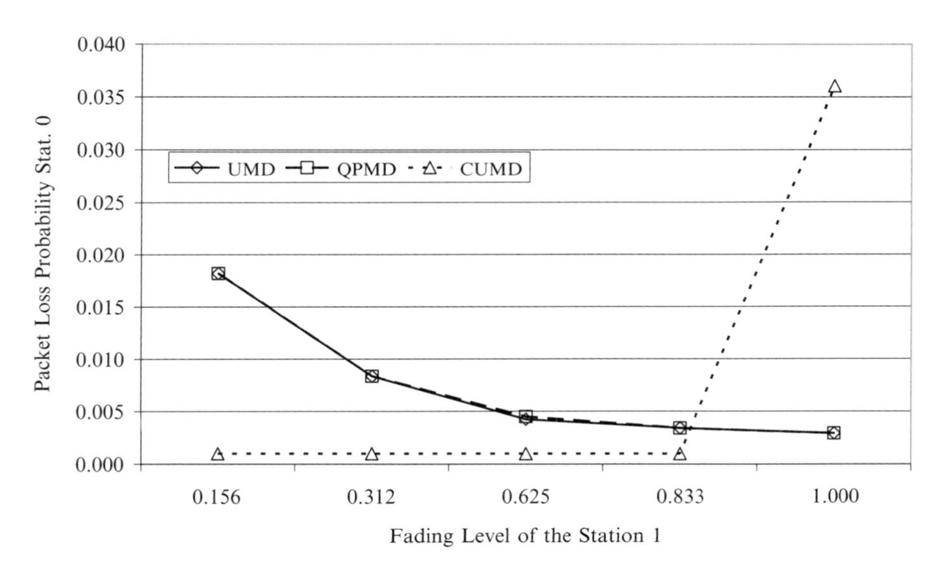

Fig. 11. Packet loss probability (Stat. 0, $\gamma_z = 0.001$).

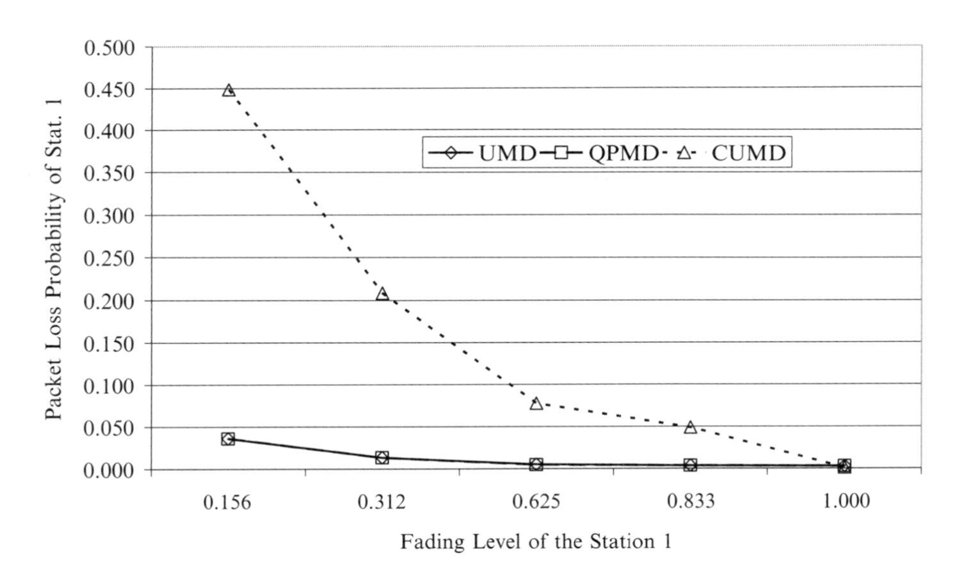

Fig. 12. Packet loss probability (Stat. 1, $\gamma_z = 0.001$).

mechanism based on the utopia point and, except for the constraint sets, which, even if acting, do not influence allocations if $\beta_1 \geq 0.625$, are the same algorithm.

UMD, being not constrained, has the only aim of minimizing the distance with the utopia point (i.e. to approach the behaviour where stations have the complete availability of channel bandwidth). It does not consider any global benefit for the network but acts in a completely competitive environment, where each station pursue its own benefit. Detailed comments about it have been reported in [2]: compared with methods where the overall benefit of the network is considered (e.g. minimizing the overall packet loss probability), faded stations are privileged, just because they are entities of the competitive problems exactly like the other stations.

QPMD changes the nature of allocation because it defines a new reference point (QoS Point), which, implicitly, contains the performance constraints. The QoS point values are the bandwidth allocations contained in Table 2. The shape of bandwidth allocations is similar to UMD but its aim is to keep minimum the quantity $\left(\|\mathbf{F}(\mathbf{C})-\mathbf{F}^{QoS}(\mathbf{C}^{QoS})\|_2\right)^2$. Being the performance requirements the same for both stations, QPMD privileges the faded station, compared to UMD. The behavior is clear in Fig. 5 and Fig. 6. The consequent packet loss probability values are shown in Figs. 7 and 8. As required by its aim, QPMD is the best scheme to have a common (among stations) way to approach performance constraints. The three approaches give the same allocations when they act in clear sky ($\beta_1 = 1$).

The commented behaviors for the three bandwidth allocation schemes hold true for the second set of tests. Performance requirements can never be matched together (Table 3). $C_1^{thr} \geq 10.24$ [Mbps], except for clear sky case. CUMD always chooses to match Station 0 requirements if $\beta_1 < 1$ because the available bandwidth allows it. Station 1 traffic is considered as a best-effort flow. The effect on the packet loss is evident in Fig. 11, where packet loss probability values of Station 0 are always below threshold, and in Fig. 12, where the values are well above the threshold. When $\beta_1 = 1$: the case is totally symmetric and both choices (either Station 0 or Station 1) guarantees minimum distance. As said in session 6.1, the choice is random (Station 1 is chosen, in this case). UMD and QMPD assigns bandwidth allocations to approach, respectively, the utopia point ($C_0 = 10.24$ [Mbps], $C_1 = 10.24$ [Mbps]) and the ideal QoS point got assigning the bandwidth values contained in Table 3 by varying β_1. QPMD behavior may be explained by observing Fig. 4.b, which reports exactly the situation in Table 3. If $\beta_1 < 1: C_1^{thr} \geq C_{TOT} = 10.24$ [Mbps] and $C_0^{thr} < C_{TOT}$. It means that the QoS Point is outside the feasible allocation region and C_1^{thr} is moving from 57.32 to 10.73 [kbps] when β_1 increases its value from 0.156 to 0.833. The projection of the QoS Point over the capacity equality constraint (14), called C_{QPMD}, in Fig. 4.b, corresponds to the projection of the utopia point in the UMD scheme (C_{MD}, in Fig. 2). That is the motivation because the same allocation is got for the two schemes. If $\beta_1 = 1$, $C_0^{thr} = C_1^{thr}$. The comments are similar to the previous case. Correspondence of UMD and QPMD is obvious in this case because the utopia point ($C_0 = 10.24$ [Mbps], $C_1 = 10.24$ [Mbps]) and the QoS point $C_0 = 8.94$ [Mbps], $C_1 = 8.94$ [Mbps] are located along the same straight line, which is orthogonal to the bandwidth constraint (14). Their projection (C_{QPMD} and C_{MD}) are obviously the same point.

8 Conclusions

The paper presents possible allocation schemes for satellite communications, aimed at guaranteeing specific Quality of Service requirements. Traffic is modelled as superposition of TCP sources. The considered framework is the Multi – Objective

Programming Optimization. In particular, the paper introduces two new techniques (CUMD and QPMD), based on the Minimum Distance mechanism, which exploit the features of MOP environment. The paper investigates the behaviour of the two schemes and compares the results with a reference MOP scheme, already published by the authors.

References

[1] S. Kota, K. Pahlavan, P. A. Leppänen, "Broadband Satellite Communications for Internet Access," Kluwer Accademic Publishers, Boston, 2004.
[2] I. Bisio, M. Marchese, "Bandwidth Allocation Strategies for TCP/IP Traffic over High Altitude Platform: a Multi-Objective Programming Approach," IEEE International Conference on Communications (ICC'06), Istanbul, Turkey, June 2006.
[3] N. Celandroni, F. Davoli, E. Ferro, "Static and Dynamic Resource Allocation in a Multiservice Satellite Network with Fading," International Journal of Satellite Communications, vol. 21, no. 4–5, pp. 469–488, July-October 2003.
[4] K. Miettinen, "Nonlinear Multi – Objective Optimization," Kluwer Accademic Publishers, Boston, 1999.
[5] I. Bisio, M. Marchese, "Analytical Expression and Performance Evaluation of TCP Packet Loss Probability over Geostationary Satellite," IEEE Communications Letters, vol. 8, no. 4, pp. 232–234, April 2004.

Carrier Pairing, a Technique for Increasing Interactive Satellite Systems Capacity. An Assessment of its Applicability to Different System Architectures

G. Gallinaro[1], R. Rinaldo[2], A. Vernucci[1]

[1]Space Engineering S.p.A. - Rome, Italy
[2]European Space Agency - ESTEC - Noordwijk, Holland

Abstract. The Carrier Pairing technique, i.e. the sharing of the same frequency band for both Forward Link and Reverse Link carriers in a star or multi-star satellite network, is here discussed. In particular a possible mechanization and its performance in a realistic reference scenario are discussed to understand the merits and limitations of that technique.

The analyses herein presented were carried out with regard to a reference scenario featuring a single Hub station per spot transmitting a standard DVB-S2 carrier with Adaptive Coding and Modulation (ACM) to User Terminals, which in turn transmit enhanced DVB-RCS signals (Turbo-Φ code, QPSK-16APSK ACM) to the Hub. Impairments caused by the non-linear satellite channel on echo canceller performance have been taken into account, also attempting to mitigate them by means of pre-distortion techniques.

1 Introduction

The Carrier Pairing technique was originally conceived with the aim to increase the spectral efficiency of interactive satellite systems comprising an Hub station and a great number of User Terminals (UTs), by allocating a common band segment to signals transmitted by the Hub and the UTs. Generally speaking, in the Forward-Link (FL, Hub → UTs) it is feasible to manage interference resulting from signals spectral overlap, thanks to the much higher level of the signal transmitted by the Hub compared to those transmitted by the UTs. In the Reverse-Link (RL, UTs → Hub) the otherwise intolerable interference caused by the Hub-transmitted signal can be mitigated, at the Hub receive side, by locally adding to the received composite signal a suitably modified replica of the signal transmitted by the Hub itself into the FL. Carrier Pairing is a known technique that has already been adopted for commercial equipment mainly intended for operation in global-coverage satellite systems.

The urgent need to improve satellite systems competitiveness is leading research to conceive solutions permitting to increase their capacity, thus increasing economy of scale and ultimately permitting to reduce service tariffs. The road being followed is that of proposing and assessing new payload architectures on the one hand (e.g. multi-beam), and, on the other hand, investigating new high-performance access solutions which Carrier Pairing is an example of. At this regard an important issue to be dealt with is the consistency between advanced payload architectures and the enhanced access solution. For instance, for the Carrier Pairing case, it would be important to assess its advantages in different system scenarios (number of beams, traffic distributions, frequency reuse factor), so as to understand to which extent the advantages stemming from the adoption of advanced architectures can be added-up to those deriving from the optimization of the access technique.

The subject paper, which is based on some of the results obtained in the course of an on-going contract awarded by ESA to Space Engineering, begins introducing briefly the Carrier Pairing concept and its possible mechanizations, and discussing the issues to be kept under control for maximizing the technique effectiveness. Then, after defining some reference system scenarios, the performance of Carrier Pairing in those scenarios is discussed, showing the applicable results of a comprehensive simulation campaign carried out in the context of the cited ESA study.

The paper is organized as follows. The next section contains a brief introduction to the Carrier Pairing techniques and the related interference cancellation scheme used at the Hub side for recovering the RL signals. Section 3 shows the Bit-Error-Rate (BER)/Frame-Error-Rate (FER) performances achievable on a non-linear satellite channel at the Hub demodulator.

Section 4 finally discusses the overall system throughput which could be achieved in a multi-beam system scenario using this technique and compares it with that achievable with a conventional approach in which separate frequency bands are utilized for the FL and RL. The comparison is done assuming that the same total bandwidth and on-board power are used in both approaches to eventually assess if Carrier Pairing is a viable choice for improving the spectral efficiency of next generation broadband multimedia satellite systems.

2 The Carrier Pairing Technique

In a satellite system implementing a conventional star network architecture (i.e. with a conventional frequency plan) we need for each link two different frequency bands: one for the up-link segment and one for the down-link segment. In the end, for a bidirectional circuit we need four frequency bands as shown in Fig. 1:

For example, assuming that the system operates at K_a-band, each of the four links may be accommodated within the bandwidth shown below:

- FL up-link (from Hub to satellite): 27.5 ÷ 28.0 GHz
- FL down-link (from satellite to UTs): 19.7 ÷ 20.2 GHz

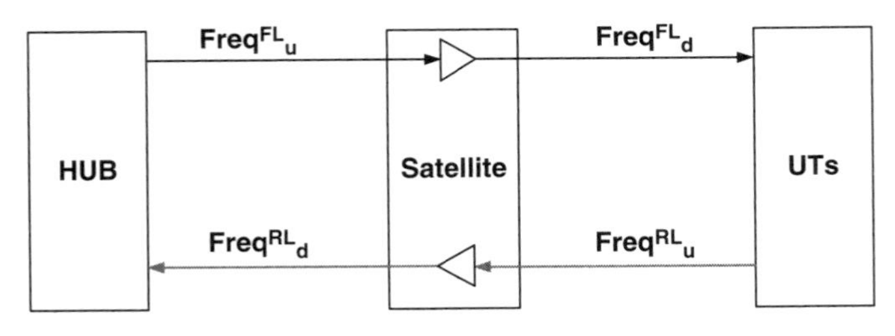

Fig. 1. Forward/reverse link frequencies in conventional systems.

- RL up-link (from UTs to satellite): 29.5 ÷ 30 GHz
- RL down-link (from satellite to Hub): 18.3 ÷ 18.8 GHz

To reduce the occupied system bandwidth it is possible to share the same bandwidth for FL and RL up-link as well as for FL and RL down-link (see Figs. 2 and 3).

With this approach only two frequency bands are required instead of four. There is thus no distinction between UT beams and Hub beams. As a consequence, a Hub can only serve a single beam.

The on-board High-Power Amplifier (HPA) is operated in multi-carrier mode as it amplifies both the FL and RL carriers. In practice the HPA amplifies a single wide-band, high-power, FL carrier plus a multitude of low-power, narrowband, RL carriers.

As evident from Fig. 3, the required power density of the FL carrier is much higher than that of the RL carriers due to the different antenna merit factor (G/T) of Hubs and UTs.

The required relative power density of the FL carriers with respect to the RL carriers is approximately equal to the ratio of the G/T between the Hubs and the UTs.

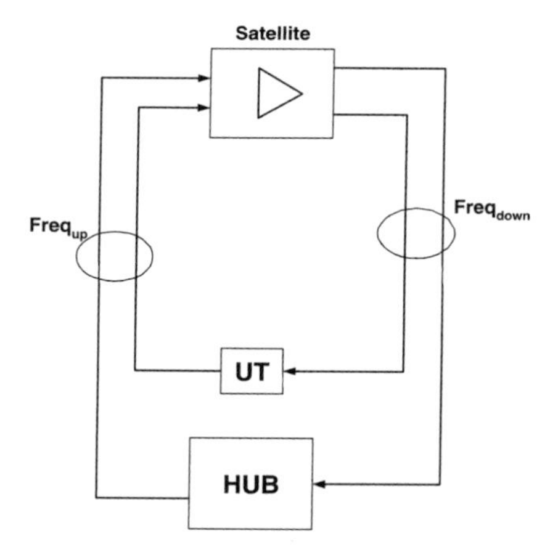

Fig. 2. The carrier pairing concept.

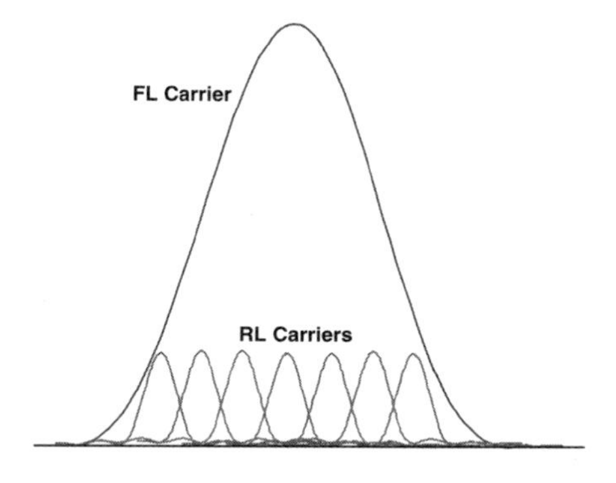

Fig. 3. Signal spectrum with carrier pairing.

Table 1. Assumed hub and UT RF parameters.

Parameter	Hubs	UTs
Saturated EIRP	44.5 dBW	81.7 dBW
Antenna Gain (Transmit/Receive)	45.1 dBi / 41.4 dBi	61.0 dBi / 57.5 dBi
HPA Saturated Power	1 W	120 W (for 4 carriers)
Post-HPA Loss	1 dB	2. dB
Minimum Operational OBO	2 dB	2.5 dB
Pre-Low-Noise Amp Losses	0.5 dB	0.5 dB
Receiver Noise Figure	2.5 dB	2. dB
Clear Sky G/T	17. dB/k	33.9 dB/k

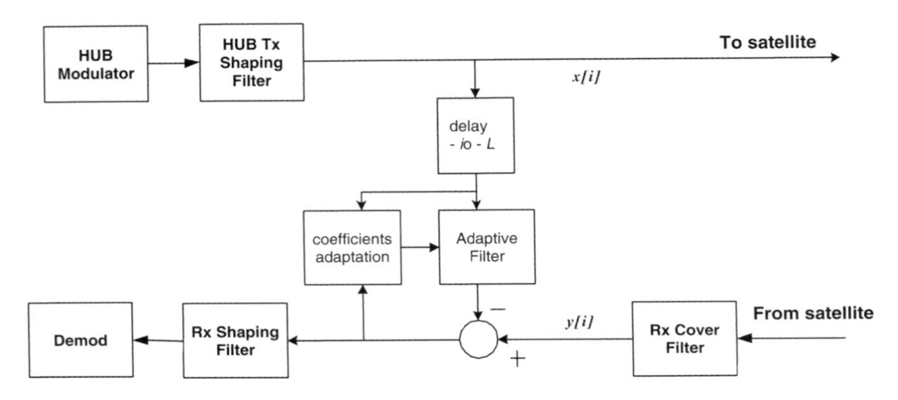

Fig. 4. FL interference cancellation with an adaptive filter (Echo Canceller).

In a typical system scenario, for example, the clear sky G/T of the Hubs and UTs can be respectively 33.9 dB/k and 17.0 dB/k (see for example Table 1).

Hence, the power density difference between the FL carrier and the RL carriers may be in the order of 17 dB. Actually, because there is some difference in the E_b/N_o requirement of FL and RL, due to the larger codeword length which is possible in the FL, such difference could be a little lower (e.g. 12 or 13 dB).

Such a power density ratio will allow the UTs to demodulate and decode the FL carriers without any special processing. Even advanced Adaptive Coding and Modulation (ACM) schemes, like the recently standardized DVB-S2, can be used on the FL although operating modes with the highest spectral efficiency would not be possible due the RL carrier interference floor.

On the other hand, the Hub, in order to successfully demodulate and decode the RL carriers, has to cancel its own transmitted signal. That signal being known at the Hub, a conventional echo canceller can be used as shown in Fig. 4.

The adaptive filter may be implemented through the LMS (Least Mean Square) algorithm trying to minimize the Mean Square Error (MSE) between the recovered signal (received signal minus the echo signal estimated by the adaptive filter) and the reference signal (i.e. the Hub FL signal before transmission to the satellite).

The required adaptive Filter length is depending on:

- the degree of accuracy of bulk round trip delay estimation
- the memory of the channel

The adaptive filter can be split into two independently adapted filters in case of operation at two samples/symbol, obtaining a hardware complexity reduction of a factor of two. Indicating with $\mathbf{f}^{(k)}$ ($k = 0$ or 1) the two filters, the adaptation rule for computing the coefficients of the filter at iteration $i+1$ is:

$$\mathbf{f}^{(k)}[i+1]=\mathbf{f}^{(k)}[i]+\frac{\mu}{\left|\mathbf{x}_i\right|^2}\mathbf{x}_i\left(y^{(k)}[i]-\left(\mathbf{f}^{(k)}[i]\right)^H\mathbf{x}_i\right)^*$$

Where $y^{(k)}[i]$ is the received signal at iteration i (the even or the odd sample depending on the k value) and $\mathbf{x}_i$ is the vector of reference signal samples within the filter memory at iteration i.

The LMS adaptation step μ has to be optimized in order to guarantee good performance and algorithm stability.

3 Waveform Simulation Results

A waveform simulation campaign was undertaken to understand the potential performance of Carrier Pairing also taking into account actual satellite non-linearities. The simulation scenario was composed by a high data rate carrier (FL) and several low-data rate carriers (RL) within the same bandwidth (see Fig. 3).

The satellite has been modelled as the cascade of an Input MUltipleXer (IMUX) and a non-linear amplifier.

The satellite Output MUltipleXer (OMUX) was not simulated explicitly because we were interested in the demodulation of the low rate RL carriers whose bandwidth is much smaller than that of a typical satellite OMUX. Anyway, if desired, the frequency response of the OMUX may be conglobated in the receiver cover filter.

Thermal noise may be added on both Up and Down-Link. However, only results with the uplink noise are shown here because the down-link noise contribution was shown to be irrelevant for the demodulation of the RL carriers at the Hub side as the Hub G/T is high enough to make the up-link the limiting factor here.

The general block diagram of the simulator is shown in Fig. 5.

The FL signal structure in the simulator was assumed conforming with the Digital Video Broadcasting – Satellite v. 2 (DVB-S2) signal specifications [2]. Both Quaternary Phase Shift Keying (QPSK) and 16-Amplitude/Phase Shift Keying (APSK) modulations were considered for the DVB-S2 signal to test the performance of the echo canceller with both constant and non-constant FL signal envelope as difference in envelope statistics may produce different losses in a non-linear channel.

For the RL, a signal structure emulating that of the DVB-Return Channel via Satellite (DVB-RCS) was used [1]. Modulation was thus QPSK. However, the recently proposed turbo-Φ code was used instead of the currently specified DVB-RCS turbo codes due to its higher performance and flexibility [4]. Simulations with the DVB-RCS standard convolutional code were also performed.

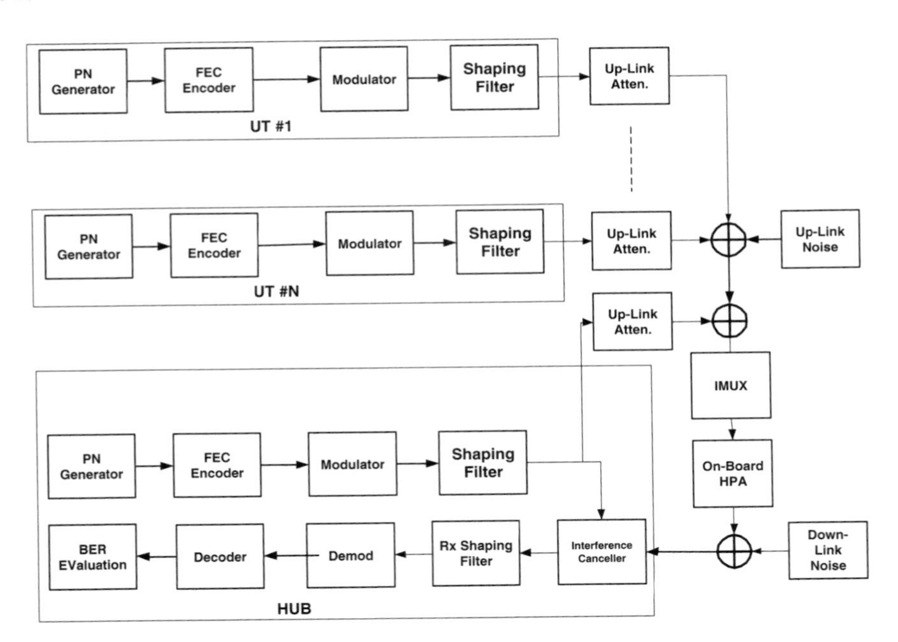

Fig. 5. Carrier pairing simulator general structure (Legenda: FEC = Forward Error Correction, PN = Pseudo Noise, RX = Receive, TX = Transmit).

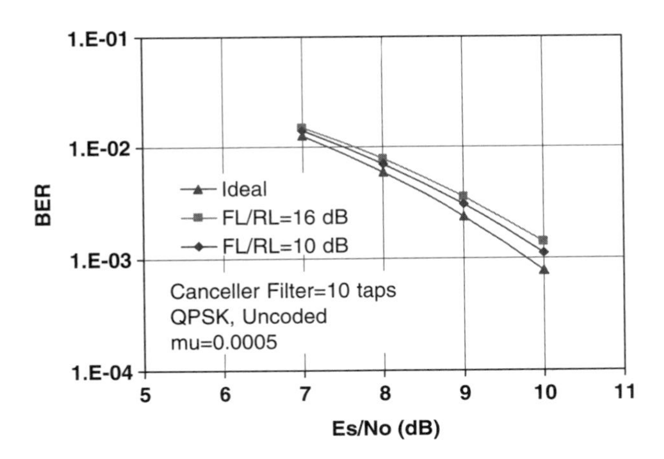

Fig. 6. Echo canceller performance, uncoded case.

In linear channel, interference cancellation is practically perfect, even with a short filter (few symbols), provided that the bulk delay compensation of the reference signal is correct, i.e. the maximum delay error (plus the channel memory) is less than the adaptive filter impulse response length.

The adaptive filter coefficients appears stable, even in very long simulations, apart for the case where the reference signal delay error is at the limit of the impulse response length. In Fig. 6 the echo canceller performance are shown in the linear and uncoded case, for different ratios of FL/RL power. The capability to almost perfectly cancel the interference in such linear conditions is apparent.

In order to analyze the effects of the on-board HPA non-linearity, the linearized Travelling Wave Tube Amplifier (TWTA) model specified in Annex H of the DVB-S2 specifications [2] was used.

As expected, the performance degradation is strongly dependent on the TWTA operating Input Back-Off (IBO). In Figs. 7 and 8 simulation results are shown for the case where a rate 1/2 convolutional FEC code is used and the FL/RL power ratio is either 20 dB or 10 dB.

BER degradations at IBO = 10 dB is negligible. At IBO = 7 dB, however, degradations start to become significant. It is, in fact, not possible, in a non-linear channel, to completely cancel the FL carrier interference using a linear canceller.

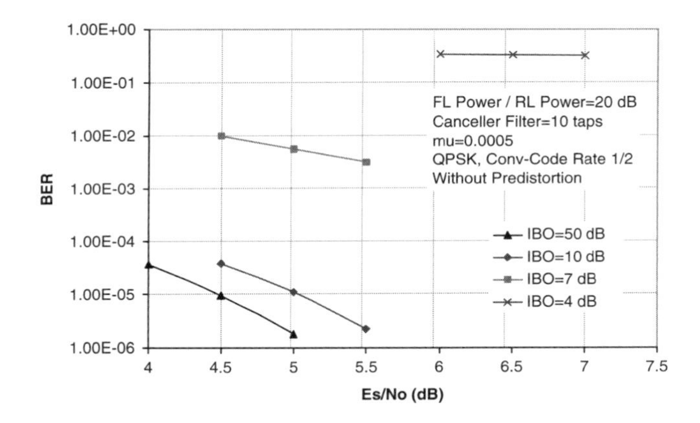

Fig. 7. Echo canceller performance with TWTA non-linearity.

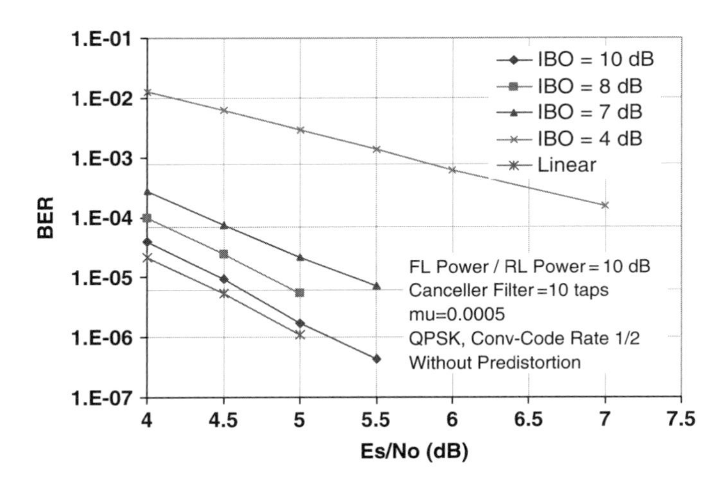

Fig. 8. Echo canceller with TWTA and F/R = 10 dB.

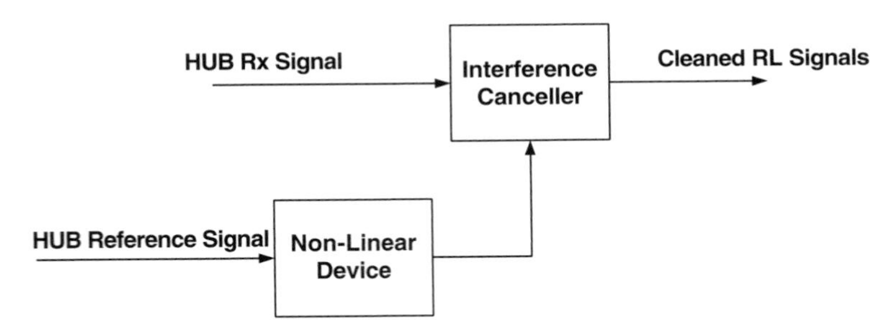

Fig. 9. Predistorter structure.

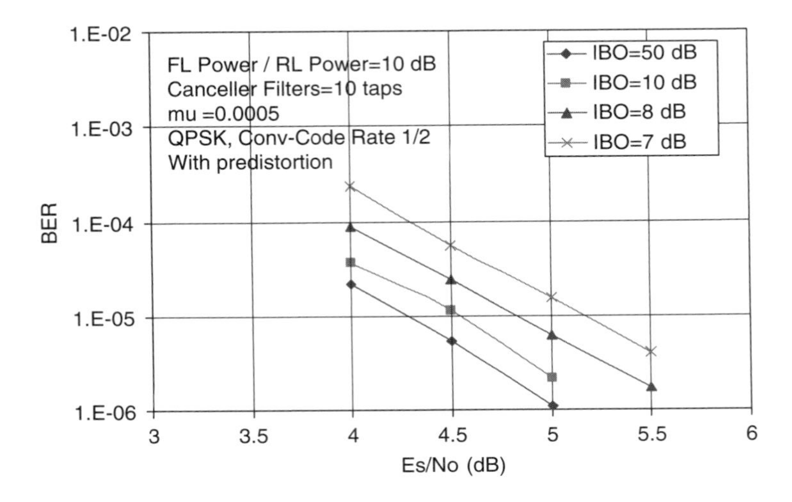

Fig. 10. Echo canceller performance with predistortion (FL/RL: 10 dB).

A pre-distorter replicating the HPA non-linearity for the local reference signal can be used before the canceller to try compensating for the non-linearity (see Fig. 9).

A waveform predistorter with 2 samples/symbol (Fractionally Spaced Predistorter) was thus tested. Improvements were however limited due to:

- the non-linearity is with memory (due to the IMUX filter presence) while the assumed pre-compensator is without memory
- the effects of the RL signal in the non-linear channel cannot be neglected.

Some performance improvements were obtained only for the highest ratio between FL and RL total signal power (greater than 10 dB). However, for a FL/RL power ratio (FL/RL) equal to 10 dB, a performance improvement of only 0.2 dB (@BER = 1E-5) at IBO = 7 dB was obtained with the assumed rate ½ convolutional FEC code (see Fig. 10).

Figures 11 and 12, respectively show the BER and FER in the same conditions as in Fig. 10 but with a rate ½ Turbo-Φ code. A DVB-RCS packet of 1504 symbols (1504 information bit) was assumed.

The effects of using the 16APSK modulation on the Forward Link carrier has been also considered as the resulting envelope fluctuation of the FL carrier may have an impact on the performance of the RL signals. In particular, the BER performance obtained with the echo canceller when 16APSK is used on the FL is shown in Fig. 13.

Furthermore, Fig. 14 compares the obtained RL carrier BER/FER performances when either 16APSK or QPSK are used on the FL carrier. The comparison is shown for the case of IBO = 7 dB and FL/RL power ratio of 16 dB. It appears that 16APSK on the FL produces a loss which is about 0.1 dB greater than the one obtained with a QPSK carrier on the FL.

Figure 15 summarizes the achievable performances on RL carrier demodulation for different FL/RL power ratio. Finally, in Fig. 16 the effect of the predistorter is shown for IBO = 6 dB and F/R = 16 dB.

On the basis of the obtained results the optimum OBO can be evaluated with respect to the global losses introduced in the system. The optimum OBO was found to be about 2 dB (corresponding to an IBO of about 7 dB for the considered TWTA) and the corresponding total loss about 3 dB.

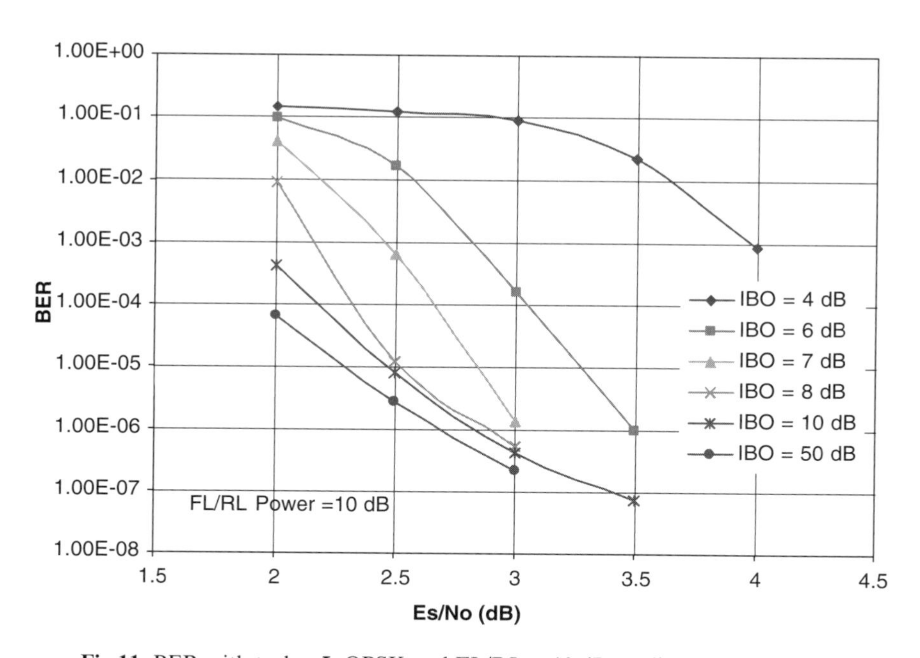

Fig.11. BER with turbo *Φ*, QPSK and FL/RL = 10 dB predistortion assumed.

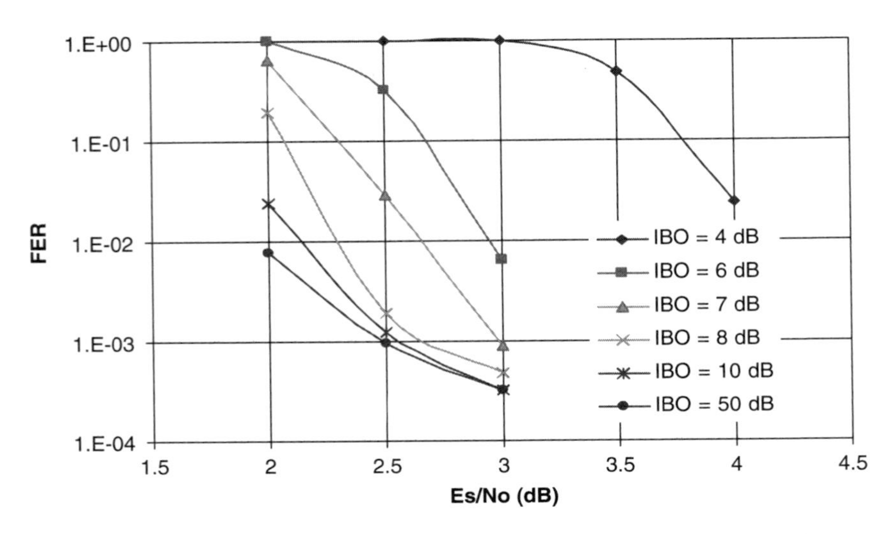

Fig. 12. FER with turbo Φ, QPSK and F/R = 10 dB predistortion assumed.

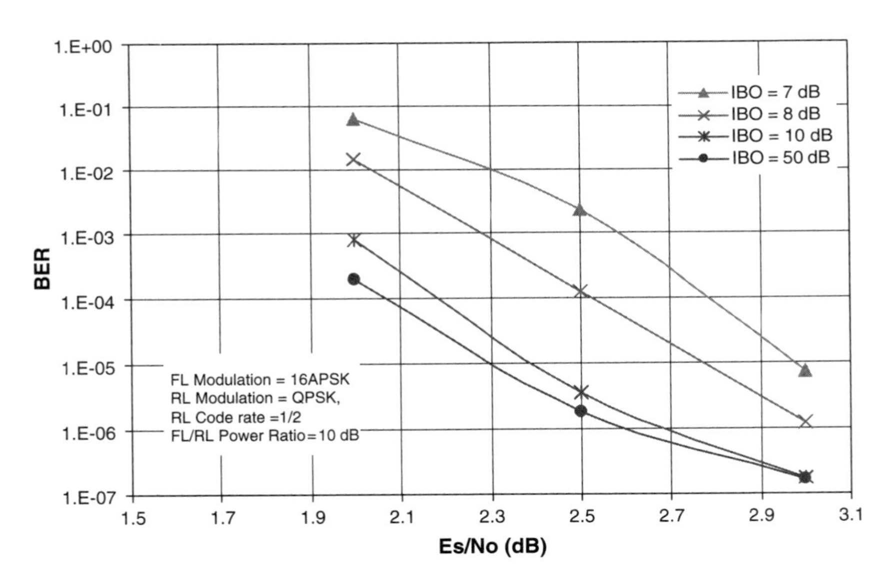

Fig. 13. BER On RL carriers with interference from A 16APSK FL carrier with FL/RL = 10 dB. Rate ½ Turbo-Φ FEC code.

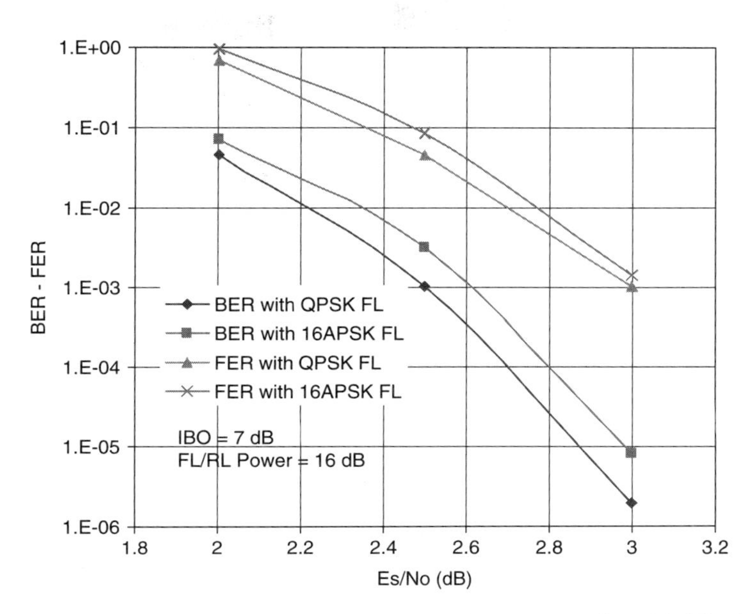

Fig. 14. Performance of RL carriers (IBO = 7 dB, FL/RL power ratio = 16 dB) with either QPSK or 16APSK interfering carrier on the forward link. Rate ½ Turbo-$\boldsymbol{\Phi}$ FEC Code.

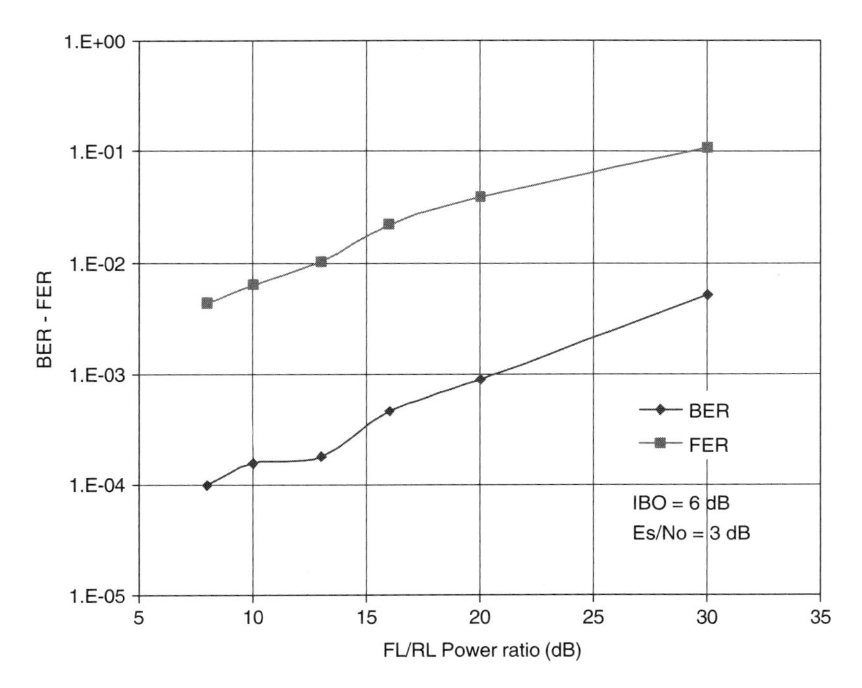

Fig. 15. RL performance for different FL/RL power ratio, Es/No = 3 dB, IBO = 6 dB. QPSK modulation on the FL carrier. Rate ½ Turbo-$\boldsymbol{\Phi}$ FEC code.

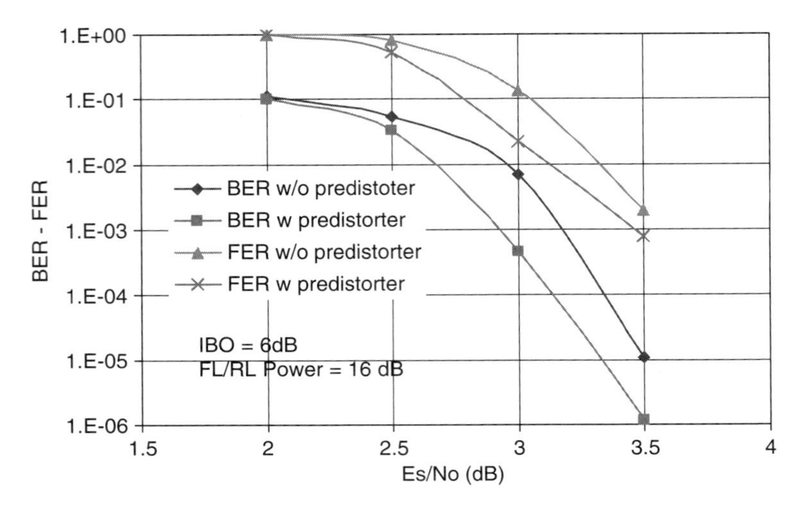

Fig. 16. Predistorter performance For QPSK FL, IBO = 6 dB and F/R = 16 dB. Rate ½ turbo-Φ FEC Code.

4 System Performances

The advantages provided by the Carrier Pairing technique with regard to the overall satellite system spectral efficiency will be more or less important depending on the considered system configuration. Ideally, one should compare the cost per transmitted bit for each of such configurations, something which however is not trivial at all to perform. We therefore took a pragmatic approach; in particular, we designed a K_a-band reference satellite system according to current best practice and evaluated its spectral efficiency with and without Carrier Pairing.

Figure 17 shows the antenna coverage of the assumed reference system (user link: FL down-link + RL up-link), having a total of 88 spots. Each spot has a beamwidth of approximately 0.5° (corresponding to an antenna gain of about 47 dBi at beam edge). A payload with such a great number of beams, although challenging, is actually within the capability of current technology. Moreover we assumed the adoption of a conventional frequency-reuse scheme based on a three-colour pattern (i.e. splitting the available bandwidth in three segments and appropriately assigning band segments to the spots).

The quantitative simulation results herein reported were obtained assuming a ground segment composed by Hubs and UTs whose Radio Frequency (RF) characteristics have already been shown in Table 1. Such characteristics are in line with those of typical Hubs and UTs. Similarly, characteristics of the onboard transponders are given in Table 2.

In the Carrier Pairing case, the number of 25-MBaud FL carriers (each having a bandwidth of about 31 MHz) in a spot ranges from 1 to 5, depending on the beam traffic load. Conversely, the number of RL carriers in a spot was assumed to be proportional to the FL spot bandwidth, with a proportionality factor equal to the ratio of the FL carrier bandwidth to the RL carrier bandwidth.

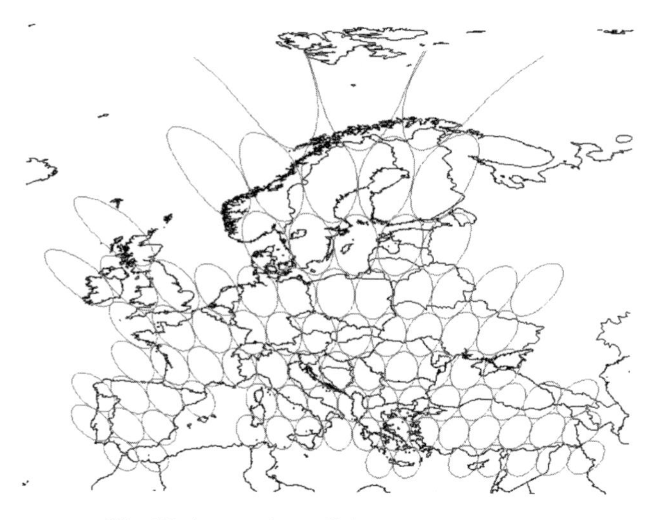

Fig. 17. Assumed user link antenna coverage.

Table 2. Satellite repeater characteristics – carrier pairing case.

HPA type	special design for operation at different saturation power levels with similar efficiency
HPA Maximum RF Saturated Power	65 W
Total RF Saturated Power	4615 W
Nominal Operational OBO	3 dB
Average TWTA Eff. @3 dB OBO	27.7%
HPA DC Power	8.3 KW
Maximum HPA Bandwidth	325 MHz
Spot Beamwidth	0.5 deg
Antenna Gain (EOC)	47.3 dBi
Post-HPA Loss	2.5 dB
Receiver Noise Figure	2.5 dB
Pre-Low-Noise Amp Losses	1.5 dB

In each spot there will just be a single HPA that amplifies all FL and RL carriers, therefore the HPA saturation power will be shared among the various FL and RL carriers. The HPA is operated at an RF saturation power level of (13 * B/31) W, where B is the total bandwidth (in MHz) allocated to the spot (e.g., for a 93-MHz spot, carrying three 25-MBaud FL carriers plus the corresponding number of RL carriers, the HPA would be operated at an RF saturation power of 39 W). The HPA design shall be such that the same physical device can be operated at different saturation power levels attaining similar efficiency at all levels.

For the reference system not using Carrier Pairing, the same total RF onboard power was assumed but separate FL and RL transponders (i.e. HPAs) were used. Each FL transponder has an RF power slightly lower than that of the Carrier Pairing case, as some power has to be allocated to the RL transponders.

As to the OBO, in the non Carrier Pairing case the FL transponder OBO was dependent on the number of carriers per HPA. For beams where a single FL carrier was allocated an OBO of 1 dB was considered (also for compatibility with 16APSK modulation). For multicarrier HPAs the nominal OBO was 1.6 dB, 2.3 dB, 2.6 dB or 2.8 dB depending whether the number of carriers were 2, 3, 4 or 5.

For the Carrier Pairing case, taking into account the waveform simulation results, an OBO in the order of 2 dB was found the best choice for operation with a single FL carrier (plus the associated number of RL carriers). However, taking into account that a single HPA may also be utilized by multiple FL carriers, the optimal OBO may actually have to be greater. For system performance assessment we assumed a conservative value of 3 dB independently of the number of FL carriers per beam. The ACM operational modes and the required E_s/N_oS for the FL are shown in Table 3.

The required E_s/N_o also includes an additional 0.5 dB margin for ACM operation (apart from the lowest mode).

For the RL, an ACM enhanced DVB-RCS access was assumed with Turbo-Φ coding. Table 4 shows the physical layer modes and required $E_S/(N_o + I_o)$.

Also in this case, an additional 1.5 dB margin has been taken into account for all modes but the most protected one, as safeguard against errors in ACM adaptation. Finally, for the Carrier Pairing case, performance of the RL ACM modes have been degraded by 0.5 dB with respect to those shown in Table 4 (used for the conventional system) to account for the uncompensated residual FL carrier interference.

A three-colour frequency reuse scheme was firstly used also for the Carrier Pairing system.

A synthesis of the throughput and availability figures obtainable on the FL and RL for different repartitions between FL and RL of the on-board power is given in Table 5.

Power repartition refers to the split of the HPA output RF power between FL and RL carriers. The transponder being operated in an almost linear region, the same

Table 3. FL ACM modes and morresponding required E_s/N_o for the non-precoded mode. for the precoded case figures are 0.5 dB lower.

Modulation	Code rate	Req. E_s/N_o (dB)
QPSK	1/4	−1.31
QPSK	1/3	0.3
QPSK	2/5	1.24
QPSK	1/2	2.54
QPSK	3/5	3.77
QPSK	2/3	4.64
QPSK	3/4	5.57
QPSK	5/6	6.72
8PSK	3/5	7.92
8PSK	2/3	9.04
8PSK	3/4	10.33
8PSK	5/6	11.77
16APSK	3/4	13.01
16APSK	5/6	14.41

Table 4. RL ACM modes and corresponding required E_s/N_o.

Modulation order	Coding rate	Efficiency (bps/Hz)	Theor. E_b/N_o (dB)	Theor. E_s/N_o (dB)	Req. E_s/N_o (dB)
4	1/3	0.66	1.3	−0.46	1.55
4	1/2	1.00	1.15	1.15	3.1
4	3/5	1.20	1.66	2.45	4.5
4	2/3	1.33	2.07	3.32	5.3
4	3/4	1.50	2.68	4.44	6.4
4	6/7	1.71	3.78	6.12	8.1
8	2/3	2.00	3.86	6.87	9.2
8	3/4	2.25	4.85	8.37	10.7
8	4/5	2.40	5.38	9.18	11.5
16	3/4	3.00	5.70	10.47	13.5
16	5/6	3.33	6.77	12.00	15.0
16	9/10	3.60	7.85	13.41	16.4

Table 5. Performance of carrier pairing for three different power repartition hypotheses.

	Power repartition FL / RL		
	4 to 1	7 to 1	10 to 1
FL Availability	99.77%	99.78%	99.76%
FL Throughput	14.8 Gbit/s	18.9 Gbit/s	20.7 Gbit/s
RL Availability	99.24%	38.97%	8.9%
RL Throughput	15.4 Gbit/s	5.03 Gbit/s	–

power repartition is assumed at the transponder input, i.e. any small signal suppression effect is assumed negligible.

Unfortunately, for FL to RL power allocation ratios of about 10 dB or higher, the RL carriers suffer too much interference caused by the other co-frequency beams. It shall, in fact, be stressed that, whilst it is possible to cancel-out almost all FL interference from the same beam, cancelling out the interference caused by FL carriers from other beams is not possible (at least with a simple echo canceller).

Such RL performance with a 3-colour frequency reuse scheme would have been probably predictable looking at the Carrier-to-Interference ratio (C/I) distribution with a three-colour frequency reuse scheme (see, at this regard, Figs. 18 and 19). It appears that the worst case up-link C/I is in the order of 11 dB. For a power ratio between FL and RL of 10 to 1, this implies that C/I on RL is 10 dB worse than it would appears from Figs. 18 and 19 below. This would lead to a significant penalization of the RL performance.

A power ratio of 10 dB is thus only useful if an asymmetric traffic is expected between FL and RL.

For symmetric traffic load a FL/RL power repartition of 4 to 1 would be required to get an acceptable RL throughput. A total throughput of 14.8 Gbit/s and

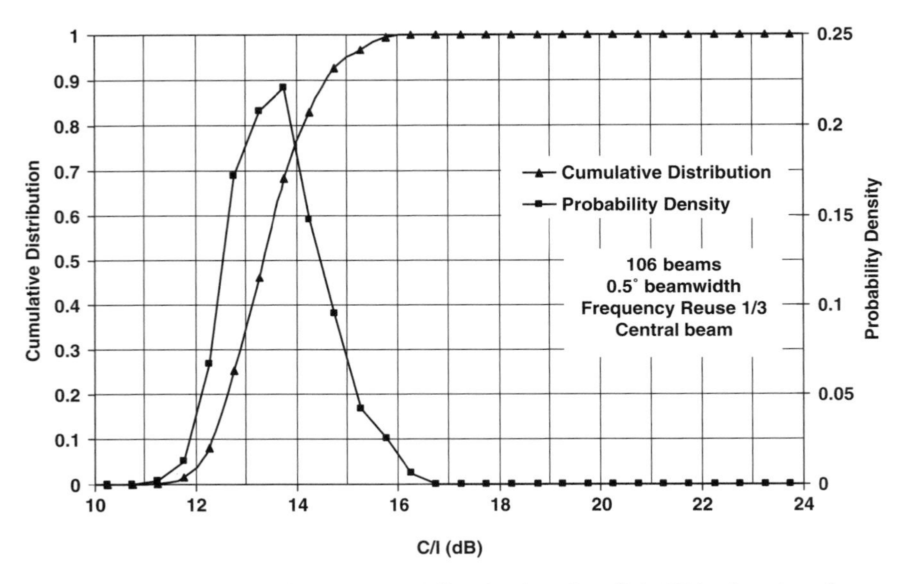

Fig. 18. Cumulative distribution and probability density of up-link C/I in the selected coverage with 0.5° beamwidth. Three-colour pattern frequency reuse. Statistics for central beam. Perfect power control.

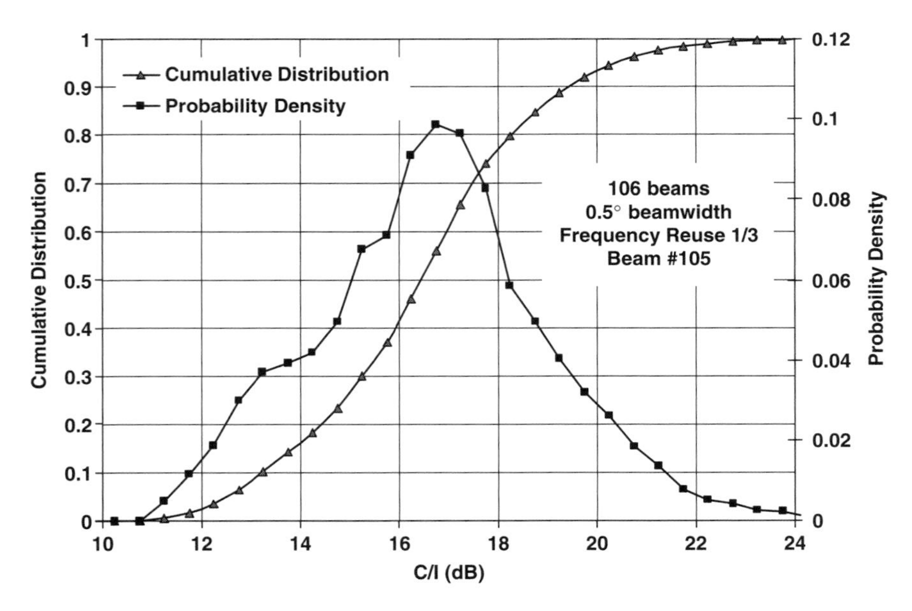

Fig. 19. Cumulative distribution and probability density of up-link C/I in the selected coverage with 0.5° beamwidth. Three-colour pattern frequency reuse. Statistics for most peripheral beam. Perfect power control.

15.4 Gbit/s in fact results for that repartition case respectively for the FL and RL. These numbers have to be compared against 14.4 Gbit/s and 18.0 Gbit/s which would be achievable respectively for the FL and RL of a conventional system using separate FL and RL transponders (and frequency bands).

There is a small improvement of the FL capacity, in exchange for a reduction in the RL throughput.

Availability is instead slightly reduced (particularly for the RL) with respect to the performance figures achieved in the reference scenario (99.9% and 99.77%).

In conclusions, Carrier Pairing seems not to be particularly attractive in the present context, at least for symmetric (FL to RL) traffic. The following considerations are however to be done.

In our evaluation we have assumed that the RL carriers are fully used with a fill-factor for the Time Division Multiple Access (TDMA) frame of 100%. This in practice never happens. So we could design the system with a nominal power ratio between FL and RL, e.g. 4 to 1, but the actual interference experienced by the FL carrier would be somewhat smaller due to the fact that, statistically, not all RL carriers are simultaneously active. At this regard, to control the RL latency in packet applications, enough free capacity should be available to exploit the free-capacity assignment mechanism in DVB-RCS. Whilst unused free capacity assignments are a complete waste in traditional systems, in this case there is a partial compensation with an increase of the FL efficiency which can be exploited to further reduce the FL to RL nominal power allocation ratio. However, with this approach interference on the FL carrier could be quite unpredictable especially when a small number of beams is considered.

In conclusion, it is our feeling that, apart for cases with very unbalanced FL to RL traffic, Carrier Pairing may find useful usage mainly when a few beams needs to operate in Carrier Pairing mode (either because the coverage is based on few beams or because there are a few "hot spots" in the coverage).

For example Table 6 shows the results when only 8 out of 88 beams at the center of the coverage are used in Carrier Pairing mode, whilst the other beams are used for RL only.

The FL throughput is that achieved in the 8 beams used in Carrier Pairing mode (to be added to the throughput provided in the normal FL transponders). The RL throughput is the sum of the throughputs of all 88 RL transponders, including the 8 ones used in Carrier Pairing mode. For the transponders not used in Carrier Pairing mode the characteristics of the conventional RL transponders are used.

Table 6. Performance using carrier pairing in 8 hot spots. A FL/RL power allocation ratio of 7 To 1 was used.

FL Availability	99.793%
FL Throughput	1.2 Gbit/s
RL Availability	99.648%
RL Throughput	18.1 Gbit/s

5 Conclusions

The Carrier Pairing technique is already known to significantly improve the performance of current-generation single-beam K_u-band broadband satellite systems.

An analysis was then carried out to determine whether, and to which extent, Carrier Pairing can also help improving the spectral efficiency of next-generation multi-beam broadband satellite systems operating at K_a band. Analysis results indicate that, in that context, Carrier Pairing can be a useful approach when the traffic distribution is uneven, it allowing to augment the capacity of (a limited number of) "hot spots". Conversely, trying to adopt Carrier Pairing in all the beams of the coverage, in systematic manner, may not be expected to significantly improve the overall system spectral efficiency. As a matter of fact, intra-beam interference would then become the limiting factor for a system featuring a fairly high number of spots.

Finally, in presence of asymmetric FL / RL traffic, Carrier Pairing grants more degrees of freedom when setting the FL / RL power ratio.

References

[1] ETSI EN 301 790, "Digital Video Broadcasting (DVB); Interaction channel for satellite distribution systems, V1.3.1, 2002

[2] ETSI EN 302 307 v.1.1.1, "Digital Video Broadcasting; Second generation framing structure, channel coding & modulation systems for Broadcasting, Interactive Services, News Gathering and other broadband satellite applications, V1.1.1., 2005

[3] M Debbah, G. Gallinaro, R. Muller, R. Rinaldo, A. Vernucci, "Interference Mitigation for the Reverse-Link of Interactive Satellite Networks," submitted to Globecom 2006.

[4] S. Benedetto, R. Garello, G. Montorsi; C Berrou, C. Douillard et al, "MHOMS: High-Speed ACM Modem for Satellite Applications", IEEE Wireless Comm., April 2005.

Reconfigurability for Satellite Terminals: Feasibility and Convenience

Luca Simone Ronga, Enrico Del Re

CNIT, University of Florence Research Unit

Abstract. The reasons for the growing interest in reconfigurability is motivated and supported by a technology evolution of signal processing components, such as analog-to-digital and digital-to-analog converters, signal processors and FPGAs, available in the market with ever increasing performances and lower power consumptions. Reconfigurability however is expensive, in terms of development costs, power consumption, computational efficiency of devices, so not all radio systems may benefit from this innovative design. The selection of convenience areas is a tough task. Each communication context, satellites being one, has to be properly analyzed before making the decisions. We try in this paper to analyze the potentials of Software Radio for space applications, where large investments but also large risks are present.

1 The Silent Revolution

Some years ago, a silent revolution has initiated in the design of radio devices. Driven by the anticipated enormous potentials, the transformation of signal processing functions into software modules, which is at the basis of the *Software Radio* concept, has gained more and more attention from radio industry, operators and research institutions [1]. The reasons for this interest is motivated and supported by a technology evolution of signal processing components, such as analog-to-digital and digital-to-analog converters, signal processors and FPGAs, available in the market with ever increasing performances and lower power consumptions. ADC market is characterized by a constant growth of performances in terms of processing speed and dissipated power: an example is the Analog Device AD9461 ADC operating at 130 Msamples/s with 16 bits of resolution and 2.4 W of power consumption, or the ADS5546 from Texas Instruments producing 190 Msamples/s with 14 bits of resolution and 1.1 W of power. If we target an hand-held device, a more constrained energy consumption is required at frontend, so we could chose the AD9215 at 105 Msamples/s with 10 bits of resolution which dissipated at most 145 mW.

Data acquisition is fundamental to digital signal processing, but when high sample streams are produced by the front-ends, suitable processing cores must be present, to fully exploit the potential of software radio concept. FPGA, DSP and GPP are the leading processing solutions for digital radio implementation. Fast computing devices are present in the market, but the design of a digital radio change substantially when a fast reconfiguration capability is required: single mode signal processing can be optimized at low level on the selected set of signal processors, but if several standards have to be accomplished by the same device, abstraction of functions will reduce the effect of optimization efforts, requiring powerful devices for relatively simple functions.

Reconfigurability is expensive, in terms of development costs, power consumption, computational efficiency of devices, so not all radio systems may benefit from this innovative design.

The selection of convenience areas is a tough task. Each communication context, satellites being one, has to be properly analyzed before making the decisions. If in the mid-term, reconfiguration and software radio are likely to be widely adopted by manufacturers also for space related products, in the near-term the injection of this new design paradigm may appear objectionable.

We try in the following paragraphs to analyze the potentials of Software Radio for space applications, where large investments but also large risks are present.

2 Terminals and Services

Satellite systems differ substantially from terrestrial ones in terms of number and diffusion of adopted standards, type of offered services, and latency of technology innovations. An accurate analysis must be carried out to evaluate the effective impact of software radio technology on space segments.

Satellite terminals may be roughly classified depending on a set of common characteristics:

***Adopted satellite access standard*:** receiving only DVB-S/S2, interactive DVB-RCS, S-UMTS or proprietary access;
***Power and size availability*:** battery powered hand-held, battery powered portable, AC powered fixed terminal;
***Access modes*:** single standard access, multi-standard access with user selection, multi-standard access with automatic selection;
***User mobility*:** stationary, slow intra-spot motion, fast inter-spot motion.

Under this incomplete classification, we can assert that software signal processing can be successfully applied to stationary AC powered receiving only fixed terminals. This is a feasible application of software radio, but it is not convenient, since the benefits from reconfiguration in this case are marginal. A portable or hand-held terminal will get a significant benefit in terms of usage and offered services from the ability to switch between different access modes. In this case the technological effort is consistent for the limitations on power and terminal size, but the convenience is high.

The main issues affecting the design of new generation satellite terminals can be simplified in three key points: the terminal size and portability, the accessed bandwith and the integration with other communication systems.

A terminal with reduced size, with the ability to access satellite services on the palm of one hand is one of the most complex challenges in the communication field. The main problem is characterized by the available power available at the terminal, which poses a serious budget link problem. In some scenarios link cooperation techniques [3, 4, 5] may represent the solution. Improvements are also obtained with the evolution of capacity approaching channel coding techniques (i.e. Turbo and LDPC alone or in combination). Moreover the ability to exploit all available service coverages (satellite, terrestrial, with or without fixed infrastructure) is another key resource. All the cited capabilities are possible through the adoption of reconfigurable architectures for terminals and systems, so in this case the benefit resulting from software radio is high.

As concerns the accessed bandwidth, the main limiting factors are power, processing complexity and spectrum. SDR platforms do not provide significant improvements in terms of processing complexity, and in some cases available complexity is even worse with DSP and FPGA. Spectrum can be better exploited by the adoption of more efficient techniques, but in the complex the SDR technology is not beneficial to bandwidth.

The situation changes with respect to the last issue: the integration with other communication systems. SDR provide an abstraction pillow between connection and transport, promoting the convergence of services at network layer (IP, IPv6). The ideal SDR platform accesses multiple communication standards, though radio-frequency sections are reasonably duplicated to avoid unnecessary waste of processing resources. The contribution from SDR technology is high.

3 Signal Processing Solutions and Evolutions

Signal processing in satellite air interfaces are usually represented as a cascade of processing blocks implemented into ASIC chipsets. The reconfiguration capability can be obtained with two main approaches: parametrization and SW reconfiguration. The former consists in a set of control registers passed to the various signal processing devices, in order to slightly modify their behavior. The latter implies a true software radio architecture where the processing is represented by segments of code for a single processing core.

SR concept by Mitola [6] is obtained by reducing the "parametrized" part to RF sampling and expanding the SW part to the rest of the access device.

Real SDR implementations consist of a variable balance between the two approaches.

Satellite reconfigurable SDR terminals will be probably designed by replicating antennas and RF sections (due to the wide frequency segment of satellite bands) controlled by parameters, while baseband processing can be easily implemented in software. The resultung general architecure, depicted in Fig. 1, can provide all the desired fetures: multistandard, multichannel, multiservice and multiband access to satellite services.

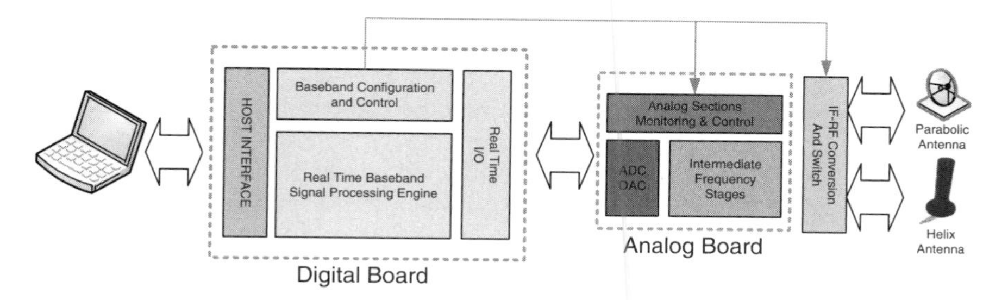

Fig. 1. Implemented SDR architecture for transportable satellite terminal.

Software reconfiguration levels enabled by SDR technology are:

- reconfiguration at commissioning
- with downtime
- on a per call (session) basis
- per timeslot.

The first kind of reconfiguration, provided by the manufacturers, defines the property of the device based on pricing profiles. It is issued once in the lifetime of the terminal, and allows opimization of the production lines, at the expense of a moderate waste of resources for low cost terminals.

The reconfiguration with downtime is performed to correct bugs and install upgrades of the terminal features. In some cases it prevents the factory recalls, which may reveal expensive especially in the case of large numbers of sold devices. Can be performed some times during the lifetime, it is operated by users or support centers after directives from the manufacturers.

The most dynamic reconfiguration levels, one on a per session basis and one during the sessions are the most attractive features expected from SDR platforms. They are characterized by a continuous service, and enable several service features. They are mainly operated by users or automated by procedures. The control of this two reconfiguration levels is supervised or instantiated by service operators, by adaptive decisional processes derived from the observation of communication and traffic parameters (i.e. congestion, load, link quality).

All the cited reconfiguration levels are possible with SDR architecture, but the last two levels require a consistent technological effort in order to be implemented with reasonable final costs for the device.

The reconfiguration is possible through software architectures, whose reference in the literature is represented by SCA (Software Communication Architecture) developed in the framework of the US'Army leaded JTRS project [7]. The Joint Tactical Radio System family of radios will range from low cost terminals with limited waveform support to multi-band, multi-mode, multiple channel radios supporting advanced narrowband and wideband waveform capabilities with integrated computer networking features. These radios shall conform to open physical and software architectures. The JTRS will develop a family of affordable, high-capacity tactical radios to provide both line-of-sight and beyond-line-of-sight C4I capabilities to the warfighters. This family of radios will cover an operating spectrum from 2 to 2000 MHz, and will be capable of transmitting voice, video and data.

Following this rationale, a series of research initiatives have been issued to prove the technical feasibility of software radio concepts to satellite communications with added value to offered services. As an example we can consider the context of fast deployable overlay networks. In this case the transportable satellite terminal acts as a service gateway between a local (wireless or wired) terrestrial network and a global satellite backbone. The ideal terminal is a transportable multi-standard terminal, able to select the suitable pair of satellite and terrestrial standards to bridge together. A prototype of this terminal has been designed and partially implemented at CNIT laboratories. The functional block diagram is in Fig. 2.

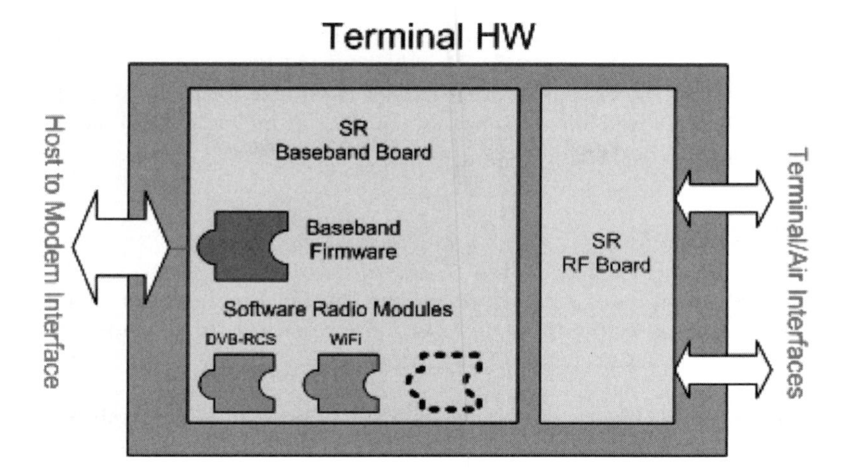

Fig. 2. Software-defined radio satellite terminal functional scheme.

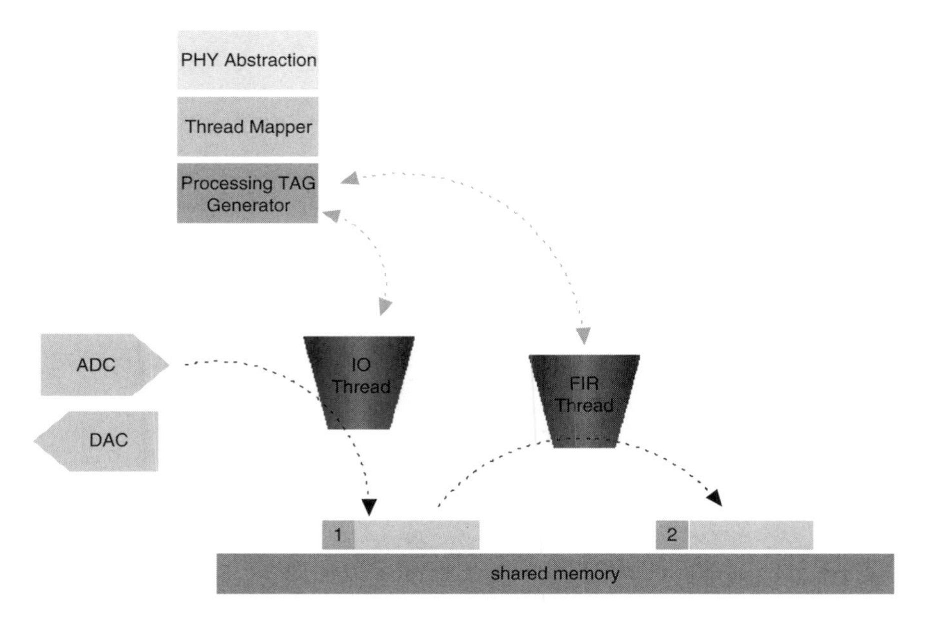

Fig. 3. Parcel processing for SDR computing.

The software definition of signal processing sections will also allow new approaches to signal processing design, removing the limitations imposed by real hardware devices [6]. The most relevant impact of this issue is the adoption on radio devices of processing schemes used by computing. Parallel processing, threads, pipelining are only a few examples of these schemes. Figure 3 shows a novel computing scheme for SDR satellite terminals called *parcel processing*. With this scheme a shared memory is accessed by parallel processing threads implemented on different hardware devices. The processing flow is regulated by labels attached to signal portions (e.g. parcels). This processing method is suitable for multi-hardware software radio terminals and allows complex architectures mixing general purpose and specialized signal processing devices.

4 Securing Devices

The software definition of complex processing functions represents a great opportunity but also exposes the terminal to new dangerous security treads. The most relevant security issues related to SDR platforms are:

Alteration of R-CFG file: event that occurs when an undesired entity substitutes the true configuration file with another one where malicious code is present. Securing the radio definition code downloaded from satellite implies a strong authentication of the source and a shared key mechanism to secure the download phase. In addition a tamper-proof HW is required, to be authenticated by the service provider.

Violation of user's privacy: usually managed by mechanism in upper layers, with the exception of the replacement of MAC address with a dynamical *Temporary Mobile Identifier* (TMI) to address the SDR device.

Terminal cloning: obtained through copies of R-CFG code on anonymous HW. The only solution is to insert a unique identifier on HW in order to authenticate the overall (SW+HW) terminal. This is a difficult task, however, on SDR platform where GPPs are employed.

HW tampering and cloning: is somewhat related to the previous item, and occurs when generic clones of HW platforms are released. A solution may be represented by the trusted computinh [8] (TC) paradigm, but it is still a controversial issue. TC in the personal computer field is considered a limiting factor to the freedom of the market.

5 Adding Intelligence to Communication Devices

The natural evolution of a dynamic HW satellite terminal is the adoption of intelligent decisional processes implemented in the terminal itself. When a complete cognitive cycle is adopted, we refers to it as *cognitive radio* [2].

The main objective of this features is the smart, dynamical distributed selection of communication resources through sensing and learning. The main advantages in the satellite context is similar to those in the terrestrial wireless: the coexistence of a primary communication service, like DVB-S downstream, with a secondary service without any perceived degradation in the performances of the primary service.

Cognitive radio features for the satelite terminal requires an additional technological effort in terms of integration with sensors (environmental, spectrum and terminal status) and the development of context-aware services. Moreover, the latencies experienced in GEO system represent heavy limitations to the processes involved in the cognitive cycle.

6 Conclusions

Satellite Terminal reconfigurability is a great opportunity for manufacturers and service operators both in terms of production efficiency and implementation of new services. The main enabling technology for reconfiguration is represented by software radio, which can improve size and mobility of terminals and their integration with other communication systems. The adoption of SDR technology on the design of mobile satellite terminals represents the most challenging and evolutionary scenario, at the same time, it is also a consistent technological effort for the scientific community and manufacturers. Many technological aspects still remain to be investigated in several areas: cognitive processes, security, and signal processing design.

References

[1] E. Del Re Ed. *Software Radio Technologies and Services*. Springer, 2001.
[2] Haykin, S. *Cognitive radio: brain-empowered wireless communications*, Selected Areas in Communications, IEEE Journal on, Volume 23, Issue 2, Feb 2005 Page(s):201–220.
[3] Sendonaris A., Erkip E., and Aazhang B. "User cooperation diversity - part I: system description". *IEEE Transactions on Communications*, vol. 51, pp. 1927–1938, Nov. 2003.
[4] Sendonaris A., Erkip E., and Aazhang B. "User cooperation diversity - part II: Implementation Aspects and Performance Analysis". *IEEE Transactions on Communications*, vol. 51, pp. 1939–1948, Nov. 2003.
[5] Ribeiro, A. and Giannakis G. "Fixed and Random Access Cooperative Networks". *EURASIP NEWSLETTER*, invited article - submutted January 18, 2006.
[6] Joseph Mitola, III *Software Radio Architecture: Object-Oriented Approaches to Wireless Systems Engineering* Wiley, November 2000.
[7] Joint Tactical Radio System *http://jtrs.army.mil/*
[8] Trusted Computing *https://www.trustedcomputinggroup.org*

Link Cooperation Technique for DVB-S2 Downlink Reception with Mobile Terminals

Luca Simone Ronga, Enrico Del Re, Fabrizio Gandon

CNIT, University of Florence Research Unit
Via di Santa Marta 3
50139 Florence Italy

Abstract. Direct reception of DVB-S (2) satellite signals from mobile terminals equipped with non directive antennas is becoming of great interest among manufacturers and operators. Low orbit constellations are technically preferred for mobile terminal reception due to the reduced path loss. Economical issues however, have recently redirected the interest to medium and geostationary constellations, eventually assisted by high altitude platforms. The satellite power is limited by technology and the maximum allowable mass of satellites. This work explores the opportunity of application of link cooperation techniques for downlink reception of DVB-S (2) bitstreams.

1 Introduction

DVB-S(2) is the second generation system for Broadcasting, Interactive Services, News Gathering and other broadband satellite applications [1,2]. This system gets advantages from the most recent developments of channel coding LDPC, joined with several modulation orders (QPSK, 8-PSK, 16-APSK and 32-APSK). The possibility to change the modulation and coding parameters for each frame (VCM) and the ability to change these parameters according to the channel (ACM), are the main new system characteristics.

Direct reception of DVB-S2 satellite signals from mobile terminals, equipped with non directive antennas, is becoming of great interest among manufacturers and operators. Low orbit constellations are technically preferred for mobile terminal reception due to the reduced path loss. Economical issues however, have recently redirected the interest to medium and geostationary constellations, eventually assisted by high altitude platforms. Since the satellite power is limited by technology and the maximum allowable mass of satellites, downlink EIRP is a limited resource which can be increased at the expense of coverage, by reducing the spot dimensions [3]. Even in the latter case, a sufficient C/N value cannot be reached by the receiver handset for the correct reception of the DVB-S(2) downstream. Recently, a new class of methods called *cooperative communication* has been proposed [4,5,6], that enables single-antenna mobiles in a multi-user environment to share their antennas and generate a virtual multiple-antenna transmitter that allows them to achieve transmit/receive diversity. The mobile wireless channel suffers from fading, meaning that the signal attenuation can vary significantly over the course of a given transmission. Transmitting/receiving independent copies of the signal generates diversity and can effectively combat the deleterious effects of fading. In particular, spatial diversity is generated by transmitting/receiving signals from different locations, thus allowing independently faded versions of the signal at the receiver. Cooperative communication generates this diversity in a new and interesting way.

The main cooperation strategies are Detect and Forward [4,5], Amplify and Forward [6] and Selective Forward [7]. The considered cooperation scheme in this paper is Amplify and Forward (AF) [8].

2 System Model

The adopted cooperation scenario is depicted in Fig. 1. The main operating parameters are reported in Table 1.

The basic idea of AF strategy [8] is that around a given terminal, there can be other single-antenna terminals which can be used to enhance diversity by forming a virtual (or distributed) multiantenna system (see Fig. 1) where the satellite signal is received from the active terminal and a number of cooperating relays. The cooperating terminals retransmit the received signal after amplification. The AF strategy is particularly efficient when the cooperating terminals are located close to the active one so that the cooperative links ($c(1)$,$c(2)$,$c(3)$) are characterized by high

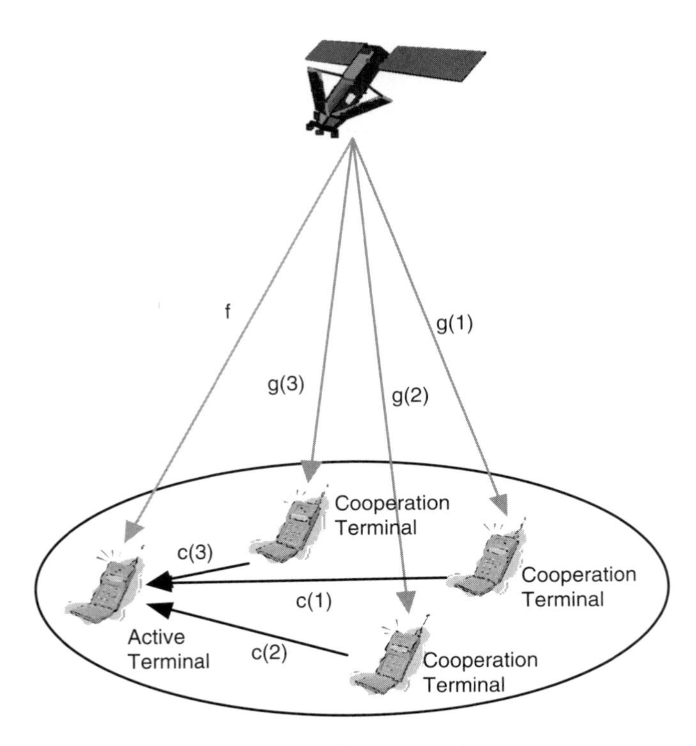

Fig. 1. Downlink satellite cooperation scenario.

Table 1. Main operational parameters.

d_{sat}	36000	[km]	satellite terminal distance
d_{coop}	10	[km]	cooperative terminal
L_{sat}	–205.34	[dB]	satellite terminal path loss
L_{coop}	–118.5	[dB]	cooperative terminal path loss
B_{sat}	36	[MHz]	transpoder bandwidth
P_{sat}	70	[dBW]	satellite power
P_{max}	250	[mW]	cooperative terminal maximum power
G/T_{Rx}	–24	[dB/K]	handheld receiver G/T
T_{sys}	290	[K]	system temperature
F_c	2000	[MHz]	cooperation channel frequency
F_d	11750	[MHz]	downlink channel frequency

signal-to-noise ratios and the link from the satellite to the active terminal (f) is comparable with the links from the satellite to cooperating devices. AF requires minimal processing at the cooperating terminal but it needs a consistent storage capability of the analog received signal. As in [8] we consider the amplification factor A relationship given by

$$A_i^2 = \frac{P_{max}}{P_{sat}\,|\,g(i)\,|^2 + N} \tag{1}$$

where P_{sat} is the satellite downlink power and P_{max} the cooperative terminal maximum power; M is the number of cooperating terminals, $g(i)$ the i-th link pathloss, $N = KT_{sys}B_{sat}$ the noise spectral density at the earth terminals (see Table 1). With this choice we obtain an expression of the resulting C/N on the active terminal.

$$\frac{C}{N} = \gamma f + \sum_{i=1}^{M} \frac{\gamma g_i \gamma c_i}{\gamma g_i + \gamma c_i + 1} \tag{2}$$

By assuming that all of the cooperating terminals have the same characteristics and the cooperative channels are similar we can simplify the previous expression in

$$\frac{C}{N} = \gamma f + M\,\frac{\gamma g \gamma c}{\gamma g + \gamma c + 1} \tag{3}$$

Furthermore we can consider $\gamma f = \gamma g$ so the variables concerning the channel become two (γf and γc).

$$\frac{C}{N} = \gamma f \left(1 + M\,\frac{\gamma c}{1 + \gamma f + \gamma c}\right) \tag{4}$$

The previous expression becomes (*see Appendix B*)

$$\frac{C}{N} = \frac{P_{sat}|f|^2}{N}\left(1 + M\,\frac{A^2|\,c\,|^2}{1 + A^2|\,c\,|^2}\right) \tag{5}$$

so the signal-to-noise ratio depends on $\gamma_z = f(P_{sat}, A, M, f, c, N)$.

3 Link Budget Consideration

As we can see in (5), AF cooperation can provide some advantages:

- *C/N* improvement at M growth with all other parameters fixed;
- *C/N* improvement depending on the choice of A and P_{sat} with M, d_{coop} and F_c fixed (see Fig. 2);
- *C/N* improvement with variable L_{coop} and M with P_{sat} and A fixed (see Fig. 3);
- P_{sat} decreasing (spot area coverage expansion) at M growth for a fixed *C/N*;

The target is to try to get a value of the (5) such to guarantee the fruition of the standard DVB-S2 services, a fact that was not realizable using only one mobile. Figure 2 shows the limit of *C/N* improvement due to choice not to sorpass the P_{max} constraint

$$A^2 \leq \frac{P_{max}}{P_{sat}|f|^2 + N} \tag{6}$$

and the amplification factor range where it is convenient to work to obtain performances gain ($A \approx 110 - 125$ dB). We chose $M = 10$, $d_{coop} = 10$ km and $F_c = 2$ GHz as

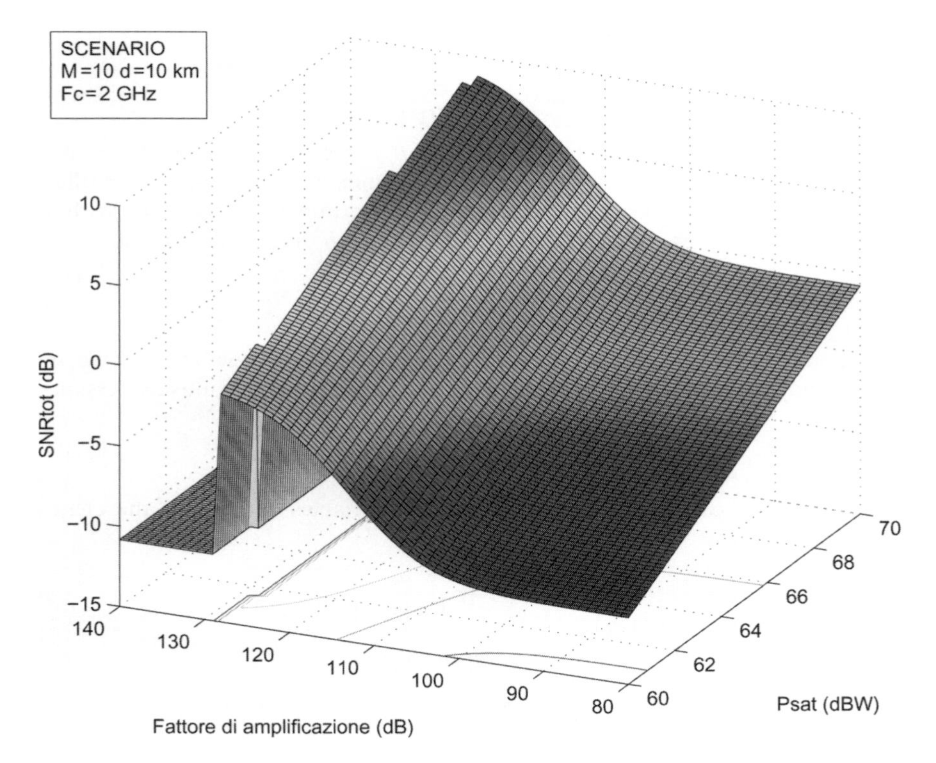

Fig. 2. Receiver SNR vs cooperative terminal amplification factor (A) and satellite tx power (P_{sat}).

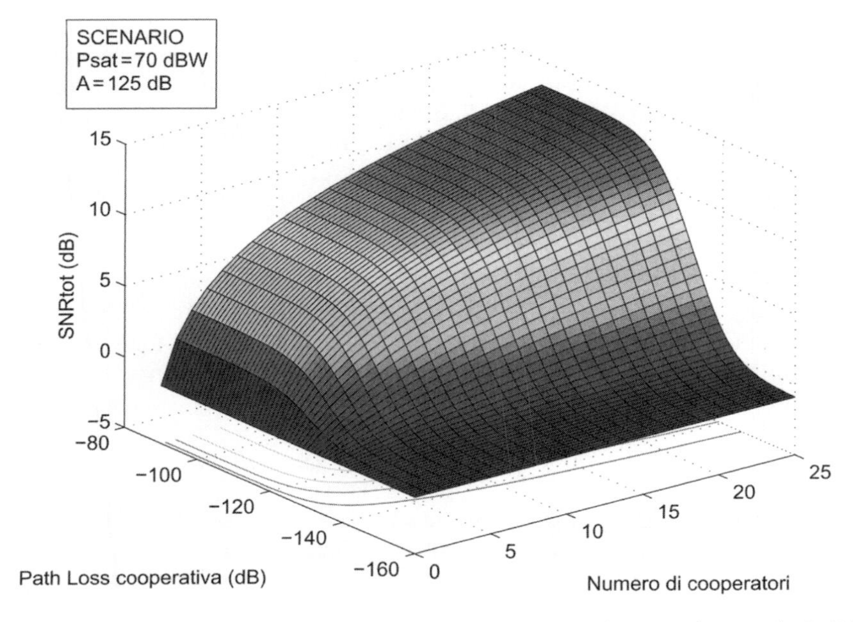

Fig. 3. Receiver SNR vs cooperation link loss (L_{coop}) and number of cooperating terminals (M).

Table 2. Required *C/N* with 7.2 Mbaud in downlink.

Modulation	useful Mb/s	Eb/No (dB)	C/N (dB)
QPSK 1/2	7.2	1.05	0.08
QPSK 2/3	9.52	1.89	2.13
QPSK 3/4	10.71	2.31	3.07
QPSK 5/6	11.91	2.99	4.21
QPSK 8/9	12.72	3.73	5.23
8-PSK 2/3	14.26	3.65	5.65
8-PSK 3/4	16.04	4.43	6.94
8-PSK 5/6	17.85	5.41	8.38
16-APSK 3/4	21.36	5.49	9.24
16-APSK 4/5	22.79	6.03	10.07
16-APSK 5/6	23.76	6.42	10.63

fixed variables. That amplification factor range depends on the quality of cooperative links and it assume lower values decreasing L_{coop}. This last dependence is better shown in Fig. 3, where the amplification factor is set to its maximum allowable value not violating the P_{max} constraint. In this figure. we can notice the *C/N* improvement as M and L_{coop} decrease. The required *C/N* for the transmission modes in DVB-S2 standard [2] are reported in Table 2. As we can see, the AF cooperation strategy with A = 125 dB, P_{sat} = 70 dBW, B_{sat} = 9 MHz gives the chance to use the modulations QPSK, 8-PSK and 16-APSK in the downlink (the required values are under the surface of Fig. 3). So for a given configuration of cooperators (link quality L_{coop} and number M) a specific subset of DVB-S2 compliant modulations can be adopted.

All the results in this section have been derived from theoretical considerations. In particular we considered AWGN satellite channel and the coefficients f and g representing the satellite and cooperation path losses. In the next section a more realistic scenario is considered, with a cluster of satellite terminals with channels modeled with Corazza-Vatalaro model [9,10,11].

4 Cluster Performance

The Corazza-Vatalaro channel is a combination of a Rice and a long-normal factors, with shadowing affecting both direct and diffused components. The p.d.f. of the multiplicative fading coefficient $pcv(r)$ is:

$$pCV(r) = \int_0^{+\infty} \frac{1}{v}\, pRice\left(\frac{r}{v}\right) pLognormal(v)\, dv \tag{7}$$

where r is the received signal envelope and v is the mean power of the directed component. This model has been implemented in Simulink as shown in Fig. 4. The CVchannel block is part of the complete system of Fig. 5 which represents cooperation environment of Fig. 5 under the hypothesis of 10 cooperators. Starting from the left, the model represents the downlink signal available at the satellite whose power is $EIRP_{sat}$. After *CVchannel* and *Free Space Path Loss 205 dB* blocks (top of Fig. 5) the signal is received from the active terminal. The other 10 block chains model the cooperation links. The signals coming from cooperators and active path is combined at the active terminal radio stages (adder block in the model), then demodulated and revealed. It is worth noting that the path-loss value indicated (118 dB) derives from the choice to use a cooperation frequency $F_c = 2$ GHz and a distance $d_c = 10km$.

It is not considered for the moment the fading effect on the cooperative links, as expected in environments characterized by limited distances (within 10 km) and good visibility among terminals.

The model has been simulated with a time resolution equal to $1/2B_{sat} = 1/14.8MHz$, with B_{sat} being the bandwidth of the modulated QPSK signal ($FEC = 1/2$) considering an useful data rate of 7.2 Mb/s (Table 2). The resulting BER versus E_b/N_o curves for different configurations have been plotted.

The first graph in Fig. 6 shows the performances of QPSK and 8-PSK modulations with a Corazza-Vatalaro channel characterized by a Rice factor $R = 20$.

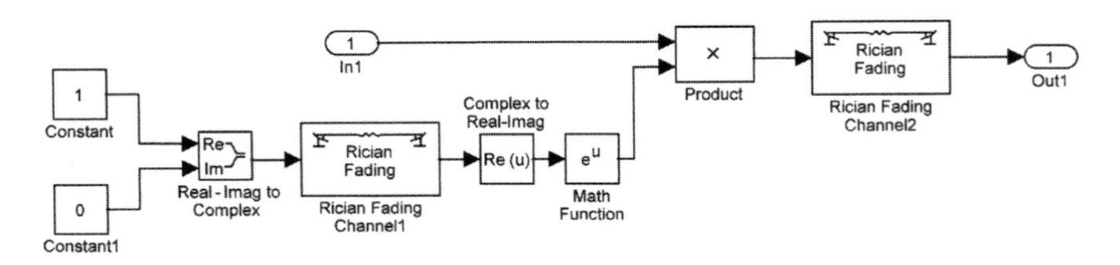

Fig. 4. Satellite CV channel model implementation.

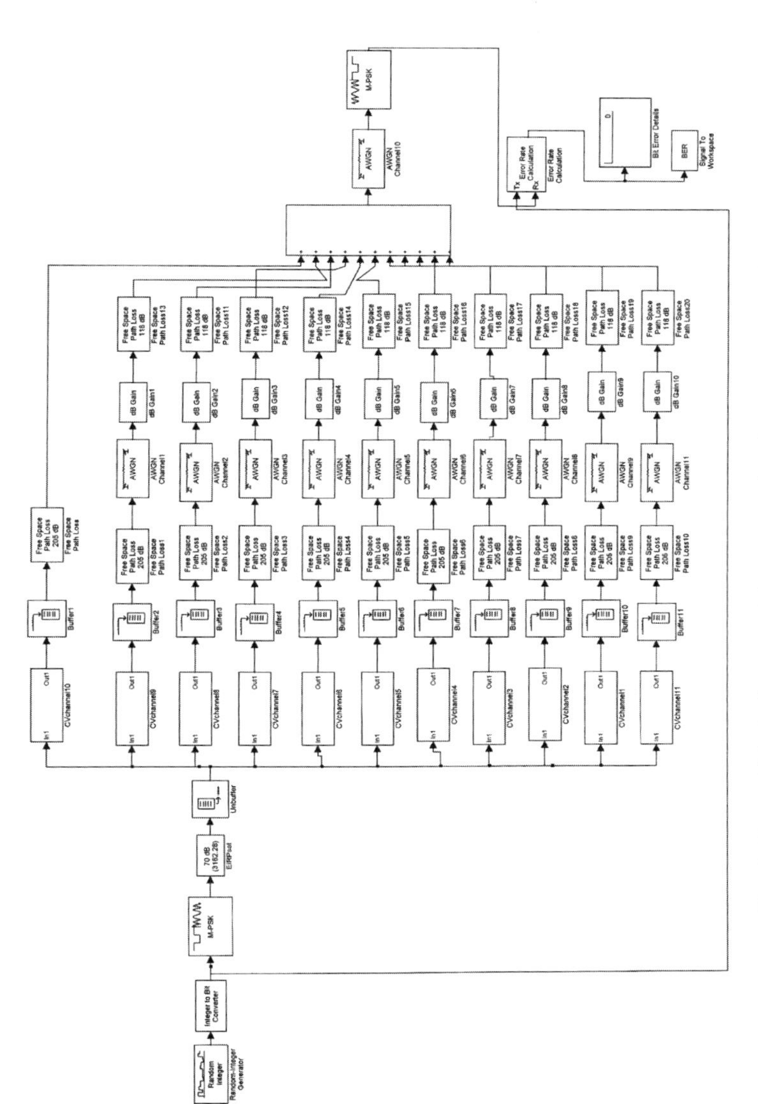

Fig. 5. General block diagram of AF cooperation for DVB-S2 downlink.

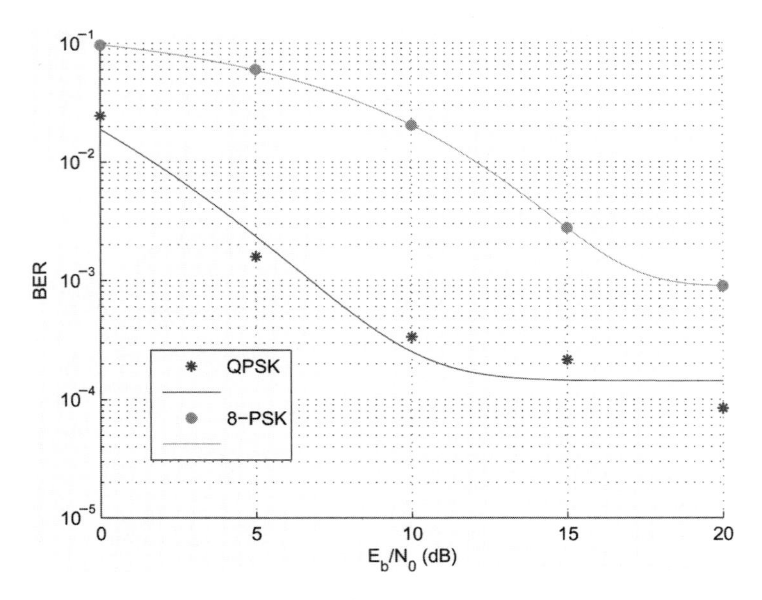

Fig. 6. QPSK and 8PSK with rice factor K = 20 and 10 cooperators.

The performances show an significant improvement, in terms of error probability, in comparison to the case in absence of cooperation for the same modulations (QPSK and 8PSK). QPSK shows a $BER = 10^{-4}$ for $E_b/N_0 = 20dB$, while for the 8PSK gives a $BER = 10^{-3}$ to parity of E_b/N_0. Moreover 8PSK performances become sensibly worse with smaller values of R due to the reduction of the deterministic component of the ricean channel which result in heavy fluctuations of the signal. In the graph of Fig. 7 three conditions of shadowing are considered:

- R = 20 correspondent to very light shadowing values;
- R = 15 representing an intermediate value;
- R = 10 with significant shadowing values.

The curves of Fig. 8 show the advantages deriving from the use of the cooperation AF strategy considering the QPSK modulation. We can see how the performances improve as the number of cooperators increase: on the top of the figure is represented the situation in the absence of relays, then follow the performances with 5, 10 and 15 cooperators. The comparison has been issued choosing a Rice factor $R = 1$; the results ($BER < 10^{-2}$) are acceptable for the channel coding techniques present in the DVB-S2 standard.

By varying the Rice factor R we obtain the results shown in Fig. 9 where QPSK performances with heavy shadowing ($R = 0.6$), medium shadowing ($R = 1$) and light shadowing ($R = 4$) are compared. For R = 4 the performances are close to the target ($BER = 10^{-4}$), while for $R = 0.6$ the BER values are higher then target resulting unacceptable for DVB-S2.

It is worth noting that in these simulations all the handset share the same Rice factor R, modeling the situation where the consumers cooperators all work under

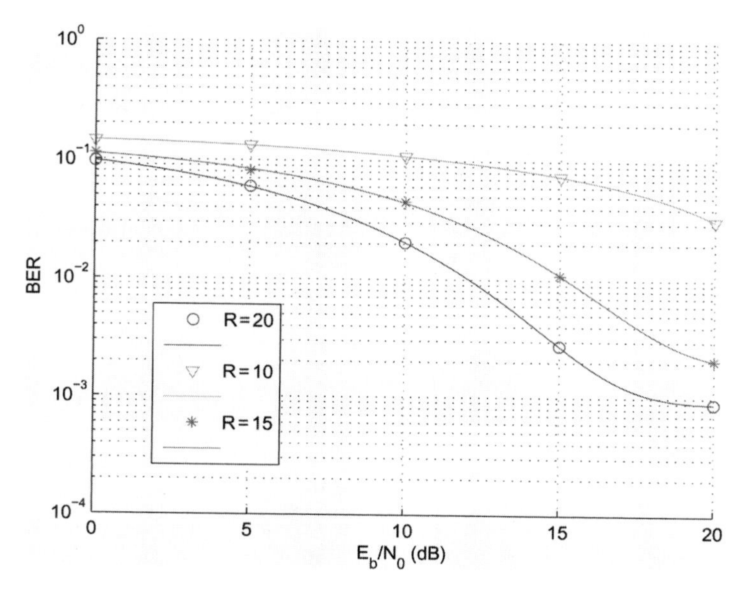

Fig. 7. 8PSK with variable R = 10;15;20 and 10 cooperators.

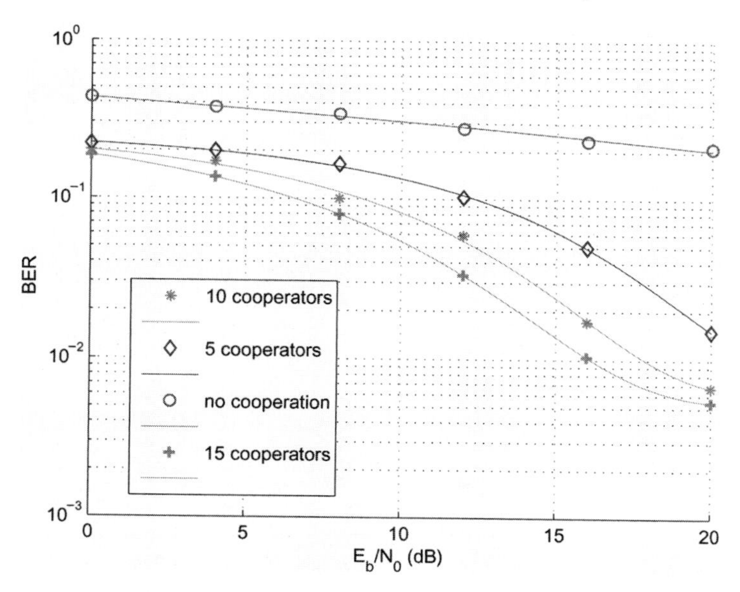

Fig. 8. QPSK with cooperator number variable M = 0;5;10;15 with R = 1.

homogeneous operational conditions. By considering a less critical situation, where only a subset of cooperating terminals are subject to heavy shadowing, we can see (Fig. 10) that the performances improve. Figure 10 shows the BER in the case of 50% of the handset are in heavy shadowing ($R = 0.6$) while the remaining ones have $R = 1$.

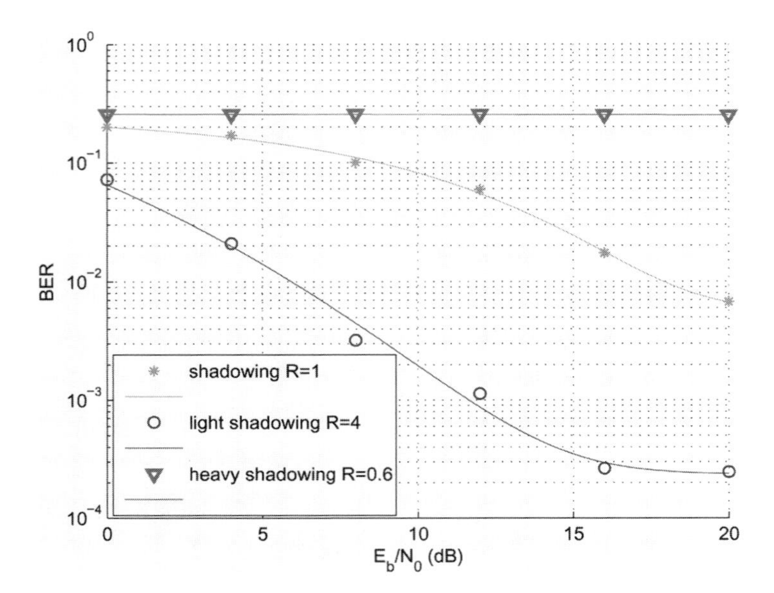

Fig. 9. QPSK for variable R = 0.6;1;4 and 10 cooperators.

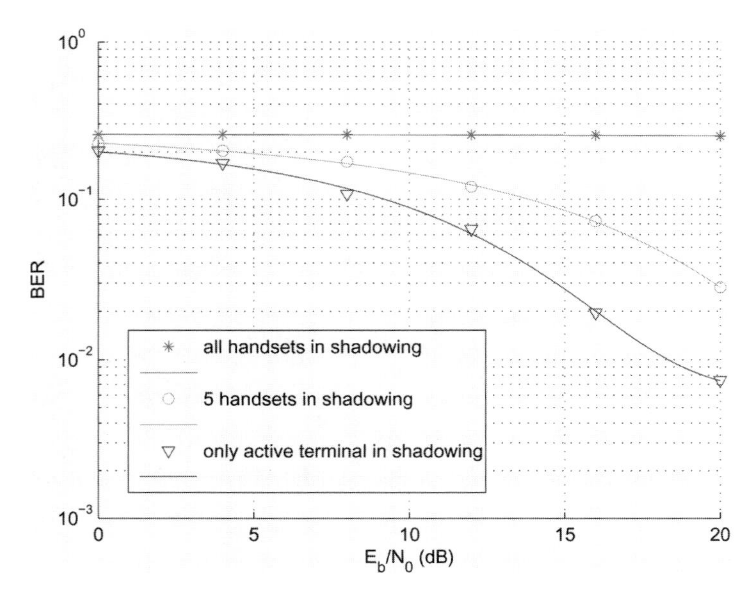

Fig. 10. QPSK varying handset number in shadowing for R = 0.6.

5 Conclusions

This paper shows a possible solution to the problem of the extension to the mobility (direct receipt on mobile terminals, equipped with non directive antennas) of a satellite transmission DVB-S2. The idea is to build a cooperation among a set of mobile terminals, in a way that the signal received by each single device is the result of the composition of more replicas of the same signal sent by other cooperating devices.

The choice of the adopted link cooperation method (Amplify and Forward) has been suggested by the satellite operational context (Fig. 1), characterized by unbalanced link strengths and limited complexity available at cooperators.

Link budget analysis shows that by choosing feasible system parameters (satellite spot power, co-operation amplification factor, number of co-operating terminals, terminal power dedicated to co-operation) we obtain signal-to-noise ratios compatible with DVB-S down-link profiles for up to 16-APSK constellations.

Under a more realistic scenario, where all the cooperators are independently faded accordingly to the Corazza-Vatalaro channel model, high order modulations are still possible in presence of favorable propagation conditions.

Link cooperation enables the reception of DVB-S2 services from handheld terminals when a cluster of cooperating users is present. This is a common context when professional users are involved (emergency rescue teams, tactical scenarios).

In the case of personal communications, the link cooperation technique may be offered as an option to overcome reception limitation, so the single subscriber has the possibility to choose if to participate to the cluster or not. This model allows the user to retain the control over the power resources of its terminal.

References

[1] ETSI TR 102 376 V1.1.1 (2005–02) "Digital Video Broadcasting (DVB): User guidelines for the second generation system for Broadcasting, Interactive Services, News Gathering and other broadband satellite applications (DVB-S2),".

[2] ETSI EN 302 307 V1.1.1 (2004–06) "Digital Video Broadcasting (DVB); Second generation framing structure, channel coding and modulation systems for Broadcasting, Interactive Services, News Gathering and other broadband satellite applications.".

[3] Richharia M. "Mobile Satellite Communications: Principles and Trends". *Addison-Wesley*, ISBN 0 201 33142 X

[4] Sendonaris A., Erkip E., and Aazhang B. "User cooperation diversity - part I: system description". *IEEE Transactions on Communications*, vol. 51, pp. 1927–1938, Nov. 2003.

[5] Sendonaris A., Erkip E., and Aazhang B. "User cooperation diversity - part II: Implementation Aspects and Performance Analysis". *IEEE Transactions on Communications*, vol. 51, pp. 1939–1948, Nov. 2003.

[6] J. Nicholas Laneman "Cooperative Diversity in Wireless Networks: Algorithms and Architectures". *PhD Thesis*, Massachusetts Institute of Technology, Cambridge MA, Sep. 2002.

[7] Ribeiro, A. and Giannakis G. "Fixed and Random Access Cooperative Networks". *EURASIP NEWSLETTER*, invited article - submitted January 18, 2006.

[8] Ribeiro, A. and Cai X. and Giannakis G. “Symbol Error Probabilities for General Cooperative Links”. *IEEE Transactions on wireless communications*, 4(3): 1264–1273, 2005.
[9] F. Vatalaro, G. Corazza “Probability of Error and Outage in a Rice-Lognormal Channel for Terrestrial and Satellite Personal Communications”. *IEEE Transactions on Communications*, vol. 44, pp. 921–924, Aug. 1996.
[10] F. Vatalaro, G. Corazza, C. Caini, C. Ferrarelli, “Analysis of LEO, MEO, and GEO Global Mobile Satellite Systems in the Presence of Interference and Fading”. *IEEE Journal on selected areas in communications*, vol. 13, pp. 291–300, Feb. 1995.
[11] G. Corazza, F. Vatalaro “A Statistical Model for Land Mobile Satellite Channels and Its Applications to Nongeostationary Orbit Systems”. *IEEE Transactions on vehicular technology*, vol. 43, pp. 738–741, Aug. 1994.

Broadband Mobile Satellite Services: The Ku-band Revolution

Antonio Arcidiacono, Daniele Finocchiaro, Sébastien Grazzini

EUTELSAT SA, 70 rue Balard, 75015 Paris (France)
Phone: +33-1.5398.4747, Fax: +33-1.5398.3181
e-mail: aarcidiacono@eutelsat.fr, dfinocchiaro@eutelsat.fr, sgrazzini@eutelsat.fr

Abstract. In this paper we analyze the emergence of new system architectures and products that are able to deliver broadband services into mobile environments using Ku-band. In particular, we analyze the latest technological solutions that have been developed to cope with the stringent requirements of a mobile environment. Such solutions have brought broadband to environments such as business jets, commercial aircrafts, trains, and cars, which are today the new frontiers where broadband MSS can be offered.

1 Introduction

After 25 years where mobile satellite services (MSS) have been delivered using L-band infrastructures, with the arrival of broadband services it is evident than the scarce frequency resource in L-band is a structural limit to the scalability of such services. At the same time, Ku-band satellites have been until recently positioned to cover land masses only, and therefore they did not seem well suited to offer MSS that, by nature, require coverage over land and sea masses. But recent evolution in the market demand for ubiquitous broadband services, with service quality comparable to those today available via ADSL or fibre networks, has generated a new business opportunity for satellite operators and satellite service providers. Consequently, market offers have appeared, starting from regional maritime services in Ku-band, evolving to global broadband services for aeronautical, and more recently maritime, scenarios, from the launch of the "Connexion by Boeing" offer [6] to the provision of GSM services onboard cruise ships [7].

These offers have initially suffered of the typical problems of new technologies, with relatively bulky and expensive equipment, but have opened a new opportunity for developing and deploying advanced system architectures and new products tailored to the different market opportunities.

In this paper we analyze the emergence of new system architectures and products that are able to deliver broadband services into mobile environments using Ku-band. In particular, we analyze the latest technological solutions that have been developed to cope with the stringent requirements of a mobile environment. Such solutions have brought broadband to environments such as business jets, commercial aircrafts, trains, and cars, which are today the new frontiers where broadband MSS can be offered.

Two main topics are analyzed: antenna systems and modulation and coding techniques. Each mobile environment requires particular solutions for these topics, tailored to the specific physical conditions, and compatible with a sound business model. Moreover, a global network architecture can optimize costs by sharing the same infrastructure among different commercial offers.

We show different antenna solutions, demonstrating that this is the key element that impacts on the service efficiency, and as such it needs to be tailored specifically for each mobile environment. We present solutions, commercially deployed or still under study, and ranging from phased-array flat antennas, to reflector(s)-based antennas, to other innovative concepts.

Concerning modulation and coding, the latest spread spectrum technologies are presented, where state-of-the-art CDMA-based systems are coupled with ACM (Adaptive Coding and Modulation) techniques to achieve maximum efficiency.

We discuss the application of "multipath routing" technology, where the satellite link is coupled in a synergic way with other wireless links, so as to augment the availability of the service in situations where satellite alone could not be sufficient.

The global network architecture must in these cases take into account other networks, other than the satellite one. It is important to offer a seamless continuous service to the end users, even in the presence of shadowing or link switching. Thus, new problems have been studied and solved to guarantee the interaction of satellite networks with terrestrial wireless networks.

This paper is organized as follows. In Section 2 the main advantages of Ku-band, when compared to L-band, are presented. In Section 3 the main issues to be solved are presented in a general fashion. In Section 4 a brief presentation of advanced technologies, applicable in different scenarios, is done. Sections 5 to 7 present the different mobile environment and their specificities. Finally, some conclusions are drawn.

2 From L-Band to Ku-Band

Historically, connectivity services for mobile vehicles have been delivered through the use of satellites transmitting in L-band (out of which only a few tens of MHz are assigned to satellite use from regulatory authorities), beginning with Marisat in 1976. Targeted to telephone communication at first, these services have evolved towards IP connectivity and, more recently, the delivery of IP broadband. The use of L-band gives important benefits, such as small onboard antenna size and little or no attenuation due to rain.

However, the amount of L-band available, and more specifically the portion allocated to MSS, is limited. Moreover, frequency reuse due to different orbital slots is extremely limited.

Broadband applications require a much greater amount of bitrate for the final user than normally available. One technical solution has been to create small spots of coverage, so that the same frequency can be re-used in different spots, thus increasing the total amount of available bandwidth for 'unicast' communication (while 'broadcast' type of communication is penalized by the multi-spot approach).

In the past, some operators failed to run a successful business out of their L-band technology (Globalstar, Iridium). Other operators succeeded in creating a large base of users (Inmarsat, Thuraya). The advantages for L-band operators are:

- small terminals (even handheld) as the antenna size is small and just coarse pointing is required;
- the coverage is wide, over seas and lands;
- the signal is not attenuated by rain, and less prone to shadowing due to trees than Ku-band;
- there is a large installed base (vessels, aircrafts).

On the other hand, L-band suffers from the lack of bandwidth, and this has consequences on the costs to the users. For example, the cost of a minute of Inmarsat communication can range from several Euros to tens of Euros. These costs are hardly compatible with a 'broadband' user experience at reasonable prices.

To definitely overcome the problems due to the scarcity of L-band, the only choice is to move to a higher frequency band. Ku-band (frequencies between 11 and 14 GHz, out of which 2+2 GHz assigned to satellite use) is an ideal candidate to offer broadband services. Although only a part of the overall Ku spectrum is usable in a mobile environment (in particular, only 500 MHz – from 14 to 14.5 GHz – can be used in the uplink direction from a mobile vehicle), bandwidth can be augmented by frequency reuse at different orbital positions.

In the recent years, on the top of TV distribution which is its principal application today, Ku-band has succeeded to provide broadband services to fixed users, either with one-way (asymmetric) or two-way (symmetric) data flow. The cost of the user terminals has become affordable, and spectrum utilization is efficient enough, so as to allow good prices to be proposed to final customers.

EUTELSAT has been the first operator to deploy the use of Ku-band for mobile communication in Europe with the Euteltracs product. Euteltracs is targeted to localization and messaging for terminals installed on trucks. After 15 years of operation, Euteltracs counts today more than 30000 active terminals over all Europe.

The natural evolution is today the commercial deployment of broadband services in Ku-band towards all mobile environments: boats, trains, aircrafts, cars. . . Building upon the past experience, and thanks to recent developments in antenna and modern design, the costs of a service in Ku-band are now only a fraction of those in L-band. Thus, new markets are reachable (see Fig. 1 for a graphical summary).

Ku-band still has some limitations when compared to L-band, and in particular:

- bigger antennas, that require precise pointing and tracking while in motion;
- need to extend coverage over oceans for the aeronautical and maritime case;

We will see in the rest of this paper how these problems have been solved.

3 Issues in Mobile Broadband

Current applications and services require data rate in the range of up to 2 Mbit/s for the return link and several Mbit/s for the forward link, taking into consideration the average number of potential user's in an aircraft, a train or a vessel. These figures are certainly going to increase, as the number of IP-based applications is ever increasing, and reaches a wider audience each day. Thanks to the explosion of

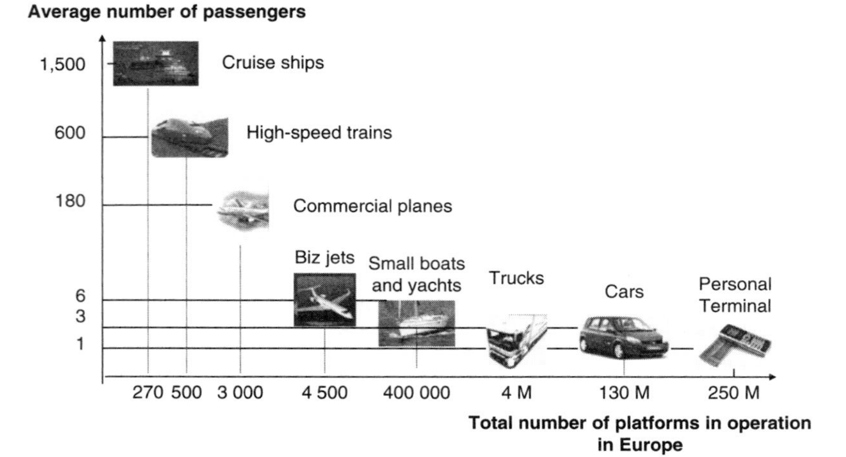

Fig. 1. Addressable market for different mobile environments.

DSL-like connection systems, users at home have become familiar with bandwidth-demanding applications. In order to be attractive from a commercial point of view, the service offered in the mobile environment should at least allow for the same type of applications that the user already has at his home or office, at not exceedingly high costs.

The main issue to be solved in a mobile broadband scenario is the one of cost. In fact, while classical telephony needs could be satisfied with a few allocated kbits, newer broadband applications are much hungrier in bandwidth. Even simple web browsing requires peaks of (at least) hundreds of kbits in order to allow the download of a web page in a reasonable time. Moreover, users are becoming acquainted with broadband at low price at home (DSL offers with Mbits connectivity are usually in the range of tens of € per month), and expect a comparable experience in a situation of mobility. Users feel connectivity as something that should be reasonably 'cheap', while they are ready to pay more for premium content (e.g., top movies in a Video-on-Demand offer) or advanced services.

It is thus fundamental to optimize the cost of the system at the maximum possible extent. Restricting the discussion to the connectivity domain, the main cost drivers are:

- the total cost of the installation of the equipment on the mobile vehicle. This is usually dominated by the cost of the antenna system; however, the cost of the other parts, and of the installation work itself can be significant too;
- the cost of the satellite segment. This is related to how performing is the modern in total bits/sec/Hz (i.e. with respect to bandwidth utilization) and how the same bandwidth is shared between multiple users or mobiles;
- the cost of the ground segment. This is due to cost of material and installation, and operating costs due to maintenance and operations of the ground equipment.

While a more detailed discussion will be done for each particular type of mobile, some general considerations on the optimization of costs are given here below.

Optimization of the Installation Costs

- the installation must be simple and must take little time (in particular, some type of mobiles, e.g. trains, simply cannot be immobilized for long periods);
- there must be a fairly limited number of equipments to be installed – putting more software packages on the same hardware simplifies both installation and future maintenance;
- as far as possible, the antenna must be efficient, simple, reliable and reasonably cheap.

Optimization of Satellite Segment Costs

- the antenna and modern must have good performance on the satellite link;
- in particular, a good protocol must be used in the two directions (for example DVB-S2 for the forward link);
- most importantly, the antenna must be as efficient as possible (gain, sensitivity, interference profile) within the limitations of the specific mobile environment;
- new techniques of 'VCM' (Variable Coding and Modulation) or 'ACM' (Adaptive Coding and Modulation) must be used, where suitable, to optimize satellite segment performance with respect to a single mobile at any specific moment;
- the same satellite segment should be shared by a large amount of users/vehicles, of course with a reasonable QoS strategy, to share the costs.

Optimization of Ground Segment Costs

- a centralized hub/NOC allows to share operational costs among several services;
- the hub/NOC should be highly automated and not rely too much on human intervention.

From a more technical point of view, the main issue to be solved is that of satellite interference. In fact, mobile vehicles usually require small or low-profile antennas. This type of antenna has a large beamwidth, thus has the double problem:

- in reception, it grabs noise coming from satellites adjacent to the one used;
- in emission, it interferes with adjacent satellites. The amount of allowable interference is limited by ETSI regulation.

While an accurate choice of the satellite segment to be used can help in smoothing these problems, the antenna design plays a major role. It is thus important that, within the size constraints coming from the vehicle, the antenna has a good gain and a good interference profile (if the antenna interferes too much, it will be forced by regulation to transmit at lower EIRP, thus at lower bitrate). Also the tracking algorithm and mechanics are important not to loose some dB due to poor tracking accuracy.

The modem also plays an important role in bounding adjacent satellite interference in the return link, in that some coding techniques (in particular spread spectrum) allow using a higher bitrate while remaining within the ETSI regulation mask.

Another problem related to low-profile antennas is the equivalent surface at low elevation angles. Even a fairly large low-profile antenna offers a limited surface to a satellite at low elevation. Thus, the performance of the antenna will be limited and the satellite link cannot be used at its optimum. Thus, low-profile antennas should be used only when necessary, and their profile be accurately designed for the specific vehicle.

4 Technological Solutions

4.1 Satellite Link Optimization for Mobile Use

Satellite communications have been used for some years now, introducing communication standards such as DVB-S and modulation and coding techniques such as TDMA. Today, one of the major issues in Broadband Mobile Satellite Services is to adapt these modulation and coding techniques to a mobile environment.

In satellite communications, the physical layer spectral efficiency depends mainly on satellite coverage and antennas performances. In our case, we can assume an EIRP in the range 44–54 dBW, and a G/T in the range 0–10 dB/K.

In the railway scenario, trains are generally operating on land masses well covered by Ku-band satellites, which implies that we have good performances. However the antenna must often be small and/or low-profile, interference with adjacent satellites should therefore be constrained within the limits imposed by radio regulations.

In the maritime and aeronautical scenario, we must take into account that boats and aircrafts can be either at the centre or at the edge of a beam. Thus, the satellite link has varying characteristics, and should be optimized for each situation.

For big boats, it is often possible to install big antennas that improve the overall performance and limit interference with adjacent satellites. However, this is not possible for smaller boats and aircrafts.

There are two major issues to solve in terms of the satellite link:

- interference generated to, and received from, adjacent satellites;
- optimisation of the link for different coverage scenarios (centre or edge of the beam).

We will see in next sections that the first issue can be solved using spread spectrum for the return link, and the second one using ACM (adaptive coding and modulation) or VCM (variable coding and modulation) for the forward link.

4.2 Spread Spectrum and CDMA

Spread spectrum technology uses wide band, noise-like signals. Spread spectrum transmitters use similar transmits power levels to narrow band transmitters. Because spread spectrum signals are so wide, they transmit at a much lower spectral power density, measured in Watts per Hertz, than narrowband transmitters. As a consequence, the interference generated on adjacent satellites operating at the same frequencies, due to the use of small antennas with a large beamwidth, is greatly reduced.

For this reason, spread spectrum is the technology of choice when working with small or low-profile antennas, as is the case in almost all mobile environments.

One of the foreseen spreading techniques is Code Division Multiple Access (CDMA). In these systems all users transmit in the same bandwidth simultaneously. The frequency spectrum of a data-signal is spread using a code uncorrelated with that signal. As a result the bandwidth occupancy is much higher then required.

4.3 Adaptive Coding and Modulation

Most existing satellite systems use QPSK modulation with concatenated Convolutional and Reed-Solomon error correction codes, as per the DVB standard. In addition, the RF links are run using fixed margins, with the size of the margin dependent upon the maximum attenuation predicted for the service area. One way to increase the system data throughput is to manage each terminal separately and have each one use the highest possible level of modulation in combination with the highest possible code rate as the instantaneous link conditions allow. As link conditions fade for each individual terminal, the modulation level and code rate is changed to maintain BER requirements. This technique, called Dynamic Link Assignment (DLA), significantly increases average information throughput per unit bandwidth. In addition to these increases in capacity, the technique may lower satellite and terminal EIRP requirements. These gains allow the service cost to be reduced.

Additionally, DLA provides a significant advantage for spot beam systems. In the typical spot-beam system, the satellite antenna gain from edge of coverage to beam centre varies by about 13 dB. As a result, the satellite EIRP varies by the same amount over the beam coverage. Dynamic links as opposed to fixed links allows each terminal to automatically sense where it is in the spot beam and to operate at as high a modulation level and code rate as possible.

This also significantly increases the transponder data throughput.

In scenarios such as aeronautical and maritime one, taking the "worst case" modulation greatly reduces the overall performance. On the other hand, optimizing the modulation parameters for each aircraft according to its current position allows extending the area of service towards the edges of the spot, while at the same time fully exploiting the good link parameters at the centre of the spot.

In other scenarios, such as the railway one, the VCM solution can be sufficient. According to the fact that trains do not travel through the whole coverage map, it is acceptable to fix once and for all a modulation for each vehicle. However, VCM allows to multiplex on the same transponder data intended for different mobiles with different parameters.

Figure 2, taken from [1], shows the advantages of using VCM/ACM techniques in the case of the DVB-S2 protocol. If the coding and modulation parameters are fixed (such as in DVB-S), then there is a global C/N and bitrate so that all mobiles with the required C/N (or better) will receive the specified bitrate. By using VCM, different

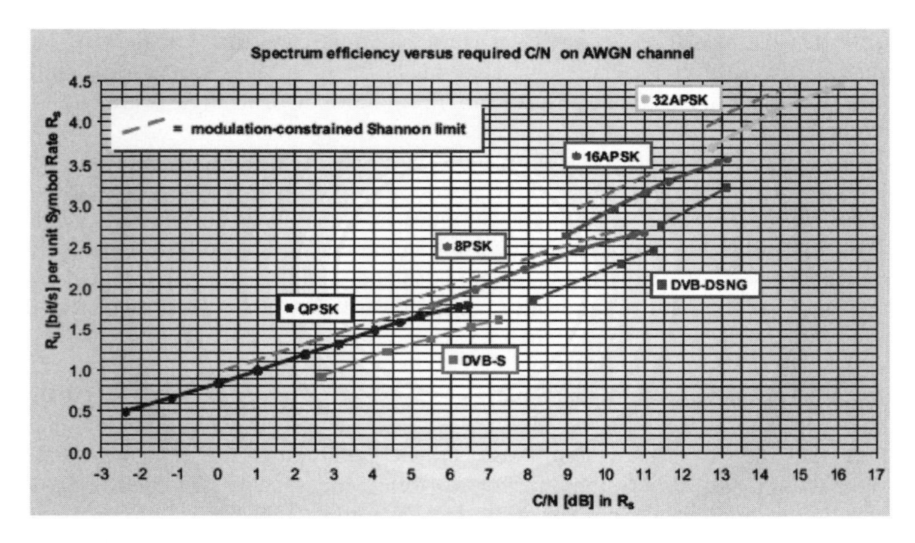

Fig. 2. Required C/N versus spectrum efficiency in different modes of DVB-S2 (Source: [1]).

mobiles can be preconfigured so as to optimize the link for their own C/N: for example mobiles at the centre of the spot may receive a higher bitrate, or consume less bandwidth. By using ACM, this adaptation is dynamic, thus taking into account the movement of the mobile or changes in the weather conditions.

By simply working with QPSK and 8PSK modulation, DVB-S2 allow a variation of 13 dB for the C/N of terminals sharing the same multiplex.

Figure 3 illustrates the possible performance gain due to ACM, by showing what represent a 13 dB variation of EIRP on the EUTELSAT Atlantic Bird 2 coverage. The smaller area corresponds to 54 dBW of EIRP, i.e. the best possible coverage. The wider area corresponds to 41 dBW (cut at an elevation of 10 degrees, the minimum at which a mobile antenna can operate). Aircrafts in the centre of the smaller area and aircrafts within the wider marked area can share the same DVB-S2 multiplex, each one optimizing the satellite resource.

By using ACM/VCM techniques, it is possible to extend the service to areas with worse coverage, while at the same time maintaining good performances at the centre of the spot. In this way, the problem of coverage over seas, typical of Ku-band spots, is vastly reduced.

4.4 TCP Acceleration and Quality of Service

In order to offer connectivity to mobiles, the most reasonable choice is to offer an "IP pipe" capable of transporting any type of IP traffic. Today the TCP/IP protocol suite is capable of handling a large range of application and services, in a flexible and powerful way. However, some care must be taken in the case of mobile broadband services.

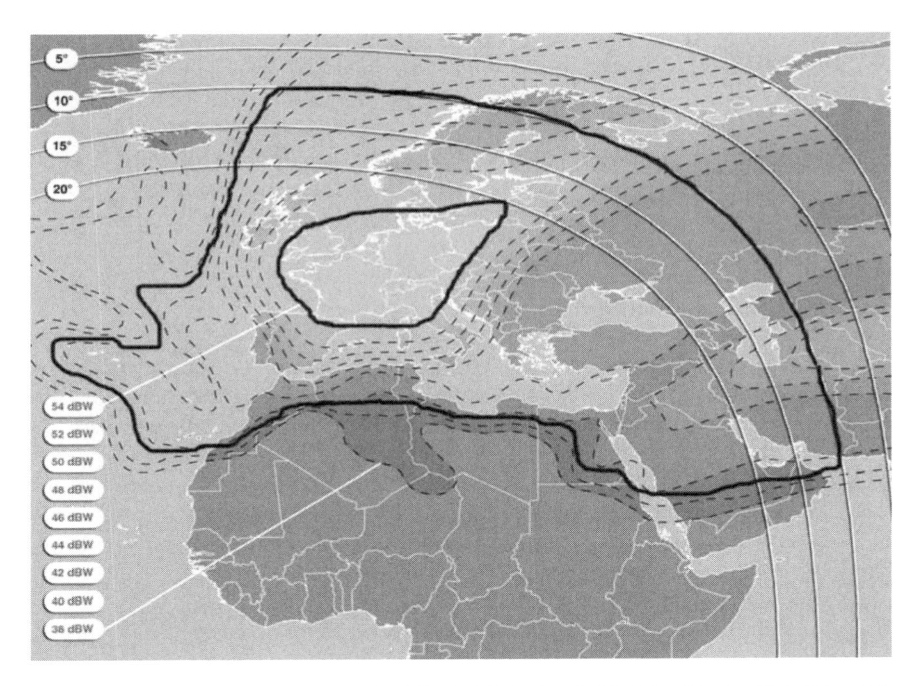

Fig. 3. Downlink european coverage of EUTELSAT Atlantic Bird 2, with the 54 dBW and the 41 dBW contours.

First of all, the performance of the TCP protocol over the satellite link is limited by the delay of the link and the different nature of errors. Even if the bandwidth is available, TCP is not able to use it fully.

TCP acceleration is a well known technique to augment the performance of the TCP/IP protocol stack over links with peculiar characteristics, such as the satellite. A number of protocols that implement this technique have been proposed, see [3] for a survey.

Another issue is the management of the available bandwidth, in particular in the case of congestion, in order to guarantee a fair quality of service to each user in the system. The amount of bandwidth to be allocated for a particular user is in part determined by the commercial offer: premium users may receive a *guaranteed* bandwidth, or simply a higher priority, or access to particular services such as Voice-over-IP (VoIP). Then the technical constraints apply: amount of bandwidth available at the terminal level, according to the implemented VCM/ACM scheme, and so on. Given the nature of current satellite networks, the main problems to be addressed to guarantee a fair quality of service are:

- the 'forward' and 'return' link are often implemented in rather different way. So the bandwidth management must be applied in a coordinated manner to both directions;

- modems and hubs often implement some kind of bandwidth management at the *terminal* level. The system must be able to treat appropriately the quality of service offered to *single users*, that may share or not a terminal.

In order to optimize bandwidth costs, it is very likely that the satellite link is almost always full. So it is very important to accurately manage congestion as a 'normal' situation, and be able to offer a good service under this condition.

4.5 The Network Architecture

An architecture for the Mobile Broadband Satellite Services does not reduce to a hub and a number of modems. Other elements, integrated into a coherent network architecture, are needed to provide the service while matching the business constraints.

The NOC, in collaboration with onboard servers, must be able in particular to:

- share the same system among the maximum possible number of terminals/users (even on different type o mobiles);
- apply suitable bandwidth management at user level;
- manage different type of links and the transparent handovers;
- when applicable, manage different satellite spots, in principle served by hub located in different places.

5 The Maritime Case

Compared to trains and aircraft, the maritime environment is the one with least constraints. For example, the vessel is more stable than trains (roll rates are slower, vibration is less), however the antenna steering range may exceed the range of trains, which is typically unproblematic since space, size and weight are not critical issues. This has allowed the deployment of some broadband commercial services in Ku-band:

- DSAT MARITIME services offer a cost-effective solution available from medium bit-rate communications (64 kbit/s) to broadband applications (up to 2Mbit/s). It provides a hub-less, fully meshed multi-services network. A central Network Control Centre is operated by EUTELSAT. Service is in full operation in Ku-band on EUTELSAT W1, W2 and W3 satellites. Data transmission includes: point to point, point to multipoint; broadcast, streaming media; interconnection of LANs, Internet Access; interconnection of PBX, Public phone with pre-paid card.
- DSTAR MARITIME service delivers cost-effective real broadband IP services (up to 2 Mbit/s) for "always-on" connectivity. Data transmission allows for a full range of applications, including broadband Internet access, e-mail and fax, video-on-demand, voice over IP and videoconferencing (Fig. 4).

As a recent application on Ku-band broadband services, WINS (a joint venture between Maltasat International and Skylogic, EUTELSAT's broadband affiliate)

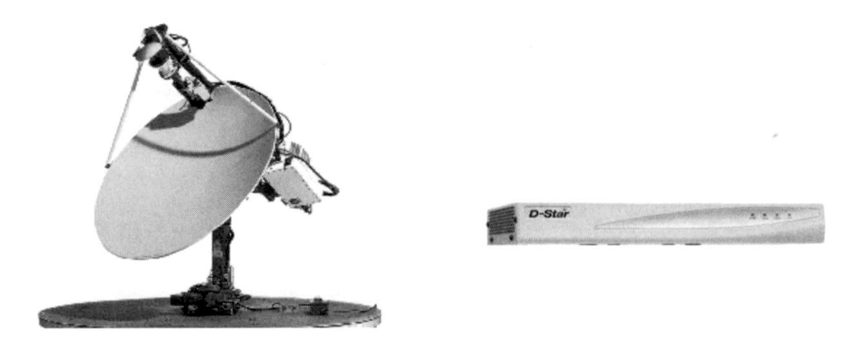

Fig. 4. The antenna used for DSAT, and the modem used for D-STAR maritime services.

offers GSM telephony and broadband services on luxury cruise ships. See [7] for more information.

The evolution in the maritime scenario is to provide service to smaller boats, such as yachts. The cost of the service (both equipment and bandwidth) is once again the key point to be tackled. We expect that in the near future new 2-way system will be developed at lower price and size, so that installation is possible on smaller boats.

Today a typical solution for small boats is to use an asymmetric link, where the Ku-band is used only in the 'forward' link (hub to mobile). For the return link, the choice can be made among terrestrial (e.g. GPRS) or other satellite (e.g. L-band) solutions. The advantage is that a receive-only antenna system in Ku-band in much cheaper that an equivalent 2-way antenna system. Moreover, many boats already have a TVRO antenna installed, that could be used also for broadband reception. On the other hand, the return channel is usually less demanding in bandwidth (as it is mostly used for TCP acknowledgment), so the higher cost of the GPRS or L-band link has a limited impact. Of course the trade-off depends also on the expected use of the system.

If the boat moves over long routes, the use of ACM is recommended, in order to optimize satellite segment utilization according to the exact location of the boat.

6 The Railway Case

In this setting, the availability of a terrestrial infrastructure allows for the use of terrestrial (even if wireless) links, such as GPRS or Wi-Fi. Even if some of these links are narrowband, they can be used as return channel in an asymmetric traffic flow, where broadband is received by satellite.

However, the train presents different problems (such as temporary link loss due to tunnels, or obstacles in the satellite line-of-sight; interference from the power lines; vibrations) that must be overcome. Different EU-funded projects have tackled this problem (for example FIFTH, and, more recently, MOWGLY [5]).

The provision of Internet and multimedia services on-board trains is nowadays one of the crucial challenges which characterise the competition among railway operators. These operators are planning to include in their offerings to passengers an effective solution for accessing Internet – both for business and entertainment purposes – to enjoy digital TV, to distribute "railway operator" specific contents and so on. Satellite-based systems seem to be particularly suited to providing broadband connectivity to trains' passengers. Following (or, somehow, generating) this new trend, several projects proposing technical, satellite-based solutions for multimedia services provision in the railway environment have been recently undertaken around the world.

The railway environment has a lot of challenges that must be tackled to provide a good service. In principle, trains can be served using terrestrial wireless technology. For example, existing GPRS or UMTS networks can be used, but the connection speed can hardly be considered broadband when shared among a large number of users. Or, technologies such as Wi-Fi and, more recently, Wi-Max have been proposed; in this case, the costs of deployment of an ad-hoc network are very high, except for short lines (e.g. commuter trains near a big city).

A better solution is to integrate satellite and terrestrial wireless in a hybrid system that can take the advantage of both technologies. Ku-band satellite can offer broadband almost everywhere in the countryside, and terrestrial wireless networks can complement satellite coverage in shadowed zones (tunnels, stations, urban zones, or areas shadowed by trees or hills). The onboard equipment shall implement a 'multipath-routing' technology capable of:

- selecting, with the appropriate advance, the most appropriate communication link. This selection can be done based on a-priori knowledge of the availability of the links, on real-time monitoring of link quality, on economic considerations (required bandwidth and associated cost).
- switching the data flow from a link to another. The switch should, as far as possible, be transparent to the end user, so at least preserve the active TCP sessions. This switching can be between two different type of link (vertical handover) or between two cells of a cellular network (horizontal handover).

The routing technology above is a key point for the final quality of service as perceived by the end user. A number of companies have proposed solutions to this problem, that are currently under trial or operation. Some solutions are based on the use of GPS localization of the train, coupled with a geographical database constantly updated, which allows the router, for example, to take appropriate measures way before the train enters into a tunnel.

An implementation of this switching functionality is the "Smart Router", whose logical decomposition is shown in Fig. 5, developed by Rockwell Collins France and MBI within the MOWGLY European project.

The antenna choice, as always, is of great importance. One must consider that each type of train can accommodate a different size of antenna, also according to the tracks that the train will follow. Some double-deck trains only allow 8 cm or so for the height of the antenna. Single-deck trains in UK have quite stringent requirements too (up to 25 cm). Other trains can allow heights of 45 cm or even more.

These requirements have driven the industry to the development of specialized low-profile antennas. These antennas can be either 'classical' single or dual reflector

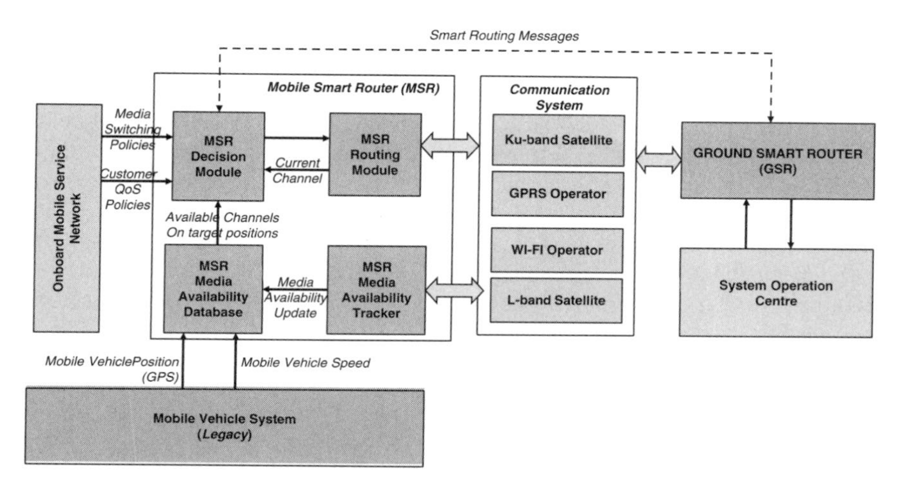

Fig. 5. Logical decomposition of the "Smart Router" (courtesy of rockwell collins francs / MBI).

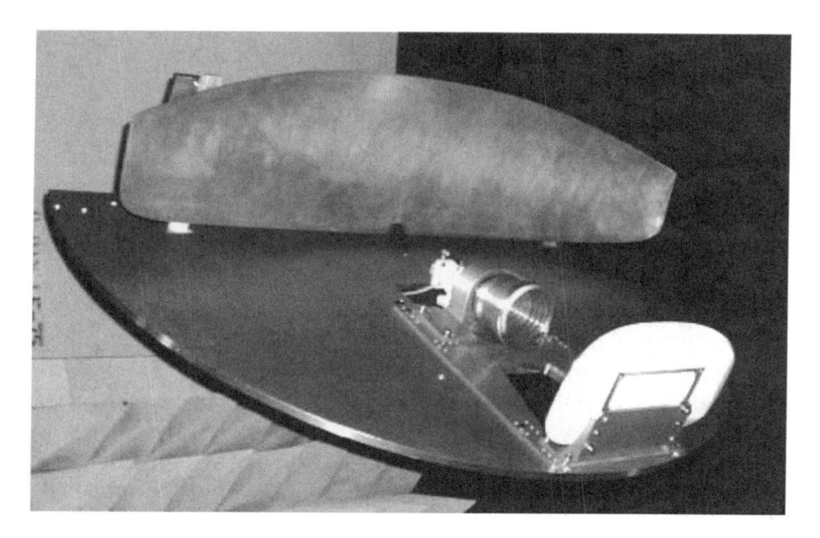

Fig. 6. A prototype antenna suitable for the pendolino trains (courtesy of teleinformatica& sistemi – patent pending).

antennas, or 'flat' antennas based on phased arrays technology. In any case the problem to solve is to have a high equivalent surface in the direction of the satellite, while maintaining a reduced height.

Two antennas dedicated for train applications are being developed within the MOWGLY European project [5], of which EUTELSAT is a partner. The first antenna, shown in Fig. 6, is an offset dual-reflector Gregorian antenna developed by Teleinformatica&Sistemi and installed on a tracking platform by ORBIT.

The antenna has a 80 × 20 cm main reflector and achieves a G/T of 9 dB/K and an EIRP of 43 dBW. ORBIT also provides a refined pointing system based on the use of a narrowband receiver, able to discriminate the target satellite from adjacent ones. This antenna is already in use in a few precommercial trials in Europe.

The second antenna is a flat antenna with a 80 × 15 cm phased array, mechanical steering in azimuth, and electronic steering in elevation. Both antennas are expected to be used in on-field trials before the end of 2006 coupled with DVB-S2.

In real commercial cases, the ideal solution is to have the highest possible antenna for the particular train. In fact, a low-profile antenna has intrinsically poorer performances (in gain and sensitivity), and a poorer interference pattern; as a consequence, a lower bitrate can be achieved and at a higher cost. The antenna size also impacts on the modem. In fact, a small antenna with a poor interference pattern requires a spread spectrum technology to minimize interference while maintaining good performance.

The modem must also be able to cope with the frequent fast fading typical of trains, for example shadowing due to electric poles. On TGV lines, for example, these poles create about 20 dB attenuation during 6 ms, periodically every 600 ms. Appropriate strategies have been implemented both on the forward and the return link, in order not to loose the signal lock, and to limit the number of lost packets (using e.g. FEC, buffering).

7 The Aeronautical Case

In the case of aircrafts, satellite communications have no concurrence in practice (terrestrial solutions have been studied and tried in the past, but without success). Today, all aircrafts have L-band based communication equipment, used at least for emergency and administrative matters. However, the use of L-band for services to customers (e.g. telephony) has proved to be too costly to have a commercial success. Although the L-band offer has evolved towards more performing systems (e.g. B-GAN), broadband offers will likely remain more costly than Ku-band offers.

A commercial service in Ku-band has been started in recent years by Connexion By Boeing (CBB). This service, started for commercial aircrafts, is now in operation on a few commercial airlines. The end user price is about $30 for Internet connection on a long-haul flight. However, initial investments have been so high that it is now difficult to recover from them.

This is a list of the main existing aeronautical solutions:

- Inmarsat Swift 64 provides passengers in the corporate jet market access to the Internet with speeds of 64 kbps. The service is distributed to seats with conventional IP cabling/routing systems.
- Connexion™ by Boeing (CBB) is available to the private business jet market and was launched on commercial aircrafts by Lufthansa in May 2004. It is based on the use of Ku-band. CBB provides real-time, high-speed Internet access to air travellers in flight. The business model for this service is far from being proven. It has been announced that Boeing might shut down Connexion after U.S. airlines have failed to show an interest, and turn a profit in 6 years.

- I-4/B-GAN™: with its fourth generation of satellites, the Inmarsat I-4, Inmarsat built a "Broadband Global Area Network" that is operational since 2005 with bit rates up to 432kbit/s.
- TV reception: reception of live television via Ku-band satellites is offered to airlines. Solutions for analogue DirectTV or digital DVB-S based systems are available by using the same satellite beam as the one used for home satellite television. Such systems are therefore only available above continent.

A broadband service targeted to business jets is offered by ARINC. Skylogic, an EUTELSAT affiliate, operates the broadband access service for this system through its European hub, so that US customers can seamlessly roam in the European coverage area. This system uses spread spectrum technology developed by Viasat in its Arclight product. Some important features of this system are the following:

- spreading occurs over the full transponder bandwidth. Consequently, the interference on adjacent satellites is greatly reduced with respect to narrow-band systems, thus allowing mobile terminals to transmit at higher speed (currently up to 512 kbps);
- the forward and return channel are superimposed in the same transponder, using PCMA (Paired Carrier Multiple Access). This allows to open the service on a new area by allocating just one transponder for the service (while other system, such as the one by CBB, require at least two separate transponders for the two directions);

The antenna size for the ARINC service is as small as 30 cm, and it is usually accommodated in the tail of the aircraft. The complete onboard system is shown in Fig. 7 below.

The system also allows dynamic bandwidth allocation, so that satellite resource is assigned only when a terminal is using it. Consequently, the satellite segment can be dimensioned with respect to the maximum number of terminals simultaneously

Fig. 7. The complete onboard system for the ARINC service on business jets: the viasat modem, the RANTEC tail-mount antenna, and the antenna control unit (courtesy of Viasat).

active, which, in the case of business jets, is much smaller than the total number of terminals.

In aeronautical applications, there are strong requirements on the onboard equipment, related either to the security of the aircraft (fire, vibration, shocks, . . .) and to the extreme environmental conditions (temperature, attitude changes, . . .). Equipment to be placed on aircraft is thus more costly than in other environments, and the installation itself can be quite complicated.

The size and weight of the system are important. In particular, the size of the antenna, and of its radome, directly impact on the drag of the aircraft. Added to the total weight of the equipment, this translates into more fuel consumption (or less accommodated voyagers), thus into higher operating costs.

In order to optimize bandwidth usage, it is important to use ACM technologies. In fact, during a long flight, the satellite link characteristics can vary considerably (in power and sensitivity of the transponder).

Handover between different coverage areas can also occur. In this case, the continuity of the connections has to be guaranteed by the NOC, in collaboration with some onboard equipment.

8 Conclusions

The use of Ku-band is bringing a revolution in the field of MSS services, as it allows real broadband to be provided on mobiles at reasonable prices. This is the result of intrinsic properties of Ku-band, and of recent developments in technology and its economic factors. Today a Ku-based solution can be deployed on almost any mobile, and further evolutions are to be expected to fill the remaining gaps.

In this paper we have discussed the different issues related to a broadband offer in Ku-band. One of the key elements is the antenna, which must be tailored to the specific environmental conditions. However, we have shown other elements that significantly contribute to the overall performance and economics of the system.

References

[1] Alberto Morello and Vittoria Mignone, "DVB-S2 – Ready for lift off", EBU Technical Review, October 2004.
[2] Bruce R. Elbert, *The Satellite Communications Applications Handbook*, 2nd ed., Artech House, 2004.
[3] Y. Fun Hu, Gérard Maral, Erina Ferro (eds.), *Service Efficient Network Interconnection via Satellite*, Wiley, 2002.
[4] M. Richharia, *Mobile Satellite Communications – Principles and Trends*, Addison-Wesley, 2001.
[5] MOWGLY – Mobile Wideband Global Link sYstem, web site, *http://www.mowgly.org*
[6] Connexion By Boeing, *http://www.connexionbyboeing.com*
[7] WINS, *http://www.winssystems.com*

Flower Constellations for Telemedicine Services

M. De Sanctis[1], T. Rossi[1], M. Lucente[1], M. Ruggieri[1], C. Bruccoleri[2],
D. Mortari[2], D. Izzo[3]

[1]University of Rome "Tor Vergata",
Dept. of Electronic Engineering Via del Politecnico 1, 00133 Rome-Italy
Tel: +39 06 7259 7258, Fax: +39 06 7259 7455,
e-mail: ruggieri@uniroma2.it
[2]Texas A&M University.
[3]European Space Agency (ESA).

Abstract. Flower constellations are a particular set of satellite constellations where every satellite covers the same repeating space track. When the flower constellations are visualized on an Earth centred earth fixed reference frame, the relative orbits shows flower-shaped figures centered on the Earth. In this paper the shape and the position of a particular flower constellations has been designed for the provision of telemedicine services. Once that performance metrics of the constellation have been defined and the service targets have been identified, the performance of the flower constellation have been compared with the well known polar and Walker constellations. The particular properties of the flower constellations allow an optimized coverage of a list of targets. It was found that the flower constellations provide better performance in terms access availability and mean access time.

1 Introduction

The Flower Constellations (FCs) constitute an infinite set of satellite constellations characterized by axial-symmetric dynamics. They have been discovered [1,2] on the way to the generalization of the concept of some existing satellite constellations. The dynamics of a FC identify a set of implicit rotating reference frames on which the satellites follow the same closed-loop relative trajectory. In particular, when the constellation axis of symmetry is chosen to be the planet's spin axis, then one of the implicit rotating reference frames coincides with a planet fixed reference frame, and, as a result, all the satellites will follow the same relative trajectory (repeating space track or compatible orbits). These relative trajectories constitute a continuous, closed-loop, symmetric pattern of flower petals.

FCs are potentially suitable for deep space observation systems, for global/regional navigation, for distributed space systems (interferometry, single and multiple-point observing systems), for telecommunications, as well as for other applications. FCs have been designed for Earth's global navigation, giving much better navigation performance (in terms of GDOP, ADOP, and coverage) than the existing Global Positioning System (GPS), Global Navigation Satellite System (GLONASS), and GALILEO constellations, using the same number of spacecraft or achieving the same performance using fewer satellites [3,4].

In this paper, we deal with telemedicine applications which are a particular type of telecommunication service which can exploit real-time and/or store-and-forward applications.

For this type of service we can identify a specific number of locations involved in providing and accessing to the service. There are several telemedicine providers which exploit existing satellite systems for the provision of this telecommunication service. Most of the sites interested in accessing telemedicine services are located in rural areas and, hence, satellite systems are the most suitable choice for the platform service.

However, none of these satellite systems are specifically designed for telemedicine services. Furthermore they do not provide a direct connection between the service suppliers and the service customers.

In this work we design a specific FC for the provision of telemedicine services with the following features:

- near continuous coverage of a list of targets interested in providing and accessing the service;
- direct connection of service suppliers and service customers via satellites and Inter Satellite Links (ISLs);
- maximum Round Trip Time of 200 ms.

This paper is organised as follows. Section 2 introduces telemedicine services and propose a list of service targets. Section 3 discusses the theory and the design of FCs and Walker constellations; the design of the two constellations is optimised for the coverage of the targets identified in Section 2. Performance comparison of the proposed constellations is shown in Section 4, while conclusions are drawn in Section 5.

2 Telemedicine Service

In this work, the design of a FC has been applied to a particular application of telecommunication: telemedicine. Telemedicine services enable the communication and sharing of medical information in electronic form, and thus facilitate access to remote expertise. This type of service is important for large and scarcely populated countries where there is a lack of health care facilities. Immobile patients should not be required to travel long distances to receive diagnosis and medical assistance.

The objective of this work is to build up a satellite constellation able to provide medical consultancies from advanced hospitals located in Europe or USA (service suppliers) to rural and/or remote areas in Africa or Asia (service customers) via a pure satellite communication network; hereafter service suppliers and customers will be referred as "targets" for the satellite constellations.

The identification of the service customer targets on Earth has been performed with respect to high demographic density and lack of medical infrastructures and terrestrial communications networks. Most of the service customers are located in rural areas while service suppliers are located in metropolitan areas. In Table 1 all the identified targets are listed. We should highlight that such targets are only a sample of possible locations that could be interested to such service.

Table 1. Telemedicine service Earth targets.

Service suppliers	Latitude	Longitude
Fucino	42.80	13.13
New York	40.71	−74.00
Houston	29.76	−95.30
Seattle	47.60	−122.33
Los Angeles	34.05	−118.24
Service customer	**Latitude**	**Longitude**
Lanzhou (China)	35.96	104.89
Lusambo (Congo)	−5.69	23.73
Baliuag (Philippines)	14.62	120.97
Mekar (Indonesia)	−6.18	106.63
Beroga (Malaysia)	3.16	101.71
Musawa (Nigeria)	12.00	8.31
Youngsfield (South Africa)	−33.93	18.46

We assume that the service customers are provided with fixed or portable terminals that are used by medical/emergency teams wanting to connect to the service suppliers in order to request any of the several services that are listed in the following:

- qualified medical assistance by using teleconference or instant messaging;
- analysis of clinical data (electrocardiogram, radiological data, etc.);
- monitoring of vital parameters (blood pressure, pulse, oximetry, respiration, etc.);
- access to medical information from digital libraries;
- follow a continuing education course (tele-medicine learning).

In order to allow such services being provided, the satellite constellation must be designed for real-time applications and store-and-forward applications.

After the identification of several service suppliers and service customers, the satellite constellations will be designed with the objective to meet the following requirements:

- Near-continuous coverage of all the service customers.
- Interconnection via Inter Satellite Links (ISLs) of all the service customers and at least one service suppliers.
- Round Trip Time (RTT) less than 200 ms in order to allow interactive applications.
- Total number of satellites lower than 10, in order to keep system cost low.

3 Constellation Design

3.1 Flower Constellation Design

The FC design methodology has been extensively described in [1–5]; a summary of the main underlying principles is briefly presented in this Section for the benefit of the reader.

The name Flower Constellations has been chosen because of the compatible orbit relative trajectories in the Earth-Centred Earth-Fixed (ECEF) reference frame resemble flower petals.

A FC is a set of spacecrafts characterized by the same repeating space track, a property obtained through a suitable phasing scheme. In general FCs are characterized by 6 integer parameters:

- N_p: number of petals;
- N_d: number of days to repeat the space track;
- N_s: number of satellites;
- F_n: phase numerator;
- F_d: phase denominator;
- F_h: phase step;

of which the first two define the semimajor axis (or the period), whereas the latter three define the satellite distribution along the relative path. Four more orbital parameters, eccentricity, inclination, argument of perigee, and Right Ascension of Ascending Node (RAAN) of the first orbit (e, i, ω, Ω_0, respectively), define the orbit shape and orientation. The number of orbits is determined by the N_p parameter and all the orbits have identical shape, inclination, and argument of perigee. They are only rotated in RAAN to obtain an even distribution about the central body.

The choice of a suitable phasing scheme is critical to reveal the most interesting dynamics obtainable by FCs. The chosen phasing scheme is a function of Ω_k, RAAN of kth orbit, $M_k(t_0)$ the mean anomaly at the initial time and the phasing parameters F_n, F_d, and F_h. The chosen scheme is designed to guarantee that every satellite is placed in a position compatible with the repeating space track constraint.

The FC approach provides great flexibility and interesting dynamics that reveal the presence of the so called *secondary path*, a relative motion of the satellites resulting in intriguing motion patterns that can be exploited to obtain useful properties as those described in [2,3, and 5]. More details and insights and interesting properties are provided in Refs. [6,7].

Once that a list of service suppliers and service customers has been defined, by using an optimization process based on a Genetic Algorithm (GA) a FC has been designed.

The problem has been decomposed in two steps: the first is finding an orbit with a ground track that allow the observation of all the sites, the second is to find a distribution of satellites along the track that provides a good time access to the ground targets. The first step has been approached using a GA for a single satellite track, whereas for the second FCs have been utilized to extend the solution achieved in the first step to a FC of 8 satellites.

Since the GA methodology is well known and has been extensively studied, its theory will not be reviewed here; the interested reader can found ample documentation in [8,9,10].

The cost function utilized to guide the GA optimization process is designed to maximize the dwell time over each target. Since the computational load to evaluate the correct dwell time is too demanding, the dwell time is here maximized by minimizing the satellite ground relative velocity. In order to provide a preference for satellites passing over the site, the relative velocity is weighted by the ratio of the angular displacement of the target site from the sub-satellite point direction (λ) with respect to the antenna field of view. These considerations yield to the following penalty function:

$$\min J_{obs} = \sum_k \alpha_k \left(1 + \frac{\lambda_k}{\theta_{FOV}}\right) | \nu_{rel} |$$

where α_k is the relative weight of the k-th site with respect to the other target sites. Therefore, by keeping the ground track spacecraft velocity $\mathbf{v}_{rel}$ as low as possible above target site we indirectly increase the dwell time over the site itself. The angle ϑ_{FOV} is the half field of view of the on board antenna, assumed to be pointed at nadir, and α_k is the weight of the kth site, with $k = 1..N$, with N being the number of target sites.

In order to run the GA, the design space (i.e. the chromosome) must be defined. The following parameters have been encoded in the chromosome string:

$$e \varepsilon (0,1), i \varepsilon [0,\pi], \omega \varepsilon [0,2\pi],$$
$$\Omega \varepsilon [0,2\pi], t(k) \varepsilon [0,t_f], k = 1..N$$

The semi-major axis of the desired orbit is chosen by the designer. The times $t(k)$ represent the instants in time in which the kth site is accessed by the spacecraft and t_f is the final time of the simulation. The extension of the design space caused by the introduction of these time array seems counterintuitive at first. It is however justified by the need of keeping the computation time manageable: while propagating the orbit and compute these access times analytically seems to be the obvious and correct solution, this latter approach requires propagation of the orbit and a search process that slows down each GA iteration considerably. The introduction of the access times array in the design space instead allows for a quick evaluation of the cost function during each GA iterations, thus allowing a bigger population and more evaluations to be completed to achieve an improved solution. The coarse solution obtained through the GA is then refined using a gradient method search.

Some of the design variable can be kept fixed as, for instance, was done with the inclination that was kept fixed at the critical value of 63.4 degrees. The critical inclination has been chosen in order to release the need to control the drift of the perigee due to the perturbations.

Once the optimization process has been completed a FC with the parameters listed in Table 2 has been created matching the FC parameters to the ground track resulted from the optimization.

The repeating ground track of this FC is shown in Fig. 1, while the relative orbit in a ECEF (Earth Centered Earth Fixed) system is shown in Figs. 2 and 3.

It can be noticed that six equally-spaced petals are placed above the locations of service providers and suppliers.

3.2 Walker Constellation Design

Walker constellations are the classical type of satellite constellation. The design of a Walker constellation is easier with respect to the design of a Flower Constellation since the number parameters is lower. A Walker constellation is characterized by

Table 2. Flower constellations parameters.

Number of Petals (N_p)	6
Number of Days (N_d)	1
Number of Satellites (N_s)	8
Phase Numerator (F_n)	1
Phase Denominator (F_d)	8
Phase Step (F_h)	0
Inclination (i)	63.4°
Perigee Height (h_p)	3000 km
Argument of Perigee (ω)	0^0
$RAAN_0$	243.77°
M_0	168°

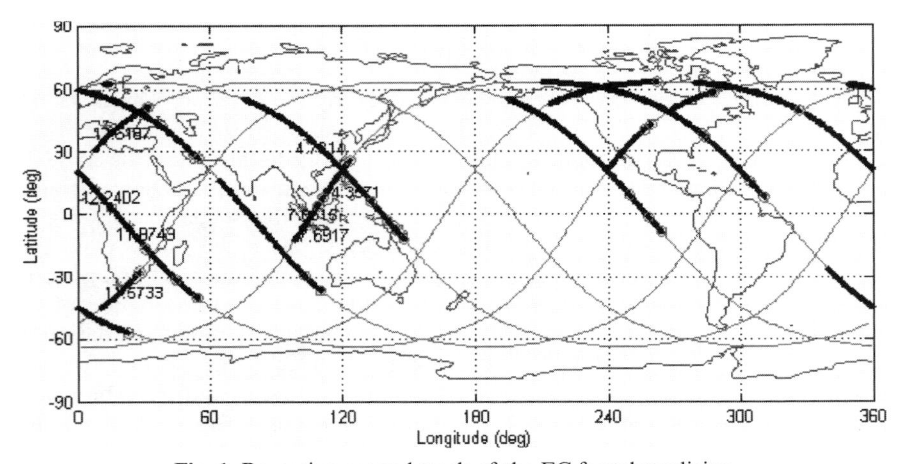

Fig. 1. Repeating ground track of the FC for telemedicine.

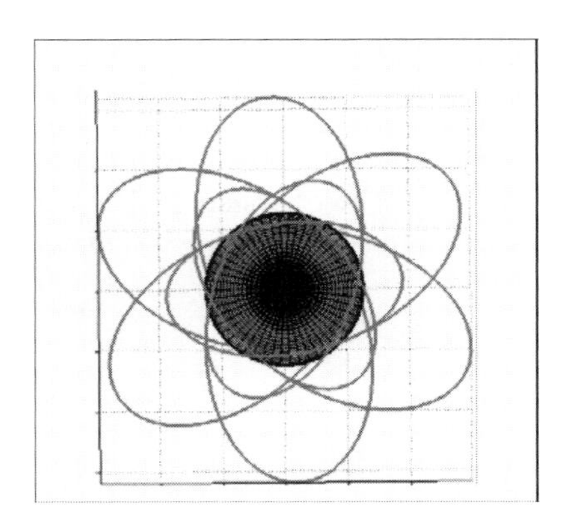

Fig. 2. Relative orbit of the FC for telemedicine (view from the pole).

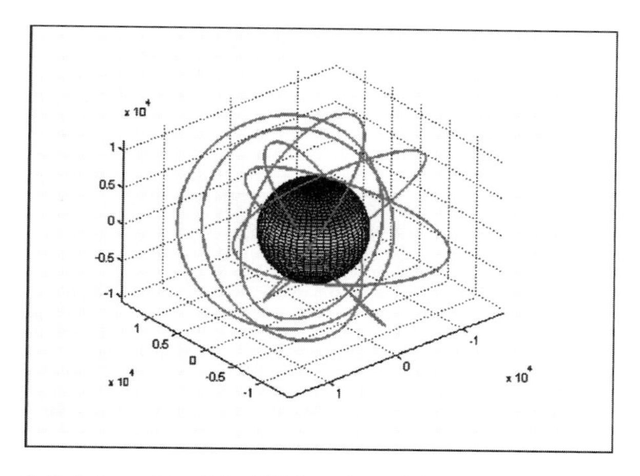

Fig. 3. Relative orbit of the FC for telemedicine (orthographic view).

Table 3. Walker constellations parameters.

Number of Satellites (t)	8
Number of Planes (p)	4
Satellites per Plane ($s = t/p$)	2
Inter Plane Spacing (f)	1
RAANspread	180°
Inclination (i)	45°
Orbit Height (h)	6,841 km

three integer parameters t, p, f and three real parameters RAANspread, h, i. The parameters t, p, f define respectively the number of satellites, the number of orbit planes and the relative spacing between satellites in adjacent planes. The parameters RAANspread, h and i define the orbit height, the inclination and the constraint on the maximum spreading of the RAAN for the satellites.

The Walker constellation is not designed to show repeating ground track. This means that the satellites belonging to a Walker constellation covers all the longitudes with the passing of time. On the other hand, the latitudinal coverage can be restricted to the equatorial regions by lowering the inclination. After the identification of the target with the highest latitude we have set the orbital inclination to 45 degrees. The number of satellites has been set to 8 as it is for the FC, while we have set the orbit altitude to 6,841 km in order to compare an elliptic orbit constellation (the FC) and a circular orbit constellation (the Walker constellation) with the same average satellite altitude. Furthermore, we had three choices on the number p of orbital planes (8, 4 and 2) and we will show the results achieved with the best configuration. Performing several evaluation of the coverage for different RAANspread, we have also found that a RAANspread = 180 degrees is the best choice.

The parameters of the Walker constellation are listed in Table 3.

4 Performance Comparison

We are going to compare the FC and the Walker constellation in terms of the following performance metrics:

- access time (%) between service providers and customers.
- mean access time between service providers and customers.

As previously described targets have been selected in order to model service providers and service customers.

Telecommunications chains (using only the satellite constellation network) between all service providers and customers have been created, selecting all usefull links (using 1, or none-ISL); afterwards for every chain time access periods have been avaluated and overlapped (in time) in order to establish the complete time-continuous coverage.

The temporal coverage, evaluated in terms of percentage of the propagation time is shown in Fig. 4.

The propagation time for the FC is set to one day (Nd = 1) while the propagation time for the Walker constellation is set to one month.

With respect to this performance metric the bar chart in Fig. 3 shows that the FC provides better performance for five customers by seven, giving an overall chain access of 59% versus 55% provided by the Walker constellation.

The mean chains access time which is the second performance evaluation metric is shown in Fig. 5; it can be observed that, for most of the customers (five), the FC provides better results compared with the ones offered by the Walker constellation.

As a matter of fact the Walker constellation provides fractioned accesses with a mean time duration of 1220 sec. with respect to the mean time duration of 1912 sec. offered by the FC; this results means very frequent handovers for the service provided by the Walker constellation.

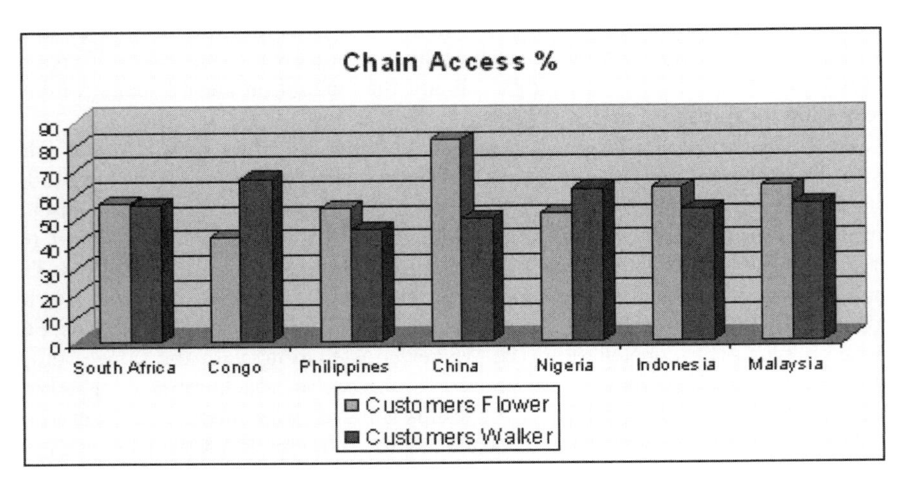

Fig. 4. Chain access availability percentage performance comparison for all service customers.

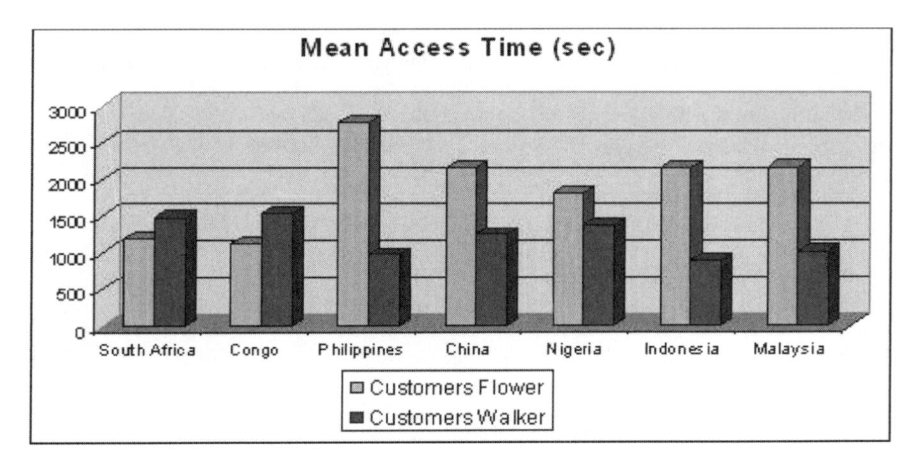

Fig. 5. Mean access time performance comparison for all service customers.

From this analysis it can be deduced that a FC provides better performance for telemedicine services in terms of availability and quality.

5 Conclusions

In this paper we designed a FC which is optimized to cover a set of targets interested in providing and accessing telemedicine services. However, the proposed optimized design can be used for a more general class of communication services and a different set of targets. It has been found that the features of the FC allow a specialised coverage of a list of targets thus providing better performance in terms of access availability with respect to a classical Walker constellation, at the expenses of a more complex design.

Acknowledgement

This work has been carried out in the frame of the European Space Agency – Ariadna extended study: "The Flower Constellation Set and its Possible Applications".

References

[1] Mortari, D., Wilkins, M.P., and Bruccoleri, C. "The Flower Constellations," The Journal of the Astronautical Sciences, Special Issue: The John L. Junkins Astrodynamics Symposium, Vol. 52, Nos. 1 and 2, January-June 2004, pp. 107–127.

[2] Wilkins, M., Bruccoleri, C., and Mortari, D. "Constellation Design Using Flower Constellations," Paper AAS 04–208 of the 2004 Space Flight Mechanics Meeting Conference, Maui, Hawaii, February 9–13, 2004.

[3] Park, K., Wilkins, M., Bruccoleri, C., and Mortari, D. "Uniformly Distributed Flower Constellation Design Study for Global Positioning System," Paper AAS 04–297 of the 2004 Space Flight Mechanics Meeting Conference, Maui, Hawaii, February 9–13, 2004. Submitted to the ION Journal of Navigation.
[4] Park K., Ruggieri, M., and Mortari, D. "Comparisons between GalileoSat and Global Navigation Flower Constellations," 2005 IEEE Aerospace Conference, March 5–12, 2005, Big Sky, Montana.
[5] Abdelkhalik, O. and Mortari, D. "The Two-way Orbits," 2005 IEEE Aerospace Conference, March 5–12, 2005, Big Sky, Montana.
[6] Mortari, D. and Wilkins, M.P. "The Flower Constellation Set Theory. Part I: Compatibility and Phasing," Submitted to the IEEE Transactions on Aerospace and Electronic Systems.
[7] Wilkins, M.P. and Mortari, D. "The Flower Constellation Set Theory. Part II: Secondary Paths and Equivalency," Submitted to the IEEE Transactions on Aerospace and Electronic Systems.
[8] Goldberg, D.E. "Genetic Algorithm in Search, Optimization and Machine Learning", Addison Wesley, 1989.
[9] http://en.wikipedia.org/wiki/Evolutionary_algorithm, From Wikipedia, the free encyclopedia.
[10] Wolpert, D.H., Macready, W.G. (1995), No Free Lunch Theorems for Search, Technical Report SFI-TR-95-02-010 (Santa Fe Institute).

Analysis of the Robustness of Filtered Multitone Modulation Schemes Over Satellite Channels

Andrea M. Tonello, Francesco Pecile

DIEGM – Università di Udine
Via delle Scienze, 208 - 33100 Udine - Italy
Phone: +39 0432 558042, Fax: +39 0432 558251, e-mail: tonello@uniud.it

Abstract. In this paper we analyze the performance of Filtered Multitone (FMT) modulation systems in satellite channels. FMT is a generalized OFDM scheme that deploys sub-channel shaping filters. It is a spectral efficient scheme that can support orthogonal multiuser transmission. Herein, we carry out a detailed analysis to understand whether it yields increased robustness, compared to OFDM, in typical LEO satellite channels. We consider an asynchronous multiuser scenario that introduces carrier frequency offsets, time offsets, and time/frequency channel selectivity. The analysis that we present in this paper allows to benchmark the multitone system and understand how robust it is to frequency selective time-variant fading and carrier Doppler shifts. Quasi-closed form expressions in the case of rectangular frequency domain pulses and raised cosine pulses are derived for the signal-to-interference power ratio at the receiver. It is found that the sub-channel spectral containment of the FMT system can yield increased performance compared to OFDM.

1 Introduction

In this paper we analyze the performance of multicarrier modulation [1] based architectures in satellite time-frequency selective channels. We consider low earth orbit (LEO) channels where the satellite elevation angle seen by a given user changes continuously. This, together with the terminal movement, introduces channel time variations. The satellite movement also induces high carrier Doppler shifts [2], [3]. Furthermore, multipath propagation, and thus frequency selective fading, can be present in urban areas. Multicarrier modulation and in particular orthogonal frequency division multiplexing (OFDM) has been proposed as an attractive modulation technique for LEO satellite communications [2]. In this paper, we investigate the performance of a more general multicarrier scheme that is referred to as Filtered Multitone (FMT) modulation. FMT is a discrete time multicarrier system that deploys sub-channel shaping pulses (Fig.1) [4]. OFDM can be viewed as an FMT scheme that deploys rectangular time domain filters. OFDM is also referred to as discrete multitone modulation (DMT). FMT modulation has been proposed for spectral efficient transmission over broadband frequency selective channels both in wireline [4] and in wireless scenarios [5]. It has interesting properties in terms of spectral efficiency, efficient implementation, and capability to support orthogonal multiuser transmission [5]. The design of the sub-channel filters, and the choice of the sub-carrier spacing in an FMT system aims at subdividing the spectrum in a number

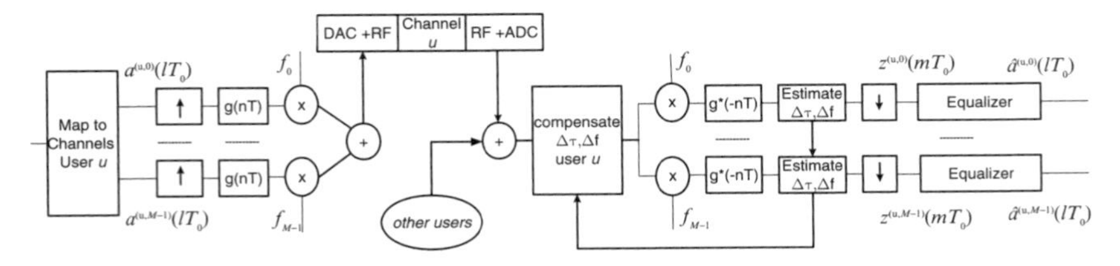

Fig. 1. Multiuser FMT system model with highlighted transmitter and receiver of user *u*.

of sub-channels that do not overlap in the frequency domain, such that we can avoid the ICI and get low ISI contributions. In an OFDM system the insertion of a cyclic prefix longer than the channel time dispersion is such that the ISI and ICI are eliminated, and the receiver simplifies to a simple one-tap equalizer per sub-channel.

Although multitone systems are robust to the channel frequency selectivity, they are sensitive to carrier frequency offsets, phase noise [6], and channel fast time variations [7], [8]. In [9] we have studied the performance limits of FMT modulation and we have shown that FMT can provide both frequency and time diversity gains when optimal multi-channel equalization is used. However, if complexity is an issue, it is likely that linear single channel equalizers are used [10]. In this paper we consider a multiuser FMT system where users are multiplexed via the assignment of a number of available sub-channels. In particular, we consider the uplink scenario where users are asynchronous. Our objective is to determine how robust the multiuser system is to users' time offsets, carrier frequency offsets, and channel time/frequency selectivity. The unified analysis that we carry out in this paper allows to evaluate the performance of both the FMT and the OFDM system.

This paper is organized as follows. In Section 2 we describe the multiuser FMT architecture. We particularize the description for the presence of time-frequency offsets only, and time-frequency channel selectivity only. In Section 3 we analyze the signal and interference power at the receiver outputs and we specialize the results to the FMT and OFDM cases. In Section 4 we report a performance comparison between FMT and OFDM. Finally, the conclusions follow.

2 Multiuser FMT System Model

An asynchronous multiuser FMT architecture is depicted in Fig. 1. The complex baseband transmitted signal $x^{(u)}(nT)$ of user u is obtained by a filter bank modulator with prototype pulse $g(nT)$ and sub-channel carrier frequency $f_k = k/(MT)$, $k = 0, \ldots, M-1$, with T being the transmission period

$$x^{(u)}(nT) = \sum_{k=0}^{M-1} x^{(u,k)}(nT) \tag{1}$$

$$x^{(u,k)}(nT) = \sum_{m \in \mathbb{Z}} a^{(u,k)}(mT_0)\, g(nT - mT_0)\, e^{j2\pi f_k nT}, \tag{2}$$

where $a^{(u,k)}(mT_0)$ is the k -th sub-channel data stream of user u that we assume to belong to the M-PSK/M-QAM constellation set and that has rate $1/T_0$ with

$T_0 = NT \geq MT$. If the sub-carrier spacing $f_k - f_{k-1}$ is larger than $1/T_0$ the scheme is referred to as non-critically sampled FMT, otherwise if $f_k - f_{k-1} = 1/T_0$ it is referred to as critically sampled FMT. In ideal FMT the prototype pulse has impulse response $g(nT) = \text{sinc}\,(nT/T_0)$. In this case a frequency guard equal to $f_G = 1/MT - 1/NT$ exists between sub-channels. A practical choice for the prototype pulse is to use a root-raised-cosine pulse. It is interesting to note that (2) allows to represent also a cyclically prefixed (CP) OFDM signal when the sub-carrier spacing is $1/MT$ and the prototype pulse is defined as $g(nT) = \text{rect}\,(nT/T_0)$. The interpolation factor N is chosen to increase the frequency separation between sub-channels, thus to minimize the amount of inter-carrier interference (ICI) and multiple access interference (MAI) at the receiver side. A possible efficient implementation of the transmitter that is based on polyphase filtering is described in [4]. The discrete time signal is digital-to-analog converted, RF modulated, and transmitted over the air. Distinct FMT sub-channels can be assigned to distinct users. In this case, the symbols are set to zero for the unassigned FMT sub-channels:

$$a^{(u,k)}(mT_0) = 0 \; for \; k \notin K_u, \tag{3}$$

where K_u denotes the set of M_u sub-channel indices assigned to user u.

At the receiver, after RF demodulation, and analog-to-digital conversion, the discrete time received signal can be written as

$$y(iT) = \sum_{u=1}^{N_U} \sum_{k=0}^{M-1} \sum_{n \in \mathbb{Z}} x^{(u,k)}(nT)\, g_{CH}^{(u)}(iT - nT; iT) + \eta(iT), \tag{4}$$

where N_U is the number of users, $g_{CH}^{(u)}(nT; mT)$ is the channel impulse response and $\eta(iT)$ is the additive Gaussian noise with zero mean contribution. Then $y(iT)$ is passed through an analysis filter bank with prototype pulse $h^{(u,k)}(nT)$. The sampled output at rate $1/T_0$ corresponding to user u and subchannel k is

$$z^{(u,k)}(lT_0) = \sum_{i \in \mathbb{Z}} y(iT)\, h^{(u,k)}(lT_0 - iT). \tag{5}$$

Defining in a convenient way the channel impulse response $g_{CH}^{(u)}(nT; mT)$ and the pulse $h^{(u,k)}(nT)$, we can analyze the effects of time and frequency offsets and of channel time-frequency selectivity as it is shown in the next sections.

2.1 Time and Frequency Offsets

In this section we consider only the effects of time and frequency offsets. Thus, we define

$$g_{CH}^{(u)}(iT - nT; iT) = e^{j2\pi\Delta_f^{(u)} iT}\, \delta\left(iT - nT - \Delta_\tau^{(u)}\right), \tag{6}$$

where $\Delta_\tau^{(u)}$ is the time offset and $\Delta_f^{(u)}$ is the carrier frequency offset of user u.

Note that we assume the time/frequency offsets to be identical for all sub-channels that are assigned to a given user.

Substituting (6) into (4) we can write the received signal as follows

$$y(iT) = \sum_{u=1}^{N_U} \sum_{k=0}^{M-1} \sum_{n \in \mathbb{Z}} x^{(u,k)}\left(iT - \Delta_\tau^{(u)}\right) e^{j2\pi\Delta_f^{(u)} iT} + \eta(iT). \tag{7}$$

We consider a single user based FMT receiver (Fig. 1) where we first acquire time/frequency synchronization with each active user. Then, for each user, we compensate the time/frequency offsets, we run FMT demodulation via a bank of filters that is matched to the transmitter bank, and we sample the outputs at rate $1/T_0$. In formulas we have

$$h^{(u,k)}(nT)=h(nT+\Delta_\tau^{(u)})e^{j2\pi(f_k+\Delta_f^{(u)})(nT+\Delta_\tau^{(u)}-lT_0)} \tag{8}$$

and substituting (8) into (6) we can write the sub-channel output as

$$\begin{aligned} z^{(u,k)}(lT_0) &= \sum_{i\in\mathbb{Z}} y(iT)h(lT_0-iT+\Delta_\tau^{(u)})e^{j2\pi(f_k+\Delta_f^{(u)})(-iT+\Delta_\tau^{(u)})} \\ &= \sum_{u'=1}^{N_U}\sum_{k'=0}^{M-1}\sum_{l\in\mathbb{Z}} a^{(u',k')}(mT_0)g_{EQ}^{(u',k'),(u,k)}(lT_0;mT_0)+\eta^{(u,k)}(lT_0), \end{aligned} \tag{9}$$

where $\eta^{(u,k)}(lT_0)$ is the sequence of filtered noise samples, and where $g_{EQ}^{(u',k'),(u,k)}(lT_0;mT_0)$ is the multi-channel impulse response.

We can rewrite (9) as follows

$$\begin{aligned} z^{(u,k)}(lT_0) &= \sum_{l\in\mathbb{Z}} a^{(u,k)}(mT_0)g_{EQ}^{(u,k),(u,k)}(lT_0;mT_0) \\ &\quad + ICI^{(u,k)}(lT_0)+MAI^{(u,k)}(lT_0)+\eta^{(u,k)}(lT_0), \end{aligned} \tag{10}$$

where we highlight the fact that the sub-channel filter output of index k may suffer from ISI, ICI (from the sub-channels of index $k' \neq k$ that are assigned to user u), and MAI (from all sub-channels that belong to the other users) as a consequence of frequency overlapping sub-channels, time/frequency offsets, and channel time/frequency selectivity. When the sub-channels do not overlap, e.g., with ideal root-raised-cosine pulses with appropriate sub-carrier spacing, and the carrier frequency offsets of all users do not exceed half the frequency guard the ICI and MAI components are zero. Some intersymbol interference over each sub-channel may be present and can be counteracted with sub-channel equalization.

In the uplink, multiuser OFDM is severely affected by time misalignments and carrier frequency offsets. This is due to the fact that in conventional OFDM, sub-channels exhibit *sinc* like frequency response, therefore their orthogonality can be easily lost in the absence of precise synchronization [2]. In an asynchronous multi-user environment, increased robustness and better performance can be obtained with filtered multitone (FMT) modulation architectures where the sub-channels are shaped with appropriate frequency concentrated pulses as proved above [5].

2.2 Channel Time-Frequency Selectivity

In this section we consider only the effects of time-frequency selectivity of the channel. We model the baseband channel with a discrete time-variant filter $g_{CH}^{(u)}(nT;mT)$ that comprises the effect of the DAC and ADC stages as follows

$$g_{CH}^{(u)}(iT-nT;iT)=\sum_{p\in \mathrm{p}} \alpha_p^{(u)}(iT)\delta(iT-nT-pT), \tag{11}$$

where the time-variant tap amplitudes $\alpha_p(nT) p \in \mathbf{P} \subset \mathbb{Z}$ are stationary complex Gaussian. Under the WSSU isotropic scattering model [3] they have uncorrelated quadrature components, with zero mean, correlation

$$r_{p,p'}(nT) = E\left[\alpha_p(mT)^* \alpha_{p'}(mT + nT)\right] = \delta_{p,p'} \Omega_p J_0(2\pi f_D nT), \tag{12}$$

and power-spectral density $R_{p,p'}(f) = \delta_{p,p'} rep_{1/T}\left\{R_p(f)\right\}$ with

$$R_p(f) = \begin{cases} 0 & |f| > f_D \\ \Omega_p/(\pi f_D)\left(1 - (f/f_D)^2\right)^{-1/2} & |f| < f_D \end{cases} \tag{13}$$

Substituting (11) into (4) we obtain

$$y(iT) = \sum_{u=1}^{N_U} \sum_{k=0}^{M-1} \sum_{p \in \mathrm{P}} \alpha_p^{(u)}(iT) x^{(u,k)}(iT - pT) + \eta(iT). \tag{14}$$

At the receiver we consider the following filter

$$h^{(u,k)}(nT) = h(nT) e^{j2\pi f_k (nT - lT_0)} \tag{15}$$

and substituting (15) into (5) we can write the $\hat{k}$-th sub-channel filter-bank output as

$$\begin{aligned} z^{(\hat{k})}(lT_0) &= \sum_{i \in \mathbb{Z}} y(iT) e^{-j2\pi f_{\hat{k}} iT} h(lT_0 - iT) \\ &= \sum_{k=0}^{M-1} \sum_{m=-\infty}^{\infty} a^{(k)}(mT_0) g_{EQ}^{(k,\hat{k})}(lT_0; mT_0) + \eta^{(\hat{k})}(lT_0), \end{aligned} \tag{16}$$

where $g_{EQ}^{(k,\hat{k})}(lT_0; mT_0)$ the equivalent impulse response between the input sub-channel k and output sub-channel $\hat{k}$.

It follows that the output in the absence of noise is

$$z^{(k)}(lT_0) = a^{(k)}(lT_0) g_{EQ}^{(k,\hat{k})}(lT_0; lT_0) + ISI^{(k)}(lT_0) + ICI^{(k)}(lT_0), \tag{17}$$

where the first term represents the useful data contribution, the second additive term is the ISI contribution, the third term is the ICI contribution. In this case MAI is not present because for simplicity of the analytical evaluation we consider a single user scenario. An in-depth analysis of the signal over interference power ratio is treated in the next sections.

3 Analytical Evaluation of the Interference in Time-Frequency Selective Channels

Our objective is to determine the robustness of the system to the channel time and frequency selectivity as a function of the design parameters. To do so we evaluate the power of the interference components. The analysis is quite general and applies both to FMT and OFDM. The results have a practical relevance because allow to understand the sources of interference as a function of the design parameters. In the following we assume the data symbols to be i.i.d. with zero mean, and average power $M_a^{(k)} = E\left[|a^{(k)}(mT_0)|^2\right]$.

First, it should be noted that

$$z^{(\hat{k})}(lT_0)=\sum_{k=0}^{M-1} z^{(k,\hat{k})}(lT_0)+\eta^{(\hat{k})}(lT_0) \tag{18}$$

where

$$z^{(k,\hat{k})}(lT_0)=\sum_{m=-\infty}^{\infty} a^{(k)}(mT_0)g_{EQ}^{(k,\hat{k})}(lT_0;mT_0) \tag{19}$$

is the contribution of the data stream transmitted on sub-channel k to the filter output of index $\hat{k}$. Further, the average power of (19) equals

$$M_z^{(k,\hat{k})}=E\left[|z^{(k,\hat{k})}(lT_0)|^2\right]=M_a^{(k)}\sum_m E\left[|g_{EQ}^{(k,\hat{k})}(lT_0;mT_0)|^2\right], \tag{20}$$

where the second equality holds with independent zero mean data symbols.

With the WSSU scattering tapped delay line channel model we can write

$$\begin{aligned} M_z^{(k,\hat{k})}=M_a\sum_m\sum_p\sum_i r_{\alpha_p}(iT)g^{(k)}(iT+i'T-pT+lT_0-mT_0) \\ \times h^{(\hat{k})}(-iT-i'T)g^{(k)*}(i'T-pT+lT_0-mT_0)h^{(\hat{k})*}(-i'T). \end{aligned} \tag{21}$$

It should be noted that if we fix $k=\hat{k}$ in (21), and we isolate the term that corresponds to $m=0$ we obtain the average signal power $S^{(k)}=M_a^{(k)}\ E\left[|g_{EQ}^{(k,\hat{k})}(lT_0;lT_0)|^2\right]$, while the sum of all other terms yields the ISI power $M_{ISI}^{(k)}=E\left[|ISI^{(k)}(lT_0)|^2\right]$. On the contrary, the total power of the ICI can be obtained as $M_{ICI}^{(k)}=\sum_{k\neq\hat{k}} M_z^{(k,\hat{k})}$.

To proceed we define the following function

$$gh^{(k,\hat{k})}(iT;sT)=g^{(k)}(iT-sT)h^{(\hat{k})}(-iT) \tag{22}$$

and we can rewrite (21) as

$$M_z^{(k,\hat{k})}=\frac{M_a^{(k)}}{T}\sum_m\sum_p\sum_i r_{\alpha_p}(iT)c_{gh}^{(k,\hat{k})}(iT;pT+mT_0-lT_0), \tag{23}$$

where the autocorrelation of function (22) is defined as

$$c_{gh}^{(k,\hat{k})}(iT;sT)=T\sum_{i'} gh^{(k,\hat{k})}(iT+i'T;sT)gh^{(k,\hat{k})*}(i'T;sT). \tag{24}$$

The expression (23) is general and can be particularized for a certain choice of the sub-channel pulse as shown in the next sub-sections. In some cases it is convenient to calculate it in the frequency domain as

$$\begin{aligned} M_z^{(k,\hat{k})}&=\frac{M_a^{(k)}}{T}\sum_m\sum_p\sum_i r_{\alpha_p}(iT)\int_{-1/2T}^{1/2T} C_{gh}^{(k,\hat{k})}(f;pT+mT_0-lT_0)e^{j2\pi fiT}df \\ &=\frac{M_a^{(k)}}{T^2}\sum_m\sum_p\int_{-1/2T}^{1/2T} R_{\alpha_p}(-f)C_{gh}^{(k,\hat{k})}(f;pT+mT_0-lT_0)df \end{aligned}, \tag{25}$$

where we have used the discrete-time Fourier transforms $C_{gh}^{(k,\hat{k})}(f;sT)=T\sum_n c_{gh}^{(k,\hat{k})}(nT;sT)e^{-j2\pi fnT}$ and $R_{\alpha_p}(f)=T\sum_n r_{\alpha_p}(nT)e^{-j2\pi fnT}$.

The first transform can be written as

$$C_{gh}^{(k,\hat{k})}(f;sT)=\left|GH^{(k,\hat{k})}(f;sT)\right|^2, \tag{26}$$

where $GH^{(k,\hat{k})}(f;sT)$ is the discrete-time Fourier transform of the function (22)

$$GH^{(k,\hat{k})}(f;sT)=rep_{1/T}\left[(G^{(k)}(f)\,e^{-j2\pi fsT})*H^{(\hat{k})}(-f)\right] \tag{27}$$

and $G^{(k)}(f)=F\left[g^{(k)}(t)\right], H^{(k)}(f)=F[h^{(k)}(t)]$.

3.1 Results for the FMT Case

In FMT the receiver filter-bank is matched to the transmitter filter-bank, i.e., $h^{(k)}(nT)=g^{(k)^*}(-nT)$. Thus, $H^{(k)}(f)=F\left[g^{(k)^*}(-nT)\right]=G^{(k)^*}(f)$.

To proceed we need to define the prototype pulse. In the following subsections we obtain results when we deploy a sinc pulse, and a root-raised cosine pulse.

3.1.1 FMT with Sinc Prototype Pulse. If we consider a sinc prototype pulse (rectangular frequency domain pulse)

$$g(nT)=\mathrm{sinc}\left(\frac{nT}{T_0}\right) \tag{28}$$

$$G(f)=T_0\,rep_{1/T}\{rect(fT_0)\} \tag{29}$$

we obtain that

$$M_z^{(k,\hat{k})}=\frac{M_a^{(k)}T_0^4}{T^2\pi f_D}\sum_m\sum_p\Omega_p\int_{-f_D}^{f_D}\frac{(|f+f_{\hat{k}}-f_k|-1/T_0)^2}{\sqrt{1-(f/f_D)^2}}$$
$$\times\mathrm{sinc}^2\left(\left(\|f+f_{\hat{k}}-f_k\|-\frac{1}{T_0}\right)(pT+mT_0)\right)df \tag{30}$$

assuming $f_D\leq 1/MT$ (smaller than the sub-carrier spacing). It can be shown that (30) can be computed also in the time-domain as follows

$$M_z^{(k,\hat{k})}=\frac{M_a^{(k)}N^3}{\pi^2}\sum_m\sum_p\sum_i r_{\alpha_p}(iT)\,e^{j2\pi(f_{\hat{k}}-f_k)iT}$$
$$\times\frac{\mathrm{sinc}\left(\frac{2(p+mN)}{N}\right)-\mathrm{sinc}\left(\frac{2i}{N}\right)}{i^2-(p+mN)^2}. \tag{31}$$

Now, the signal power can be obtained by isolating the term in (30) or (31) of index $m=0$

$$S^{(k)}=\frac{2M_a^{(k)}T_0^4}{T^2\pi f_D}\sum_p\Omega_p\int_0^{f_D}\frac{(f-1/T_0)^2}{\sqrt{1-(f/f_D)^2}}\,\mathrm{sinc}^2\left(\left(f-\frac{1}{T_0}\right)pT\right)df. \tag{32}$$

The power of the sub-channel ISI is

$$M_{ISI}^{(k)}=\frac{2M_a^{(k)}T_0^4}{T^2\pi f_D}\sum_{m\neq 0}\sum_p\Omega_p\int_0^{f_D}\frac{(f-1/T_0)^2}{\sqrt{1-(f/f_D)^2}}$$
$$\times\mathrm{sinc}^2\left(\left(f-\frac{1}{T_0}\right)(pT+mT_0)\right)df. \tag{33}$$

The total power of the ICI (assuming sub-channel data streams with identical power M_a) is

$$M_{ICI}^{(k)} = \sum_{k \neq k} M_z^{(k,k)} = \frac{2M_a T_0^4}{T^2 \pi f_D} \sum_m \sum_p \Omega_p \int_{f_G}^{f_D} \frac{(f - f_G)^2}{\sqrt{1-(f/f_D)^2}} \times \text{sinc}^2\big((f - f_G)(pT + mT_0)\big)\, df. \quad (34)$$

Now, it should be observed that if $f_D \leq f_G$, (34) is always zero for k $\neq \hat{k}$. Therefore, ICI is not present. Otherwise, if $f_G < f_D \leq 1/MT$ only two adjacent sub-channels can generate ICI. This is a very interesting aspect of FMT which exhibits no ICI when band limited pulses are used and a frequency guard larger than the maximum Doppler is used between sub-channels. Clearly, fast fading can introduce some ISI because it distorts the received sub-channel pulse as we will discuss in more detail in the following. If the channel is flat and static then there is neither ISI nor ICI, i.e., the system is orthogonal.

3.1.2 FMT with Root-Raised Cosine Prototype Pulse. Another possibility is to use root-raised cosine pulses

$$g(nT) = \text{rrcos}\left(\frac{nT}{T_0}\right) \quad (35)$$

$$G(f) = T_0\, rep_{1/T}\{\text{RRCOS}(fT_0)\} \quad (36)$$

where

$$\text{rrcos}(t) = \text{sinc}\left(\alpha t + \frac{1}{4}\right)\frac{\sin\left(\pi\left(t - \frac{1}{4}\right)\right)}{4t} + \text{sinc}\left(\alpha t - \frac{1}{4}\right)\frac{\sin\left(\pi\left(t - \frac{1}{4}\right)\right)}{4t} \quad (37)$$

$$\text{RRCOS}(f) = \begin{cases} 1 & 0 \leq |f| < f_1 \\ \cos\left(\frac{\pi}{2}\frac{|f| - f_1}{\alpha}\right) & f_1 \leq |f| < f_2. \\ 0 & |f| > f_2 \end{cases} \quad (38)$$

with $f_1 = 0.5\,(1 - \alpha)$ and $f_2 = 0.5\,(1 + \alpha)$. Note that α is the roll-off factor of the filter. With this pulse the computation of (26) requires some cumbersome algebra. For space limitation we don't report herein the full analytical results but just the graphical comparison in Section 4.

3.2 Results for the OFDM Case

Now, we consider a CP-OFDM scheme for which the subchannel filters are $g^{(k)}(iT) = \text{rect}\,(iT/T_0)\, e^{j2\pi f_k iT}$ and $h^{(k)}(iT) = \text{rect}\,(-iT/T_1)\, e^{j2\pi f_k iT}$ where we have $T_0 = NT$, $T_1 = MT$, and $f_k = k/(MT)$. Using, the power of the interference seen by sub-channel $\hat{k}$ reads

$$M_z^{(k,k)}(lT_0)=M_a^{(k)}\sum_m\sum_p\sum_{i=0}^{M-1}\sum_{i'=0}^{M-1}r_{\alpha_p}(iT-i'T)e^{j2\pi(f_k-f_k)(iT-i'T)}$$
$$\times\text{rect}\left(\frac{iT-lT_0+mT_0+pT}{T_0}\right)\text{rect}\left(\frac{i'T-lT_0+mT_0+pT}{T_0}\right). \quad (39)$$

Thus, the power of the useful term is

$$S^{(k)}=M_a^{(k)}\sum_p\sum_{i=0}^{M-1}\sum_{i'=0}^{M-1}r_{\alpha_p}(iT-i'T)\,\text{rect}\left(\frac{iT+pT}{T_0}\right)\text{rect}\left(\frac{i'T+pT}{T_0}\right). \quad (40)$$

The power of the sub-channel ISI is

$$M_{ISI}^{(k)}=M_a^{(k)}\sum_{m\neq 0}\sum_p\sum_{i=0}^{M-1}\sum_{i'=0}^{M-1}r_{\alpha_p}(iT-i'T)$$
$$\times\text{rect}\left(\frac{iT+mT_0+pT}{T_0}\right)\text{rect}\left(\frac{i'T+mT_0+pT}{T_0}\right). \quad (41)$$

The power of the ICI is (assuming sub-channel data streams with identical power M_a)

$$M_{ICI}^{(k)}=M_{TOT}^{(k)}-(S^{(k)}+M_{ISI}^{(k)}), \quad (42)$$

where the total power is

$$M_{TOT}^{(k)}=MM_a\sum_m\sum_p r_{\alpha_p}(0)\sum_{i=0}^{M-1}\text{rect}^2\left(\frac{i'T+mT_0+pT}{T_0}\right). \quad (43)$$

4 Frequency Selective Fading and Flat Fading

From the general expressions that we have obtained in the previous section we can evaluate the signal-to-interference power ratio:

$$SIR^{(k)}=\frac{S^{(k)}}{M_{ISI}^{(k)}+M_{ICI}^{(k)}}. \quad (44)$$

For ease of understanding we consider first a multipath channel with quasistatic fading, then a time-variant flat fading channel. The multipath channel has power delay profile $\Omega_p \sim e^{-pT/(\gamma T)}$. We truncate the channel at –20 dB obtaining N_P taps, and we normalize its power to one. The time-variant channel has the temporal correlation defined in (12).

4.1 FMT and OFDM Comparison in Frequency Selective Static Fading

Let us assume the channel to be quasi-static but frequency selective. Then, we can elaborate further the formulas and specialize the results as follows.

4.1.1 FMT with Sinc Prototype Pulse

$$S_{FMT}^{(k)}=M_a^{(k)}N^2\sum_{p=0}^{N_P}\Omega_p\,\text{sinc}^2\left(\frac{p}{N}\right), \quad (45)$$

$$M_{ISI-FMT}^{(k)} = M_a^{(k)} N^2 \sum_{m \neq 0} \sum_{p=0}^{N_p} \Omega_p \,\mathrm{sinc}^2\left(\frac{p+mN}{N}\right). \tag{46}$$

The total power of the ICI is always zero (assuming frequency confined pulses).

4.1.2 FMT with Root-Raised Cosine Prototype Pulse

$$S_{FMT}^{(k)} = M_a^{(k)} N^2 \sum_{p=0}^{N_p} \Omega_p \,\mathrm{rcos}^2\left(\frac{p}{N}\right) \tag{47}$$

$$M_{ISI-FMT}^{(k)} = M_a^{(k)} N^2 \sum_{m \neq 0} \sum_{p=0}^{N_p} \Omega_p \,\mathrm{rcos}^2\left(\frac{p+mN}{N}\right) \tag{48}$$

The total power of the ICI is always zero (assuming frequency confined pulses).

4.1.3 FMT with Rect Prototype Pulse (OFDM). For the CP-OFDM system, the power of the useful term, the ISI and the ICI read as follows

$$S_{OFDM}^{(k)} = M_a^{(k)} \left(\sum_{p=0}^{min\,(N-M,\,N_p)} \Omega_p M^2 + \sum_{p=N-M+1}^{min\,(N-1,\,N_p)} \Omega_p (N-p)^2 \right) \tag{49}$$

$$M_{OFDM-ISI}^{(k)} = M_a^{(k)} \sum_{m \neq 0} \sum_{p=0}^{N_p} \Omega_p \sum_{i=0}^{M-1} \sum_{i'=0}^{M-1} \mathrm{rect}\left(\frac{iT+mT_0+pT}{T_0}\right) \mathrm{rect}\left(\frac{i'T+mT_0+pT}{T_0}\right) \tag{50}$$

$$M_{OFDM-ICI}^{(k)} = M_a \sum_{m} \sum_{p=0}^{N_p} \Omega_p \sum_{i'=0}^{M-1} \left[\mathrm{rect}\left(\frac{i'T+mT_0+pT}{T_0}\right) \times \left(M - \sum_{i=0}^{M-1} \mathrm{rect}\left(\frac{iT+mT_0+pT}{T_0}\right)\right)\right]. \tag{51}$$

In particular, as it is well known, when the channel is shorter than the CP ($Np \leq \mu = N - M$) the useful power is $S^{(k)} = M_a^{(k)} M^2 \sum_{p=0}^{N_p} \Omega_p$, while the ISI and ICI are zero.

In Fig. 2A we report the SIR as a function of the normalized delay spread for the FMT system while in Fig. 2B we consider the CP-OFDM system. The plot shows that the FMT architecture is robust to channel frequency selectivity. Indeed the CP-OFDM system maintains the orthogonality for channels shorter than the CP. But for channels longer than the CP it also suffers as a result of ISI and ICI.

4.2 FMT and OFDM Comparison in Flat Fast Fading

Now, let us assume the channel to be flat but time-variant. Elaborating further the formulas we obtain the following particular results.

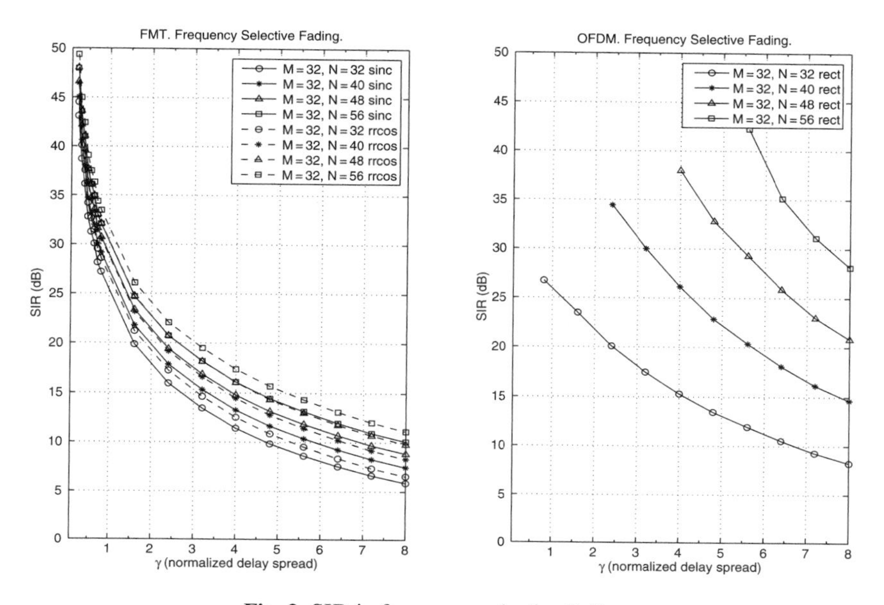

Fig. 2. SIR in frequency selective fading.

4.2.1 FMT with Sinc Prototype Pulse

$$S^{(k)}_{FMT} = \frac{2M^{(k)}_a \Omega_0 N^4}{\pi}\left(\frac{(f_D T)^2 \pi}{4} - \frac{2 f_D T}{N} + \frac{\pi}{2N^2}\right), \tag{52}$$

$$M^{(k)}_{FMT-ISI} = \frac{M^{(k)}_a N^2 \Omega_0}{\pi^2} \sum_{m=1}^{\infty} \frac{1 - J_0(2\pi f_D m T_0)}{m^2}. \tag{53}$$

The total power of the ICI equals (assuming sub-channel data streams with identical power M_a)

$$M^{(k)}_{FMT-ICI} = \frac{2M_a T_0^4 \Omega_0}{T^2 \pi f_D} \sum_m \int_{f_G}^{f_D} \frac{(f - f_G)^2}{\sqrt{1-(f/f_D)^2}} \operatorname{sinc}^2\big((f - f_G) m T_0\big)\, df. \tag{54}$$

Note that (54) is zero when the sub-channels are separated more than the maximum Doppler.

4.2.2 FMT with Root-Raised Cosine Prototype Pulse. In the case of a root-raised cosine filter the useful and ISI term have complicated expressions that we don't report for space limitations. They are a combination of Bessel and StruveH functions [11]. For the ICI term instead we can't obtain a closed form, but as the sinc case we have that $M^{(k)}_{FMT-ICI}$ is zero if the sub-channels are separated more than the maximum Doppler.

4.2.3 FMT with Rect Prototype Pulse (OFDM). For the CP-OFDM system the power of the useful term, the ISI and the ICI read as follows

$$S_{OFDM}^{(k)} = M_a \Omega_0 \sum_{i=0}^{M-1} \sum_{i'=0}^{M-1} J_0(2\pi f_D T(i-i')), \tag{55}$$

$$M_{OFDM-ICI}^{(k)} = M_a \Omega_0 \left(M^2 - \sum_{i'=0}^{M-1} \sum_{i=0}^{M-1} J_0(2\pi f_D T(i-i')) \right). \tag{56}$$

The total power of the ISI is always zero. Note that (56) is identical to the one reported in [8]. In Fig. 3A we report the SIR as a function of the normalized delay spread for the FMT system, with sinc and root-raised cosine, while in Fig. 3B we consider the CP-OFDM system. The plot shows that the OFDM scheme is more robust than the ideal FMT scheme for large Doppler spreads. If we use root-raised cosine with roll-off $\alpha = 0.2$ we can improve the performance of the FMT. For moderate Doppler the performance of FMT can be improved also with sub-channel equalization.

Finally, we point out that when equalization is used in the FMT scheme we can exploit the sub-channel time-frequency diversity while the one tap equalizer in the OFDM scheme does not allow to pick any diversity gain [9]. As an example, we report in Fig. 4a comparison of bit-error rate performance in fast fading between FMT and OFDM that has been obtained via simulation. The OFDM scheme uses 128 tones while the FMT scheme uses 32 tones with a 11 taps sub-channel equalizer. The sub-channel filter in the simulation is a truncated root-raised-cosine pulse and 4-PSK modulation is used.

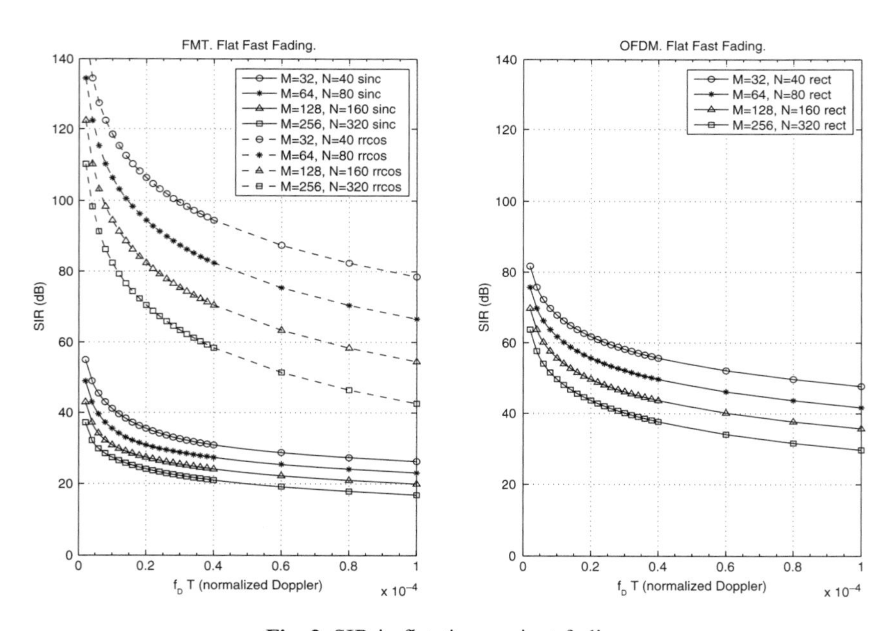

Fig. 3. SIR in flat time variant fading.

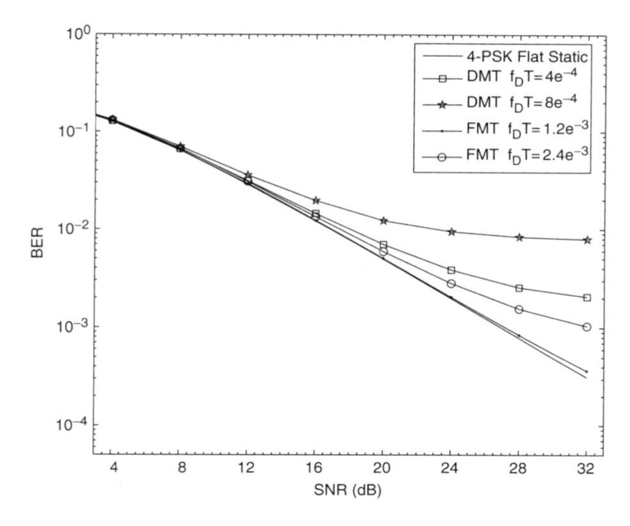

Fig. 4. BER performance in time variant fading for OFDM and FMT.

The figure shows that when equalization is used FMT has superior BER performance than OFDM.

5 Conclusions

We have presented an analysis of multiuser FMT in LEO satellite channels and in particular we have considered the effect of time-frequency offsets and time-frequency channel selectivity. We have obtained quasi-closed form expressions for the signal-to-interference power ratio in both FMT and OFDM systems. The sub-channel spectral containment of FMT yields increased robustness to the ICI and ISI compared to OFDM.

References

[1] Z. Wang, G. Giannakis, "Wireless multicarrier communications", *IEEE Signal Proc. Mag.*, pp. 29–48, May 2000.

[2] L. Wei, C. Schlegel, "Synchronization requirements for multi-user OFDM on satellite mobile two-path Rayleigh fading channels", *IEEE Trans. on Comm.*, Feb/March/April 1995 pp. 887–895.

[3] N. Sagias, A. Papathanassiou, P. T. Mathiopoulos, G. Tombras, "Burst Timing Synchronization for OFDM-Based LEO and MEO Wideband Mobile Satellite Systems", *Proc. 7th International Workshop on DSP Techniques for Space Communications*, Sesimbra, Portugal, October 2001.

[4] G. Cherubini, E. Eleftheriou, S. Ölçer, "Filtered multitone modulation for very high-speed digital subscribe lines", *IEEE JSAC*, pp. 1016–1028, June 2002.

[5] A. Tonello, "Asynchronous multicarrier multiple access: optimal and sub-optimal detection and decoding", *Bell Labs Technical Journal* vol. 7 n. 3, pp. 191–217, 2003.
[6] L. Tomba, W. A. Krzymien, "Effect of carrier phase noise and frequency offset on the performance of multicarrier CDMA systems", *Proc. of IEEE ICC 96, Dallas*, pp. 1513–1517, June. 1996.
[7] M. Speth, S. A. Fetchel, G. Fock, H. Meyr, "Optimum receiver design for wireless broad-band systems using OFDM-Part I", *IEEE Trans. on Commun.*, vol. 47, no. 11, pp 1668–1677, November 1999.
[8] G. L. Stuber, Principles of Mobile Communications, Kluwer, 1996.
[9] A. Tonello, "Performance limits for filtered multitone modulation in fading channels", *IEEE Trans. on Wireless Comm.*, vol. 4, pp. 2121–2135, Sept. 2005.
[10] N. Benvenuto, S. Tomasin, L. Tomba, "Equalization methods in DMT and FMT systems for broadband wireless communications", *IEEE Trans. on Commun.*, vol. 50, no. 9, pp. 1413–1418, Sept. 2002.
[11] I.S. Gradshteyn, I.M. Ryzhic, *Table of Integrals, Series, and Products*, Academic Press, USA, fourth edition, 1983.

VeRT Prototype Architecture and First Trials Campaign Results

Viviana Artibani, Gianluca Graglia, Giovanni Guarino

Navigation & Integrated Comms Directorate
Alcatel Alenia Space Italia SpA
Via Saccomuro 24, 00131 Rome, Italy

Abstract. VeRT (Vehicular Remote Tolling) is a research program cofunded by GALILEO Joint Undertaking in the framework of the Activity C first call: Introduction of GALILEO services Using EGNOS
GJU/03/118/issue2/OM/ms, 2003 31st July.

The program belongs to the GJU activities aiming at the APPLICATIONS RESEARCH coming from the EGNOS and GALILEO employment, which are competitive with respect to the only GPS system usage. VeRT belongs to the "Road Applications" area and in particular it focuses on research of applications, sustainable from the economic and social point of view, either for enhancing the "Road Tolling" services or for introducing new added value services to be employed in the "mobility" field only by employing GALILEO.

The VeRT programme aims at exploiting the capabilities offered by Galileo to provide new applications in the road sector, by means of the EGNOS system for demonstration purposes; specifically, an extended service concept of road tolling: it is indicated as "remote tolling", such to cover basic tolling service, but also additional "pay-per use services" on motorways transport, as well as in urban environment (parking and access to restricted zones) or in both the environments as the pay per use insurance.

The higher level of accuracy and integrity provided by GALILEO (respect to the GPS-only) represents an essential condition connected with the services.

From a general point of view, the users must be sure that they are paying for the offered service they are using, thus the right positioning of the users and the guarantee of the service provided assumes an important role in the service architecture.

The adoption of a satellite based toll and road services facility in the infrastructural approach, opens the way to further commercial opportunities.

This aspect is taken into right account within the VeRT project: a market analysis has been conducted in which revenue mechanism for the service provision has been estimated by analysing market opportunities and defining the opportune marketing strategies.

1 Prototype Architecture Features

Alcatel Alenia Space Italia is involved as leader of the VeRT overall system architecture design (Fig. 1) with particular focus on the navigation aspects related to the VeRT services.

The key factors that supported the design of the VeRT prototype are the following (1):

- VeRT is based on a grounded technology, improving a reusable Infrastructure
  The IT infrastructure is mainly "component based", where a component represent "a functional block" of the system. The main basic functionalities of the proposed system thus use a set of re-configurable and distributed set of components that basically allow to deploy a Service Control Centre, to equip the necessary on-board terminal with positioning and multimedia messaging capabilities, use of digital maps, etc. This type of infrastructure facilitates the addition of new/improved components, and to interface with external systems (either UT or Service centres).
- VeRT is composed of a Service Control Centre
  The Service Control Centre is the core part of the system; based on a modular and scalable infrastructure it acquires data from User Terminals, databases, in

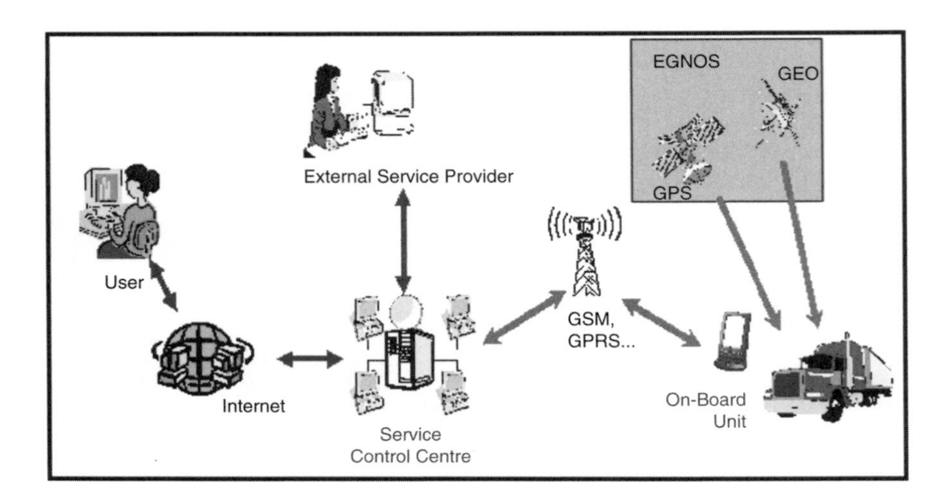

Fig. 1. VeRT Prototype overall architecture.

order to process heterogeneous information by means of specific data fusion capabilities. Communication technologies such as WiFi, GSM, GPRS and Satellite, integrated with powerful elaboration units and information sources links, allow all the components of the services value chain, from the services provider till to the end user, to co-operate on the collection and exploitation of the information.
- VeRT is deployed in standard (on-board) User Terminals

  The User Terminals are basically composed by available, and continuously evolving, handheld/tablet devices such as PDAs connected with GPS/EGNOS receiver (which is foreseen to be embedded in the near future) and having communications (WiFi, GSM, GPRS), visualisation (multimedia MMI) and internet applications facilities. Advanced services, based on EGNOS/GALILEO infrastructure, such as those foreseen for the "on the road" users, need to rely on "cooperative" User Terminals, as they can exchange (through several communication channels) with the SCC, geo referenced and multimedia information, necessary for the provision and the exploitation of the services.

Here after it is reported the Physical Generic Configuration of the VeRT Service Prototype (Fig. 2):

As general consideration, it has to be taken into account that from a functional point of view, the Application Server, the System Control Centre and the DB Server represent the VeRT IT Infrastructure (Application Side) in charge of managing and generating the VeRT Services; while the HMI and the Service Control Centre represent the Service Side of the VeRT Infrastructure in charge of providing a suitable VeRT System Interface for the Final Users (both private and business users) (2).

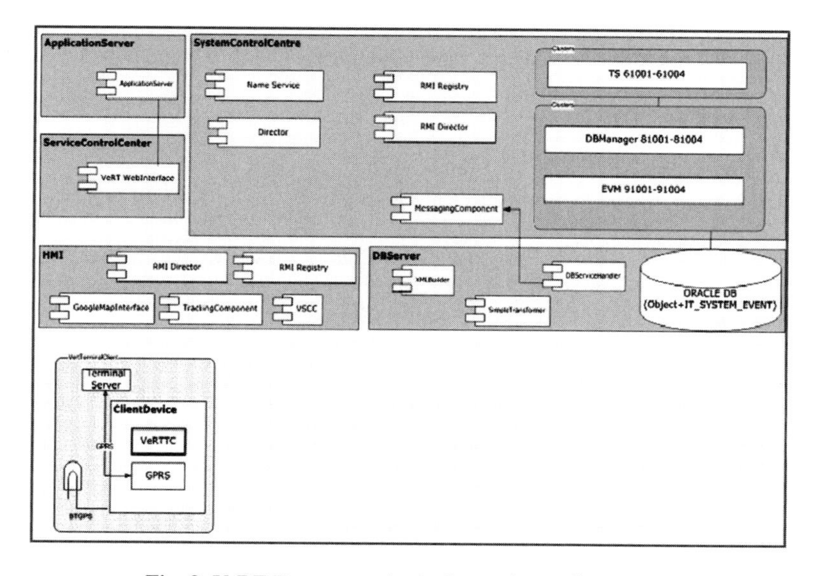

Fig. 2. VeRT Prototype physical generic configuration.

2 Reference Market

The Road Domain offers significant opportunities for a profitable use of GALILEO an a maximisation of the consequent social benefits. The key objectives are:

- To define high added-value services for responding to the market demands within the regulatory environment which supports and stimulates the market
- To extend the focal point from "Tolling" to "Telematics"
- To concentrate the system not on a stand-alone service but to aim at a poli-functional system

During the VeRT Phase A, it has been conducted a user requirements classification with the scope of highlighting for each user requirement the added value level due to EGNOS/GALILEO employment with respect to the GPS only (3). From this analysis it appears evident how the incoming of the EGNOS/GALILEO capabilities has a different impact over the services offered by the VeRT system. This has been expressed by dividing the services in three main groups following the priority of the EGNOS/GALILEO added-value level:

Primary Service Group

- Automatic Fee Collection (AFC) Services Class
- Safety and Emergency Management (SEM) Services Class
- LTZ Access Management (LTZ) Services Class
- Parking Access Management (PAM) Services Class

Secondary Service Group

- Risk Management (RIM) Services Class
- Routing Management (ROM) Services Class

Ancillary Service Group

- Traffic Management (TRM) Services Class
- Fleet & Freight Management (FFM) Services Class

As well explained also in the Business plan document (4), the basic package should be composed by services with an high penetration in the mass market but with a quite limited impact of the EGNOS/GALILEO added value as:

- Traffic Information / Routing Information / Point of Interest Provision
  At the same time the service bundle should be completed with "Add-ons" which have limited revenues potential but the EGNOS/GALILEO employment has a strong impact on these services as:
- Road Tolling / Parking Payment / LTZ Access
  So, remarking the VeRT Project scope of demonstrating the feasibility of new services in the road domain by the introduction of EGNOS/GALILEO, it is

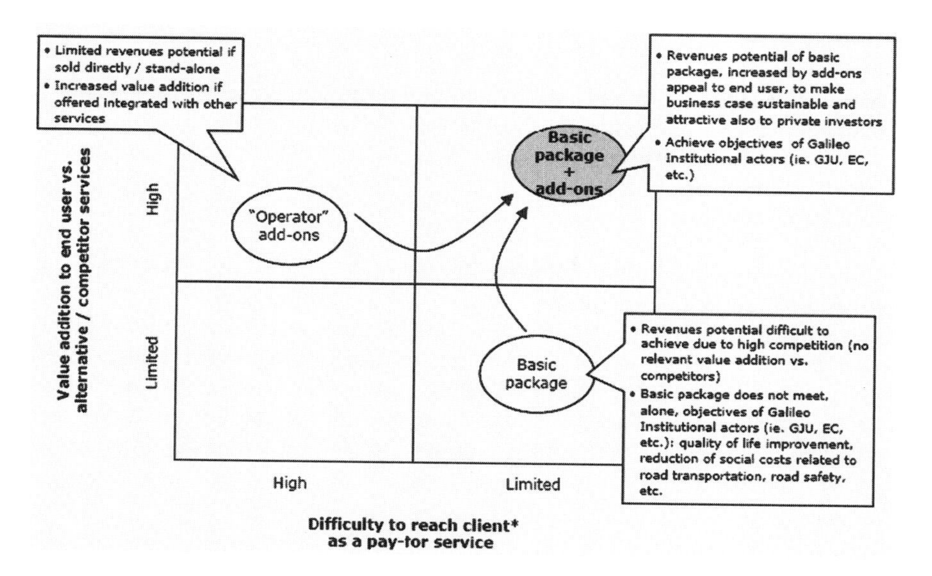

Fig. 3. VeRT Offer structure strategy.

important to clarify that it is necessary to provide to the clients a bundle of services, some of which could not take advantages by the EGNOS/GALILEO introduction, for penetrating the mass market (Fig. 3).

In conclusion the focus should be shifted from VeRT as "Vehicular Remote Tolling" to VeRT "Vehicular Road Telematics"

3 VeRT Trials Campaign Scenarios

Due to the Market Analysis outputs, it has been decided to implement and demonstrate all the services foreseen in the proposal phase, representing both the services where the GALILEO (today by means of EGNOS) introduction provides an high added-value and the services necessary to penetrate the market as the General Telematics Services (Info provision, Routing).

During July 2006 it has been conducted an intensive Trails Campaign with the objective of testing the developed prototype both for Motorway and Urban Services. The objectives of this Trials Campaign was to validate the VeRT System, verifying that the service data generated at Service Control Centre Level were coherent with the routes established in the Demonstration Plan and followed during the Trials Execution.

Alcatel Alenia Space Italia has contributed deeply to the trials campaign with its personnel involved during the VeRT prototype definition and development phase by participating to the trials design preparation and execution, ensuring the consistency of the collected data for performing the trails result analysis.

The trials locations were the Toscana-Ligure Motorway for the Motorway Services Test Execution and the City of Turin for the Urban Services Test Execution (5).

Motorway Scenario

It has been decided to divided into two different campaigns the Motorway and the Urban Services. More specifically, the following services:

- Road Tolling
- Emergency Service

have been performed and tested in the Motorway environment of SALT (Società Autostrade Liguria Toscana; Fig.4).

For the Road Tolling service, the OBU recognised that the user was entering/exiting a motorway and communicated during these events (entering/exiting) all the necessary information to the SCC. For the trials execution, the presence of DSRC radio-beacons in the Motorway Entrance and Exit has been considered not relevant, taking into account that the OBU did not interact with the DSRC equipment.

Urban Scenario

The services performed and tested in urban environment were the following:

- LTZ Access
- Location Based Services
- Parking Services

The tests have been executed in the city of Turin. Here after it is reported a figure with the Turin LTZ which has been used as test field (Fig. 5):

Conceptually the access to LTZ is very similar to the access to a highway. The big difference consists in the different environments for the provision of the two

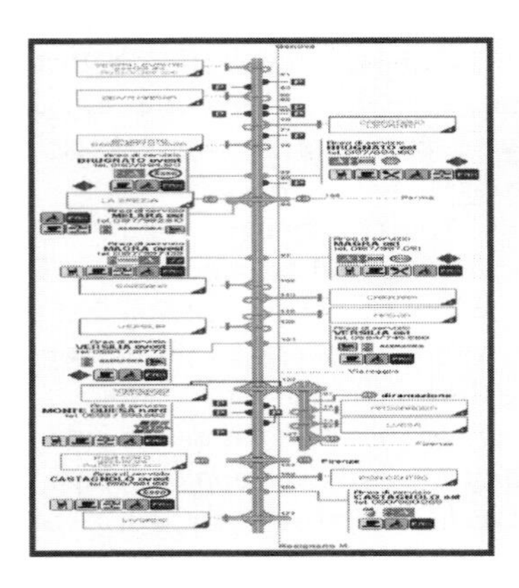

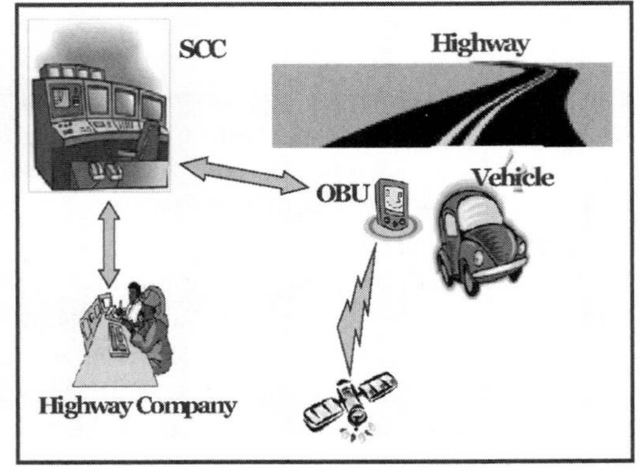

Fig. 4. Motorway scenario (SALT motorway).

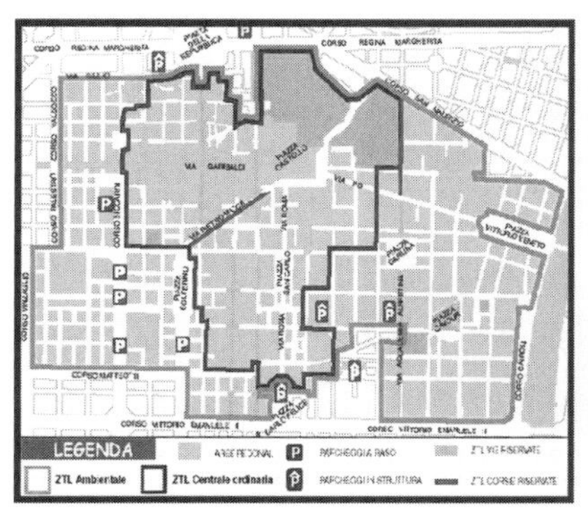

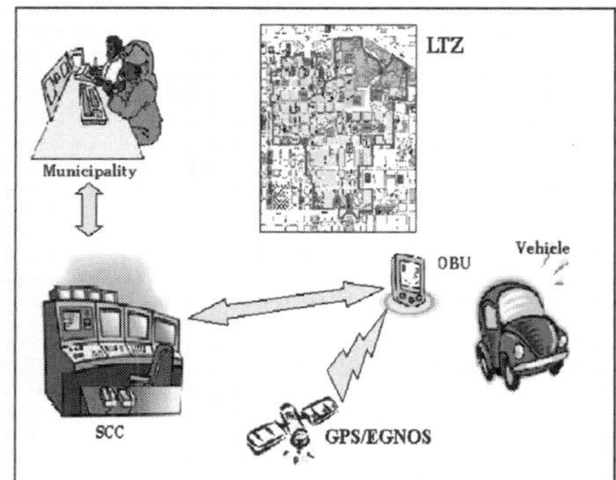

Fig. 5. Urban scenario (turin city).

services. In case of access to a highway the environment is substantially rural and there are no problems regarding the access identification for two reasons: very good navigation satellite visibility and road network quite simple. In case of access to an LTZ these two conditions are typically not satisfied. In fact, due to the urban canyon the visibility of the navigation satellite is very penalised and the multipath is increased, while the road network, especially in the historical centres, is very complex and this can represent a difficulty for the map matching software module.

In the Centre of Turin DSRC beacons (using the Telepass technology) are present at the access point of the LTZ. For the LTZ Service, the OBU recognised that the user is entering/exiting an LTZ and communicated during these events (entering/exiting) all the necessary information to the SCC. For the trials execution, the presence of DSRC radio-beacons in the LTZ gates has been considered not relevant, taking into account that the OBU did not interact with the DSRC equipment.

Regarding the parking services, the case of on-street parking lots includes parking along public roads where the Municipality has given the permission for parking. Usually the parking management is done from the municipality itself or by a private company. In general the payment is anticipated and done by means of parking machine spread along the roads. A typical environment for on-street parking is reported Fig. 6.

The off-street parking lots includes structure under or over the floor level closed by barrier (Fig. 7).

These structures can be planned and built either by the Municipality or by medium long term private concessionaire. The latest generation is completely automatic for control and toll management. The payment can be made at automatic cash points or with monthly or annual prepaid subscription. The fee due is calculated according to the hourly rate for the time stopped in the car park. Usually it is defined a maximum parking time. The access to the parking is possible by using a magnetic card or a ticket.

Fig. 6. On-street parking.

Fig. 7. Off-street parking.

Motorway and Urban Scenarios

Among the services provided by the VeRT Prototype there are some that can be considered not related to a specific environment but usable in both motorway and urban context. The services performed and tested in both the motorway and urban scenarios are the following:

- Pay per Use Insurance
- Fleet Management

4 VeRT Prototype Configuration

For the Tests execution it has been used as User Terminals for the installation of the VeRT System Software two different Mobile Hardware Platforms (6). In this way it has been possible to test from the user point of view the usability of the VeRT System on different devices. In fact, the VeRT System has the scope to answer to the requirements of two different macro-typologies of Final Users: Private Users and Business Users. For this reason it has been considered opportune the use of two different mobile-device on which testing the VeRT System, taking into account the different requirements and service typologies connected to the two users categories to which VeRT intends to answer. Both the devices use the same GPS/EGNOS mass-market receiver.

The following hardware configurations for the VeRT mobile equipment have been used during the tests depending on the VeRT user typology:

During the tests execution, two EGNOS Processing Units, already used in the EGNOS Signal Measurements Survey, have been used in parallel to the VeRT System in order to acquire (by means of the GPS/EGNOS L1/L2 receiver + antenna, also already used in the EGNOS Signal Measurement Survey) the GPS/EGNOS raw data, having the possibility of generating a Measurement Chain in order to evidence the behaviour of the VeRT System by performing in parallel a GPS/EGNOS Data acquisition.

In fact, the VeRT System produces as outputs all the parameters related to each of the implemented Services. So in case of Road Charging, the final outputs were:

- Entrance Gate
- Exit Gate

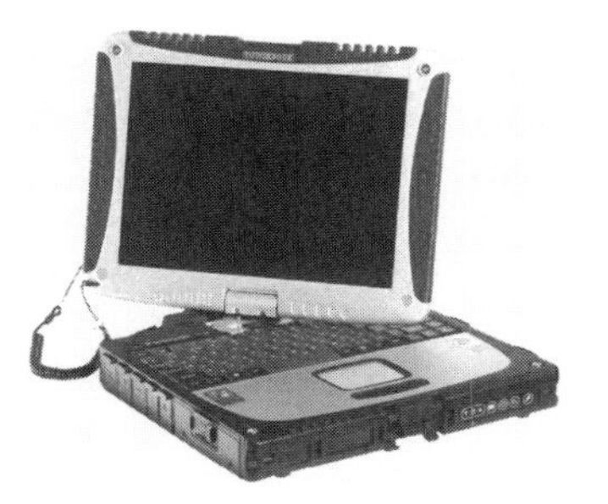

VeRT professional equipment.

VeRT equipment for private user.

- Entrance Date/Time
- Exit Date/Time
- Due Fee

In order to evaluate the compliance of the VeRT System respect to the expected goals, it has been necessary to install the VeRT System SW on suitable mobile devices which have been used during the test execution for the generation (by communicating the relevant data to the Service Control Centre collocated in the NEXT premises and in charge for the data processing) of the relevant parameters for the different VeRT Services under test.

The VeRT Mobile Equipment has been used in parallel with other equipment constituting the so-called Measurement Chains in charge for the Positioning Data acquisition used as inputs by the VeRT Mobile Equipment and not available otherwise as outputs from the VeRT System (Fig. 8).

These data will be used, during the results analysis phase, for evaluating their impact on the correct identification of relevant events Equipment (LTZ Access, Motorway Entrance, . . .) performed in the VeRT Mobile Equipment, which constitute the basis for the service parameters generation performed by the VeRT System. From an operative point of view, for the VeRT Centralise Infrastructure ready for the demonstration phase has been chosen the following configuration (Fig. 9):

- Application Server: it has been installed on a Server physically residing at Developer Company Premises
- System Control Centre: it has been installed on a Server physically residing at Developer Company Premises
- DB Server: it has been installed on a Server physically residing at Developer Company Premises
- HMI: it has been installed on a Server physically residing at Final Operator's Premises (Turin Municipality for Urban tests, SALT premises for Motorway tests)

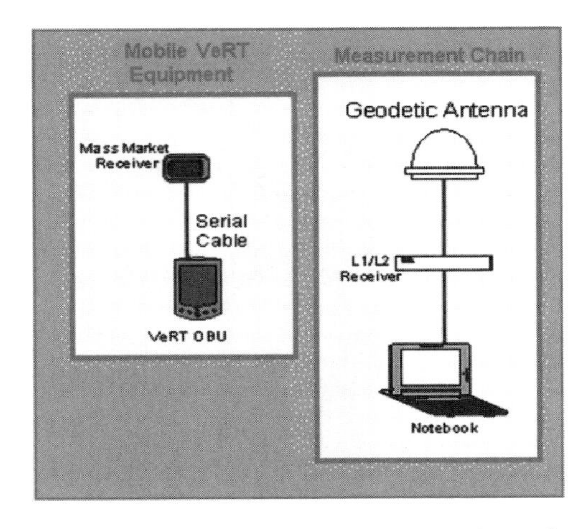

Fig. 8. VeRT User terminal and measurement chain configuration.

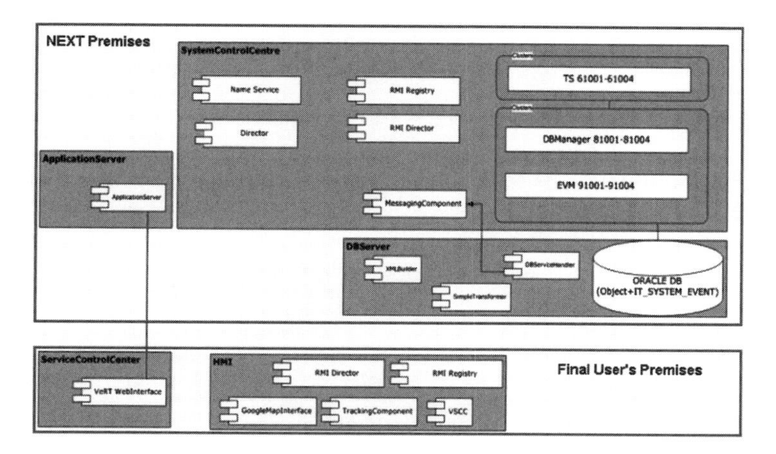

Fig. 9. VeRT centralised infrastructure configuration.

- Service Control Centre: it has been installed on a PC physically residing at Final Operator's Premises (Turin Municipality for Urban tests, SALT premises for Motorway tests)

5 EGNOS Preliminary Trials Campaign

Before the beginning of the VeRT Prototype Tests Execution, it has been conducted a preliminary EGNOS Trials Campaign during June 2006. The survey of the EGNOS SIS measurement had the scope of measuring, with a suitable data acquisition campaign, the EGNOS SIS in order to define both journeys with a low risk to have a lack of EGNOS signal and journeys characterised by bad environmental conditions (urban canyon, orientation, masking angle, etc.).

The fundamental elements which constitute the system used for the acquisition of the navigation signals of EGNOS were:

- Base Station
- Rover Station

The Base Station received data from EGNOS space segment during all the Site Survey Transfer time of Rover Station. The collected data have represented a reference for post-evaluate the different cases in which the rover station were during the transfer in urban environment. A base_station.dat log file has been saved and used to match data with rover_station.dat log file.

The Rover Station "surfed" around the city in selected zones and routes of interest collecting all needed data (described in paragraph "Data Format") for post data processing. This data were afflicted by all disturb components of first and second order:

- Urban canyon effects
- Foliage Coverage
- Multipath
- Electro Magnetic Interferences (tram, Electric backbones, Radio Repeaters . . .)

In the following figures the main elements composing both the Base Station and Rover Station:

Universal amplified GPS antenna.

GPS/EGNOS L1/L2 receiver.

The BS and the RS have been equipped with the same type of antenna for receiving the GPS/EGNOS SIS and the same kind of receiver, which acquires the antenna signals with a great number of navigation data contained in different strings with specific protocols. The reason in choosing the same equipment both for the BS and the RS is for guaranteeing the same behaviour in the data acquisition chain, permitting a consistent and effective comparison of the two parallel data acquisition.

Figure 10 shows the complete schemes of the configuration of the two measurement chains used in the site survey.

It is compulsory that the Operators present at the Base and Rover Station has to synchronise the operations for the data acquisition.

The data have been saved and a laptop (in the Base Station) and a laptop (in the Rover Station), and then post-processed from a software which provides the following parameters for the evaluation of the performances obtained during the test campaigns:

- The power of the EGNOS SIS, in terms of SNR (Signal to Noise Ratio); with these information it is possible to obtain a measure of the availability and continuity of the EGNOS system.

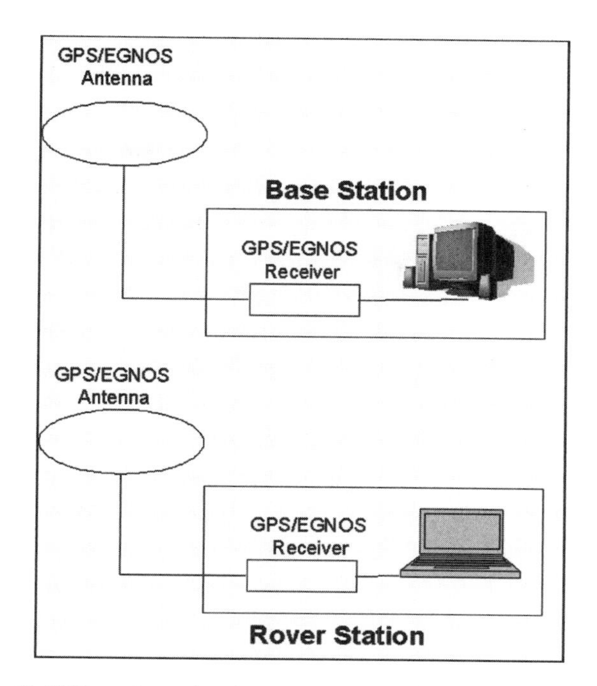

Fig. 10. HW configuration for the EGNOS SIS measurement chains.

- The position of the BS and RS in terms of latitude, longitude and altitude with reference to the WGS-84 (World Geodetic System 1984). The data of the BS provide an idea of the accuracy through an statistic evaluation of the dispersion of the positioning values during the temporal interval necessary to the RS to cover an established route; for the mobile station the positioning data are useful to have an idea of the positions collected by the receiver of the RS during the travel.
- The BS and the RS horizontal and vertical positioning error in terms of latitude, longitude and altitude standard deviation; these parameters provide an estimation of the accuracy of the positioning.
- The RTCA (Radio Technical Commission for Aeronautics) sent by the satellites. There are 16 different message types, identified by integer between 0. . .63, which specify the differential correction and other information in a defined period.

References

[1] VeRT Team, VeRT-DD03 Preliminary Design of the Service Prototype, December 2004.
[2] VeRT Team, VeRT-DD04 Service Prototype Detailed Design, April 2005.
[3] VeRT Team, VeRT-DD02 Preliminary Design of the Final System, November 2004.
[4] Bain Company, VeRT Cost Benefit Analysis, November 2004.
[5] AAS-I, VeRT-DD05 Service Prototype Validation Plan, June 2006.
[6] AAS-I, VeRT Procurement Technical Note, May 2006.

Chapter V

Perspectives in Satellite Communications

ISI – The Integral SatCom Initiative Towards FP7

Giovanni Emanuele Corazza

DEIS/ARCES - University of Bologna
e-mail: giovanni.corazza@unibo.it

Abstract. The paper addresses the Integral Satcom Initiative, a European Technology Platform devoted to the entire field of Satellite Communications. The scope, status, and future plans of ISI are outlined in the paper.

1 Introduction

As the worldwide telecommunications network is evolving fast, and historically separate sectors are converging into a single competition arena, there is an apparent need to coordinate efforts in the field of satellite communications, in order to maximize its chances for business consolidation and growth. This is precisely the purpose of the Integral Satcom Initiative (ISI), a Technology Platform which is being set up in Europe in view of the upcoming 7th Framework Programme (FP7) of the European Union (EU), and of the implementation of the European Space Policy, in collaboration with the European Space Agency (ESA). This perspective article highlights the ISI founding principles and its present status.

The European Commission (EC) is working towards the implementation of FP7, which will run from 2007 to 2013, and is expected to have extremely significant impact and structuring effects for Europe, with focus over sectors of recognized strategic relevance [1]. Two very important sectors in FP7 will be Information and Communications Technologies (ICT) in the Information Society [2], and Space & Security, which is in relation to the implementation of the European Space Policy [3], in cooperation with ESA [4].

Within FP7, a new element has been introduced, known as Technology Platform. Technology Platforms (TPs) are industry-led initiatives intended to define research and development (R&D) priorities and timeframes. They focus on areas of significant economic impact and high societal relevance, where there is high public interest and scope for genuine value added through European and International response. The initial work of a TP focuses on the preparation of a Strategic Vision document, a Strategic Research Agenda and the mobilization of the necessary critical mass of research and innovation effort. More information regarding TPs can be found in [5].

A pioneering example of platform is the Task Force on Advanced Satellite Mobile Systems (ASMS-TF), launched in 2001 under the auspices of EC and ESA [6]. In fact, the ASMS-TF represents the interests of a broad industrial community, and works actively on R&D, standardization, regulatory matters, and commercial operations. Indeed, the ASMS-TF was instrumental in the launch of the Integral Satcom Initiative.

2 ISI: Scope, Rationale, and Governance

ISI is an industry-led action forum designed to bring together all aspects related to satellite communication (satcom). In fact, ISI addresses broadcasting, broadband, and mobile satellite communications, as well as their convergence and integration into the global telecommunication network infrastructure, in support of all forms of space communication and exploitation.

ISI is designed as an open platform, embracing all relevant and interested private and public stakeholders. ISI intends to collaborate and cooperate with the European Commission, the European Space Agency, the EU and ESA Member States and Associated States, the National Space Agencies, International Organizations, User fora, and other TPs. ISI fosters international cooperation under a global perspective, and already enumerates participants from outside of Europe. See Fig. 1 for a pictorial view of the scope of ISI.

ISI is determined to contribute significantly to several EU and ESA policies, in order to promote European industrial competitiveness, growth and employment in a sustainable way, in synergy with National priorities. Representative EU sectors of interest include ICT, Space, Security, Transport, Development, and Environment. Specific policy initiatives of interest include i2010 [7], the European Space Policy, and in general all those initiatives which can benefit from the existence of an efficient satellite communications infrastructure, or which are aimed at the development of innovative satellite services and technologies.

The rationale of ISI is based on the fact that satellite communications constitute a strategic sector for Europe, with significant economic impact and high societal relevance. Satcoms are instrumental for European and International broadcasting, mobile communications, broadband access, bridging the digital divide, safety, crisis management, disaster relief, and dual use applications.

In fact, satellites provide both direct access to, and the backbone of European and Worldwide digital information broadcast networks, as well as interactive and subscription TV services, mobile services to ships, aircrafts and land-based users, and

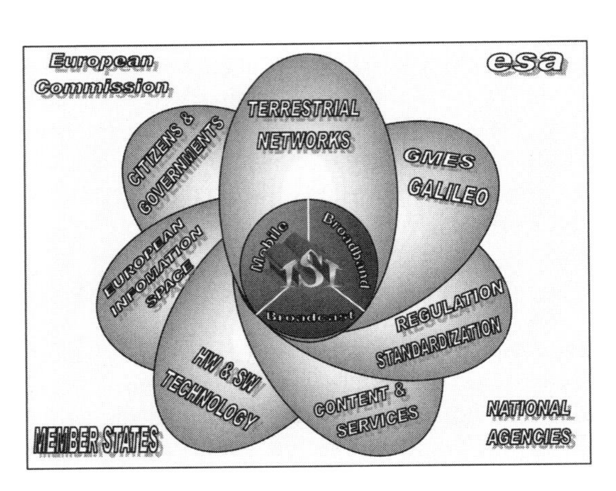

Fig. 1. A pictorial view of ISI scope and inter-relationships.

data distribution within business networks. Satellites are also a key element in the Internet backbone, and enable both broad and narrow-band Internet access services from remote and rural locations. Satellite services provide an essential component of disaster relief activities worldwide, offering reliability, instant and long-term availability, over very wide areas. In addition to civil applications, the unique coverage advantages of satellite systems position them as key players for risk and crisis management for institutional, government and defense applications.

ISI works towards the convergence and integration of satellite and terrestrial networks, both fixed and mobile, considering all interworking and interoperability aspects. ISI supports the development of applications and services according to a user-centric approach, to enable all citizens to become full members of the knowledge-based society. ISI addresses the integration of satellite communications with navigation, Earth observation, and Air Traffic Management systems. Specific attention is devoted to Galileo and GMES. Data relay systems and the use of Unmanned Aerial Vehicles are in the scope of ISI, as well.

3 ISI: Success Factors

ISI intends to be instrumental in achieving and maintaining European leadership and competitiveness in all of the above-mentioned fields, fostering the entire industrial sector, and maximizing the value of related research and technology development. In practice, there are several concrete targets that must be achieved for satcoms:

- Identify user requirements and market segments
- Reduce costs (terminals, networks, tariffs, licenses, etcetera)
- Develop open standards
- Secure spectrum allocation and sharing rules
- Harmonize the regulatory framework
- Coordinate efforts while preserving competition
- Contribute to the user centric services trends
- Accompany the convergence move
- Integrate satcoms with Terrestrial networks, Galileo, GMES, Air Traffic Management systems, Security programmes
- Reduce time to market for competitiveness

4 ISI: Regulation and Standardization

One of the main priorities of ISI is to contribute to the harmonization of the European and International regulatory framework for satellite communications, helping in the removal of barriers. ISI works for the allocation of sufficient spectrum for all satellite communication applications and services. ISI favors the consideration of a regulatory framework for complementary ground components (CGC).

ISI promotes open standards and international standardization approaches. Indeed, ISI fosters wide adoption of common standards to enlarge markets, reduce costs and tariffs, facilitate interoperability and roaming, and ensure fair competition for the benefit of citizens, user communities and governments.

5 ISI: Constituency and Documents

ISI embodies the critical mass required to pursue the above objectives considering short term priorities, medium-term evolutions, and long-term strategic directions. ISI Participants are increasing rapidly. Presently, there are more than 120 institutions from 24 different Countries: Austria, Belgium, Bulgaria, Finland, France, Germany, Greece, Hungary, Israel, Ireland, Italy, Luxemburg, Norway, Poland, Portugal, Romania, Russia, Slovenia, Sweden, South Korea, Spain, Switzerland, United Kingdom, USA. The full list of entities can be found in [8].

The ISI participating entities have worked towards the production of Strategic Vision Document [9], which has been signed by the highest company or institution representatives, and the Strategic Research Agenda [10]. This document defines the priorities for R&D in the satcoms sector for the medium-and long-term.

References

[1] http://www.cordis.lu/fp7
[2] http://europa.eu.int/information_society
[3] http://europa.eu.int/comm/space
[4] http://www.esa.int
[5] http://www.cordis.lu/technology-platforms
[6] http://www.asms-tf.org
[7] http://europa.eu.int/information_society/eeurope/i2010
[8] http://www.isi-initiative.eu.org
[9] ISI Strategic Vision Document, November 2005.
[10] ISI Strategic Research Agenda, January 2006.

Diversity Reception Over Correlated Ricean Fading Satellite Channels

Petros S. Bithas, P. Takis Mathiopoulos

Institute for Space Applications and Remote Sensing,
National Observatory of Athens, Metaxa & Vas. Pavlou Street,
15236 Athens, Greece
e-mail: pbithas@space.noa.gr, mathio@space.noa.gr

Abstract. In this paper, the performance of switch and stay combing (SSC) diversity receivers operating over correlated Ricean fading channels, is studied. By representing the bivariate Ricean distribution as infinite series the probability density function (PDF) of the SSC output signal-to-noise ratio (SNR) has been derived. Capitalizing on this PDF the moments of the output SNR and the corresponding cumulative distribution function (CDF) have been also obtained. Furthermore, an analytical expression for the moments generating function (MGF) is derived and by employing the MGF-based approach the average bit error probability (ABEP) for several modulation schemes is studied. Finally, various performance evaluation results, such as average output SNR (ASNR), amount of fading (AoF), outage probability (P_{out}) and ABEP are presented.

1 Introduction

The mobile radio channel is particularly dynamic due to multipath propagation, which has a strong negative impact on the average bit error probability (ABEP) of any modulation technique [1]. Diversity is a powerful communication receiver technique used to compensate for fading channel impairments. The most important diversity reception methods employed in digital communication receivers are maximal-ratio combining (MRC), equal gain combining (EGC), selection combining (SC) and switch and stay combing (SSC) [2]. Among them, SSC diversity is the least complex and can be used in conjunction with coherent, non-coherent and differentially coherent modulation schemes. It is well known that in many real life communication scenarios the combined signals are correlated [2]. A typical example for such signal correlation exists in small size mobile terminal units where the distance between diversity antenna is small. Due to this correlation between the signals received at the diversity branches, there is degradation in the diversity gain.

As far as the Ricean distribution is concerned, it is often used to model propagation paths consisting of one strong direct line-of-sight (LoS) signal and many random reflected and usually weaker signals. Such fading environments are typically observed in microcellular and mobile satellite radio links [2]. Especially for mobile satellite communications, the Ricean distribution is used to accurately characterize the mobile satellite channel, for the single-state [3] and the clear state [4]. Moreover, in [5] it was depicted that the Ricean-K factor characterizes the land mobile satellite channel during unshadowed periods.

The technical literature concerning diversity receivers operating over correlated fading channels is quite extensive, e.g., [6–9]. In [6] expressions for the outage probability (P_{out}) and average bit error probability (ABEP) of dual SC with correlated Rayleigh, were derived either in closed-form or in terms of single integrals. In [7] the cumulative distribution functions (CDF) of SC, in correlated Rayleigh, Ricean and Nakagami-*m* fading channels were derived in terms of single fold integral and infinite series expressions. In [8] the ABEP of dual-branch EGC and MRC receivers was obtained in correlative Weibull fading. More recently, in [9] the performance of MRC in nonidentical Weibull fading channels with arbitrary parameters was evaluated.

Past work concerning the performance of SSC operating over correlated fading channels can be found in [10–13]. One of the first attempts to investigate the performance of SSC diversity receivers operating on independent and correlated Ricean fading channels was made in [10]. However, in this reference only a integral representation for the ABEP using non-coherent frequency shift keying (NCFSK), has been obtained. In [11] the performance of SSC diversity receivers was evaluated and optimized for several channel conditions, including different fading channels and unbalanced branches fading correlation. However, the research reported in [11] was restricted to correlated Nakagami-*m* fading conditions. In [12] the moment generating function (MGF) of SSC is derived in finite integral representation and has been also applied to correlated Nakagami-*m* channel. Recently in [13] analytical results for the performance of non ideal referenced based SSC for *M*-ary digitally modulated signals in correlated Nakagami-*m* fading channels were derived. Although previous research on generalized fading channels is rather extensive, the problem of analyzing the performance of SSC over correlated Ricean fading channel has not been adequately addressed. The main difficulty for this is the complicated form of the bivariate probability density function (PDF) derived in [10] and the absence of an alternative expression for the multivariate distribution. On the contrary as it will agreed in our paper by using a PDF of the form derived in [14], the most important statistic metrics of SSC output signal to noise ratio (SNR) can be obtained.

The remainder of this paper is organized as follows. In Section 1 the system and channel model are presented. In Section 2 the statistics and the most important performance metrics of the SSC output SNR are derived while in Section 3 some numerical evaluation results are given.

2 System and Channel Model

Let us consider a dual-branch SSC diversity receiver operating over a correlated Ricean fading channel. The baseband receiver signal at the ℓth (ℓ = 1, 2) input branch is $\zeta_\ell = sh_\ell + n_\ell$, where s is the complex transmitted symbol, $E_s = E\langle |s|^2 \rangle$ is the transmitted average symbol energy, where $E\langle \cdot \rangle$ denotes expectation and $|\cdot|$ absolute value, h_ℓ is the complex channel fading envelope, with its magnitude $R_\ell = |h_\ell|$, and n_ℓ is the additive white Gaussian noise (AWGN) with single-sided power spectral density N_0.

The instantaneous SNR per symbol at the ℓth input branch is $\gamma, = R_\ell E_S/(2N_0)$ and the corresponding average SNR per symbol at both input branches is

$\overline{\gamma} = \Omega E_s/N_0$, where $\Omega = E\left\langle R_\ell^2\right\rangle$. Using a similar procedure as for deriving [14, eq. (9)], the joint PDF of γ_1 and γ_2 can be obtained as

$$f_{\gamma_1,\gamma_2}(\gamma_1,\gamma_2) = \sum_{\substack{i,h=0\\ \upsilon_1+\upsilon_2+\upsilon_3=i}}^{\infty} \mathcal{A}\exp\left[-\beta_1(\gamma_1+\gamma_2)\right] \times\left(\mathcal{B}\gamma_1^{\beta_2-1}\gamma_2^{\beta_3-1}+C\overline{\gamma}^{-1}\gamma_1^{\beta_2-1/2}\gamma_2^{\beta_3-1/2}\right) \quad (1)$$

where

$$\mathcal{A}=\frac{2^{\upsilon_3+2h-1}(1+K)^{1+\beta_4}\rho^{2h}K^i\exp\left(-\frac{2K}{1+\rho}\right)}{\sqrt{\pi}\overline{\gamma}^{1+\beta_4}(1-\rho^2)^{1+2h}\upsilon_1!\upsilon_2!\upsilon_3!i!(1+\rho)^{2i}},$$

$$\mathcal{B}=\frac{[1+(-1)^{\upsilon_3}]\,\Gamma[h+(1+\upsilon_3)/2]}{\Gamma[h+1+\upsilon_3/2]\,\Gamma(1+2h)},$$

$$C=\frac{[-1+(-1)^{\upsilon_3}]\,2\rho(1+K)\Gamma(1+h+\upsilon_3/2)}{(\rho^2-1)\Gamma(2+2h)\,\Gamma[h+(3+\upsilon_3)/2]},$$

$$\beta_1=\frac{(1+K)}{(1-\rho^2)\overline{\gamma}},\ \beta_2=\upsilon_1+\frac{\upsilon_3}{2}+h+1,$$

$$\beta_3=\upsilon_2+\frac{\upsilon_3}{2}+h+1,\ \beta_4=i+2h+1$$

where $\Gamma(\cdot)$ is the Gamma function [15, eq. (8.310/1)], K is the Ricean factor, defined as the ratio of the specular signal power to the scattered power, ρ is the correlation coefficient between γ_1 and γ_2. In [14] was proved that the infinity series expression in (2) converge always and converge rapidly.

3 Statistics and Performance Analysis

In this section the most important statistical metrics of SSC diversity receivers and a detailed performance analysis for these receivers operating over correlated Ricean fading channels will be presented.

3.1 Probability Density Function (PDF)

Let γ_{ssc} be the instantaneous SNR per symbol at the output of the SSC and γ_τ the predetermined switching threshold. By using [11], the PDF of γ_{ssc}, $f_{\gamma ssc}(x)$, can be obtained as follows

$$f_{\gamma ssc}(x) = \begin{cases} r_{ssc}(x), & x\leq\gamma_\tau \\ r_{ssc}(x)+f_{Rice}(x), & x>\gamma_\tau \end{cases} \quad (2)$$

where

$$f_{Rice}(x) = \frac{1+K}{\overline{\gamma}}\exp(-K)\exp\left[-\frac{(1+K)}{\overline{\gamma}}x\right]I_0\left(2\sqrt{\frac{K(K+1)}{\overline{\gamma}}}x^{1/2}\right)$$

with $I_0(\cdot)$ being the zeroth-order modified Bessel function of the first kind [15, eq. (8.406)]. Moreover, using [11, eq. (70)], [15, eq. (3.351)], $r_{ssc}(x)$ can be obtained as

$$r_{ssc}(x) = \sum_{\substack{i,h=0 \\ \upsilon_1+\upsilon_2+\upsilon_3=i}}^{\infty} \mathcal{A} \exp(-\beta_1 x) x^{\beta_2 - 1/2} \times \left[\frac{\mathcal{B}}{\sqrt{x}\,\beta_1^{\beta_3}} \gamma(\beta_3, \beta_1 \gamma_\tau) + \frac{\mathcal{C}}{\overline{\gamma}\beta_1^{\beta_3+1/2}} \gamma(\beta_3 + 1/2, \beta_1 \gamma_\tau) \right] \quad (3)$$

where $\gamma(\cdot,\cdot)$ is the lower incomplete Gamma function [15, eq. (8.350)].

3.2 Cumulative Density Function (CDF) and Outage Probability (P_{out})

Using [16, eq. (20)], the CDF of γ_{ssc} can be obtained as

$$F_{\gamma ssc}(x) = Pr(\gamma_\tau \le \gamma_1 \le x) + Pr(\gamma_2 < \gamma_\tau \text{ and } \gamma_1 < x). \quad (4)$$

After some manipulations, (4) can be expressed in terms of CDF's as

$$F_{\gamma ssc}(x) = \begin{cases} F_{\gamma_1,\gamma_2}(x, \gamma_\tau), & x \le \gamma_\tau \\ F_\gamma(x) - F_\gamma(\gamma_\tau) + F_{\gamma_1,\gamma_2}(x, \gamma_\tau), & x > \gamma_\tau \end{cases} \quad (5)$$

with

$$F_\gamma(x) = Q_1\left[\sqrt{2K}, \sqrt{\frac{2(1+K)}{\overline{\gamma}}x}\right]$$

where $Q_1(\cdot)$ is the first order Marcum-Q function [2, eq. (4.33)]. In (5) $F_{\gamma_1,\gamma_2}(x, \gamma_\tau)$ can be derived by using [15, eq. (3.352)] as

$$F_{\gamma_1,\gamma_2}(x, \gamma_\tau) = \frac{\mathcal{A}}{\beta_1^{\beta_2+\beta_3}} \Big[\mathcal{B}\gamma(\beta_2, \beta_1 x)\gamma(\beta_3, \beta_1 \gamma_\tau) + \frac{\mathcal{C}}{\beta_1}\gamma(\beta_2 + 1/2, \beta_1 x)\gamma(\beta_3 + 1/2, \beta_1 \gamma_\tau) \Big]. \quad (6)$$

Clearly, the probability that the output SNR falls below a given threshold (γ_{th}), P_{out}, can be obtained for γ_{ssc} as

$$P_{out}(\gamma_{th}) = F_{\gamma_{ssc}}(\gamma_{th}). \quad (7)$$

3.3 Moments Generating Function (MGF) and Average Bit Error Probability (ABEP)

Based on (2) the MGF for γ_{ssc}, $\mathcal{M}_{\gamma_{ssc}}(s)$, defined in [17, eq. (5.62)], can be expressed in terms of two integrals as

$$\mathcal{M}_{\gamma_{ssc}}(s) = \int_0^\infty \exp(-sx) r_{ssc}(x)\,dx + \int_{\gamma\tau}^\infty \exp(-sx) f_{Rice}(x)\,dx = \mathcal{I}_1 + \mathcal{I}_2. \quad (8)$$

Using [15, eq. (3.381/4)], $\mathcal{I}_1$ can be solved as

$$\mathcal{I}_1 = \mathcal{A}\left[\frac{\Gamma(\beta_2)}{(\beta_1+s)^{\beta_2}}\mathcal{B}\beta_1^{\beta_3}\gamma(\beta_3,\beta_1\gamma_\tau) + \mathcal{C}\beta_1^{\beta_3-1/2}\frac{\Gamma(\beta_2+1/2)}{(\beta_1+s)^{\beta_2+1/2}}\gamma(\beta_3+1/2,\beta_1\gamma_\tau)\right]. \tag{9}$$

Setting $\psi = \sqrt{2x\left(\frac{1+K}{\overline{\gamma}}+s\right)}$ and using again [2, eq. (4.33)], $\mathcal{I}_2$ can be solved as

$$\mathcal{I}_2 = Q_1\left(\sqrt{\frac{2K(1+K)}{1+K+\overline{\gamma}s}},\sqrt{\frac{2(1+K+\overline{\gamma}s)\gamma_\tau}{\overline{\gamma}}}\right) \times \exp\left(\frac{K(1+K)}{1+K+\overline{\gamma}s}\right)\frac{(1+K)\exp(-K)}{1+K+\overline{\gamma}s}. \tag{10}$$

Using (8), (9) and (10) and based on the well-known MGF approach [2, 18], the ABEP for several coherent and non-coherent modulation schemes can be obtained.

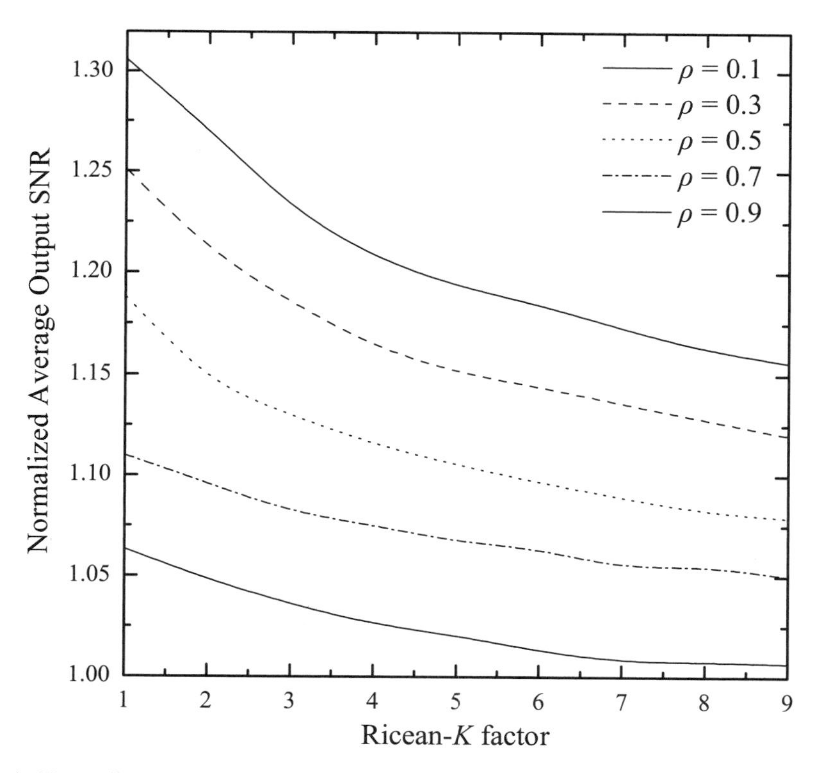

Fig. 1. Normalized average output SNR (ASNR) versus the Ricean-K factor for several values of ρ.

3.4 Moments, Average Output SNR (ASNR) and Amount of Fading (AoF)

Based on (2) the moments for γ_{ssc}, $\mu_{\gamma ssc}(n)$, defined in [17, eq. (5.38)], can be expressed in terms of two integrals as

$$\begin{aligned}\mu_{\gamma_{ssc}}(n) &= \int_0^\infty x^n r_{ssc}(x)\,dx + \int_{\gamma_\tau}^\infty x^n f_{Rice}(x)\,dx \\ &= \mathcal{I}_3 + \mathcal{I}_4.\end{aligned} \tag{11}$$

Using [15, eq. (3.381/4)], $\mathcal{I}_3$ can be easily solved as

$$\begin{aligned}\mathcal{I}_3 = \mathcal{A}\Bigg[&\mathcal{B}\beta_1^{-\beta_3}\gamma(\beta_3, \beta_1\gamma_\tau)\,\frac{\Gamma(n+\beta_2)}{\beta_1^{n+\beta_2}} \\ &+ \frac{\mathcal{C}\gamma(\beta_3+1/2, \beta_1\gamma_\tau)}{\beta_1^{\beta_3+1/2}}\,\frac{\Gamma(n+\beta_2+1/2)}{\beta_1^{n+\beta_2+1/2}}\Bigg].\end{aligned} \tag{12}$$

Setting $\phi = \sqrt{2\frac{1+K}{\overline{\gamma}}x}$ in $\mathcal{I}_4$ and after some straight forward mathematical manipulations yields

$$\mathcal{I}_4 = \frac{\overline{\gamma}^n \exp(K)}{2^n(1+K)^n} Q_{2n+1,0}\left(K, \sqrt{\frac{2(1+K)\gamma_\tau}{\overline{\gamma}}}\right) \tag{13}$$

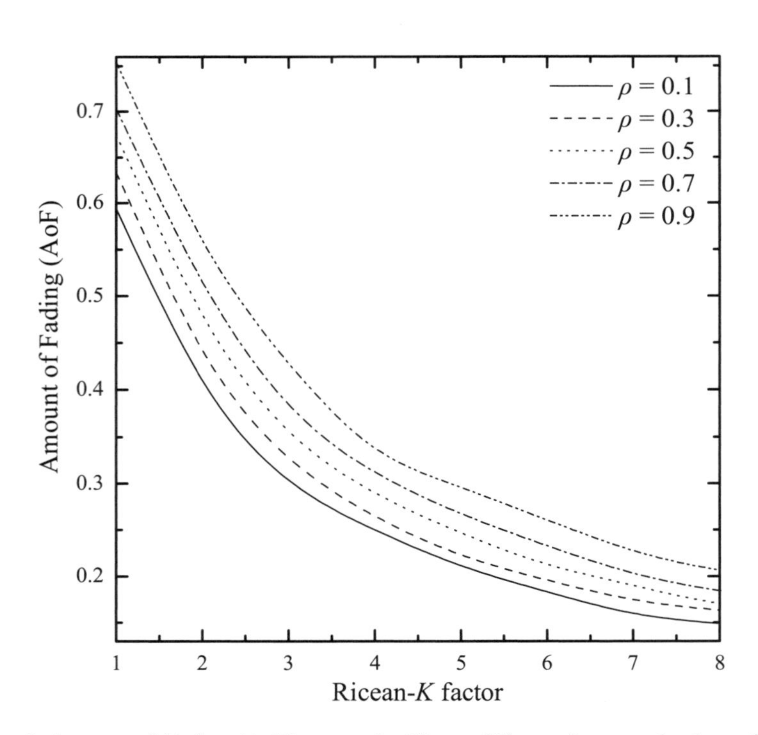

Fig. 2. Amount of Fading (AoF) versus the Ricean-K factor for several values of ρ.

where $Q_{m,n}(\cdot,\cdot)$ is the Nuttal Q-function defined in [2, eq. (4.104)]. The ASNR, $\overline{\gamma}_{out}$, is a useful performance measure serving as an excellent indicator for the overall system fidelity. The AoF, defined as AoF $\triangleq \mathrm{var}(x)/\overline{\gamma}^2$, is a unified measure of the severity of the fading channel [2], which can be expressed in terms of first- and second-order moments of γ_{ssc} as

$$\mathrm{AoF} = \frac{\mu_{\gamma_{ssc}}(2)}{\mu_{\gamma_{ssc}}(1)^2} - 1. \tag{14}$$

Both these important performance metrics can be derived by using (11), (12) and (13).

4 Numerical Performance Evaluation Results

In this section, by using the previous mathematical analysis, performance evaluation results obtained by means of numerical techniques are presented. These results include different correlated Ricean fading conditions and several modulation formats.

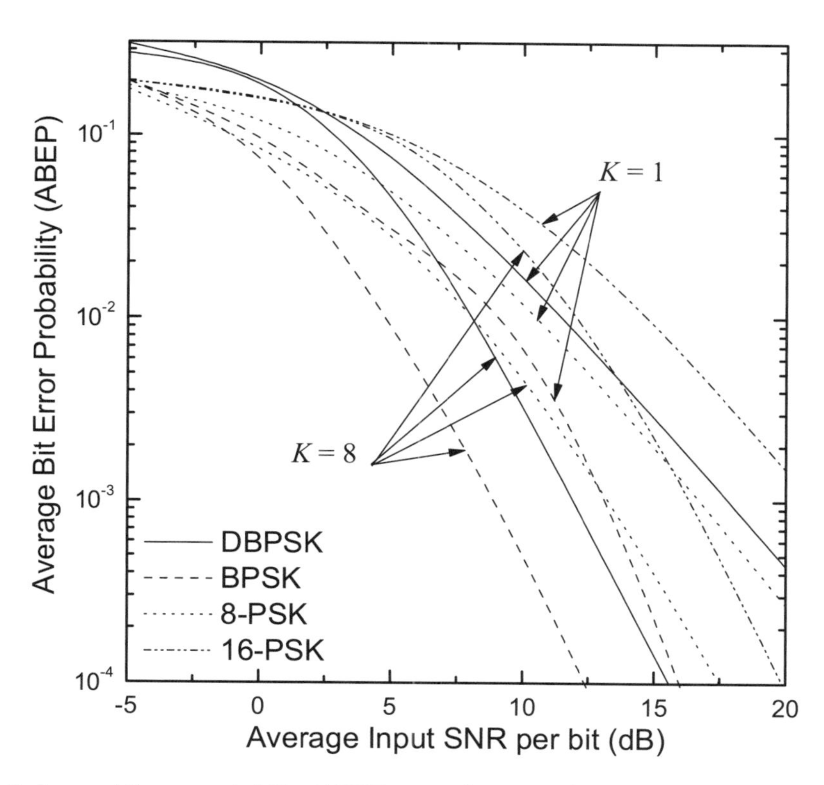

Fig. 3. Average bit error probability (ABEP) versus the average input SNR per bit for DBPSK and M-PSK (M = 8, 16) signaling formats for different values of the Ricean-K factor.

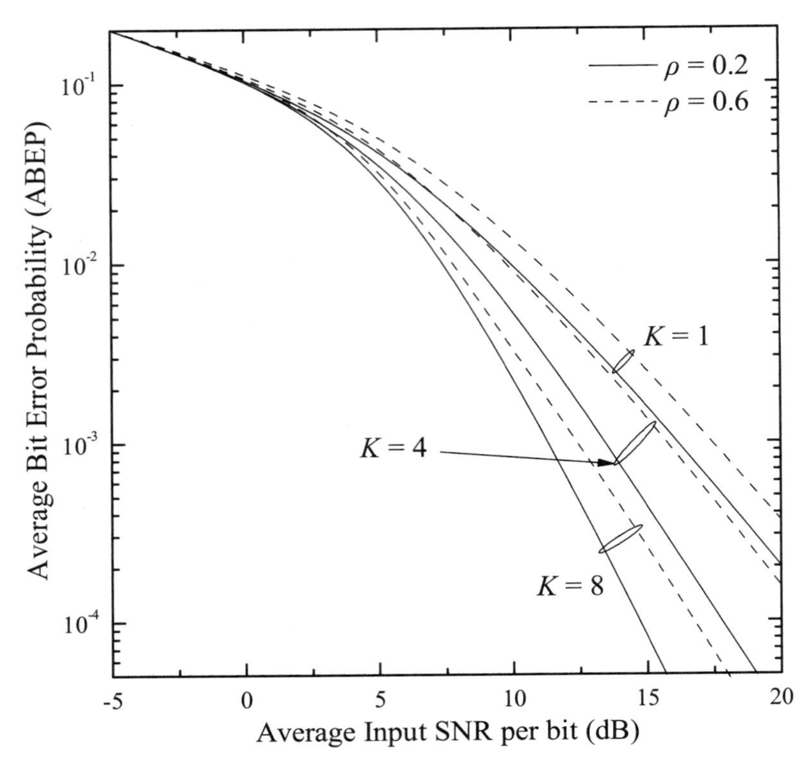

Fig. 4. Average bit error probability (ABEP) versus the average input SNR per bit for 16-QAM signaling format for different values of the Ricean-K factor and ρ.

In Figs. 1 and 2 the ASNR and AoF are plotted as a function of the Ricean-K factor for several values of the correlation coefficient ρ. In Fig.1 as K and/or ρ increase the ASNR decreases, which means that the diversity gain lessens. In Fig. 2 as K increases and/or ρ decreases the AoF lessens.

In Figs. 3 and 4 the ABEP is plotted as a function of the average input SNR per bit, i.e., $\overline{\gamma}_b = \overline{\gamma}/\log_2 M$, for several values of K. In Fig. 3 the differential binary phase shift keying (DBPSK), binary phase shift keying (BPSK) and M-ary phase shift keying (M-PSK) (gray encoding signals are considered) signaling formats are employed. It is depicted that as K increases the ABEP improves and the BPSK has the best performance. In Fig. 4 considering the 16- quadrature amplitude modulation (QAM) scheme, it can been obtained as K increases and/or ρ decreases the ABEP lessens.

Finally, in Fig. 5 the P_{out} is plotted as a function of the normalized outage threshold $\gamma_{th}/\overline{\gamma}_b$ for several values of K and ρ. As K increases and/or ρ decreases the P_{out} lessens.

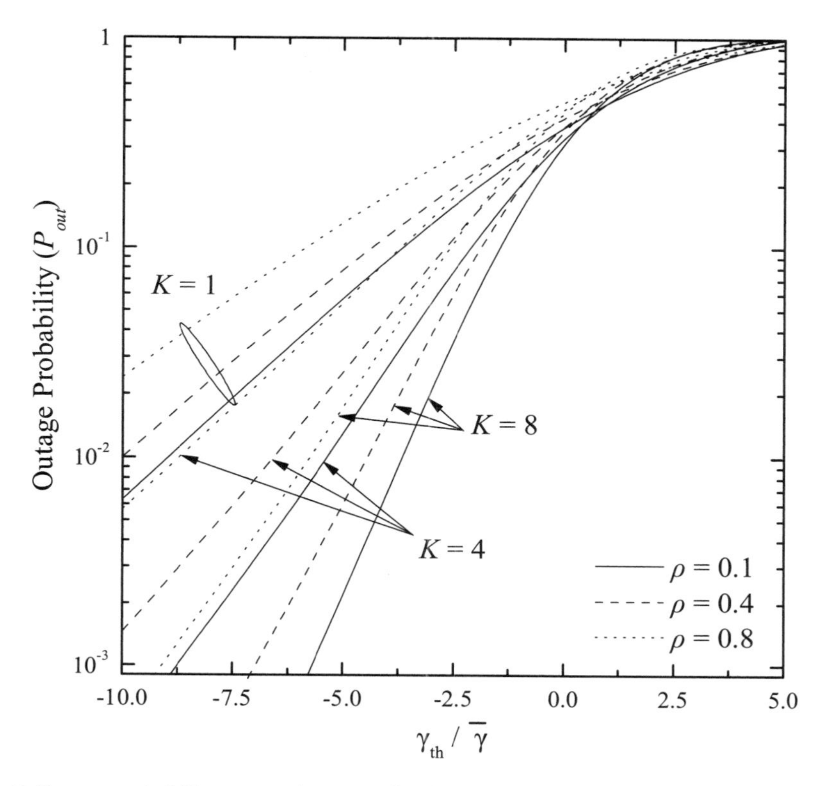

Fig. 5. Outage probability versus the normalized average input SNR for several values of the Ricean-K factor and ρ.

Acknowledgments

This work has been performed within the framework of the Satellite Network of Excellence (SatNEx-II) project (IST-027393), a Network of Excellence (NoE) funded by European Commission (EC) under the FP6 program.

References

[1] T. S. Rappaport, *Wireless Communications*. New Jersey: Prentice Hall PTR, 2002.

[2] M. K. Simon and M.-S. Alouini, *Digital Communication over Fading Channels*, 2nd ed. New York: Wiley, 2005.

[3] G. E. Corazza and F. Vatalaro, "A statistical model for land mobile satellite channels and its applications to nongeostationary orbit system," *IEEE Trans. Veh. Technol.*, vol. 43, no. 3, pp. 738–742, Aug. 1994.

[4] H. Wakana, "A propagation model for land-mobile-satellite communication," in *Proc. IEEE Antennas and Propagation Sociaty International Symposium*, vol. 3, June 1991, pp. 1526–1529.
[5] E. Lutz, D. Cygan, M. Dippold, F. Dolainsky, and W. Papke, "The land mobile satellite communication channel-recording, statistics, and channel model," *IEEE Trans. Veh. Technol.*, vol. 40, no. 2, pp. 375–386, May 1991.
[6] M. K. Simon and M.-S. Alouini, "A unified performance analysis of digital communications with dual selective combining diversity over correlated Rayleigh and Nakagami-*m* fading channels," *IEEE Trans. Commun.*, vol. 47, no. 1, pp. 33–43, Jan. 1999.
[7] Y. Chen and C. Tellambura, "Distribution functions of selection combiner output in equally correlated Rayleigh, Rician, and Nakagami-*m* fading channels," *IEEE Trans. Commun.*, vol. 52, no. 11, pp. 1948–1956, Nov. 2004.
[8] G. K. Karagiannidis, D. A. Zogas, N. C. Sagias, S. A. Kotsopoulos, and G. S. Tombras, "Equal-gain and maximal-ratio combining over nonidentical Weibull fading channels," *IEEE Trans. Wireless Commun.*, vol. 4, no. 3, pp. 841–846, May 2005.
[9] M. H. Ismail and M. M. Matalgah, "Performance of dual maximal ratio combining diversity in nonidentical correlated Weibull fading channels using padé approximation," *Electron. Letters*, vol. 32, no. 19, pp. 1752–1754, Sept. 1996.
[10] A. A. Abu-Dayya and N. C. Beaulieu, "Switched diversity on microcellular Ricean channels," *IEEE Trans. Veh. Technol.*, vol. 43, no. 4, pp. 970–976, Nov. 1994.
[11] Y. C. Ko, M.-S. Alouini, and M. K. Simon, "Analysis and optimization of switched diversity systems," *IEEE Trans. Veh. Technol.*, vol. 49, no. 5, pp. 1813–1831, Sept. 2000.
[12] C. Tellambura, A. Annamalai, and V. K. Bhargava, "Unified analysis of switched diversity systems in independent and correlated fading channels," *IEEE Trans. Commun.*, vol. 49, no. 11, Nov. 2001.
[13] G. Femenias, "Reference-based dual switch and stay diversity systems over correlated Nakagami fading channels," *IEEE Trans. Veh. Technol.*, vol. 52, no. 4, 2003.
[14] D. A. Zogas and G. K. Karagiannidis, "Infinite series representations associated with the bivariate Ricean distribution and their applications," *IEEE Trans. Commun.*, vol. 53, no. 11, Nov. 2005.
[15] I. S. Gradshteyn and I. M. Ryzhik, *Table of Integrals, Series, and Products*, 6th ed. New York: Academic Press, 2000.
[16] A. A. Abu-Dayya and N. C. Beaulieu, "Analysis of switched diversity systems on generalized-fading channels," *IEEE Trans. Commun.*, vol. 42, no. 11, pp. 2959–2964, Nov. 1994.
[17] A. Papoulis, *Probability, Random Variables, and Stochastic processes*, 2nd ed. McGraw-Hill, 1984.
[18] N. C. Sagias and G. K. Karagiannidis, "Gaussian class multivariate Weibull distributions: theory and applications in fading channels," *IEEE Trans. Inform. Theory*, vol. 51, no. 10, pp. 3608–3619, Oct. 2005.

Application of Long Erasure Codes and ARQ Schemes for Achieving High Data Transfer Performance Over Long Delay Networks

Tomaso de Cola*, Harald Ernst°, Mario Marchese*

*CNIT - Italian National Consortium for Telecommunications Genoa Research Unit, University of Genoa, Via Opera Pia 13, 16145, Genoa, Italy
e-mail: tomaso.decola@cnit.it, mario.marchese@cnit.it
°DLR - Deutschen Zentrum für Luft- und Raumfahrt, German Aerospace Center Institute for Communications and Navigation, 82234 Wessling, Germany
e-mail: harald.ernst@dlr.de

Abstract. The extension of telecommunication frontiers towards deep space scenarios has opened new horizons in the way of designing network infrastructures and has raised the necessity of developing novel communication paradigms, more suited to this specific environment. Under this view, given the hazardous operative conditions (very large latencies and frequent link disconnections), in which the exchange of data among remote peers has to be performed, the exploitation of TCP-based transmission schemes does not offer satisfactory results. On the contrary, the use of erasure coding schemes and more appropriate Automatic Repeat reQuest(ARQ) schemes, available within the Packet Layer Coding-based and the Consultative Committee for Space Data Systems (CCSDS) protocol architectures respectively, assure better performance results.

In this paper, the adoption of erasure codes within CCSDS protocol stack is considered and its effectiveness is evaluated with respect to ARQ-based transmission schemes available within the CCSDS File Delivery Protocol. Finally, the combined use of erasure codes and ARQ schemes is proposed in order to improve the overall performance.

1 Introduction

Since the end of eighties, the exploration of space and the proliferation of scientific experiments have shown, on the one hand, the necessity of reliable telecommunication infrastructures and, on the other hand, have revealed the shortcomings deriving from the use of TCP-based protocols. In particular, the large latencies experienced by typical deep space environments negatively affect the TCP performance because of its transmission paradigm based on a feedback scheme [1]. In this perspective, the features offered by the Consultative Committee for Space Data Systems (CCSDS) recommendations in terms of suspending and resuming capabilities are an effective resource to assure reliable data communication over space networks. Moreover, the support of highly efficient ARQ schemes available within the CCSDS File Delivery Protocol (CFDP) helps improve the overall data communication performance in terms of both throughput and loss recovery effectiveness. Although the use of CCSDS protocols has revealed its powerful abilities in recovering from consistent information losses and tolerating long disconnection periods, it is not completely able to properly exploit the channel bandwidth while recovery operations are performed. From this point of view, the adoption of erasure schemes and hence the Transport Layer Coding approach [2] would be beneficial for its recovery capabilities even in presence of bursty information losses. Starting from the aforementioned issues, this paper analyses the use of the Packet Layer Coding approach within the CFDP implementation and hence proposes a combined use of erasure coding and ARQ schemes for improving the overall performance.

The remainder of the paper is organized as follows. The state of the art and the related works carried out in the area of deep space communications are envisaged in Section 2, while Section 3 addresses the peculiarities of such scenario by introducing the transmission channel model based on discrete Markov chains. The CCSDS protocol architecture, the Packet Layer Coding approach and the issues regarding the joint use of ARQ schemes and erasure codes are shown in Section 4. The investigation completes in Section 5, where the performance analysis of the different CCSDS configurations is shown; in Section 6 the conclusions are drawn.

2 Background

Over last years, the scientific community has made strong efforts for designing appropriate protocols and architectures able to guarantee reliable data communication over space networks. From the standardization point of view, relevant contributions have been provided by the CCSDS institution together with the Delay Tolerant Network [3] (formerly known as InterPlanetary Internet Project) working group within IRTF (Internet Research Task Force). In this perspective, it is worth mentioning the CCSDS File Delivery Protocol, which is able to tolerate long disconnection periods and to react properly to information losses thanks to suspending/resuming capabilities and efficient ARQ schemes, respectively.

Furthermore, the study of alternative mechanisms, based on erasure coding schemes and aimed at guaranteeing reliable communications deserves a particular attention. In particular, it is worth mentioning the work carried on within the Reliable Multicast Transport IETF working group, addressed to the design of protocol architectures able to support multicast communications over wireless links, by means of erasure codes implemented over the transport layer. From this standpoint, the advantages offered by the long erasure codes, and in particular by Low Density Parity Check codes (LDPC). Under this view their adoption over the transport layer is identified as Transport Layer Coding and proposed in [2]. Further considerations about the software complexity issues, arising from LDPC implementations, and the related performance are addressed in [4].

Beside the standardization activities performed within CCSDS and IETF, a special attention has to be paid to relevant protocol solutions, implemented at different layers of the OSI protocol stack, as proposed in [5]. In particular, TP-Planet protocol implemented at the transport layer, emerges as a promising solution.

Finally, this work takes the CCSDS File Delivery Protocol (CFDP) as reference and applies the Transport Layer Coding approach for improving the overall data communication performance over space networks.

3 The Deep Space Environment

3.1 The Reference Scenario

To better capture the environment peculiarities and hence properly study protocol implementations able to counteract the hazardous conditions in which the data communication is achieved, the following scenario is assumed:

- Two remote stations, placed on the Earth and on a remote planet (e.g. Mars or Moon), communicating each other through specialised protocol stacks, based on the CCSDS protocol architecture and, in particular, implementing the CCSDS File Delivery Protocol (CFDP).
- Two satellites orbiting around Earth and the remote planet, respectively, guarantee the end-to-end path, by acting as relay nodes.
- The long-haul link connecting the two satellites is actually the deep space link, which is the focus of this work.

The whole scenario is depicted in Fig. 1.

3.2 The Deep Space Link

The strong impairments introduced by the deep space links, such as deep fading periods, blackout events and variable propagation delays, have to be properly taken into account while designing transmission schemes suited to the space environments. Under this view, the necessity of adopting a transmission channel model, able to

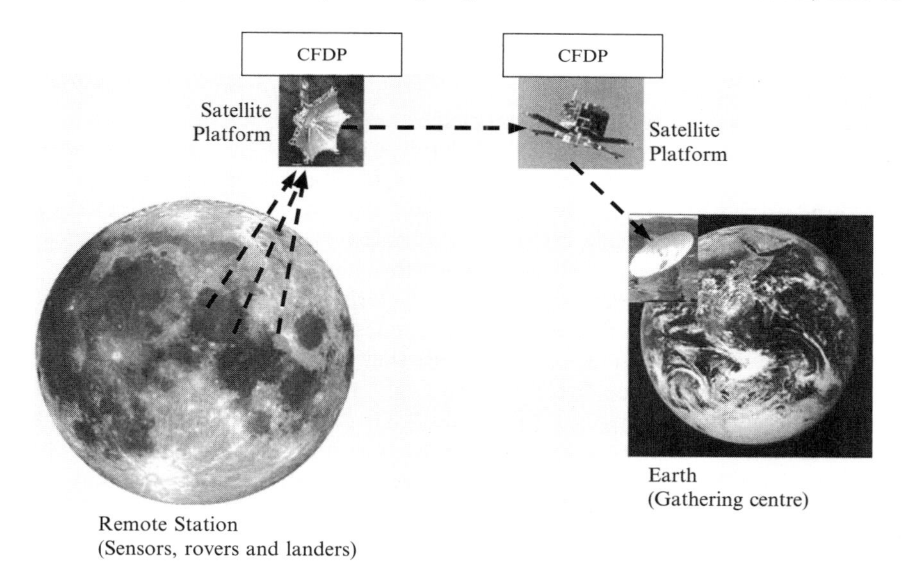

Fig.1. The reference scenario.

capture the main peculiarities of the physical link, is hence straightforward. Given the high number of factors characterizing the transmission channel dynamics, some simplifications are introduced. Firstly, the propagation delay, whose variability is due to the relative motion of planets, is assumed constant. Secondly, the blackout events are neglected, since, in general, data communications in space scenarios are scheduled in advance, through ephemerides calculations.

On the basis of these considerations, the main aspects that have to be properly taken into account concern the relative motion of the satellite platforms, the multipath fading effects due for instance to the solar flares and other hostile radiofrequency conditions. Hence, the adaptation of common models employed for characterizing the wireless transmission channel is an appropriate solution. In particular the use of Discrete-Time Markov Chains (DTMC) for representing the channel behaviour has been envisaged; in detail, the use of first order Markov chains with 4 states is proposed.

The transition between two arbitrary consecutive states, i and j, is ruled by the transition probability matrix $P = \{p_{i,j}\}$. On the other hand, the steady-state probability of being in the ith state is denoted as π_i, where $i \in \{0,1,2,3\}$. Each state accounts the channel reliability by means of the Bit Error Ratio (BER). In practice, a BER value equal to BER_i is assigned to the ith state; for consecutive states, the following inequality holds: $BER_i < BER_j$ $\forall$i,j $\in$ $\{0,1,2,3\}$, with $i < j$.

In particular, BER is the bit error rate measured at the receiver side, by taking into account the employment of forward error codes, applied at the lower layers.

Finally, under the hypothesis that the Markov chain is embedded at the start of each packet transmission, the mean permanence time τ_i within the ith state can be expressed as the bit duration time (i.e. the reciprocal of the channel bandwidth, here indicated as B_w) divided by one minus the probability of being in the same state, namely $(1-p_{ii})$. This yields:

$$\tau_i = \frac{1}{B_w \cdot (1 - p_{ii})}$$

To fully evaluate the impact of corrupted bits on the transmission performance, it is also necessary to provide a statistical characterization of the packet loss process. Under this view, the use of the GAP error length model is promising. In practice, error and error-free gaps are defined as occurrences of consecutive successful and unsuccessful received packets, respectively. It is immediate to realize that the length of error-free and error gaps differently affect the channel reliability.

Finally, a set of parameters suitable to characterize the deep space link behaviour has been identified. From this point of view, states 0 and 3 are expected to strongly impact on the DTMC channel model because they correspond to the best and worst operative conditions, respectively. Consequently, the permanence times τ_0 and τ_3 have been considered. Besides, also the steady-state probability of being in state 3, namely π_3, has been analysed. Finally, in order to account also for the impact of the other states on the channel dynamics, π_2 values have been taken into account.

Operatively, the values of permanence times τ_0 and τ_3, as well as the steady state probabilities π_2 and π_3 have been fixed within each tests; hence, the lengths of error-free and error gaps have then been evaluated through MATLAB™.

The whole channel model is sketched in Fig. 2.

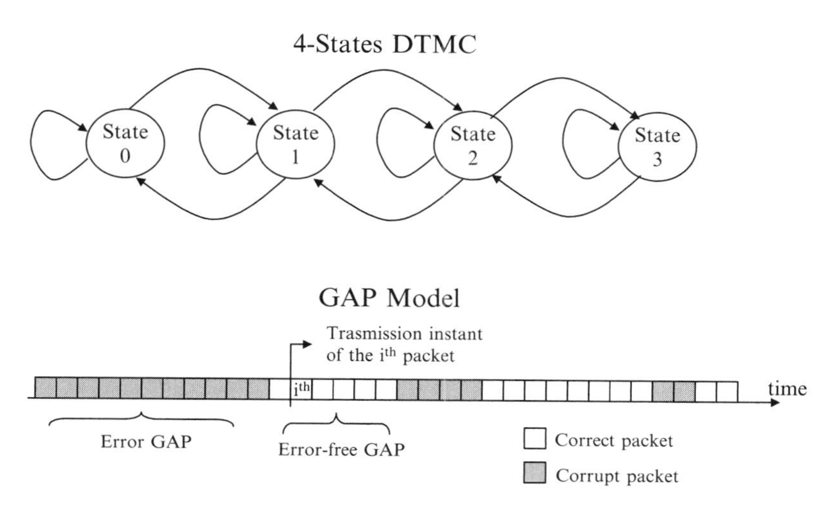

Fig. 2. The 4-states DTMC and the GAP model.

4 The Proposed Protocol Architecture

4.1 The CCSDS File Delivery Protocol (CFDP)

The CCSDS File Delivery Protocol supports file transfer operations in space environments. In facts, CFDP transmitting entity assembles data into PDUs, identified in the following as CFDP blocks, whose payload can carry up to 65536 bytes, while the header length is assumed here of 20 bytes. Once the data transmission is completed, an End-of-File notification (EOF PDU) is transmitted to the other side, which will be responsible of issuing an ACK PDU to acknowledge the receipt of the EOF PDU. The reliability issues are addressed by the CFDP entity in dependence of the operating mode in which it is configured, either acknowledged or unacknowledged. In the latter, no specific options for assuring the communication reliability are implemented: protocols acting in the lower layers are responsible for that (if necessary). On the other hand, when CFDP operates in acknowledged mode, the communication reliability is assured my means of negative acknowledgments (NAK, issued by the receiving CFDP entity) and requiring the retransmission of the missed data PDUs. Once the loss of a data block is detected, four different algorithms rule the recovery mechanism: immediate, prompted, asynchronous and deferred, which differ for the time in which the recovery procedure is invoked. More in detail, "immediate" consists in performing the information loss detection as soon as new CFDP blocks are received. Differently, "deferred" checks that the data communication has terminated correctly, when an End Of File (EOF) block, issued by the sender, is received. As far as "prompted" is concerned, it allows starting the recovery procedure in dependence of explicit requests notified by the transmitting entity. Finally, "asynchronous" triggers the recovery phase on the basis of asynchronous events such as externals commands and actions (e.g. made by human operators). In particular, in this work, the acknowledged, running the deferred algorithm and the unacknowledged mode have been analysed.

Finally, a particular note has to be dedicated to the suspending and resuming features provided by CFDP. In particular, when the protocol entity is configured for operating in "extended operations", it is able to suspend the transmission on the basis of the notifications, indicating the unavailability of the transmission medium, issued by the lower layer protocols. Afterwards, the data blocks are temporally stored in the local CFDP buffer; the transmission is resumed again once positive notifications about the channel availability are provided.

4.2 Proposed CFDP Improvements

In this work, CFDP working in both acknowledged and unacknowledged modes is investigated. The proposed CFDP improvements regard the use of erasure coding schemes, aimed at guaranteeing reliable exchange of data also when the communication is achieved in very hazardous conditions. In practice, two protocol proposals have been conceived, namely "CLDGM" and "CLDGM-deferred", whose description follows.

CLDGM. It concerns the integration of erasure coding schemes into CFDP protocol when running in unacknowledged mode, by applying the Transport Layer Coding approach as shown in [2]and [6]. In practice, the adoption of LDGM codes, derived from the Low Density Parity Check codes, is envisaged for their ability of protecting the communication against bursty data losses. In facts, the integrated scheme works as follows: CFDP aggregates different data blocks, split them into k information "packets", and hence encode them into n packets, exploiting a LDGM generator matrix. The necessity of merging several CFDP PDUs is that LDGM performance strictly depends on the number of information packets (k): higher k is, more effective is the encoding procedure. Operatively, for the sake of performance, k has been set to 1000. Moreover, it is straightforward that the LDGM performance strictly depends on the ratio among the number of encoded packets and the total number of generated packets, referred in the following as code-rate. In particular, in this work, code-rate values ranging from 0.125 up to 0.875 have been considered and, for the sake of the completeness, block and packet sizes varying from 1024 to 65536 bytes and 128 to 1024 bytes, respectively have been considered in order to characterize the impact of link errors on the overall performance. In the following this approach will be referred as *CLDGM* (which stands for CFDP with LDGM codes). Finally, for the sake of the clarity, only the configurations offering the best performance are shown in the results.

CLDGM-Deferred. The second approach combines the use of NAK PDUs with LDGM codes in order to allow data retransmission when LDGM effectiveness is not sufficient. On the other hand, the design of this scheme has taken into account the necessity of aggregating several CFDP blocks as well as the time spent in retransmitting the aggregated CFDP blocks, that can get lost. In practice, the integration of LDGM codes within CFDP follows the implementation adopted in the CLDGM proposal. In particular, the number of encoding packets (k) has been fixed to 1000 in order to avoid the retransmission of too big CFDP aggregated blocks. The deferred issuance of NAK PDUs, on the other hand, conforms the CFDP specification. Code-rate and packet sizes have been varied, during the tests, within the same intervals as CLDGM. This proposal will be referred in the following as *CLDGM-deferred*. For the sake of the completeness, the two proposals have been compared with CFDP working in the following configurations:

- acknowledged mode, with deferred NAK. This scheme is indicated in the performance analysis as *CFDP-deferred*;
- unacknowledged mode, with extended operations. In this case, the *a priori knowledge* of the transmission medium availability help achieve reliable communications without necessity of either data retransmissions or employment of erasure codes. In practice, the transmission of a new block of data is performed whenever the transmission channel is experiencing error-free GAP periods. This solution is actually an "ideal solution" and has been taken into account in order to assess the effectiveness of the other solutions. This scheme is indicated in the following as *CFDP-extended*.

5 The Performance Analysis

5.1 The Testbed

The investigation has been focused on the transfer of data between two remote peers, implementing a full CCSDS stack. For the sake of the analysis, the transfer of 100 Mbytes has been considered. As for the protocol solutions investigated, different configurations in terms of code rate, block size as well as packet size have been considered, as mentioned in Section 4.2. In practice, tests with CLDGM have been carried out by varying the code rate of LDGM codes and the size of encoded packets. In fact, code rate has been ranging from 0.125 up to 0.875, while packet size from 128 up to 1024 bytes. Tuning properly these parameters is a necessary step to achieve satisfactory performance results, because on the one hand low values of code rate and packet sizes make the communication more robust even against strong signal degradations. On the other hand, a reduced number of redundancy packets along with bigger packets allows filling the satellite link bandwidth. In general performance also should depend on the CLDGM block size, because setting a fixed amount of data to be transmitted implies an increasing overhead once the block size is reduced. In our tests, however, CLDGM blocks have been merged together in order to form a unique large bit vector of 1 Mbytes to allow efficient LDGM encoding. For CFDP-deferred, only the packet and CFDP block size impact on the performance. In this perspective, packet size has been varied between 128 and 1024 bytes, while CFDP block size from 1024 up to 65536 bytes. It is straightforward that transmitting large CFDP blocks in correspondence of short packets is highly inefficient because of the increased overhead. On the other hand, when larger packets are assumed, the overhead is lower but the impact of link errors may affect seriously the performance. Finally, CLDG-deferred have been tested by considering the same parameters as CLDGM.

The tests are accomplished through a simulation tool designed for the aim. A number of runs sufficient to obtain a width of the confidence interval less than 1% of the measured values for 95% of the cases has been imposed.

As far as the deep space transmission medium is concerned, the forward-link bandwidth is set to 1 Mbit/s, while the reverse link has availability for 1 Kbit/s. The propagation delays in the reverse and forward directions are equal and ranging from 0.250s to 200s for each experiment. The states within the DTMC model assume BER values equal to 10^{-8}, 10^{-6}, 10^{-4} and 10^{-2}, for states 0, 1, 2, and 3, respectively (Fig. 2). Moreover, the steady state probability π_2 and π_3 has been fixed as well as the average permanence times τ_0 and τ_3 within states 0 and 3 in order to evaluate the effectiveness of the proposals. In particular four case studies have been identified, in dependence of τ_0 and τ_3 values, in order to show the different impact of bursty losses on the communication reliability:

- Scenario 1: τ_0 = 20s and τ_3 = 5s
- Scenario 2: τ_0 = 60s and τ_3 = 5s
- Scenario 3: τ_0 = 5s and τ_3 = 20s
- Scenario 4: τ_0 = 5s and τ_3 = 60s

From this picture it is immediate to see that higher values of τ_0 imply longer sequences of correct packets and hence more reliable data communications; on the other hand, higher values of τ_3 determine several losses because in state 3, a BER of 10^{-2} is set.

5.2 The Metrics

The probability of missing a CFDP block, indicated as Loss Probability (P_{loss}) and defined as one minus the ratio among the transmitted and received blocks, is the performance metric together with the real use of the channel, indicated as Effective Throughput. The latter is measured as the product of ($1 - P_{loss}$) and the ratio of the Transfer Size and the Transfer Time evaluated as the time elapsed from the transmission of the first bit and the reception of the last one. Transfer Size is measured in [bit], Transfer Time in [s] and Bandwidth in [bit/s].

In facts:

$$P_{loss} = 1 - \frac{\text{Received Blocks}}{\text{Transmitted Blocks}}$$

$$\text{Effective Throughput} = (1 - P_{loss}) \cdot \frac{\text{Transfer Size}}{\text{Transfer Time}} \cdot \frac{1}{\text{Bandwidth}}$$

In order to characterize the different performance constraints of the traffic transported through CFDP blocks, five classes of service have been introduced, A, B, C, D, E, presenting different constraints in terms of the maximum P_{loss} and Transfer Time acceptable. Actually, three thresholds for P_{loss}, namely P_{loss_1}, P_{loss_2}, P_{loss_3}, and equal to 0.025, 0.05 and 0.15, respectively, are chosen. As regards the constraints on

Table 1. Classes of service and related performance constraints.

Class of service	Delivery time	Loss probability
A: spacecraft location data and classes of telemetry data updates.	$< T_1 = 2\ T_{min}$	$< P_{loss\_2}$
B: critical instrument status notification or urgent remote control commands.	$< T_1 = 2\ T_{min}$	$< P_{loss\_3}$
C: measurements, planet's surface images.	$< T_2 = 4\ T_{min}$	$< P_{loss\_3}$
D: periodic notifications bulks of data sent on a best-effort basis.	$< T_2 = 4\ T_{min}$	$< P_{loss\_2}$
E: other file transfers.	any	$< P_{loss\_1}$

the Transfer Time, taking as reference the minimum time, T_{min}, required to accomplish the whole transfer of data (equal to the ratio between the Transfer Size and the Bandwidth, plus twice the propagation delay, required to complete successfully the data transfer by means of acknowledgment issuance), two thresholds T_1 and T_2 have been set. The whole classification is shown in Table 1.

5.3 The Results

Scenario 1 (τ_0 = 20s, τ_3 = 5s). In this configuration, since the average time spent in state 0 is much longer than state 3, the error gaps have a moderate length. Consequently Loss Probability requirement has no great impact on all the tests. In particular, since loss probabilities less or equal to 0.05 are experienced for classes A and B as well as C and D, the investigation (reported in Fig. 3) is addressed to the performance exhibited by classes A, D and E.

In general, it is possible to see that CFDP-extended, which represents an ideal protocol solution, outperforms the other proposals because of its capabilities of transmitting data when the channel is reliable. As far as class A is concerned, CLDGM achieves the best performance results independently of the propagation delays. In fact, the Effective Throughput measured for CLDGM ranges from 0.85 to 0.45 as propagation delays vary from 0.25s to 200s. CLDGM-deferred too performs efficiently in the case of delays ranging from 0.25s to 50s, achieving performance results very close to CLDGM one: from 0.85 down to 0.72, while CLDGM gives 0.78 for 50s. For larger delays, the recovery phase is longer and, consequently, Effective Throughput drops to 0.45. Finally, CFDP-deferred shows poor performance, and, in particular, in the case of 100s and 200s, it is unable to match the performance requirements. As far as class D is concerned, CLDGM provides the best performance results from 0.855 to 0.69 as delay varies from 0.25 to 100s. Once the delay raises up to 200s, CLDGM-deferred gives more satisfactory results because combining erasure codes and retransmission procedures help reduce the loss recovery operations. In fact, Effective Throughput achieves 0.52 for 200s, while CLDGM does not come over 0.45.

Finally, with class E, given the total relaxation of the delay constraint combined with the severe loss constraint and with the limited length of the error gaps, CLDGM and CLDGM-deferred present the best results, ranging from 0.85 to 0.45, and from 0.85 to 0.52, respectively.

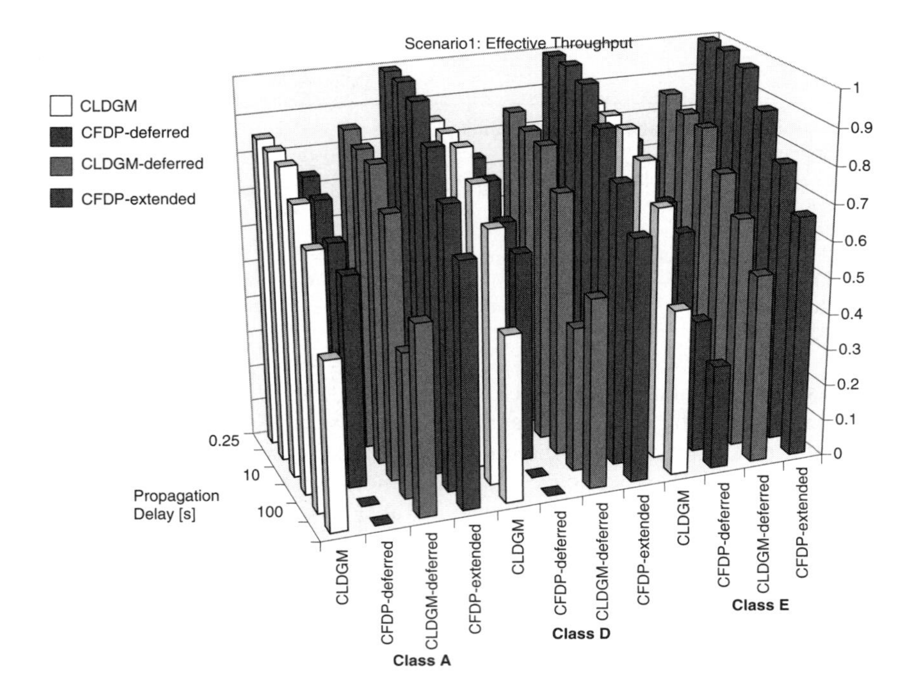

Fig. 3. Scenario 1: the overall performance.

Scenario 2 (τ_0 = 60s, τ_3 = 5s). In this study case, the effect of link errors on the overall performance is even more limited since the mean time spent in state 0 is much longer than in state 3. The discussion of the results, shown in Fig. 4, can be limited to classes A, D and E since identical performance results are offered by classes A and B, and C and D, respectively. In facts, CFDP-extended guarantees the highest effective-throughput values (for each class), ranging from 0.988 to 0.673 as the propagation delay is varied from 0.25s to 200s. As for class A, all the solutions present very similar results. In particular, CLDGM-deferred gives the best performance results, ranging from 0.88 to 0.55, as delay varies from 0.25s to 200s. As far as class D is concerned, CLDGM-deferred and CLDGM confirm performance observed for class A. In fact, they show results ranging from 0.865 to 0.381 and from 0.868 to 0.464, respectively. Finally, as in scenario 1, CLDGM and CLDGM-deferred provide the best results for class E, ranging from 0.86 to 0.46 and from 0.88 to 0.55 (for delay varying from 0.25s to 200s), respectively.

Scenario 3 (τ_0 = 5s, τ_3 = 20s). The longer permanence in state 3, if compared to state 0, implies an increased length of error gaps and hence less effective results are expected. For the sake of simplicity, the investigation does not take class D into account, since it does not add further information with respect to class C evaluation. In practice, apart from CFDP-extended that exhibits the most satisfactory results (effective throughput of 0.90–0.602), the other three solutions present performance results strictly dependent of the service classes, as shown in Fig. 5. In more detail,

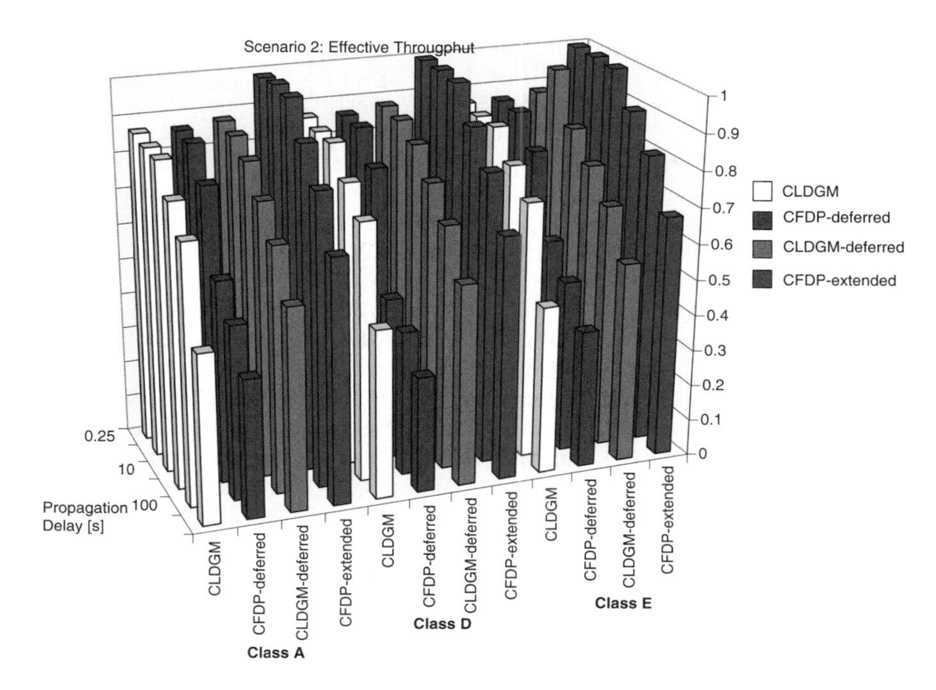

Fig. 4. Scenario 2: the overall performance.

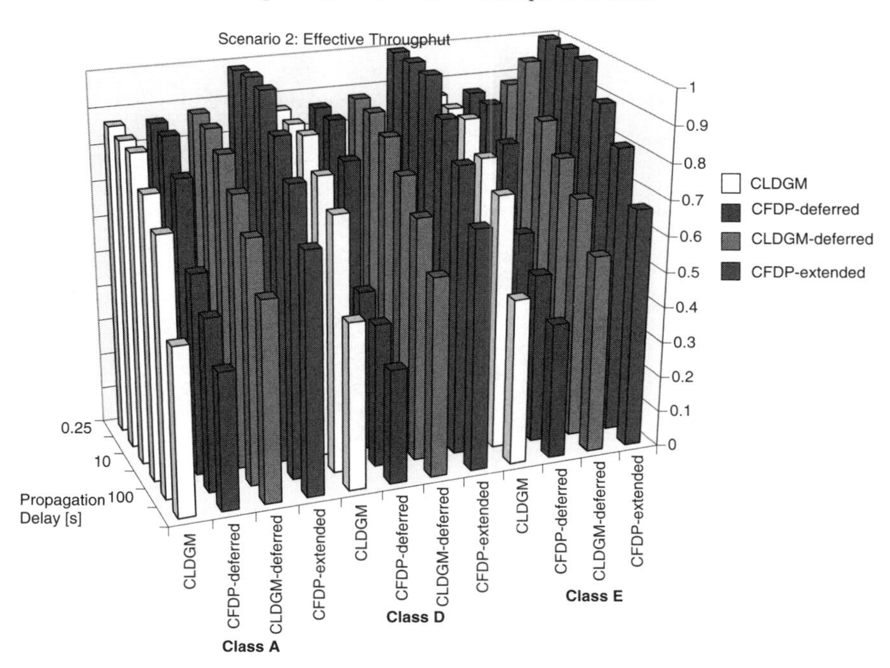

Fig. 5. Scenario 3: the overall performance.

for class A, CLDGM-deferred achieves the best performance results, varying from 0.69 to 0.38. As for class B, CLDGM-deferred again provides satisfactory results (0.68–0.59) as delay ranges from 0.25s to 50s. Once delay further increases, CLDGM performs better, giving Effective Throughput of 0.56 and 0.37 for 100s and 200s, respectively. Class C results show that CLDGM-deferred and CFDP-deferred prove to be powerful, achieving performance ranging from 0.69 to 0.38 in both cases. Finally, the strict constraints on Loss Probability of class E can be efficiently matched by CLDGM-deferred, achieving performance from 0.69 to 0.38.

Scenario 4 (τ_0 = 5s, τ_3 = 60s). As the permanence in state 3 gets longer (60s), the length of error gaps increases accordingly, severely affecting the global performance. In this case, class A, C, D, and E provides meaningful results, while results for class B are aligned with class A ones. In particular, from Fig. 6, it is possible to realize that, apart from CFDP-extended that behaves almost ideally, CFDP-deferred and CLDGM-deferred completely outperform CLDGM, since the long error runs cannot be only counteracted by erasure code application. In more detail, as far as class A is concerned, CFDP-deferred is the most promising for delays lower than 200s (Effective Throughput: 0.86–0.42). In the case of 200s, CLDGM-deferred is more performing, achieving 0.4. As for classes C and D, again CFDP-deferred ensures high performance results ranging from 0.86 to 0.25 in both cases. It is worth noting that CLDGM is unable to match class D performance constraints because of the strict constraints on Loss Probability. Finally, for class E, CFDP-deferred is efficient when

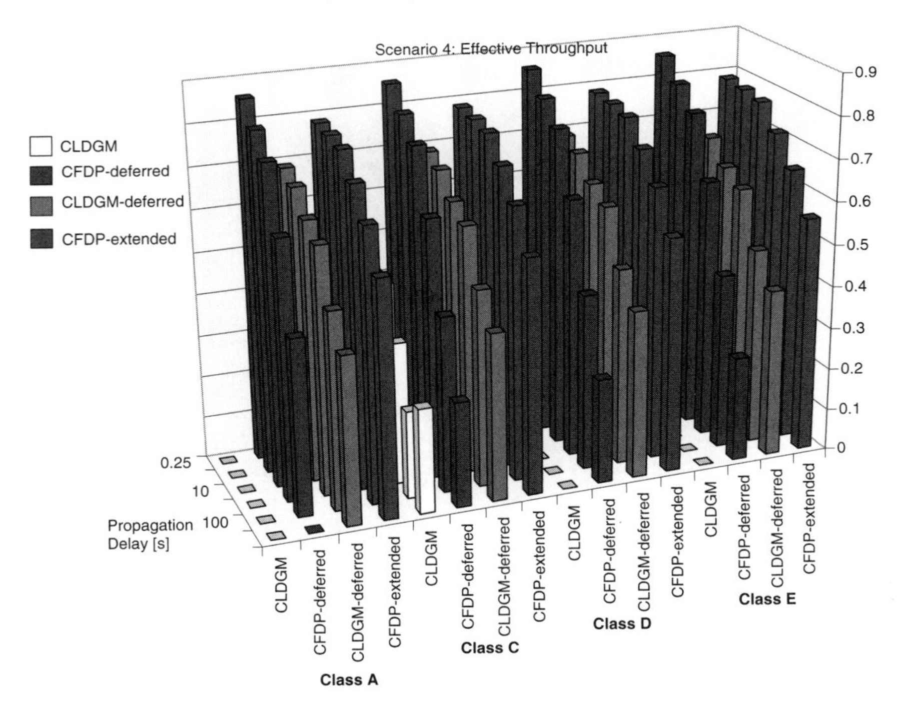

Fig. 6. Scenario 4: the overall performance.

the delay is lower or equal to 50s, achieving performance ranging from 0.86 to 0.61. Otherwise, in presence of larger delays, CLDGM-deferred offers better performance values, corresponding to 0.47 and 0.4 for delay of 100s and 200s, respectively.

5.4 Comparison

In order to assess more deeply the performance and the effectiveness provided by the investigated results, CFDP-deferred, CLDGM and CLDGM-deferred are compared with CFDP-extended, by introducing the Efficiency (%), expressed as the ratio between the effective throughput achieved by the above solutions (indicated as CFDP variants) and CFDP-extended one:

$$\textit{Efficiency}\,(\%) = \frac{\text{Effective Throughput (CFDP variants)}}{\text{Effective Throughput (CFDP} - \text{extended)}} \cdot 100$$

The analysis of classes A, D, and E, when the propagation delay is 100s, follows.

From Fig. 7, it is possible to see that when class A constraints have to be satisfied, CLDGM-deferred offers the best efficiency results when strong impairments (scenarios 4 and 3) are introduced by the channel. Actually, the combined use of erasure codes and retransmissions allows achieving the best performance results, corresponding to 72.30% and 62.87% for scenarios 4 and 3, respectively. On the other hand, when minor losses are exhibited, it is CLDGM that offers the most satisfactory efficiency results, equal to 88.5% and 90.06% for scenarios 2 and 1, respectively. As for

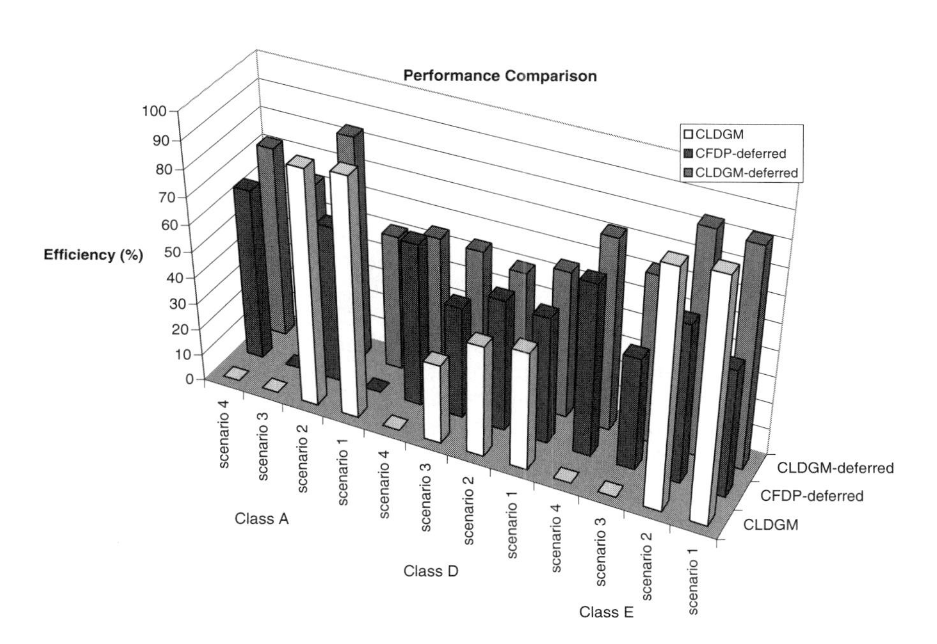

Fig. 7. Performance comparison (Efficiency %).

class D and E, which require loss probabilities lower than 0.05 and 0.025 respectively, CLDGM-deferred always behaves better than CFDP-deferred, presenting efficiency values ranging from 84.05% to 55.30% and from 82.58% to 52%, respectively. Finally, in scenarios 1 and 2, when the class E requirements have to be satisfied, CLDGM-deferred performance is very close to CLDGM. In facts, CLDGM-deferred achieves results from 82.58% to 84.05% for scenarios 1 and 2 respectively, while CLDGM achieves 90.0% and 88.5%.

6 Conclusions

This work has been devoted to the design of novel protocol solutions, based on the CCSDS File Delivery Protocol (CFDP), to achieve data communications over long-delay networks. Two proposals CLDGM and CLDGM-deferred have been introduced in this work and deeply investigated with respect to CFDP-deferred and CFDP-extended. The performance analysis, carried out for different scenario configurations, has identified CLDGM together with CLDGM-deferred as promising solutions, able to match the specific constraints of five classes of service. In particular, CLDGM, thanks to the powerful LDGM erasure codes, offers very satisfactory results in scenarios 1 and 2, where moderate losses are experienced. CLDGM-deferred, in these cases, is less efficient even if its behaviour is very satisfying. On the other hand, the adoption of CLDGM-deferred is "mandatory" when "almost reliable" data communications have to be carried out in very hazardous conditions, such as in scenarios 3 and 4. It allows considering CLDGM-deferred as an efficient solution, whose application is very wide.

References

[1] D. C. Lee, W. B. Baek, "Expected file-delivery time of deferred NAK ARQ in CCSDS file-delivery protocol," IEEE Transactions on Communications, vol. 52, no. 8, Aug 2004, pp. 1408–1416

[2] H. Ernst, L. Sartorello, S. Scalise, "Transport layer coding for the land mobile satellite channel," 59th IEEE Vehicular Technology Conference (VTC'04), Milan, Italy, May 2004, pp. 2916–2920

[3] S. Burleigh, A. Hooke, L. Torgerson, K. Fall, V. Cerf, B. Durst, K. Scott, H. Weiss, "Delay-tolerant networking: An approach to interplanetary internet," IEEE Communications Magazine, vol. 41, no. 6, Jun 2003, pp. 128–136

[4] V. Roca, "Design, Evaluation and Comparison of Four Large Block FEC Codecs, LDPC, LDGM Staircase and LDGM Triangle, plus a Reed Solomon Small Block FEC Codec," INRIA, Rapport de Recherche, June 2004, http://www.inrialpes.fr/planete

[5] I. F. Akyildiz, Ö. B. Akan, C. Chen, J. Fang, W. Su, "The state of the art in interplanetary internet", IEEE Communications Magazine, vol. 42, no. 7, Jul 2004, pp. 108–118

[6] T. de Cola, H. Ernst, M. Marchese, "Joint Application of CCSDS File Delivery Protocol and Erasure Coding Schemes over Space Communications," IEEE International Conference on Communications (ICC 2006), Instanbul, Turkey, June 2006

Interconnection of Laboratory Equipment via Satellite and Space Links: Investigating the Performance of Software Platforms for the Management of Measurement Instrumentation

Luca Berruti, Franco Davoli, Stefano Vignola, Sandro Zappatore

National Inter-university Consortium for Telecommunications (CNIT) DIST-University of Genoa Research Unit,– Via Opera Pia 13 – 16145 Genova - Italy
Phone: +390103532990, Fax: +3935322154, e-mail: luca.berruti@cnit.it, franco.davoli@cnit.it, stefano.vignola@cnit.it, sandro.zappatore@cnit.it

Abstract. In the recent years, systems devoted to access remote laboratories through a networking infrastructure have been actively studied, and a number of specific software platforms, able to manage the instrumentation, have been proposed and implemented. Nevertheless, the problems involved in controlling remote devices on board satellite laboratories or in accessing remote equipment via satellite links are still not sufficiently investigated.

The paper briefly introduces the architecture developed within the Labnet project, especially as concerns its core software component, whose aim is providing unified access to heterogeneous equipment for a multiplicity of users.

Since the use of an earth-to-space link can possible affect the overall performance of systems based on a TCP-IP suite, a number of tests has been carried out, in order to evaluate the effectiveness and robustness of the proposed solution. Furthermore, the results regarding the performance of the Labnet platform are discussed and compared with those achieved by exploiting the facilities offered by a commercial and very popular software package.

1 Introduction

The recent years have seen an increasing use of satellite networks, especially those based on geo-stationary spacecrafts, owing to the offer of low cost terminals, and to new bandwidth availability in Ka band. The commercial exploitation of the Ka satellite band fostered the development of new packet-based multimedia applications, and the porting of a number of functionalities, initially devised for terrestrial networks. Through the last five years, the Italian National Consortium for Telecommunications (CNIT) has undertaken several projects, aimed at deploying a satellite network [1], as well as at providing a platform for remote laboratory management [2, 3] and at performing distance learning activities in higher education [4]. Still on the satellite side, CNIT has been actively involved in SatNEx, a European Network of Excellence (NoE) in satellite communications [5, 6], where research integration, training and dissemination activities are carried out, involving, among other tools, the use of satellite platforms. Recently, the new NoE SatNEx II has been launched, a follow-up of the previous one, which will also include experimental activities over satellite networks.

Although distance learning and training can fruitfully exploit services of a remote measurement platform, the possibility of remotely driving experiments represents an issue of great interest and relevance for many sectors, ranging from medicine and biology to mechanics and telecommunications. For instance, one can imagine a biological analysis test bench on-board a space laboratory, or a set of bio-medical devices mounted on mobile first aid and rescue units, to be used in case of earthquakes and hydro-geological disasters. In these scenarios, tele-measurement platforms can provide researchers on the earth a means to directly investigate phenomena in the absence of gravity, and medical staff a sophisticated infrastructure to carry out diagnoses, as well as environmental biotic and chemical analysis. The former case obviously requires an earth-to-space communication link, while in the latter a satellite link often represents the only available and reliable channel to connect the mobile units to an operative center.

Programmable equipment, remotely controllable devices, and fast/efficient interfaces constitute only the basis of a remote measurement platform. In general, one has to deal with issues concerning the possible heterogeneity of the application environments and of the instruments, the resource sharing, the efficient storage and transmission of data captured and of controls, needed for properly setting all the elements included in a remote laboratory. Hence, the software portion of such a platform plays a very important and critical role, as it must provide data and service abstraction so that end users can effectively receive data collected by the instruments, and send commands and possibly other data to them, thus preserving the realistic aspects of the experiment. In the literature, several architectures have been proposed to remotely control and manage measurement instrumentation [7–12].

As concerns commercial products, the Labview® suite by National Instruments represents one of the most widely employed software packages to remotely pilot instrumentation. In spite of its simplicity of use, no source code is available, and some strategies and communication protocols are partially unknown, limiting the implementation of new functionalities, the software portability and scalability.

Although a wide variety of architectures were proposed, few appear to be oriented to remote measurements for a distance learning experimental environment [13–15] and, for the most part, no support (e.g., multicast) is provided for an efficient and simultaneous dissemination of the measured data to a potentially large group of users. Furthermore, the application scenarios commonly involve the use of local area networks or terrestrial communication channels. The aspects concerning the exploitation of satellite links for accessing remote laboratories and measurement equipment are not sufficiently well focused and considered. The present paper addresses these problems and provides some experimental investigation.

The paper is organized as follows. Section 2 describes an ad-hoc designed software platform, which was originally developed within the LABNET project [3]. In the third Section, a number of tests, aimed at evaluating the performance of the system, are shown and the related results are discussed. The goal is to highlight the effects of a satellite link on the performance. Furthermore, a comparison between the behaviour of the proposed platform and a commercial one is presented, in order to prove the effectiveness of the solution devised. Finally, in the last Section, the conclusions are drawn.

2 LABNET Server Architecture

An efficient tele-laboratory system requires a Supervising Central Unit (SCU), whose task is to control and monitor the instrumentation involved in any experiment. The SCU decouples the user(s) (and the user software tools) from any issue related to the commands/controls and the communication protocols specific of each equipment. In other words, the SCU plays the role of an "interface" between the *inner* laboratory space (i.e., "the domain" of the real instruments and of their specific rules and protocols) and the *outer* laboratory space (i.e., the user space, the "domain" of abstracted instruments and standardized protocols).

Within the LABNET project, an ad-hoc SCU, called LabNet-Server (LNS), was proposed and its development started. Through the last four years, the Labnet Tele-Laboratory Architecture has been continuously evolving, and a number of versions of the LNS were released, in order to better meet the new requirements of the laboratory system. The current LNS is able to address several crucial concerns, such as: i) the intrinsic heterogeneity of the application environments and of the instruments, ii) the software portability and scalability, iii) the level of flexibility, and iv) the capability of efficiently exploiting IP-based satellite channels, as well as of multicasting the data gathered from the instrumentation for an efficient use of the transmission resources. All these aspects, although quite relevant, are not sufficiently well focused, and often neglected, in some products available on the market. Figure 1 illustrates the LNS architecture and shows its main components.

Owing to the LNS, the facilities of the remote laboratory can be exploited by means of any common Internet browser, which communicates with the LNS. The latter maintains a real-time database, which grants access to data gathered from the

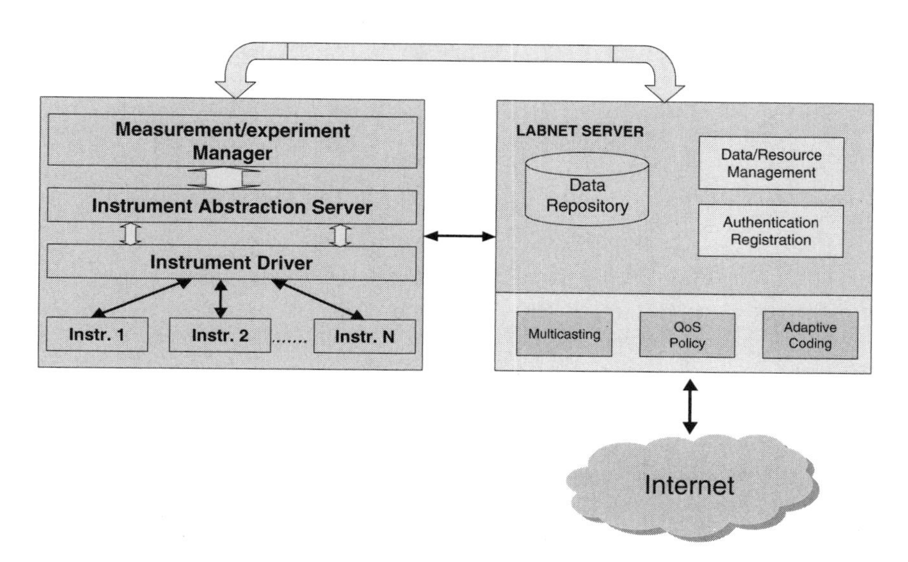

Fig. 1. Overall software architecture of the LABNET server.

instrumentation. In order to access the laboratory, a user must contact the Labnet Web-site and choose an experiment. In turn, the Web server uploads the client with a proper archive of Java applets and the LNS starts an experiment session, setting-up any kind of actions required for the correct execution of all the measurements.

The Java applets carry out the actual communication between the LNS and the user stations. Data exchanged can be roughly divided into two groups: the first one consists of commands toward the server, and then to the instrumentation on the field; the second one is related to data gathered by the LNS from the instruments and addressed to the clients. In order to assure a good level of interaction, the LNS can adopt the most suitable data coding, QoS strategy and, if needed, it can enable multicast transmission to save bandwidth. In this manner, the LNS hides the languages/environments specific to the laboratory test bench from the final user. It is just a task of the LNS to communicate with the proper experiment manager that actually operates the data exchange, by exploiting the services offered by the Instrument Abstraction Server (IAS). The IAS eventually communicates with and controls the physical devices through a set of drivers.

Finally, a set of ancillary modules completes the LNS in order to i) grant access only to registered users, who possess the proper access rights; ii) schedule the resources among competing users.

3 Performance Evaluation

A significant number of tests has been performed on the LNS with a number of clients, connected via a real satellite link. The tests were aimed at i) evaluating the efficiency of the LNS in terms of the possible delay and jitter of data packets observed at the receiver end, and ii) comparing the effectiveness of the proposed software platform with the "Data Socket Server (DSS)", a component of the Labview® package. The DSS, similarly to the LNS, stores and publishes data gathered from instrumentation for any possible use of the client stations. The choice of the DSS is motivated by the fact that Labview is a well-known package, very popular in the fields of tele-measurements and instrumentation remote control.

The experimental set-up, involving the satellite link, is depicted in Fig. 2.

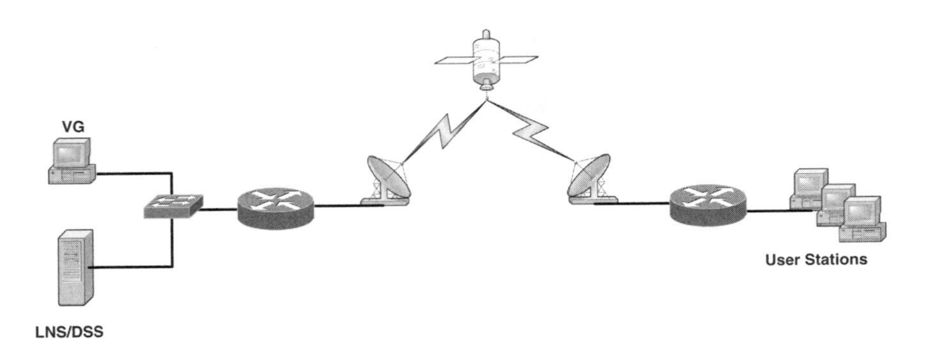

Fig. 2. The experimental set-up.

The "variable generator" (VG) plays the role of an experiment manager: It (quasi-) synchronously produces a set of data packets, conveying a group of 60 variables (52 of them scalars of 4 bytes each, 8 one-dimensional arrays of 1024 bytes each). When the LNS is in use, the VG periodically generates a single packet, containing all the variables. Upon receiving the packet, the LNS decodes it, updates the values of the referenced variables in its repository, and, finally, for each variable, it generates as many data packets as the number of client stations connected (we have chosen not to adopt the multicast option in the LNS for fairness of comparison). Therefore, if v is the number of variables, and the client stations are c, the LNS periodically sends v*c data packets on the satellite link. In the set-up centred on the DSS, the VG periodically updates 60 variables (of the same type and size as in the previous case) in the repository maintained by the DSS. The latter, in turn, notifies the clients the new values of the 60 variables. As the internal working mechanisms of the DSS are undocumented, the number and the role of packets involved in the overall process is not well known.

Since the total net payload (consisting of variables generated at the VG) is the same in both cases, the possible differences in performance can be attributed to the different protocols, data storing, retrieving, and forwarding strategies adopted by the LNS and the DSS. For all our tests, we employed a real satellite link (DVB-RCS like) in Ka band, exploiting the Skyplex processor onboard Hot Bird 6. Although the gross bandwidth amounts to 2 Mbit/s, the net capacity measured at the IP layer never exceeds 1.2 Mbit/s, owing to various overheads. The modems/routers at the earth stations are produced by Viasat, which seems to adopt a FIFO queue at the IP layer, and the number of user stations participating in all the tests is three.

In the set-up centred on the LNS, the operating system was Linux (kernel v. 2.6.11); in the one involving the DSS, Windows XP Pro sp2 was used. All the PCs employed were Fujitsu-Siemens Scenic P300 VKM266, and the switch was a CISCO Catalyst 2900xl.

Specifically, for each experimental set-up described above, the VG quasi-synchronously produces, every D seconds, a set of 60 variables, as already specified. Hence, the total net payload amounts to 8400 bytes. The "variable generation time" D (i.e., the time interval between the generation of two consecutive sets of variables) was set in different experiments at 1000, 500, 350, and 300 ms, respectively. Although the dispatching and publishing mechanisms of the DSS are not well known, the performances of the LNS and DSS can be jointly evaluated by considering the jitter of the "variable" transit time, viz the time a variable waits to be notified to a client station since its generation at the VG.

Let us consider a generic variable within the set periodically produced by the VG. Since the latter is implemented as a task in the user space, the generation time of the variable is itself affected by a certain level of jitter, owing to the intrinsic, non-real-time nature of the operating system. Therefore, the generic instant t_k when the k-th variable is produced can be written as

$$t_k = kD + \varepsilon_{tx} \tag{1}$$

where D is the (theoretic) "variable" time, and ε_{tx} is a random variable expressing the uncertainty about t_k. The algorithm used by the VG assures that t_k has zero mean.

The user station receives the k-th variable at the instant T_k given by

$$T_k = t_k + o + \varepsilon = kD + \varepsilon_{tx} + o + \varepsilon \quad (2)$$

where o is a fixed time offset due to the physical transmission (about 560 ms at our latitudes), and ε is a random variable, expressing the uncertainty related to the time spent in i) the queuing process at the VG, ii) the LNS / DSS to perform their tasks (packet decoding, data archiving, dispatching, . . .), iii) the modem/router queues, iv) the switch processor on-board the satellite, and v) the de-queuing/de-assembling process at the receiver end. It should be remarked that, in the case of the LNS, the first and last mentioned processes are quite tiny (the LNS protocol is based on UDP); thus, ε is essentially due to all the remaining contributions. On the contrary, when the DSS is in use, it is impossible to determine the different time contributions of ε. In both cases, ε represents the total uncertainty due to all the active processes involved in advertising that a variable has changed.

The notification instant of the (k+1)-th variable, T_{k+1}, is given by

$$T_{k+1} = (k + 1)D + \varepsilon'_{tx} + o + \varepsilon' \quad (3)$$

Then, the time interval between the arrivals of two consecutive variables is

$$\Delta T = D + \alpha_{tx} + \beta \quad (4)$$

where $\alpha_{tx} = \varepsilon'_{tx} - \varepsilon_{tx}$ is a zero mean random variable, whose variance σ_α^2 is twice that of ε_{tx}, and $\beta = \varepsilon' - \varepsilon$.

As α_{tx} and β are independent and α_{tx} has zero mean, then

$$E[\beta^2] = E[(\Delta T - D)^2] - E[\alpha_{tx}^2] = E[(\Delta T - D)^2] - \sigma_\alpha^2 \quad (5)$$

Since

$$E[\beta^2] = E[(\varepsilon' - \varepsilon)^2] = E[(\varepsilon')^2] + E[(\varepsilon)^2] - 2E[\varepsilon' \varepsilon] \quad (6)$$

under the assumption that ε' and ε are independent,

$$\begin{aligned} E[\beta^2] &= E[(\varepsilon')^2] + E[(\varepsilon)^2] - 2E[\varepsilon']\,E[\varepsilon] \\ &= 2\left\{E[(\varepsilon)^2] - E[\varepsilon]^2\right\} = 2\sigma_\varepsilon^2 \end{aligned} \quad (7)$$

where σ_ε^2 is the variance of ε.

Combining Eq. (5) with Eq. (7) gives

$$\sigma_\varepsilon^2 = \tfrac{1}{2}\left\{E[(\Delta T - D)^2] - \sigma_\alpha^2\right\} \quad (8)$$

Hence, in a laboratory platform managed by the LNS or DSS, the RMS of the time needed to advertise a user station that a variable at the VG has changed can be estimated by

$$RMS(\varepsilon) = \sigma_\varepsilon = \sqrt{\tfrac{1}{2}\left(E[(\Delta T - D)^2] - \sigma_\alpha^2\right)} \quad (9)$$

In practice, by evaluating the variance of the "variable" time computed over all the user stations, viz $E[(\Delta T - D)^2]$, and by computing the variance of the "variable" time at the VG, viz σ_α^2, it is possible to estimate the standard deviation of ε, that is the root mean square of the time the variable needs, "passing-through" the LNS or the DSS, to be notified, via a satellite link, to a user station. Finally, the overall variance of the time a general variable needs to reach a user station since its generation at the VG is computed by averaging the RMS (ε) over all the variables involved in the experiment.

Besides the experimental set-up previously mentioned (see Fig. 2), other two quite similar set-ups have been exploited, in order to carry out altogether three groups of tests, for both the LNS and the DSS. The first group is centred on the set-up sketched in Fig. 2 and includes the actual satellite link. In the second group of tests, the client stations are connected to the LNS (or DSS) via a terrestrial link, whose bandwidth (1.2 Mbit/s) amounts to the average bandwidth available on the satellite link. In this case, the experimental set-up includes two CISCO routers, back-to-back connected by means of two synchronous serial ports. In order to operate under conditions as similar as possible to those of the previous case, the queues of the router interfaces were set to FIFO. Eventually, in the last group of tests, the client stations are directly connected to the LNS (or DSS) by means a high speed (100 Mbit/s) LAN: no routers and satellite links are employed. In this manner, as the latencies owing to the communication channels can be neglected, we simply evaluate the jitter of the transit time inserted by the LNS (or DSS). In other words, ε used in equation (2), here represents a random variable, expressing the uncertainty related to the time spent by the LNS (or DSS) to perform their tasks.

The results achieved by the tests are summarized in Tables 1 and 2. The former shows data related to the LNS; the latter reports data obtained with the DSS. The columns labelled RMS refer to the Root Mean Square of the delay jitter (i.e., the difference between the expected and the actual variable transit time, viz the time a variable needs to reach a client since its arrival to the LNS). For every test, the values have been calculated by averaging the RMS of the delay jitter over all the variables involved in the experiment and, for each variable, over 10 repeated sets of 1000 transmissions.

Table 1. Estimated packet loss and RMS of data packet transit time vs packet time (i.e., the time interval between two successive packet departures at the VG) for a LAN, a terrestrial and satellite link, respectively, when the LNS is used to access the laboratory.

	LAN		Terrestrial		Satellite	
Variable Time [ms]	Loss [%]	RMS [μs]	Loss [%]	RMS [μs]	Loss [%]	RMS [μs]
1000	0	70 ± 2	0	139 ± 71	0	16615 ± 70
500	0	72 ±3	0	170 ± 78	0	15980 ±240
350	0	75 ± 4	0	258 ± 90	0	11030 ± 530
300	0	71 ± 5	1.6	14141 ± 78	2.3	9120 ± 212

Table 2. Estimated packet loss and RMS of data packet transit time vs packet time (i.e., the time interval between two successive packet departures at the VG) for a LAN, a terrestrial and satellite link, respectively, when the DSS is used to access the laboratory.

	LAN		Terrestrial		Satellite	
Variable Time [ms]	Loss [%]	RMS [μs]	Loss [%]	RMS [μs]	Loss [%]	RMS [μs]
1000	0	16750	0.2	103000	60	–
500	0	14500	25	189000	82	–
350	1.3	24500	60	–	96	–

The Tables report the results organized according to the group of tests the data belong to. Specifically, the group named "Satellite" refers to the tests involving the actual satellite link; the group named "Terrestrial" refers to the second group of tests above described, which permit to investigate the behaviour of the LNS (or DSS) when a terrestrial link (1,2 Mbit/s) is in use; finally, the group named "LAN" is related to tests performed on a high speed LAN. Some comments regarding the Tables are in order. As regards Table 1, the number of repetitions assures that the range specified for each value is characterized by a confidence level of 99%. The RMS values reported in Table 2 have a tolerance of 20% with a confidence level of 95%. Furthermore, whenever the variable losses exceeded 30%, we have preferred to omit the corresponding RMS for two reasons: i) owing to the significant variable losses, there are too few data in order to compute a stable and reliable RMS value; ii) such high values of loss are often associated to instabilities in the processes controlling the DSS: indeed, the DSS crashed during several tests.

The results highlight that the LNS performance is almost the same in the case of LAN and a terrestrial link, while a satellite link yields higher RMS values. Although the RMS values in this latter case are significantly higher than those measured in the other cases, the RMS values never exceed 3% of the variable time. Moreover, no loss is present for variable times of 1000, 500, 350 ms. The losses at 300 ms, both in the case of terrestrial and satellite link, are due to the queue length, inadequate to completely allocate room for the data bursts associated to the transmission of packets from the LNS.

On the contrary, the performance of the DSS, especially as concerns the packet loss, dramatically decreases when a satellite link is in use. Comparing the columns of Table 2 related to the terrestrial and satellite links, highlights how the propagation delay, inherent to the satellite link, strongly affects the overall performance of a tele-measurement platform centred on the DSS. Furthermore, the DSS appears unable to manage bursts of variables, whose inter-arrival times are less than 350 ms.

There are a number of reasons for the different behaviour of the LNS and the DSS, and it is not simple to motivate them. The DSS uses TCP as a transport protocol, whose performance may be negatively affected by the presence of a large bandwidth-delay product, whereas the LNS relies on UDP. Moreover, the mechanism of bandwidth allocation, which controls all the satellite modems/routers, seems to further reduce the efficiency of TCP. It is quite difficult to motivate the behaviour of the DSS in the absence of information regarding its internal structure. Therefore, the DSS appears more suitable to manage asynchronous controls and data within networks characterized by high bit-rates.

4 Conclusions

The paper has presented a possible extension of the tele-laboratory system designed within the Labnet and related projects. Specifically, the attention has been addressed to evaluate how an earth-to-space link can affect the overall efficiency of the supervising central unit, the software component that plays an important role in the entire system. To this aim, a number of tests have been carried out, whose results prove the effectiveness of the proposed solution. Furthermore, a comparison with a very popular commercial software package has highlighted that the devised platform appears more suitable to be exploited in all those contexts that include communication channels characterized by high delay-bandwidth products. Further work is in order to investigate whether some performance improvements might be achieved, by implementing the software within a multi-thread environment.

Acknowledgements

This work was partially supported by the Italian Ministry of Education, University and Research (MIUR) within the PRIN-CRIMSON project, and by the European Commission under the SatNEx Network of Excellence.

References

[1] F. Davoli, G. Nicolai, L. S. Ronga, S. Vignola, S. Zappatore, A. Zinicola, "A Ka/Ku Band Integrated Satellite Network Platform for Experimental Measurements and Services: the CNIT Experience", *Proc. 11th Ka and Broadband Commun. Conf.*, Rome, Italy, Sept. 2005, pp. 715–723.

[2] F. Davoli, S. Vignola, S. Zappatore, "A Multimedia Laboratory Environment for Generalized Remote Measurement and Control", *Proc. 2002 Tyrrhenian Internat. Workshop on Digital Communications (IWDC 2002)*, Capri, Italy, Sept. 2002, pp. 413–418.

[3] F. Davoli, G. Spanò, S. Vignola, S. Zappatore, "LABNET - Towards Remote Laboratories with Unified Access", *IEEE Trans. Instr. Meas.*, Oct. 2006 (to appear).

[4] G. Mazzini, A. Ravaioli, C. Fontana, P. Toppan, O. Andrisano, L. S. Ronga, S. Vignola, "The Teledoc2 Project: a Heterogeneous Infrastructure for International E-Learning", in F. Davoli, S. Palazzo, S. Zappatore, Eds., *Distributed Cooperative Laboratories - Networking, Instrumentation, and Measurements*, Springer, New York, NY, 2006, pp. 475–496.

[5] M. Werner, A. Donner, R. Sheriff, F. Hu, R. Rumeau, H. Brandt, G. Maral, M. Bousquet, B. Evans, G. Corazza, "SatNEx – The European Satellite Communications Network of Excellence", *Proc. 59th IEEE Vehicular Technology Conf. (VTC 2004-Spring)*, Milan, Italy, May 2004, vol. 5, p. 2842.

[6] B. G. Evans, "SatNEx – A European Network of Excellence in Satellite Communication", *Internat. J. Satell. Commun.Network.*, vol. 23, no. 5, p. 263, Sept./Oct. 2005.

[7] F. He, F. Wang, W. Li, X. Han, J. Liu, "Object Request Brokers for Distributed Measurement", *IEEE Comp. Appl. in Power*, vol. 14, no. 1, pp. 50–54, Jan. 2001.

[8] D. Grimaldi, L. Nigro, F. Pupo, "Java-based Distributed Measurement Systems", *IEEE Trans. Instr. Meas.*, vol. 47, no. 1, pp. 100–103, Feb. 1998.

[9] L. Nigro, F. Pupo, "A Modular Approach to Real Time Programming Using Actors and Java", *Proc. 22nd IFAC/IFIP Workshop on Real Time Programming (WRTP-97)*, Lyon, France, Sept. 1997, pp. 83–88.

[10] M. Bertocco, F. Ferraris, C. Offelli, M. Parvis, "A Client-Server Architecture for Distributed Measurement Systems", *IEEE Trans. Instr. Meas.*, vol. 47, no. 5, pp. 1143–1148, Oct. 1998.

[11] J. W. Overstreet, A. Tzes, "An Internet-Based Real-Time Control Engineering Laboratory", *IEEE Contr. Syst. Mag.*, vol.19, no. 5, pp. 19–34, Oct. 1999.

[12] S. You, T. Wang, R. Eagleson, C. Meng, Q. Zhan, "A Low Cost Internet-Based Telerobotic System for Access to Remote Laboratories", *Artificial Intelligence in Engineering*, vol. 15, no. 3, pp. 265–279, July 2001.

[13] A. Ferrero, S. Salicone, C. Bonora, M. Parmigiani, "ReMLab: A Java-Based Remote, Didactic Measurement Laboratory", *IEEE Trans. Instr. Meas.*, vol. 52, no. 3, pp. 710–715, June 2003.

[14] X. F. Yuan, J. G. Teng, "Interactive Web-Based Package for Computer-Aided Learning of Structural Behavior", *Comp. Appl. in Engineering Education*, vol. 10, no. 2, pp. 79–87, Oct. 2002.

[15] T. Murphy, V. G. Gomes, J. A. Romagnoli, "Facilitating Process Control Teaching and Learning in a Virtual Laboratory Environment", *Comp. Appl. in Engineering Education*, vol. 10, no. 3, pp. 121–136, Jan. 2003.

A Common Representation of QoS Levels for Resource Allocation in Hybrid Satellite/Terrestrial Networks

Laura Rosati[1], Gianluca Reali[2]

[1]German Aerospace Center (DLR), Institute of Communications and Navigation
P.O. Box 1116, Oberpfaffenhofen, Germany
Phone: +49 8153 282801, Fax: +49 8153 281871, e-mail: laura.rosati@dlr.de
[2]University of Perugia, Department of Electronic and Information Engineering (DIEI)
via G. Duranti 93, 06125 Perugia, Italy
Phone: +39 075 5853651, Fax: +39 075 5853654, e-mail: gianluca.reali@diei.unipg.it

Abstract. The future hybrid satellite/terrestrial networks are challenged to provide different end-to-end Quality of Service (QoS) classes to the users. In particular, it is expected that QoS routing procedures are implemented, in order to decides the best route to the destination and optimize resource distribution through the path. This fact arises the question of a common representation of the QoS levels such that they can be compared and combined. It was shown in the literature how they can be mapped into a single parameter, the so-called virtual delay. We present and investigate the computational complexity of two approaches to distribute the virtual delay over the communication path (i.e., allocating the resources) in order to guarantee a QoS performance requested by the user minimizing the cost of the provided service. The former consists of a joint routing/resource allocation optimization. The latter assumes that the selection of the path is correctly performed and models the resource allocation problem as an equality constrained convex minimization problem.

1 Introduction

ITU-T defines Quality of Service (QoS) as "the collective effect of service performance which determine the degree of satisfaction of a user of the service" [1]. The IETF has proposed QoS architectures to provide guaranteed service level to different applications over *terrestrial* networks. These architectures include Integrated Services (IntServ) [2], Differentiated Services (DiffServ) [3] and MultiProtocol Label Switching (MPLS) [4].

Also the future global *satellite* networks will likely use some on-board switching techniques enabling the provision of service level guarantees. In [5] a QoS framework for satellite IP networks including requirements, objectives and mechanisms is described. In [6] different combinations of buffer management policies to be adopted in satellite systems are presented to guaranty the QoS required by the user.

In this paper we investigate the issue in a *hybrid satellite/terrestrial* environment. The research challenges and technology advances needed to accomplish the integrated format are presented in [7]. The space segment is expected to operate in the future in collaboration with the terrestrial component in order to provide a complementary rather than an alternative service. In particular it is expected that the

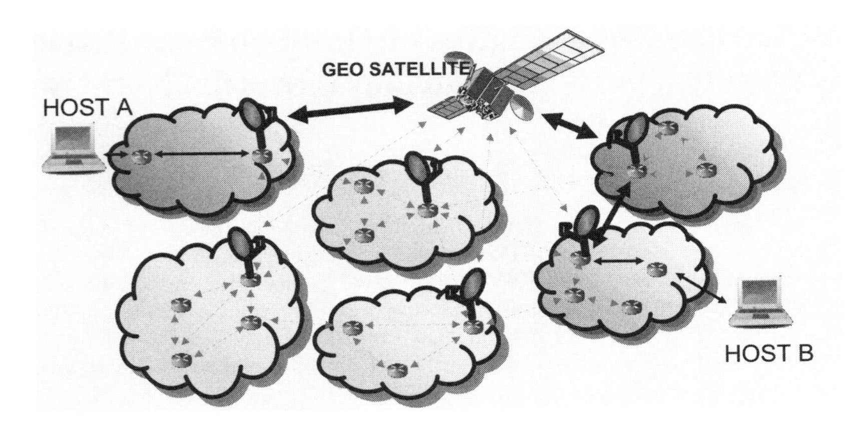

Fig. 1. Hybrid satellite/terrestrial scenario.

satellite will be no longer seen as a component of an alternative routing path but as a part of unique (really integrated) system.

We assume that the system accepts user's requests made in terms of QoS parameters. To achieve a seamless integration, each component of the heterogeneous network is challenged to be "end-to-end QoS-aware". In particular, it is expected that QoS routing procedures are implemented, in order to define for each communication flow the best source-destination path (as depicted in Fig. 1), verify if the route has sufficient resources the QoS requested by the user and optimize resource allocation through the path. In particular, since radio resources are costly and scarcely available and allowance has to be made for housekeeping procedures, the optimization of link layer is of paramount importance.

In this paper we assume that the global network is composed of administratively independent domains, which can be either a terrestrial or a satellite component. Each domain implements admission control and resource allocation functions to provide different levels of IP QoS. In order to accomplish this task, for instance, the satellite can use Bandwidth on Demand (BoD) techniques, allocating an amount of bandwidth to the user on the basis of the QoS requested by the user itself. This approach can be combined with non-uniform *traffic based* bandwidth allocation among the beams constituting the satellite coverage region (for instance beam hopping techniques [10]) in order to enhance the performance of the system in terms of bandwidth efficiency.

The provision of QoS-oriented services in hybrid networks arises the question of a common representation of the QoS levels such that they can be compared and combined. The problem of the mapping of the quality of the network support into a single parameter was faced in [14, 15, 16].

The QoS parameters can be classified as service parameters and network parameters. The service parameters are those that can be defined at call level, and may be negotiated between users and network. Typical service parameters are the transfer delay, the delay jitter, and the packet loss probability. All other parameters that influence the QoS are called network parameters. For instance the network resilience, which characterizes the intrinsic quality of the network, is not specific for individual flows, and is a typical network parameter. The service parameters were used to define the virtual delay d.

An intuitive, although partial, rationale of this model comes from the observation that both delay jitter and loss probability may be traded with delay. Indeed, the delay jitter could be reduced by using a playout buffer at the network egress at the cost of an additional for queuing delay. Hence, a possible model used for describing the actual delay jitter is the equivalent queueing delay. A similar approach may be used for packet loss probability due to buffer overflow. Consequently, a given network service with specific guarantees on delay jitter and packet losses may be modeled as an equivalent service with a given virtual delay, without any delay jitter or losses. The interested reader can find in [14, 16] further details concerning the computation of the virtual delay from the service parameters.

In summary, it is assumed that a virtual end-to-end delay d is computed from service parameters by summing up the actual edge-to-edge delay and the virtual components representing the QoS parameters. A low virtual delay value indicates a good service and vice versa. In principle the sum could be weighted according to whatever criteria, such as customer sensitivity. If we transfer an information unit from a point A to a point B crossing N domains, with guaranteed delays $d_1 \ldots d_N$, then the total end-to-end virtual delay is

$$d = \sum_{i=1}^{N} d_i. \tag{1}$$

When a customer wants to transfer information, it specifies the service quality desired, thus a total virtual end-to-end allowed delay d_{max} is associated with it. The virtual delays d_i of each crossed domain are chosen such that they satisfy the end-to-end constraints, i.e.,

$$\sum_{i=1}^{N} d_i \le d_{max} \tag{2}$$

In fact, we will consider in the paper

$$\sum_{i=1}^{N} d_i = d_{max} \tag{3}$$

in order to avoid waste of resources. The number of commodity units may be determined by using a function $f(d)$, where d is the virtual delay. Since such function gives the number of commodity units associated with the information transmission, in [17, 18] it was used to define a pricing strategy for guaranteed network services, depending on both the actually used and the reserved network resources. The end-to-end price of the network support for performance guaranteed services is given by the sum of the single tariffs charged by the domains involved in the end-to-end transfer; analytically,

$$F(\underline{d}) = \sum_{i=1}^{N} \xi_i f(d_i) \tag{4}$$

where $F(\underline{d}): \mathbb{R}_+^N \rightarrow \mathbb{R}_+$ and, for each crossed domain i, d_i is the guaranteed delay; $\xi_i \in \mathbb{R}_+$ is the price of one commodity unit, which might also depend on the network parameters; $f(d_i): \mathbb{R}_+ \rightarrow \mathbb{R}_+$ is the cost function that associates a measure with the transfer of each information unit, expressed in commodity units. In principle, each domain could select such function arbitrarily. In particular, this selection will reflect the fact that the satellite bandwidth cost is higher than the terrestrial domains bandwidth cost.

Our task is to determine an algorithm which minimizes the cost of the service provided to users. Essentially it is necessary to solve two problems. The former is to find the end-to-end path through independent domains (*routing problem*), the latter is to distribute the virtual delay over the communication path according to the constraint of Eq. 3 (*resource allocation*).

In this paper we investigate two approaches to distribute the virtual delay over the path. The former is a Minimum Price (M-P) routing algorithm [17, 19] which consists of a *joint* routing / resource allocation optimization. In the following it will be referred to as "joint algorithm". It can be seen as a cross-layer technique involving the network and the link layer in the protocol stack. Some approaches present in the literature investigating the Simultaneous Routing and Resource Allocation (SRRA) problem are listed below. In [20] a controller allocates power and schedules the data to be routed over the links in reaction to channel state and queue backlog information. The same topic was investigated in [21] for a scenario constituted of a multibeam satellite down-link which transmit data to ground locations over time-varying channels. The SRRA problem was formulated in [22, 23] as a convex optimization problem over the network flow variables and the communication variables. The aim of this work was extended in [24] to also address transmission scheduling.

The latter algorithm presented in this paper assumes that the selection of the path is correctly performed and faces the resource allocation problem distributing the total allowed virtual delay over the domains involved, such that the total cost for accessing the service over the selected path is minimized. In the following this approach will be referred to as "disjoint algorithm".

The structure of the paper is as follows. In Section 2 we describe the notation, some mathematical definitions and the network model. In Section 3 we recall some known results on M-P routing algorithms and shows the estimation of the computational complexity of the joint optimization approach. In Section 4 we describe our approach to the problem (i.e., the disjoint algorithm). In particular we model the cost of the service as a convex function of the virtual delays, thus the problem becomes a convex optimization one. In Section 5 we present the disjoint optimization complexity analysis. The relevant simulation results are showed in Section 6. In Section 7 we drive the conclusions of the work.

2 Notation and Definitions

- $\mathbb{R}$: the set of real numbers.
- $\mathbb{R}^n$: the set of real n-vectors ($n \times 1$ matrices).
- $\mathbb{R}^{1xn}$: the set of real n-row-vectors ($1 \times n$ matrices).
- $\mathbb{R}^{mxn}$: the set of real $m \times n$ matrices.
- $\mathbb{R}_+$: the set of nonnegative real numbers, i.e., $\mathbb{R}_+ = \{x \in \mathbb{R} | x \geq 0\}$.
- $\|x\|$: norm of $x \in \mathbb{R}^n$.

- $\| x \|_2$: the euclidean norm of $x \in \mathbb{R}^n$, $\| x \| = (x_1^2 + \ldots + x_n^2)^{1/2}$.
- $x \preceq y$: (if x and y are vectors) component-wise inequality: $x_i \leq y_i \forall i$.
- *dom f*: domain of function f.
- $f: A \to B$: f is a function on the set $dom\, f \subseteq A$ into the set B.
- $f'(x)$: *first derivative* of a differentiable function $f: \mathbb{R} \to \mathbb{R}$ evaluated at x.
- $f''(x)$: *second derivative* of a twice differentiable function $f: \mathbb{R} \to \mathbb{R}$ evaluated at x.
- $f'''(x)$: *third derivative* of a three times differentiable function $f: \mathbb{R} \to \mathbb{R}$ evaluated at x.
- $\nabla f(x)$: *gradient* of a differentiable function $f: \mathbb{R}^N \to \mathbb{R}$ at x : $(\nabla f(x)_i) = \frac{\partial f}{\partial x_i}$ evaluated at x.
- $\nabla^2 f(x)$: *Hessian* of a twice differentiable function $f: \mathbb{R}^N \to \mathbb{R}$ at x:

  $(\nabla^2 f(x)_i) = \frac{\partial^2 f}{\partial x_i \partial x_j}$ evaluated at x.

- A function $f: \mathbb{R}^N \to \mathbb{R}$ is *convex* if *dom f* is a convex set and if $\forall x, y \in dom\, f$ and $\forall \theta \in \mathbb{R}$ with $0 \leq \theta \leq 1$, we have

$$f(\theta x + (1-\theta) y) \leq \theta f(x) + (1-\theta) f(y). \tag{5}$$

- A function $f: \mathbb{R}^N \to \mathbb{R}$ is *strongly convex* on S if there exists an $m_{SC} > 0$ and an $M_{SC} > 0$ such that

$$m_{SC} I \preceq \nabla^2 f(d) \preceq M_{SC} I \quad \forall d \in S. \tag{6}$$

- A convex function $f: \mathbb{R} \to \mathbb{R}$ is *self-concordant* if $\forall d \in dom\, F$

$$|f'''(d)| \leq f''(d). \tag{7}$$

- A function $f: \mathbb{R}^N \to \mathbb{R}$ is *self-concordant* if it is self-concordant along every line in its domain, i.e., if the function $\tilde{f}(t) = f(d + tv)$ is a self-concordant function of $t\, \forall d \in dom\, f$ and $\forall v$.

2.1 Network Model

We model the topology of a data network by an undirected graph. In this model a collection of n nodes, labeled by $i = 1 \ldots n$, may send, receive, and relay data across m communication links. We label the node by integers $\xi_i \geq 0$, which represent the cost of each commodity unit per time unit.

3 Existing Routing Algorithms and Complexity Analysis

In this Section, Q denotes the number of paths connecting the source and the destination. Since the number of such paths typically grows exponentially with the number n of nodes of the network, we will express Q as $O(a^n)$, where a is a constant. To determine the computational complexity of the joint algorithm, we compute the total number of floating-point operations (flops) to be executed in the worst case, as a function of various problem dimensions, by neglecting all terms, except the dominant ones. A generic algorithm is said to run in $O(f(n))$ time if for some numbers c and n_0, the processing time of the algorithm is at most $cf(n)\ \forall n \geqslant n_0$.

3.1 M-P Routing Algorithm Based on Min-Plus Convolutions

The joint algorithm is based on the use of min-plus convolutions. The general structure of the cost of each path is:

$$g1,i(d_{1i}) = \min_{0 \leq d \leq d_{1i}} \left[\xi_i f_i(d) + g1,i-1\,(d_{1i}-d)\right] = \xi_1 f_1 * \xi_2 f_2 * \cdots * \xi_i f_i \qquad (8)$$

For detailed information on min plus algebra, the reader should refer to [25, 26]. For the sake of clearness we quote briefly from [17, 19] the steps which constitute the joint algorithm:

- Starting from the source, all the departing inter-domain paths that do not create loops are considered.
- A metric is associated to each path, obtained by computing the min plus convolutions of the cost functions of all domains of the path, computed in the range [0, d_{max}].
- If M_m paths converge towards the same input port of the generic m-th domain, they are compared. Since the maximum allowed delay is d_{max}, for each delay value in the range [0, d_{max}], only the path relevant to the minimum cost function survives, and all other paths are discarded.
- At the destination domain, the values of the cost functions of the paths survived, computed at d_{max}, are compared, and the cost function corresponding to the minimum value is selected.
- Finally, the maximum delay d_{max} is distributed over the selected domains.

3.2 Joint Optimization Complexity Analysis

Since the domain of the cost functions are meaningful only between 0 and d_{max}, this holds also for their convolution. Therefore each convolution may be restricted to the meaningful range. It turns out that the cost of each step is constant, denoted as C_{M-P}.

Fact 1: In the worst case the total cost of the joint algorithm is

$$C_{M-P}\, O\,(a^n n). \qquad (9)$$

Proof: In the worst case, the network is very dense, that is every node is adjacent to every other node, and every path is composed of up to n domains; further, at each domain, we have to compare the relevant cost function, defined in the range [0, d_{max}], with that of the an different paths which could converge towards the domain. In summary, in the worst case we must compute $O(a^n)$ min-plus convolutions. Then, we must compare the values of the cost functions associated with the Q paths. This step requires $Q-1$ inequalities, i.e., $Q-1$ flops (a number negligible which does not influence the asymptotical computational complexity of the algorithm). At the end of the algorithm, once we have found the optimum path, we can find the optimal distribution of the virtual delays in few flops (also in this case this number is negligible with respect to the total computational complexity of the algorithm). On the basis of the previous observations, we can estimate that in the worst case the total cost of the joint algorithm is given by Eq. 9.

3.3 Observations

A joint optimization in the terms ξ and $f(d)$ characterizes the joint algorithm. Despite this complexity, the relevant algorithms could make sense in operation if the number of involved domains is low (e.g., regional or national). On the contrary, if the number of considered domains is large (world-wide communications), a different solution is necessary. In [17] Authors have also proposed a simpler approach by means of discretization of the domain of the cost functions. This way the possible values of the virtual delay belong to a discrete and finite set, which simplifies the routing algorithm at the detriment of the flexibility in allocating the virtual delay values. Another approach assumes that the selection of the path is correctly performed. Note that the setting of the term ξ will reflect the fact that the price of one commodity unit processed by the satellite is higher than the correspondent price for a terrestrial domain. As a consequence a solution with a low computational complexity may be found by considering the terms ξ_i only, which is the cost of each commodity unit in the i-th domain. We stress that in general this approach cannot *guarantee* the optimum solution, but rather a satisfactory solution with a convergence time much lower than the one obtainable by facing the general problem. Thus, the path could be found considering only the term ξ. This type of problem is known in literature as shortest path problem, specifically based on a metric that is the amount of money per commodity unit per time unit ξ. Due to this metric, in this work we will use the term *cheapest* path instead of shortest path. Then the problem becomes an optimization one; we have to find the virtual delays $d_1 \ldots d_N$ (with $\sum_{i=1}^{N} d_i \leq d_{\max}$), so as to minimize $F(d)$. It is worth noting that we do not care about the specific technique used within domains to guarantee QoS. We only need an abstract edge-to-edge description, which is the virtual delay and its related cost function.

4 M-P Routing Through Disjoint Optimization

In this Section we investigate the disjoint algorithm. As discussed in Section 3.3, we note that the source-destination paths could be defined on the basis of the commodity price. This way we can pre-select a number of candidate solutions over the same physical path, characterized by the lowest commodity price. After this, the best solution (i.e., virtual delay distribution over the path) is determined. We observe that this approach cannot guarantee the best solution over the network if the cost functions of the domains are very different. A good and satisfactory solution is found any way. On the contrary, if the cost functions are identical, it is trivial to show that the path that has the minimum commodity price include the best solution, found by the subsequent optimization. Between the two extremes, it is expected that if the cost function are similar, even if not identical, the proposed approach can either provide either the best solution or a very good one, very close to the optimum.

Once we have defined the cheapest path between two domains A and B, our task is to find the virtual delays $d_1, \ldots, d_N$ associated with the domains $1, \ldots, N$ of the path so as to minimize the total network cost with QoS guarantees. We present a general approach which can be applied to a general choice of $F(\underline{d})$. We assume that $F(\underline{d})$ is twice continuously differentiable, and strongly convex with constants m_{sc} and M_{sc}.

4.1 Equality Constrained Minimization Problem

Once defined the cheapest path, from Eq. 4 and Eq. 3, the disjoint algorithm can be formulated as follows:

$$\begin{aligned} &\text{minimize } F(\underline{d}) = \sum_{i=1}^{N} \xi_i f(d_i) \\ &\text{subject to } A\underline{d} \;\; = d_{\max}, \end{aligned} \tag{10}$$

where $\underline{d} = (d_1, \ldots, d_N)$ are the optimization variables, $F(\underline{d})\text{:}\mathbb{R}_+^N \rightarrow \mathbb{R}_+$ is convex and $A = (1 \ldots 1)$ with $A \in \mathbb{R}^{1 \times N}$. The physical dimension of $\underline{d}$ is the time, thus $\underline{d} \geq 0$.

We denote $F!$ as the optimal value of this problem, i.e., $F! = \mathit{inf}\{F(\underline{d}) \mid A\underline{d} = d_{\max}\}$. A point $\underline{d}! \in \mathit{dom}\, f$ is optimal for (10) if and only if there is a $\nu! \in \mathbb{R}$ such that

$$\begin{aligned} &A\underline{d}! = d_{\max}, \\ &\nabla F(\underline{d}!) + A^T \nu! = 0 \end{aligned} \tag{11}$$

hence, solving the equality constrained optimization problem (10) is equivalent to finding a solution of the equations (11), called Karush-Kuhn-Tucker (KKT) equations, which are a set of $N + 1$ equations in the $N + 1$ variables $\underline{d}!$, $\nu!$. There are several general approaches for equality constrained problems. Below we discuss feasible descent methods.

We assume that a suitable starting point $\underline{d}(0)$ is available. This point must be feasible, i.e., $A\underline{d}(0) = d_{\max}$, and $\underline{d}(0) \in \mathit{dom}\, F$. These methods are called *descent* because all iterates $\underline{d}(k)$ are feasible and $F(\underline{d}(k + 1)) < F(\underline{d}(k))$, except when $\underline{d}(k)$ is optimal. The outline of a general feasible descent method for equality constrained minimization is as follows:

Feasible descent method for equality constrained minimization: given a starting point $\underline{d}(0) \in \mathit{dom}\, F$, $A\, \underline{d} = \alpha_{\max}$ *repeat*

- *Determine a feasible descent direction* v, $Av = 0$
- *Line search. Choose a step size* $t_{ls} > 0$
- *Update.* $\underline{d} := \underline{d} + t_{ls}\, v$

until the stopping criterion is satisfied.

We adopt the same line search for all the descent methods, thus the step which differentiates the different descent methods is the determination of the feasible descent direction v. The *projected gradient method* uses as search direction the Euclidean projection of the negative gradient $-\nabla F$ on the set of feasible directions; *the steepest descent method* uses a step of unit norm that gives the largest decrease in the linear approximation of F; the step used in *the Newton method* is defined as the quantity that must be added to d to solve the problem when the quadratic approximation is used in place of F. The iterates $v(k)$ that converge to an optimal dual variable $v!$, i.e., satisfy

$$\lim_{k \to \infty} \nabla F(\underline{d}(k)) + A^T v(k) = 0. \tag{12}$$

The stopping criterion for a feasible descent method generally has the form

$$\left\| \nabla F(\underline{d}(k)) + A^T v(k) \right\| \le \eta \tag{13}$$

where η is small. This is justified in [27].

4.1.1 Line Search. Different kinds of line search exist [29, 31]. Below we show the results obtainable by using the *backtracking* line search. It depends on two constants α_{ls}, β_{ls} with $0 < \alpha_{ls} < 0.5$, $0 < \beta_{ls} < 1$. The step length t_{ls} is chosen to approximately minimize F along the ray $\{d + t_{ls}\, v \mid t_{ls} \ge 0\}$:
Backtracking line search:
given a descent direction v for F at $\underline{d} \in$ *dom F*
$t_{ls} := 1.$
while $(F(\underline{d} + tv) > F(\underline{d}) + \alpha_{ls}\, t_{ls} \nabla F(\underline{d})^T v)$
$t_{ls} := \beta_{ls}\, t_{ls}.$
end

The line search is called backtracking since it starts with unit step size and then reduces it by the factor β_{ls} until the stopping condition $F(\underline{d} + tv) \le F(\underline{d}) + \alpha_{ls}\, t_{ls} \nabla F(\underline{d})^T v$ holds.

4.1.2 Projected Gradient Method and Steepest Descent Method. Any norm $\|\cdot\|$can be bounded in terms of the euclidean norm, i.e., there exists a constant $\gamma \in (0, 1]$ such that $\|\underline{d}\| \ge \gamma \|\underline{d}\|_2$. It can be shown [27] that:

$$F(\underline{d}(k) - F!) \le c_{lin}^{k} (F(\underline{d}(0) - F!)) \tag{14}$$

where $c_{lin} = \min\{2m_{sc}\,\alpha_{ls},\ 2\,\alpha_{ls}\,\beta_{ls}\,m_{sc}/M_{sc}\} < 1$ for the gradient method, and $c_{lin} = 1 - m_{sc}\,\alpha_{ls}\,\gamma^2 \min\{1,\ \beta_{ls}\,\gamma^2/M_{sc}\} < 1$ for the steepest descent method. Thus, in the gradient and in the steepest descent method, $F(\underline{d}(k))$ converges to $F!$ at least as fast as a geometric series with an exponent that depends on the condition number bound M_{sc}/m_{sc}. In the terminology of iterative methods, the convergence is said at *least linear*.

4.1.3 Newton's Method. We replace the objective with its second order Taylor approximation near $\underline{d}$, to formulate the problem:

$$\text{minimize } \tilde{F}(\underline{d} + v) = F(\underline{d}) + \nabla F(\underline{d})^T v + \frac{1}{2} v^T \nabla^2 F(\underline{d}) v \tag{15}$$

$$\text{subject to } A(\underline{d} + v) = d_{max} \tag{16}$$

with variable v. This is a (convex) quadratic minimization problem with equality constraints, and can be solved analytically. The Newton step v_{nt} is defined as the quantity that must be added to d to solve the problem when the quadratic approximation is used in place of F.

Another alternative is provided by a family of algorithms for unconstrained optimization called quasi-Newton methods [30, 29, 31]. These methods require less computational effort to form the search direction, but since they share some of the strong advantages of Newton methods, such as rapid convergence near $d!$, we will not consider them. The interested reader can find more details about Newton's method in [30, 31]. Among the different feasible descent methods, we have decided to focus our attention on Newton's method for the reasons, experimentally verified, shown in below.

4.1.4 Comparison of the Feasible Descent Methods. Concerning the gradient and the steepest descent methods can be observed:

- The choice of backtracking parameters α_{ls}, β_{ls} has a noticeable but not dramatic effect on convergence.
- The methods often exhibit approximately linear convergence, i.e., the error $F(\underline{d}(k)) - F!$ converges to zero as a geometric series.
- The convergence rate depends greatly on the condition number of the Hessian, that is the ratio of value the largest and the lowest eigenvalues.

About the Newton method we can say that it has several very important advantages over gradient and steepest descent methods:

- It scales well with problem size.
- The good performance of Newton's method is not dependent on the choice of algorithm parameters. In contrast, the choice of the norm for the steepest descent plays a critical role in its performance.
- Its convergence is rapid in general, and quadratic near $\underline{d}!$.
- Once the quadratic convergence phase is reached, a number close to 6 iterations, if the value of η is between 10^{-6} and 10^{-7}), are required to find a very accurate solution.

For the above reasons, in the following we will focus our attention on the Newton's method computational complexity.

5 Disjoint Optimization Complexity Analysis

5.1 Complexity of the Cheapest Path Problems

The disjoint algorithm assumes that the path selection is correctly performed. Routing in Internet is mainly performed by means of two mechanisms: Border Gateway Protocol (BGP) [32] for inter-domain routing and Open Shortest Path First (OSPF) [33] is used for intra-domain routing. In this Section we investigate the computational complexity that the routing process would require if implemented as discussed in Section 3. A cheapest path problem consists in finding a path of minimum cost from a specific source node to another specified sink node, assuming that each link has an associated cost. Hence, in this phase we do not consider any virtual delay, but associate the network connections with the commodity price ξ only, and find the minimum cost path by using known algorithms. The network flow literature [34] typically classifies algorithmic approaches for solving cheapest path problems into two groups: label setting and label correcting. As regards the label setting algorithms we have considered a simple implementation of them, Dijkstra algorithm, then two versions [34] of it: Dial's implementation and R-HEAP implementation. We have also used a special implementation of label correcting algorithm that requires polynomial time. We denote $max\ \{\xi_i, i = 1,\ldots,n\}$ as ξ_{max}. Depending on the values assumed by n, m, and ξ_{max} we can select the one which gives the best performance. About the three different versions of Dijkstra algorithm shown in [34], we can say that the original $O(n^2)$ one has the optimal running time for fully dense networks (with at least n^2 arcs). A potential disadvantage of this Dial's scheme, as compared to the original one, is that ξ_{max} may be very large, thus requiring large storage and increased computational time. The R-heap implementation runs in $O(m + n \log \xi_{max})$ time units. Using more sophisticated data structures, it is possible to reduce this bound to $O(m + n\sqrt{\log n})$, which is a linear time algorithm for all but the sparsest classes of shortest path problems. Label setting has the most attractive worst-case performance but practical experience has shown that label correcting is fairly more efficient. In the following, we will express the computational complexity of a cheapest path problem as

$$O\left(S(n,m,\xi_{\max})\right). \tag{17}$$

5.2 Complexity of the Newton's Method

The Newton's method for solving an equality constrained minimization problem is constituted of the 4 main steps described in 4.1. We assume that the cost of the first step is negligible. The other steps are repeated until the stopping criterion is satisfied, thus to obtain the total computational time we must multiply the running time of these steps by the number of iterations. If $F(\underline{d})$ is strongly convex and the value of η is between 10^{-6} and 10^{-7}, the number of iterations is upper bounded by [27]:

$$6+(\alpha_{\mathrm{ls}}\beta_{\mathrm{ls}}\min\{1,3(1-2\alpha_{\mathrm{ls}})\})\frac{m_{\mathrm{sc}}^3}{M_{\mathrm{sc}}^2 L_1}(F(\underline{d}(0))-F!) \tag{18}$$

If $F(\underline{d})$ is self-concordant, for the same η values the number of iterations is upper bounded by [27]:

$$\frac{20-8\alpha_{\mathrm{ls}}}{\alpha_{\mathrm{ls}}\beta_{\mathrm{ls}}(1-2\alpha_{\mathrm{ls}})^2}(F(\underline{d}(0))-F!)+6 \tag{19}$$

This expression depends only on the line search parameters α_{ls} and β_{ls}. If, for example, we take $\alpha_{\mathrm{ls}} = 0.1$ and $\beta_{\mathrm{ls}} = 0.9$ the previous expression becomes $334(F(\underline{d}(0)) - F!) + 6$. The entire line search can be carried out at an effort comparable to simply evaluating F, thus the step which requires the largest computational time is the computation of the Newton step.

Fact 2: The number of flops required by the Newton step for our problem may be expressed in the form:

$$N_{W-f} = t_{\mathrm{H}} + t_{\mathrm{g}} + 5N. \tag{20}$$

where t_{H} and t_{g} are the times necessary to calculate $H = \nabla^2 F(\underline{d})$ and $g = \nabla F(\underline{d})$.

Proof: In this Section we describe methods that can be used to compute the Newton step, i.e., to solve the KKT system

$$\begin{bmatrix} H & A^T \\ A & 0 \end{bmatrix} \begin{bmatrix} v \\ w \end{bmatrix} = \begin{bmatrix} -g \\ 0 \end{bmatrix} \tag{21}$$

where $H = \nabla^2 F(\underline{d})$ and $g = \nabla F(\underline{d})$, for v (and w). A simple straightforward approach is to solve the KKT system which is a set of $N + 1$ linear equations in $N + 1$ variables. The KKT matrix is symmetric and positive definite. By eliminating v from the KKT system and solving for ω we obtain the reduced equations:

$$w = -(AH^{-1}A^T)^{-1}AH^{-1}g \tag{22}$$

$$v = H^{-1}(-A^T w - g) \tag{23}$$

which give us an alternate method for computing v and w, constituted of the following steps:

1) Form $H^{-1}A^T \in \mathbb{R}^{N \times 1}$ and $H^{-1}g \in \mathbb{R}^{N \times 1}$. We have:

$$H = diag(\nabla^2 F(\underline{d})_{11} \ldots \nabla^2 F(\underline{d})_{NN}) \tag{24}$$

and

$$H^{-1} = diag\left((\nabla^2 F(\underline{d})_{11})^{-1} \ldots (\nabla^2 F(\underline{d})_{NN})^{-1}\right). \tag{25}$$

The term of the i-row of $H^{-1} A^T$ is $(\nabla^2 F(\underline{d})_{ii})^{-1}$. The term of the i-row of $H^{-1} g$ is: $(\nabla^2 F(\underline{d})_{ii})^{-1} \nabla F(\underline{d})_i$.

2) Form $S = -AH^{-1} A^T$ We note that $S \in \mathbb{R}$. We have:

$$S = -\sum_{i=1}^{N} (\nabla^2 F(\underline{d})_{ii})^{-1} \tag{26}$$

3) Form $w = S^{-1} A (H^{-1} g)$. We note that $w \in \mathbb{R}$. We have:

$$w = S^{-1} \sum_{i=1}^{N} \left((\nabla^2 F(\underline{d})_{ii})^{-1} \nabla F(\underline{d})_i\right). \tag{27}$$

4) Form $v = H^{-1} (-A^T w - g)$. We note that $v \in \mathbb{R}^{N \times 1}$. The term of the i-row of v is:

$$\left(w - \nabla F(\underline{d})_i)\right) (\nabla^2 F(\underline{d})_{ii})^{-1} \tag{28}$$

Now we estimate the total computational complexity associated with each step:

- We define t_{H} and t_{g} as the time necessary to calculate H and g, respectively. To compute H^{-1} we need N divisions, thus N flops. We need N flops also to calculate $H^{-1} g$, while $H^{-1} A^T$ does not need any further flop since it is constituted of the elements of H^{-1}.

- To form S we have to sum up the N terms $\neq 0$ of H^{-1}, thus we need $N - 1$ flops.
- To form w, we have to sum up the N terms $\neq 0$ of $H^{-1}\, g$ and then multiply for S^{-1}, thus we need $N + 1$ flops.
- To form v, we have to calculate N–1 sums and N divisions, thus we need $2N$–1 flops.

Thus, the number of flops of the step is

$$N_{W-f} = t_{\mathrm{H}} + t_{\mathrm{g}} + 5N. \tag{29}$$

In summary, if the cost function is self-concordant, $\alpha_{\mathrm{ls}} = 0.1$, and $\beta_{\mathrm{ls}} = 0.9$, the worst-case complexity can be expressed as:

$$(334\,(F(\underline{d}(0) - F!)) + 6)\,(t_{\mathrm{H}} + t_{\mathrm{g}} + 5N). \tag{30}$$

If we take into account also the routing algorithm, then from Eq. 17 we have that the total computational complexity of the disjoint algorithm is:

$$(334\,(F(\underline{d}(0) - F!)) + 6)\,(t_{\mathrm{H}} + t_{\mathrm{g}} + 5N) + O(S(n, m, \xi_{\max})), \tag{31}$$

that is polynomially bounded.

6 Simulation Results

Now we make some consideration about the choice of $f(d)$. This is a cost function, thus it must be chosen such that $f(d) \geq 0 \forall d \in dom\ f$; it is meaningful if it is monotonic nondecreasing since the cost has to decrease with the value of the virtual delay ($f'\,(d) \leq 0$); moreover, in order to use convex optimization algorithms, it is desirable for it to be convex ($f''\,(d) \geq 0$), in particular strongly convex or self-concordant. Since these properties are preserved by sums, we can conclude that the same properties are valid for Eq. 4.

The use of strongly convex and self-concordant functions is important since for these classes of functions we can determine an upper bound to the number of iterations of the algorithm. Any case, we have experimentally verified that this number is typically pretty much lower than this bound. In addition, the feasible methods described in Section 4.1 can be applied also to functions which are not strongly convex or self-concordant with good results from the point of view of the processing time. We considered the following function:

$$f(d) = e^{-xd + k} \tag{32}$$

with $x \in \mathbb{R}_+ - \{0\}, k \in \mathbb{R}$ and $dom\ f = \mathbb{R}$, as cost function associated with the satellite domain. The physical dimension of $\underline{d}$ is the time, thus $\underline{d} \succeq 0$. We have $f(d) > 0$, $f'\,(d) < 0$ and $f''(d) > 0 \forall d \in \mathbb{R}_+$.

We considered the following function:

$$f(d) = -\log\left(\frac{d}{d_{\max}}\right) + \frac{m}{2}d^2 \tag{33}$$

with m $\in \mathbb{R}_+ - \{0\}$ and $dom\ f = \mathbb{R}_+ - \{0\}$, as cost function associated with the terrestrial domains.

Since $d < d_{max}$, we have $f(d) > 0 \forall d \in dom\ f$; if $d < \sqrt{\frac{1}{m}}$ then $f'(d) < 0$. We can note that $f(d)$ is strongly convex $\forall d \in dom\ f$. As a consequence, if we assume that each path comprises at least one terrestrial domain, also $F(d)$ is strongly convex, thus efficient algorithms can be computed to minimize it without verifying if it is self-concordant.

We considered a network constituted of one satellite and three groups of terrestrial domains. The first group, constituted of 10 domains, implements the cost function (33) with $m = 1$; the second one, constituted of 20 domains the same cost function with $m = 3$, the third one, constituted of 30 domains adopts $m = 5$. The satellite implements the cost function (32) with $x = 2$ and $k = 1$. The values of ξ are: 10 for the satellite, 1 for the first group of terrestrial domains, 2 for the second and 3 for the third one.

We implemented by using MATLAB the three feasible optimization methods described in Section 4 setting η to 10^{-6} and assuming as starting feasible point $\underline{d}(0) = \{1, . . . ,1\}$ (the sum of the virtual delays is constrained to be 61). We used the backtracking line search setting $\alpha_{ls} = 0.1$ and $\beta_{ls} = 0.9$. In Table 1 the main simulation parameters are listed.

Table 1. Simulation parameters.

	x	k	m	ξ	$d_i(0)$
Satellite	2	1	/	10	1
Terrestrial Domains (Group 1)	/	/	2	1	1
Terrestrial Domains (Group 2)	/	/	3	2	1
Terrestrial Domains (Group 3)	/	/	5	3	1

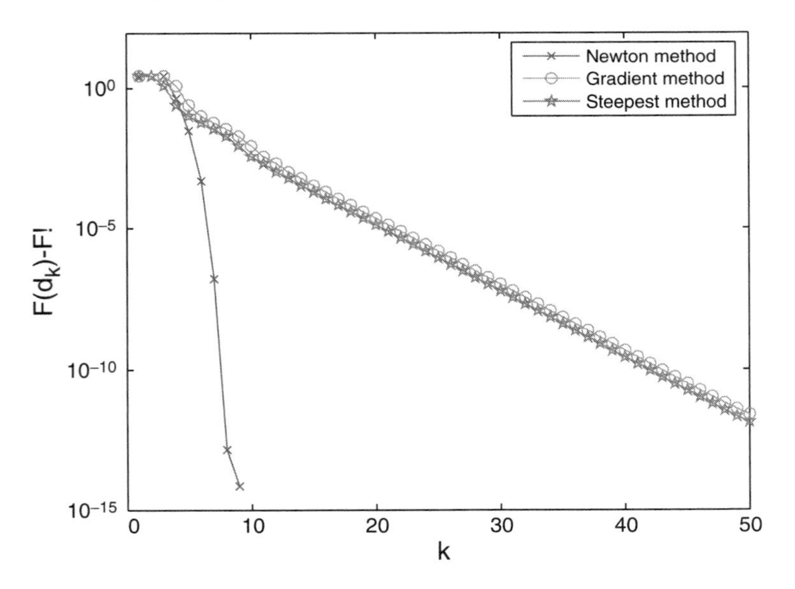

Fig. 2. Error $F(d_k) - F!$ versus iteration number.

In Fig. 2 we plotted, for each algorithm iteration k, the difference between the value of the cost function F evaluated at $d^{(k)}$ and the optimal value $F!$. It can be appreciated that the gradient method and the steepest descent method exhibit approximately linear convergence, i.e., the error $F(\underline{d}(k)) - F!$ converges to zero as a geometric series. In general the convergence of the Newton method is more rapid, and quadratic near $F!$.

7 Conclusions

In this paper we have faced the M-P routing problem, which consists in finding the cheapest path among different independent domains, which charge users for IP network services with guaranteed QoS performance. We have investigated two solutions. The former is based on min-plus convolutions and involves the link and the network layer of the protocol stack. We showed that the computational complexity due to this approach is *exponentially* bounded. The latter formulates the M-P problem as an equality constrained convex problem which can be solved by means of a feasible descent method. We showed the definitions, the mathematical aspects and we focused our attention on the characteristics of the convergence of 3 types of feasible descent methods. We showed that the computational complexity due to this approach is *polynomially* bounded.

Acknowledgment

This work has been partially funded by the European Community under the 6-th Framework Programme IST Networks of Excellence "SatNEx" (contract No. 507052) and "SatNEx II" (contract No. 027393).

References

[1] "Communications quality of service: a framework and definitions," ITU-T Recommendation SG-12 G.1000, 2001.
[2] R. Braden, D. Clark, and S. Shenker, "Integrated Services in the Internet Architecture: An Overview," IETF RFC 1633, 1994.
[3] S. Blake, D. Black, M. Carlson, E. Davies, Z. Wang, and W. Weiss, "An architecture for Differentiated Services," IETF RFC 2475, 1998.
[4] E. Rosen, A. Viswanathan, and R. Callon, "Multiprotocol label switching architecture," IETF RFC 3031, Jan 2001.
[5] S. Kota and M. Marchese, "Quality of service for satellite IP networks: a survey," *International Journal of satellite communications and networking*, vol. 21, pp. 303–349, 2003.
[6] N. Courville, "QoS-oriented traffic management in multimedia satellite systems," *International Journal of satellite communications and networking*, vol. 21, pp. 367–399, 2003.
[7] F. Daoud, "Hybrid satellite/terrestrial networks integration," *Computer Networks*, vol. 34, pp. 781–797, 2000.
[8] X. Maufroid, R. Rinaldo, and R. C. Garcia, "Analysis of Beam Hopping Techniques in Future Multi-Beam Broadband Satellite Networks," in *Proceedings 23rd International*

Communication Satellite Systems Conference (ICSSC) and 11th Ka and Broadband Communications Conference, Rome, Italy, 2005.
[9] D. K. Okello and M. Kaplan, "Adaptive Beam Allocation for Multimedia Ka-band Satellite Networks," in *Proceedings 59th Vehicular Technology Conference (VTC)*, Milan, Italy, 2004.
[10] L. Rosati and G. Reali, "On Traffic-demand Based Multi-Beam Bandwidth Allocation in Future Satellite Networks Using Beam-Hopping Techniques," in *Proceedings of Advanced Satellite Mobile Systems (ASMS)*, Herrsching, Germany, 2006.
[11] F. Chiti, R. Fantacci, T. Pecorella, L. Giacomelli, M. Poggesi, and F. Poggianti, "A resource allocation scheme based on traffic prediction for DVB-RCS systems," in *Proceedings 59th Vehicular Technology Conference (VTC)*, Milan, Italy, 2004.
[12] F. D. Priscoli and A. Pietrabissa, "Load-adaptive bandwidth-on-demand protocol for satellite networks," in *Proc. of the 41st IEEE Conf. on Decision and control*, Las Vegas, USA, 2002, pp. 4066–4071.
[13] L. Chisci, R. Fantacci, F. Francioli, and T. Pecorella, "Multi-terminal dynamic bandwidth allocation in GEO Satellite Networks," in *Proceedings 59th Vehicular Technology Conference (VTC)*, Milan, Italy, 2004.
[14] D. Di Sorte, M. Femminella, and G. Reali, "A Novel Approach to Charge for IP Services with QoS support," *Journal of Network and Systems Management*, vol. 8, no. 11, 2004.
[15] D. Di Sorte, M. Femminella, and G. Reali, "A QoS Index for IP Services to Effectively Support Usage-based Charging," *IEEE Communications Letters*, vol. 8, no. 11, 2004.
[16] N. Blefari-Melazzi, D. Di Sorte, M. Femminella, and G. Reali, "Theoretical analysis of a virtual delay based tariff model," in *IEEE International Conference on Communications (ICC)*, Anchorage, USA, 2003.
[17] N. Blefari-Melazzi, D. Di Sorte, and G. Reali, "Inter-domain Routing Algorithms that Maximize Users. Benefit in an Internet Business Model," in *Proceedings of IEEE Globecom 2002*, Taipei, Taiwan, 2002.
[18] D. Di Sorte, M. Feminella, G. Reali, and S. Zeisberg, "Network Service Provisioning in UWB Open Mobile Access Networks," *IEEE Journal on Selected Areas in Communications (JSAC)*, vol. 20, no. 9, 2002.
[19] D. Di Sorte and G. Reali, "Minimum Price Inter-Domain Routing Algorithm," *IEEE Communications Letters*, vol. 6, no. 4, 2002.
[20] M. J. Neely, E. Modiano, and C. E. Rohrs, "Dynamic Power Allocation and Routing for Time Varying Wireless Networks," *IEEE/ACM Transactions on Networking*, vol. 11, no. 1, 2003.
[21] M. J. Neely, E. Modiano, and C. E. Rohrs, "Power Allocation and Routing in Multibeam Satellites with Time-Varying Channels," *IEEE/ACM Transactions on Networking*, vol. 11, no. 1, 2003.
[22] L. Xiao, M. Johansson, and S. Boyd, "Simultaneous Routing and Resource Allocation via Dual Decomposition," *IEEE Transactions on Communications*, vol. 52, no. 7, pp. 1136–1144, 2004.
[23] M. Johansson, L. Xiao, and S. Boyd, "Simultaneous Routing and Resource Allocation in CDMA wireless data networks," in *IEEE International Conference on Communications*, Anchorage, Alaska, 2003.
[24] M. Johansson and L. Xiao, "Scheduling, Routing and Power Allocation for Fairness in Wireless Networks," in *Proceedings 59th Vehicular Technology Conference (VTC)*, Milan, Italy, 2004.
[25] F. Baccelli, G. Cohen, G. J. Olsder, and J. P. Quadrat, *Synchronization and linearity, an algebra for discrete event systems*. New York: Wiley, 1992.
[26] C. S. Chang, "Deterministic traffic specification via projections under the min-plus algebra," in *Proceedings of IEEE INFOCOM*, New York, USA, 1999.

[27] S. P. Boyd and L. Vandenberghe, *Convex Optimization*. Stanford University: EE 364 Course Reader, 2001.
[28] J. Jahn, *Introduction to the Theory of Nonlinear Optimization*. Springer, 1996.
[29] J. Hiriart-Urruty and C. Lemarechal, *Convex analysis and Minimization algorithms*. Springer-Verlag, 1993.
[30] M. S. Bazaraa, H. D. Sherali, and C. M. Shetty, *Nonlinear programming. Theory and algorithms. Second edition*. Wiley, 1993.
[31] D. P. Bertsekas, *Nonlinear Programming, second edition*. Athena Scientific, 1995.
[32] Y. Rekhter and T. Li, "A border gateway protocol 4 (BGP-4)." RFC 1771.
[33] J. Moy, "OSPF version 2." RFC 2328.
[34] R. K. Ahujaa, J. L. Magnanti, and J. B. Orlin, *Network flows*. Prentice Hall, 1993.

Broadband Satellite Communication in EHF Band

Franco Provenzale, Maurizio Tripodi, Daniela A. Vasconi

Selex-Communications, *a Finmeccanica Company*
Via dell'Industria, 4, 00040, Pomezia, Italy
Phone: +3991091644, Fax: +3991091480, e-mail: maurizio.tripodi@selex-comms.com
Phone: +390957576404, e-mail: franco.provenzale@selex-comms.com,
Phone: +3991091887, e-mail: danielaadriana.vasconi@selex-comms.com

Abstract. The continuous increase in Satellite resources demand, especially in Government and Military missions, led to both efficiently use the Satellite Bandwidth/Power resources, as well as differentiating frequency Bands.

The attractiveness of the highest frequencies in terms of reduced antenna dish size, maintaining terminals with good throughput performances, induced us to develop an EHF DVB-RCS Terminal for Government and Military use. This paper describes the Terminal and Network architecture, the overall performances obtained and the future developments planned in this area.

1 Introduction

The EHF Band is going to be more and more used for Military Satellite application even in Europe, being widely diffused in US.

The Italian Defence Satellite (SICRAL) is one of the first in Europe equipped with an EHF Transponder (44 GHz Up and 20 GHz down) for Telecommunication purposes. The Demo Network that has been set up has several objectives:

- To demonstrate the effectiveness of the EHF band for an IP-switched network based on the DVB-RCS (Digital Video Broadcasting – Return Channel Via Satellite) open standard
- To validate the link analysis in the 44 GHz band
- To validate the Satellite coverage
- To gather elements of propagation statistics at that frequency
- To gather elements of traffic statistics supported by the network

This paper will present the collected results based on the above objectives and will provide a synthesis on the conclusions.

2 Elements of the Trial Network

Figure 1 provides a general overview of the network topology.

The network structure is composed, essentially, by three main Satellite resources in addition to the Terrestrial Back-Bone. The real demo Network was composed

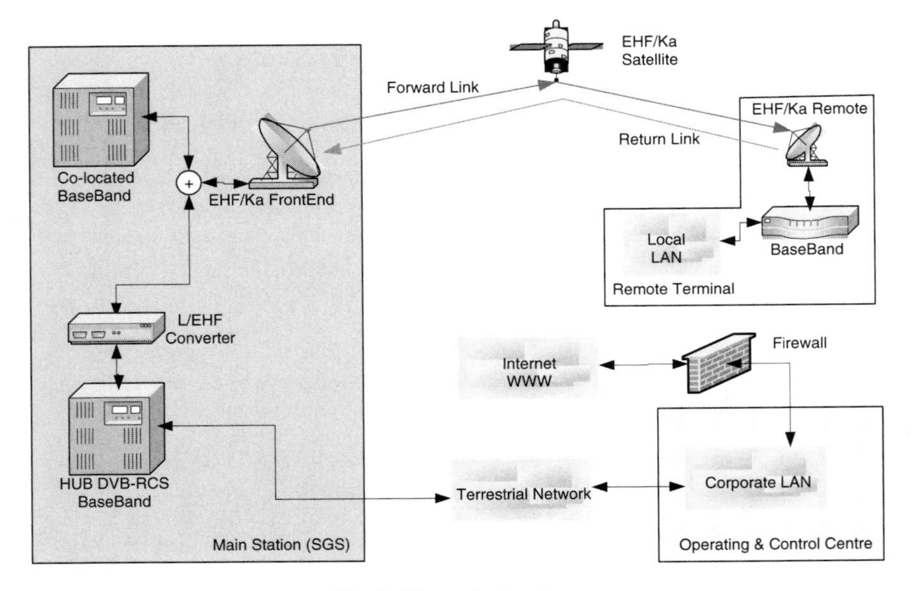

Fig. 1. Network structure.

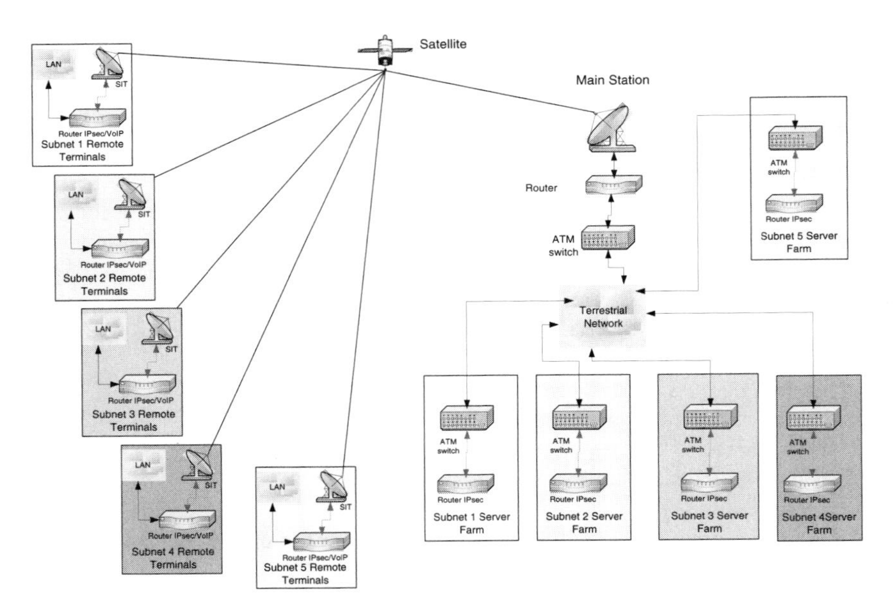

Fig. 2. Sub-networks details.

by 12 remote Terminals, structured in five sub-networks (Fig. 2). Each of these sub-networks was virtually separated by IPsec devices.

Figure 3 illustrates the detailed structure of one of these sub-networks: the IPsec traffic and the GRE tunnel are showed up. The IPsec traffic is exchanged between

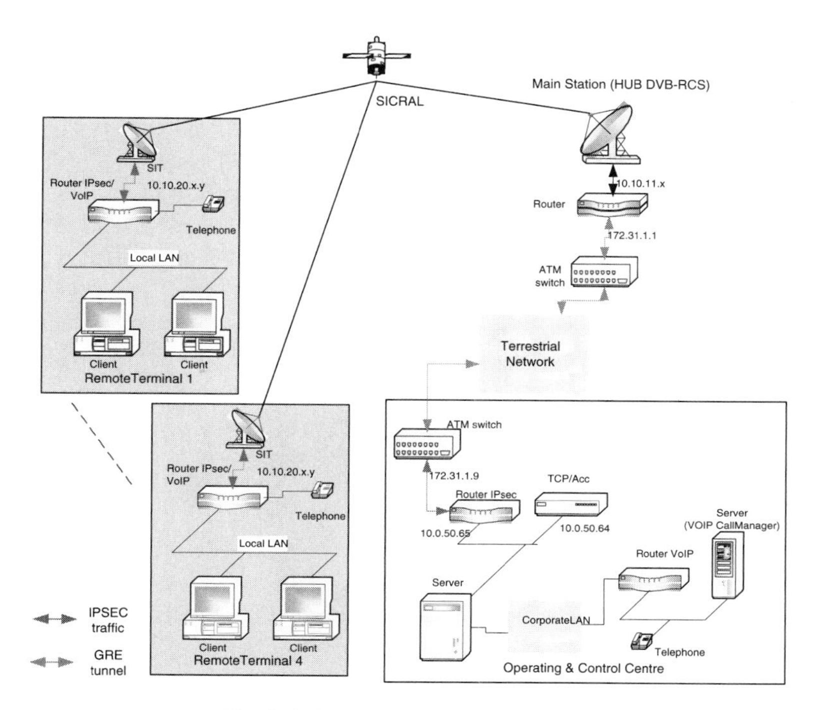

Fig. 3. Sub-network detailed structure.

the remote Terminals and the Operating & Control Centre, passing through the GRE tunnel created in the Main Station.

2.1 The Satellite

SICRAL is Italy's first military satellite, launched in 2001. It has been designed to operate in EHF, SHF and UHF frequency bands with fixed and mobile terminals.

The EHF Transponder of the SICRAL Satellite has been used. The Satellite provides a domestic coverage over Italy and a partial coverage over the east-cost of Europe faced to the Adriatic Sea, in this frequency band.

Figure 4 shows the distribution of the remote Terminals on the territory. The Terminals have been placed in order to cover the edges of the Satellite coverage and to provide a significant set of trials.

Thanks to Italian MoD, enough bandwidth/power has been used for the Trial network to simulate a much bigger network (typically made by more than 500 terminals).

In particular an 8 Mbps Forward Channel and an aggregate of 1.5 Mbps on the Return Channels have been used.

Fig. 4. Terminal distribution on territory.

2.2 The HUB Station

The HUB Station has been based on the available RF Subsystem including the 4.2m Antenna and the dedicated HUB base band. The RF subsystem has been designed to provide UpLink Power Control (UPPC) feature.

UpLink Power Control provides a compensation of the rain fade effects in satellite communications, just increasing the transmission power. In EHF band, the atmospheric effects are a severe problem considering that the attenuation may be many dB high even at moderate rain rates.

The following algorithm is under evaluation to compensate the up link power fading (Fig. 5). It calculates the EHF HPA attenuation to be applied.

At the beginning, an operator inserts the expected beacon levels at clear sky, B_{cc} [dBm] in the Monitor & Control system (M&C).

The Antenna Control Unit (ACU) periodically provides the misured beacon level B_{mis} [dBm] at 20 GHz to the M&C.

The M&C periodically calculates the attenuation level A_{20} [dB] at 20 GHz and the attenuation level A_{44} [dB] at 44 GHz (up link), according to the following formulas:

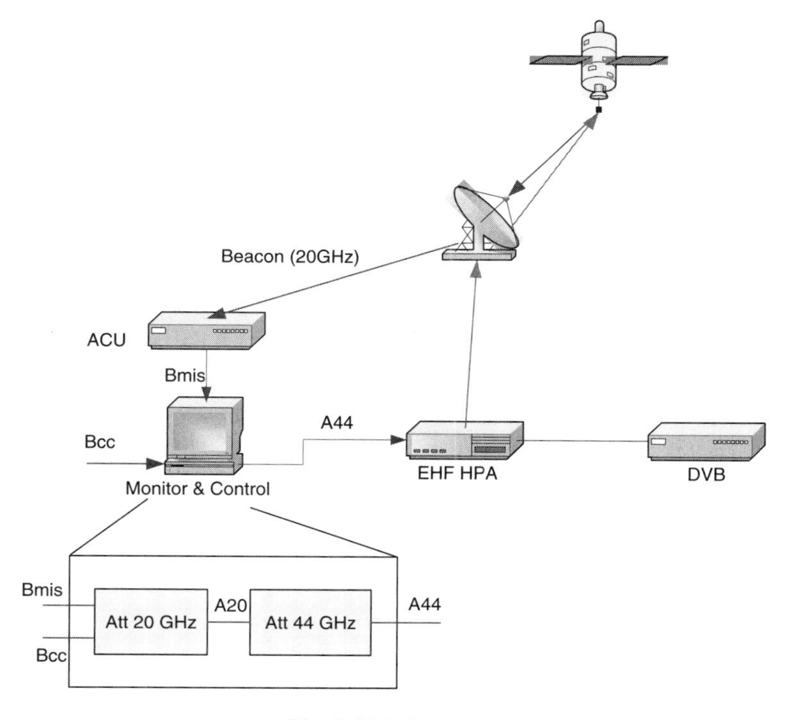

Fig. 5. UPPC scheme.

$$A_{20} = B_{cc} - B_{mis} \tag{1}$$

$$A_{44} = 4 * k_{44}^{*}(A_{20}/4^{*} k_{20})^{\wedge}(\alpha 44/\alpha 20) \tag{2}$$

where:

$$k_{44} = 0.4236, k_{20} = 0.08488, \alpha_{44} = 0.9102, \alpha_{20} = 1.0914 \tag{3}$$

Using (2) the following estimated values of A_{44} are obtained, while the intermediate values can be calculated by linear interpolation.

A_{20} [dB]	A_{44} [dB]
0.1	0.6
0.3	1.5
0.5	2.3
1	4.2
1.5	5.8
2	7.4
3	10.4
4	13.3
5	16

(4)

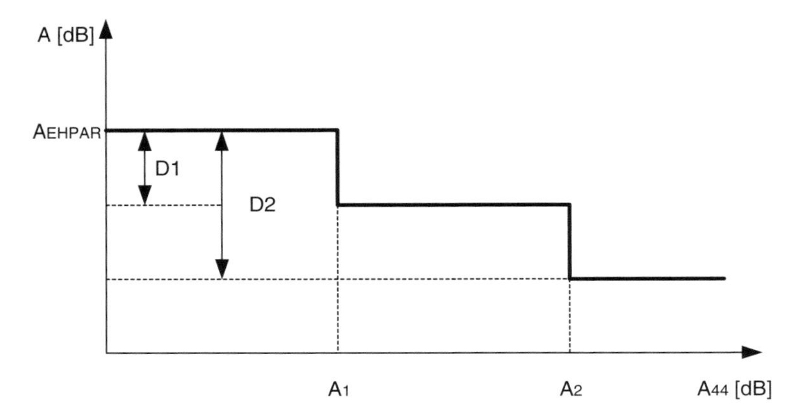

Fig. 6. EHF HPA variable attenuation.

The EHF HPA attenuation A [dB] is then set according to the following rule (Fig. 6), in order to compensate the up link fading:

$$
\begin{array}{ll}
A = A_{EHPAR} & \text{if } A_{44} \leq A_1 \\
A = A_{EHPAR} - D_1 & \text{if } A_1 \leq A_{44} \leq A_2 \\
A = A_{EHPAR} - D_2 & D_2 > D_1, \text{if } A_{44} \geq A_2
\end{array}
$$

where:

A_{EHPAR} = EHF HPA attenuation in down link rain conditions and up link clear sky (default value 4.7 dB)

A_1 = medium rain threshold, settable parameter (default value 4.7 dB)

A_2 = high rain threshold, settable parameter (default value 7 dB)

D_1 = medium rain EIRP increased value, settable parameter (default value 2 dB)

D_2 = high rain EIRP increased value, settable parameter (default value 4 dB)

Another possible UPPC solution, under evaluation at present, is shown in Fig. 7.

The ACU periodically provides the misured beacon level B_{mis} [dBm] at 20 GHz to the M&C. The down link attenuation A_{dw} can be calculated from this value.

An FDMA modem can be used in loop mode to get the misured E_b/N_0 and to calculate the total attenuation A_{tot}, which is a function of both down link and up link attenuations:

$$A_{tot} = f(A_{up}, A_{dw}) \tag{6}$$

From the knowledge of A_{tot} and A_{dw}, the up link attenuation A_{up} at 44 GHz can be calculated.

2.3 The Terminal

The Remote Terminal key subsystems are the 1.2m carbon fibre Antenna subsystem, the 31 dBm SSPA at 44 GHz and the Indoor DVB-RCS Modem. The design of the terminal was focused on the overall performances, but primarily on the EIRP, but

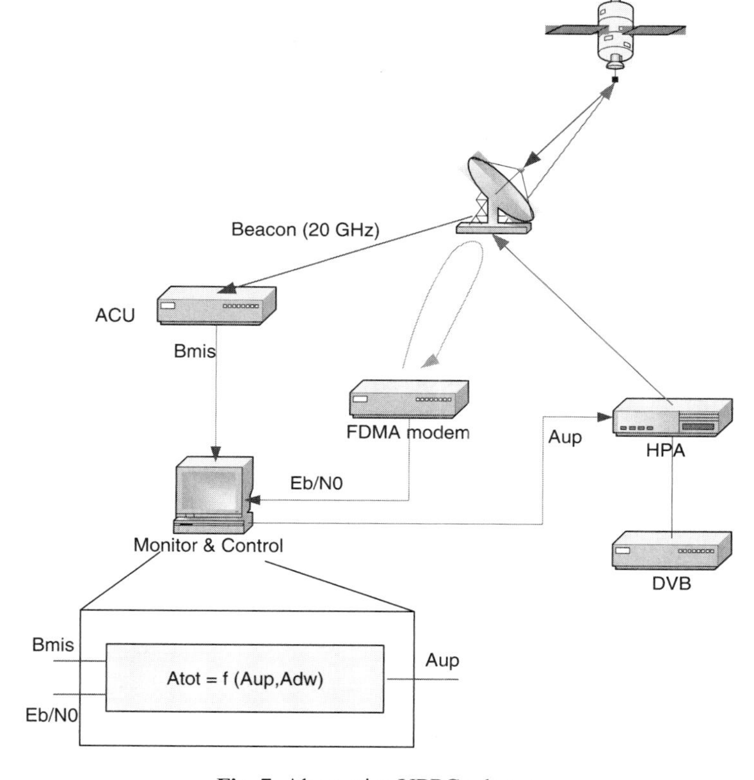

Fig. 7. Alternative UPPC scheme.

maintaining the cost as low as possible. The terminal has been designed to mitigate the heavy propagation fades, by means of two main features: enough EIRP margin and Dynamic Rate Assignment (DRA). This last feature will be explained in the next paragraph.

2.4 DVB-RCS Network Solutions

The EHF Demo network is an IP-switched network based on the DVB-RCS open standard for broadband satellite communications. It provides a two way unbalanced link between Hub and remote Terminals (Forward link from Hub to Terminals and Return link from Terminals to Hub).

The Forward link is a standard DVB-S (Digital Video Broadcasting over Satellite) broadcast channel, based on QPSK modulation, Reed Solomon and convolutional code for Forward Error Correction (FEC), Time Division Multiplexing (TDM) access and MPEG2 Transport Streams for carrying data (user traffic, DVB-S signalling data, DVB-RCS signalling data, network operator monitor & control data.

The Return link is based on a MF-TDMA (Multiple Frequency – Time Division Multiple Access) access scheme, QPSK modulation, and Reed Solomon and convolutional code or Turbo coding FEC, MPEG2 or ATM burst profiles.

A protocol stack for Forward and Return links is provided in Fig. 8 and a simplified scheme of DVB-RCS principles of operation is illustrated in Fig. 9.

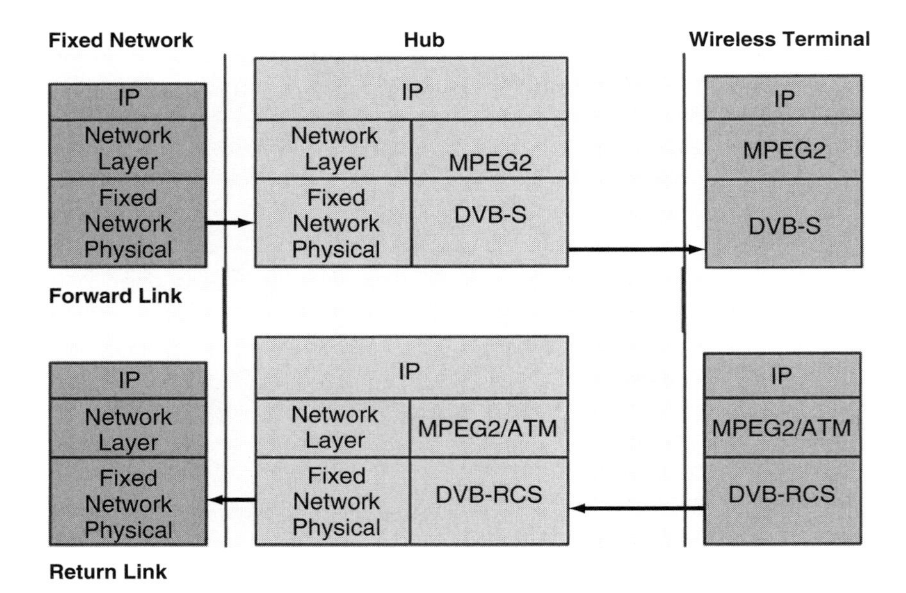

Fig. 8. Protocol stacks (forward and return links).

DVB-RCS Principles

PCR = Program Clock Reference
BTP = Burst Transmission Plan
RQ = Request

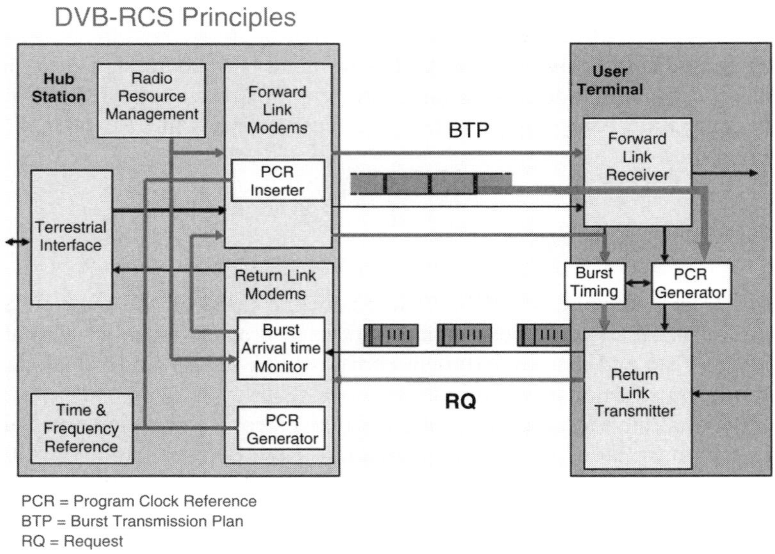

Fig. 9. DVB-RCS principles.

2.4.1 Dynamic Rate Assignment (DRA). The Return link access scheme can operate in two different ways: fixed or dynamic slot MF-TDMA. In fixed slot MF-TDMA, the bandwidth and duration of successive slots are fixed.

In order to reduce the effects of EHF propagation fading, the MF-TDMA slots are dynamically assigned. This means that the Terminal can change frequency, bit rate, FEC-rate and burst length from burst to burst, depending on the link properties (Fig. 10) or the congestion on the network (DRA).

All the Terminals are equipped with a fast frequency hopping feature to achieve the DRA.

2.4.2 TCP Over Large Delay Network. A satellite network is characterized by a large delay (about 550–600 ms round-trip delay) and this aspect limits the TCP speed. Many solutions have been adopted to overcome this problem.

One solution is to introduce a TCP accelerator server in the Hub and a TCP accelerator agent in each Terminal. The TCP accelerator ends the standard TCP protocol sending an "ACK" message to the terrestrial side and uses a satellite optimised protocol over satellite (Fig. 11). Using this solution, a very high speed on FTP download is achieved (tens of Mbps).

The problem with this kind of solution is related with security. The encryption of TCP headers makes impossible to use a TCP accelerator.

For this reason, the above solution has been rejected, while the solution adopted during the trials consists in modifying the TCP Receive Window Size.

The TCP bandwidth depends on the Receive Window Size and the Satellite Round Trip Delay.

With a default window of 64 Kbytes and a Satellite Round Trip Delay of 600 ms, the maximum possible bandwidth is little less than 900 Kbps, but increasing the

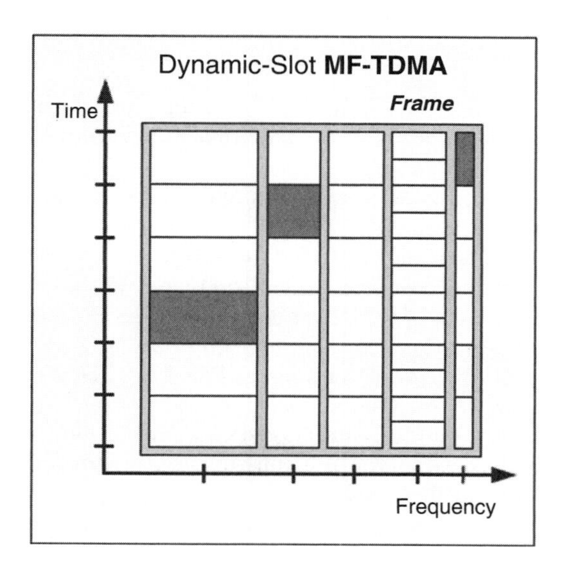

Fig. 10. Dynamic slot MF-TDMA scheme.

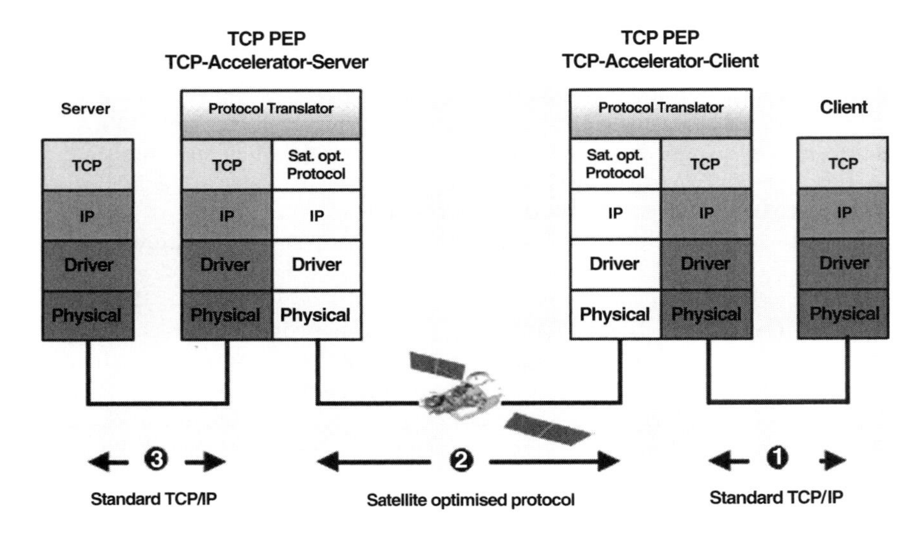

Fig. 11. TCP acceleration.

window up to 256 Kbytes with the same Satellite Round Trip Delay the maximum bandwidth is almost 3.5 Mbps.

3 Trials Results

The results obtained during the trials are provided in the following sections.

3.1 Link Results

The following figures illustrates how the Return link carrier type varies, trying to follow and to compensate the E_s/N_0 changes, thus implementing the Dynamic Rate Assignment.

When the E_s/N_0 decreases, the system tries to overcome the worse link conditions and it selects lower bit rate carriers.

Figures 12–14 show the results obtained for Site 1, which is located in the middle of Italy and of the satellite coverage, while Figs. 16 and 17 shows the data of Site 2, which is located beyond the edge of the satellite coverage. Both the sites were under variable weather conditions.

The meteorological observed conditions of Site 1 are given in Fig. 15, those of Site 2 are given in Fig. 18.

A steady fair weather, on the contrary, gives a quite constant value of E_s/N_0 and a constant value of the carrier type as a consequence, as shown in Fig. 19 of Site 3, which is located in the north of Italy.

The analysis of the traffic data during a period of more than three months provides the Service Interruption graphic of Fig. 20.

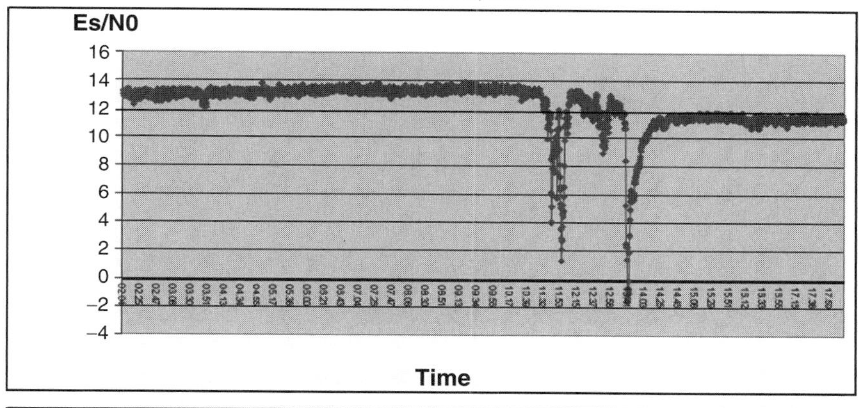

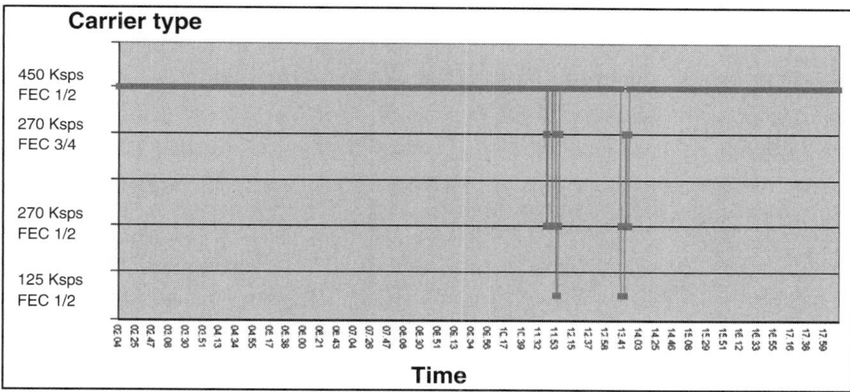

Fig. 12. Site 1 link statistic.

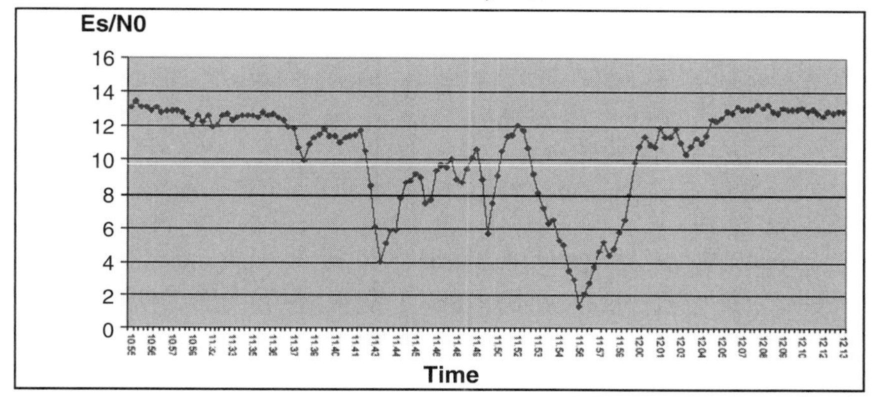

Fig. 13. Site 1 link statistic (details – first section).

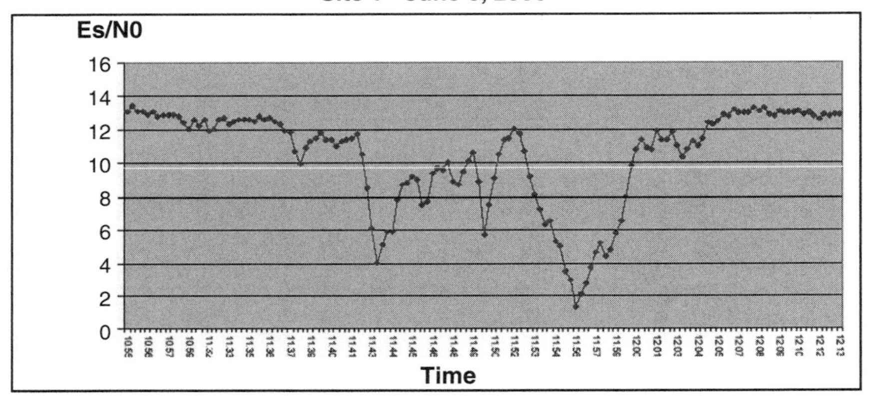

Fig. 13. (*Continued*) Site 1 link statistic (details – first section).

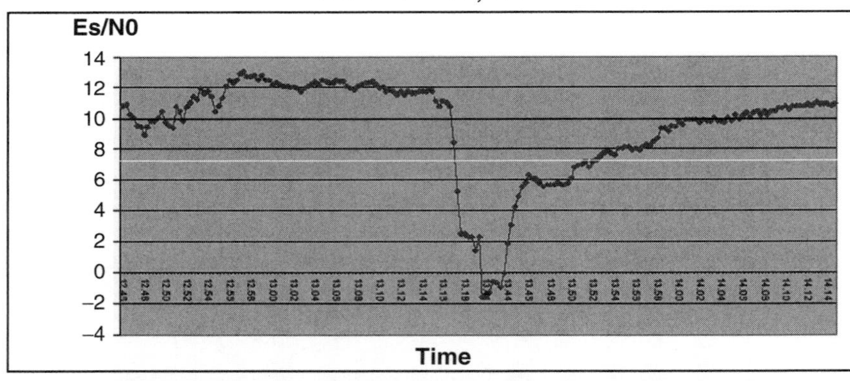

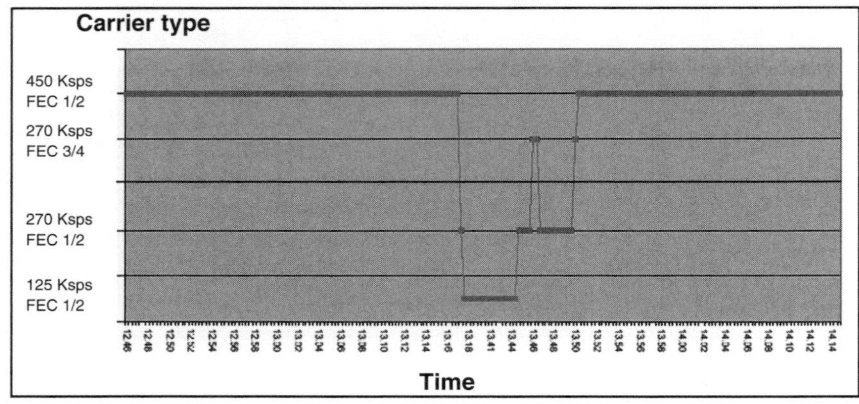

Fig. 14. Site 1 link statistic (details – second section).

Site 1

June 6, 2006

UTC	MSLP hPa	TEMP °C	RELH %	WIND knots	VIS	SKY	SIGWX
07.50	1020	19	68	SSE-8	Good	Overcast	-
08.20	1021	19	68	S-7	Good	Overcast	-
08.50	1020	19	68	SSE-9	Good	Overcast	Light rain shower
09.20	1020	19	72	SSE-8	Moderate	Overcast	Thunderstorm
10.20	1021	16	72	NE-6	Moderate	Overcast	Thunderstorm
11.20	1021	14	82	WSW-6	Moderate	Overcast	Thunderstorm
11.50	1021	14	82	N-8	-	Overcast	Light thunderstorm rain
12.20	1020	14	82	NNE-4	Moderate	Overcast	Thunderstorm
13.20	1021	14	82	N-8	Poor	Very cloudy	Rain shower
13.50	1021	13	82	NNE-11	Poor	Very cloudy	Rain shower
14.20	1020	13	82	NE-8	Poor	Overcast	Light rain shower
14.50	1020	14	76	VAR-3	Good	Overcast	-
15.20	1020	15	77	S-5	Moderate	Overcast	-
15.50	1019	16	67	SW-4	Good	Overcast	-
16.20	1019	17	67	W-6	Good	Scattered clouds	-

Fig. 15. Site 1 meteorological information

Site 2 - June 2-3, 2006

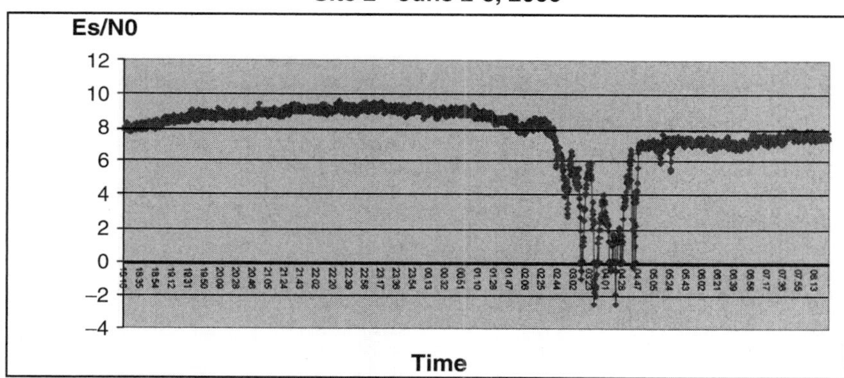

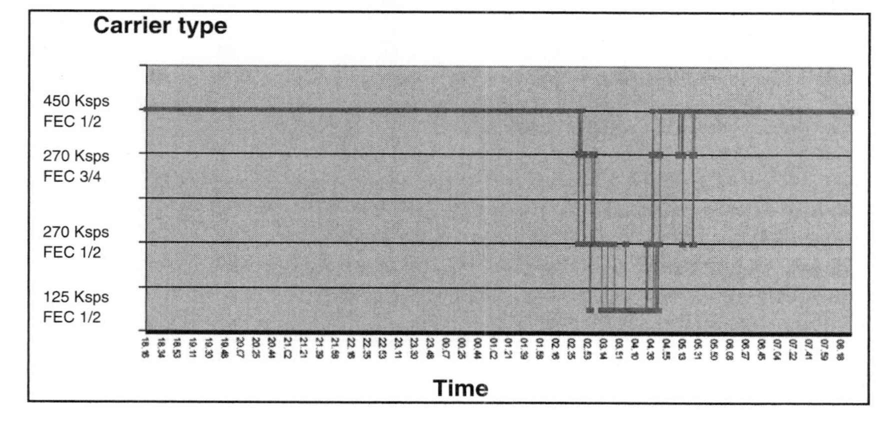

Fig. 16. Site 2 link statistic.

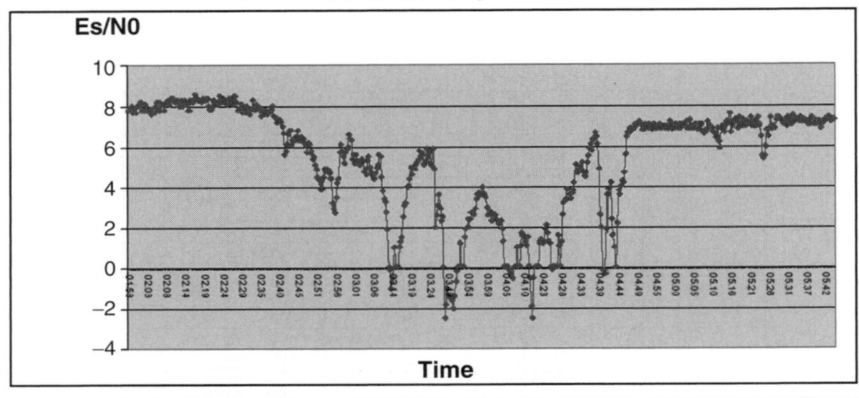

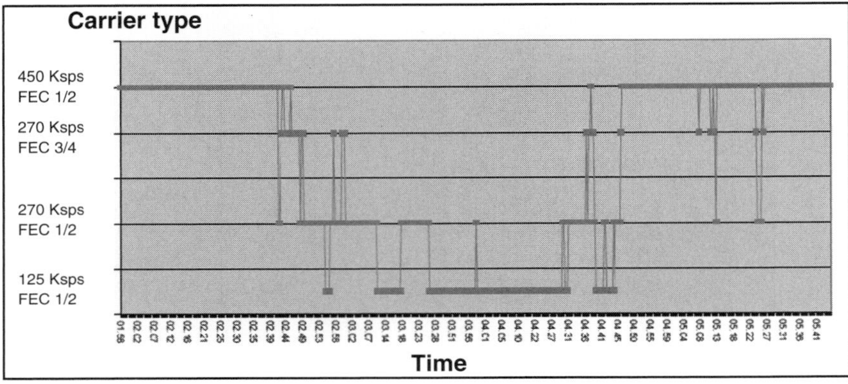

Fig. 17. Site 2 link statistic (details).

Site 2 neighbourhood

June 3, 2006

UTC	MSLP hPa	TEMP °C	RELH %	WIND knots	VIS	SKY	SIGWX
04.00	1011	14	100	W-8	Moderate	Overcast	Heavy thunderstorm rain
05.00	1012	12	100	SSW-8	Moderate	Overcast	Heavy thunderstorm rain
06.00	1011	12	100	E-10	-	Overcast	-
08.00	1010	18	93	SE-12	Good	Scattered clouds	-

Fig. 18. Site 2 neighbourhood meteorological information.

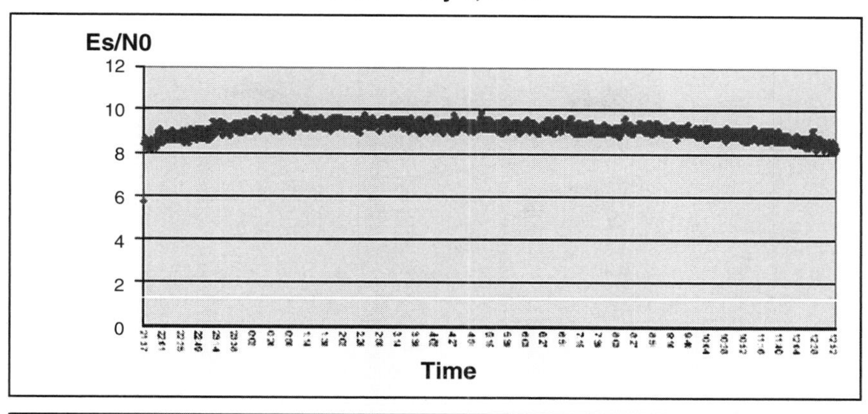

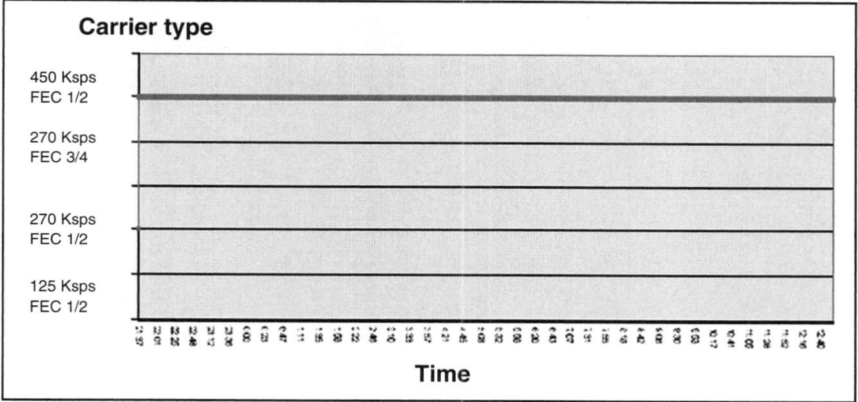

Fig. 19. Site 3 link statistic.

Four sites have been considered, which are located in different areas covered by the satellite, characterized by different meteorological conditions and landscape.

The total amount of Service Interruptions, expressed in hours and minutes, and the percentage of Service Availability of each site is provided in the following table.

It must be noticed that the Service Availability is at least more than 99%, considering that the observation window includes spring months, which are usually characterized by large amounts of precipitation.

Observation Window March 3 - June 9, 2006

	Location	Service interruptions (hh:mm)	Service availability (%)
Site 1	South	3:15	99.86%
Site 2	North	8:56	99.61%
Site 3(*)	Northwest	19:12	99.17%
Site 4	Middle	9:07	99.60%

(*) rainy zone

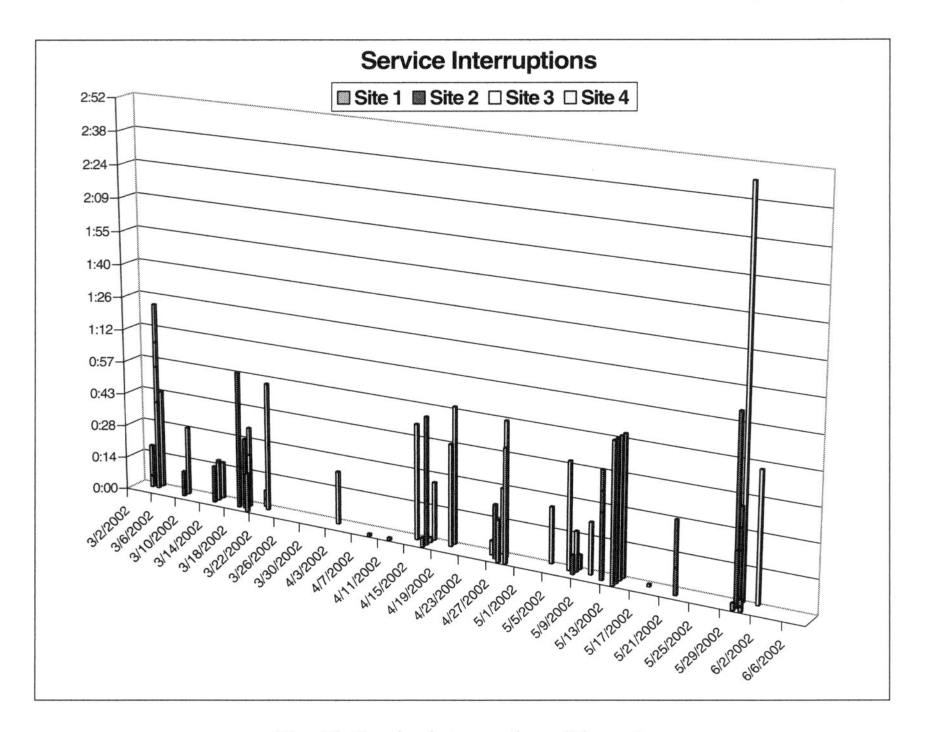

Fig. 20. Service interruptions (hh:mm).

3.2 Throughput Analysis

Several tests on throughput performances have been carried out during the trial period: in the following table a synthesis of the main results is provided.

Throughput Performances

	TCP standard		TCP enhanced	
Application	Forward (Download)	Return (Upload)	Forward (Download)	Return (Upload)
FTP	850 Kbps	248 Kbps	1,3 Mbps	250 Kbps
Internet Download	800 Kbps	240 Kbps	1,3 Mbps	250 Kbps

The Table shows the performances obtained with various protocols (application) with both standard TCP parameters and optimized TCP (using the TCP Receive Window Size method). To be noted that the Return link throughput is limited by the Satellite Channel Rate (in this case limited to 300 Kbps), while the Forward link is limited by the Receive Window Size and Satellite Round Trip Delay, as described above.

3.3 Traffic Analysis

Three QoS (Quality of Services) categories for assignment capacity in the return link have been treated:

- CRA
- VoIP bandwidth on demand (BoD VoIP)
- Best effort (BE) bandwidth on demand (BoD BE)

During the trials several traffic traces have been captured: Figure 21 shows an example of the observation window of the peak capacity requested by a single terminal during a window of several days.

The traffic analysis performed is particularly important to gather information useful to dimension the overall Network throughput for each of the services provided.

4 Conclusions and Future Steps

Taking into account the link results, the Service Availability statistics and the throughput analysis, the trials results confirm the validity and the potentiality of the EHF band for an IP-switched network based on the DVB-RCS open standard.

Some improvements on this type of technology applied in Military environment are requested to better fulfil the specific user's requirements and particularly:

- To fit a non-homogeneous network (based on different Terminal size and capacity). There is the need to have more Forward links at different bit rates: it is clear that very small Terminals cannot manage too much high bit rates.

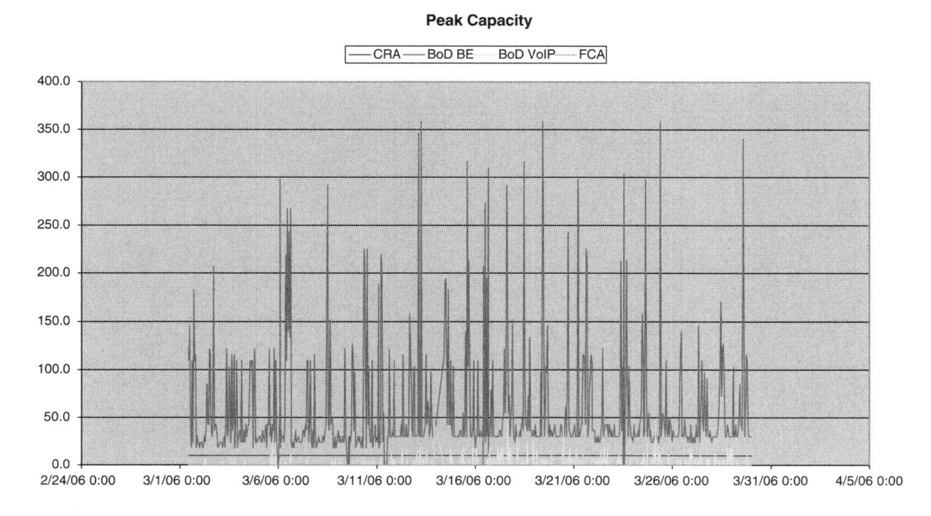

Fig. 21. Traffic peak capacity (Kbps).

- To introduce the meshed topology for peer-to-peer direct connection. If two Terminals have to exchange data (a typical example consists in a VoIP connection), a double satellite hope is necessary at the moment, passing through the Hub and with all the related disadvantages. A meshed topology will eliminate double hope, just connecting together the two Terminals.
- To have a communicating On-the-Move Terminal.

References

[1] EN 301 790 V1.2.2 (2000–12), "Digital Video Broadcasting (DVB); Interaction channel for satellite distribution systems, ETSI and EBU, 2000.
[2] EN 300 421 V1.1.2 (1997–08), "Digital Video Broadcasting (DVB); Framing structure, channel coding and modulation for 11/12 GHz satellite services, ETSI and EBU, 1997.
[3] Nera Satcom AS, "DVB-RCS Nera SatLink Gateway User Manual", Rev. F, Mar 2005.
[4] V. Jacobson, R. Braden, D. Borman, "RFC 1323 – TCP Extensions for High Performance", May 1992.
[5] T. Gagliano, "Documento di Progetto per l' MCS dei Terminali Trasportabili EHF", AIR-0049/01 DI, Rev. 1, Selex Communications SpA, Jan 2006.
[6] "Propagation in non-ionized media", CCIR Report 721–3, Annex to Vol. V, 1990.

Iterative Demapping and Decoding for DVB-S2 Communications

Simone Morosi, Romano Fantacci, Enrico Del Re, Rosalba Suffritti

Department of Electronics and Telecommunications, University of Florence,
Via di S. Marta, 3, Firenze, 50139, Italy
Phone/Fax: +39 055 4796485
e-mail: morosi@lenst.det.unifit.it, fantacci@lenst.det.unifit.it,
delre@lenst.det.unifit.it, suffritti@lenst.det.unifit.it

Abstract. In this paper an original detection strategy for Satellite Digital Broadcasting communications is definited: particularly, we consider the DVB-S2 system, which is proposed as a development of the DVB systems and exploits iterative decoding and higher order modulation; these features allow the derivation of advanced detectors which are based on an iterative demapping and decoding approach. The adoption of this strategy approaches permits a remarkable performance gain and an improvement of the system throughput.

1 Introduction

Recently, the use of satellites in telecommunications has had a growing importance. The satellites are essential for linking users at large distance or in cases where cable connection is unpractical or uneconomic. Moreover, their use is essential in case of mobile users spread over a large area, in particular for aeronautical and maritime communications. Satellite systems give the opportunity of serving efficiently not uniformly distributed and asymmetric traffic due to the flexibility obtained by multiple and redirectable beams and to the possibility of dynamic reconfiguration of the resources. Finally, they allow to extend the range of terrestrial fixed and mobile networks. Satellite systems are mainly used in broadcasting and telephony, where they are especially suitable and efficient. Presently, the interest is focused on voice transmission, high definition video and picture transmission, wide band access to data, etc.

The DVB-S2 (*Digital Video Broadcasting - Satellite - Second Generation*) standard [4], proposed by DVB project[1], is a system aimed at providing a variety of satellite applications, higher power and bandwidth efficiency. The DVB-S2 system was introduced to improve the performance which is obtained by DVB systems; it is very flexible and its main characteristics are the following: a flexible input stream adapter, suitable for operation with single and multiple formats of the input streams, a powerful FEC system based on LDPC (Low-Density Parity Check) codes concatenated

[1] The Digital Video Broadcasting Project (DVB) is an industry-led consortium of over 270 broadcasters, manufacturers, network operators, software developers, regulatory bodies and others in over 35 countries committed to designing global standards for the universal delivery of digital television and data services.

with BCH codes, a wide range of code rates and 4 constellations optimized for non-linear trasponders. In the case of interactive and point-to-point applications, the ACM (Adaptive Coding Modulation) is adopted to optimize channel coding and modulation on a frame-by-frame basis. This technique provides more versatile and robust communications and a dynamic link adaptation to propagation conditions, targeting each individual receiving terminal.

The introduction of higher order modulation schemes permits to analyze the feasibility of a receiver which is based on the Iterative Detection and Decoding. In this paper, in order to ease the application of the Turbo principle, a well-known Turbo Code system, which has been previously considered in the DVB-RCS standard [5], has been introduced and tested. The proposed system permits to achieve remarkable results with a moderate complexity increase.

This paper is organized as follows. In Section II the system model is presented and the Turbo principle is described. Moreover the "tail-biting" technique and the "duo-binary" Turbo Codes are introduced and analysed. In Section III the behavior of the proposed receivers is discussed while in Section IV the simulation results are presented. The concluding remarks are given in Section V.

2 System Model

The successful proposal of Turbo codes [1] suggests the idea of an iterative (Turbo) processing techniques in the design of Satellite Digital Broadcasting communications. The considered Turbo Codes are composed of two Recursive Systematic Convolutional (RSC) codes connected by an interleaver.

Moreover, considering the flexibility of format of input stream and the wide range of code rates of DVB-S2 system, another structure is considered (see Figs. 1 and 2): while the former is used when the code rate is lower than 1/2, the latter is introduced for rates which are higher than 1/2. The two structures are different for the number of sistematic streams which are transmitted: while in the former case the data stream is only one, when the coding rates is higher than 1/2 the sistematic streams are two,

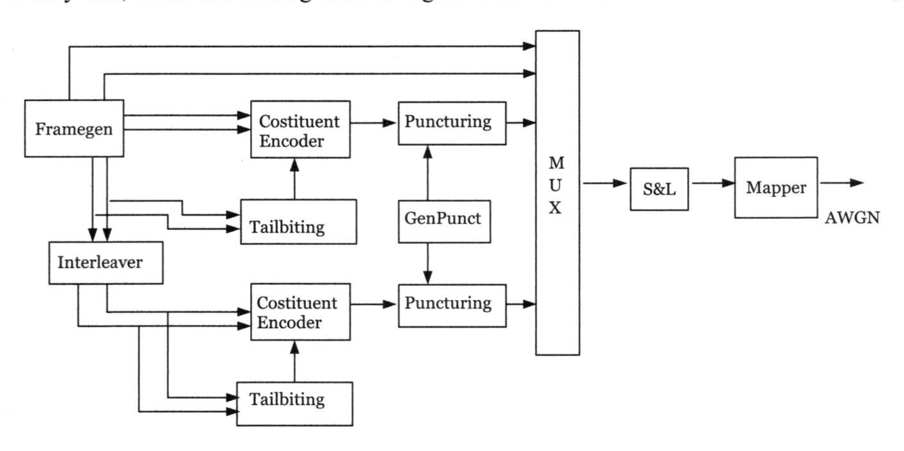

Fig. 1. Transmitter for DVB-S2 like system with rate >1/2.

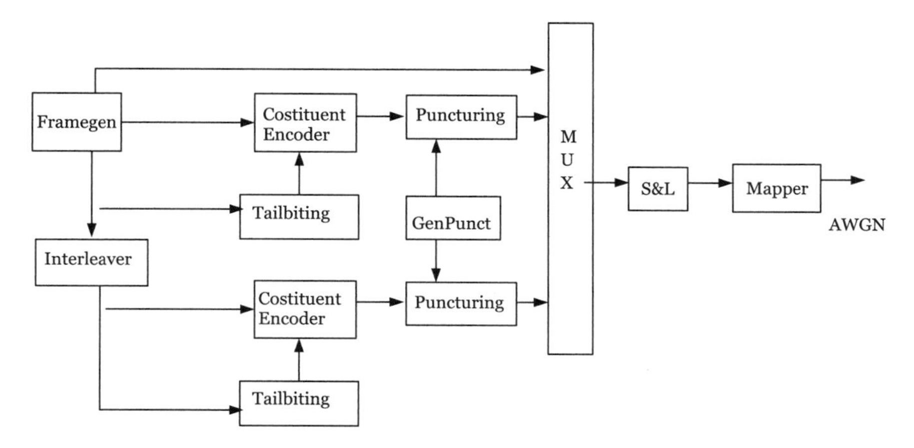

Fig. 2. Transmitter for DVB-S2 like system with rate < 1/2.

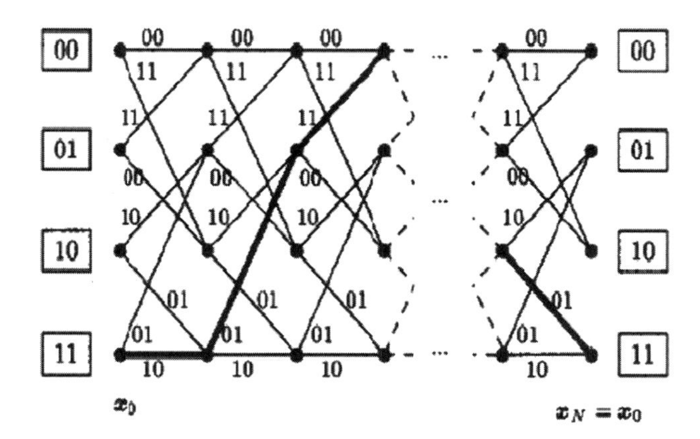

Fig. 3. Tail-biting technique.

that is the data input is a couple of bits. Two different Parallel Concatenated Convolutional Code (PCCC) encoders are so devised, one is a classical convolutional binary code while the other is a "'double-binary'" convolutional code [3]. Since the natural coding rate of the turbo encoder is equal to 1/2, a regular puncturing is performed at the output of the costituent encoders in order to obtain higher code-rate values. The costituent encoders call for the *tail-biting termination* technique so that the code trellis can be viewed as a circle (Fig. 3).

This transforms a convolutional code into a block code allowing any state of the encoder as the initial state and enconding the sequence so that the final state of the encoder is equal to the initial state [6]. There is no need to force the encoder back to the all zero-state by appending a block of tail bits to the information vector (zero termination): therefore the rate loss which is present in classical convolutional codes is avoided [7]. In the following, we describe a rate-k_0/n_0, $k_0 < n_0$, convolutional encoder as a device which generates the n_0-tuple

$$\mathbf{v}_t = \left(v_t^{(1)}, v_t^{(2)}, \ldots, v_t^{(n_0)}\right) \tag{1}$$

of code bits at time t given the k_0-tuple

$$\mathbf{u}_t = \left(u_t^{(1)}, u_t^{(2)}, \ldots, u_t^{(k_0)}\right) \tag{2}$$

The state of the encoder at time t is denoted by $\mathbf{x}_t = (x_t^{(1)}, x_t^{(2)}, \ldots, x_t^{(m)})$ where m is the number of memory elements of the encoder.

The correct initial state, which permits to fulfill the tail-biting boundary condition $\mathbf{x}_0 = \mathbf{x}_N$ can be calculated using the state-space representation

$$\mathbf{x}_{t+1} = \mathbf{A}\mathbf{x}_t + \mathbf{B}\mathbf{u}_t^T \tag{3}$$

$$\mathbf{v}_t^T = \mathbf{C}\mathbf{x}_t + \mathbf{D}\mathbf{u}_t^T \tag{4}$$

of the encoder, where $\mathbf{A}$ is the $(m \times m)$ state matrix, $\mathbf{B}$ denotes the $(m \times k_0)$ control matrix, $\mathbf{C}$ is the $(n_0 \times m)$ observation matrix, and $\mathbf{D}$ denotes the $(n_0 \times k_0)$ transition matrix.

The complete solution of (3) is given by the superposition of the zero-input solution $\mathbf{x}_t^{[zi]}$ and the zero-state solution $\mathbf{x}_t^{[zs]}$. The zero-input solution $\mathbf{x}_t^{[zi]}$ is the state achieved after t cycles if the encoding started in a given state $\mathbf{x}_0$ and all input bits are zero, whereas the zero-state solution $\mathbf{x}_t^{[zs]}$ is the resulting state at time t if the encoding started in the all-zero state $\mathbf{x}_0 = \mathbf{0}$ and the information word $\mathbf{u} = (\mathbf{u}_0, \mathbf{u}_1, \ldots, \mathbf{u}_{t-1})$ has been input. If we demand that the state at time $t = N$, that is after N cycles, is equal to the initial state $\mathbf{x}_0$, we obtain, after working (3), the equation

$$(\mathbf{A}^N + \mathbf{I}_m)\mathbf{x}_0 = \mathbf{x}_N^{[zs]} \tag{5}$$

where $\mathbf{I}_m$ denotes the $(m \times m)$ identity matrix.

Provided the matrix $(\mathbf{A}^N + \mathbf{I}_m)$ is invertible, the correct initial state $\mathbf{x}_0$ can be calculated from the zero-state response $\mathbf{x}_N^{[zs]}$. Therefore the encoding process can be divided into two steps:

- Firstly, the zero-state response $\mathbf{x}_N^{[zs]}$ for a given information word $\mathbf{u}$ is determinated. The encoder starts in the all-zero state $\mathbf{x}_0 = \mathbf{0}$; all $N\,k_0$ information bits are input while the output bits are ignored. After N cycles, the encoder is in the state $\mathbf{x}_N^{[zs]}$. We can calculate the corresponding initial state $\mathbf{X}_0$ using (5) and initialize the encoder consequently.
- The second step is the actual encoding. The encoder starts in the correct initial state $\mathbf{x}_0$ found in the first step; the information word $\mathbf{u}$ is input, and a valid codeword $\mathbf{v}$ is calculated.

In the following, we investigate the "Duo-binary Turbo codes" theory [8], [9], that is a generalization of the classical Binary Turbo Codes theory. In this case, generally, each component code rate was equal to 1/2 so that the resulting turbo code had a natural coding rate of 1/3. Obtaining higher rates involved puncturing the redundancy part of the encoded sequence. In cases where code rates higher than 1/2 are needed, especially when the turbo code is associated with high level modulation schemes, a better solution than puncturing involves component encoders able to encode 2 bits at the same time, that is "Duo-binary RSC encoders".

This structure can lead to improve the global performance in comparison with the original turbo codes, especially for block coding. In practice, for the construction of duo-binary turbo codes, we consider a parallel concatenation of two identical RSC encoders with 2-bits word interleaving (see Fig. 4). In order to encode data blocks without any rate loss or performance degradation due to direct truncating, the principle of circular trellises (tail-biting termination) is adopted. Figure 5 shows the structure of the costituent encoder that receives in input a couple of bits that is it said to be working at "symbol-level". As for the interleaver, the permutation function is performed on two levels:

- a *inter-symbol permutation* that modifies the order of symbols. This rule is almost regular. It is based on faint vectorial fluctuations around the locations given by regular permutation.
- a *intra-symbol permutation* that reverses bits in some prearranged couples.

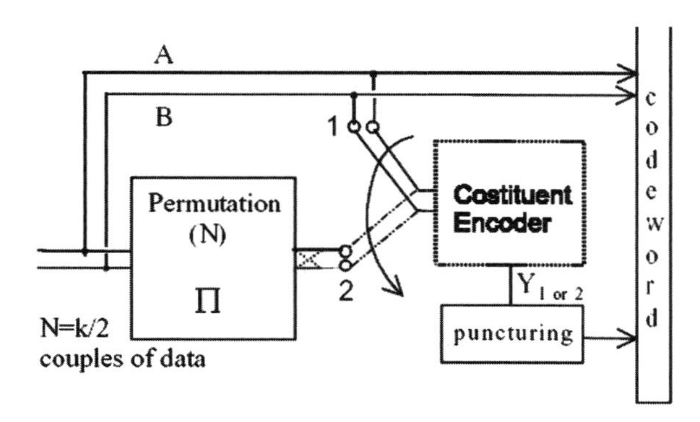

Fig. 4. Duo-binary turbo encoder.

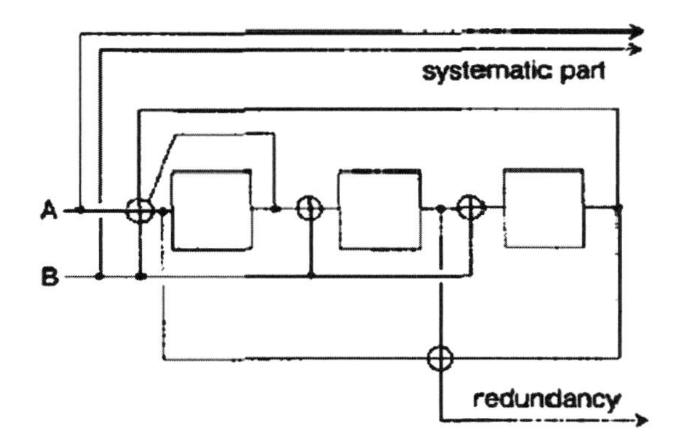

Fig. 5. 8-state quaternary recursive systematic convolutional (RSC) code with generators 15,13.

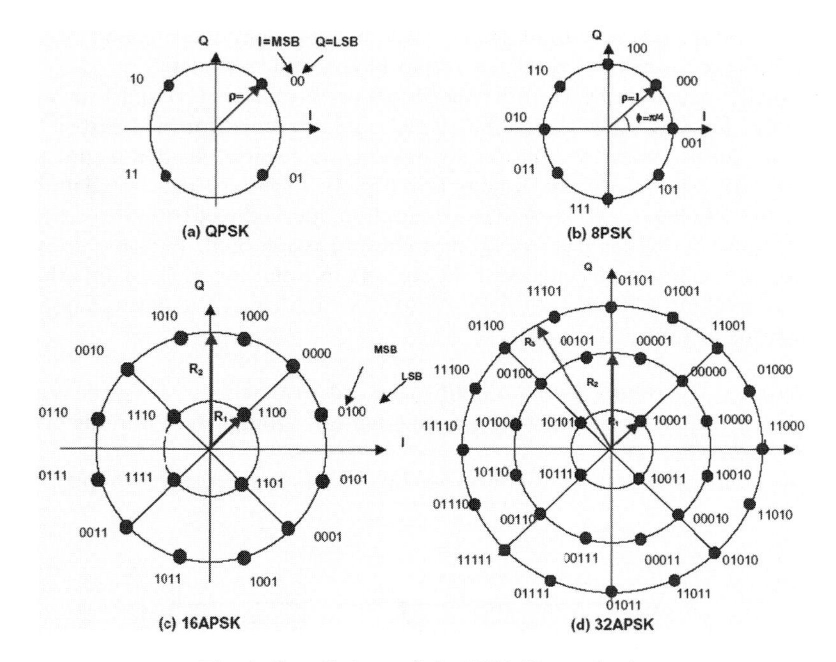

Fig. 6. Costellations of the DVB-S2 standard.

The encoded sequence, both in the structure for code rates higher than 1/2 and in that for code rates lower than 1/2, is multiplexed and it is sent to the mapper. The DVB-S2 standard provides for 4 costellations: the classical QPSK and 8PSK, and the 16APSK (amplitude and phase shift keying) and the 32APSK which allow to minimize the effects of the non-linear distorsion due to non-linear trasponders. To minimize these effects, new advanced pre-distortion methods are adopted in the up-link station. These modulations are based on concentric "rings" of equi-spaced points: 16 APSK is composed by a inner 4PSK surrounded by an outer 12PSK while 32APSK is composed by a inner 4PSK, by a 12PSK ring in the middle and by an outer 16PSK (see Fig. 6).

3 The Proposed Receivers

As for the transmitter, also for the receiver we propose two structures, one for code rates higher than 1/2 and the other for code rates lower than 1/2. Both structures rely on the iterative turbo decoding, which is based on the MAP algorithm. For every structure the canonical hard demapping, the soft demapping and the iterative soft demapping have been realized so as to show the improvements due to use of the different techniques (see Figs. 7, 8, 11, and 12).

In the case of Soft Demapping, the receiver requires soft information about reliability of both the systematic and the parity bits. The complex channel symbols are demapped by a log-likelihood ratio calculation for each of the M coded bits

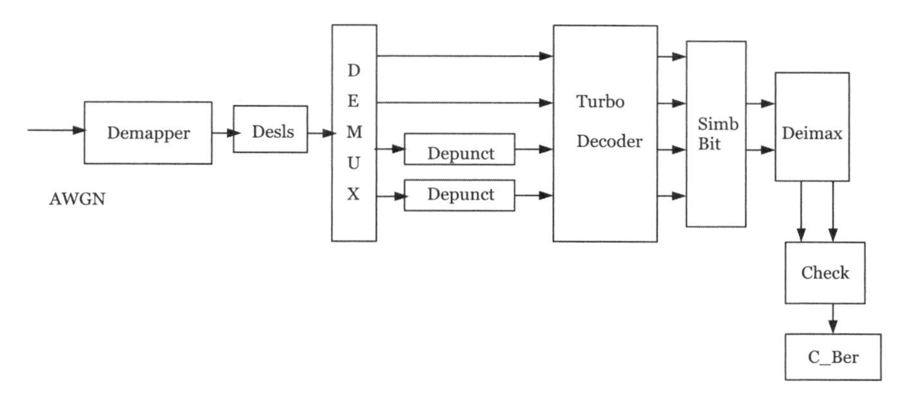

Fig. 7. Receiver with hard or soft demapping and iterative decoding for DVB-S2 like system with rate >1/2.

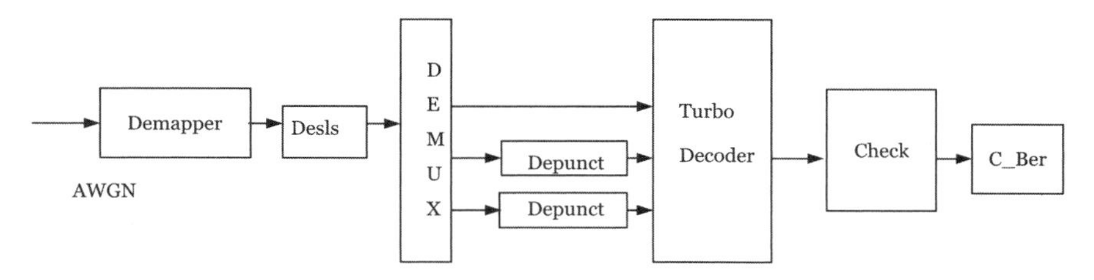

Fig. 8. Receiver with hard or soft demapping and iterative decoding for DVB-S2 like system with rate <1/2.

per symbol. Generally, the concept of Log Likelihood Ratios (LLRs) is useful to simplify the passing of information from one component decoder to the other in the iterative turbo decoding [10]. The LLR of a data bit c_i is denoted as $L(c_i)$ and is defined as the logarithm of the ratio of the probabilities of the bit taking its two possible values, ie:

$$L(c_i) \stackrel{\Delta}{=} \ln \frac{P(c_i = 1)}{P(c_i = 0)} \tag{6}$$

The sign of the LLR $L(c_i)$ of a bit c_i indicates whether the bit is more likely to be 1 or 0, and the magnitude of the LLR gives an indication of how likely it is that the sign of the LLR gives the correct value of c_i.

This concept proves useful also for the Soft Demapping. In this case the Demapper produces the conditional LLRs, $L(c_i \mid \mathbf{z})$, $i = 1, \ldots, M$, which can be calculated using the following expression [2]:

$$L(c_i \mid z) = \ln\left(\frac{P(c_i = 1 \mid \mathbf{z})}{P(c_i = 0 \mid \mathbf{z})}\right) = L_{ap}(c_i) + \tag{7}$$

$$\ln\left(\frac{\frac{1}{\sigma^2}\sum_{j = bin1\,(i)} \exp\left(\left\|\mathbf{z} - y_j\right\|^2 + \sum_{k = other\,(j)} L_{ap}(c_k)\right)}{\frac{1}{\sigma^2}\sum_{j = bin0\,(i)} \exp\left(\left\|\mathbf{z} - y_j\right\|^2 + \sum_{t = other\,(j)} L_{ap}(c_t)\right)}\right)$$

where:

- $\sum_{j\ =\ bin1\ (i)}$ is referred to the symbols with the ith considered bit equal to 1.
- $\sum_{j\ =\ bin0\ (i)}$ is referred to the symbols with the ith considered bit equal to 0.
- $\mathbf{z}$ is the matched filter output.
- y_j is the jth considered symbol.
- $\sum_{k\ =\ other\ (j)}$ is referred to the other bits of the considered symbol that are equal to 1.
- $\sum_{t\ =\ other\ (j)}$ is referred to the other bits of the considered symbol that are equal to 0.

These values are called the *a posteriori* log-likelihood ratios.

After the Demapper, these soft values are passed to the iterative decoder composed by the two component decoders that are connected by the interleavers as shown in Figs. 9 and 10. At each new iteration, the iterative structure permits to allow an additional information to the first decoder in order to obtain a more accurate set of soft outputs, which are then used by the second decoder as *a priori* information. The structure for code rates lower than 1/2 works in the logarithmic domain while that for code rates higher than 1/2 works in the domain of the probabilities. In the latter case, since the structure uses the "Duo-binary Turbo codes" and so it works at "symbol level", the MAP algorithm is modified for producing the

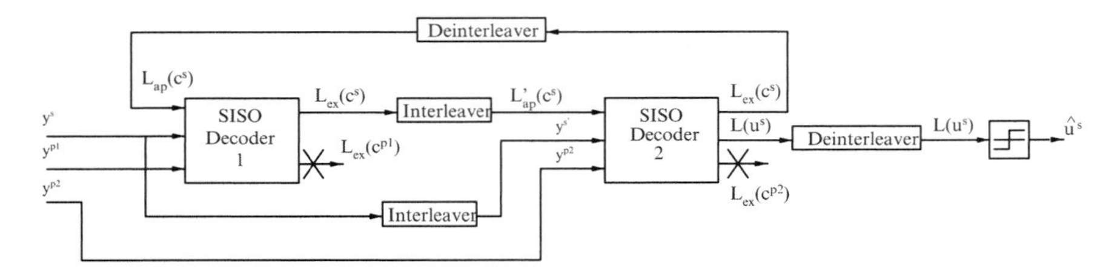

Fig. 9. Binary turbo decoder.

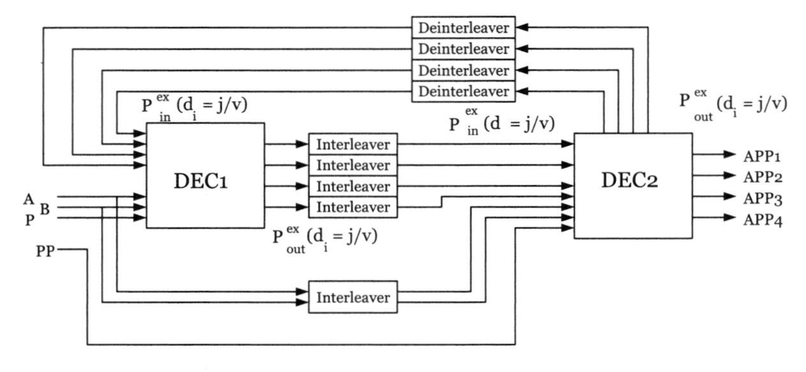

Fig. 10. Duo-binary turbo decoder.

a posteriori probability of each sistematic symbol. Moreover, also in the decoder, the permutation function of the interleavers and of the deinterleavers is performed on two levels.

In case of Iterative Soft Demapping, the decoding algorithm is properly modified to produce also the *extrinsic* information about the parity bits. The *extrinsic* information is obtained subtracting from the *a posteriori* information, the *a priori* information and the received systematic channel input. This relation is valid for sistematic and non-sistematic coded bits. After a fixed number of turbo decoder iterations, through the deinterleaver, the *extrinsic* information of coded bits at the output of the turbo decoder are fed back to the input of the soft demapping as the *a priori* information, $L(c_i)$, for the next receiver iteration, as shown in Figs. 11 and 12. The demapper can utilize the a priori information received from the decoder and calculate improved *a posteriori* values, $L(c_i \mid \mathbf{z})$, which are passed as *extrinsic* values to the decoder for further iterative decoding steps [2].

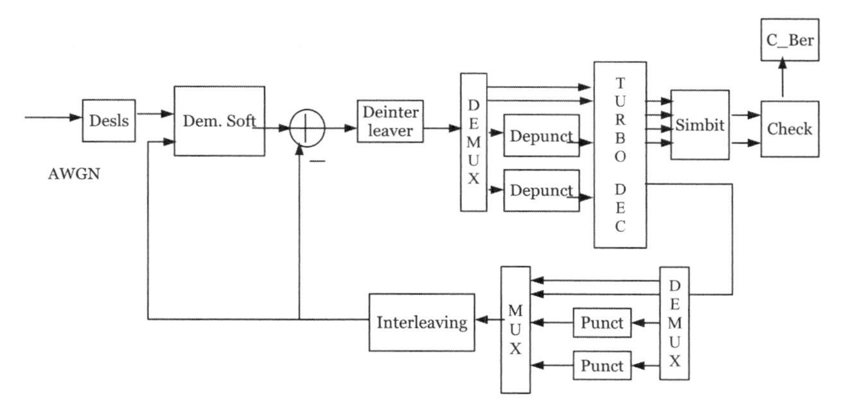

Fig. 11. Receiver with iterative soft demapping and iterative decoding for DVB-S2 like system with rate > 1/2.

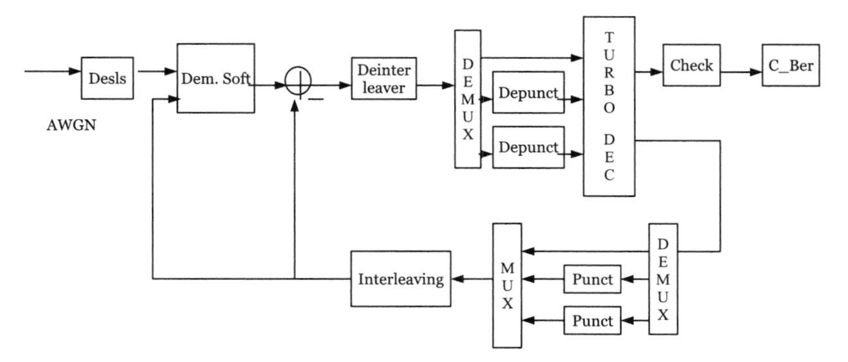

Fig. 12. Receiver with tterative soft demapping and iterative decoding for DVB-S2 like system with Rate < 1/2.

4 Simulations Results

In order to demonstrate the effectiveness and the performance of the proposed receivers, a set of computer simulations was performed.

For all simulations we used the *Log-MAP* Algorithm as decoding algorithm in the Turbo Decoder, since it permits to achieve very good performance but with a complexity lower than the MAP algorithm. The performance of the proposed receivers has been analyzed in a synchronous AWGN channel.

As shown before, we considered two structures: the former where code rates are lower than 1/2 and the latter where code rates are higher than 1/2. In the following we will separate these two cases also for simulation results.

In Fig. 13 we report the performance in terms of Bit Error Rate (BER) in case of QPSK modulation, rate equal to 1/3, soft demapping and iterative decoding. The performance has been analyzed for different values of the number of iterations of the turbo decoder. As the signal to noise ratio (SNR) grows, the value of BER decreases and the gain is more evident as the number of the iterations passes from 2 to 4, to 8 and to 10 iterations. The optimum value of the number of the decoding iterations results to be equal to 8. This occurs because, beyond this value, the additional information which can be obtain from the extrinsic information is lower.

In Fig. 14 we show the performance in terms of Bit Error Rate (BER) in case of QPSK modulation, rate equal to 1/3, iterative soft demapping and iterative decoding. In this case, we report the results as the total number of iterations changes.

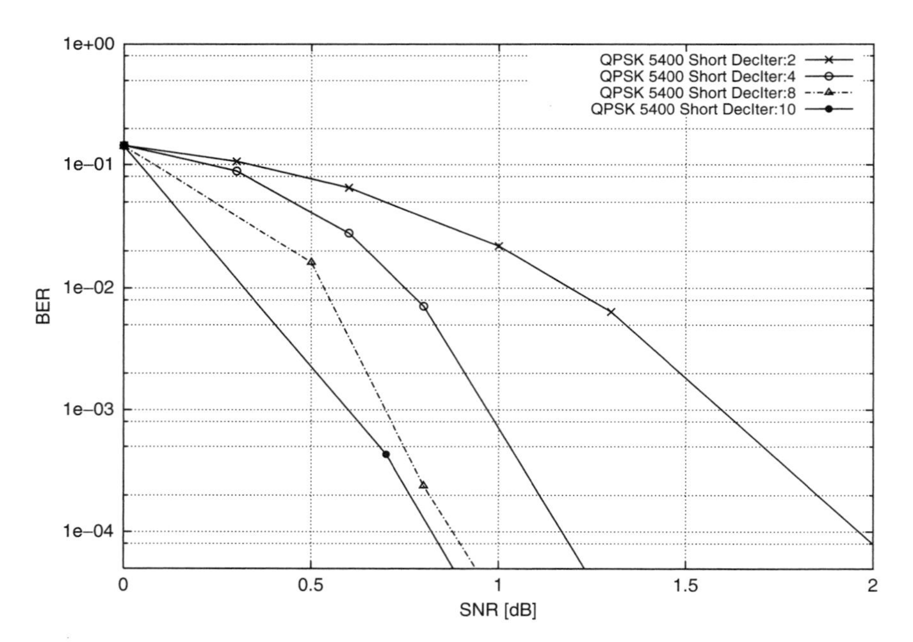

Fig. 13. Performance in terms of bit error rate (BER) for QPSK, rate 1/3, in case of soft demapping and iterative decoding.

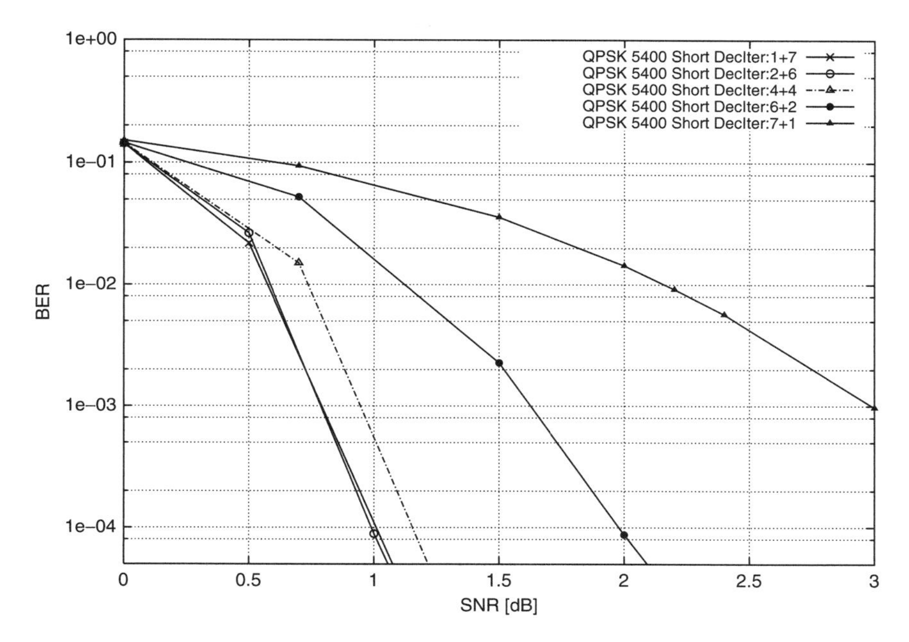

Fig. 14. Performance in terms of bit error rate (BER) for QPSK, rate 1/3, in case of iterative soft demapping and iterative decoding: 1 + 7, 2 + 6, 4 + 4, 6 + 2, 7 + 1 iterations.

As "total number" we mean the sum of the number of iterations of the turbo decoder before the outer demapping iteration and of the ones after the external iteration. We can observe which is the best strategy for realizing the iterative demapping. Indeed, the best results are obtained when few turbo iterations are performed before the external iteration of demapping and many turbo iterations are carried out after the iteration of demapping.

In Figs. 15 and 16 we report the performance comparison between hard demapping, soft demapping and iterative soft demapping in terms of Frame Error Rate (FER) and of net throughput for QPSK modulation with Gray mapping and code rate equal to 1/3. The considered total number of iterations is equal to 8. The soft demapping, both iterative and non-iterative, provides much better performance than the hard demapping. In this case no advantage is due to the iterative demapping in comparison with the non-iterative approach: this result depends on the mapping strategy which has been adopted and on the number of demapping iterations. The Gray mapping determines the independence of the bits in phase and in quadrature and so it is not possible to take advantage of the joint information of the bits belonging to the same symbol.

In the following we analyze some simulation results for the structure where the code rates are higher than 1/2. As shown before, this structure uses the duo-binary turbo codes: hence it was necessary to modify the decoding algorithm and the soft demapper.

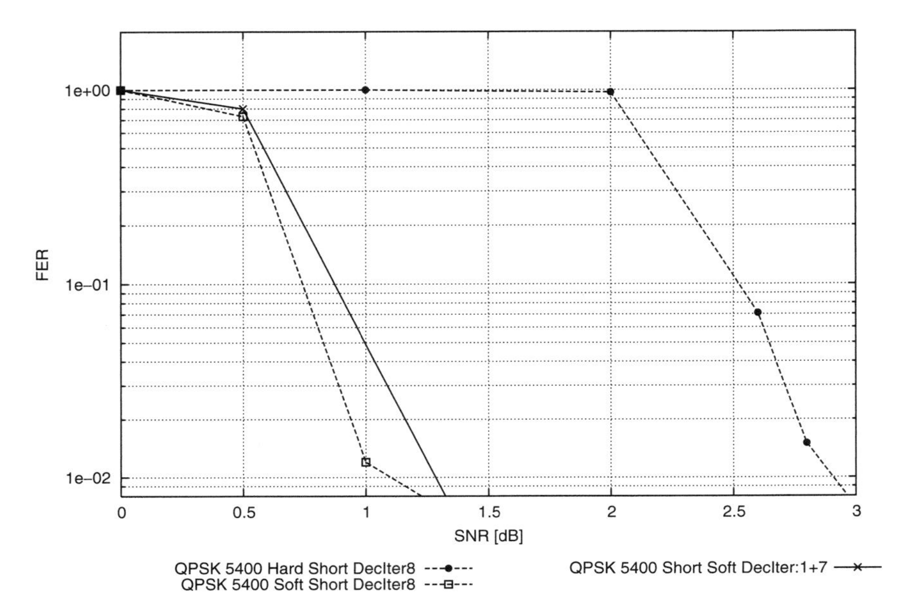

Fig. 15. Performance comparison in terms of frame error rate (FER) for QPSK, rate 1/3, between hard demapping, soft demapping and iterative soft demapping.

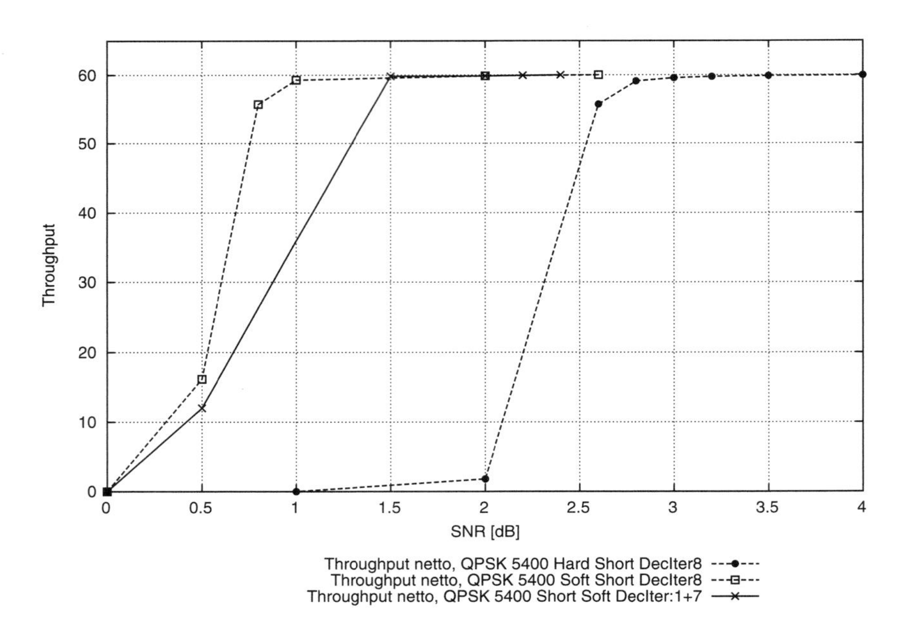

Fig. 16. Performance comparison in terms of net throughput for QPSK, rate 1/3, between hard demapping, soft demapping and iterative soft demapping.

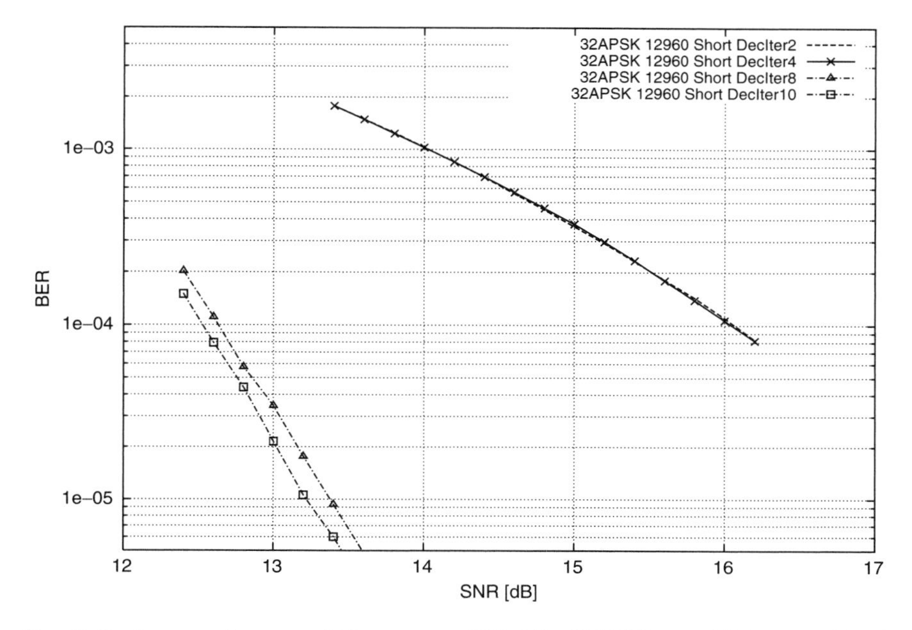

Fig. 17. Performance in terms of bit error rate (BER) for 32APSK, rate 4/5, in case of hard.

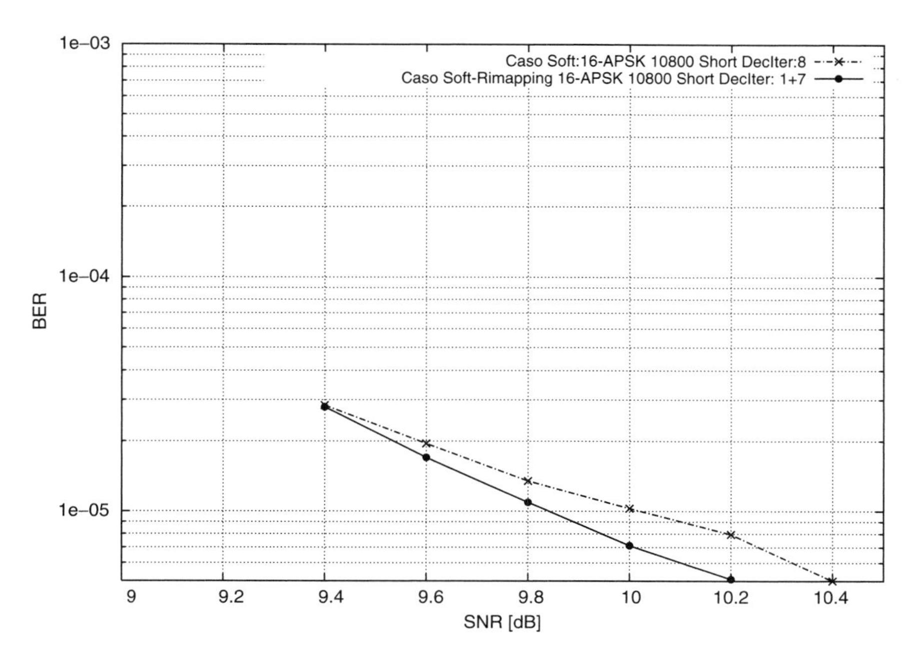

Fig. 18. Performance comparison in terms of bit error rate (BER) for 16APSK, rate 2/3, between soft demapping and iterative decoding.

In Fig. 17 we report the performance in terms of Bit Error Rate (BER) for 32APSK modulation, code rate equal to 4/5, hard demapping and iterative decoding. As in Fig. 13, as the signal to noise ratio grows, the BER decreases more quickly as the number of the iteration of the turbo decoder increases.

Finally, in Fig. 18, we show the comparison in terms of Bit Error Rate between non-iterative soft demapping and iterative soft demapping for 16APSK modulation and code rate equal to 2/3. In this case the iterative soft demapping behaves better than the non-iterative one, i.e., the iterative demapping produces some advantages. Since the 16APSK modulation does not present a Gray mapping it allows to exploit the joint information of bits belonging to the same symbol.

5 Concluding Remarks

In this paper we have proposed a novel detection strategy for satellite digital broadcasting systems. The advanced detectors which have been described, are based on an iterative decoding and on iterative and non-iterative demapping approach. The adoption of the soft demapping has presented a remarkable performance gain in comparison with the canonical hard strategy in all cases which have been analyzed. The advantages of the iterative soft demapping depend on the mapping strategy which is choosen and on the number of demapping iterations. The best performance is obtained when a non-Gray approach is considered.

References

[1] C. Berrou, A. Glavieux and P. Thitimajshima "Near Shannon limit error-correcting coding and decoding: Turbo codes" *Proc. 1993 Int. Conf. on Communications (ICC '93)*, Washington DC.
[2] S. ten Brink, J. Speidiel and Yan Ran-Hong "Iterative Demapping and Decoding for multilevel modulation" *IEEE Global Telecomunications Conference*, Vol. 1, pp. 579–584, 1998.
[3] C. Berrou and C. Douillard "Turbo Codes With Rate-m/(m+1) Costituent Convolutional Codes" *IEEE Transactions on Communications*, Vol. 53, No. 10, pp. 1630–1638, October 2005.
[4] DVB-S2, *Second generation framing structure, channel coding and modulation system for Broadcasting, Interative Services, News Gathering and other broadband satellite applications*, ETSI EN 302 307, v. 1.1.1, 2005.
[5] DVB-RCS, *Digital Video Broadcasting (DVB); Interaction channel for satellite distribution system*, ETSI EN 301 790, v. 1.3.1, 2003–03.
[6] C. Weiss, C. Bettstetter "Code Construction and Decoding of Parallel Concatenated Tail-Biting Codes" *IEEE Transactions on Information Theory*, 2001.
[7] C. Weiss, C. Bettstetter, S. Riedel and D.J. Costello "Turbo Decoding with Tail-Biting Trellises" *IEEE*, 1998.
[8] C. Berrou, M. Jézéquel, C. Douillard and S. Keroudan "The advantages of nonbinary turbo codes" Proc.Information Theory Workshop, Cairns, Australia, pp. 61–63, September, 2001.
[9] C. Berrou and M. Jézéquel "Non binary convolutional codes for turbo coding" Electronics Letters, Vol. 35, pp. 39–40, January, 1999.
[10] J. P. Woodard and L. Hanzo "Comparative study of Turbo Decoding Techniques: an Overview", ed. IEEE, 2000.

New Perspectives in the WAVE W-Band Satellite Project

A. Jebril[1], M. Lucente[1], T. Rossi[1], M. Ruggieri[1], S. Morosi[2]

[1]University of Rome "Tor Vergata", Dept. of Electronic Engineering
Via del Politecnico 1, 00133 Rome-Italy
Tel: +39 06 7259 7258, Fax: +39 06 7259 7455, e-mail: ruggieri@uniroma2.it
[2]University of Firenze, Dept. of Electronics and Telecommunications
Via di S. Marta, 3, 50139 Firenze-Italy
Tel: +39 055 4796 485, Fax: +39 055 472858, e-mail: morosi@lenst.det.unifi.it

Abstract. In 2004 ASI funded phase-A of the WAVE satellite mission, a feasibility study to design and develop a W-band geostationary payload. The aim was to perform experimental studies of the W-band channel and possible utilization in satellite data communications and data-relay services. The next phase of this project will address the experimental-only nature of W-band and the strategic relevance of making investments towards the development of the payload by performing preliminary experiments, such as the Aero-WAVE mission. The Aero-WAVE mission is presented as a basic test to provide the necessary measures leading to the development of the primary mission into GEO orbit. Aero-WAVE is a scientific payload that operates in W-band. It will be embarked on-board a stratospheric platform or High Altitude Platform (HAP) in order to perform a first test of the channel behaviour at an altitude of about 20 km. The stratospheric platform is a high altitude aircraft known as the M-55 Geophysica, a Russian aircraft well known in the atmospheric research environment and the European scientific community since 1996. The M-55 Geophysica will be equipped with the payload instrumentation to carry out the measurements. These measurements and the resulting data provided throughout the Aero-WAVE mission will give good indications of the channel behaviour at different weather conditions and at different locations within the GEO-WAVE coverage.

1 Introduction

During the last few years, the growth of innovative multimedia services have led towards the need to explore higher and higher frequency bands such as W-band (75–110 GHz). This unexplored frequency range could satisfy and improve the advanced and innovative services allowing high-volume data transfers. Therefore, it is becoming more and more necessary to exploit this frequency band which can be considered as the new frontier of satellite communications. Nonetheless, W-band potentialities are still unused due to the unknown atmospheric channel behaviour at these high frequencies since no satellite mission of either a scientific or commercial nature has performed any space experimentations at W-band frequencies.

Italy, through the Italian Space Agency (ASI), is one of the first countries that have made an effort towards the exploitation of W-band. It has financed two satellite projects; DAVID (DAta and Video Interactive Distribution) and WAVE

(W-band Analysis and VErification), in addition to other terrestrial and balloon-based tests and proposals towards the study of the W-band satellite channel.

2 WAVE A2: New Developments

W-band is recently considered as a "technological frontier"; thus representing the true challenge towards which the research community and the industry should concentrate their efforts.

The study of phase-A has been carried out [1, 2] to determine the feasibility of W-band to be used for the development of a telecommunications payload in a GEO orbit mission to provide the following:

- an experimental study on the propagation in W-band;
- preliminary experimentations towards the commercial utilization of the channel as a data relay.

This study has resulted in the feasibility of a W-band payload in both technical and commercial terms. It also showed that the development of a payload for telecommunications in W-band would be of a fundamental strategic importance from both technological and scientific points of view.

The second phase of the WAVE project A2, that is expected to start by the beginning of 2007, will carry out in parallel two main studies as depicted in Fig. 1; the first is a demonstrative study concerning the development of an experimental payload on-board a High Altitude Platform (HAP), referred to as Aero-WAVE, in addition to the development of a small nano-satellite platform. The second is a pre-operative study on the GEO-orbit outlined in the previous phase, and a payload in LEO-orbit on a dedicated micro-satellite platform. All the studies related to the development of the project WAVE-A2 phase will be focused on the following objectives:

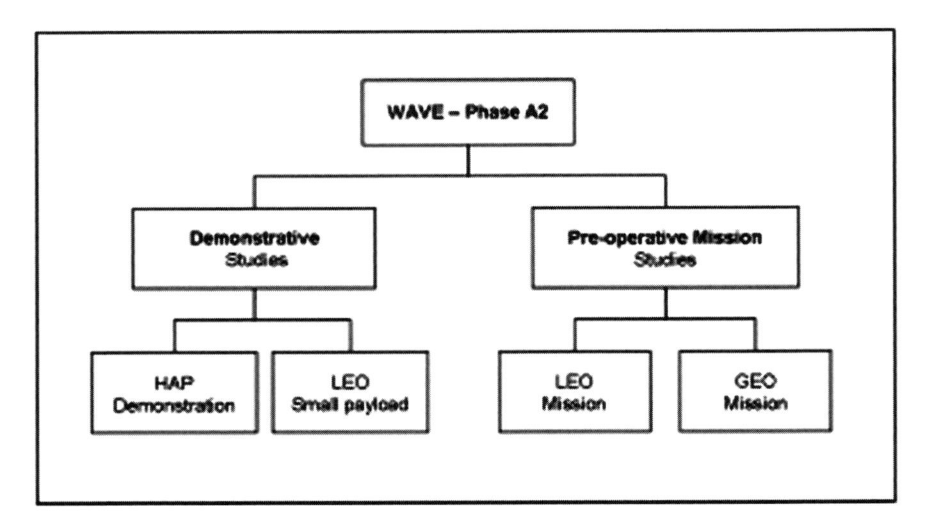

Fig. 1. The WAVE-A2 scheme.

- detailed definition of the proposed missions and of their payload;
- definition of the configuration and of the specifications of the whole system;
- analysis of the architecture TX/RX front-ends and of the antenna;
- evaluation of possible launching platforms and preliminary study of boarding;
- evaluation of the strategic technological elements and risk analysis;
- preliminary identification study of the earth segment (TT&C, Payload Control Centre).

3 The W-Band Experiment

The W-band experiment at this phase of the WAVE consists in the installation of a W-band payload with two RF transmitting channels on-board the Geophysica M-55 aircraft. The M-55 aircraft has been developed in Russia to be used as a high-altitude reconnaissance aircraft in operation since August 1988. Since January 2002, the M-55 Geophysica is supervised by the European Economic Interest Group (EEIG) Geophysica, providing services to the international scientific community [3]. The M-55 aircraft is currently the only subsonic aircraft in Russia and Europe performing long endurance flight at altitudes up to 21 km with the capability to perform long endurance high altitude flights (at altitudes of 17–21 km) providing services to a number of applications such as:

- scientific research of upper atmosphere layers;
- geomagnetic research;
- ecological monitoring;
- mapping and making thematic maps of land and water surfaces in different spectral ranges;
- retransmission of electromagnetic signals and localization of distress signals in global rescue system.

The scientific instrumentation is installed in special bays on-board the M-55 aircraft and operates during the flight in fully automatic mode. At cruising altitude the speed remains constant due to the flight altitude not exceeding 270 km/h. The flight radius of action can reach up to 1600 km.

The data transmission section operates at 94 GHz, and with a 92 GHz beacon, both used to provide the necessary signal reception experiments in W-band at the fixed earth station in Spino d'Adda and at the transportable station around Rome. The received signals will provide the necessary elements to analyze the channel behaviour at different weather conditions. Two flight campaigns are proposed; the first will be over the area around Rome where the transportable earth station will be used. This will give a good opportunity to provide different measurements in different locations. The second campaign is over the area of Spino D'Adda fixed station. The aircraft altitude for both campaigns will be around 20 km and the duration of each flight is about 4–5 hours. The flight route of all campaigns will have a diameter of about 10 km with an aircraft roll angle of 22°. The aircraft displacement data are shown in Table 1.

Assuming that the earth station is located perpendicularly below the center of the aircraft circulation orbit, and the on-board data transmission antenna has a beamwidth of 1.04°, the ground footprint has a diameter of 470.42 m. Therefore the

Table 1. M-55 aircraft flight data.

Cruising altitude	Turning radius	Angle of roll	Roll stability rate
20 km	10.76 km	22°	± 0.2°

corresponding elevation angle of both earth stations will be 61.72°. The on-board antenna will have an angle of 230.28° with respect to the yaw axis of the aircraft movement. The data transmission experiments are based on the transmission of a known bit stream saved in the on-board memory using a different data rate and different fixed coding-modulating scheme for every flight. The main goal at this stage is to counteract the large dynamic range of the expected attenuation by transmitting various rates of coded signals and the use of bit error control so that the power margins provision is traded for bandwidth and coding complexity. Therefore, in order for the W-band measurements to approach as much as possible the time-variant capacity provided by the W-band link and the strict requirements of the overall link quality and low Bit Error Rate (BER) values (BER$< 10^{-11}$), variable coding rates and hence the yielding variable information rates should be used. This can be accomplished by utilizing the flexibility of the rate-compatible punctured convolutional (RCPC) coding for better error correction adopted previously by the DAVID project [4][5]. This coding technique allows an encoder/decoder pair to change code rates without the need to make major changes to the hardware.

4 The Payload

The proposed payload design has been carried out considering the use of existing proved hardware to minimise failure risks, easy integration and test procedure. In this payload, it is possible to use solid-state power amplifiers (SSPA) given that there is no need for higher power transmission at this stage since the transmission distance to the HAP is relatively small. A 100 mW device would be suitable for the data transmission section, while a unidirectional antenna is proposed for the beacon in order to carry out the propagation experiment even in poor M-55 stability conditions. The beacon required power can be reached combining two or more 100 mW SSPA stages. In the up-conversion section, a 82 GHz master oscillator is used, based upon a reference source at 100 MHz. This reference source is also used to generate a 2 GHz signal from which, by means of multipliers, the 10 and 12 GHz signals are obtained. The first part of the payload is an on-board memory containing the bit stream to be transmitted followed by the modulation section in which the signal will be modulated by a 12 GHz Ku-band carrier signal obtained from the 82 GHz master oscillator previously proposed for the DAVID mission [5]. This modulated signal is then up-converted to the required W-band 94 GHz frequency using the 82 GHz master oscillator. The modulated signal is then amplified using a solid-state power amplifier, filtered and then sent through the antenna. Another important part is the beacon which will generate a 92 GHz sinusoidal signal to be transmitted for power measurements. To measure the power of the sinusoidal signals, the beacon must send a sufficiently stable signal. With the proposed minimal configuration, it is appropriate to use also radiometric equipment, in order to perform the so-called "bias removal", specifically to estimate the "zero dB" attenuation level. This measurement technique, derived and validated in previous propagation experiments at lower frequencies (Olympus, Italsat) allows the calibration of the channel power, considering the clear-air attenuation due to presence of gases in atmosphere, which also affects the signal transmitted by beacon. At the receiver end, the signal is demodulated and the resulting baseband bit stream is compared with that of the originally transmitted (saved in the on-board memory) in order to determine the erroneous bits and hence, the BER value.

5 Simulation Results

Since all the atmospheric phenomena are strongly interdependent [6], the total attenuation is evaluated by combining the calculated attenuation values of each atmospheric effect separately. This implies a combination of the cumulative distributions of all contributions in a suitable way. The ITU recommendations; ITU-R Rec. P676-5 [7], ITU-R P840-3 [8] and ITU-R P618-7 [9] present a general method to calculate the total attenuation $A_T(p)$ from each individual attenuation contribution at a fixed value of percentage probability of time. The total attenuation, A_T (dB), represents the combined effect due to gases, clouds, rain and tropospheric scintillation. Given that the probability level p is fixed:

$$A_T(p) = A_O(p) + A_{WV}(p) + \sqrt{\left[A_R(p) + A_C(p)\right]^2 + A_S^2(p)} \quad (1)$$

where:

$A_O(p)$, attenuation due to oxygen
$A_{WV}(p)$, attenuation due to water vapour
$A_R(p)$, attenuation due to rain
$A_C(p)$, attenuation due to clouds
$A_S(p)$, attenuation due to tropospheric scintillation

Based on the above equations, the first estimates of the additional attenuation performed for the feasibility study of the payload at 94 GHz are presented below. The simulations were performed considering the two earth stations of Spino D'Adda and Rome at various elevation angles between 90° (best case) and 45° (worst case). Table 2 illustrates the estimated attenuation levels for the earth stations of Spino D'Adda and Rome at the minimum and maximum elevation angles 45° and 90°.

Figure 2 shows the simulation results of the percentage probability of time (x-axis dB) as a function of the total attenuation values (y-axis %) for various elevation

Table 2. Total attenuation values for Spino D'Adda and rome at 94 GHz (1% probability).

	Elevation angle		
Location	45∞	61.7∞	90∞
Spino D'Adda	13.72	12.19	11.69
Rome	14.92	13.33	13.04

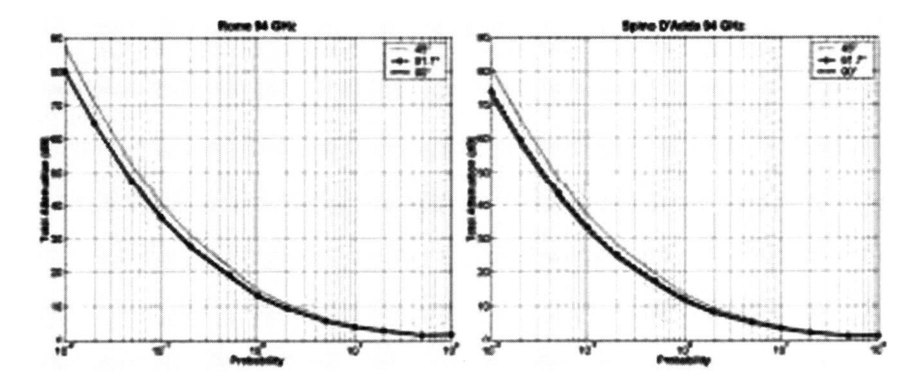

Fig. 2. Total attenuation estimated at 94 GHz (Rome and Spino D'Adda).

angles at the earth stations in Rome and Spino D'Adda at 94 GHz. It can be seen that the total attenuation values are significantly reduced for probabilities more than 5% of the time. The simulation was carried out using the available climatic data at Spino d'Adda and Rome. This means that it is feasible to establish the W-band link for limited availability for more than 95% of the time.

6 Future Perspectives in W-Band

In January 2004, a national directive has defined the new frontiers for the mankind with a "Vision for Space Exploration". The first step of this new challenge is the re-exploration of the Moon, focusing thereafter towards the Mars planet and other planets of the Solar System. At present, only a worldwide effort would afford to collect the necessary scientific, technological, human resources to face this huge project. The required expertise and know-how is very widespread, ranging from propulsion to communication systems up to the effects on astronauts living in space for long time. In this context, the use of terrestrial applications of W-band and scientific and technological results from WAVE experience could have a considerable advantage since no significant atmospheric effects are present outside the Earth.

A number of scenarios are foreseen for the return on the moon missions as shown above in Fig. 3, including the possible use of one or more W-band GEO (and possibly LEO) satellites to work as a data relay between rovers or base stations on the moon and the corresponding earth stations [10].

A likely scenario for the Moon mission could foresee the possible use of a number of landers and rovers on the surface in different regions, supported by one or more orbiters. Orbiters perform scientific observations with different instruments (active and/or passive), and support landers and rovers creating a communication infrastructure (data, telemetry, commands, etc.) between Moon and Earth. This could include many interesting applications that can be envisaged where W-band is the main actor, offering a key contribution in space exploration missions concerning:

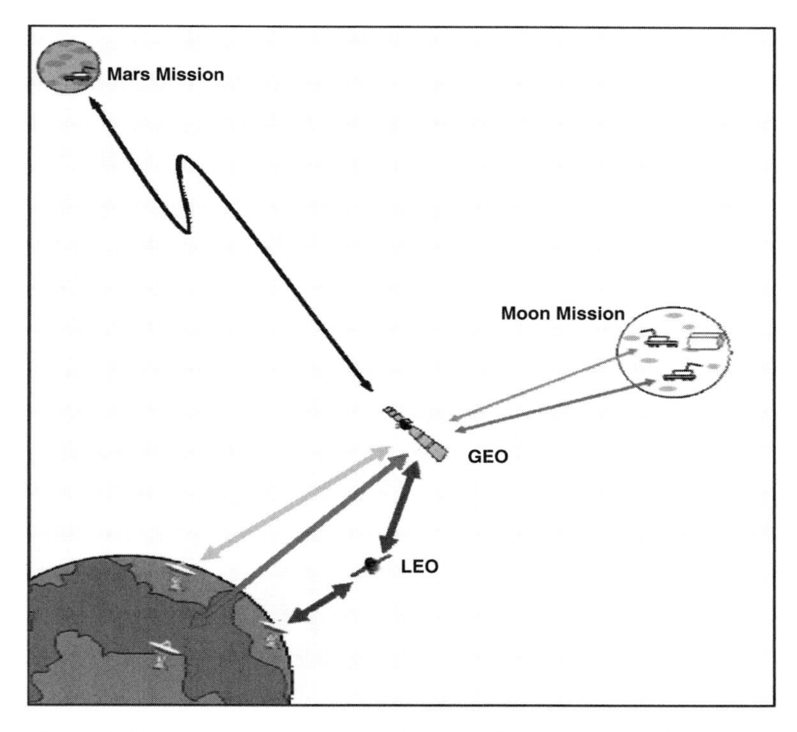

Fig. 3. A foreseen scenario for the exploration of the moon and planet mars.

- *Communications*—a W-band full space communication network can be used to exchange communications between orbiters and landers and/or rovers on the surface, between orbiters and ISS (International space Station) and/or other DRS (Data Relay Satellite) GEO satellites
- *Surface Survey*—orbiters can be provided with W-band radars for remote sensing, in order to produce high resolution surface images and to get multiple benefits such as searching and mapping for ice in polar regions and in darkest craters, very helpful in the creation of permanent bases on the Moon; and also identification of geological and geomorphologic features.

However, since the potential benefits of the W-band technology are affected by atmospheric attenuation in case of terrestrial applications, W-band can be fully exploited in space applications. Such a vision can be applied also to future exploration missions to Mars and beyond. Any long range space mission will be based on the use of CEV (Crew Exploration Vehicles), orbiters, landers, rovers, USV (Unmanned Space Vehicles).

Deploying W-band communication links between these vehicles will provide reliable, accurate, high rate, safe, secure data transmission and some additional advantages such as:

- Reduced mass and volume of components
- Small size of antennas

- Small antenna beamwidth
- Low power consumption

High resolution radar imaging from orbiters and satellites of potential landing and settlement areas (polar regions, crater bottoms) represents a fundamental preliminary step in the colonisation process of the Moon and Mars. USV can be used in the thin atmosphere of Mars for aerial radar survey and mapping of much broader surfaces than with wheeled rovers. In particular, Lunar orbiters and satellites offer favourable support for W band radar applications:

- absence of atmosphere
- much lower orbital speed compared to Earth (1:5)
- very low distance from surface
- static environment (no time constraints)

Under these circumstances, W-band radar imaging can offer the required resolution, at much more favourable conditions compared to medium frequency SAR (Synthetic Aperture Radar) solutions, resulting in a potential optimization of the synergy between radar and communication hardware.

We can also envisage applications related to safety-on-trip using high resolution radar for:

- detection and mapping of small debris
- rendezvous and docking operations

Moreover, energy beams for power transmission from the power plant (nuclear or solar) to users (ground based, orbiting or flying) could improve high focusing, pencil beam capability with smaller antennas.

The choice of W-band for the above applications might be preferred over the use of lower band frequencies considering the following advantages:

- availability of the technology and electronic components operating in W-band.
- reduced mass and size of components
- broad bandwidth and high data rates.

7 Conclusion

In this paper, an overview of the Aero-WAVE test is presented. A manned HAP known as the M-55 Geophysica aircraft will be used to perform a number of propagation experiments, transmitting an experimental W-band signal over the area around Rome and Spino D'Adda.

A description was provided on the payload to be delivered, and the main aspects of the experiments and the simulation results of the expected total attenuation levels at both earth stations are provided. These results show that the W-band link can be feasibly established to provide broadband communication services for 95% of the time, based on the available climatic and data attenuation values for each individual atmospheric effect. The results obtained during the Aero-WAVE tests are expected to help in the characterization of the W-band channel behaviour in different weather conditions at both, Spino D'Adda and Rome.

References

[1] A. Jebril, M. Lucente, T. Rossi, M. Ruggieri, "WAVE – A New Satellite Mission in W Band", IEEE Aerospace Conference 2005, Big Sky, Montana, March 2005.
[2] A. Bosisio, P. Cambriani, V. Dainelli, A. Jebril, M. Lucente, S. Morosi, A. Pisano, L. Ronzitti, T. Rossi, M. Ruggieri, L. Scucchia, V. Speziale, "The WAVE Mission Payload", IEEE Aerospace Conference 2005, Big Sky, Montana, March 2005.
[3] High-altitude M-55 Geophysica aircraft Investigators Handbook, Myasishchev Design Bureau, Third edition, Russia 2002.
[4] M. Ruggieri, S. De Fina, M. Pratesi, A. Salomè, E. Saggese, C. Bonifazi, "The W-band Data Collection Experiment of the DAVID Mission", Special Section of IEEE Transactions on AES on "The DAVID Mission of the Italian Space Agency", vol. 38, no. 4, pp. 1377–1387, Oct. 2002.
[5] M. Ruggieri, S. De Fina, A. Bosisio, "Exploitation of the W-band for High Capacity Satellite Communications", IEEE Transactions on Aerospace and Electronic System Vol. 39, No. 1, Jan. 2003.
[6] J. Sanders, "Rain attenuation of millimeter waves at 5.77, 3.3 and 2.2 mm", IEEE Transactions on Antennas and Propagation, Ap-23, pp. 213–220, 1975.
[7] ITU-R Rec. P676-5 (Attenuation by atmospheric gases).
[8] ITU-R P840–3 (Attenuation due to clouds and fog).
[9] ITU-R P618-7 "Propagation data and prediction methods required for the design of Earth-space telecommunication systems".
[10] "The Next Steps in Exploring Deep Space", Executive Summary, A Cosmic Study by the International Academy of Astronautics, Jul. 2004.

HAP-LEO Link Communication Systems Based on Optical Technology

Silvello Betti, Valeria Carrozzo, Elisa Duca, Federica Teodori

Università degli Studi di Roma "Tor Vergata" – Dipartimento di Ingegneria Elettronica
Via del Politecnico, 1 – 00133 Rome, Italy
Phone: +390672597447, Fax: +39725974335
e-mail: betti@ing.uniroma2.it, carrozzo@ing.uniroma2.it,
duca@ing.uniroma2.it, fede.teodori@fastwebnet.it

Abstract. One among the most recent configurations in satellite architectures, is enhanced with respect to the traditional ones, by the presence of HAPs (High Altitude Platforms). Those balloons are located at stratospheric altitudes, and make possible to separate the link from LEO to ground into two segments: the former crosses the highest levels of atmosphere and can be developed by optical technology, the latter is more sensitive to scattering and absorption, since it crosses lower atmosphere, so it must be carried out in RF domain where more consolidated technology can be used. The insertion of HAPs in the global architecture enhances the system performance in terms of LEO-Ground link capacity, connectivity and flexibility; in addition the wide band-width offered by optics makes more and more effective those improvements. Even better results can be obtained if a proper coding system is designed.

1 Scenario Description

For a HAP-LEO link an optical communication system based on 1550 nm technology is proposed. This wavelength ensures a wide capacity and the consolidated technology used in terrestrial links. In addition EDFA (Erbium Doped Amplifiers) optical amplifiers work in that band, offering some important improvements in terms of system performance. The optical communication subsystem should be designed starting from the knowledge of the operative scenario. In the case we consider, data stream is mainly transmitted from LEO to HAP: LEO can collect information during its orbit, for example from high resolution Earth Observation imageries. Instead of transferring data to a Ground Station with a RF link in the visibility window, it is possible to increase the downloading effectiveness by locating a HAP just above the destination Ground Station [1,3,4]. As a matter of fact, the HAP location at 20 km allows to avoid the most limitative atmosphere layers making possible to carry out an optical link from LEO to HAP (Fig. 1). With a 10 Gbit/s link, even in a very short time window for download it is possible to transfer up to some Terabit per day. HAP can transmit stored data to Ground Station until the next time LEO becomes visible to HAP and it starts its downloading again. Between two consecutive accesses there is a time delay, that sometimes is some hours long, depending on the HAP latitude. That means HAP has a quite long time to empty its memory, so the link to Ground does not require a wide band [5]. During this silent delay, HAP

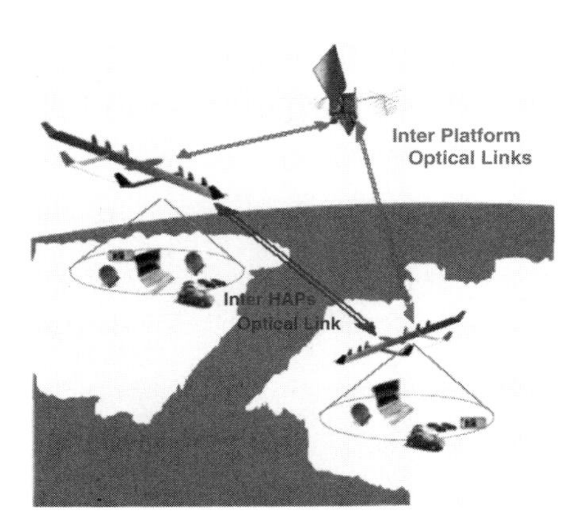

Fig. 1. HAP-Satellite link scenario: by integration of optical and RF technologies it is possible to enhance system performance.

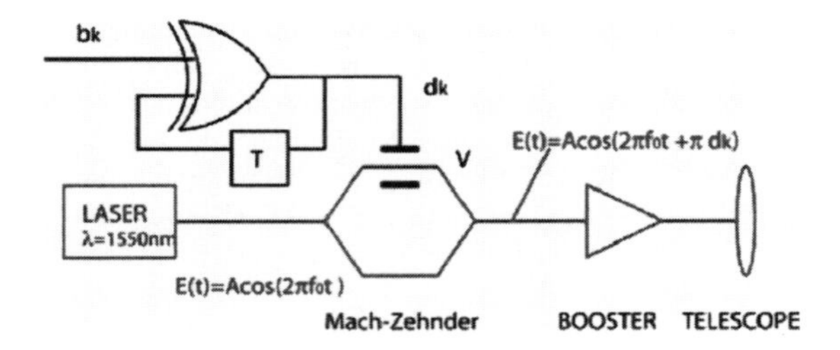

Fig. 2. LEO optical transmitter. The laser wave is externally modulated by a Mach-Zehnder interferometer that is electrically driven by a differential encoded stream.

can transmit data even to another HAP exploiting optical technology, as well. That can improve the downloading speed and the network connectivity.

2 Transmitter

The transmitter subsystem foresees the presence of a semiconductor laser that generates an optical CW (Continuous Wave) signal at the desired wavelength. One can design the system assuming that each LEO can transmit on its own optical carrier in order to reduce interference and to increase the global capacity. The phase modulation is carried out by the use of a differentially encoded stream in the electrical domain, as shown in Fig. 2. Such electrical stream can be used as driver voltage for the Mach-Zehnder modulator, that creates a phase modulation in the optical domain on the CW.

The incoming optical field is equally splitted inside the modulator, along one branch the driving voltage changes the refractive index, thus induces a phase shift that affects the outgoing field. Thus, a differential encoded phase can be assigned to the field that exhibits amplitude variation according to the phase value. Notice that the modulator can change the phase with a finite response time, such a parameter will affect the received eye diagram. A deeper explanation will be given in the proper section.

The optical field is then amplified by a booster (EDFA – Erbium Doped Fibre Amplifier) and transmitted by a telescope, whose diameter (D_t) gives a measurement of the transmitter gain by the following law [6]:

$$G_t = \frac{\pi^2 D_t^2}{\lambda^2} \tag{1}$$

The transmitted field can be written as:

$$E(t) = A\cos\left\{\omega_0 t + [\varphi(t)]\right\} \tag{2}$$

where it has been assumed that the phase undergoes continuous changes from bit to bit, in a transition time that is within the range 3–6 ps.

3 Receiver

On the receiver hand, a telescope focuses the optical power on the first optical band pass filter, whose goal is to reduce the background noise. It worth to observe that the proposed scheme is well suited for a WDM communications system, where each HAP can download simultaneously streams from different LEOs that use different optical carrier. In that case, the optical received spectrum can be represented as in Fig. 3, where the 4th channel is dropped by a direct filtering [7].

Between the background filter and the selective filter is placed an optical preamplifier (EDFA), that increases the receiver performance in terms of electrical signal to noise ratio [13]. The desired channel is then selected, it is divided into two component by a 3dB splitter, each component feeds one of two branches of the interferometer. That configuration implements the differential demodulation scheme in the optical domain. In one branch, the field undergoes a delay equal to the bit duration (T_b), then the components are coupled again in another coupler that sums the field again.

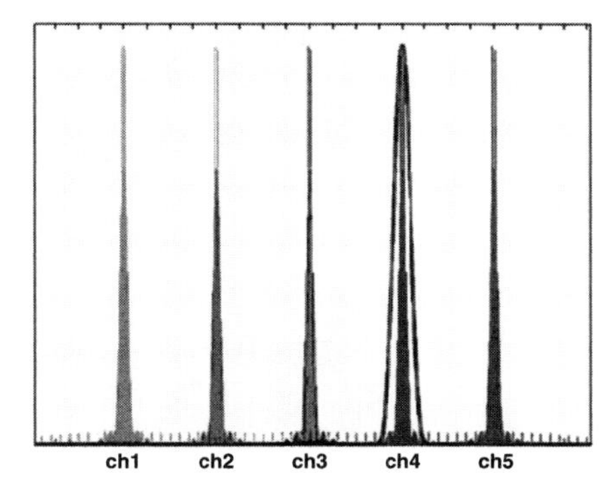

Fig. 3. From a WDM comb, the desired signal is selected by a direct filtering whose spectral shape is represented with the black line.

Thus, the photodetector receives the following optical field:

$$E_{pin}(t)=\frac{1}{\sqrt{2}}E(t)+\frac{1}{\sqrt{2}}E(t-T_b) \tag{3}$$

then, the current is generated by a beating of the signal that gives:

$$I(t)=\frac{E_b}{T_b}\left\{1+cos\left[\Delta\theta_k+\varphi_N(t)\right]\right\} \tag{4}$$

where φ_N is the phase noise and the phase variations can be calculated as follows:

$$\Delta\theta_k=\omega_0 T_b+\pi\left(a_k-a_{k-1}\right) \tag{5}$$

notice that the first term in previous equation is related to the laser chirp, and it can be controlled in the receiver end chain inside the demodulator phase controller; the second term is the phase variation that contains information.

As shown in Fig. 5, the eye diagram is similar to a RZ signalling. That is a consequence of the differential encoding. As it has been explained before, each phase variation is carried out, on the transmitter modulator, in a finite time interval, creating a high level ("1") on the demodulated signal. Thus, the field imaginary component becomes nonzero in correspondence of such phase variations, and the received current undergoes a "return to zero" dynamics. This pattern effect is only present when a sequence of "1" is demodulated, that is related to a sequence of bit to bit variations.

Another scheme is proposed for the same system. It is based on a balanced receiver: the current is obtained by separating the demodulated field in a 3dB splitter, whose branches end into a photodetector, as shown in Fig. 4. After the 3dB coupler the optical field on each branch is:

$$E_{pin}(t)=\frac{1}{\sqrt{2}}E(t)+\frac{1}{\sqrt{2}}E(t-T_b) \tag{6}$$

$$E_{pin}(t)=\frac{1}{\sqrt{2}}E(t)-\frac{1}{\sqrt{2}}E(t-T_b) \tag{7}$$

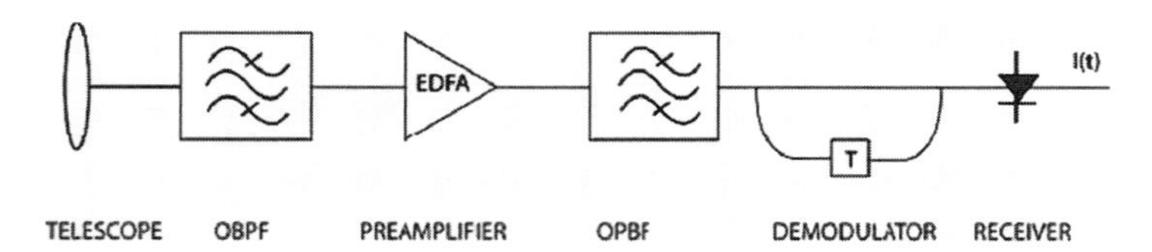

Fig. 4. (a) DPSK receiver scheme with a non balanced photodetector.

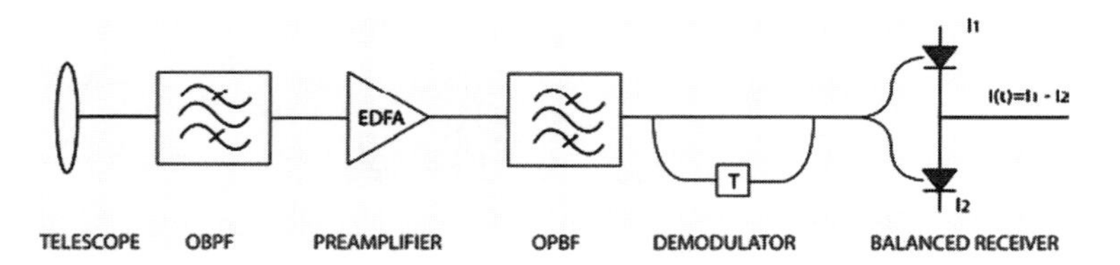

Fig. 4. (b) Scheme for DPSK signalling, based on balanced receiver.

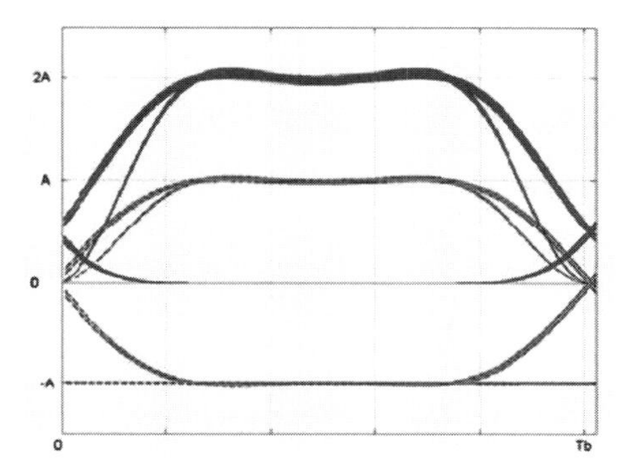

Fig. 5. Eye diagram for the current obtained by a non balanced receiver in solid line, and by a balanced receiver in dashed line.

resulting from the scattering matrix for the coupler:

$$[S]=\frac{1}{\sqrt{2}}\begin{pmatrix}1 & 1\\ 1 & -1\end{pmatrix} \tag{8}$$

The detected current exhibits a dynamics from positive and negative values, as it can be observed from the eye diagram in Fig. 5.

$$I(t)=I_1(t)-I_2(t)=\frac{E_b}{T_b}cos(\theta_k)cos\left[\varphi_N(t)\right] \tag{9}$$

By now we have considered a phase noise term in the received field. It worth to point that, this noise can be generated by the receiver itself, pointing error [8] on the transmitter side, wrong Doppler correction [9,10]. All these sources contribute to make the phase noise arise, with a distribution we have supposed to be uniform with zero mean over a narrow range of some degrees.

4 Doppler Compensation

The Doppler effect arises when the source or destination is moving with respect to its counterpart. Many techniques to suppress the phenomenon or reduce its impact on the system performance have been proposed [10]. An interesting solution is to use a direct filtering locked on the desired channel, by exploiting the deterministic nature of Doppler phenomenon. As a matter of fact, if the HAP is supposed to be placed within a spot of few squared kilometres, the relative distance between source and destination can be "a priori" known by calculation from LEO orbit geometric parameters. Thus, it is possible to predict with a negligible error, the Doppler shift that affects each optical carrier, by solving:

$$\Delta f(t)=f_0\bullet\frac{v(t)}{c}cos[\alpha(t)] \tag{10}$$

where f_0 is the optical carrier, $v(t)$ is the relative speed and $\alpha(t)$ is the angle between the velocity vector and signal propagation direction.

When a new link is established, LEO approaches HAP until the minimum range is reached, thus the frequency shift in the signal spectrum is positive for the first half time of each access, then it becomes negative. For the proposed scenario it is possible to upper bound the shift within a 10 GHz band, centred at the optical carrier. This value can be relevant if the transmitted data rate is low, i.e. few Gbit/sec. In that case the narrow band filter after the preamplifier can not increase the receiver effectiveness as it would be if there was no Doppler shift. For that reason, when a low data rate link is established, a frequency controller on the receiver end should be implemented, in order to lock the narrow band filter at the received signal spectrum and change its central frequency with the Doppler shift fluctuations (Fig. 6).

This approach is somehow expensive in terms of software complexity, but the enhancement of the received OSNR is relevant. Another case is when the transmitter data rate is very high, i.e up to 40 Gbit/sec. In that instance the signal bandwidth can reach 80 GHz, and the optical filter bandwidth increasing due to Doppler shift is quite negligible.

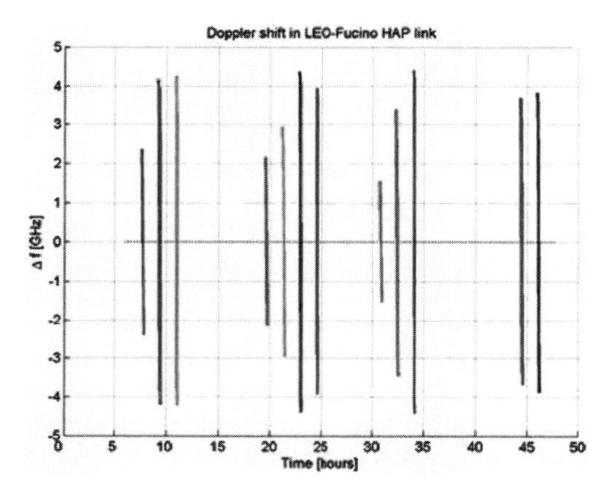

Fig. 6. Doppler shift for an optical link LEO-HAP (on Fucino), at 1550 nm.

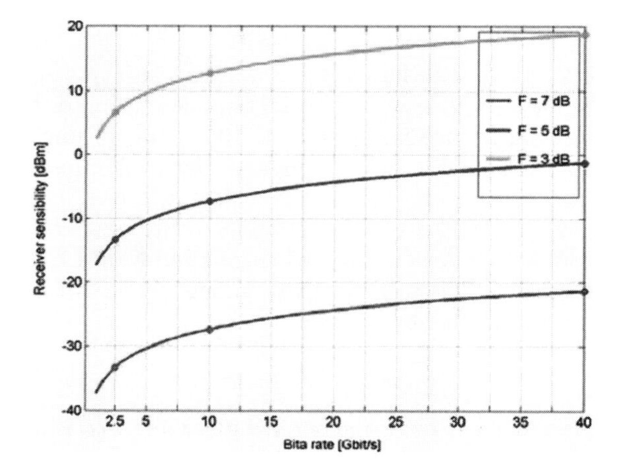

Fig. 7. Receiver sensitivity increase for different noise figure (F), with the signal data rate.

5 System Performance

The LEO-HAP link performance are limited in terms of Optical Signal to Noise Ratio on the receiver end. That parameter is affected by different impairments, as: pointing error, HAP random motion due to wind, high background noise and so on.

A first approach to improve system performance is related to the correct device and hardware choice. For example, higher diameter lenses guarantee a higher gain both in the receiver and the transmitter end.

But this leads to the decrease of the spot size as well, with the consequence of a more complicated pointing procedure. The availability of a wide spot reduces the time required to carry out the connection establishment, this for the higher probability to find an intersection of the uncertain area of the target and the transmitter spot.

The use of wavelength around 1550 nm gives a partial compensation for that reduction of the beam width. The choice of the transmitter and receiver lenses diameters has to be made taking into account all these implications: the increase of the available gain can not always compensate the time consumption required to scan the uncertain area where the target can be found if a narrow beam is used. Another factor the spot size depends on is the link range, as shown in Fig. 8. As a matter of fact, a LEO-HAP link is rather short, if compared with other scenarios, (i.e., LEO GEO link), and just a section of stratosphere is crossed by the optical beam, a very high gain for the transmitter telescope is not required. This reduction of the lenses leads to a wider optical spot, occurrence that can simplify the pointing procedure for the relative reduction of the uncertain area related to the spot size at the link range [11]. The receiver embarked on HAP payload, has been already described in the receiver section. A telescope focuses the received beam on a background filter. This first filter has an important role: it reduces the impact of the stars and Sun noise field on the signal bandwidth. Its importance becomes more relevant for the proposed scheme, where the optical preamplifier could be saturated by a strong incoming noise making its performance worst. The preamplifier can reduce the system Bit Error Rate by the increase of receiver sensitivity. A second optical filter is designed to reduce the impact of Doppler on the signal spectrum [12,10]. The presence of an

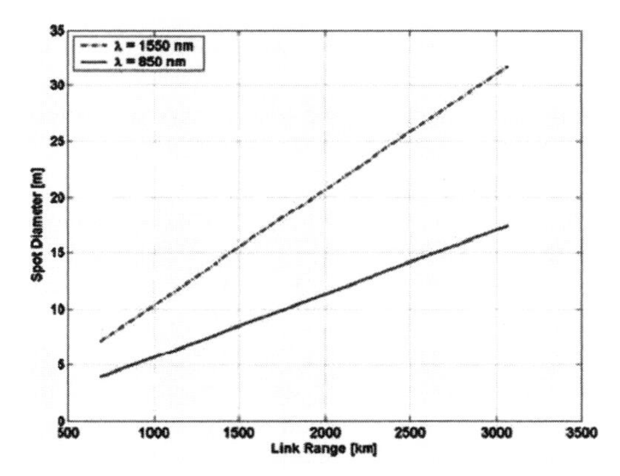

Fig. 8. Spot diameter increase with range link for a 15 cm diameter lens.

optical preamplifier is relevant for the improvement in the receiver sensibility, that can be written as:

$$P_{in} = R \cdot hf \cdot F \left[\gamma^2 + \gamma \sqrt{\frac{\Delta f_{opt}}{R}} \right] \tag{11}$$

where R is the data rate, F is the preamplifier noise figure, γ is the optical signal to noise ratio and Δf is the optical filter bandwidth. The dependence on the data rate has been graphically shown in Fig. 7.

The high speed offered by optical technology increases the available capacity of a LEO-HAP link to some Terabit per day. A 10 Gbit/sec link has been considered. Then the actual connection window has been calculated. In particular, from geometrical parameter obtained via simulation with AGI-STK, the theoretical connection window can be easily calculated considering as start time the moment HAP and LEO become visible each other. Actually, the pointing procedure must be completed before the communication link can be established, in addition sometimes the distance is too high to ensure a connection that satisfies the required BER.

The BER calculation have been implemented assuming no phase error affects the signal, even if that hypothesis can be easily removed with no large lost in performance. In particular, if a limited phase noise with uniform distribution is assumed, it has been proved that the increase of the receiver sensitivity is about 1 dB [9]. The BER can be found by solving [14]:

$$P_e = Q_1(a, b) - \frac{1}{2} e^{-\frac{(a^2 + b^2)}{2}} \cdot I_0(ab) \tag{12}$$

where Q_1 is the Marcum Q function and I_0 is the first kind modified Bessel function of order zero. The terms a and b are related to the phase error:

$$a = \sqrt{SNR(1 - cos\theta)}$$
$$b = \sqrt{SNR(1 + cos\theta)}.$$

Within the actual connection windows, it could be calculated the overall amount of information the link can transmit. That leads to start investigations in new applications for satellite network: at first, data relay becomes more effective since hard memory is now enormous, if compared to the RF link capacity. A huge hard memory can be embarked on HAP and it can be used to storage information downloaded from LEO at 10 Gbit/sec in the connection window. On the other hand, an RF link from HAP to Earth can download all data even with a reduced bandwidth.

This architecture offers a wide band link also to places located at middle latitude, where the LEO-HAP connection time is shorter, for examples Rome. In particular three different locations have been considered: Fucino (Rome), Shadnagar (India) and Poker Flat (Alaska). For these latitudes the available capacity for each day has been calculated and the results are reported in Fig. 9. The most effective site is Poker Flat, with a very high latitude if compared to Fucino and Shadnagar. Anyway the presence of an optical link from LEO to HAP ensures even at low latitude, as for Shadnagar, a capacity of some Terabit/day.

At Fucino the capacity reaches value of 10 Terabit/day if the transmitted power is 10 dBm. The link becomes even more effective if the power increases. In particular, in Fig. 10 has been shown the trend of link efficiency with respect to the optical transmitted power. It worth to notice that 10 dBm corresponds to a low efficiency, that

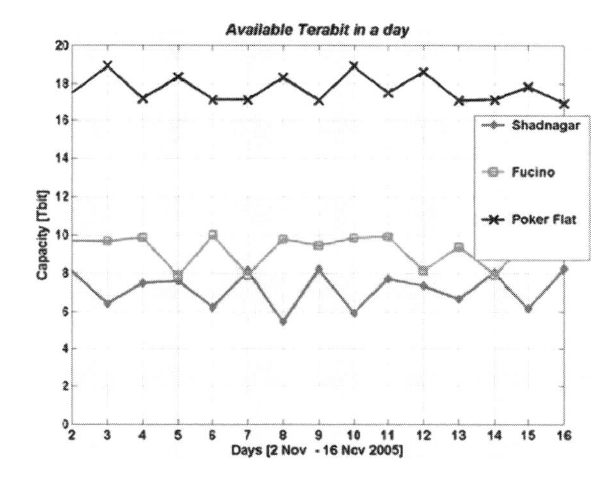

Fig. 9. Maximum available capacity for an optical link, operating at 1550 nm with 10 dBm transmitted power. Different results have been found for the considered ground stations: Poker Flat, Shadnagar and Fucino.

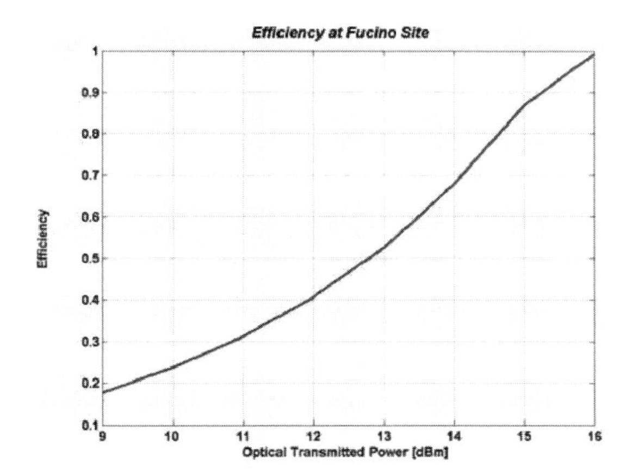

Fig. 10. Link efficiency with the increase of the optical power.

means a wide part of time intervals where LEO and HAP are visible is not actually used for transmission because the link does not satisfy the quality constraints, or the visibility is too short to let the pointing procedure end successfully.

5.1 Coding Effects on System Performance

To improve the performance of Hap-Leo link we have considered two different kinds of coding (block and convolutional) and two techniques of decision (hard and soft). The results have been obtained considering SNR available at Fucino site for a scenario period of two days.

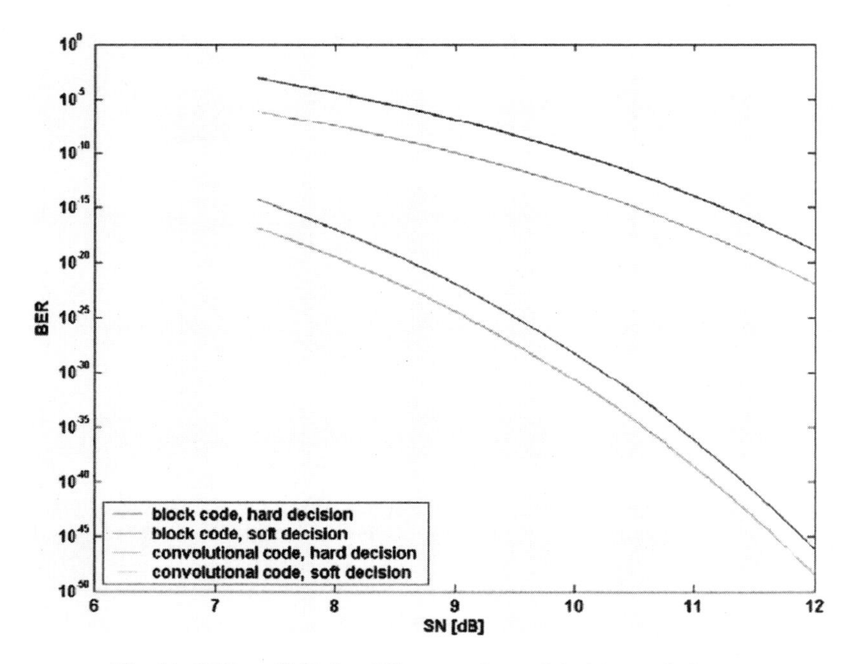

Fig. 11. BER vs. SNR for different codes and decision techniques.

The use of codes requires a lower OSNR at the receiver end for a given BER. In particular, the optical power requested shows a higher decrease with convolutional techniques compared with the block ones, as can be seen in Figs. 11 and 13. It is also clear that soft decision involves a lower required power than hard decision, for an assigned Bit Error Rate and a specified code.

The plots of Fig. 11 were obtained considering codes with equal correction power and with equal coding rate ($R_c = 1/2$). Bit error probability for the codes has been calculated from [2]:

$$\left.\begin{aligned} P_{e\_HD} &\leq \frac{2t+1}{n}(2^k-1)\left(\sqrt{4P_e(1-P_e)}\right)^d \\ P_{e\_SD} &\leq \frac{2^k-1}{2}\, erfc\left(\sqrt{\frac{E_b}{N_0}}\right) \end{aligned}\right\} \text{Block codes} \tag{13}$$

$$\left.\begin{aligned} P_{e\_HD} &\leq 2^d P_e^{d/2} \\ P_{e\_SD} &\leq \frac{1}{2} e^{-d(E_b/N_0)} \end{aligned}\right\} \text{Convolutional codes} \tag{14}$$

where t is the number of errors that can be corrected at the receiver end, and d represents the maximum number of errors detectable; P_e is given by

$$P_e = \frac{1}{2}\, erfc\left(\sqrt{\frac{E_b}{N_0}}\right) \tag{15}$$

for On–Off–Keying (OOK) modulation, or by equation (12) for DPSK modulation.

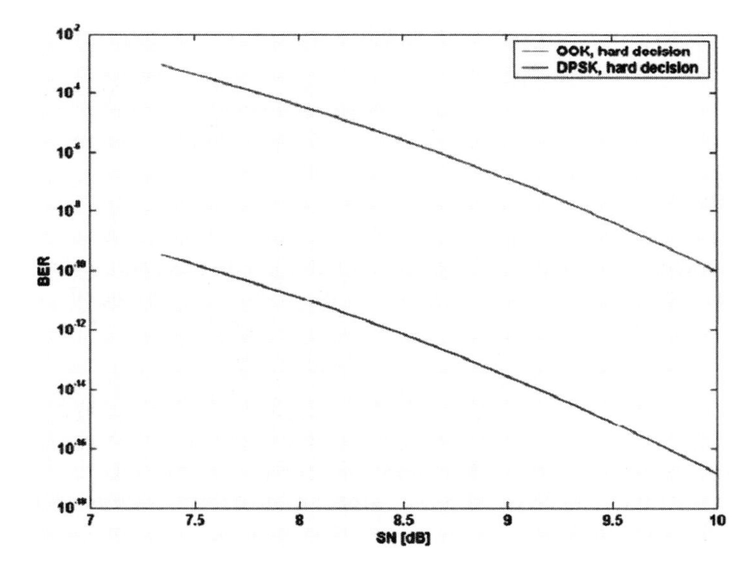

Fig. 12. BER vs. SNR for different modulation format.

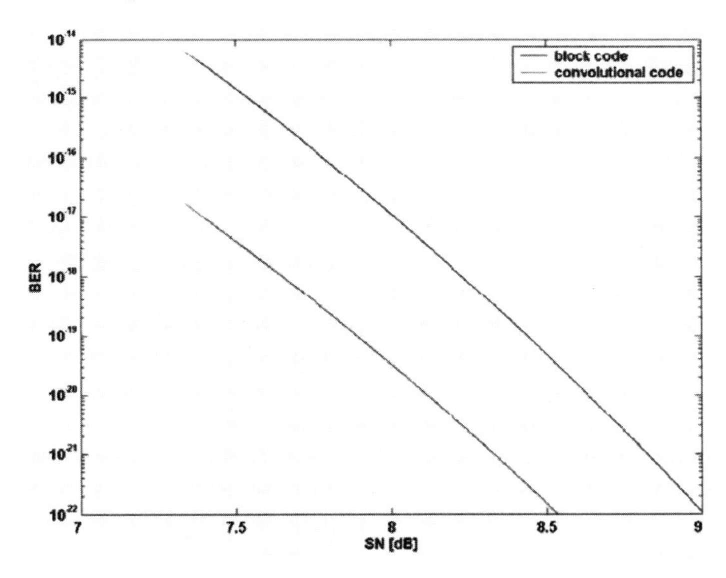

Fig. 13. BER vs. SNR for block/convolutional codes, with soft decision.

The performance of OOK and DPSK in terms of BER against SNR in case of hard decision technique is reported in Fig. 12. We can see that an OOK modulated signal grants to reach, for a given SNR, a BER lower than that achievable with a DPSK modulated signal.

In Fig. 13 the case of soft decision is shown for block and convolutional codes. It is worthwhile to point out that, in case of soft decision, the Bit Error

Rate is independent of modulation scheme, as can be deduced from equations (13) and (14).

The reported plots in this Figure confirm that convolutional codes have better performance than block codes, in term of BER.

6 Conclusions

By the cooperation of RF and optical technology it is possible to develop high performance satellite network. The increase in the communication effectiveness is related to the possibility of data dump even for medium latitude and to the enlarge of downloadable volumes. These features give a contribution to the future earth observation systems. In particular, STK simulations have shown the possibility to carry out some optical links from LEO to HAP located above low latitude sites such as Shadnagar or Fucino, with very actractive performance in terms of available capacity per day. Those calculations have been performed considering some constraints on the minimum received power (optical sensibility) in order to guarantee the required Bit Error Rate.

References

[1] M. Antonini, E. Cianca, A. De Luise, M. Pratesi, M. Ruggieri, "Stratospheric Relay: Feasibility of New Satellite-High Altitude Platforms Integrated Scenarios", IEEE Aerospace Conference, March 8–15, 2003, Proc. no. 3, pp. 1211–1219.

[2] S. Benedetto, E. Biglieri, V. Castellani, "Digital Trasmission Theory". 1987 Prentice Hall, Inc., 9.

[3] G. Hyde, B. Edelson, "Laser satellite communications: current status and directions". 1997 – Space Policy.

[4] W. Leeb, Laser space communication systems, technologies and applications. 2000 The review of laser engineering, 28, 804–808.

[5] C. Fragale, A. Jebril, M. Lucente, M. Pratesi, T. Rossi, M. Ruggieri, "White Paper on Signal Propagation in the Frequency Range Greater than 60 GHz for Satellite Links", Sympotic 2004, Bratislava, 24–26 Oct 2004.

[6] R.M. Gagliardi, S. Karp "Optical Communications", Wiley Series in Telecommunications and Signal Processing – J.G. Proakis, Series Editor.

[7] G. Lenz, B.J. Eggleton, C.R. Giles, C.K. Madsen, R. E. Slusher "Dispersive Properties of Optical Filters for WDM Systems" – IEEE Journal of Quantum Electronics, Vol 34, No8, August 1998.

[8] G. Baister, P. Gatenby, "Pointing acquisition and tracking for optical space communications" Electronics and Communications Engineering Journal, 271–280, 1994.

[9] S. Arnon, N. Kopeika, Performance limitation of free space optical communication satellite networks due to vibrations: direct detection digital mode, 1997 IEEE Optical Engineering.

[10] C. Svec, T. Shay, "Wide dynamic range doppler-shift compensation for space-borne optical communications". IEEE, Photonics technology letters, 16, 2004, 260–262.

[11] S.G. Lambert, W.L. Casey "Laser Communications in Space", Artech House Publishers, 1995.

[12] S. Betti, V. Carrozzo, E. Duca "Optical Intersatellite System Based on DPSK Modulation" II International Symposium on Wireless Communication Systems, September 5–7, 2005 – Siena (Italy), pp. 817–821.
[13] G.P. Agrawal, "Fiber-Optic Communication System", Wiley series in microwave and optical engineering, John Wiley 1997 and sons, inc., New York, 2nd edn.
[14] S. Haykin, "Communication systems", John Wiley and Sons Inc., 4th edn.

Integrated Broadband Wireless Network

Massimo Celidonio[1], Dario Di Zenobio[1], Giovanni Nicolai[2]

[1]Fondazione Ugo Bordoni, Rome, Italy
Tel: +39 06 54801, e-mail: celi@fub.it, dizen@fub.it
[2]Aersat SpA, Rome, Italy
Tel: +39 066482111, e-mail: g.nicolai@alice.it

Abstract. Telecommunication systems based on wireless technology (Wi-Fi, WiMax, Sat DVB-RCS, etc.) offer more than others systems, the opportunity to provide a wide range of broadband services and applications. There are some circumstances where these systems contribute to solve the problem of need of communications such as, for example, for communication restoration during emergencies and catastrophic events as well as for communication services deployment in rural suburban areas typically affected by "digital divide".

This paper presents a proposal for an unique integrated wireless network system, mainly based on TCP-IP protocol, which provides a wide range of applications like voice, video, internet browsing, data access, etc.

A specific section of the paper is dedicated to WiMax technology and, to evidentiate the performances of this innovative technology, some results of the trials carried out in Italy in 2005–2006 by Fondazione Ugo Bordoni (FUB) and Ministry of Communications, are reported.

1 Communication Network Architecture

The main characteristics of the proposed integrated satellite and terrestrial communication network system (see Fig. 1), capable to provide broadband services for fixed and mobile users with high availability are:

- it allows the interoperability among existing terrestrial wireless networks (2G, 3G, WiFi/WiMax, Tetra, IP, etc.)
- it includes star/mesh satellite networks operating with transparent or regenerative transponders based on hub gateways interconnected with terrestrial networks.

1.1 Wireless and Satellite Network Integration

To give the possibility, for large communities of users, to access to IP applications and services on broadband networks, the Wireless Broadband Network (WBN) Architecture shown in Fig. 2 has been considered. The main blocks of this architecture are:

- Wireless Access and Distribution Network (WADN)
- Satellite and Wireless User Terminals (SWUT)
- Satellite Network Access Point (SNAP)

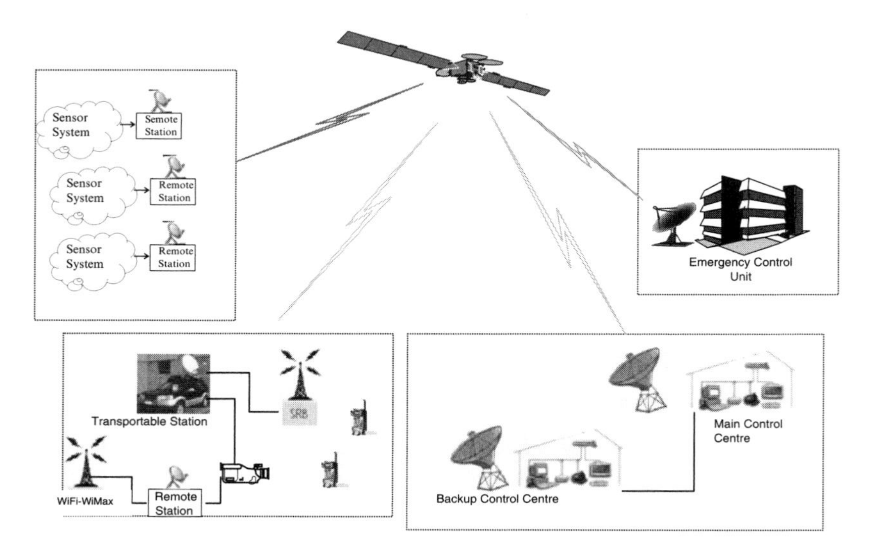

Fig. 1. Communication network bock diagram.

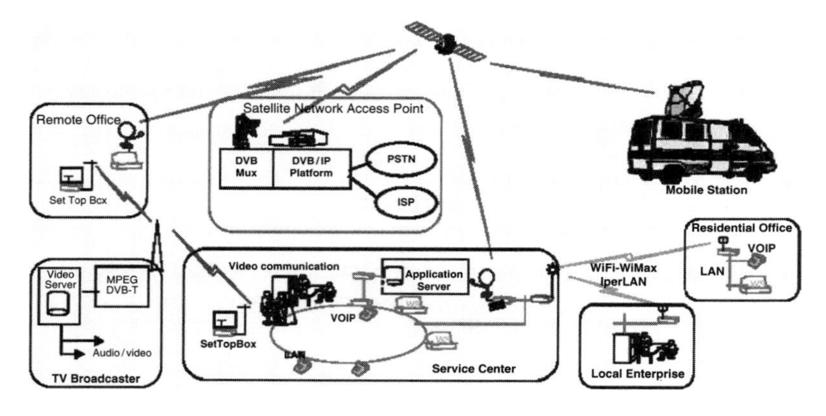

Fig. 2. Satellite and wireless network architecture.

- Mobile Integrated Station (MIS)
- Service Centre (SC)

A detailed description of these blocks will be given in the next paragraphs.

2 Some Applications and Services for the WBN

Typical applications for the WBN, described in the present paper, are

- communication restoration during emergency events;
- contribute to reduce the digital divide.

To better understand the peculiarities of these two applications a more detailed analysis is reported.

2.1 Applications and Services for Communication Restoration During Emergency Events

As far as the catastrophic events are concerned they are usually managed according to the following phases:

Phase 1 – during event: human and technical resources engaged for first aid;
Phase 2 – after event: communication resources restoration and management.

In particular, in phase 2, the restoration of the communication interoperability between involved areas is a primary objective. A multi purpose wireless IP network is able to guarantee a wide range of services, exactly where they are essential, in very short times. Typical applications and services are:

- IP Audio/Video communications by satellite via a mobile terminal between emergency areas and a central telecommunication gateway;
- IP Audio/Video communication in the emergency areas via WiFi/WiMax access points integrated in the mobile terminal and/or via TETRA and GSM networks;
- Localization and positioning management (closed and open loop). Maps transfer service performed by the mobile terminal;
- Human Health Monitoring HHM (life parameters) involved in hazard operations;
- Internet browsing, database access, file transfer and file sharing;
- Hospital Receptivity Information (HRI).

2.2 Applications and Services to Reduce the Digital Divide

The digital divide is still present in industrialized countries where the regional topography (rural and mountain areas) does not allow fast communication infrastructures such as fibre optic or ADSL technologies, capable to provide broadband services. In these cases the broadband service demand can be granted by the use of wireless technologies which do not require large investment cost and long times for the implementation.

The following wireless technologies could be applicable for the above mentioned cases:

- DVB RCS satellite technologies constituting the network backbone between rural areas and the tlc gateway;
- Wireless Local Loop, WiFi, HiperLan and WiMax technologies for the broadband service local distribution;
- Terrestrial Digital TV for downlink channel distribution

For example, the DVB RCS satellite terminals, operating in the Ku band, can provide data access up to 38 Mbps in download and up to 4 Mbps in upload.

This data rate capability can allow the implementation of the following services:

- Administrative Management Services
- Emergency Services
- Service distribution to the user demand via WiFi-WiMax technologies

The broadband data services shall be based on a unique standard protocol TCP/IP and the useful data rate could be: 50 Mbps for Base Station and 10 Mbps for User Terminal.

3 Wireless Network Elements

The Main Wireless Network Elements of a Broadband Telecommunication Network are the following:

- Satellite Network Access Point (SNAP)
- Wireless Access and Distribution Network (WADN)
- Satellite and Wireless User Terminals (SWUT)

The functions of these Network Elements are described in the following.

3.1 Satellite Network Access Point (SNAP)

An emergency or rural satellite network shall be a mixed of different architecture/technologies in order to be flexible during the deployment to the different circumstances. Transparent and/or re-generative satellite could be used jointly with star and/or mesh network technologies for ground/user terminals.

The main satellite networks elements of a SNAP will be:

- Satellite Network Control Centre (SNCC);
- Satellite Network Traffic Station (SNTS) usually named Hub Station;
- Telco Network Gateways (TNG) co-located with the SNTS.

The SNCC and the SNTS are normally co-located in the Satellite Operator Centres and provide three main functions:

- The Satellite Network Control
- The Traffic Management
- The connections between the User Terminals (UT) and the Telco Network Gateways (TNG).

The SNCC and SNTS do not need to be allocated in the Recovery areas but these are existing facilities providing a fast access to the TLC Network via the Service Network Elements (SWUT, MIS and SC).

The standardized technology for SNCC-SNTS stations is the DVB RCS providing a DVB IP download channels bit rate of 38 Mbps and an upload return channels max bit rate of 4 Mbps. The DVB IP channel is a TDM carrier while the RCS access is performed in TDMA in order to allow the connection to the SNCC-SNTS Station for hundreds of User Terminals.

3.2 Wireless Access and Distribution Network (WADN)

The WADN is the access and distribution network we propose and it is based on a very attractive solutions of WiFi and WiMax technologies. A more detailed description of these technologies is reported in the following sections.

3.3 Using WiMax Technology

While many technologies currently available for fixed broadband wireless can only provide line of sight (LOS) coverage or, alternatively, have good building penetration performance but small coverage area, the WiMax technology has been optimized to provide excellent, large, non line of sight (NLOS) coverage. WiMax's advanced technology, just following the preliminary technical features declared by manufacturing companies, provides the best performances in any of the above mentioned link conditions – large coverage distances of up to 50 kilometres under LOS conditions and typical cell radii of up to 5 miles/8 km under NLOS conditions. This results have been partially confirmed by the results obtained during the trials carried out in Italy during 2005–2006 by FUB and Ministry of Communications in collaboration with the principal manufacturing companies of this technology which are reported in the last paragraph of this paper.

3.4 Network Elements

The Standard BS will be Equipped With:

- Basic WiMax implementation;
- Standard RF output power for a lower cost BS.

The full featured BS will be equipped with:

- Higher RF output power in respect to the standard BS;
- Tx/Rx diversity combined with space-time coding and MRC reception;
- Sub-channelling;
- ARQ.

Both the standard and full-featured base stations can be WiMax compliant. It is important to understand that there are a number of options within WiMax that give operators and vendors the ability to build networks that best fit their application and business case.

3.5 WiMax/WiFi Network Connection Module with Gateway Capabilities

WiFi hot spots are being installed worldwide at a rapid pace. One of the reasons which obstacles hot spot growing is the lack of availability of high capacity, cost-effective backhaul solutions. The problem could be overcame with the use of WiMax technology which could also help in fill up the coverage gaps among WiFi hot spot coverage areas.

WiMax offers a useful throughput better than 60 megabytes per second (Mbps) in a 20-MHz channel. This is more efficient if compared with the 25 Mbps available in the standards 802.11 a or g. The quality of service of WiMax is also superior, thanks notably to more sophisticated control over transmission, which guarantees a good level for the services offered. Thus, there will be limited latency delays, which prove to be essential, for example, in transmitting Voice over IP. It also provides better security levels as, the final equipment, must be identified before being connected to the network.

Furthermore WiMax technology should be substantially less costly if compared with other ones proposed today under proprietary systems. This is explained by the peculiar charactericts of the devices as well as with the competition among suppliers once the standard has been adopted.

3.6 WiFi Ground Signal Distribution Point

The WiFi (802.11b/g) radio signal is mainly a line-of-sight signal, meaning it doesn't bounce off things like walls, ceilings, or even the atmosphere, as some radio signals do. But it can get through some opaque objects like walls, ceilings, and floors under certain conditions.

Those certain conditions depend on how electrically dense the barriers are. A typical drywall between offices isn't normally a problem, but conduit, plumbing, or steel studs can increase the density. Going through several of these walls can defeat the signal quickly.

4 Service Network Elements

The Main Service Network Elements of the proposed Broadband Telecommunication Network are the following:

- Satellite and Wireless User Terminal (SWUT)
- Mobile Integrated Station (MIS)
- Service Centre (SC)

The functions of these Service Network Elements are described here below.

4.1 Satellite and Wireless User Terminals (SWUT)

The satellite terminal is a two way, bandwidth-on-demand broadband VSAT. The forward channel provides a total capacity of 60 Mbps, while return channel to the hub can transmit up to 4 Mbps. For increase return channel throughput and reduce the satellite bandwidth employed to transmit, it use DVB-RCS system.

The main features of a satellite terminal are shown in Table 1.

Table 1. Satellite and wireless terminal feature.

Return Channel
Format: MF-TDMA
Symbol Rate: 156, 312, 625 Ksps, 1, 25, 2, 5 Msps
Turbo coding: DVB-RCS compliant
Modulation: QPSK
Forward channel
Format: DVB/MPEG-2 TS, DVB-MPE for IP data
Symbol rates: 2, 5 to 36 Msps
FEC: DVB compliant R/S and Convolutional
Modulation QPSK
Performance
Protocols: TCP/IP, UDP/IP, IGMP, RIP, 1&2, IP QoS support
TCP accelerator

4.2 Mobile Integrated Station (MIS)

The concept of a Mobile Integrated Station MIS is to bring communication broadband services in the Recovery areas either for emergency or telemedicine applications. A typical Mobile Integrated Station is shown in Fig. 3. The MIS station is normally equipped with:

- Satellite Access to the SNCC-SNTS Station and to the TNG
- Wireless Access for Service Distribution for local users
- Audio/Video Recording and Encoding equipment
- VoIP gateway

Typical offered applications are:

- Voice over IP VoIP
- Audio/Video Communications
- Emergency and Telemedicine Services
- News Distribution for Emergency

Fig. 3. Mobile integrated station layout.

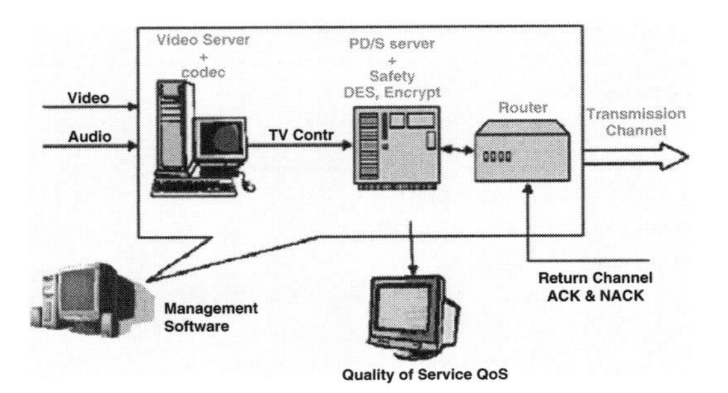

Fig. 4. Regia centre.

4.3 Service Centre (SC)

The Service Centre is normally co-located in the Service Coordination Centres and allows the Access to standard DVB IP platforms capable to provide a wide range of applications. Typical offered applications are:

- Voice over IP VoIP
- Web TV e Business Television
- Distant Learning
- Emergency Services for Telemedicine
- Web Hosting
- Application Hosting
- News Distribution

Furthermore the Service Centre could include a Regia Centre (Fig. 4) where the Live Events Contributions via SNG Mobile Station can be collected and provided to local Broadcaster for Information Dissemination.

5 Some Results of the Italian Wimax Trials

The recent development of broadband wireless technologies has raised the attention of Italian ICT operators in respect of the new WiMax technology, able to allow the implementation of broadband wireless networks in point-to-multipoint topology with metropolitan extension.

For this reason the Italian Ministry of Communication (MINCOM) has authorized a technological trial to be carried out in different Italian Regions and Cities. The trial will involve different WiMax manufacturing companies as well as the official Italian resellers and is planned to be concluded at the end of June 2006. FUB, R&D company supporting the MINCOM, will lead and monitor all the trials.

The trials have been allocated in the 3.4–3.6 GHz band that has been identified in Europe as the core-band to be licensed to offer Wimax services in the future.

The instances of authorized trials have been 53, and these have been conducted in Piemonte, Valle d'Aosta, Sicilia, Sardegna, Abruzzo and in the cities of Roma, Milano, Parma and Arezzo.

The participation of the main manufacturers (Siemens, Alcatel, Ericsson, Nortel, Marconi Communication, Selex, etc.), have made possible to test a large number of WiMax devices available on the market.

Different conditions of propagation, including Line-Of-Sight (LOS) or Non-Line-Of-Sight (NLOS), as well as a variety of scenarios, such as urban, mountainous, rural, coastal, etc., have been considered. The purpose has been the evaluation of the existing technology and the verification of their employment to solve the complex problem of "digital divide".

The network architectures considered cover a large selection of cases including the point-to-multipoint configuration and the backhauling configuration where two or more base-stations are connected in a chain to establish a transport backbone network.

A considerable number of system parameters have been evaluated during the trials. The maximum transmit power (EIRP) allowed for the trial has been fixed at 36dBm. This value has been fixed in agreement with the Ministry of Defence that is, actually, the licensed Italian entity to utilize the 3.4–3.6 GHz frequency band.

Some test results of mixed trial conducted in suburban and rural areas are shown in Table 2.

Table 2. Sample test results.

a) Experimental results from Marconi Communications

Measurements in a mixed urban/rural environment (distance BS-SS: 2,95Km/condition NLOS)

Frame size (bytes)	Throughput (Kbps)	Latency RTD (ms)	Jitter (ms)
1518	120,95	77	5,2 uplink
1518	1021	66	4,1 downlink

Measurements in a mixed urban/rural environment (distance BS-SS: 6,65Km/condition LOS)

Frame size (bytes)	Throughput (Kbps)	Latency RTD (ms)	Jitter (ms)
1518	417,85	46	9,4 uplink
1518	7749	33	4,1 downlink

b) Experimental results from Alcatel

Measurements in urban environment (distance BS-SS: 1,2Km / condition NLOS)

Frame size (bytes)	Throughput (Kbps)	Latency RTD (ms)	Jitter (ms)
1500	4764	32	4,1 uplink
1500	9946	32	2,2 downlink

Measurements in extra-urban environment (distance BS-SS: 6,5Km / condition NLOS)

Frame size (bytes)	Throughput (Kbps)	Latency RTD (ms)	Jitter (ms)
1500	4864	32	3 uplink
1500	9958	31	1,9 downlink

6 Conclusion

The architectures for satellite and wireless recovery networks capable to carry out broadband applications are day by day evolving due to three main factors:

- The service convergence to IP protocols
- The wireless evolution of WiMax technologies
- The satellite evolution of DVB RCS technologies

The possibility to plan Recovery Networks for Emergency Communications and Digital Divide Reduction is nowadays a reality allowed by the wireless technology convergence and it could be carried out in short time and without large investment costs.

References

[1] "*WiMax Tecnology for LOS and NLOS environments*", WhitePaper, WiMax Forum, Aug. 2004;
[2] "*Satellite and Wireless Network Evolution for Broadband Applications*" - 11th Ka Band Conference & ICSSC - September 28, 2005 Rome
[3] "*Applications and Services for Security and Emergency Management*" - 10th Ka and Broadband Communication Conference - September 30, 2004 Vicenza

Unscented Filtering for LEO Satellite Orbit Determination

Stefano Lagrasta

Telespazio S.p.A., Via Tiburtina, 965 - 00156, Rome (Italy)
Tel: +39 06 4079 6315, Fax: +39 06 40999438
e-mail: stefano_lagrasta@telespazio.it

Abstract. The specific interest of Telespazio towards spacecraft radio-assisted navigation does not only concern with the traditional "user segment", but the development as well of ground segment components able to support the space missions with an accurate orbit determination.

Spacecraft platforms already exist, and are planned to be launched, including on board GPS receivers, thus providing the capability of processing related raw data to attain a quite precise reconstruction of the satellite dynamics.

In the proposed application, it is assumed that measurements z_k are provided of a non-linear dynamic system that represents the orbit motion of a LEO spacecraft; the observation samples originate from a dual-frequency (L1/L2) GPS receiver available on-board.

The described solution allows to estimate the state vector x, that includes at least position and velocity of the satellite, with a sequential processing which avoids the calculation of any Jacobian or design matrix and thus suppresses the need of reconstructing any derivatives of plant model equations. It is attained through the use of the so called "Unscented Filter" (UF).

The proposed algorithm is tested with real data, consisting of raw navigation observables and post-processed reference orbits available from Champ satellite mission. CHAMP (CHAllenging Minisatellite Payload) is a German small satellite characterised by near polar, low altitude orbit for geoscientific and atmospheric research and applications.

Among the other problems, a critical aspect to be faced is that one of "outliers" detection and elimination from the orbit determination process; the selection of "proper" signals to be treated is performed in the frame of the sequential calculation, so that only a single "pass" is performed on available raw data. This means that the overall approach is a good candidate as well for a real-time determination of the state vector.

As a conclusion, the UF algorithm has proven to be a working and robust solution, even starting from a bad initial approximation of the state vector, whilst this is typically not the case for an Extended Kalman Filter (EKF).

1 Navigation Observables

The raw read-outs from a navigation receiver equipment consist mainly of the so called *code range* ($\tilde{\rho}$) and *carrier range* ($\tilde{\phi}$) observables.

1.1 Code Range

The raw code-range measurement, that is achieved after correlating the "local replica" of a PRN code with the signal from a space vehicle (SV), can be modelled as follows:

$$\tilde{\rho}(T_{ET}) = c \cdot (T_{ET}(t_R) - T_{SV}(t_E)) + \text{noise} + \text{multipath} + \ldots \tag{1}$$

where:

- $T_{ET}(t_R)$ is the End Terminal receiver clock time measurement, at "receipt" of RF signal, when co-ordinate reference time is t_R
- $T_{SV}(t_E)$ is the space Vehicle clock time measurement at the "emission" of RF signal, when co-ordinate reference time is t_E

The time instants t_E, t_R are related to an "absolute" reference time scale, reproducing the ideal terrestrial time on the surface of Earth equi-potential geoid. The difference:

$$\tau = t_R - t_E \tag{2}$$

is the real "light travel time" of the navigation signal. On the other side T_{ET}, T_{SV} depend upon accuracy of local (receiver and navigation payload) clock oscillators, and relativistic effects.

1.2 Clock Time Offsets

One has:

$$T_{SV}(t_E) = t_E - \Delta t_{REL1}(t_E) + \varepsilon_{SV} \tag{3}$$

$$T_{ET}(t_R) = t_R + \Delta t_{REL3}(t_R) + \varepsilon_{ET} \tag{4}$$

where:

ε_{ET} and ε_{SV} are respectively the End Terminal and the Space Vehicle clocks "time errors", that describe "how well" the real clock is capable to implement its time measurement function

- $\Delta t_{REL1}(t)$ and $\Delta t_{REL3}(t)$ are is a relativistic terms, due to the fact that the SV and the ET are moving with respect to the surface of geoid

Denoting with vectors $\underline{R}_{SV}$, $\underline{R}_{ET}$, respectively the positions of SV and ET in Inertial, Earth-Centred (ECI) axes, one can demonstrate that:

$$\Delta t_{REL1} = \text{bias} + 2 \cdot \underline{R}_{SV}^{T} \cdot \dot{\underline{R}}_{SV} / c^2 \tag{5}$$

$$\Delta t_{REL3} = \text{bias} - 2 \cdot \underline{R}_{ET}^{T} \cdot \dot{\underline{R}}_{ET} / c^2 \tag{6}$$

thus, relativistic time offsets are composed by a bias and a time-varying term. For the sake of simplicity, the bias part of the relativistic effect will be assimilated to the time errors ε_{ET} and ε_{SV}. In the case of a purely Keplerian orbit motion, note that the following identity holds:

$$\underline{R}^T \cdot \dot{\underline{R}} = e \cdot \sqrt{\mu \cdot a} \cdot \sin(E) \tag{7}$$

with: μ = Earth gravity constant, e = satellite orbit eccentricity, a = satellite orbit semi-major axis, E = actual eccentric anomaly = E(t). One sees that the time-varying part of the relativistic effect is mainly due to the eccentricity of satellite orbit, which produces a radial velocity component.

1.3 Propagation Time

The signal propagation time depends upon the real distance ρ between SV and ET, as well as on delays due to both the atmosphere (Δt_{IONO}, Δt_{TROPO}) and the gravity gradient; the latter implies a new relativistic effect, denoted as Δt_{REL2}. One can write:

$$\tau = (t_R - t_E) = \rho\,(t_R, \tau)\,/\,c + \Delta t_{REL2} + \Delta t_{IONO} + \Delta t_{TROPO} \tag{8}$$

$$\text{with: } \rho\,(t_R, \tau) = \|\underline{R}_{SV}(t_R - \tau) - \underline{R}_{ET}(t_R)\| \tag{9}$$

so that τ appears implicitly defined, being in both terms of the previous equation. As already said, Δt_{REL2} (t) is a relativistic term, accounting for the geometry of the light path in the Earth gravitation field, that is characterised by a gradient. The following estimate can be considered:

$$\Delta t_{REL2}(t_R, \tau) = \mu \cdot \frac{(1+\gamma_{PPN})}{c^3} \cdot \ln\left(\frac{R_{ET}(t_R) + R_{SV}(t_R - \tau) + \rho\,(t_R, \tau)}{R_{ET}(t_R) + R_{SV}(t_R - \tau) - \rho\,(t_R, \tau)}\right) \tag{10}$$

In the previous formula, γ_{PPN} is the "Post-Newtonian Parameter" of General Relativity.

The delay due to light travel through the "troposphere", intended here as the "neutral" part of the atmosphere, is defined as Δt_{TROPO}; the delay due to the variable refraction index caused by free electrons is named as "ionospheric" delay, Δt_{IONO}.

1.4 Changing Reference Frame, From Inertial to Earth Fixed

As already pointed out, vectors $\underline{R}_{SV}$, $\underline{R}_{ET}$ denote respectively the positions of SV and ET in Inertial (ECI) axes. With vectors $\underline{r}_{SV}$, $\underline{r}_{ET}$, the positions are intended to be expressed in Earth- Centred and Fixed (ECF) axes.

It is important to account for the following inequality, which is not immediately obvious, due to the fact that we routinely consider distances invariant w.r.t. the reference frame that is considered:

$$\|\underline{R}_{SV}(t_R - \tau) - \underline{R}_{ET}(t_R)\| \neq \|\underline{r}_{SV}(t_R - \tau) - \underline{r}_{ET}(t_R)\| \tag{11}$$

The inequality is due to the fact that the two vectors involved are not referred to the same time instant.

First, a time-dependent attitude matrix **A** exists, bringing from ECF to ECI frame:

$$\underline{R}_{SV}(t) = \mathbf{A}(t) \cdot \underline{r}_{SV}(t) \text{ with: } \mathbf{A}^T(t) \cdot \mathbf{A}(t) = \mathbf{A}(t) \cdot \mathbf{A}^T(t) = \mathbf{I} \tag{12}$$

A multiplying "delta-matrix" brings from $\mathbf{A}(t_R)$ to $\mathbf{A}(t_R - \tau)$:

$$\mathbf{A}(t_R - \tau) = \mathbf{A}(t_R) \cdot \Delta\mathbf{A}(t_R, \tau) \tag{13}$$

and thus:

$$\underline{R}_{SV}(t_R - \tau) - \underline{R}_{ET}(t_R) = \mathbf{A}(t_R) \cdot (\Delta\mathbf{A}(t_R, \tau) \cdot \underline{r}_{SV}(t_R - \tau) - \underline{r}_{ET}(t_R)) \tag{14}$$

due to the ortho-normality of an attitude matrix, $\|\mathbf{A}\| = 1$ and thus:

$$\rho(t_R, \tau) = \| \Delta\mathbf{A}(t_R, \tau) \cdot \underline{r}_{SV}(t_R - \tau) - \underline{r}_{ET}(t_R) \| \tag{15}$$

It might be further shown that the scalar product $(\underline{R} \cdot \underline{R})$ between position and linear velocity is independent on the reference system, i.e.:

$$\underline{R}^T \cdot \dot{\underline{R}} \equiv r^T \cdot \dot{\underline{r}}, \text{ assuming: } \underline{R}(t) = \mathbf{A}(t) \cdot r(t) \tag{16}$$

It can be demonstrated that $\Delta\mathbf{A}$ results, with quite good approximation, to be a "pure" rotation about z-axis, of the angle accumulated by the Earth rotation in the time interval τ:

$$\Delta\mathbf{A}(t_R, \tau) \cong \Delta\mathbf{A}(\tau) = \begin{Vmatrix} \cos(\tau \cdot \omega_0) & \sin(\tau \cdot \omega_0) & 0 \\ -\sin(\tau \cdot \omega_0) & \cos(\tau \cdot \omega_0) & 0 \\ 0 & 0 & 1 \end{Vmatrix}$$

$$\omega_0 > 0, \ \omega_0 \cong 7.29212 \ 10^{-5} [\text{rad/s}] \tag{17}$$

1.5 Code Range, "Iono Free" Derived Expression

In the case of a LEO satellite, Δt_{TROPO} is negligible. The same cannot be affirmed concerning the ionospheric delay. Actual models for Δt_{IONO} available from GPS navigation message (Klobuchar) are not sufficiently accurate. In the first order approximation, Δt_{IONO} is inversely proportional to the second power of carrier frequency:

$$\Delta t_{IONO} \propto 1 / f^2 \tag{18}$$

this suggests the possibility of linear combinations between raw measurements taken at L1 and L2, allowing to "suppress" the ionospheric delay component. The following is named as a "iono free" observable:

$$\tilde{\rho}_{IF} = \frac{f_{L1}^2}{f_{L1}^2 - f_{L2}^2} \cdot \tilde{\rho}_{L1} - \frac{f_{L2}^2}{f_{L1}^2 - f_{L2}^2} \cdot \tilde{\rho}_{L2} \tag{19}$$

where $\tilde{\rho}_{L1}$, $\tilde{\rho}_{L2}$ are code-range measurements taken on GPS L1 and L2 signal frequencies. Accounting for all previous formulae, the expression for the "new", iono-free raw code range measurement takes now the form:

$$\begin{aligned} \tilde{\rho}_{IF}(t_R) &= \rho(t_R, \tau) + c \cdot (\varepsilon_{ET} - \varepsilon_{SV}) + \dots \\ &+ c \cdot (\Delta t_{REL1} + \Delta t_{REL2} + \Delta t_{REL3} + \Delta t_{TROPO}) + \text{noise} + \text{multipath} \\ &\text{with: } \rho(t_R, \tau) = \left\| \Delta\mathbf{A}(t_R, \tau) \cdot \underline{r}_{SV}(t_R - \tau) - \underline{r}_{ET}(t_R) \right\| \end{aligned} \tag{20}$$

The noise level on derived code-range measurement $\tilde{\rho}_{IF}$ appears in general *amplified* with respect to the noise on original code-range raw observables, being:

$$\sigma^2[\tilde{\rho}_{IF}] = \left(\frac{f_{L1}^2}{f_{L1}^2 - f_{L2}^2}\right)^2 \cdot \sigma^2[\tilde{\rho}_{L1}] + \left(\frac{f_{L2}^2}{f_{L1}^2 - f_{L2}^2}\right)^2 \cdot \sigma^2[\tilde{\rho}_{L2}] \quad (21)$$

1.6 Carrier Range

The carrier-phase observable, converted from the very "basic" read-out into units of length (multiplying cycles by wavelength λ), is modelled as:

$$\begin{aligned}\tilde{\phi}(T_{ET}) &= \rho(t_R, \tau) + c \cdot (\varepsilon_{ET} - \varepsilon_{SV}) + \ldots \\ &+ c \cdot (\Delta t_{REL1} + \Delta t_{REL2} - \Delta t_{IONO} + \Delta t_{TROPO}) + \lambda \cdot N \end{aligned} \quad (22)$$

where "N" -the so called "initial ambiguity" - is an integer term, with relative sign, not a-priori known, that remains constant, unless carrier tracking is lost. Carrier range measurements on L1 and L2 have different ambiguity terms.

2 Structure of the Estimation Model

Measurement formulae, which explicit the link between raw observables and LEO orbit state vector components, are strongly non-linear: the raw navigation data are essentially "distance" (ranging) magnitudes; this obviously implies that Euclidean norm operators figure out, working on vector differences.

State vector dynamics equations are also nonlinear in the case of an orbiting spacecraft, equipped with a GPS receiver: force models are nonlinear, and – if working in ECF coordinates- apparent accelerations are to be introduced. By the way, models of high complexity are deployed in Flight Dynamics Systems developed to support the mission of a real satellite.

A fundamental point of strength is the application of the *Unscented Filtering* (UF) to implement the state estimation, which allows to maintain the original, intrinsic non-linear structure of the equations one has to deal with. There is no linearization process within the UF; models are treated in their original format. The following reference equations are considered as an abstract system; state vector reconstruction is our goal:

$$\dot{\underline{x}}(t) = f(\underline{x}(t), \underline{u}(t), t) + \underline{w}(t) \quad (23)$$

$$\underline{z}_k = h(\underline{x}_k, \underline{u}_k, t_k) + \underline{v}_k \quad (24)$$

$$\underline{x}_k = \underline{x}(t_k),\ \underline{u}_k = \underline{u}(t_k),\ \underline{z}_k = \underline{z}(t_k),\ t_k = t_0 + k^*\Delta T,\ k = 0,1,2, \ldots$$

In the above equations and figures, $\underline{v}$ and $\underline{w}$ are noise vectors. The corresponding UF calculation scheme is recalled for commodity of the reader in the appendix of this article; we remind here that the idea of UF, as a robust alternative to the Extended Kalman Filter (EKF) [1,2,3,4,5], was conceived and developed in recent times by Simon J. Julier and Jeffrey K. Uhlmann for the Robotics research Group at Engineering Science Dept. of the University of Oxford [6,7,8,9,10,11].

It is now necessary to precise the equations of system dynamics, i.e.: function $f(\cdot)$, and of system observation, that is $h(\cdot)$.

2.1 System Dynamics

The "cause" of LEO satellite motion are real forces; when using a GPS sensor, it is very convenient to operate in Earth-fixed co-ordinates, and not in inertial ones. The ECF system is non-inertial and thus, apart from projecting their resultant in this frame, additional "apparent" accelerations are to be taken into account. In what follows, vector $\underline{r}$ denotes the LEO satellite C.o.G. coordinates.

The force model can be very (quite arbitrarily) complex; assuming, for the sake of simplicity, to consider only the J2 gravity force (quadrupole moment) component, forgetting here other forces and related parameters (e.g.: solar pressure and air drag, that might be otherwise estimated), the resultant $\underline{f}$ of real forces acting on satellite C.o.G., projected in ECF frame, is:

$$\frac{1}{m}\cdot\underline{f}=g_r\cdot\frac{1}{r}\cdot\underline{r}+g_z\cdot\begin{bmatrix}0\\0\\1\end{bmatrix} \tag{25}$$

$$g_r\doteq-\frac{\mu}{r^2}\left[1+\frac{3}{2}\cdot J_2\cdot\left(\frac{a_E}{r}\right)^2\cdot(1-5\cdot\sin^2\phi_c)\right] \tag{26}$$

$$g_z\doteq-\frac{3\mu}{r^2}\cdot J_2\cdot\left(\frac{a_E}{r}\right)^2\cdot\sin\phi_c,\ \sin\phi_c=\frac{r_x}{r} \tag{27}$$

The state vector components must include the receiver clock offset ε_{ET} that means a minimum dimension of 7 elements, i.e.:

$$\underline{x}=[\underline{r}^T,\underline{v}^T,c\cdot\varepsilon_{ET}]^T \tag{28}$$

being $\underline{v}$ the (relative) ET linear velocity in ECF co-ordinates.

Accounting for apparent accelerations in ECF reference frame, the following non-linear dynamics model function f(·) is obtained:

$$\frac{d}{dt}\underline{r}=\underline{v} \tag{29}$$

$$\frac{d}{dt}\underline{v}=-\underline{\omega}\wedge(\underline{\omega}\wedge\underline{r})-2\cdot\underline{\omega}\wedge\underline{v}+g_r\cdot\frac{1}{r}\cdot\underline{r}+g_z\cdot\begin{bmatrix}0\\0\\1\end{bmatrix} \tag{30}$$

$$\frac{d}{dt}(c\cdot\varepsilon_{ET})=0 \tag{31}$$

with g_r, g_z computed according to previous formulae and $\underline{\omega}$ is the absolute angular velocity of Earth, assumed constant; $\underline{\omega}$ is intended to be given in ECF axes, consisting of a mean vector over the available observation period or more sophisticated interpolations of the effective angular speed.

2.2 System Observation

The measurement vector $\underline{z}_k$, available at receiver time T_k, depends upon which kind of raw "GPS observable" is considered. In the case of code range, taking into account the definition of state variables, the following simplified description applies, where $\underline{r}_j$ is the position of j-th space vehicle in view of the terminal:

$$\underline{z}_k = \begin{bmatrix} h_1(\underline{r}, \underline{v}, \varepsilon_{ET}) \\ h_2(\underline{r}, \underline{v}, \varepsilon_{ET}) \\ \dots \\ h_m(\underline{r}, \underline{v}, \varepsilon_{ET}) \end{bmatrix} + \underline{v}_k \tag{32}$$

$h_j(\underline{r}, \underline{v}, \varepsilon_{ET})$ = "iono-free" linear combination of code range measurements on L1 and L2, available from the j-th GPS satellite in view; j = 1,2,. . .,m

$$h_j(\underline{r}, \underline{v}, \varepsilon_{ET}) = \left\| \underline{\tilde{r}}_j - \underline{r}_{ET} \right\| + c \cdot (\varepsilon_{ET} - \varepsilon_j) + \dots$$
$$+ c \cdot (\hat{\Delta}t_{REL1} + \hat{\Delta}t_{REL2} + \hat{\Delta}t_{REL3}) \quad j = 1, 2, \dots, m \tag{33}$$

$$\underline{\tilde{r}}_j = \Delta A(\tau_j) \cdot \underline{r}_j(T_k - \tau_j - \varepsilon_{ET}) \quad j = 1, 2, \dots, m \tag{34}$$

One has, with good approximation:

$$\underline{r}_{ET} \cong \underline{r} + \mathbf{M} \cdot \underline{p} \tag{35}$$

where p is the position of the LEO GPS antenna phase centre in "local orbit frame" coordinates, coincident with body-fixed frame, if neglecting the small satellite attitude control errors; the transform matrix **M** can be built up as follows:

$$M = [u1, u2, u3]T \tag{36}$$
$$\underline{u}_1 = \underline{u}_2 \times \underline{u}_3$$
$$\underline{u}_2 = (\underline{v}_I \times \underline{r}) / \| \underline{v}_I \times \underline{r} \|, \text{ with: } \underline{v}_I = \underline{v} + \underline{\omega} \times \underline{r}$$
$$\underline{u}_3 = -\underline{r} / r$$

The question is, now, that τ_j is implicitly defined; given $\underline{r}$, $\underline{v}$ and ε_{ET}, one must loop to refine its value, iterating the following substitution or performing a similar sequential solution approach:

$$\tau_j \leftarrow 1/c \cdot \left\| \Delta A(\tau_j) \cdot \underline{r}_j(T_k - \tau_j - \varepsilon_{ET}) - \underline{r}_{ET} \right\| + \dots$$
$$+ c \cdot (\hat{\Delta}t_{REL2} + \hat{\Delta}t_{IONO}), j = 1, 2, \dots, m \tag{37}$$

3 Using Real Test Data

3.1 Outliers Detection

When using GPS raw data from a real orbiting receiver, detection and rejection of "wrong data" becomes a crucial matter.

Whatever it is the algorithm adopted to solve for orbital position, ome must be sure that navigation are "consistent" in their numerical values, according to what is expected.

Measurements that do not satisfy this constraint are to be considered "wrong" and suppressed from subsequent calculation process.

It both code and carrier range measurements are available on L1 and L2 GPS channels, the "Narrow Lane minus Wide Lane" test for consistency may be adopted.

The principle can be sketched as follows: form a linear combination "C" based on $\tilde{\rho}_{L1}$, $\tilde{\rho}_{L2}$, $\tilde{\phi}_{L1}$, $\tilde{\phi}_{L2}$, so that resulting output is approximately a constant (unless a "problem" occurs), perturbed just by noise; then, apply a pseudo-derivation filter, to detect any unexpected, sudden variations in "C". The proposed linear combination is as follows:

$$C = NL/\lambda_{WL} - WL \tag{38}$$

where:

$$NL = k_1 \cdot \tilde{\rho}_{L1} + k_2 \cdot \tilde{\rho}_{L2}, \quad WL = \tilde{\phi}_{L1}/c - \tilde{\phi}_{L2}/c_{L2}$$
$$\lambda_{WL} = c/(f_{L1} - f_{L2}), k_1 = f_{L1}/(f_{L1} + f_{L2}), \quad k_2 = f_{L2}/(f_{L1} + f_{L2});$$
$$f_{L1} = 1575.42 \times 10^6 \text{ [Hz]}, \quad f_{L2} = 1227.60 \times 10^6 \text{ [Hz]}$$

One can demonstrate that:

$$C \cong N_{L1} - N_{L2} + \text{multipath} + \text{noise} \tag{39}$$

where N_{L1}, N_{L2} are the "ambiguity" terms on carrier phase measurements. If the receiver tracking loop on carrier phase does not fail *and* the four measurements $\tilde{\rho}_{L1}$, $\tilde{\rho}_{L2}$, $\tilde{\phi}_{L1}$, $\tilde{\phi}_{L2}$ are consistent, then C is very close to a constant (it only suffers from code range measurements noise).

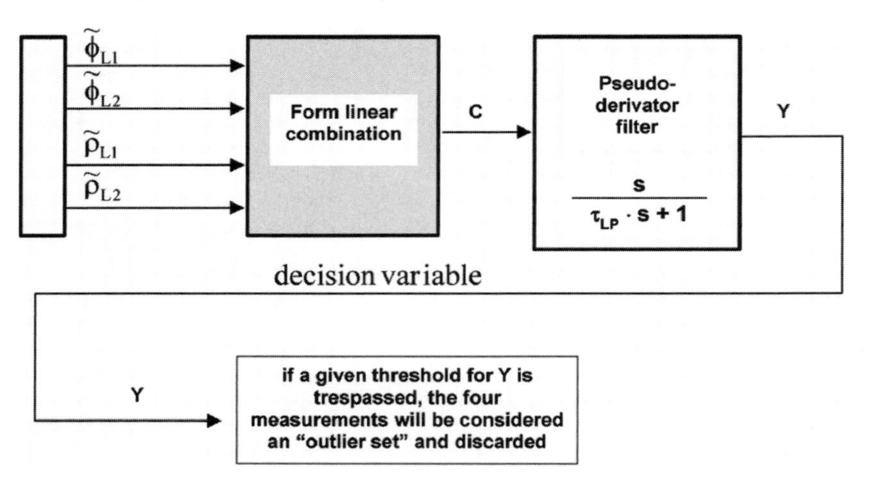

The pseudo-derivation filter is properly implemented by the following recursive equation:

$$Y(t_k) = \frac{1}{\tau_{LP}} \cdot \left(C(t_k) - s(t_k)\right) \tag{40}$$

$$s(t_{k+1}) = Ad \cdot s(t_k) + (1 - Ad) \cdot C(t_k), Ad = \exp(-dt/\tau_{LP}) \tag{41}$$

τ_{LP} is the "time constant" of the filter, whilst dt is the sampling time of the discretized algorithm. It will be necessary to re-initialise the calculation process, in order to avoid "false alarms", by imposing:

$$s(t_k) = C(t_k) \tag{42}$$

1) at first computation step of a whole data set
2) every time that visibility status changes for a single satellite, so that it is acquired at time t_k, whilst it was not in view at t_{k-1}

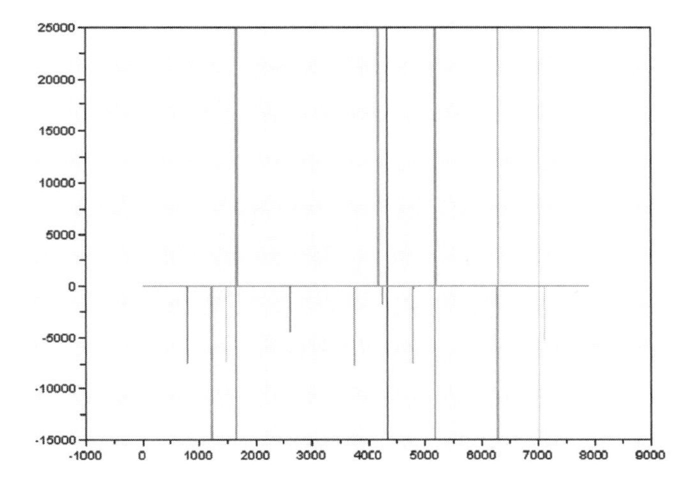

3.2 Run Results

A long raw CHAMP GPS receiver data set (about 24 hours, collected on February 7th, 2003) was processed with the UF technique described herein, applying the consistency check method provided in the previous section to exclude outliers. No precise IGS orbits were used; just the ephemeris info which is broadcast by GPS itself. Purpose was to set up and verify robustness of the overall calculation scheme, not to attain high precision. For positioning calculation, noisy iono-free code range measurements were formed as the (only) available input to the UF.

Significant spikes in the outlier detection output Y appeared; that means: a lot of data sets were unsuitable for orbit determination, as shown in the following figure for all satellites in view over the time window.

On the other side, when measurements were consistent, Y exhibited a low value, like it was expected, as shown here below.

The "scalar distance" between CHAMP ECF position computed by UF and the reference NASA/JPL/GFZ orbit is provided in the following picture; most of samples are accurate estimates, with a 3D separation below 5 metres, which is considered acceptable, under the working hypotheses.

4 Conclusions

The application of the Unscented Filtering (UF) approach has been presented, to solve the navigation positioning equations for a LEO satellite, without any initial or intermediate linearization step and in a robust configuration, eventually suitable for real-time processing.

The UF has proven to be a working algorithm even starting from a bad initial approximation of the state vector, whilst this is typically not the case for an Extended Kalman Filter (EKF). Moreover, UF is applicable to arbitrarily complex exogenous force models, more sophisticated than the "test" one presented in this article.

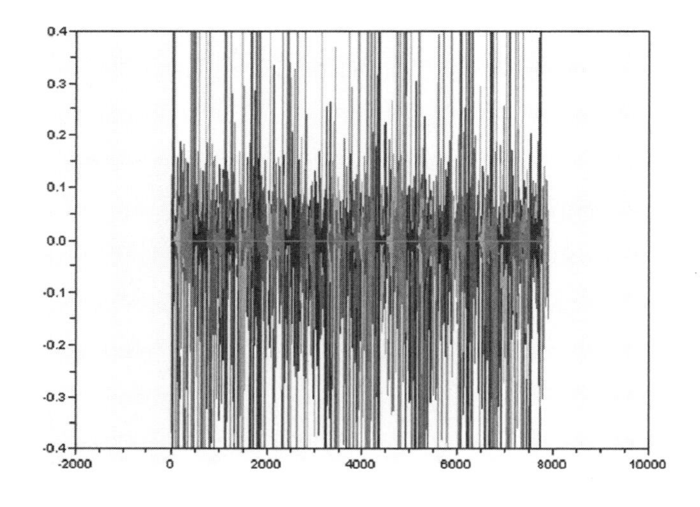

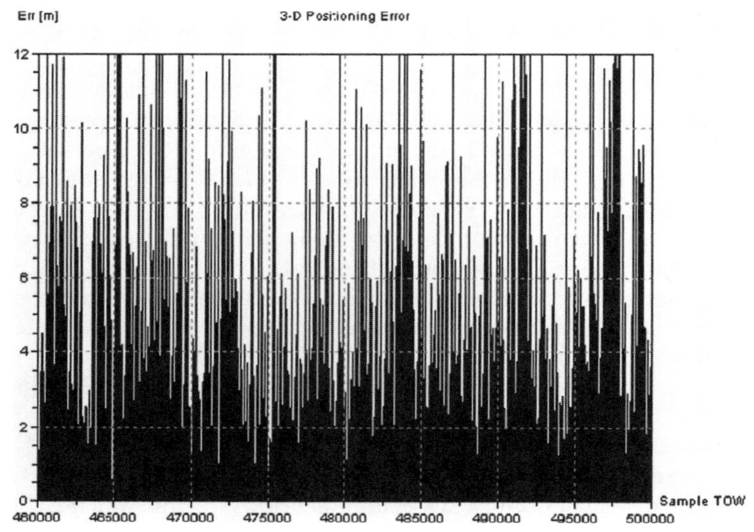

The approach has been successfully tested on real GPS data and comparison orbit available by GFZ / NASA / JPL from Champ LEO satellite mission. Refinements are required to attain a better accuracy, by smoothing inputs with carrier ranges and switching to the use of IGS orbits for GPS satellites.

5 Appendix

The "Unscented filter" is a promising sequential algorithm, aiming at estimating the state vector of a dynamic system, especially suited for the case of non-linear systems, with output measurements and state evolution perturbed by exogenous disturbances.

As such, its reference application domain is the same as the EKF; now, the fundamental difference between EKF and Unscented Filter (UF) is the fact that the latter is fully non-linear, i.e.: it does not require the calculation of any Jacobian or design matrix from the derivatives of plant model equations (1), (2), where the "noise" vector w(t) accounts for unknown perturbation inputs and for not-modelled dynamics.

It is assumed that measurements $z_k = z(t_k)$ provided by the dynamic system are available at the discrete sampling instants $\{t_k\}$ and perturbed by $v_k = v(t_k)$, a noise, realistically related to the operation of some sensor equipment;

w(t) and $v(t_k)$, considered both as zero mean, white and uncorrelated gaussian processes, appear characterised respectively by the power spectral density **Q** and the covariance matrix **R**.

The UF algorithm is centred on the so called "sigma points": a set of (2n + 1) vectors of state space, being n the dimension of state x(t). At step "k" of the sequential processing, the above mentioned set is denoted as $\{\chi_k(i), i = 0, \dots, 2n\}$. For system initialisation, step k = 0, the "sigma points" $\{\chi_0(i)\}$ are computed from the assumption of an initial configuration $\mathbf{P}_0$ for the state error covariance matrix, so that the following calculation scheme is applicable for *any* step "k":

$$\mathbf{Aux} = (n + \tilde{k}) \cdot [\mathbf{P}_k + \Delta T \cdot \mathbf{Q}_k] \quad (43)$$

$$\mathbf{M} = \sqrt{\mathbf{Aux}} \quad (44)$$

$$S \doteq \| 0 | -M | +M \| \quad (45)$$

$$\chi_k(i) = \hat{x}_k + \sigma_k(i), \; i = 0, \dots, 2n \quad (46)$$

where σ_k (i) = i-th column of matrix **S**

Above, $\tilde{k}$ is a derived parameter, related to a couple of tuning elements $\bar{\alpha}$, $\bar{k}$, whose selection is critical for a correct behaviour of the algorithm; for choosing the scalars $\bar{\alpha}$, $\bar{k}$, the original work of Juliers, Uhlmann [6,7,8,9,10,11] should be addressed. Applicable values for the problems described in this paper can be:

$$\tilde{k} = \bar{\alpha}^2 \cdot (n + \bar{k}) - n, \quad \bar{k} = 2, \bar{\alpha} = 1 \quad (47)$$

Let us now recall the overall UF processing phases foreseen at step "k", arranged for the application case described in this paper.

Phase "1" = completion of last step prediction phase

Generate:

- the predicted state vector $\hat{x}_{\bar{k}}$ for current step
- a predicted value $\hat{z}_{\bar{k}}$ for the innovation process
- a predicted state estimation error covariance matrix $\mathbf{P}_{\bar{k}}$
- the auxiliary covariances $\mathbf{PH}_k$ and $\mathbf{PXH}_k$

by decomposing computation in the sub-steps:

$$\hat{x}_{\bar{k}} = \sum_{i=0}^{2n} w_i \cdot \chi_k(i) \quad (48)$$

$$\hat{z}_{\bar{k}} = \sum_{i=0}^{2n} w_i \cdot \zeta_k(i) \tag{49}$$

$$\zeta_k(i) = h(\chi_k(i), u_k, t_k), \quad i = 0, \ldots, 2n \tag{50}$$

with weights $\{w_i\}$ defined as:

$$w_i = \begin{cases} = \dfrac{\tilde{k}}{n+\tilde{k}} & \text{if } i = 0 \\ = \dfrac{0.5}{n+\tilde{k}} & \text{if } i = 1, \ldots, 2n \end{cases} \tag{51}$$

and:

$$\mathbf{P}_{\bar{k}} = \sum_{i=0}^{2n} e\chi_k(i) \cdot e\chi_k(i)^T \cdot \tilde{w}_i \tag{52}$$

$$\mathbf{P}_{\mathbf{H}k} = \sum_{i=0}^{2n} e\zeta_k(i) \cdot e\zeta_k(i)^T \cdot \tilde{w}_i \tag{53}$$

$$\mathbf{P}_{\mathbf{XH}k} = \sum_{i=0}^{2n} e\chi_k(i) \cdot \zeta_k(i)^T \cdot \tilde{w}_i \tag{54}$$

with weights $\{\tilde{w}_i\}$ defined as:

$$\tilde{w}_i = \begin{cases} = \dfrac{\tilde{k}}{n+\tilde{k}} + (1 - \bar{\alpha}^2) & \text{if } i = 0 \\ = w_i & \text{if } i = 1, \ldots, 2n \end{cases} \tag{55}$$

and:

$$e\chi_k(i) = \chi_k(i) - \hat{x}_{\bar{k}}, \quad i = 0, \ldots, 2n \tag{56}$$

$$e\zeta_k(i) = \zeta_k(i) - \hat{z}_{\bar{k}}, \quad i = 0, \ldots, 2n \tag{57}$$

Phase "2" = correction of available predictions, using current measurement read-out
Compute:

$$\hat{x}_k = \hat{x}_{\bar{k}} + K_k \cdot (z_k - \hat{z}_{\bar{k}}) \tag{58}$$

$$P_k = P_{\bar{k}} - K_k \cdot PZ_k \cdot K_k^T \tag{59}$$

with:

$$PZ_k = PH_k + R_k \tag{60}$$

$$K_k = PXH_k \cdot (PZ_k)^{-1} \tag{61}$$

Phase "3" = generation of predictions for next step (k+1)
Re-compute the set of sigma points $\{\chi_k(i), i = 0, \ldots, 2n\}$, by applying rules (A.1), . . ., (A.4);
Now, use each new "sigma point" point $\chi_k(i)$ as initial (Cauchy) condition to integrate the (non-linear) state vector dynamics, as represented by equation (1.1), from time t_k to instant t_{k+1}, getting for each the "predicted projection" $\chi_{k+1}(i)$ applicable at next step "k+1":

$$\chi_{k+1}(i) = \tilde{x}(t_{k+1}), \tag{62}$$

obtained by integrating the dynamics:

$$\dot{\tilde{x}}(t) = f(\tilde{x}(t), u(t), t), \quad i = 0, \ldots, 2n$$

with the initial condition: $\tilde{x}(t_k) = \chi_k(i)$

References

[1] A. Gelb, editor. "Applied Optimal Estimation". The MIT Press, 1974

[2] A.H. Jazwinski. "Stochastic Process and Filtering Theory" Academic Press, 1970

[3] P.S. Maybeck. "Stochastic Models, Estimation and Control" (Vol.1 & 2). Academic Press, 1982

[4] K.S. Miller, D.M. Leskiw. "An Introduction to Kalman Filtering with Applications". Robert E. Krieger Publishing Company Malabar, Florida, 1987

[5] G. Welch, G. Bishop. "An Introduction to the Kalman Filter". Department of Computer Science, University of North Carolina at Chapel Hill, NC 27599–3175. http://www.cs.unc.edu/~welch

[6] S.J. Julier and J.K. Uhlmann. "A General Method for Approximating Nonlinear Transformations of Probability Distributions", 1994

[7] S.J. Julier and J.K. Uhlmann. A Consistent, Debiased Method for Converting Between Polar and Cartesian Coordinate Systems". The Proceedings of Aereosense: The 11th International Symposium on Aereospace/Defense Sensing, Simulation and Controls, Orlando, Florida. SPIE, 1997. Acquisition, Tracking and Pointing XI

[8] S.J. Julier. Process Models for the Navigation of High-Speed Land Vehicles. PhD thesis, Robotics Research Group, Department of Engineering Science, University of Oxford, 1997

[9] S.J. Julier, J.K. Uhlmann and H.F. Durrant-Whyte. "A New Approach for the Nonlinear Transformations of Means and Covariances in Linear Filters". IEEE Transactions on Automatic Control, 1996

[10] S.J. Julier, J.K. Uhlmann and H.F. Durrant-Whyte. "A New Approach for Filtering Nonlinear Systems". The proceedings of the American Control Conference, Seattle, Washington, 1995, pages 1628–1632

[11] J.K. Uhlmann. "Simultaneous map building and localisation for real time applications". Technical report, University of Oxford, 1994. Transfer thesis

[12] A. Kleusberg. "Analytical GPS Navigation Solution". Technical report, Department of Geodesy and Geoinformatics - University of Stuttgart

Index

Signals and Communication Technology

(continued from page ii)

Digital Television
A Practical Guide for Engineers
W. Fischer ISBN 3-540-01155-2

Speech Enhancement
J. Benesty (Ed.)
ISBN 3-540-24039-X

Multimedia Communication Technology
Representation, Transmission
and Identification of Multimedia Signals
J.R. Ohm ISBN 3-540-01249-4

Information Measures
Information and its Description in Science
and Engineering
C. Arndt ISBN 3-540-40855-X

Processing of SAR Data
Fundamentals, Signal Processing,
Interferometry
A. Hein ISBN 3-540-05043-4

Chaos-Based Digital Communication Systems
Operating Principles, Analysis Methods,
and Performance Evaluation
F.C.M. Lau and C.K. Tse
ISBN 3-540-00602-8

Adaptive Signal Processing
Application to Real-World Problems
J. Benesty and Y. Huang (Eds.)
ISBN 3-540-00051-8

Multimedia Information Retrieval and Management Technological
Fundamentals and Applications D. Feng,
W.C. Siu, and H.J. Zhang (Eds.)
ISBN 3-540-00244-8

Structured Cable Systems
A.B. Semenov, S.K. Strizhakov, and
I.R. Suncheley
ISBN 3-540-43000-8

UMTS
The Physical Layer of the Universal Mobile
Telecommunications System
A. Springer and R. Weigel
ISBN 3-540-42162-9

Advanced Theory of Signal Detection
Weak Signal Detection in Generalized
Obeservations
I. Song, J. Bae, and S.Y. Kim
ISBN 3-540-43064-4

Wireless Internet Access over GSM and UMTS
M. Taferner and E. Bonek
ISBN 3-540-42551-9

Printed in the United States of America.